Carbohydrate Chemistry

Volume 15

Part II

A Specialist Periodical Report

Carbohydrate Chemistry

Volume 15
Part II

Macromolecules

A Review of the Literature Published during 1981

Senior Reporter
J. F. Kennedy, *University of Birmingham*

Reporters
D. P. Atkins, *University of Birmingham*
I. M. Morrison, *Hannah Research Institute, Ayr*
C. M. Sturgeon, *University of Edinburgh*
R. J. Sturgeon, *Heriot-Watt University, Edinburgh*
C. A. White, *University of Birmingham*

The Royal Society of Chemistry
Burlington House, London W1V 0BN

ISBN 0-85186-152-0
ISSN 0576-7172

Printed in Great Britain by
Whitstable Litho Ltd., Whitstable, Kent

Preface

This Report, the fifteenth of this Title, is, like its predecessors, a comprehensive summary of publications, both primary and secondary, presented in a concise and readable form.

In spite of the supportive comments we have received for this Title from readers in various parts of the world, reference must be made to the 1985 decision of the Royal Society of Chemistry to cease publication of 'Carbohydrate Chemistry' Part II in its present form. This decision has been based on insufficient income to support the high cost of production. Such a cessation was hinted at in the preface to Volume 14 as being a possibility. It is regrettable not only since the carbohydrate macromolecular chemistry may no longer be available in such an accessible form but also since so much effort has been put into the change to the production of camera-ready copy despite the problems associated with the complexities of carbohydrate names and structures and abbreviated forms thereof.

In the production of this volume I have been supported once again by colleagues who have unstintingly contributed to many volumes of this Title and by Drs C. A. White and D. P. Atkins who joined in the producton of this volume to ease the work-load. I again very gratefully acknowledge the considerable effort put in by every reporter, and the help and advice they have given.

Finally, the editorial and other assistance provided by Mrs R.H. Pape, Assistant Editor, Books, at the Royal Society of Chemistry in the production of this Report is much appreciated.

April 1986 JOHN F. KENNEDY

Contents

Abbreviations

The following abbreviations have been used:

Chemicals and Chemical Groups

ADP	adenosine 5'-diphosphate
AMP	adenosine 5'-phosphate
ATP	adenosine 5'-triphosphate
CDP	cytidine 5'-diphosphate
CMP	cytidine 5'-phosphate
DEAE	N,N-diethylaminoethyl
DMF	N,N-dimethylformamide
DMSO	dimethyl sulphoxide
DNA	deoxyribonucleic acid
GDP	guanosine 5'-phosphate
RNA	ribonucleic acid
THF	tetrahydrofuran
TMS	trimethylsilyl
UDP	uridine 5'-diphosphate

Physicochemical Methods

c.d.	circular dichroism
d.s.c.	differential scanning calorimetry
e.s.r.	electron spin resonance
g.c.-m.s.	gas-liquid chromatography-mass spectrometry
g.l.c.	gas-liquid chromatography
g.p.c.	gel permeation chromatography
h.p.l.c.	high-performance liquid chromatography
i.r.	infrared
m.s.	mass spectrometry
n.m.r.	nuclear magnetic resonance
o.r.d.	optical rotatory dispersion
t.l.c.	thin-layer chromatography
u.v.	ultraviolet

1
Introduction

BY J. F. KENNEDY

The objectives of the Part II Report have remained in the area of providing a summary of the literature of carbohydrate-containing and carbohydrate-directed macromolecules, but since the inception of 'Carbohydrate Chemistry' fourteen volumes and therefore fourteen years ago, the coverage has necessarily changed. This change has often been nothing more than keeping up to date with improvements in the scientific understanding of the macromolecules concerned. However, since what have come to be called biotechnological processes - i.e. the life aspects of molecules - have become a driving force in today's scientific thinking, so in this Title we have continued to cite applications of the macromolecules as well as to report on the basic science.

The introduction to Volume 14 Part II[1] was necessarily comprehensive and set the scene for subsequent volumes of this Report. In it, the following aspects were covered:

1. Scope and Coverage of the Report,
2. Organization, Nomenclature, and Use of the Report,
3. Significant Advances in Macromolecular Carbohydrate Chemistry,
4. Conclusions and Readership,

and readers of this volume are advised to consult the previous one to derive maximum benefit.

March 1982 saw the publication of the 100th volume of Carbohydrate Research - a notable landmark. This publication, very early on in its lifetime, became a primary publication channel for important carbohydrate work, particularly academic, and it is pleasing to see that the journal looks easily set for another one hundred volumes. The sister journal, Carbohydrate Polymers, which publishes articles more in the fields of application and industrial research, has a long way to go to catch this up, because it is a much younger journal. Nevertheless there is scope for both journals to continue successfully with the continual significant advances in both academic and industrial carbohydrate chemistry.

A special issue of Carbohydrate Research[2] was published to honour Professor Sumio Umezawa and his work on antibiotics, particularly in the field of sugar-containing antibiotics.

References

1. J.F. Kennedy, in 'Carbohydrate Chemistry', Part II (Macromolecules), ed. J.F. Kennedy (Specialist Periodical Reports), The Royal Society of Chemistry, London, 1983, Vol. 14, p. 1.
2. T. Tsuchiya, Carbohydr. Res., 1982, 109, 1.

2
General Methods

BY R. J. STURGEON

1 Gas-Liquid Chromatography

An improved g.l.c. method for the simultaneous determination of alditol acetates of neutral and amino-sugars has been reported.[1] Short retention times and baseline separation between D-glucose and D-galactose were recorded. The method has been used for the analysis of sialoglycoproteins from bone.

A chiral polysiloxane stationary phase has been used in the g.l.c. analysis of mixtures of neutral and amino-sugars.[2] Complete separation of the alditol acetates of the neutral sugars, including 3-O-methyl- and 4-O-methyl-D-glucitols, was accomplished. Capillary g.l.c. separation of monosaccharides as their alditol acetates has been reported.[3]

2-Amino-2-deoxy-D-galactose and 2-amino-2-deoxy-D-glucose, after reduction with sodium borohydride to the corresponding amino alcohols, and conversion to the trifluoroacetyl derivatives, have been separated by g.l.c. and detected using a nitrogen-specific flame thermionic detector.[4] The method may be applied to the determination of amino-sugars in glycoconjugates.

The quantitative g.l.c. analysis of sucrose in the presence of the oximes of D-glucose and D-fructose has been reported using a buffered oximation reagent.[5] No hydrolysis of the sucrose takes place during the oximation procedure, nor do the reagents affect the subsequent silylation of the disaccharide.

1-Methyl-imidazole has been used as a solvent and a catalyst for the preparation of aldononitrile acetates of aldoses.[6] Existing methods have been modified for the analysis of glycoses, amino-sugars, and Smith degradation products by using g.l.c. determination of their aldononitrile acetate derivatives.[7]

Monosaccharides have been converted to produce one derivative for each aldose, as detected by g.l.c., in a sequence which involves the formation of aldoximes, their reduction with borane to the corresponding aminopolyols, and subsequent conversion to the *N*-ethoxycarbonyl-*O*-trimethylsilyl derivatives.[8] Ketoses produce two diastereomers upon derivatization. The formation of side and decomposition products is negligible.

The monosaccharide composition of glycoconjugates has been determined by g.l.c.-m.s. of the *O*-trimethylsilyl ethers after liberation of the component sugars, *N*-deacetylation, and deamination of amino-sugars and neuraminic acid.[9] 6-Deoxyhexoses, hexose, and pentoses, derived from cardiac glycosides, have been estimated by g.l.c.-m.s. of their *O*-trimethylsilyl ethers.[10,11] Monosaccharides may also be separated as *O*-trimethylsilylalditols on fused-silica capillary columns.[12] These derivatives have been found to offer a convenient alternative to the *O*-acetate derivatives for the quantitation of acid hydrolysates of polysaccharides. High resolution g.l.c. separations of alditol acetates on fused-silica wall-coated open tubular columns have been achieved.[13]

A differential g.l.c. method for determination of uronic acids in carbohydrate mixtures has been developed.[14] In a two-step procedure, neutral sugars are first determined by g.l.c. of their aldononitrile acetates, before the uronic acids in turn are reduced and converted to the same derivatives. The relative proportions of *D*-mannuronic acid and *L*-guluronic acids in alginic acid have been determined in a g.l.c. method.[15] After lowering the viscosity of the alginic acid by limited acid hydrolysis, carboxyl groups are first esterified by reaction with 1-ethyl-3-{3-(dimethylamino)-propyl}-carbodi-imide, then reduced with sodium borohydride, and the resulting mixture of hexosans is converted by acid hydrolysis to monosaccharides. The monosaccharides in turn are reduced with sodium borohydride to hexitols which can be analysed as the butane boronic esters.

New chiral stationary phases for the g.l.c. separation of the enantiomers of amines, amino-alcohols, hydroxy-acids, and monosaccharides have been described.[16] The separation of monosaccharide enantiomers as the *O*-trifluoroacetates is achieved on a stationary phase by attachment of *L*-valyl-S-α-phenyl ethylamide to the functionalized cyanoethyl side chains of the polysiloxane stationary phase. Urinary polyols have been estimated by g.l.c. of their acetates.[17] Interference in the chromatographic patterns due to monosaccharides was overcome by forming methyloxime-acetate derivatives of reducing sugars.

2 Column and Ion-exchange Chromatography

Macroporous microspherical cellulose has been used as a stationary phase in the chromatography of water-soluble high and low molecular weight carbohydrates.[18] Distribution coefficients of the polymer molecules depend on the hydrodynamic diameters of the polymer molecules. The hydrodynamic behaviour of reduced glycopolypeptides has been studied by gel filtration in guanidine hydrochloride in conjunction with h.p.l.c.[19] Even though carbohydrate-rich glycopeptides may occasionally yield an underestimated molecular weight value, the method appears to be useful for the rapid estimation of molecular weights of simple polypeptides.

The separation of 2-amino-2-deoxy-D-glucose from 2-amino-2-deoxy-D-galactose in the routine amino-acid analysis of plant glycoproteins has been improved by the use of longer columns on the amino-acid analyser.[20] An improved chromatographic separation of sugars and sugar alcohols on cation-exchange resins (Ca^{2+}) is reported.[21] The addition of small amounts of triethylamine to the eluant catalyses the mutarotation of reducing sugars, resulting in reduced peak widths without affecting the elution times. The separation of strongly basic ion-exchange resins for the separation of alditols from monosaccharides has previously been reported as a method for the estimation of the degree of polymerization of neutral oligo- and poly-saccharides. In this procedure, the alditols were reported not to bind to the resin, and were estimated after derivatization by g.l.c. It has now been shown that in a number of cases alditols do bind to the resins, but their elution from the resin is influenced by methanol, ammonium acetate, and ammonium formate.[22] Cation-exchange resins in the silver form retain oligosaccharides to a greater extent than the calcium forms of the resins, resulting in a greater number of oligosaccharides being separated.[23]

3 Thin-layer Chromatography

A continuous-flow t.l.c. system has been developed for the identification and measurement of urinary carbohydrates impregnated on paper.[24] Muramic acid, 2-amino-2-deoxy-D-glucose, 2-amino-2-deoxy-D-galactose, and their corresponding aminodeoxyalditols have been separated by two-dimensional t.l.c. as their 2,4-dinitrophenyl derivatives.[25] Some of the factors influencing the derivatization and separation of these sugars are reported.

4 High-pressure Liquid Chromatography

A review of the different types of silica and mobile phases that are used in the h.p.l.c. of sugars has been published.[26] Methods of column preparation, pre-column derivatization, and detection methods are discussed.

Chromatography on boronic acid silica reduces the separation time of diol-containing substances without loss of resolution when compared to less rigid polymers such as cellulose and polymethacrylic acid.[27] This stationary phase has been proposed for the separation of substances such as catecholamines and glycoproteins. Proteoglycans from bovine nasal, bovine articular and rat chondrosarcoma cartilage have been analysed by h.p.l.c. on a silica-based material bonded with an amide phase.[28] The biochemical integrity of the proteoglycans is retained during this procedure.

High-performance size-exclusion chromatography has been used to obtain data for hydrodynamic molecular radii of dextrans.[29] Glycoprotein hormones and glycoprotein allergens from plant and animal origin have been analysed by size-exclusion chromatography on the TSK series of modifield silicas under h.p.l.c. conditions.[30]

Silica gel chemically modified with aminopropyl groups, or silica gel with 1,4-diaminobutane present in the mobile phase, is effective in the separation of neutral oligosaccharides by h.p.l.c.[31] The method has been used successfully in the purification of neutral oligosaccharides with different degrees of branching and derived from dolichol-linked oligosaccharide intermediates.

Oligosaccharides (up to d.p. 12) obtained from acid and enzymic hydrolysis of starch have been separated by reversed-phase h.p.l.c.[32]

A method for the rapid separation of anionic oligosaccarides has been developed using an anion-exchange column which offers superior speed, resolution, and yield when compared with other ion-exchange methods or high-voltage paper electrophoresis.[33] Baseline resolution of anionic oligosaccharides which differ in their net negative charge is achieved, regardless of whether the charge is imparted by neuraminosyl or phosphate moieties.

A h.p.l.c. procedure for the separation and isolation of neuraminosyl-oligosaccharides has been developed using an isocratic system to give baseline separation of neuraminic acid, 3′- and 6′-neuraminosyl-lactose.[34] A complex mixture of oligosaccharides obtained by alkaline borohydride degradation of human, bronchial mucous glycoproteins has been separated by h.p.l.c. on bonded primary amine packing using a linear gradient solvent system.[35] The separation of isomeric oligosaccharides is achieved, suggesting that the total number of hydroxyl groups, as well as their configuration, determines the retention time of each oligosaccharide.

A method has been developed for separation of reduced, neutral oligosaccharides by h.p.l.c.[36] D-Glucitol and reduced D-gluco-oligosaccharides containing between one and twenty D-glucosyl residues have been separated. The method has also been applied to the analysis and preparative isolation of glycoprotein-derived oligosaccharides obtained by enzymic release by endoglycosidases or by chemical release by hydrazinolysis. Subnanomolar quantities of oligosaccharides may be detected when sodium borotritide is used in the reduction step.

The urinary excretion of isomeric chondroitin sulphates has been monitored by h.p.l.c. of the unsaturated disaccharides produced by digestion with chondroitin sulphate lyases.[37] A systematic variation in the composition of the h.p.l.c. mobile phase was used for the selection of the optimal conditions for separation. Endo-β-D-2-acetamido-2-deoxy-D-glucanase activity has been assayed using the dansyl derivative of the L-asparaginyl oligosaccharide (D-Man*p*)$_5$-(D-Glc*p*NAc)$_2$-L-Asn as the substrate.[38] Analysis of the product, dansyl 2-acetamido-2-deoxy-D-glucosyl-L-asparagine, has been achieved by a reverse-phase h.p.l.c. method. The lower limit of detection is 0.1 nmol of dansyl glycopeptides. Reducing sugars have been reacted with dansyl hydrazine prior to their separation by h.p.l.c.[39] Fluorescent detection provides a sensitivity of 10 pmol per injection for the monosaccharide components of glycoproteins.

Numerous alkyl and aryl D-glucosides, D-galactosides, D-mannosides, D-glucosiduronic acids, and their acetyl and benzoyl derivatives have been separated by h.p.l.c.[40] A rapid isocratic h.p.l.c. method for the estimation of neuraminic acid in serum has been described.[41] Separation is achieved on a cation-exchange resin using 0.003M sulphuric acid as the mobile phase. Reversed-phase h.p.l.c. has been used in the separation of some methyl glycosides.[42] In the aldohexose series, the pyranosides elute before furanosides, but in the aldopentose series the furanosides elute first.

Mono- and di-saccharides and polyhydric alcohols have been separated by h.p.l.c. on silica columns modified with tetra-ethylenepentamine.[43] The method, which offers a high degree of resolution, requires low operating temperatures.

Tryptic glycopeptides from viral sources have been chromatographed using reverse-phase ion-pair partition chromatography with a hydrophilic pairing agent, phosphoric acid.[44] An h.p.l.c. method for the analysis of reduced, unsaturated disaccharides derived from the enzymic digestion and sodium borohydride reduction of chondroitin sulphate, dermatan sulphate, heparan sulphate, and heparin has been described.[45] The method avoids the possibility of obtaining anomeric forms of the unsaturated disaccharides.

An improved separation of monosaccharides by h.p.l.c. has been achieved using porous microparticle carbohydrate columns.[46] Fourteen oligosaccharylpyrophosphoryl dolichols, precursors of the L-asparagine-linked oligosaccharides of glycoproteins, have been separated by h.p.l.c. on silica gel.[47] Some of the resolved products were shown to retain their activity as substrates in enzyme-catalysed reactions. Monosaccharides have been separated as their borate complexes by h.p. anion-exchange chromatography.[48] The sugars are detected fluorimetrically after separation with 2-cyano-acetamide. Monosaccharide residues of glycoproteins, after conversion to 2-aminopyridyl derivatives, have been purified by ion-exchange chromatography and then separated by h.p.l.c. on reversed phase columns.[49] Although it was difficult to detect mono- and di-saccharide derivatives because of interference from 2-aminopyridine and other fluorescent materials, higher oligosaccharides can be separated.

Improvements in a detector for liquid chromatography based on optical activity of the components allow 100 ng of mono- or di-saccharide to be detected.[50] Simultaneous determination of six carbohydrate components of human urine, which is injected after deionization, has been accomplished. Improvements have been made to the sensitivity of refractive-index detectors by eliminating noise caused by poor temperature control of the detector and eluant.[51]

The use of 2-cyanoacetamide in post-column detection of aldoses has been evaluated.[52] The condensation reactions obtained by heating carbohydrates with 3-methyl-2-benzothiazolinone hydrazone in alkali containing 2-methoxyethanol have been described.[53]

Monosaccharide aldoses and ketoses and disaccharides produce chromogens which can be measured at 390 nm. The potential in using the method for post-column detection of carbohydrates separated by h.p.l.c. is discussed.

5 Electrophoresis

The accuracy of molecular weight estimates for glycoproteins has been evaluated by SDS-pore gradient electrophoresis of glycoproteins of known molecular weight and known amino-acid and carbohydrate composition.[54] A comparison was made of Tris-glycine and Tris-borate-e.d.t.a. buffer systems with and without reduction by 2-mercaptoethanol. Although both the gradient gel system and the formation of borate complexes with the individual carbohydrate moieties of the glycoprotein contribute to the precision of the molecular weight estimate, it is the thermodynamic and hydrodynamic properties of the individual glycoprotein which determine its mobility on gel electrophoresis.

Lectins have been used for the detection of glycoproteins transferred to nitrocellulose sheets after electrophoretic separation on polyacrylamide gels.[55] Fluorescein-derivatized asialoglycopeptides have been resolved by polyacrylamide gel electrophoresis.[56] When the method is used in conjunction with specific exoglycosidase digestion of the glycopeptides, a linear correlation is observed between the relative electrophoretic mobility and the number of monosaccharide residues removed by the enzymes.

D-Glucuronic acid and L-iduronic acid may be separated by electrophoresis on Titan III cellulose acetate plates.[57] The method has been applied to the analysis of uronic acids in chondroitin 4- and 6-sulphates and dermatan sulphate.

Analytical isotachophoresis has been used in the estimation of neuraminic acid.[58] Under the conditions of assay neuraminic acid appears as a trough between two u.v. impurity components of the buffer. The trough length is proportional to the amount of neuraminic acid injected into the system.

Electrophoretic and chromatographic methods for separating carbohydrates have been reviewed in the latest edition of a treatise on the chemistry and biochemistry of carbohydrates.[59]

6 Analytical Methods

Modifications of the Park-Johnson ferricyanide submicromethod for the assay of reducing groups in carbohydrates have been reported.[60] Oxalic acid is used as a solvent for ferric ferrocyanide in place of sodium dodecyl sulphate, which gives turbid solutions.

When aldoses are determined iodometrically in the presence of magnesium oxide, no over-oxidation takes place with the minimum formation of iodate ions.[61] Colorimetric analyses of carbohydrates have been carried out on a liquid scintillation counter over a wider range of concentrations than can be measured with a conventional light spectrophotometer.[62] The use of alkaline solutions of 4-hydroxybenzoic acid hydrazide for the determination of reducing sugars in extracts of plant tissues has been evaluated.[63]

An enzymic microassay for lactose, with a sensitivity range of 12.5-500 nmol per sample, has been developed.[64] The method is based upon the hydrolysis of lactose to D-glucose and D-galactose by β-D-galactosidase followed by the colorimetric assay of the free D-glucose using the D-glucose oxidase procedure.

A simple detection method for D-galactose and D-galactose 1-phosphate in blood-impregnated filter paper has been developed as a screening test for galactosaemia.[65] When D-galactose (or D-galactose 1-phosphate after treatment with phosphatase) is oxidized by D-galactose dehydrogenase, the resulting NADH may be measured fluorimetrically. In a similar approach the microdetermination of D-galactose and D-galactose 1-phosphate in dried blood spots has been achieved.[66] D-Galactose is determined fluorimetrically with D-galactose dehydrogenase whilst the D-galactose 1-phosphate is determined using the same procedure after hydrolysis of the phosphate ester with alkaline phosphatase. In Nelson's Somogyi method for the determination of reducing sugars, the arsenomolybdate reagent has been replaced by the Folin-Ciocalteu phenol reagent.[67] The method compares favourably with the original method in stability of colour and reproducibility. A new type of multilayer film analytical element, consisting of a spreading layer, a blocking layer, an enzyme layer (composed of D-glucose oxidase, peroxidase and dye), and a transparent layer, has been developed for the determination of D-glucose in blood.[68] After a spot of whole blood in dropped on the film, the D-glucose concentration is determined without further manipulations.

Sulphydryl reagents such as cysteine and glutathione interfere with the D-glucose oxidase assay for D-glucose.[69] This interference can be eliminated by use of N-ethylmaleimide. D-Glucose has been estimated enzymically using a commercial analyser after elution from blood spotted on to filter paper.[70] An oxygen-stabilized enzyme electrode has been designed for the analysis of D-glucose in fermentation broths.[71]

In a new enzymatic method for the determination of D-glucose in human sera, the hydrogen peroxide produced by D-glucose oxidase is measured from the change in absorbance due to oxidation of NAD(P)H in the presence of catalase, aldehyde dehydrogenase, and a high concentration of ethanol.[72]

Several advantages of this method, as compared with other D-glucose oxidase methods coupled with leuco dyes, are recorded. The estimation of D-glucose in haemolysed blood samples using the hexokinase—D-glucose 6-phosphate dehydrogenase method has been described.[73]

Two new systems have been developed for the differential measurement of pH in small volumes.[74] The apparatus is capable of following the pH change caused by hexokinase-catalysed reaction of D-glucose and ATP, when the sugar concentration is in the range 1-18 mM. The system can be used to monitor D-glucose concentrations in biological fluids. Optimal reaction conditions and kinetic properties of D-glucose dehydrogenase from Bacillus megaterium have been studied in order to develop a method for serum D-glucose determination.[75] The K_m value of the enzyme for D-glucose is influenced by the pH of the medium and by ionic strength. Suitable conditions for the use of D-glucose dehydrogenase in rate assay and end-point assay methods were identified. D-Glucose concentrations determined in serum using these methods show good correlation with those obtained with the hexokinase method. D-Glucose dehydrogenase has been covalently attached to nylon tubing.[76] The activity and pH optimum of the bound enzyme depends on the transition metals used as spacers between the enzyme and the support.

A fluorimetric method for the measurement of α-L-fucopyranose released from glycoproteins by α-L-fucosidase has been reported.[77] The L-fucose is oxidized with L-fucose dehydrogenase and NAD, and the NADH produced is used in conjunction with diaphorase to convert the nonfluorescent dye resazurin to the highly fluorescent product resorufin.

L-Fucose has been measured radiometrically using a method capable of measuring quantities of the sugar as low as 25 pmol.[78]

The assay couples L-fucose dehydrogenase to L-glutamate dehydrogenase and L-glutamate decarboxylase to yield $^{14}CO_2$ which is derived from {1-^{14}C}-α-ketoglutarate. L-Fucose dehydrogenase has been used previously in fluorescence assays of the sugar, but the method is limited when the sample to be assayed contains endogenous fluorescent substances, commonly observed in hydrolysed glycopeptides and membrane preparations at low operating pressures on columns having a relatively lengthy life and high linear sample capacity.

The phloroglucinol method for the estimation of D-xylose in urine has been assessed with respect to the interference from low levels of D-glucose.[79]

A micromethod for the determination of free or glycoprotein-bound neuraminic acid on the basis of oxidation with periodic acid and reaction with thiobarbiturate has been described.[80] A modification of the thiobarbituric acid technique for the determination of neuraminic acid has been developed.[81] Elimination of interference caused by 2-deoxy-D-ribose and several other sugars, without loss of neuraminic acid, is achieved by pH-dependent extraction of the chromogen. The neuraminic acid level in sialoglycoconjugates has been measured in a new enzymic method after release from glycoproteins with neuraminidase.[82] Neuraminic acid is cleaved to release pyruvate using N-acetylneuraminic acid aldolase, followed by oxidation of pyruvate oxidase to produce hydrogen peroxide which is measured colorimetrically by the peroxidase—4-chlorophenol—4-amino-antipyrine method. Correlation of results with those obtained using a chemical method of analysis was reported. Free neuraminic acid split from red-blood-cell receptors to Newcastle disease virus has been determined during simultaneous elution and haemolysis of the cells.[83] The procedure depends on the presence of neuraminidase activity on the preadsorbed virions.

Fluorescamine {4-phenylspiro(furan-2-(3H),1′-(3′H)-isobenzofuran)-3,3′-dione} may be applied to the quantitative and qualitative analysis of amino-sugars.[84] The method has been applied to the analysis of the product of acid hydrolysis of chitin. Methods have been developed for the identification of the glycosidic linkages between amino-sugars and L-asparagine, L-serine, and L-threonine in glycoproteins.[85] After alkaline elimination in the presence of sodium borotritide, acid hydrolysis, and dansylation, the sugar components of O-glycosyl linkages are identified after separation of the dansyl hexosaminitols by thin—layer electrophoresis. The linkage compound *N*^{4}-(2-acetamido-2-deoxy-β-*D*-glucopyranosyl) hydrogen *L*-asparaginate is isolated from the *N*-glycosidic linkage region of glycoproteins by partial acid hydrolysis and identified as its dansyl derivative.

The mannitol levels in serum samples from dogs after mannitol infusion have been measured using the spectrophotometric measurement of the initial rate of NADH formation when the sugar is oxidized by a bacterial mannitol dehydrogenase preparation.[86]

7 Structural Methods

A comprehensive review of physical methods used for structural analysis of carbohydrates has been published.[59] The methods include high-resonance n.m.r. spectroscopy, m.s., polarimetry, u.v. and i.r. spectroscopy, and *X*-ray diffraction.

N.m.r. Spectroscopy.——A review dealing with ^{13}C n.m.r. spectroscopy of polysaccharides has been published.[87]

The primary structures of polysaccharides and their derivatives have been investigated using ^{13}C magic-angle spinning-crosspolarization n.m.r. spectroscopy.[88] Although the resolution of the individual carbon resonances was not perfect when xanthan gum and its octylamide derivative were used, the technique provided considerably more detail than the corresponding ^{13}C n.m.r. studies in an equivalent period of time. A ^{13}C n.m.r. analysis method has been described for the identification of carbohydrate residues in *D*-galactosides and *D*-glucosides.[89] Assignments can be made to the anomeric configuration and to the ring size of the sugars. ^{13}C N.m.r. spectroscopy is a practical method for following the kinetics of enzymic digestion of individual carbohydrate residues of glycopeptides and for determining the structure of the products of partial digestion.[90] In a study of the hydrolysis of hen ovalbumin glycopeptides by α-*D*-mannosidase, the method showed that some α-*D*-mannopyranosyl-(1 → 3)-*D*-mannopyranosyl linkages are cleaved at a faster rate than the corresponding (1 → 6) linkages.

The resolution-enhanced 500 MHz ^{1}H n.m.r. spectra of three *N*^{4}-oligo-(*D*-mannopyranosyl)-*L*-asparagines can be interpreted in terms of the complete primary structures of these compounds.[91] The proposed structures are mainly based upon interpretation of the sets of chemical-shift values of H-1*s* and H-2*s* of constituting *D*-mannopyranosyl residues. The data are sensitive to the type and configuration of the glycosidic linkage of the D-mannopyranosyl residue and to the position of that sugar in the chain.

Mass Spectrometry.— Carbohydrates, after separation by h.p.l.c., have been measured using a mass detector.[92] The solvent is evaporated after nebulization in a heated column producing finely divided solute particles which pass through a light beam. Light scattered from the particles is detected by a photomultiplier and amplified on a chart recorder. The mass detector, when compared with refractive-index detection, showed a ten-fold increase in sensitivity, improved stability, and an ability to detect products separated by gradient elution.

By careful selection of derivatives and analysis conditions, m.s. has been applied to the microscale sequencing of a dodecasaccharide derived from a glycolipid.[93] The structural analysis by m.s. of some synthetic fully methylated L-rhamno-oligosaccharides has been reported.[94] A number of partially methylated aldoses have been separated by g.l.c. on capillary columns as their trimethylsilylated diethyl dithioacetal derivatives.[95] Single peaks for each derivative are obtained and m.s. is used in the identification of peaks. Thirteen methyl ethers of methyl N-acetyl-N-methyl-α-D-neuraminate methyl glycoside have been identified by g.l.c.-m.s. after partial methylation of methyl N-acetyl-α-D-neuraminate methyl glycoside.[96] Chemical-ionization (methane) m.s. has been used to enhance the sensitivity of detection of partially methylated alditol acetates of neutral and amino-sugars.[97] The mass spectra of 52 partially methylated and acetylated methyl glycosides of D-galactose, D-mannose, D-glucose and 2-acetamido-2-deoxy-D-glucose have been determined.[98] A scheme for the rapid determination of the position of the methyl and acetyl residues is proposed.

Field-desorption m.s. has been used to analyse oligosaccharides containing five to fourteen hexose units, without prior derivatization.[99] The molecular weight of the oligosaccharide can be determined by means of the abundant quasi-molecular ions of the type MNa^+, MH^+, MNa_2^{2+}, and MNa_3^{3+}. The method has the potential for structural elucidation of oligosaccharides and their reaction products up to a range of 3.5×10^3 mass units. Neuraminic acid isolated from erythrocyte ghosts has been estimated by methane chemical-ionization m.s. of the O-trimethylsilyl methyl glycoside using phenyl 2-acetamido-2-deoxy-(3,4,6-tri-O-trimethylsilyl)-α-D-glucopyranoside as internal standard.[100] Nanogram levels of the sugar can be detected in the presence of other contaminating substances.

The anomeric configuration of a 2-acetamido-2-deoxy-D-glucopyranosyl residue linked to a secondary hydroxyl group of another sugar residue has been established from g.l.c.-m.s. data obtained when such a disaccharide is successfully treated with periodate, sodium borohydride, and acetic anhydride.[101]

Miscellaneous Methods.——A modification of the Hakomori methylation reagent has been reported, using potassium hydride in place of sodium hydride.[102] The preparation of the potassium methylsulphenylmethide is faster and may be performed at ambient temperature. Significantly fewer impurities in the reaction products are also observed.

Internal 8-O-substituted N-glycolylneuraminic acid residues of methylated saccharides undergo de-N-acylation whereas the terminal fully methylated N-glycolylneuraminic acid residues do not under the conditions of methanolysis.[103,104] This difference in the stability of N-acyl groups has been observed in polysialosyl chains comprising either N-glycolylneuraminic acid or N-acetylneuraminic acid.

On methylation and subsequent saponification, 2-O-(4-O-methyl-α-D-glucopyranosyluronic acid)-D-xylose and 4-O-(α-D-galactopyranosyluronic acid)-D-xylose give four and two methyl glycosides, respectively.[105] Investigation of their structures using the Hakomori methylation procedure showed that the 4-O-substituted uronic acid residues afford β-elimination products in addition to the permethylated derivatives. The aldobiouronic acid with an unsubstituted 4′-OH group gave the products of direct methylation.

The hydrazinolysis-nitrous acid deamination of glycopeptides leads to the specific cleavage of 2-acetamido-2-deoxy-D-glucosyl linkages.[106] The 2,5-anhydro-D-mannose-containing oligosaccharides thus obtained are reduced with sodium borohydride, methylated, and analysed by g.l.c.-m.s.

Partially methylated alditol acetates have been separated by g.l.c. on glass capillary columns.[107] Some characteristics of the column and the application to structural analysis of polysaccharides are reported. Methylation techniques in the structural analysis of glycoproteins and glycolipids have been reviewed.[108]

Kinetic data for eleven common methyl glucopyranosides in unbuffered sodium periodate have been reported.[109] The results indicate the extent of variations in reactivity that can arise from differences in configuration, especially at the anomeric centre, and illustrate the role of intermediate hemiacetals in determining overall reaction rates. The site of substitution of peripheral branches on the core D-mannosyl residues of complex-type carbohydrate units of glycoproteins have been studied by Smith degradation.[110] By using glycopeptides of known structure (1) it is shown that the site of attachment of the (1 → 4)-linked, third branch to the D-mannose core in tri- and tetra-antennary structures may be identified. The critical factor is the concentration of acid used for the cleavage of the periodate-oxidized product. The results support the concept that in glycoproteins from most sources the (1 → 4)-linked branch is attached to the (1 → 3)-linked D-mannosyl residue of the core portion in the complex-type structures.

The possible conformations of neuraminic acid have been analysed using semi-empirical potential functions.[111] In solution, neuraminic acid may exist preponderantly in two different conformations which differ in the orientation of the terminal hydroxymethyl group of the glycerol side chain. The present model is consistent with ^{1}H- and ^{13}C-n.m.r. spectroscopic data, but differs from earlier models. C.d. spectra have been recorded for methyl α- and β-D-glycosides of neuraminic acid.[112] The Cotton effects around 220 nm arise from the n → π* transition of the carboxyl group and are negative for α-D-linked glycosides and positive for β-linked glycosides.

The results of X-ray diffraction analysis of the crystalline conformation of homo- and regular hetero-D-glucan chains have been used to establish the hypothesis that the glycosidic linkage type is the determinant of polysaccharide conformation.[113] In this respect, polysaccharides are more like synthetic polymers than proteins or nucleotides. In the latter it is variations in the substituents which are responsible for the conformational diversity. The continuing compilation of crystal structures of carbohydrates, nucleosides, and nucleotides has been published.[114]

An improved method for the determination of the molecular weights of oligosaccarides by a combination of paper electrophoresis and h.p.l.c. of their 2-aminopyridyl derivatives has been reported.[115] The electrophoretic mobilities of the derivatives are independent of linkage positions, anomeric configurations, and the presence of acetamido groups, up to molecular weights of 1.7×10^{3}.

β-D-Glucans, but not their oligo- or mono-saccharide

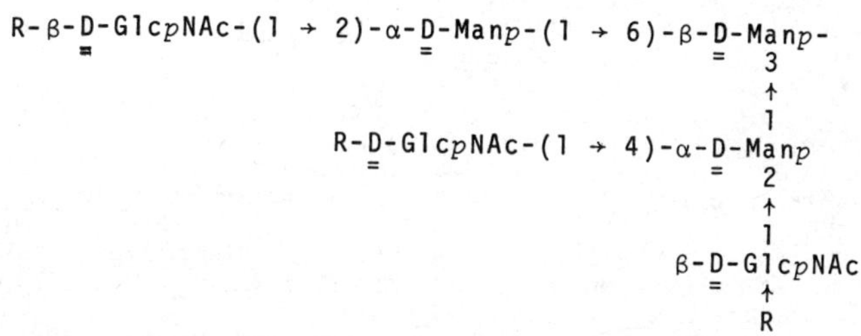

(1) R = other glycosyl residues

degradation products, interact with Congo Red.[116] Radial diffusion of β-D-glucanases into a substrate-bearing gel slab has been used as a sensitive method for assay of the enzymes, by measuring the clearing zone formed by enzyme action on the substrate.

The use of glycosyltransferases in assessing oligosaccharide structure has been reviewed.[117]

References

1 T.Anastassiades, R.Puzic, and O.Puzic, J.Chromatogr., 1981, 225, 309.
2 C.Green, V.M.Doctor, G.Holzer, and J.Oro, J.Chromatogr., 1981, 207, 268.
3 J.Klok, E.H.Nieberg van Velzen, J.W.Leeuw, and P.A.Schenck, J.Chromatogr., 1981, 207, 273.
4 T.Shinohara, J.Chromatogr., 1981, 207, 262.
5 K.J.Schäffler and P.G.Morel du Boil, J.Chromatogr., 1981, 207, 221.
6 C.C.Chen and G.D.McGinnis, Carbohydr.Res., 1981, 90, 127.
7 S.H.Turner and R.Cherniak, Carbohydr.Res., 1981, 95, 137.
8 H.Frank, H.J.C.das Neves, and E.Bayer, J.Chromatogr., 1981, 207, 213.
9 I.Mononen, Carbohydr.Res., 1981, 88, 39.
10 B.Kopp, J.Jurenitsch, and W.Kubelka, J.Chromatogr., 1981, 210, 291.
11 J.Jurenitsch, B.Kopp, I.Gabler-Kolacsek, and W.Kubelka, J.Chromatogr., 1981, 210, 337
12 A.G.W.Bradbury, D.J.Halliday, and D.G.Medcalf, J.Chromatogr., 1981, 213, 146.
13 R.Oshima, A.Yoshikawa, and K.Kumanotani, J.Chromatogr., 1981, 213, 141.
14 J.Lehrfeld, Anal.Biochem., 1981, 115, 410.
15 H.S.Prihar, B.K.Bugashetti, and D.S.Feingold, Anal.Biochem., 1981, 114, 294.
16 W.A.König, I.Benecke, and S.Sievers, J.Chromatogr., 1981, 217, 71.
17 J.N.Mount and M.F.Laker, J.Chromatogr., 1981, 226, 191.
18 Y.A.Eltekov, N.M.Strakhova, J.Kàlal, J.Peška, and J.Stamberg, J.Polym.Sci., Polym.Symp., 1980, 68, 247.
19 N.Ui, J.Chromatogr., 1981, 215, 289.
20 I.E.P. Taylor, Phytochemistry, 1981, 20, 2769.
21 L.A.T.Verhaar and B.F.M.Kuster, J.Chromatogr., 1981, 210, 279.
22 M.Tanaka, Carbohydr.Res., 1981, 88, 1.
23 H.D.Scobell and K.M.Brobst, J.Chromatogr., 1981, 212, 51.
24 J.R.Alonso-Fernàndez, M.D.Bóveda, C.Parrado, J.Peña, and J.M.Fraga, J.Chromatogr., 1981, 217, 357.
25 M.J.Talieri, N.Kilic, and J.S.Thompson, J.Chromatogr., 1981 206, 353.
26 L.A.T.Verhaar and B.F.M.Kuster, J.Chromatogr., 1981, 220, 313.
27 M.Glad, S.Ohlson, L.Hansson, M.O.Mansson, and K.Mosbach, J.Chromatogr., 1981, 200, 254.
28 E.R.Schwartz, J.Stevens, and D.E.Schmidt, Anal.Biochem., 1981, 112, 170.
29 M.E.Himmel and P.G.Squire, J.Chromatogr., 1981, 210, 443.
30 D.H.Calam and J.Davidson, J.Chromatogr., 1981, 218, 581.
31 S.J.Turco, Anal.Biochem., 1981, 118, 278.
32 N.W.H.Cheetham, P.Sirimanne, and W.R.Day, J.Chromatogr., 1981, 207, 439.
33 J.U.Baenziger and M.Natowicz, Anal.Biochem., 1981, 112, 357.
34 M.Le Bergh, P.Koppen, and D.H.van den Eijnden, Carbohydr.Res., 1981, 94, 225.
35 A.Boersma, G.Lamblin, P.Degand, and P.Roussel, Carbohydr.Res., 1981, 94, C7. C7.
36 S.J.Mellis and J.U.Baenziger, Anal.Biochem., 1981, 114, 276.
37 G.J.L.Lee and H.Tieckelmann, J.Chromatogr., 1981, 222, 23.
38 H.Iwase, T.Morinaga, Y.T.Li, and S.C.Li, Anal.Biochem., 1981, 113, 93.
39 W.F.Alpenfels, Anal.Biochem., 1981, 114, 153.
40 K.Yoshida, K.Kotsubo, and H.Shigematsu, J.Chromatogr., 1981, 208, 104.
41 H.K.B.Silver, K.A.Krim, M.J.Gray, and F.A.Salinas, J.Chromatogr., 1981, 224, 381.

42 N.W.H.Cheetham and P.Sirimanne, J.Chromatogr., 1981, 208, 100.
43 D.L.Hendrix, R.E.Lee, J.G.Baust, and H.James, J.Chromatogr., 1981, 210, 45.
44 M.C.Kemp, W.L.Hollaway, R.L.Prestidge, J.C.Bennett, and R.W.Compans, J.Liquid Chromatogr., 1981, 4, 587.
45 G.J.L.Lee, D.W.Liu, J.W.Pav, and H.Tieckelmann, J.Chromatogr., 1981, 212, 65.
46 M.T.Yang, L.P.Milligan, and G.W.Mathison, J.Chromatogr., 1981, 209, 316.
47 G.B.Wells, S.J.Turco, B.A.Hanson, and R.L.Lester, Anal.Biochem., 1981, 110, 397.
48 S.Honda, M.Takahashi, K.Kakehi, and S.Ganno, Anal.Biochem., 1981, 113, 130.
49 S.Hase, T.Ikenaka, and Y.Matsushima, J.Biochem.(Tokyo), 1981, 90, 407.
50 J.C.Kuo and E.S.Yeung, J.Chromatogr., 1981, 223, 321.
51 S.I.M.Johncock and P.J.Wagstaffe, Analyst, 1980, 105, 581.
52 S.Honda, M.Takahasi, Y.Nishimura, K.Kakehi, and S.Ganno, Anal.Biochem., 1981, 118, 162.
53 S.Honda, Y.Nishimura, H.Chiba, and K.Kakehi, Anal.Chim.Acta, 1981, 131, 293.
54 J.F.Poduslo, Anal.Biochem., 1981, 114, 131.
55 W.F.Glass, R.C.Briggs, and L.S.Hnilica, Anal.Biochem., 1981, 115, 219.
56 R.D.Poretz and G.Pieczenik, Anal.Biochem., 1981, 115, 170.
57 I.Miyamoto and S.Nagase, Anal.Biochem., 1981, 115, 308.
58 E.Weiland, W.Thorn, and F.Bläker, J.Chromatogr., 1981, 214, 156.
59 M.I.Horowitz, in 'Carbohydrates: Chemistry and Biochemistry', 2nd Edition, ed. W.Pigman and D.Horton, Academic Press, New York, 1980, Vol.1B, p.1445.
60 M.Porro, S.Viti, G.Antoni, and P.Neri, Anal.Biochem., 1981, 118, 301.
61 H.S.Isbell and H.L.Frush, Carbohydr.Res., 1981, 92, 131.
62 I.M.Morrison and R.E.Brice, Carbohydr.Res., 1981, 98, 237.
63 M.J.Koziol, Anal.Chim.Acta, 1981, 128, 195.
64 D.P.Kotler, A.R.Tierney, and N.S.Rosenweig, Anal.Biochem., 1981, 110, 393.
65 H.Misuma, H.Wada, M.Kawakami, H.Ninomiya, and T.Shohmori, Clin.Chim.Acta, 1981, 111, 27.
66 Y.Fujimura, S.Ishii, M.Kawamura, and H.Naruse, Anal.Biochem., 1981, 117, 187.
67 C.Hatanaka and Y.Kobara, Agric.Biol.Chem., 1980, 44, 2943.
68 A.Ohkubo, S.Kamei, M.Yamanaka, F.Arai, M.Kitajima, and A.Kondo, Clin.Chem., 1981, 27, 1287.
69 N.Haugaard, J.Cutler, and M.R.Ruggieri, Anal.Biochem., 1981, 116, 341.
70 R.Taylor and C.Pennock, Clin.Chem., 1981, 27, 1624.
71 S.O.Enfors, Enzyme Microb.Technol., 1981, 3, 29.
72 F.Heinz and T.W.Beushausen, J.Clin.Chem.Clin.Biochem., 1981, 19, 977.
73 H.Schlebusch, M.Sorger, E.Munz, A.C.Kessler, and W.Zwez, J.Clin.Chem.Clin.Biochem., 1980, 18, 885.
74 A.Mosca, G.Gossi, M.Luzzana, L.R.Bernardi, W.S.Friauf, R.L.Berger, H.P.Hopkins, and V.Carey, Anal.Biochem., 1981, 112, 287.
75 M.Sugiura, S.Hayakawa, Y.Ito, and K.Hirano, Chem.Pharm.Bull., 1981, 29, 146
76 E.Bissè, A.Scholer, and D.J.von der Schmitt, J.App.Biochem., 1981, 3, 176.
77 M.A.Cohenford, J.C.Urbanowski, and J.A. Dain, Anal.Biochem., 1981, 112, 76.
78 D.S.Grove and G.S.Serif, Anal.Biochem., 1981, 111, 122.
79 J.P.Straub, Clin.Chem., 1981, 27, 198.
80 R.Kattermann and R.Krieger, J.Clin.Chem.Clin.Biochem., 1981, 19, 31.
81 J.Roboz, M.Suttajit, and J.G.Bekesi, Anal.Biochem., 1981, 110, 380.
82 K.Sugahara, K.Sugimoto, O.Nomura, and T.Usui, Clin.Chim.Acta, 1981, 108, 493.
83 B.Rivetz, M.A.Lipkind, E.Shichmanter, and E.Bogin, Experientia, 1980, 36, 370.
84 A.C.Chen and R.T.Mayer, J.Chromatogr., 1981, 207, 445.
85 D.C.Farwell and A.S.Dion, Anal.Biochem., 1981, 113, 423.
86 C.H.Blomquist, B.D.Snyder, and W.G.Neihaus, J.Clin.Chem.Clin.Biochem., 1981, 19, 139.
87 P.A.J.Gorin, Adv.Carbohydr.Chem.Biochem., 1981, 38, 13.
88 L.D.Hall and M.Yalpani, Carbohydr.Res., 1981, 91, Cl.
89 R.C.Beier, B.P.Mundy, and G.A.Strobel, Can.J.Chem., 1981, 58, 2800.
90 E.Berman and A.Allerhand, J.Biol.Chem., 1981, 256, 6657.
91 H.van Halbeek, L.Dorland, G.A.Veldink, J.F.G.Vliegenthart, J.C.Michalski, J.Montreuil, G.Strecker, and W.E.Hull, FEBS Lett., 1980, 121, 65.
92 R.MacRae and J.Dick, J.Chromatogr., 1981, 210, 138.

93 M.E.Breimer, G.C.Hansson, K.-A.Karlsson, H.Leifler, W.Pimlott, and B.E.Samuelsson, FEBS Lett., 1981, 124, 299.
94 V.Kováčik, P.Kováč, and A.Lipták, Carbohydr.Res., 1981, 98, 242.
95 S.Honda, M.Nagata, and K.Kakehi, J.Chromatogr., 1981, 209, 299.
96 C.Bruvier, Y.Leroy, J.Montreuil, B.Fournet, and J.P.Kamerling, J.Chromatogr., 1981, 210, 487.
97 R.A.Laine, Anal.Biochem., 1981, 116, 383.
98 B.Fournet, G.Strecker, Y.Leroy, and J.Montreuil, Anal.Biochem., 1981, 116, 489.
99 M.Linscheid, J.Dangona, A.L.Burlingame, A.Dell, and C.E.Ballou, Proc.Natl.Acad.Sci.USA, 1981, 78, 1471.
100 J.Ashraf, D.A.Butterfield, J.Järnefelt, and R.A.Laine, J.Lipid Res., 1980, 21, 1137.
101 S.R.Sarfati, Carbohydr. Res., 1981, 94, 236.
102 L.R.Phillips and B.A.Fraser, Carbohydr.Res., 1981, 90, 149.
103 S.Inoue and G.Matsumura, FEBS Lett., 1980, 12, 33.
104 S.Inoue and M.Iwasaki, Biochem.Biophys.Res.Comm., 1980, 93, 162.
105 K.Shimizu, Carbohydr.Res., 1981, 92, 65.
106 G.Strecker, A.Pierce-Cretel, B.Fournet, G.Spik, and J.Montreuil, Anal.Biochem., 1981, 111, 17.
107 N.Shibuya, J.Chromatogr., 1981, 208, 96.
108 H.Rauvala, J.Finne, T.Krusius, J.Karkkainen, and J.Jarnefelt, Adv.Carbohyr.Chem.Biochem., 1981, 38, 389.
109 K.M.Aalmo and T.J. Painter, Carbohydr.Res., 1981, 89, 73.
110 T.Krusius and J.Finne, Carbohydr.Res., 1981, 90, 203.
111 K.Veluraja and V.S.R.Rao, Biochim.Biophys.Acta, 1980, 630, 442.
112 H.Ogura and K.Furuhata, Tetrahedron Lett., 1981, 22, 4265.
113 R.H.Marchessault and Y.Deslandes, Carbohydr.Polymers, 1981, 1, 31.
114 G.A.Jeffrey and M.Sundaralingam, Adv.Carbohydr.Chem.Biochem., 1981, 38, 417.
115 S.Hase, T.Ikenaka, and Y.Matsushima, J.Biochem.(Tokyo), 1981, 90, 1275.
116 P.J.Wood, Carbohydr.Res., 1981, 94, C19.
117 T.A.Beyer, J.E.Sadler, J.I.Rearick, J.C.Paulson, and R.L.Hill, Adv.Enzymol.Related Areas Mol.Biol., 1981, 52, 23.

3
Plant and Algal Polysaccharides

BY I. M. MORRISON

1 Introduction

A volume of the comprehensive treatise on the biochemistry of plants has been devoted to the structure and function of carbohydrates.[1] The proceedings of a symposium on the mechanism of saccharide polymerization and depolymerization have been published with several articles relevant to this chapter.[2] The state of macromolecular chemistry at the present time, with particular reference to the pulp and paper industry, has been reviewed.[3]

2 Starch

The structure of the hydrated amylose-iodine complex has been determined by combined methods of X-ray diffraction and stereochemical packing analysis and shown to be an orthorhombic unit cell with dimensions of a = 13.60, b = 23.42, and c (the fibre repeat) = 8.17Å.[4] Two amylose chains pass through the unit cell and iodide ions were present as an almost linear chain in the centre of the six-fold, left-handed amylose helix. Eight molecules of hydration per unit cell were located in good hydrogen-bonding positions between the amylose helices. The crystal structure of V_h amylose has been refined by similar methods to show another orthorhombic unit cell of dimensions a = 13.65, b = 23.70, and c = 8.05Å.[5] The chain conformation is a left-handed, six-fold helix with O-6 in the position *gauche* to both O-5 and C-4. The water molecules are located inside the helical channel of the amylose and in the interstitial spaces, forming an intensive hydrogen-bonded network.

Starch has been degraded by plasma and thermolysis in the ionization chamber of a mass spectrometer, and the sugars and oligosaccharides formed were characterized by direct chemical ionization mass spectrometry.[6] The technique, especially with negative ions, extends mass spectrometry to polymers.

The molecular-size distribution of modified corn starch has

been determined by size—exclusion chromatography.[7] The high performance liquid chromatography of malto-oligosaccharides has been reported.[8]

The degree of mechanical damage imparted to starch granules during flour milling has been correlated with various absorbancies in the near—infrared reflectance spectrum.[9] The wavelengths of importance corresponded to overtones and combinations of vibration frequencies due to free and hydrogen-bonded hydroxy groups in the starch.

A resonance Raman spectroscopic study has been made of the blue colouring of iodine/iodide in amylose.[10] Similar spectra were obtained regardless of the KI and I_2 concentrations, degree of polymerization, and excitation wavelengths, but the relative intensities of the lines changed. It was concluded that the basic unit changed from I_6^{2-} to I_{10}^{2-} through I_8^{2-} with decreasing KI concentration. A simple straight-line relationship has been obtained between average chain length of linear amyloglucans and extent of iodine staining.[11] The wavelength of maximal optical absorbance, the maximal absorbance, and the iodine-binding capacity per chain are all linearly related to the chain length.

The sorption of water and water-soluble alcohols by starch granules in aqueous suspension has been studied, and potato and corn starches absorb 33 and 28% water, respectively, at pH 7.[12] The method was not suitable for cationic starches. Alcohols were also absorbed in large amounts. The kinetics of adsorption of 2,3-dialdehydostarch can be expressed by an exponential equation which is applicable for energetically uniform heterogeneous cellulose surfaces.[13] The activation energy did not depend on the initial concentration of starch in solution but increased linearly with the increased amount of starch adsorbed.

Some physicochemical properties of heat-moisture treatment of starches have been determined.[14] The water-binding capacity and enzyme susceptibility increased, while the swelling power decreased. The starches with the highest moisture content before heating had the lowest swelling powers. The same treatments had adverse effects on the functional properties and baking potential of the starches.[15]

An enzymatic and an acidic hydrolysis procedure have been compared for the analysis of starch in feeds and digesta.[16] Comparable results were obtained for purified starches and feeds with high starch contents. Only in treated materials like digesta with more than 17% cellulose did problems arise due to hydrolysis of cellulose by the acidic treatment. The enzymes do not need to be highly purified to be used in starch determination even in the presence of cellulose and hemicelluloses.[17]

A series of n.m.r. spectra of whole seeds of various types have been obtained by using cross-polarization magic-angle spinning techniques.[18] Selected signals provided a means of comparing the protein content relative to the starch content within a group of seed varieties.

The multi-branched nature of amylose samples from several plant sources has been revealed by methods for estimating reducing and non-reducing residues.[19] The isoamylase of a *Pseudomonas* partially split the branch linkage in potato amylose. The concurrent action of pullulanase from *Aerobacter* and β-amylase from sweet potato hydrolysed the amylose completely.

An enzymatic method of determining the A and B chains in amylopectin has led to a ratio of 1:1, not 2:1 as previously suggested.[20] Partial debranching with pullulanase gave results consistent with earlier suggestions that A chains are predominantly and selectively removed by this enzyme.

The changes in physicochemical properties of starches isolated from maize, barley, and triticale grains which had been allowed to sprout for various lengths of time have been determined.[21] The water-binding capacity initially decreased but then increased with increasing time whereas swelling power decreased and solubilities and enzyme susceptibilities both increased with time.

A procedure adapted for the rapid analysis of total starch and amylose has been used on a series of 37 barley genotypes.[22] The amylose content was not related to either starch content or grain yield.

Investigations of the starch granules of wheat throughout their development have suggested that there are two distinct populations of granules which are initiated at separate stages of kernel development.[23] This theory was supported by results using a Coulter counter as well as light and electron microscopy and chemical analysis.

In studies on the characterization of legume starches, the fine structures of amylose and amylopectin have been determined by the use of pullulanase, β-amylase, and glucoamylase.[24] The results indicated that there were some 1→6 substituted α-D-glucopyranosyl residues in amylose. Similar studies were carried out on acid-treated starches.[25] A fast hydrolysis of amorphous and gel phases occurred with a slow hydrolysis of crystalline regions. The results were consistent with the structure being a cluster type.

In attempts to separate the starch and protein fractions of legumes, the starch fraction was found to contain between 7.7 and 20.1% protein while the protein fraction contained 0.0-4.6% starch.[26]

During germination of dry beans at 22°C, the total starch, amylose, and amylopectin contents all declined over the first 6-day period when the amylose:amylopectin ratio also declined.[27]

The solubility and swelling characteristics of the starch from the Great Northern bean have been found to be both temperature and pH dependent.[28] The solubility was highest at pH 6.0. Oxidation increased the solubility as a function of temperature whereas acetylation decreased the solubility. Acetylation and oxidation both reduced the swelling while addition of free fatty acids reduced the viscosity and raised the gelatinization temperature. The gelatinization temperature of the starch from the winged bean was 60-70°C and it exhibited single-stage swelling and low solubility.[29] It was very soluble in DMSO and had an amylose content of 38.5%. The gelatinized starch was more susceptible to α-amylase and glucoamylase than the native starch.

The morphological and physicochemical properties of starch from mature green and ripe bananas have been determined.[30] The granules were irregularly shaped with spheroid and elongated forms predominating. Other parameters were similar to those of the starch from mung beans. The starch from mature, unripe breadfruit has been isolated and characterized by the use of pullulanase and β-amylase followed by gel-permeation chromatography.[31] The granules were 10-20 μm in size with an amylose content of 18.2% and β-amylolysis limit of 58%. The debranched amylopectin had chains of 15 and 38-45 residues, respectively, in the ratio 4.4:1.

The relationships between contents of dry matter, starch, total fermentables, and specific gravity in cassava storage roots have been determined.[32] A curvilinear relationship of the form $\underline{y}^{-1} = \underline{a} + \underline{b}$ was found to exist between dry matter and starch content and a similar one for specific gravity and starch.

The native starch from fermented cocoa beans has been isolated by solvent extraction and flocculation of non-starch material at pH 2.7.[33] The egg-shaped granules had a mean size of 4.4 μm, an amylose content of 30.4%, a gelatinization temperature of 52.5—68°C and a limiting viscosity number of 166.2 ml g^{-1} in 1M KOH.

Kuzu starch contains 21% amylose and 124 mg kg^{-1} phosphorus.[34] The β-amylolysis limits of the amylose and amylopectin were 76 and 57, respectively, with 20.5 being the average chain length of the amylopectin. The granules were of the Ca crystalline type.

A non-aqueous procedure, using glycerol and 3-chloro-1,2-propanediol, has been developed for the isolation from maize of starch granules with associated metabolites.[35] Electron microscopy failed to demonstrate the presence of the amyloplast's membrane but it may be dried onto the granule surface. Soluble metabolites were found which may derive from the stroma of amyloplasts.

The distribution of starch components of maize (Zea mays) and their properties have been investigated by gel filtration after debranching and by exhaustive hydrolysis of the granules with glucoamylase.[36] Based on the results, the relationship between genotypes and properties of starch components was discussed. For example, introduction of the amylose-extender (ae) gene produced amylopectins of longer average chain lengths. The starch and phytoglycogen fractions from normal and sugary (su) maize have been compared, and negligible differences were observed between 2 inbred lines and 2 su conversions of the lines studied.[37] The phytoglycogen was shown to be distributed in a soluble fraction and a particulate fraction which also contained some amylose. These observations are consistent with a pattern of conversion of starch into phytoglycogen in su endosperm.

The starch of wild rice (Zizania aquatica) has been isolated and is a very small granule (2-7 μm) of angular and polygonal shape.[38] It has an A-type pattern on X-ray diffraction but an amylose content of only 2%. Other properties were compared with those of wheat starch.

Normal- and waxy-type starches have been isolated from two types of Amaranthus hypochondriacus and found to be round granules (1 μm diameter) with similar X-ray diffraction diagrams to other A-types.[39] The amylose contents were 14 and 0%, respectively, while the amylopectins were similar to those from rice and maize starches.

The elephant yam (Amorphophallus campanulatus) has yielded a starch which exhibited single-stage swelling and moderate solubility in water but low resistance to solubilization in

DMSO.[40] The amylose content was 24.5%. The amylolytic susceptibility of the native starch was low but that of the gelatinized starch was high.

Oxidation of malto-oligosaccharides has yielded the series of oligomers with aldonolactones at the reducing end position and these give diamides with alkylenediamines in 30-75% yield.[41] The effect of these derivatives on the enzymatic activity of potato phosphorylase was determined.

The digestion of hydroxypropyl wheat starches by porcine pancreatic α-amylase has been examined.[42] The digestibility of the gelatinized starches decreased with increasing degree of substitution, and the action pattern favoured a non-repetitive attack whereas the digestibility of the native starches increased with extent of modification. The hydrolysates from the above digestion of starches with degrees of substitution from 0-0.17 were separated into total digest, oligo- and polysaccharide fractions and further analysed.[43] Increasing the degree of substitution caused a reduction in the reducing value, an increase in the average degree of polymerization, and a greater proportion of oligosaccharides in the hydrolysate. The oligosaccharide fraction had double the hydroxypropyl content of the undigested starch. The oligosaccharide fraction was further analysed by h.p.l.c. on a μ-Bondapak^{R} carbohydrate column.[44] The proportion of oligosaccharides with three and four D-glucose residues increased and those with five D-glucose residues decreased as the degree of substitution increased.

The changes in molecular weight of amylose and amylopectin during thermal dispersion in water have been followed by paper chromatography.[45] Grain disaggregation and molecular disintegration of amylopectin proceeded more rapidly than in amylose. The effect of sodium hydroxide was more marked on amylose than amylopectin.

Aqueous suspensions of starch at neutral pH values were heated by microwave energy in sealed tubes.[46] The colour of the hydrolysates darkened with increasing time and pressure. The total acidity increased and oligosaccharides in the range monomer to octamer were determined.

The amounts of glyceraldehyde, dihydroxyacetone, and 2-hydroxy-1,3-propanedial induced in maize starch by γ-radiation have been determined by a g.c. procedure, and the influence of several parameters such as dose and temperature has been investigated.[47]

This is the first report of 2-hydroxy-1,3-propanedial from irradiated starch.

The influence of the polysaccharide structure on the oxidation process with Ce^{4+}, Mn^{3+}, and V^{5+} ions has been studied, and it was confirmed that the rates were highest for compounds containing α-glycol groups.[48] The nature of the functional groups and the conformation of the macromolecular unit exert an influence on the oxidation rate.

An enzymatic procedure has been developed to measure starch in cereal products.[49] The precision of the test is ± 1.5% and is applicable to a wide range of products.

The amylases from pearl millet have an action pattern similar to those of other cereal α-amylases but are able to hydrolyse wheat starch more readily than millet starch.[50]

The immature wheat endosperm aleurone with seed coat and embryo detached has produced considerably less α-amylase activity than immature whole or de-embryonated wheat kernels.[51] The isoenzyme composition of the incubated endosperm aleurone was unique suggesting that the seed coat may contain factors required for normal α-amylase isoenzyme synthesis.

An enzymatic assay for the determination of α-amylase in serum has been developed which employs maltoheptaose as substrate and a coupled enzymatic indicator system.[52] H.p.l.c. was used to establish the action pattern of hydrolysis of the maltoheptaose. The stoichiometric equations for the action of α-amylase alone and α-amylase and α-glucosidase together were derived.

The multiple—attack model for the kinetics of hog pancreatic α-amylase action has been analysed.[53] The model explains the dependence of $\underline{V}$ and $\underline{Km}$ on the degree of polymerization of substrates. The distribution in the degree of polymerization of the hydrolysis products has been used to calculate the number of unitary movements of the enzyme during the single enzyme—substrate meeting.

A mixture of β-amylase and pullulanase from *Bacillus cereus* has been used to produce maltose from a variety of starches which had previously been liquefied by the α-amylase of *Bacillus subtilis*.[54] The maximum yield was 87-88% obtained at pH 6.0-8.0, 55°C, and 30h hydrolysis of a 5% solution. Similar results were obtained if acidic conditions were used for the liquefaction.

Two inbred lines of rye (*Secale cereale*), whose kernels had only 1-3% of the β-amylase activity of normal lines, have been investigated using an anti-wheat β-amylase immune serum which cross-reacted with the rye enzyme.[55] The results indicated that the reduced activity was due neither to the presence of an inhibitor nor to the production of inactive enzymes.

An inducible cell-bound glucoamylase, which degrades cyclodextrins, has been isolated from a Flavobacterium species.[56] The final degradation product in all cases was D-glucose with small amounts of maltose. Amylopectin and glycogen were very poor substrates.

The enzymes of starch metabolism in the developing grains of a high-lysine barley mutant have been determined.[57] The activity of ADP-D-glucose:starch synthetase was lower while those of UDP-D-glucose-pyrophosphorylase and starch phosphorylase were higher than in normal barley.

The structural changes in starch molecules during malting of barley have been investigated by comparing a normal malt where 14% of the starch had been lost with an over-modified malt where 68% of the starch had been lost.[58] In the normal malt, 14% of the large granules were eroded but the small granules remained intact whereas 38% of the large granules were eroded in the over-modified malt with a marked reduction in the small granules. Malting resulted in an apparent increase in the amylose content. There was no apparent difference in the amylopectin structure between the two malts.

A reduction in light intensity during starch synthesis in developing wheat grains has reduced the starch synthetase I activity but not the starch phosphorylase activity.[59] The low starch production was not due to an insufficient supply of assimilates.

[^{35}S]-Methionine has been incorporated into β-amylase from germinating rice seeds.[60] The specific radioactivity of the enzyme derived from scutellum was significantly greater than that from the endosperm. These results indicate that, at the onset of germination, β-amylase is synthesized in the scutellum and only at a later stage is an inactive, latent form of the enzyme activated in the endosperm. This latter form becomes dominant during the later stages of germination.

The role of the recessive amylose-extender (ae) allele in the biosynthesis of maize starch has been investigated.[61] Multiple forms of starch synthetase were observed, some of them appearing at the same time as the rapid increase in amylose content. The presence of modified amylopectin synthesized by the enzymes which were affected by the ae allele was suggested. Debranching of these amylopectins followed by analysis of the products was attempted to confirm this theory, but the results were inconclusive.[62]

Starch granule-bound D-glucan synthetase has been isolated from cotton leaves.[63] On further separation, amylopectin was the only D-glucan glycosidically linked with protein and this had D-glucan synthetase activity. The amylose component was free from covalently bound protein.

On germination, both in the dark and light, a DMSO-soluble D-glucan is synthesized in the cotyledons of lupins.[64] Chemical analysis indicated that it was starch-like but its chromatographic behaviour differed from that of amylopectin.

The concentrations of starch synthetase and debranching enzyme from developing seeds of Pisum sativum have been measured.[65] Primed starch synthetase activity increased from day 8 to day 14 after anthesis but then decreased by 50% at 26 days, whereas citrate-stimulated starch synthetase activity was highest at 10 days after anthesis thereafter falling to a low value. Debranching activity increased from day 8 to day 18 after anthesis. The changes in starch synthetase and debranching activity resulted from changes in the concentration of a few enzyme forms and not from the appearance of different enzyme forms.

The β-amylase activities of some soybean varieties are 10^{-2} times lower than others yet they have similar starch metabolism.[66] It was therefore concluded that starch metabolism was independent of the β-amylase activity.

The starch contents of plantains and bananas, ripened under similar conditions, have been determined with a content of 1% in ripe bananas and 9% in ripe plantains.[67] These values fell to zero and 3%, respectively, in over-ripe material.

Phosphorylase II from potato tubers has been purified and characterized.[68] No separation of primed from unprimed activity was achieved. The molecular weight was 9.6×10^4 and the native enzyme was a dimer. The reaction product was a protein-bound D-glucan which possessed long (1→4)-α-D-glucan chains.[69] Since protein was also present in amylose from potato starch grains, it is suggested that these findings may be further evidence of a common biosynthetic pathway for (1→4)-α-D-glucans from a precursor protein.

The rate of starch breakdown in isolated chloroplasts from

spinach is similar to that found in whole leaves.[70] These rates have been measured both in the dark and in light and compared with the rates of starch synthesis in the light.[71] Illumination has little inhibitory effect on breakdown at high phosphate levels but 67% inhibition was observed at low phosphate levels. Since CO_2 evolution is prevented by illumination, the oxidative pentose phosphate pathway must have been inhibited. The balance between accumulation and breakdown is very sensitive to the rate at which carbon is withdrawn from the chloroplast in exchange for inorganic phosphate.

The unusually high solution torque developed by the starch from the Hinoat variety of oats when a hot paste is cooled to 75-80°C has been abolished in the presence of acetic acid at pH 3.0.[72] Salts at 0.1M concentration reduced the torque value as did increasing sucrose concentrations.

The gelatinization of wheat starch as modified by xanthan, guar and cellulose gums has been investigated and each gum hastens the onset of initial paste viscosity.[73] The reaction seemed to be a strong association with the amylose component since the β-amylolysis limit of the amylose is reduced. The starch-xanthan gum exhibited a pseudoplastic behaviour whereas the starch-guar gum resisted shearing at low shear rates. The syneresis of curdlan gel has been repressed by the addition of starch before heating but not by other sugars.[74] The role of starch in this process was discussed.

A number of starch esters of herbicides have been prepared from activated pregelatinized starch and evaluated as slow release agents.[75] With sterile soil a picloram ester was phytotoxic for 48 days. Unmodified starch esters were hydrolysed at a slower rate than gelatinized starch esters.

A *Flavobacterium* from tap water was able to grow on tap water but the addition of starch greatly enhanced its growth.[76]

X-Ray and neutron-diffraction studies on hexa-amylose hexahydrate crystal forms A and B have given an indication of the chain-like and circular arrangement of hydrogen bonds within the molecules.[77] In the circles, homodromic and antidromic orientations of the hydrogen bonds are observed, with the homodromic circles being more stable by 8% per hydrogen bond. The changes in electronic charges on H and O atoms are greater in homodromic circles and the dipole moments are only *ca.* 3 D in the homodromic circles but 6-8 D in chain-like and antidromic arrangements.

X-Ray analysis of the hexa-amylose-3-nitroaniline (1:1) hexahydrate complex has shown that the hexa-amylose molecules are stacked along the a-axis in a head-to-tail fashion to form a channel-type structure.[78] A stereo-drawing of the packing feature of the complex with its associated water molecules was shown. Similar studies have been conducted on the hexa-amylose-benzaldehyde (1:1) hexahydrate complex and a similar packing was observed with the hexa-amylose molecule being tilted at 11.5° against the channel axis.[79] The benzene ring is inserted into the hexa-amylose ring from the secondary hydroxy side while the carbonyl group is in Van der Waals contact with the primary hydroxy side of the next hexa-amylose molecule.

A correlation between the penetration depth of the C atoms of a variety of phenols in the cavity of hexa-amylose and the changes in their ^{13}C n.m.r. chemical shifts has been investigated.[80] The formation and molecular dynamics of cycloamylose inclusion complexes with phenylalanine have been investigated by ^{13}C n.m.r. spectroscopy.[81] The influence of the cavity size was determined and the cyclohepta-amylose-phenylalanine system was most strongly coupled. The cyclohexa-amylose coupling was shallow and loose while that of the octa-amylose was deep and loose.

The induced c.d. and u.v. spectra of hepta-amylose-disubstituted benzene systems has been studied with respect to their molecular structures.[82] The dipole moment was correlated with rotational intensity and the energy-level transition calculated by molecular orbital methods.

The use of a hexa-amylose mobile phase in the separation of mono-substituted phenols by t.l.c. has been reported.[83] The R_f values depended on both the structural features of the phenolic compounds and the concentration of hexa-amylose in the mobile phase.

A selective attack of dichlorocarbene at the 4-position of phenolate ion was achieved with hexa-amylose as catalyst.[84] The mechanism was studied by ^{1}H and ^{13}C n.m.r. spectroscopy.

3 Fructans

The crystalline conformation of inulin [(2→1)-β-D-fructofuranan] has been determined by conformational analysis coupled with X-ray and electron-diffraction data.[85] The only two models possible correspond to 5-fold helices, one left-handed and the other right-handed. Steric interactions of the substituents and the exanomeric effect were believed to be responsible for the observed conformational differences.

An invertase and an inulase from germinating garlic (Allium sativum) bulbs have been separated and purified.[86] The invertase had no effect on inulin whereas the inulase, as well as hydrolysing inulin, could hydrolyse sucrose and raffinose. The enzymes had similar molecular weights (7.6×10^4). In the early stages of development the activities increased in parallel, but at later stages the invertase activity was higher.

A pathway for fructan trisaccharide synthesis from sucrose has been reported.[87]

Two non-reducing hexasaccharides (1) and (2) have been isolated from the roots of Asparagus officinalis and their structures confirmed by the usual methods.[88] A novel tetrasaccharide (3) has been isolated from Chinese milk vetch honey.[89] It is probably synthesised via erlose which is produced from sucrose in nectar.

4 Cellulose

A history of the study of cellulose structure has been published which also considered how this knowledge has been extended to synthetic polymers.[90]

A hydrate of cellulose II has been prepared and its crystal structure determined.[91] The unit cell contains disaccharide sections of two chains, and an antiparallel arrangement of adjacent chains was assumed. The chains are stacked in the same way as in cellulose II with the water molecules between the stacks, but it was not possible to position the water molecules in a regular arrangement. The treatment of cotton and rayon with ethylenediamine resulted in the transformation of cellulose IV to cellulose III as determined by X-ray analysis.[92] Thus ethylenediamine-methanol can affect lattice modifications.

The transformation of cellulose I into cellulose III in Valonia macrophysa, preformed in ethylenediamine, has maintained the external appearance of the microfibrils as shown by electron microscopy.[93] This conversion must have involved a fracturing and fibrillation of the initial crystals.

^{1}H n.m.r. spectroscopy has shown that both ramie and cotton have a unique phase structure characteristic of native

β-D-Fru*f*-(2→1)-β-D-Fru*f*-(2→1)-β-D-Fru*f*-(2→6)-α-D-Glc*p*-(1→2)-β-D-Fru*f*
1
↑
2
β-D-Fru*f*

(1)

β-D-Fru*f*-(2→1)-β-D-Fru*f*-(2→6)-α-D-Glc*p*-(1→2)-β-D-Fru*f*
1
↑
2
β-D-Fru*f*-(2→1)-β-D-Fru*f*

(2)

α-D-Glc*p*-(1→6)-α-D-Glc*p*-(1→4)-α-D-Glc*p*-(1→2)-β-D-Fru*f*

(3)

cellulose.[94] The non-crystalline component did not exhibit micro-Brownian segmental motion even in the swollen state. The phase structure was distinctly different from that of regenerated cellulose. An n.m.r. spectroscopic study of the interactions between cadoxen and cellulose has been carried out using signals from ^{1}H, ^{13}C, and ^{113}Cd.[95] A well recorded ^{13}C spectrum was obtained which could be assigned by comparison with the spectra of simpler sugars. The ^{113}Cd and ^{13}C results were not consistent with the formation of chelate-alcoholate complexes with the C-2 and C-3 hydroxy groups. Hydrogen bonding is probably the dominant interaction. It is further suggested that the metal ion serves the dual purpose of holding two amino groups in a favourable orientation to hydrogen bond with a pair of equatorial hydroxy groups on the cellulose.

The solubilization of cellulose and other plant structural carbohydrates has been achieved in a mixture of 4-methylmorpholine N-oxide and DMSO.[96] The separation and purification of homogeneous polysaccharide fractions could then be made.

Hydroxymethylcellulose has been prepared and characterized by i.r. and n.m.r. spectroscopy.[97] To achieve dissolution of cellulose, a high degree of molecular substitution was required, but once solution was achieved the molecular substitution could be lowered, the actual value of precipitation depending on the solvent.

A rapid separation of cello-oligosaccharides (degree of polymerization 2-9) has been obtained on a cation-exchange resin in 60 min with water as eluant.[98]

The cellulose content of raw and cooked vegetables has been determined by extraction procedures using detergent solutions.[99]

The u.v. spectra of extracts of hydrocellulose with the hydroxides of Li^{+}, Na^{+}, K^{+}, Ca^{2+}, Sr^{2+}, and Ba^{2+} were identical.[100] The curves of yellowing absorption coefficients against loss in weight of hydrocellulose in boiling solutions of these alkalis were linear, the slopes giving apparent molar absorption coefficients in the order $Na^{+} = K^{+} > Li^{+} > Ba^{2+} > Sr^{2+} > Ca^{2+}$. The divalent cations reduced the yield of chromogen from the 4-deoxy-D-glycero-2,3-hexodiulose released from reducing end groups of the hydrocellulose. In the same systems, greater weight losses and production of acidity were obtained with the alkaline-earth cations than alkali-metal cations when the OH^{-} ion concentration was 0.02M while the reverse was found at 1.0M.[101] The results were explained in terms of a hypothesis that divalent cations exercise their catalytic effects through enhanced ionization of the D-glucose moiety at the reducing terminus of cellulose chains.

The selectivity (viscosity at a given Kappa number) of softwood Kraft pulp has been shown to increase dramatically when soaked with SO_2-H_2O in the presence of diethylenetriaminepentamethylene phosphonic acid before oxygen bleaching in the presence of magnesium sulphate.[102] Addition of manganese(VII) sulphate to soaked pulps led to a severe loss in selectivity while to dry pulps a slight increase was obtained. Addition of the phosphonic acid to manganese(VII)-treated pulps caused a large increase in selectivity. Aqueous sulphur dioxide treatment of cellulose pulp and cotton linters dramatically reduced the degree of polymerization of the cellulose but did not affect its enzymatic digestibility or decrease its crystallinity as previously reported.[103]

The acetyl content of cellulose acetate and a variety of woods has been determined by aminolysis with pyrrolidine followed by g.c. of the resultant 1-acetylpyrrolidine.[104]

The distribution of the sulphate ester groups in the natural cellulose sulphate of Buccinum undatum has been determined by ^{13}C n.m.r.; the results were compared with those from a randomly substituted synthetic cellulose sulphate.[105] The relative simplicity of the natural cellulose sulphate spectrum indicated that this polymer consisted of blocks of disulphated (1→4)-β-D-glucan chains separated by blocks of unsulphated (1→4)-β-D-glucan chains.

A variety of sulphoalkylcelluloses have been synthesized in NaOH solutions at 65°C.[106] Increasing temperature and NaOH concentrations increased the sulphoalkylation as well as the yield of water- and alkali-soluble fractions.

The relative oxyanion formation at hydroxy groups of D-glucosides and cellulose has been reviewed.[107]

Native cellulose and pulps from a variety of hardwoods have been fractionated by the nitric acid method.[108] A regression analysis showed a direct correlation between the age of the tree and the degree of polymerization of the cellulose. The values were a useful indicator of the papermaking potential of the wood.

Cellulose powders obtained by mechanical disintegration have larger average particle sizes and broader size distributions than those obtained by hydrolytic degradation.[109]

Microcrystallization of bleached sulphate pulp from beechwood and reed increased the crystallinity index from 66 to 71% and 56 to 61%, respectively, as determined by X-ray analysis, but decreased the yield from 86 to 75% and 74 to 67%, respectively.[110] The average length of the microcrystalline pulp fibres was 55 μm compared with 220 μm for the control fibres.

By the use of polarizing microscopy, the microfibrils in the spherulites formed in the pellicle of bacterial cellulose have been shown to be radially or tangentially orientated.[111] Some of the results indicated that the initial stage of nucleation may be a function of intermediate lipid-carbohydrate compounds as in the lyotropic liquid crystal of the soap micelle.

When virulent strains of *Agrobacterium tumefaciens* bind to carrot cell walls, fibrils develop which surround the bacteria and anchor them to the cell surface.[112] The fibrils are composed of cellulose and are synthesized by the bacteria since they are produced on dead as well as living carrot cells. Within this entrapped mass, the bacteria are still able to multiply and form large clusters.

On pyrolysis of cellulose under various conditions, 82 compounds have been identified as volatile products.[113] The maximum yield in a nitrogen atmosphere was achieved at 600-650°C while in air a 50-200°C lower temperature was required. The volatile products were qualitatively similar to those found in cigarette smoke. Combined g.c. gas-phase thermal fragmentation has been developed for the separation and identification of pyrolysis products from bark polysaccharides.[114] The compounds were identified by their fragmentation pattern which was useful in demonstrating the thermal stability of the parent compounds.

Various polymer initiating systems, especially redox systems involving cerium(III) and iron(II) salts, had little effect on the thermal properties of cotton cellulose.[115]

The cellulose from *Acetobacter xylinum* has been used for visualizing the action of the cellulase enzyme system from *Trichoderma reesei* by high-resolution electron microscopy.[116] The enzymes initially bind to the cellulose ribbon, but within 10 min the ribbon is split along its longitudinal axis into bundles of microfibrils which are thinned until they are dissolved, which takes about 30 min. Separation of the various enzyme components produced less dramatic changes confirming the synergistic mode of action of these enzymes.

The cellulase from T. viride has been immobilized on Sepharose^R 4B.[117] The immobilized enzyme had a similar Km to the free enzyme, the activity increased up to 50°C and did not decrease after 4 cycles of batch use and recovery. Treatment of finely divided Pinus densiflora with the cellulase from T. viride at 40°C and pH 4.0 resulted in hydrolysis of 80-90% of the carbohydrate present.[118]

A series of mono- and oligo-saccharides were obtained from bagasse and pith by cellulase hydrolysis.[119] The presence of pentoses and their oligosaccharides in hydrolysates of alkali-treated bagasse showed that hemicelluloses had not been completely removed by this treatment.

The action of the cellulase from Penicillium funiculosum on a variety of cellulosic substrates has been investigated by scanning electron microscopy and X-ray diffraction analysis.[120] The differences were interpreted in terms of the α-cellulose content, surface area, degree of crystallinity, and crystallite dimensions of the substrate.

The kinetics of hydrolysis of ball-milled cotton linters by eight cellulase complexes from Trichoderma, Geotrichum, and Aspergillus species have been studied under steady-state and non-steady-state conditions.[121] The steady-state rate of D-glucose formation was proportional to the endo-(1→4)-β-D-glucanase content. A mechanism was proposed in which the enzyme which first attacks the insoluble substrate is this endo-D-glucanase. This contradicts the accepted view for a prehydrolytic C_1 enzyme. The C_1 role is filled by the endo-D-glucanase.

Partial acid hydrolysis has been investigated as a pretreatment to enhance the enzymatic hydrolysis of cellulosic materials.[122] Up to 100% of the potential D-glucose could be obtained, and this was due to removal of hemicellulose, reduction in the degree of polymerization, and changes in the crystal structure of the cellulose.

The purification of a cellulase isoenzyme from kidney bean abscission zones has been carried out by affinity chromatography.[123] Antibodies raised against this cellulase (pI = 9.5) did not react with the cellulase (pI = 4.5) obtained from 2,4-D-treated abscission zones.

Of 60 species of soil fungi examined, only 31 were

cellulolytic.[124] The cellulolytic activity of 25 species of pyrenomycetous fungi has been determined with 18 being able to utilize cellulose.[125]

A Bacteroides species, of human faecal origin, was able to degrade isolated cell walls of peanuts.[126]

When autospores of Oocystis solitaria were treated with Congo Red or Calcofluor White a more diffuse fibrillar material was obtained [which could be chains of (1→4)-β-D-glucan] rather than the discrete microfibrils of untreated cells.[127]

Using intact cotton fibres, it has been shown by pulse-chase experiments that (1→3)-β-D-glucans (callose) have a high rate of turnover and are likely to be intermediates in cellulose biosynthesis at the stage of secondary cell-wall formation.[128] Protoplasts from cultured soybean (Glycine max) cells have been used to study polysaccharide synthesis during the early stages of cell-wall regeneration.[129] Label from D-glucose was rapidly incorporated into cellulose but neither UDP- nor GDP-D-glucose was rapidly incorporated into (1→3)-β-D-glucans.

Calcofluor White ST can increase the rate of D-glucose polymerization into the cellulose of Acetobacter xylinum.[130] At a concentration of >0.1mM, the rate was four times the control but the assembly of crystalline cellulose I microfibrils was disrupted. It is concluded that polymerization and crystallization are cell-directed, coupled processes and that the rate of crystallization determines the rate of polymerization. Coupling must be maintained for biogenesis of cellulose I.

Addition of coumarin ($1\ g\ kg^{-1}$) to Phaseolus vulgaris leaves inhibited the synthesis of cellulose I but stimulated the synthesis of α-cellulose.[131]

A persisting cellulosic wall has been demonstrated by light microscopy in microspore mother cells during microsporogenesis in Allium tuberosum and Cyclamen persicum.[132]

Cellulose metabolism by the flagellate Trichonympha, isolated from a termite, has been studied.[133] The protozoon, in culture, harboured no endosymbiotic bacteria and produced acetate, CO_2, and H_2 from cellulose. The cellulolytic activity is therefore an inherent property.

A new method of determining the extent of binding of an ionic dye to a polyelectrolyte in solution has used the fact that dyes can adsorb to a cellulose dialysis membrane.[134]

The permeability of cellulosic walls to macromolecules may

limit the ability of enzymes to alter the properties of these walls.[135] Since proteins of molecular weight 6 x 10^4 can permeate a substantial portion of the wall, previous results may have underestimated this property.

The fermentation of cellulose by Acetivibrio cellulolyticus, Methanosarcina barkeri, and a Desulphovibrio species has been studied.[136] Only a three-component culture allowed a rapid conversion to CO_2 and CH_4 and utilized 85% of the cellulose present. This value was increased to 90% by increasing the concentrations of M. barkeri.

5 Hemicelluloses

A number of sugar residues were found in hydrolysates of the water-soluble hemicelluloses of rice brans.[137] Although L-arabinose was the major sugar residue, there was no direct evidence for the presence of an L-arabinan. The intracellular site of the synthesis of L-arabinose-containing cell-wall precursors in suspension-cultured tobacco cells has been determined as the Golgi bodies with transport being via vesicles in the low-density fraction.[138]

A D-galactan has been isolated from the leaves of Aloe barbadensis by hot-water extraction and separation from contaminating pectin, D-gluco-D-mannan, and L-arabinan.[139] The (1→4)-linked D-galactan had one branch point at O-6 for every 26 residues, and the molecular weight of the acetylated polysaccharide was 3.7 x 10^4.

The formation of monodispersed polysaccharides on Smith degradation of the acidic L-arabino-D-galactan from rapeseed (Brassica campestris) strongly suggests a molecular structure composed of regularly repeated subunits.[140] The molecular weight of the subunit was 1.2 x 10^4 and the resistance of the polysaccharide to subsequent Smith degradations suggested that the main chain was a (1→3)-D-galactan.

An L-arabino-D-galactan-protein has been released by sonication of a crude cell organelle fraction from hypocotyl tissue of bean seedlings (Phaseolus vulgaris) and suspension - cultured cells.[141] The macromolecule (mol. wt. 1.4 x 10^5) contained 90% carbohydrate and 10% protein with the major sugar residues being D-galactose, L-arabinose, and uronic acids and the major amino-acids present being L-hydroxyproline, L-serine, and L-alanine. The carbohydrate was linked *via* the first two amino-acids and also to L-threonine with the carbohydrate portion of the linkage regions containing residues of L-arabinose, D-galactose, D-glucose, and L-rhamnose.[142] The L-hydroxyproline glycosides were different from those in non-extractable cell-wall protein but have similarities with those in the wall proteins of the algae.

The changes in the L-arabino-D-galactan-proteins of the carrot have been studied during germination and growth.[143]

An extracellular L-arabino-D-galactan-protein has been obtained from suspension-cultured tobacco cells.[144] The complex (mol. wt. 2.24×10^5) contained 5.5% protein and the sugar moiety was a typical L-arabino-(1→3)-D-galactan. A β-D-glucopyranosyluronic acid residue was present at the non-reducing terminus attached to O-6 of a D-galactopyranosyl residue.

Auxin-induced changes in the sugar composition, intrinsic viscosity, and molecular-weight distribution of matrix polysaccharides in the epicotyl cell wall of *Vigna angularis* have been determined.[145] The greatest change was the reduction in L-arabino-D-galactan content, but other changes in pectin and D-xylo-D-glucan content were also observed.

The linkage regions in cell walls of rice coleoptiles have been characterized as involving O^3-D-galactopyranosyl-L-serine and N^4-(2-acetamido-2-deoxy-β-D-glucopyranosyl) hydrogen L-asparaginate.[146] The linkage region in the water-soluble L-arabino-D-galactan-protein complex from wheat endosperm has been characterized as O^4-β-D-galactopyranosyl-L-*trans*-hydroxyproline.[147] Methylation analysis of the glycopeptide linkage region from cell walls of *Chlamydomonas reinhardii* has revealed that O-glycosidically linked D-galactofuranosyl residues are present.[148]

Bacteroides thetaiotaomicron, a polysaccharide degrading bacterium from the human colon, can use larch L-arabino-D-galactan as its carbon source with yields similar to those obtained on D-glucose.[149]

The crystal structure of a regenerated form of (1→3)-α-D-glucan has been determined by X-ray diffraction analysis and stereochemical model refinement as an orthorhombic unit cell.[150] The chain conformation is almost completely extended and is very close to a 2/1 helix even though the dimer residue is the crystallographic repeat unit.

The vacuum c.d. spectrum of (1→6)-β-D-glucan has a single

positive band near 177 nm.[151] As gels formed, a negative band at 190 nm appeared followed by a blue shift in both bands with ageing.

The specificity of the interaction of direct dyes with polysaccharides has been studied by changes in solubility and in the fluorescence and absorption spectra of the dyes.[152] The strongest interactions were shown by polysaccharides with contiguous (1→4)-linked-β-D-glucopyranosyl residues, but (1→3)-β-D-glucans exhibited some complex formation. Dyes differed in their affinity. The direct dye, Calcofluor, has been used in a fluorimetric assay for (1→3)(1→4)-β-D-glucans in beer, wort, malt, and barley.[153]

The β-D-glucan content of developing and germinating kernels of several varieties of barley has been determined.[154] The D-glucan content varied considerably but did not seem to be related to the β-D-glucanase activity of the variety.

A D-glucan is present in the cotyledons of Mirabilis jalapa seeds.[155] Methylation, periodate oxidation, and graded and enzymatic hydrolysis studies have indicated 34 (1→4)- and 3 (1→3)-linkages for every 38 D-glucose residues. The branch point appears at O-2. In some places, at least three (1→3)-linked D-glucopyranosyl residues are in sequence. Both α- and β-glycosidic linkages are present.

The hemicellulose fraction accounted for 47% of the pollen tube wall of Camellia japonica and comprised essentially D-glucose residues.[156] The polysaccharide had a (1→3)-β-D-glucan chain with a degree of polymerization of 21. The pollen tube wall of C. sasanqua, C. sinensis, Tulipa gesneriana, and Lilium longiflorum also consisted mostly of D-glucose residues, and the hemicellulose fractions were essentially pure D-glucans.[157] The deposition of (1→3)-β-D-glucan during megagametogenesis in two species of Oenothera is partly in continuum with the wall and partly in the degenerating cytoplasm.[158]

The addition of β-D-glucanase before the use of α-amylase has been used to aid filtration in the measurement of fibre contents by detergent procedures.[159]

The fact that fungal D-glucans will stimulate soybeans to accumulate phytoalexins has prompted an investigation into the (1→3)-β-D-glucanases and β-D-glucosidases of soybean and their hydrolytic patterns.[160] The β-D-glucosylase I enzyme has been purified. Several lines of evidence suggest that the

(1→3)-β-D-glucanase and β-D-glucosidase activities exhibited by β-D-glucosylase I preparations are catalysed by the same enzyme. Treatment of the D-glucan elicitor from the walls of *Phytophthora megasperma* by D-glucosylase I results in extensive hydrolysis and loss of 94% of its activity.[161] It is unlikely that some other enzyme(s) may be involved in the overall reaction.

The *endo*- and *exo*-β-D-glucanases from *Zea mays* have been partially purified.[162] The *endo*-β-D-glucanase (mol. wt. 2.6 x 10^4) showed a marked preference for substrates with mixed linkages, and it probably initiated the solubilization of wall D-glucan. The *exo*-β-D-glucanase (mol. wt. 6.0 x 10^4) was required for extensive hydrolysis *in vitro*. During coleoptile development, the autolytic activity of the cell wall increased dramatically but did not parallel the growth potential of the tissue.

Conditions which induce a transmembrane electrical potential, positive with respect to the inside of the membrane vesicles, have resulted in a 4-12-fold stimulation of the activity of membrane-associated β-D-glucan synthetases in a preparation from developing cotton (*Gossypium hirsutum*) fibres.[163] The products were mainly (1→3)-β-D-glucans but some increase in (1→4)-β-D-glucan content was also found. No (1→4)-α-D-glucan was formed.

The products of β-D-glucan synthetases have been characterized using specific β-D-glucan hydrolases.[164] The β-D-glucan synthesis by particulate enzymes from suspension-cultured endosperm cell walls from ryegrass has been studied.[165]

The β-D-glucan synthetase activity of the fungus *Saprolegnia monoica* has been assayed and a product is observed which contains both 1→3- and 1→4-linked β-D-glucopyranosyl residues.[166] The presence of Mg^{2+} ions affects the structure of the product. In the absence of Mg^{2+}, only a (1→3)-β-D-glucan is formed, but as the substrate concentration is reduced and Mg^{2+} ion concentration is increased the proportion of 1→3-linked residues in the product is reduced and a pure (1→4)-β-D-glucan is formed.

High molecular weight D-glucans have been isolated from mycelial walls of *Fusarium oxysporum*.[167] The D-glucans contained both 1→3- and 1→6-linked residues in contrast to the D-glucans from *Colletotrichum lindemuthianum*, which contained 1→3- and 1→4-linked residues. However, the two types of D-glucans had very similar effects in eliciting browning and phytoalexin production in green-bean cotyledons.

An *exo*-(1→3)-β-D-glucanase derived from *Sclerotinia libertiana* has induced growth of Avena sativa coleoptiles and degraded the (1→3)(1→4)-β-D-glucans.[168] No endo-(1→4)- or endo-(1→3)-β-D-glucanase activity could be detected. The exo-(1→3)-β-D-glucanase from a Basidiomycete did not induce growth.

The metabolism of a D-glucan from the cactus Opuntia bigelovii has been studied in a desert environment.[169] The D-glucan content increased slightly during a 9-week drought period but then decreased during the irrigation period. The D-glucan is probably storage material.

The D-xylo-D-glucan from the walls of etiolated mung beans has been studied by acidic and enzymatic hydrolyses, the latter by a preparation from Aspergillus oryzae.[170] Cellobiose, cellotriose, and cellotetraose were obtained from the acid hydrolysate while isoprimeverose (4) and the pentasaccharide (5) were characterized from the enzymic hydrolysate. Similar studies using the cellulase from Trichoderma viride produced three types of oligosaccharide, hepta-, nona-, and deca-saccharides, and their structures were partially characterized.[171]

The structure of a cell-wall polysaccharide isolated from cotyledons of tora bean has been established by methylation analysis, Smith degradation, and enzymatic fragmentation.[172] The main features were those of a D-galacto-D-xylo-D-glucan where a (1→4)-β-D-glucan main chain is substituted at O-6 by D-xylopyranosyl and β-D-galactopyranosyl-D-xylopyranosyl residues. A further feature was the presence of (1→5)-α-L-arabinan side chains (degree of polymerization = 8) at some O-3 positions of the main chain.

A D-xylo-D-glucan has been isolated from the cell walls of potato (Solanum tuberosum), and the methylation analysis and fragmentation results confirmed the usual structure.[173] Some of the D-xylose residues carried further substitution at O-2 with L-arabinofuranosyl and D-galactopyranosyl residues.

The biosynthesis of the D-xylo-D-glucan in cultured soybean cells has been studied.[174] Two types were obtained from cultured cell walls (mol. wts. 1.8×10^5 and 6×10^4) and one from the culture medium, the molecular weight of which shifted from 6×10^4 to 3×10^4 during the progress of cultivation. The compositions were very similar and two kinds of oligosaccharide repeating units were present, namely a heptasaccharide (D-xylose:D-glucose 3:4) and a nonasaccharide (D-galactose:L-fucose:D-xylose:D-glucose (1:1:3:4)). Incubation of a particulate fraction from soybean α-D-Xyl*p*-(1→6)-D-Glc

(4)

α-L-Fuc*p*-(1→2)-β-D-Gal*p*-(1→2)-α-D-Xyl*p*-(1→6)-β-D-Glc*p*-(1→4)-D-Glc*p*

(5)

cells with UDP-^{14}C-D-glucose or UDP-^{14}C-D-xylose produced a labelled polymer.[175] Digestion with A. oryzae enzymes gave a labelled disaccharide characterized as isoprimeverose (4) by the usual methods. Since isoprimeverose is the smallest structural unit of D-xylo-D-glucan, the polymer can be synthesized from these nucleotide sugars. The results have been used as the basis of an assay for D-xylo-D-glucan:D-xylosyltransferase.[176] They have also been used to suggest that this transferase can be distinguished from other polysaccharide synthetase activities.

The metabolism of polysaccharides by pea stem segments treated with and without auxin has been studied by a centrifugation technique.[177] Auxin enhanced the levels of D-xylose and D-glucose in ethanol-insoluble polysaccharide, and these sugars were in a D-xylo-D-glucan. Similar studies using ethene showed that this treatment lowered the amount of D-xylo-D-glucan in ethanol-insoluble polysaccharides.[178]

The vacuum u.v.-c.d. spectra of guar, tara, and carob D-galacto-D-mannans, which had D-galactose contents of 39, 25, and 19%, respectively, have been recorded.[179] A positive band at 169 nm and a negative band at 149 nm were observed whose relative intensities increased systematically with decreasing D-galactose content for solid films. Some of the differences were attributed to conformational restriction of D-galactose residues by chain packing with a consequent increase in the c.d. intensities from these residues and a cancellation of D-mannan backbone contributions.

High—performance size-exclusion chromatography of guar gum has indicated that the molecular weight was $> 2.0 \times 10^6$.[180]

The seeds of Astragalus sinicus contain a polysaccharide which has a D-galactose:D-mannose ratio of 1:2.3 and a molecular weight of 1.6×10^4.[181] The results of methylation and Smith degradation analyses indicated that the D-galactopyranosyl residues were attached to O-6 of the main chain but that both (1→3)- and (1→4)-linkages were present in the main D-mannan chain.

Treatment of guar gum with an α-D-galactosidase has produced a polymer in which 75% of the D-galactose residues have been removed, but there was no reduction in viscosity.[182] Further treatment did reduce the viscosity and led to D-mannan-type precipitates. The interaction of guar gum with xanthan gum is increased by removal of the D-galactose residues.

The chemical structure of the D-galacto-D-mannan component of

Trypanosoma cruzi has been determined by ^{13}C n.m.r. spectroscopy and confirmed that β-D-galactofuranosyl residues were linked at the non-reducing terminus to O-3 of D-mannose residues.[183]

D-Galacto-D-mannans are the major storage reserve carbohydrate in the embryo of fenugreek but they are not mobilized during the first 24h of germination.[184]

A pure D-gluco-D-mannan has been recovered from the hot-water extract from Aloe barbadensis.[185] The methylated polysaccharide (mol. wt. 1.5×10^4) contained residues of 2,3,4,6-tetra-, 2,3,6-tri-, and 2,3-di-O-methyl-D-mannose and 2,3,6-tri-O-methyl-D-glucose in the ratios 1.3:18.3:1.2:1.0.

D-Mannose was observed in hydrolysates of the holocellulose of Panicum coloratum leaves.[186] The material containing this sugar residue was particularly resistant to extraction from the α-cellulose fraction.

The conditions for the preparation of a rigid and homogeneous gel from Konjac mannan have been published.[187]

Fractionation of cell-wall preparations from grass leaves has shown that the less dense fractions had a higher D-xylan content as well as a higher content of lignin and acetyl residues.[188]

Fully methylated D-xylan-type oligosaccharides have been analysed by mass spectrometry and the fragmentation patterns compared.[189] Within the series of di- and tri-saccharides studied, the criteria proposed permitted reliable determinations of any of the theoretically possible branching points.

The structure of the water-soluble L-arabino-D-xylan from delignified bark of Cinnamomum zeylanicum has been established.[190] The main (1→4)-β-D-xylan chain is substituted both at O-2 and O-3 with L-arabinofuranosyl and 3-O-α-D-xylopyranosyl-L-arabinofuranosyl groups.

The isolation, purification, and partial characterization of a D-glucurono-L-arabino-D-xylan, a previously unobserved component of the primary cell wall of dicotyledonous plants, have been described.[191] It represents 5% of the wall and has the usual structural features of this type of hemicellulose.

A procedure has been developed for the fractionation of the non-starchy polysaccharides of wheat bran.[192]

Determination of the activity of Ca^{2+} ions bound to carboxy groups in the D-xylan from white willow has shown that 4-O-methyl-D-glucopyranosyl uronic acid groups are concentrated in blocks where at least every second α-D-xylopyranosyl residue contains a single uronic acid residue linked by a (1→2) bond.[193] These acidic blocks alternate with blocks of unsubstituted D-xylose units which are three times as long.

The hemicellulose contents of leaves, nodes, and internodes of wheat straw were similar.[194] Ozone and sulphur dioxide at 70°C for 72 h solubilized virtually all the wheat hemicelluloses.[195] Results were also presented on the digestion of the hemicelluloses by rumen micro-organisms. Similar studies with rumen organisms have been carried out on NaOH-treated straw.[196]

The hemicellulose content of cotton straw is reduced by 50% on treatment with ozone.[197]

Auxin was found to have no effect on the structure of the L-arabino-D-xylans isolated from Avena coleoptiles.[198]

The pentosans of pearl millet (Pennisetum americanum) have been separated into four fractions and each was found to contain varying amounts of seven sugar residues.[199] Each fraction also contained protein.

A mechanism for the oxidative gelation of the water-soluble pentosan of wheat flour has been proposed that suggests an attack by addition of a protein radical to the activated double bond of a ferulic acid residue.[200]

The substrate-binding site of the endo-(1→4)-β-D-xylanase from the yeast Cryptococcus albidus has been investigated using a series of (1→4)-β-D-xylan oligosaccharides which were labelled with tritium at the reducing end.[201] The substrate binding site is composed of four subsites with the catalytic groups located in the centre. As a consequence of an assymetric distribution of negative values of affinity around the binding site, the enzyme displays a strong preference to attack near the reducing end of the substrate. The mechanism of action was found to involve pathways other than a simple hydrolytic cleavage.[202] Some D-xylosyl transfer occurred to increase the molecular size of the substrate and the transfer came from the non-reducing end of the molecule. All features of the degradation of oligosaccharides by this D-xylanase were consistent with the lysozyme-type reaction mechanism.

Particulate enzyme preparations obtained from homogenates of differentiated xylem cells from sycamore trees have catalysed the synthesis of a D-xylan from UDP-D-xylose.[203] The polysaccharide had the expected structure. Various properties of the enzyme and its requirements were determined. The specific activity of the enzyme system increased markedly as cells differentiated from the vascular cambium to the xylem.[204] The increase was closely correlated with the enhanced deposition of D-xylan occurring during the formation of secondary thickening.

The control of hemicellulose synthesis during differentiation of vascular tissue in bean callus and hypocotyl has been investigated and the D-xylan synthetase induction correlated with the induction of phenylalanine ammonia-lyase and with lignin synthesis.[205]

A cell-wall degrading agent has been isolated from Chlorella and characterized as a (1→4)-β-D-xylanase.[206]

Residual carbohydrates have been removed almost completely from milled wood lignin preparations of birchwood by treating with NaOH-dioxane.[207] Treatment with HCl-dioxane did not decrease the carbohydrate content. Studies using i.r. and n.m.r. spectroscopy indicated that a benzyl ester type of linkage existed between lignin and the carbohydrates. The lignin-carbohydrate complex from the milled wood lignin fraction of Pinus densiflora contained three sub-fractions on gel filtration.[208] Two of the fractions were homogeneous and their compositions were determined. Methylation analysis showed that the carbohydrate moiety exhibited multiple branching with the major backbone being composed of (1→4)-D-mannan chains. A water-soluble lignin-carbohydrate complex has been released from jute (Corcharus capsularis) fibre by reduction with borohydride.[209] Analysis of the D-xylans from the original and borohydride-treated fibres indicated that 34% of the acidic side chains of the D-xylan were linked to lignin by ester bonds. Lignin-carbohydrate complexes have been released from wheat straw by prolonged hot-water extraction.[210] The bound sugars were identified as D-glucose, L-arabinose, D-xylose, and D-galactose residues with uronic acids being absent. The higher molecular weight fractions were richer in D-glucose residues with the lower molecular weight ones being richer in D-xylose and L-arabinose residues.

6 Pectins

A high-performance gel-permeation chromatographic method has been developed to determine the molecular weight distribution of pectins.[211] High-methoxy, low-methoxy, and amidated pectins can be analysed within 15 min.

An investigation of oligo- and poly-galacturonic acids has been carried out by potentiometry and c.d.[212] It was shown that no Ca^{2+} ion is ever fixed in excess of stoichiometry. The c.d. spectra, in agreement with ^{13}C n.m.r. spectroscopic data, suggested that, in dilute solutions, the polymer adopts two conformations.

The gelation of pectin under conditions of low water activity has been investigated by c.d., competitive inhibition, and mechanical properties.[213] The optimum degree of esterification for pectin gelation under conditions of low water activity is 70%. The reduction in gel strength at lower degrees of esterification is not merely due to increased charge density but also results from the loss of a significant contribution which the ester group makes to the stability of the interchain junctions. Gel permeation chromatography, c.d., and viscosity measurements showed that appreciable aggregations of aqueous solutions of pectin occur at low concentrations in pectins of both low and high ester content.[214] The aggregation is independent of divalent cations. The Arrhenius slopes were greater at lower temperatures and they increase with conditions such as pH, ionic strength, concentration, and degree of esterification that are known to favour interchain association of pectin. The c.d. spectra at various pH values and in the presence of additives indicated that two modes of interchain association contribute to the network formation in calcium pectate gels.[215] The first is by interchain chelation of Ca^{2+} ions and the second is by non-ionic interchain associations analogous to those in pectin gels with low water activity. The contribution of the latter type increases on lowering the pH, as the Ca^{2+} ion binding capacity is diminished and interchain electrostatic repulsions are minimized. The mechanical properties and thermal stability of a gel are a function of the proportion of each type.

A relationship has been sought between the ability of various ions to inhibit growth and their ability to cause gelation of isolated pectins, their binding affinity for isolated cell walls, and their binding affinity for purified pectin.[216] A good correspondence was found with the second factor. Pectic gel formation was not involved in cation-induced growth inhibition. The previously determined activity coefficients of Ca^{2+} ion in oligo-D-galacturonates were plotted and extrapolated to infinity.[217] In molecular disperse solutions the binding of Ca^{2+} ions was electrostatic, whereas in aggregated molecules a chelate intermolecular binding also took place. The non-reducing terminal uronic acid residues bind Ca^{2+} ions less firmly than to the inner units of the chain.

Lemon pectin and pectic acids have been hydrolysed into acid-soluble and acid-insoluble products and separated by gel filtration.[218] The latter product had a molecular weight of 3.0—3.2 x 10^4 whereas the former could be separated into two components (mol. wts. 2.2 — 2.5 x 10^4 and < 6 x 10^3). The larger of these two components contained only (1→4)-α-D-galacturonan chains while the smaller contained residues of D-galacturonic acid and L-rhamnose as well as other neutral sugars.

Cell-wall material has been isolated from cabbage (Brassica oleracea) and methylated.[219] Hydrolysis of the product and analysis of the fragments revealed some of the structural features many of which were derived from the pectin present.

Two pectic fractions, comprising 52% of the cell wall, have been isolated from potatoes.[220] Both fractions contained a range of molecular types differing in composition. The more easily extracted fraction had a high L-arabinose:D-galactose ratio as well as a high D-galacturonic acid content. The more easily extractable fraction was probably held by Ca^{2+} ion bonds. An endo-poly-D-galacturonanase from Aspergillus japonica has been used for the isolation and characterization of cell-wall pectic substances from potato tubers.[221] Methylation analysis of the four fractions obtained by gel filtration was used to elucidate their structure. The two highest molecular size fragments had very similar compositions while the intermediate fraction had a very complex structure including D-galacturonic acid residues which were branched at C-3.

Sequential extractions of Rosa glauca cell walls showed that two different types of acidic polysaccharide were present.[222] One was extracted with chelating agents while the other remained in close association with the α-cellulose fraction.

A structurally complex pectic polysaccharide, L-rhamno-D-galacturonan I, has been isolated from the primary cell wall of suspension-cultured sycamore cells.[223] The polysaccharide (mol. wt. 2 x 10^5) has the usual pectin backbone to which is attached a wide variety of side chains containing L-arabinosyl and/or D-galactosyl residues.

Tomato pectin esterase has been used to prepare a high molecular weight pectin from citrus.[224] The product from the free enzyme had a degree of esterification of *ca.* 2% whereas an immobilized enzyme gave a value of *ca.* 19%. The molecular weight was lowered by less than 5% by the enzymatic treatments.

The steric course of the methyl esterification reaction of a D-galacturonan has been studied using *S*-adenosylmethionine and an enzyme preparation from *Phaseolus aureus*.[225] The methyl group of the methionine derivative was chiral, possessing ^{1}H, ^{2}H, and ^{3}H substituents, and it was shown that transfer of the methyl group to the oxygen of the carboxy group proceeds with inversion of the configuration of the methyl group.

Silica (1.4 — 2.3 mg g^{-1}) was found in a wide variety of pectin samples.[226]

Pectic substances were the major components of cell-wall material from pea flour and concentrates.[227] Detergent methods were shown to underestimate the pectin contents in a range of dietary fibres.[228] For example, 50% of orange pectin was recovered in the acid-detergent residue and 38% in the neutral-detergent residue. Many of the physical properties of the dietary fibre from apples were shown to be due to the high pectin content.[229] The interactions of bile acids and their conjugates with pectins of different degrees of esterification in the pH range 3.5 — 7.3 have been investigated by ^{1}H n.m.r. spectroscopy, but no molecular interactions were apparent.[230]

The pH-dependent interaction between pea cell-wall polymers possibly involved in wall deposition and growth has been investigated and there occurs, at neutral pH values, a pH- and time-dependent binding of soluble pectin, in the walls, to a heat-labile protein component of the wall.[231] The reaction is also observed in water extracts of walls. The reaction is inhibited at low pH values and by Ca^{2+} ions and is lectin-like in nature.

An elicitor of phytoalexin accumulation has been solubilized from cell walls of soybeans.[232] The active material contains residues of D-galacturonic acid, L-rhamnose, and D-xylose and is probably pectic in origin. Similar elicitors could be prepared from suspension-cultured tobacco, sycamore, and wheat cells. A pectic polysaccharide, solubilized by the action of a fungal *endo*-(1→4)-α-D-galacturonanase from sycamore cell walls, possesses activity similar to the proteinase-inhibitor inducing factor isolated from tomato leaves.[233] The synthesis and accumulation of proteinase inhibitor I in excised tomato leaves can be induced with oligosaccharides obtained by *endo*-(1→4)-α-D-galacturonanase action of a pectic polysaccharide from tomato leaves.[234] These oligosaccharides released by enzymic action at a wound or site of infection may play a hormone-like role in regulating plant defence responses in unwounded tissue some distance from the site of release.

Pectic multienzyme preparations have been separated by h.p.l.c. using both isocratic and linear-gradient conditions.[235]

Three D-galacturonanases have been isolated from carrots, two from root tissue and one from cell-suspension cultures.[236] All three enzymes were of similar size (mol. wt. 4.8×10^4) and were exo-enzymes. The only major differences were in the binding properties of the enzymes to cell-wall material.

Pectic enzymes have been extracted from haustoria of Cassytha filiformis growing on Nerium indicum and Hibiscus rosa-sinensis and both strains contained exo-(1→4)-α-D-galacturonanase and exo-pectin methylgalacturonanase activity.[237] The two strains differed markedly in the production of lyase enzymes: the Nerium strain produced endo-pectic acid lyase while the Hibiscus strain produced endo-pectin lyase.

A D-galacturonanase (mol. wt. 3.6×10^4) has been isolated from citrus fruit infected with Penicillium italicum.[238] Sodium D-galacturonan was a better substrate than citrus pectin.

The D-galacturonanase activity in seed cavity tissue from harvested cucumber fruit increases 20-fold after the fruit produces a transient burst of ethene during maturation.[239]

A single pectic enzyme, D-galacturonanase, is produced by Rhizoctonia solani in culture and during infection of cotton seedlings.[240] The enzyme (mol. wt. 4.2×10^4) has a pH optimum of 5.2.

The D-galacturonanase from Rhizopus stolonifer has been shown to be an elicitor of casbene synthetase activity in castor bean seedlings.[241] The enzyme is a glycoprotein (mol. wt. 3.2×10^4) and contains 20% carbohydrate. The principal sugar residues present are D-mannose (92%) and 2-amino-2-deoxy-D-glucose (8%) and the enzyme is an endo-hydrolase.[242] Other features of the enzyme were reported.

The active site of the endo-D-galacturonanase from tomato has been studied using oligo-D-galacturonic acids and their aldehydo derivatives.[243] The relationship of the substrate chain length to the initial and maximum reaction rates was determined. The topology of the active site was suggested from the results. The

D-galacturonanase of ripe tomatoes has been extracted in two isoenzyme forms, I and II.[244] The enzymes have different properties although their polypeptides are similar. Green fruit contains a factor which can convert form II into form I, and the amount of this factor increases during ripening.

The characteristics of the pectin methylesterase from potato have been determined and the role of the enzyme in commercial dehydrated mashed-potato production has been examined.[245] Ca^{2+} ions, but not Mg^{2+} ions, retard solubilization. Texture measurements suggested a firming effect which could be attributed to Ca^{2+} ion release during pre-cooking and stabilization of Ca^{2+} ion bridges during cooling.

The changes in pectinesterase activity and other pectin changes in mango varieties during storage and ripening have been investigated.[246] In a study of the microbiology of wetwood, the importance of Clostridia species in the degradation of pectin has been assessed.[247]

High levels of pectinolytic enzymes were produced in culture filtrates of Phoma exigua and Graphium penicillioides isolated from decaying seeds of Phaseolus aureus and Cyamopsis tetragonalobus, respectively.[248]

The susceptibility of strawberry varieties to breakdown in sulphate liquor by the endo-D-galacturonanase from Zygomycete spoilage fungi has been investigated.[249]

Fruit waste containing de-esterified pectin as a thickener has been assessed for its potential use in canned products.[250]

7 Gums and Mucilages

The carbohydrate gum from bael (Aegle marmelos) seeds has been resolved into four glycoprotein fractions.[251] One of these contained residues of D-galactose, D-glucose, L-arabinose, and L-rhamnose in the molar ratios 6:2:8:3. The linkages amongst these residues were determined as well as the anomeric configuration of the glycosyl residues. The structure of the linkage region was established as O^3-L-arabinopyranosyl-L-threonine.

A histological and a histochemical comparison of the mucilages on the root tips of several grasses has been made, and it was concluded that there were two types.[252] One is a gelatinous material originating from the root cap and the other is a firm uniformly thick mucilage overlying the columnar epidermal cells.

Reactions of the outer layer and cell wall indicated that carboxy groups were present and these were absent from the inner mucilage.

A mucous polysaccharide has been isolated from the bulbs of Narcissus tazetta.[253] It was homogeneous and contained D-mannose and D-glucose residues in the ratio 5:1 with a molecular weight of 1.19×10^5. The O-acetyl content was 22.7%, and these groups were present on C-6 and on C-2 and C-6 of most D-mannose residues. The main chain is a (1→4)-β-D-hexopyran.

The glycoprotein from neem (Azadirachta indica) gum (mol. wt. 1.3×10^4) contained 50% carbohydrate.[254] The sugar residues present were D-mannose, 2-acetamido-2-deoxy-D-glucose, D-galactose, L-arabinose, L-fucose, D-xylose, and D-glucose in the ratios 14:13:5:3:1:1:1. The linkage between the carbohydrate and the protein was an N^4-(2-acetamido-2-deoxy-β-D-glucopyranosyl) hydrogen L-asparaginate bond. Another glycoprotein from the same source contained only 19% carbohydrate.[255] This material contained the same sugar residues in the ratio 4:3:2:3:2:1:1. The linkage region was the same.

The mucilage of Opuntia ficus-indica and the partially degraded mucilage have been investigated by methylation analysis and periodate oxidation.[256] The results show that the mucilage is composed of 1,4-substituted α-D-galactopyranosyluronic acid and 1,2-substituted β-L-rhamnopyranosyl residues to which short chains of 1,6-substituted β-D-galactopyranosyl residues were attached at O-4 of all of the L-rhamnopyranosyl residues. Most of the D-galactosyl residues carry branches at O-3 while some are branched at O-4. Partial acid hydrolysis of this mucilage has yielded O-β-D-galactopyranosyl-(1→6)-D-galactose and its polymer-homologous trimer as well as fourteen oligosaccharides that contain L-arabinose residues and most of which have D-xylosyl residues at the non-reducing terminus.[257] The oligosaccharides (6) and (7) were the most abundant. Similar studies have been reported on the mucilage of O. aurantiaca with similar results.[258]

The gum exudates from some Prosopis species have been analysed by the Smith degradation procedure.[259] The degraded gums have molecular weights of 6×10^3. A simple (1→3)-D-galactan chain was obtained after only two degradations showing that the skeleton was similar to those of Acacia gums.

Autohydrolysis of the gum exudate from the Spondias dulcis tree gave a gum containing D-galactose, L-arabinose, and D-galacturonic acid in the ratio 3:3:1.[260] Three neutral and three acidic β-D-Xyl*p*-(1→5)-L-Ara*f*

(6)

β-D-Xyl*p*-(1→5)-α-L-Ara*f*-(1→5)-L-Ara*f*

(7)

oligosaccharides were obtained and their structures characterized to suggest a sequence for the repeating unit. Chromium(VI) trioxide oxidation was used to determine the anomeric configuration of the sugar residues.

8 Algal Polysaccharides

Based on chemical shifts and C-H coupling constants from model molecules, the structural conformation of aqueous agarose solutions and gels has been determined from the ^{13}C n.m.r. spectra.[261] Similar studies on the agaroses of Pterocladia species and Gracilaria secundata have been reported and all signals in the spectra were assigned to various D-galactose derivatives and agaro-oligosaccharides.[262]

The chemical composition and rheological properties of the agar isolated from Gelidium purpurascens before and after alkaline treatment have been reported.[263] The D-xylose, D-glucose, and D-glucuronic acid residues in the agar were removed on alkaline treatment together with 86% of the protein. Sequential extraction of the alga accounted for low yields of agar as losses incurred on precipitation with ethanol.

A method has been presented for the determination of the backbone conformations of polymer chains conforming to a given helical type.[264] The method is illustrated by its application to agarose.

The high – resolution ^{13}C n.m.r. spectra of slightly depolymerized alginates have been interpreted.[265] The sequence of the monomer units L-guluronate (G) and D-mannuronate (M) markedly influences the chemical shifts. Some of the individual carbon resonances were resolved into four lines, which was evidence for a dependence on the identities of the units immediately before and after them in the polymer chain. The relative intensities of the signals permitted a rapid computation of the monomeric composition (M/G ratio), the composition of end units (M/G ratio), and the monomeric sequence in terms of a complete set of four diad and eight triad frequencies. Any region with a strictly alternating M and G sequence was very short.

The cation-specific vacuum ultraviolet c.d. spectra of alginate solutions, gels, and solid films have been recorded.[266] A band at 185 nm has been assigned to carboxy groups while those at 169 and 149 nm were assigned to changes in the polymer backbone.

ι-Carrageenan has shown substantial helix formation in the presence of Li^+, Na^+, and Me_4N^+ ions as measured by optical rotation, but the polymer does not gel.[267] Under identical conditions, the polymer does gel with K^+, Rb^+, Cs^+, and NH_4^+ ions. The gel formation is in three stages. Initially the individual coils form double helices, then the double helices form domains, which finally form aggregates around the heavier Group I ions.

A detailed assignment of the ^{13}C chemical shifts of κ- and ι-carrageenan in their Na^+ and K^+ forms has been given.[268] Evidence for the conformational transition induced by temperature variation in the absence of any gel formation of κ-carrageenan is also presented. The evidence is based on both n.m.r. and optical rotation experiments.

Measurements of rigidity moduli and ^{39}K and ^{23}Na n.m.r. spectra have been used to investigate the roles played by Ca^{2+}, K^+, and Na^+ ions in ι-carrageenan gels.[269] Differences in binding were reflected in the rheological properties of the gels. Potassium ι-carrageenate gels have been prepared and studied by photon correlation spectroscopy at 633 nm.[270] The molecular and vibrational motions of the gels were interpreted in terms of their elastic properties. Below the sol-gel transition temperature, oscillatory correlation functions were found with the frequency remaining constant for two weeks. Measurements of the shear modulus for Ca^{2+}, K^+, and Na^+ ι-carrageenate gels as a function of biopolymer concentration and temperature showed that the Ca^{2+} salt appears to possess a random glass-like structure whereas the K^+ and Na^+ forms show a pseudocrystalline structure.[271] The differences were tentatively attributed to differences in the solubility of the salt forms which, through controlling the gelation temperature, moderated the kinetics of the gelation process.

The optical rotation and the conductivity of κ-carrageenans in aqueous solutions have been investigated as functions of temperature in the presence of various electrolytes.[272] The activity coefficients of Na^+ and K^+ ions have been determined and correlated with the conformation. The K^+ activity coefficient under ordered conformation is in agreement with a mechanism involving dimerization.

The use of the carrageenan from the red alga Eucheuma striatum has been investigated as a possible substitute for agar in bacteriological tests.[273]

Neocarratetraose 4-O-monosulphate β-hydrolase has been isolated from cell-free extracts of Pseudomonas carrageenovora.[274] The hydrolysis products are neocarrabiose and neocarrabiose 4-O-sulphate. The pathway for the sequential degradation of the tetrasaccharide was proposed.

λ-Carrageenan has been shown to have a hypotensive effect in the rat that was dependent on platelet stimulation.[275] Also in rats, intraperitoneal injection of ellagic acid and carrageenan reduced paw oedema induced by carrageenan itself.[276] The relationship between the anticoagulant activity and the pro- and anti-inflammatory mechanism in λ-carrageenan-induced inflammation has been investigated.[277]

The gelation behaviour of solutions of chitosan on reaction with acid anhydrides has been studied, and the effects of the time to onset of gelation and of varying the nature and concentration of the anhydride, the chitosan concentration, the temperature, and the nature of the cosolvent were all examined.[278] Gelation occurred due to the decreased solubility of the polymer molecules brought on by N-acylation, and all other parameters influenced the rate of N-acylation. The syneresis of the gels was examined. The degree of N-acylation required to initiate gelation depended on the molecular weight of the acyl anhydride and the concentrations of chitosan.[279] The energy of activation of N-acylation was determined for N-acetylation and N-hexanoylation. A mechanism was proposed for the gelation process.

An i.r. spectroscopic method has been used to determine the amide content of highly deacylated chitosans.[280] A plot of the ratio of absorbance at the amide band (1655 cm^{-1}) to the C-H stretch frequency (2867 cm^{-1}) against the degree of deacetylation was linear in the deacetylation range 90-100%.

The degree of N-acetylation of chitosan was rapidly determined by periodate oxidation and i.r. spectroscopic methods.[281] The i.r. methods gave higher results but were thought to be more reliable.

A new method of chitin determination, based on deacetylation and g.l.c. analysis of the liberated acetic acid, has been applied to a variety of chitosans.[282] The results were in good agreement with other methods even though the rate of deacetylation of some chitosans is 10 times that of others.

A highly crystalline form of α-chitin has been found in the grasping spines of the marine worm Sagitta and been examined by electron microscopy and electron diffraction.[283]

The chitin from carapace of Penaeus japonicus has been purified by a proteolytic enzyme from Pseudomonas maltophilia.[284] No deacetylation occurred and the protein content was reduced to virtually zero. The molecular weight of the enzymically prepared chitin was higher than that obtained by alkaline treatment.

Polyelectrolytic complexes have been prepared from chitosan and sodium carboxymethylcellulose.[285] None were soluble in aprotic solvents but the series prepared at high pH values were more soluble in hot formic acid than those prepared at low pH values. The former were also different from the latter in the density of carboxy sites.[286]

A ^{1}H n.m.r. method has been used to investigate the binding of methyl di-N-acetyl-β-chitobioside to wheat-germ agglutinin.[287] The linewidth broadening of the methoxy protons allowed a calculation of kinetic association constants to be obtained.

The bovine serum chitinase has been characterized as a true (1→4)-β-D-2-acetamido-2-deoxy-glucanase without any exo-β-D-2-acetamido-2-deoxy-glucosidase activity.[288] The purification and properties of the enzyme were reported.

The chitin synthesizing system of insect tissue has been reported.[289]

A study on the adsorption of heavy-metal ions by chitin phosphate and chitosan phosphate has been made.[290]

The chemical composition of the porphyran from Porphyra columbina was in good agreement with data for similar polysaccharides.[291] One fraction had an optical rotation similar to that of the porphyran from P. capensis while the optical rotation of another fraction was similar to that of the phycocolloid of P. umbilicalis. The molar composition of the latter was D-galactose:6-O-methyl-D-galactose:3,6-anhydro-L-galactose:sulphate in the ratio 1.00:0.57:0.35:0.65. Several signals in the ^{1}H n.m.r. spectra could be assigned to various anomeric protons.

A biophotoreactor, using immobilized cells of Porphyridium cruentum in a polyurethane 'mousse', has been used to produce a capsular polysaccharide sulphate which is secreted into solution.[292]

Sulphated polysaccharides have been isolated from Caulerpa species and their function in wound healing has been assessed.[293]

Fucoidanase activities have been observed in fractions from the hepatopancreas of abalone.[294] The primary product from one enzyme was L-fucose while another enzyme gave a series of neutral oligosaccharides.

Laminarin, L-fucans, and alginic acid have all been isolated from Dictyopteris plagiogramma.[295] The laminarin contained chains terminating in a D-glucose residue and those terminating in a D-mannitol residue in the ratio 3:1. The L-fucans were present in a variety of extracts and each had a slightly different ratio of L-fucose:D-xylose:D-galactose:D-mannose:D-glucuronic acid:sulphate. The structural features elucidated were similar to those from other L-fucans.

Polysaccharide material from Eucheuma spinosum appears to consist of two different components containing low and high sulphate contents.[296] The major component had the higher sulphate content while the molecular sizes of the two polysaccharides were different.

A series of acidic oligosaccharides (8) —— (20) has been obtained by graded acid hydrolysis of the methylated acidic polysaccharide associated with the coccoliths of the alga Emiliania huxleyi; the above structures were characterized.[297] The results along with methylation data from the native, carboxy-reduced, and desulphated carboxy-reduced polysaccharides have been used to give a proposed structure for this complex polymer.[298]

Glycogen accumulation in vegetative cells of the blue-green alga Anabaena is a light-dependent process.[299] The amount of glycogen produced was high during sporulation and this increased with the onset of maturation of the spores. The amino-acids L-methionine, L-tyrosine, L-glycine, and L-histidine gave the greatest enhancement of glycogen accumulation in supplemented cultures although all the amino-acids examined had a positive effect.[300]

The storage D-glucans in and isoenzyme distribution of Cyanidium caldarium have been investigated.[301] The observations propose this organism to be a primitive Rhodophytan which is an algal-bridge linking the prokaryotic blue-green and the red algae. It seems to be a true transitional organism between the nucleated and anucleated algae and is unlikely to be an endosymbiotic association of more than two cells as speculated elsewhere.

Two major arsenical constituents of the brown kelp Ecklonia radiata have been isolated and characterized as 2-hydroxy-3-sulphopropyl-5-deoxy-5-(dimethylarsenoso)-furanoside and 2,3-dihydroxypropyl-5-deoxy-5(dimethylarsenoso)-furanoside.[302]

α-D-GalpA-(1→6)-α-D-Manp-(1→3)-D-Man

(8)

D-GalpA-(1→4)-D-Galp-(1→6)-Man

(9)

D-GalpA-(1→4)-D-GalpA-(1→2/6)-Manp-(1→3)-Man

(10)

D-GalpA-(1→4)-D-GalpA-(1→3)-D-Xyl

(11)

D-GalpA-(1→2)-L-Manp6Me-(1→4)-D-GalpA-(1→2)-L-Rha

(12)

D-GalpA-(1→2)-Manp-(1→4)-D-GalpA-(1→2)-L-Rha

(13)

D-GalpA-(1→2)-L-Rhap-(1→4)-D-GalpA-(1→2)-L-Rha

(14)

D-GalpA-(1→2)-L-Manp6Me-(1→4)-D-GalA

(15)

D-GalpA-(1→2)-Manp-(1→4)-D-GalA

(16)

D-GalpA-(1→2)-L-Manp6Me-(1→4)-D-GalpA-(1→2)-L-Man6Me

(17)

D-GalpA-(1→2)-L-Manp6Me-(1→4)-D-GalpA-(1→2)-Man

(18)

D-GalpA3Me-(1→2)-L-Manp6Me-(1→4)-D-GalpA-(1→2)-L-Man6Me

(19)

D-GalpA3Me-(1→2)-L-Manp6Me-(1→4)-D-GalA

(20)

References

1 'Biochemistry of Plants. A Comprehensive Treatise', ed. J.Preiss, Academic Press, New York, 1980, Vol. 3 (Carbohydrates: Structure and Function).
2 'Mechanisms of Saccharide Polymerization and Depolymerization', ed. J.J.Marshall, Academic Press, New York, 1980.
3 H.F.Mark, Angew. Chem., Int. Ed. Engl., 1981, 20, 303.
4 T.L.Bluhm and P.Zugenmaier, Carbohydr. Res., 1981, 89, 1.
5 G.Rappenecker and P.Zugenmaier, Carbohydr. Res., 1981, 89, 11.
6 J. Metzger, Fresenius' Z. Anal. Chem., 1981, 308, 29.
7 R.G.Stone and J.A.Krasowski, Anal. Chem., 1981, 53, 736.
8 K.Kainuma and T.Ogawa, J. Chromatogr., 1981, 212, 126.
9 B.G.Osborne and S.Douglas, J. Sci. Food Agric., 1981, 31, 328.
10 T.Handa, H.Yajima, and T. Kajiura, Biopolymers, 1980, 19, 1723.
11 F.W.Fales, Biopolymers, 1980, 19, 1535.
12 J.N.BeMiller and G.W.Pratt, Cereal Chem., 1981, 58, 517.
13 M.Nedelcheva, E.Valcheva, S. Bencheva, and V.Valchev, Starch, 1981, 33, 195.
14 K.Kulp and K.Lorenz, Cereal Chem., 1981, 58, 46.
15 K.Lorenz and K.Kulp, Cereal Chem., 1981, 58, 49.
16 R.J.Kartchner and B.Theurer, J. Agric. Food Chem., 1981, 29, 8.
17 F.R.Dintzis and C.C.Harris, Cereal Chem., 1981, 58, 467.
18 D.J.O'Donnell, J.J.H.Ackerman, and G.E.Maciel, J. Agric. Food Chem., 1981, 29, 514.
19 S.Hizukuri, Y.Takeda, M.Yasuda, and A.Suzuki, Carbohydr. Res., 1981, 94, 205.
20 D.J.Manners and N.K.Matheson, Carbohydr. Res., 1981, 90, 99.
21 K.Lorenz and K.Kulp, Starch, 1981, 33, 183.
22 J.Torp, J. Sci. Food Agric., 1980, 30, 1354.
23 P.Meredith, Starch, 1981, 33, 40.
24 C.G.Biliaderis, D.R.Grant, and J.R.Vose, Cereal Chem., 1981, 58, 496.
25 C.G.Biliaderis, D.R.Grant, and J.R.Vose, Cereal Chem., 1981, 58, 502.
26 R.T.Tyler, C.G.Youngs, and F.W.Sosulski, Cereal Chem., 1981, 58, 144.
27 M.E.O.Labaneiah and B.S.Luh, Cereal Chem., 1981, 58, 135.
28 S.K.Sathe, Y.Iyer, and D.K.Salunkhe, J. Food Sci., 1981, 46, 1914.
29 S.Umadevi and D.B.Wankhede, Starch, 1981, 33, 23.
30 K.Kayisu, L.F.Hood, and P.J.Van Soest, J. Food Sci., 1981, 46, 1885.
31 P.J.Loos, L.F.Hood, and H.D. Graham, Cereal Chem., 1981, 58, 282.
32 B.A.Keating, A.R.Breen, and J.P.Evenson, J. Sci. Food Agric., 1981, 32, 997.
33 I.Geilinger, R.Amado, and H.Neukom, Starch, 1981, 33, 76.
34 A.Suzuki, A. Hizukuri, and Y.Takeda, Cereal Chem., 1981, 58, 286.
35 T.T.Y.Liu and J.C.Shannon, Plant Physiol., 1981, 67, 518.
36 Y.Ikawa, D.V.Glover, Y.Sugimoto, and H.Fuwa, Starch, 1981, 33, 9.
37 C.D.Boyer, P.A.Damewood, and E.K.G.Simpson, Starch, 1981, 33, 125.
38 K.Lorenz, Starch, 1981, 33, 73.
39 Y.Sugimoto, K.Yamada, S.Sakamoto, and H.Fuwa, Starch, 1981, 33, 112.
40 S.U.Sajjan and D.B.Wankhede, Starch, 1981, 33, 153.
41 W.N.Emmerling and B.Pfannemüller, Starch, 1981, 33, 202.
42 M.Wootton and M.A.Chaudhry, Starch, 1981, 33, 135.
43 M.Wootton and M.A.Chaudhry, Starch, 1981, 33, 168.
44 M.Wootton and M.A.Chaudhry, Starch, 1981, 33, 200.
45 M.Čeh, Č.Stropnik, V.Doleček, and S.Leskovar, Starch, 1981, 33, 45.
46 A.R.Khan, J.A.Johnson, and R.J.Robinson, Cereal Chem., 1979, 56, 303.
47 J.J.Raffi, J.-P.L.Agwel, C.M.Fréjaville, and L.R.Saint-Lèbe, J. Agric. Food Chem., 1981, 29, 548.
48 A.A.Ilin, L.S.Galbraikh, and B.P.Morin, Cellul. Chem. Technol., 1980, 14, 327.
49 M.C.Baur and R.J.Alexander, Cereal Chem., 1979, 56, 364.

50 A.Beleia and E.Varriano-Marston, Cereal Chem., 1981, 58, 437.
51 B.A.Marchylo, L.J.Lacroix, and J.E.Kruger, Plant Physiol., 1981, 67, 89.
52 E.-O.Haegele, E.Schiach, E.Rauscher, P.Lehmann, and M.Grassl, J. Chromatogr., 1981, 223, 69.
53 J.Hutney and M.Ugorski, Arch. Biochem. Biophys., 1981, 206, 29.
54 T.Yamanobe and Y.Takasaki, J. Agric. Chem. Soc. (Jpn.), 1979, 53, 77.
55 J.Daussant, B.Zbaszyniak, J.Sadowski, and I.Wiatroszak, Planta, 1981, 151, 176.
56 H.Bender, Eur. J. Biochem., 1981, 115, 287.
57 V.I.P.Batra and S.L.Mehta, Phytochemistry, 1981, 20, 635.
58 Y.Kano, N.Kunitake, T.Karakawa, H.Taniguchi, and M.Nakamura, Agric. Biol. Chem., 181, 45, 1969.
59 K.Mengel and G.K.Judel, Physiol. Plant., 1981, 51, 13.
60 K.Okamoto and T.Akazawa, Plant Physiol., 1980, 65, 81.
61 T.Baba, Y.Arai, E.Amano, and T.Itoh, Starch, 1981, 33, 79.
62 T.Baba, T.Yamamoto, Y.Arai, M.Yokota, and T.Itoh, Phytochemistry, 1981, 20, 1513.
63 C.W.Chang, Microchem. J., 1981, 26, 86.
64 H.S.Saini and N.K.Matheson, Phytochemistry, 1981, 20, 641.
65 G.L.Matters and C.D.Boyer, Phytochemistry, 1981, 20, 1805.
66 C.A.Adams, T.H.Broman, and R.W.Rinne, Ann. Botany, 1981, 48, 433.
67 J.Marriott, M.Robinson, and S.K.Karikari, J. Sci. Food Agric., 1981, 32, 1021.
68 M.N.Sivak, J.S.Tandecarz, and C.E.Cardini, Arch. Biochem. Biophys., 1981, 212, 525.
69 M.N.Sivak, J.S.Tandecarz, and C.E.Cardini, Arch. Biochem. Biophys., 1981, 212, 537.
70 M.Stitt and H.W.Heldt, Plant Physiol., 1981, 68, 755.
71 M.Stitt and H.W.Heldt, Biochim. Biophys. Acta, 1981, 638, 1.
72 D.Paton, Cereal Chem., 1981, 58, 35.
73 D.D.Christianson, J.E.Hodge, D.Osborne, and R.W.Detroy, Cereal Chem., 1981, 58, 513.
74 K.Ishida and T.Takeuchi, Agric. Biol. Chem., 1981, 45, 1409.
75 T.A.McGuire, R.E.Wing, and W.M.Doane, Starch, 1981, 33, 138.
76 D.van Der Kooij and W.A.M.Hijnen, Appl. Environ. Microbiol., 1981, 41, 216.
77 B.Lesyng and W.Saenger, Biochim. Biophys. Acta, 1981, 678, 408.
78 K.Harata, Bull. Chem. Soc. Jpn., 1980, 53, 2782.
79 K.Harata, K.Uekama, M.Otagiri, F.Hirayama, and H.Ogino, Bull. Chem. Soc. Jpn., 1981, 54, 1954.
80 M.Komiyama and H. Hirai, Bull. Chem. Soc. Jpn., 1981, 54, 828.
81 Y.Inoue and Y.Miyata, Bull. Chem. Soc. Jpn., 1981, 54, 809.
82 H.Shimizu, A.Kaito, and M.Hatano, Bull. Chem. Soc. Jpn., 1981, 54, 513.
83 W.G.Burkert, C.N.Owensby, and W.L.Hinze, J. Liq. Chromatogr., 1981, 4, 1065.
84 M.Komiyama and H.Hirai, Bull. Chem. Soc. Jpn., 1981, 54, 2053.
85 R.H.Marchessault, T.Bleha, Y.Deslandes, and J.F.Revol, Can. J. Chem., 1980, 58, 2415.
86 P.G.Bhat and T.N.Pattabiraman, Indian J. Biochem. Biophys., 1980, 17, 338.
87 R.J.Henry and B.Darbyshire, Proc. Austral. Biochem. Soc., 1980, 13, 70.
88 N.Shiomi, Phytochemistry, 1981, 20, 2581.
89 T.Takenaka and T.Echigo, J. Agric. Chem. Soc. Jpn., 1980, 54, 945.
90 H.Mark, Cellul. Chem. Technol., 1980, 14, 569.
91 D.M.Lee and J.Blackwell, Biopolymers, 1981, 20, 2165.
92 P.K.Chidambareswaran, S.Sreenivasan, N.B.Patil, and H.T.Lokhande, J. Polymer Sci., Polymer Lett., 1980, 18, 603.
93 E.Roche and H.Chanzy, Int. J. Biol. Macromol., 1981, 3, 201.
94 A.Hirai, R.Kitamaru, F.Horii, and I.Sakurada, Cellul. Chem. Technol., 1980, 14, 611.

95 A.D.Bain, D.R.Eaton, R.A.Hux, and J.P.K.Tong, Carbohydr. Res., 1980, 84, 1.
96 J.P.Joseleau, G.Chambat, and B.Chumpitazi-Hermoza, Carbohydr. Res., 1981, 90, 339.
97 T.J.Baker, L.R.Schroeder, and D.C.Johnson, Cellul. Chem. Technol., 1981, 15, 311.
98 J.Schmidt, M.John, and C.Wandrey, J. Chromatogr., 1981, 213, 151.
99 J.Herranz, C.Vidal-Valverde, and E.Rojas-Hidalgo, J. Food Sci., 1981, 46, 1927.
100 M.Lewin and I.Ziderman, Cellul. Chem. Technol., 1980, 14, 743.
101 I.Ziderman, Cellul. Chem. Technol., 1980, 14, 703.
102 B.Larsson and O.Samuelson, Cellul. Chem. Technol., 1981, 15, 237.
103 A.H.Conner, Biotechnol. Lett., 1980, 2, 439.
104 P.Månsson and B.Samuelsson, Sven. Papperstidn., 1981, 84, 15.
105 S.Hunt and T.N.Huckerby, Int. J. Biol. Macromol., 1980, 2, 389.
106 A.Ebringerova and J.Pastyr, Cellul. Chem. Technol., 1980, 14, 885.
107 S.P.Rowland, Cellul. Chem. Technol., 1980, 14, 423.
108 G.Rozmarin, A.Cernatescu-Asandei, and M.Diaconescu, Cellul. Chem. Technol., 1980, 14, 719.
109 B.Philipp and H.Frommelt, Cellul. Chem. Technol., 1980, 14, 831.
110 C.Vasiliu-Oprea, C.Negulianu, M.Popa, F.Weiner, and J.Nicoleanu, Cellul. Chem. Technol., 1980, 14, 655.
111 M.Takai, Y.Nakahara, J.Hayashi, and J.R.Colvin, Cellul. Chem. Technol., 1981, 15, 153.
112 A.G.Matthysse, K.V.Holmes, and R.H.G.Gurlitz, J. Bacteriol., 1981, 145, 583.
113 H.Sakuma, S.Munakata, and S.Sugawara, Agric. Biol. Chem., 1981, 45, 443.
114 P.Fang, G.D.McGinnis, and W.W.Wilson, Anal. Chem., 1981, 53, 2172.
115 V.Luzakova, M.Kosik, O.Minoo, and A.Blazej, Cellul. Chem. Technol., 1980, 14, 677.
116 A.R.White and R.M.Brown, Proc. Natl. Acad. Sci. USA, 1981, 78, 1047.
117 A.Marzetti, B.Focher, L.D'Angiuro, and V.Sarto, Cellul. Chem. Technol., 1981, 15, 3.
118 R.Tanaka, F.Yaku, E.Muraki, and T.Koshijima, Cellul. Chem. Technol., 1980, 14, 859.
119 B.Slutzky, A.R.Lara, and R.Lopéz-Planes, Cellul. Chem. Technol., 1980, 14, 361.
120 S.M.Betrabet, K.M.Paralikar, and N.B.Patil, Cellul. Chem. Technol., 1980, 14, 811.
121 A.A.Klesov and S.Yu.Grigorash, Biochemistry (Engl. Transl.), 1980, 45, 166.
122 D.Knappert, H.Grethlein, and A.Converse, Biotechnol. Bioeng., 1980, 22, 1449.
123 D.E.Koehler, L.N.Lewis, L.M.Shannon, and M.L.Durbin, Phytochemistry, 1981, 20, 409.
124 E.Baath and B.Soederstroem, Can. J. Bot., 1980, 58, 422.
125 T.R.Bandre and V.Sasek, Curr. Sci., 1981, 50, 229.
126 J.Dekker and J.K.Palmer, J. Agric. Food Chem., 1981, 29, 480.
127 H.Quader, Naturwissenschaften, 1981, 68, 428.
128 H.Meier, L.Buchs, A.J.Buchala, and T.Homewood, Nature (London), 1981, 289, 821.
129 A.S.Klein, D.Montezinos, and D.P.Delmer, Planta, 1981, 152, 105.
130 M.Benziman, C.H.Haigler, R.M.Brown, A.R. White, and K.M.Cooper, Proc. Natl. Acad. Sci. USA, 1980, 77, 6678.
131 R.V.S.Sundaresan and W.C.Kimmins, Ann. Bot., 1981, 47, 283.
132 N.N.Bhandari, M.Bhargava, and T.A.Geier, Ann. Bot., 1981, 48, 425.
133 M.A.Yamin, Science, 1981, 211, 58.
134 J.Gormally, J.Panak, E.Wyn-Jones, A.Dawson, D.Wedlock, and G.O. Phillips, Anal. Chim. Acta, 1981, 130, 369.
135 M.Tepfer and I.E.P.Taylor, Science, 1981, 213, 761.

136 V.M.Laube and S.M.Martin, Appl. Environ. Microbiol., 1981, 42, 413.
137 R.R.Mod, E.J.Conkerton, R.L.Ory, and F.L.Normand, Cereal Chem., 1979, 56, 356.
138 S.Kawasaka, Plant Cell Physiol., 1981, 22, 431.
139 G.Mandal and A.Das, Carbohydr. Res., 1980, 86, 247.
140 S.C.Churms, A.M.Stephen, and I.R.Siddiqui, Carbohydr. Res., 1981, 94, 119.
141 G.-J.Van Holst, F.M.Klis, P.J.M.De Wildt, C.A.M.Hazenberg, J.Buijs, and D.Stegwee, Plant Physiol., 1981, 68, 910.
142 G.-J.Van Holst and F.M.Klis, Plant Physiol., 1981, 68, 979.
143 M.A.Jermyn, Proc. Austral. Biochem. Soc., 1981, 14, 29.
144 Y.Akiyama and K.Kato, Phytochemistry, 1981, 20, 2507.
145 K.Nishitani and Y.Masuda, Physiol. Plant., 1981, 52, 482.
146 T.Hoson and S.Wada, Plant Cell Physiol., 1981, 22, 989.
147 A.Strahm, R.Amado, and H.Neukom, Phytochemistry, 1981, 20, 1061.
148 M.A.O'Neill and K.Roberts, Phytochemistry, 1981, 20, 25.
149 A.A.Salyers, R.Arthur, and A.Kuritza, J. Agric. Food Chem., 1981, 29, 475.
150 K.Ogawa, K.Okamura, and A.Sarko, Int. J. Biol. Macromol., 1981, 3, 31.
151 A.J.Stipanovic and E.S.Stevens, Int. J. Biol. Macromol., 1980, 2, 209.
152 P.J.Wood, Carbohydr. Res., 1980, 85, 271.
153 S.A.Jensen and S.Aastrup, Carlsberg Res. Commun., 1981, 46, 87.
154 N.Prentice and S.Faber, Cereal Chem., 1981, 58, 77.
155 T.K.Ghosh and C.C.V.N.Rao, Carbohydr. Res., 1981, 90, 243.
156 N.Nakamura, K.Yoshida, and H.Suzuki, Plant Cell Physiol., 1980, 21, 1383.
157 N.Nakamura and H.Suzuki, Phytochemistry, 1981, 20, 981.
158 I.N.De Halac, Ann. Bot., 1980, 46, 473.
159 N.J.L.Roth, G.H.Watts, and C.W.Newman, Cereal Chem., 1981, 58, 245.
160 K.Cline and P.Albersheim, Plant Physiol., 1981, 68, 207.
161 K.Cline and P.Albersheim. Plant Physiol., 1981, 68, 221.
162 D.J.Huber and D.J.Nevins, Planta, 1981, 151, 206.
163 A.Bacic and D.P.Delmer, Planta, 1981, 152, 346.
164 J.A.Cook, G.B.Fincher, F.Keller, and B.A.Stone, Proc. Austral. Biochem. Soc., 1979, 12, 13.
165 R.J.Henry and B.A.Stone, Proc. Austral. Biochem. Soc., 1981, 14, 49.
166 M.Fèvre and M.Rougier, Planta, 1981, 151, 232.
167 A.J.Anderson, Can. J. Bot., 1980, 58, 2343.
168 R.Yamamoto and D.J.Nevins, Physiol. Plant., 1981, 51, 118.
169 B.G.Sutton, I.P.Ting, and R.Sutton, Plant Physiol., 1981, 68, 784.
170 Y.Kato and K.Matsuda, Agric. Biol. Chem., 1980, 44, 1751.
171 Y.Kato and K.Matsuda, Agric. Biol. Chem., 1980, 44, 1759.
172 K.Ohtani and A.Misaki, Agric. Biol. Chem., 1980, 44, 2029.
173 S.G.Ring and R.R.Selvendran, Phytochemistry, 1981, 20, 2511.
174 T.Hayashi, Y.Kato, and K.Matsuda, Plant Cell Physiol., 1980, 21, 1405.
175 T.Hayashi and K.Matsuda, Plant Cell Physiol., 1981, 22, 517.
176 T.Hayashi, Y.Kato, and K.Matsuda, J. Biochem. (Toyko), 1981, 89, 325.
177 M.E.Terry, R.L.Jones, and B.A.Bonner, Plant Physiol., 1981, 68, 531.
178 M.E.Terry, B.Rubinstein, and R.L.Jones, Plant Physiol., 1981, 68, 538.
179 L.A.Buffington, E.S.Stevens, E.R.Morris, and D.A.Rees, Int. J. Biol. Macromol., 1980, 2, 199.
180 H.G.Barth and D.A.Smith, J. Chromatogr., 1981, 206, 410.
181 H.Kusunose and M.Sawamura, J. Agric. Chem. Soc. Jpn., 1980, 54, 849.
182 B.V.McCleary, R.Amado, R.Waibel, and H.Neukom, Carbohydr. Res., 1981, 92, 269.
183 P.A.J.Gorin, E.M.Barreto-Bergter, and F.S.Da Cruz, Carbohydr. Res., 1981, 88, 177.
184 D.W.M.Leung, J.D.Bewley, and J.S.G.Reid, Planta, 1981, 153, 95.
185 G.Mandal and A.Das, Carbohydr. Res., 1980, 87, 249.
186 W.D.Pitman, E.J.Soltes, and E.C.Holt, Phytochemistry, 1981, 20, 1129.
187 Y.Ota and K.Maekaji, J. Agric. Chem. Soc. Jpn., 1980, 54, 741.

188 A.H.Gordon and J.S.D.Bacon, Biochem. J., 1981, 193, 765.
189 V.Kováčik, V.Mihalov, and P.Kováč, Carbohydr. Res., 1981, 88, 189.
190 J.P.Gowda, D.C.Gowda, and Y.V.Anjaneyalu, Carbohydr. Res., 1980, 87, 241.
191 J.E.Darvill, M.McNeil, A.G.Darvill, and P.Albersheim, Plant Physiol., 1980, 66, 1135.
192 J.M.Brillouet and C.Mercier, J. Sci. Food Agric., 1981, 31, 243.
193 R.Toman, R.Kohn, A.Malovikova, and J.Rosik, Collect. Czech. Chem. Commun., 1981, 46, 1405.
194 S.H.T.Harper and J.M.Lynch, J. Sci. Food Agric., 1981, 32, 1057.
195 D.Ben-Ghedalia and J.Miron, J. Sci. Food Agric., 1981, 31, 224.
196 A.Chesson, J. Sci. Food Agric., 1981, 32, 745.
197 D.Ben-Ghedalia, G.Sheffet, and J.Miron, J. Sci. Food Agric., 1980, 30, 1337.
198 R.Yamamoto, N.Sakurai, K.Shibata, and Y.Masuda, Plant Cell Physiol., 1980, 21, 373.
199 A.V.Bailey and G.Sumrell, Cereal Chem., 1979, 56, 295.
200 R.C.Hoseney and J.M.Faubion, Cereal Chem., 1981, 58, 421.
201 P.Biely, Z.Krátký, and M.Vršanská, Eur. J. Biochem., 1981, 119, 559.
202 P.Biely, M.Vršanová, and Z.Krátký, Eur. J. Biochem., 1981, 119, 565.
203 G.Dalessandro and D.H.Northcote, Planta, 1981, 151, 53.
204 G.Dalessandro and D.H.Northcote, Planta, 1981, 151, 61.
205 G.P.Bolwell and D.H.Northcote, Planta, 1981, 152, 225.
206 G.Touet and H.G.Aach, Naturwissenschaften, 1979, 66, 525.
207 K.Lundquist, R.Simonson, and K.Tingsvik, Sven. Papperstidn., 1980, 83, 452.
208 J.I.Azuma, N.Takahashi, and T.Koshijima, Carbohydr. Res., 1981, 93, 91.
209 N.N.Das, S.C.Das, A.S.Dutt, and A.Roy, Carbohydr. Res., 1981, 94, 73.
210 Y.Vered, O.Milstein, H.M.Flowers, and J.Gressel, Eur. J. Appl. Microbiol. Biotechnol., 1981, 12, 183.
211 H.G.Barth, J. Liq. Chromatogr., 1980, 3, 1481.
212 G.Ravanat and M.Rinaudo, Biopolymers, 1980, 19, 2209.
213 E.R.Morris, M.J.Gidley, E.J.Murray, D.A.Powell, and D.A.Rees, Int. J. Biol. Macromol., 1980, 2, 327.
214 M.A.F.Davis, M.J.Gidley, E.R.Morris, D.A.Powell, and D.A.Rees, Int. J. Biol. Macromol., 1980, 2, 330.
215 M.J.Gidley, E.R.Morris, E.J.Murray, D.A.Powell, and D.A.Rees, Int. J. Biol. Macromol., 1980, 2, 332.
216 M.Tepfer and I.E.P.Taylor, Can. J. Bot., 1981, 59, 1522.
217 R.Kohn and A.Malovíková, Collect. Czech. Chem. Commun., 1981, 46, 1701.
218 H.Konno, Y.Yamasaki, and J.Ozawa, Agric. Biol. Chem., 1980, 44, 2195.
219 B.J.H.Stevens and R.R.Selvendran, J. Sci. Food Agric., 1980, 30, 1257.
220 M.C.Jarvis, M.A.Hall, D.R.Threlfall, and J.Friend, Planta, 1981, 152, 93.
221 S.Ishii, Phytochemistry, 1981, 20, 2329.
222 G.Chambat, J.-P.Joseleau, and F.Barnoud, Phytochemistry, 1981, 20, 241.
223 M.McNeil, A.G.Darvill, and P.Albersheim, Plant Physiol., 1980, 66, 1128.
224 O.Markovic, R.Kohn, and O.Luknar, Collect. Czech. Chem. Commun., 1981, 46, 266.
225 R.W.Woodward, J.Weaver, and H.G.Floss, Arch. Biochem. Biophys., 1981, 207, 51.
226 N.F.Khalil and H.J.Duncan, J. Sci. Food Agric., 1981, 32, 415.
227 R.D.Reichert, Cereal Chem., 1981, 58, 266.
228 P.S.Belo, jun. and B.O.De Lumen, J. Agric. Food Chem., 1981, 29, 370.
229 R.Gormley, J. Sci. Food Agric., 1981, 31, 392.
230 P.E.Pfeffer, L.W.Doner, P.D.Hoagland, and G.G.McDonald, J. Agric. Food Chem., 1981, 29, 455.
231 G.W.Bates and P.M.Ray, Plant Physiol., 1981, 68, 158.
232 M.G.Hahn, A.G.Darvil, and P.Albersheim, Plant Physiol., 1981, 68, 1161.

233 C.A.Ryan, P.Bishop, G.Pearce, A.G.Darvill, M.McNeil, and P.Albersheim, Plant Physiol., 1981, 68, 616.
234 P.D.Bishop, D.J.Makus, G.Pearce, and C.A.Ryan, Proc. Natl. Acad. Sci. USA, 1981, 78, 3536.
235 O.Mikeš, J.Sedláčková, L.Rexová-Benková, and J.Omelková, J. Chromatogr., 1981, 207, 99.
236 H.Konno, K.Katoh, and Y.Yamasaki, Plant Cell Physiol., 1981, 22, 899.
237 A.S.Reddy, M.Komraiah, and S.M.Reddy, Curr. Sci., 1981, 50, 283.
238 C.R.Barmore and G.E.Brown, Phytopathology, 1981, 71, 328.
239 M.E.Saltveit and R.F.McFeeters, Plant Physiol., 1980, 66, 1019.
240 L.W.Brookhouser, J.G.Hancock, and A.R.Weinhold, Phytopathology, 1980, 70, 1039.
241 S.C.Lee and C.A.West, Plant Physiol., 1981, 67, 633.
242 S.C.Lee and C.A.West, Plant Physiol., 1981, 67, 640.
243 O.Markovic and A.Slezárik, Collect. Czech. Chem. Commun., 1981, 46, 525.
244 G.A.Tucker, N.G.Robertson, and D.Grierson, Eur. J. Biochem., 1981, 115, 87.
245 K.H.Moledina, M.Haydar, B.Ooraikul, and D.Hadziyev, J. Sci. Food Agric., 1981, 32, 1091.
246 M.Ashraf, N.Khan, M.Ahmed, and M.Elahi, J. Agric. Food Chem., 1981, 29, 526.
247 B.Schink, J.C.Ward, and J.G.Zeikus, Appl. Environ. Microbiol., 1981, 42, 526.
248 M.A.S.Chary and S.M.Reddy, Curr. Sci., 1980, 49, 557.
249 J.E.Harris and C.Dennis, J. Sci. Food Agric., 1981, 31, 87.
250 C.I.Speirs, G.C.Blackwood, and J.R.Mitchell, J. Sci. Food Agric., 1980, 30, 1287.
251 P.K.Mandal and A.K.Mukherjee, Carbohydr. Res., 1981, 90, 233.
252 N.K.Miki, K.J.Clarke, and M.E.McCully, Can. J. Bot., 1980, 58, 2581.
253 M.Tomoda, M.Yokoi, A.Torigoe, and K.Maru, Chem. Pharm. Bull., 1980, 28, 3251.
254 B.R.Nayak and T.N.Pattabiraman, Indian J. Biochem. Biophys., 1980, 17, 222.
255 B.R.Nayak and T.N.Pattabiraman, Indian J. Biochem. Biophys., 1981, 18, 202.
256 D.McGarvie and H.Parolis, Carbohydr. Res., 1981, 88, 305.
257 D.McGarvie and H.Parolis, Carbohydr. Res., 1981, 94, 57.
258 D.McGarvie and H.Parolis, Carbohydr. Res., 1981, 94, 67.
259 S.C.Churms, E.H.Merrifield, and A.M.Stephen, Carbohydr. Res., 1981, 90, 261.
260 S.Basu and C.V.N.Rao, Carbohydr. Res., 1981, 94, 215.
261 F.M.Nicolaisen, I.Meyland, and K.Schaumburg, Acta Chem. Scand., 1980, B34, 103.
262 D.J.Brasch, C.T.Chuah, and L.D.Melton, Aust. J. Chem., 1981, 34, 1095.
263 J.N.C.Whyte and J.R.Englar, Phytochemistry, 1981, 20, 237.
264 S.A.Foord and E.D.T.Atkins, Int. J. Biol. Macromol., 1980, 2, 193.
265 H.Grasdalen, B.Larsen, and O.Smidsrød, Carbohydr. Res., 1981, 89, 179.
266 J.N.Liang, E.S.Stevens, S.A.Frangou, E.R.Morris, and D.A.Rees, Int. J. Biol. Macromol., 1980, 2, 204.
267 G.Robinson, E.R.Morris, and D.A.Rees, J. Chem. Soc., Chem. Commun., 1980, 152.
268 C.Rochas, M.Rinaudo, and M.Vincendon, Biopolymers, 1980, 19, 2165.
269 V.J.Morris and P.S.Belton, J. Chem. Soc., Chem. Commun., 1980, 983.
270 V.J.Morris and K.S.Fancey, Int. J. Biol. Macromol., 1981, 3, 213.
271 V.J.Morris and G.R.Chilvers, J. Sci. Food Agric., 1981, 32, 1235.
272 C.Rochas and M.Rinaudo, Biopolymers, 1980, 19, 1675.
273 E.C.Epifanio, R.L.Veroy, F.Uyenco, G.J.B.Cajipe, and E.C.Laserna, Appl. Environ. Microbiol., 1981, 41, 155.
274 M.W.McLean and F.B.Williamson, Eur. J. Biochem., 1981, 113, 447.
275 E.Deflandre and J.Damas, C. R. Seances Soc. Biol. Ses Fil., 1980, 174, 973.

276 J.Damas and G.Volon, C. R. Seances Soc. Biol. Ses Fil., 1979, 173, 637.
277 J.Damas, C. R. Seances Soc. Biol. Ses Fil., 1981, 175, 269.
278 G.K.Moore and G.A.F.Roberts, Int. J. Biol. Macromol., 1980, 2, 73.
279 G.K.Moore and G.A.F.Roberts, Int. J. Biol. Macromol., 1980, 2, 78.
280 M.Miya, R.Iwamoto, S.Yoshikawa, and S.Mima, Int. J. Biol. Macromol., 1980, 2, 323.
281 G.K.Moore and G.A.F.Roberts, Int. J. Biol. Macromol., 1980, 2, 115.
282 Z.Holan, J.Votruba, and V.Vlasáková, J. Chromatogr., 1980, 190, 67.
283 E.D.T.Atkins, J.Dlugosz, and S.Foord, Int. J. Biol. Macromol., 1979, 1, 29.
284 M.Ikeda and K.Shimahara, J. Agric. Chem. Soc. Jpn., 1978, 52, 355.
285 H.Fukuda and Y.Kikuchi, Makromol. Chem., 1979, 180, 1631.
286 H.Fukuda, Bull. Chem. Soc. Jpn., 1980, 53, 837.
287 J.-P.Grivet, F.Delmotte, and M.Monsigny, FEBS Lett., 1978, 88, 176.
288 G.Lundblad, M.Elander, J.Lind, and K.Slettengren, Eur. J. Biochem., 1979, 100, 455.
289 A.Vardanis, Biochim. Biophys. Acta, 1979, 588, 142.
290 T.Sakaguchi, T.Horikoshi, and A.Nakajima, J. Agric. Chem. Soc. Jpn., 1979, 53, 149.
291 L.H.Villarroel and A.B.Zanlungo, Carbohydr. Res., 1981, 88, 139.
292 C.Gudin and D.Thomas, C. R. Hebd. Seances Acad. Sci., 1981, 293, 35.
293 D.B.Hawthorne, T.W.Dreher, and B.R.Grant, Proc. Austral. Biochem. Soc., 1980, 13, 132.
294 T.Fujikawa, K.Koyabu, and M.Wada, J. Agric. Chem. Soc. Jpn., 1979, 53, 87.
295 E.Percival, M.A.Rahman, and H.Weigel, Phytochemistry, 1981, 20, 1579.
296 H.K.Tong, K.H.Lee, and H.A.Wong, Carbohydr. Res., 1981, 88, 168.
297 A.M.J.Fichtinger-Schepman, J.P.Kamerling, C.Versluis, and J.F.G.Vliegenthart, Carbohydr. Res., 1980, 86, 215.
298 A.M.J.Fichtinger-Schepman, J.P.Kamerling, C.Versluis, and J.F.G.Vliegenthart, Carbohydr. Res., 1981, 93, 105.
299 T.A.Sarma and S.Kanta, Z. Allgem. Mikrobiol., 1979, 19, 571.
300 T.A.Sarma and S.Kanta, Z. Allgem. Mikrobiol., 1980, 20, 653.
301 J.Seckbach and J.F.Frederick, Microbios, 1980, 29, 135.
302 J.S.Edmonds and K.A.Francesconi, Nature (London), 1981, 289, 602.

4
Microbial Polysaccharides

BY C. A. WHITE

1 Teichoic Acids

A review has summarized the biosynthesis, assembly, and location of teichoic and teichuronic acids, which includes the biosynthesis of lipoteichoic acids and their involvement in teichoic acid biosytheses.[1]

Membranes from Bacillus subtilis W23 synthesized a lipid precursor of the linkage unit that attaches teichoic acid to the cell wall.[2] It contained glycerophosphoryl-2-acetamido-2-deoxy-D-glucose, linked through an acid-labile bond to a lipid.

Glycerol-3-phosphate cytidylyltransferase, D-ribitol-5-phosphate cytidylyltransferase, and poly(ribitol phosphate) synthetase activities have been measured in cultures of Bacillus subtilis W23 as they became phosphate-starved either in batch culture or during changeover from potassium limitation to phosphate limitation in a chemostat.[3] The results indicated that repression of synthesis of all three enzymes occurred at the onset of phosphate starvation and that this was accompanied by inhibition or inactivation of glycerol-3-phosphate cytidylyltransferase and poly(ribitol phosphate) synthetase. These results show that the initial response to phosphate starvation involves more than inhibition of one enzyme as previously proposed. Syntheses of both linkage unit and poly(ribitol phosphate) are inhibited independently.

Protoplasts of Bacillus subtilis W23 readily synthesized ribitol teichoic acid from nucleotide precursors in the surrounding medium. With CDP-ribitol they made poly(ribitol phosphate), presumably attached to lipoteichoic acid carrier:[4] when CDP-glycerol and UDP-2-acetamido-2-deoxy-D-glucose were also present, a 10-fold increase in the rate of polymer synthesis occurred, and the product contained both the main chain and the linkage unit. Synthesis was inhibited by trypsin or 4-chloromercuribenzenesulphonate in the medium, and it was concluded that it occurred at the outer surface of the membrane. During synthesis, which was also achieved readily by whole cells after a brief period of wall lysis, the CMP portion of the nucleotide precursors did not pass through the membrane. No evidence was obtained for a transphosphorylation mechanism for the translocation process. It was suggested that reaction with exogenous substrates was due to temporary exposure of a protein component of the enzyme complex at the outer surface of the membrane during the normal biosynthetic cycle.

Toluenized cells of Bacillus subtilis W23 synthesized the teichoic acid, poly(ribitol phosphate), from exogenous precursors.[5] The synthesis was dependent on concomitant synthesis of the linkage unit that joins teichoic acid to peptidoglycan. Under conditions that reduced cell autolytic activity, a large proportion of the teichoic acid became linked to the cell wall, independently of peptidoglycan synthesis. The specific activity of the system was more than 30 times that of isolated membranes, so that activity could be measured readily in the cells from 2 ml of an exponential culture of bacteria.

The influence of the lectins concanavalin A and phytohaemagglutinin A and the polyelectrolytes dextran-sulphate (as its sodium salt) and diethylaminoethyl-dextran (as its hydrochloride) on bacterial transformations in Bacillus subtilis has been studied[6] in order to discriminate the target of lectins and polyelectrolytes in competent cells.

Bacillus subtilis var. niger WM was grown in continuous culture under phosphate-limited and under magnesium-limited conditions. Whole cells, cell walls, and the isolated wall polymers peptidoglycan, teichoic acid, and teichuronic acid were analysed by Curie-point pyrolysis mass spectrometry.[7] Characteristic ion peaks for the wall polymers were established and facilitated the interpretation of the mass pyrograms of walls and whole cells. The mass pyrograms of magnesium-limited cells showed the characteristic peaks for protein, peptidoglycan, and teichoic acid.

A tunicamycin-like antibiotic 24010 at a concentration of 1 μg/ml selectively inhibited the in vivo synthesis of glycerol teichoic acid of cell walls in Bacillus cereus AHU 1030.[8] Incubation of membranes of this stain with 2-acetamido-2-deoxy-D-glucopyranosyl pyrophosphorylundecaprenol and UDP-2-acetamido-2-deoxy-D-mannose led to formation of a glycolipid having a saccharide moiety identical with the cell-wall teichoic acid linkage unit (1).

The membranes also catalysed transfer of glycerol phosphate units from CDP-glycerol to this disaccharide-linked lipid. The biosynthesis of the cell-wall glycerol teichoic acid in this strain seemed to involve the disaccharide-linked lipid as an intermediate.

β-D-Man*p*NAc-(1→4)-D-GlcNAc

(1)

The presence of active, or of only reversibly inactivated, enzyme systems for the synthesis and incorporation of teichoic acid has been determined[9] by withdrawing samples of bacteria from a phosphate-pulsed chemostat culture into phosphate-rich medium that contained chloramphenicol. The increased content of wall teichoic acid following incubation of these samples was determined by chemical analysis of the recovered bacteria.

D-Alanyl-lipoteichoic acid from *Lactobacillus casei* contains a poly(glycerol phosphate) moiety that is selectively acylated with D-alanine ester residues.[10] To characterize further the mechanism of D-alanine substitution, intermediates were sought that participate in the assembly of this lipoteichoic acid. From the incorporation system utilizing either toluene-treated cells or a combination of membrane fragments and supernatant fraction, a series of membrane-associated D-{^{14}C}alanyl-lipophilic compounds was found. The assay of these compounds depended on their extractability into monophasic chloroform-methanol-water (0.8:3.2:1.0, vol/vol/vol) and subsequent partitioning into chloroform. Four lines of evidence suggested that the D-alanyl-lipophilic compounds are intermediates in the synthesis of D-alanyl-lipoteichoic acid. First, partial degradation of the poly(glycerol phosphate) moiety of D-alanyl-lipoteichoic acid by phosphodiesterase II/phosphatase from *Aspergillus niger* generated a series of D-alanyl-lipophilic compounds similar to those extracted from toluene-treated cells during the incorporation of D-alanine. Second, enzymatic degradation of the D-alanyl-lipophilic compounds by the above procedure gave D-alanyl-glycerol, the same degradation product obtained from D-alanyl-lipoteichoic acid. Third, the incorporation of D-alanine into these compounds required the same components as the incorporation of D-alanine into membrane-associated D-lipoteichoic acid. Fourth, the phosphate-induced loss of D-{^{14}C}alanine-labelled lipophilic compounds could be correlated with the stimulation of phosphatidylglycerol synthesis in the presence of excess phosphate. These experiments were interpreted to indicate that the D-alanyl-lipophilic compounds were D-alanyl-lipoteichoic acid with short polymer chains and were most likely intermediates in the assembly of the completed polymer, D-alanyl-lipoteichoic acid.

Teichuronic acid isolated from the cell walls of *Micrococcus luteus* has been examined by natural-abundance ^{13}C n.m.r. spectroscopy.[11] Proton-decoupled and proton-coupled spectra were obtained for native teichuronic acid and also after the teichuronic acid had been oxidized with periodate and reduced with borohydride. The spectra are consistent with the structure (2). Teichuronic acid synthesized *in vitro* from suitable substrates by the particulate enzyme fraction obtained from *Micrococcus luteus* yielded a ^{13}C n.m.r. spectrum which is indistinguishable from that of the native teichuronic acid, indicating a structural identity of the teichuronic acid synthesized *in vitro* with that isolated from cell walls.

→4)-β-D-Man*p*ANAc-(1→6)-α-D-Glc*p*-(1→

(2)

The membrane-bound enzymes of *Micrococcus varians* which participate in the syntheses of the teichoic acid main chain and linkage unit have been solubilized with Triton X-100 and fractionated by sucrose density-gradient centrifugation.[12] Two main fractions were obtained: a heavy fraction, containing enzymes effecting synthesis of the main chain attached to the linkage unit, which was associated with only a small amount of lipid, and a light fraction, which was rich in prenyl phosphate and catalysed only linkage-unit synthesis. The separation by density was not based entirely on polypeptide chain length, as some of the shortest chains appeared in the denser fractions and some relatively high-molecular-weight peptides occurred in the lightest fraction. High activity for linkage-unit synthesis was observed in a fraction containing only a few peptides. Addition of ficaprenyl phosphate to the enzyme preparations had no stimulatory effect. It was concluded that the enzymes for main-chain and linkage-unit syntheses form one or more fairly tightly associated complexes and that polyprenyl phosphate is an integral, firmly bound component of the complex in which the linkage unit is synthesized.

The main chain of teichoic acids can be assembled in cell-free membrane preparations by the transfer of residues from the appropriate nucleotide precursors to an incompletely characterized amphiphilic molecule, lipoteichoic acid carrier.[13] However, in the cell wall, the main chain is attached to peptidoglycan through a linkage unit which is synthesized independently. It is believed that, in these cell-free systems, lipid intermediates carrying linkage units are also able to accept residues directly from nucleotide precursors to build up the main chain. It was shown that the main chain attached to lipoteichoic acid carrier was transferred from lipoteichoic acid carrier to lipids containing the linkage unit. Thus, in these systems, there appear to be two routes to the biosynthesis of teichoic acid/linkage-unit complexes, one by direct assembly of the main chain on linkage-unit lipids and the other by transfer of the preassembled main chain from lipoteichoic acid carrier to the linkage unit. It was also shown that linkage-unit lipids from different organisms were interchangeable and that these were used for polymer synthesis by *Bacillus subtilis* 3610, in which the teichoic acid is a poly(glycerol phosphate).

The increasing use of *Staphylococcus aureus* ribitol teichoic acid, which possesses a 2-acetamido-2-deoxy-β-D-glucopyranosyl residue, in serological diagnosis especially in bacteremia and osteoarticular infections prompted the purification of this antigen.[14] In the course of the purification the wheat-germ agglutinin immobilized on Ultrogel, which has affinity for 2-acetamido-2-deoxy-D-glucopyranosyl residues, was employed to separate ribitol teichoic acid from other contaminating antigens. This study was concerned with the abnormal affinity of the 2-acetamido-2-deoxy-β-D-glucosyl ribitol teichoic acid of *Staphylococcus aureus* for the wheat-germ agglutinin immobilized on Ultrogel which is dependent on the ionic forces and the nature of the contaminants.

Native substitution with the D-alanine ester of lipoteichoic acids affects their immunological properties, the capacity to bind divalent cations, and lipoteichoic acid carrier activity.[15] The influence of the D-alanine ester on anti-autolytic activity was tested, using extracellular autolysin from *Staphylococcus aureus* and nine lipoteichoic acids with alanine / phosphorus molar ratios of between 0.23 and 0.71. The inhibitory activity, highest with D-alanine-free lipoteichoic acid, exponentially decreased with increasing alanine content, approaching zero at substitutions of greater than 0.6. Correspondingly, dipolar ionic phospholipids were not inhibitory, in contrast to negatively charged ones.

Glycosylation of lipoteichoic acid up to an extent of 0.5 did not depress inhibitory activity, and even at a degree of 0.8 the effect was comparatively small. On comparison of lipoteichoic acids from various sources, differences in lipid structures and chain lengths were without effect. The inhibitory activity drastically decreased when the glycolipid carried a single glycerophosphate residue or the hydrophilic chain had the unusual structure (3). The same results were obtained with the more complex system of autolysis of *Staphylococcus aureus* cells. It was suggested that the anti-autolytic activity of lipoteichoic acid residues in a sequence of glycerophosphate units and that the negative charges of appropriately spaced phosphodiester groups play a crucial role. The D-alanine ester effect was discussed with respect to the putative *in vivo* regulation of autolysis by lipoteichoic acid.

```
→6)-α-D-Galp-(1→6)-α-D-Galp-O-CH2
                                |
                    α-D-Galp-O-CH
                                |      O
                                CH2-O-P→
                                       OH
```

(3)

Streptococcus mutans Ingbritt was grown in a chemostat at destined dilution rates in either 0.5% D-fructose or 0.5% D-glucitol and at destined pH values in 0.5% fructose. The yield of cells was affected by the carbohydrate source, as well as by the pH, with the lowest yield being at pH 5.5 in 0.5% D-fructose.[16] D-Fructose-grown cells showed greater susceptibility to lysis by a muramidase than the corresponding D-glucose-grown cells, but there were no marked differences in the lytic susceptibilities of the corresponding cell-wall preparations or in the serological reactivities of wall lysates with antiserum to *Streptococcus mutans* Ingbritt. The greatest amounts of cellular lipoteichoic acid were obtained at high dilution rates in both D-fructose and D-glucitol, as well as at high pH values in D-fructose. The greatest amounts of extracellular lipoteichoic acid were found at low dilution rates, as estimated by rocket immunoelectrophoresis and also by haemagglutination. Three major extracellular protein components were separated by sodium dodecyl sulphate-polyacrylamide gel electrophoresis, and the effects of growth conditions on these components were determined. Results for batch-grown cultures showed that there was genotypic variation in the susceptibility of cells to lysis by a muramidase. The enhancement of lipoteichoic acid production of D-fructose and D-glucitol in batch cultures was not identical in representative strains of Streptococcus mutans serotype c, nor was the effect of D-fructose found uniformly in representative strains of the different Streptococcus mutans serotypes.

The problems caused by the non-specific binding of lipoteichoic acid and glucan-binding proteins of Streptococcus mutans to immunosorbent columns prepared from cyanogen bromide-activated Sepharose 4B have been described.[17]

Human polymorphonuclear leukocytes were shown to possess specific binding sites for lipoteichoic acid.[18] Lipoteichoic acid binding was reversible and time and temperature dependent. Scatchard plot analysis revealed an apparently single population of 6.6×10^6 lipoteichoic acid binding sites per human polymorphonuclear leukocytes with a dissociation constant of 5.6 μM. Attachment of an avirulent unencapsulated, M-negative of group A streptococci to human polymorphonuclear leukocytes was inhibited by lipoteichoic acid but not by other bacterial somatic antigens tested. Occupation of 30% of the lipoteichoic acid binding sites resulted in greater than 70% inhibition of streptococcal attachment of human polymorphonuclear leukocytes. In contrast, lipoteichoic acid failed to block attachment of Escherichia coli or antibody-coated streptococci, indicating that binding sites for Escherichia coli and the Fc portion of immunoglobulin G are distinct from those for lipoteichoic acid. Immunofluorescent studies demonstrated that lipoteichoic acid uniformly bound to polymorphonuclear leukocyte membranes for as long as 2 h at 37°C. Cross-linking of human polymorphonuclear leukocytes-bound lipoteichoic acid resulted in rapid capping of lipoteichoic receptor sites. The results suggest that lipoteichoic acid is a monovalent ligand interacting with mobile receptors in the plasma membrane of human polymorphonuclear leukocytes.

^1H and ^{13}C n.m.r. has been shown to distinguish easily between covalently and ionically associated D-alanine in lipoteichoic acids, with ^{13}C n.m.r. showing regular 1,3-linkages in the polyglycerophosphate chain and assignable resonances for those subunits carrying D-alanine and sugar residues.[19]

2 Peptidoglycans

The immunological activities of bacterial peptidoglycans have been reviewed and the structures of particular peptidoglycans and groups of peptidoglycans related to activity.[20]

Peptidoglycan monomer (D-GlcpNAc-MurNAc-L-Ala-D-isoglutamine-meso-diaminopimelic acid-D-Ala-D-Ala), labelled with ^{14}C in both the disaccharide and pentapeptide portions, was incubated with slices of mouse liver, kidney, or spleen as well as with mouse and human blood, blood cells, plasma, and serum.[21] Peptidoglycan monomer was isolated unchanged after incubations with mouse organs and blood cells. However, upon incubation with mouse or human blood, 10-50% of the peptidoglycan monomer underwent hydrolysis to the corresponding disaccharide and pentapeptide. After incubations with plasma and serum more than 90% of the {^{14}C}peptidoglycan monomer was metabolized: about 50% of the administered radioactive dose was recovered in the disaccharide unit and about 35% in the pentapeptide part. It was suggested that in blood, plasma, and serum of mouse and man an N-acetylmuramoyl-L-alanine amidase exists which splits the amide bond between the lactyl carboxyl group of the muramyl residue and the amino group of the peptide moiety in the peptidoglycan molecule.

An isotope-dilution procedure for quantitative N-terminal micro-analysis using ^{3}H-labelled 1-fluoro-2,4-dinitrobenzene and a mixture of free and acetylated ^{14}C-labelled amino acids as internal standards has been described.[22] The 2,4-dinitrophenyl derivatives are separated by polyamide thin-layer chromatography. Although primarily intended for the estimation of crosslinkage in bacterial cell-wall peptidoglycan, the method is capable of extension to other peptides. When tested on lysozyme, the procedure gave results which were in good agreement with accepted values.

The conformational energies of complexes of alternating copolymers of 2-acetamido-2-deoxy-D-glucose and N-acetylmuramic acid with hen egg-white lysozyme have been computed.[23] This involved a complete search of the conformational space at the active site of the enzyme available to these substrates and minimization of the conformational energies of the noncovalent complexes. As with the homopolymer $(D\text{-GlcpNAc})_6$, the hexasaccharide $(D\text{-GlcpNAc-MurNAc})_2$-$(D\text{-GlcpNAc})_2$ binds preferentially on the left side of the active-site cleft, involving residues such as L-Arg-45, L-Asn-46, and L-Thr-47. The alternating copolymer $(D\text{-GlcpNAc-MurNAc})_3$, however, binds with its F-site residue preferentially on the right side of the active-site cleft, involving residues such as L-Phe-34 and L-Arg-114. The lactic acid side chain prevents good binding to the F site on the left side. This result can explain the higher rate of catalysis for the cell-wall substrate (the alternating co-polymer). The relative affinities of the disaccharide D-GlcpNAc-MurNAc for all sequential pairs of sites A-F (including E and F sites on both sides of the cleft) are determined. It is found that the highest affinity of this disaccharide is for sites C and D and right-side sites E and F. The energy of the recently determined X-ray crystallographic structure of MurNAc-D-GlcpNAc-MurNAc bound to the B, C, and D sites of hen egg-white lysozyme has been minimized and found to lead a conformation quite similar to one which has been predicted previously for the trisaccharide $(D\text{-GlcpNAc})_3$. The D ring is undistorted and binds close to the surface of the active-site cleft. The structure can be extended into sites E and F by addition of two D-GlcpNAc residues, but only on the left side of the active-site cleft. This indicates that polymers bound with their D-site residues near the surface of the cleft must bind to sites E and F on the left side of the cleft, as also predicted previously.

A series of seven different UDP-N-acetylmuramyl peptide precursors of bacterial cell-wall peptidoglycan has been examined by reverse-phase high-pressure liquid chromatography.[24] Mixtures of these compounds were successfully and rapidly analysed by using Waters µBondapak C_{18} column as a stationary phase and isocratic elutions with 0.05M ammonium phosphate or formate buffers of appropriate pH. Accurate quantitation could also be readily achieved by reverse-phase high-pressure liquid chromatography. All these techniques are extremely useful for the purification of these compounds and for a wide range of biochemical studies concerning the cytoplasmic steps of the biosynthesis of peptidoglycan.

The solid-state conformational analysis of Ac-D-Ala-D-Ala-OH.H_2O, carried out by infrared absorption and X-ray diffraction, has indicated that the molecules are not extended in a regular conformation, but rather that they are partially folded,[25] the ϕ, ψ* torsional angles of the carboxyl-terminal reside in particular being in the region of the left-handed α-helix of the Ramachandran map. The acetylamino and peptide groups are found in the usual trans conformation, the latter exhibiting a deviation from rigid planarity. Only intermolecular hydrogen bonds occur in the crystal state. The solution conformational analysis, performed by infrared absorption and c.d., has revealed that the amount of intramolecular N—H···O═C hydrogen-bonded folded forms, if any, should be extremely small, even in deuteriochloroform at high dilution. In water, solvated, unordered species largely predominate.

Incubation of growing *Bacillus subtilis* with penicillin G led to the secretion of a peptidoglycan-related polymer and a nonglycan-bound pentapeptide into the culture medium.[26] The secreted polymer was isolated and characterized as a linear cell-wall glycan strand substituted predominantly by uncross-linked pentapeptide side chains. Polymer formation and secretion were most likely the result of continued synthesis and elongation of nascent glycan strands in the absence of subsequent processing by peptidoglycan transpeptidase of D-alanine carboxypeptidase enzymes. The nonglycan-bound pentapeptide L-Ala-D-iso-Glu-*meso*-diaminopimelic acid-D-Ala-D-Ala was probably formed by an N-acetylmuramoyl-L-alanine amidase active on the peptide side chains of the uncross-linked polymer. The uncross-linked peptidoglycan polymer was shown to be a good subtrate for penicillin-sensitive D-alanine carboxypeptidase purified from membranes of *Bacillus subtilis*, *Bacillus stearothermophilus*, and *Escherichia coli*. D-Alanine release was not, however, coupled to the cross-linking of peptide side chains, suggesting that these enzymes do not function as peptidoglycan transpeptidase *in vivo*. No transpeptidase or D-alanine carboxypeptidase activity was detected in mixtures of high-molecular-weight penicillin-binding proteins from *Bacillus subtilis*, *Bacillus stearothermophilus*, or *Staphylococcus aureus*. Possible reasons for the inability to demonstrate these activities are discussed. In addition, an N-acetylmuramoyl-L-alanine amidase activity which copurifies with penicillin-binding proteins with *Bacillus subtilis*, *Staphylococcus aureus*, and *Escherichia coli* was partially characterized.

Cell-wall polymers were measured both in the cells and in the cell-free medium of samples from steady-state chemostat cultures of *Bacillus subtilis*, growing at various rates under magnesium or phosphate limitation.[27] The presence of both peptidoglycan and anionic wall polymers in the culture supernatant showed the occurrence of wall turnover in these cultures. Variable proportions of the total peptidoglycan present in the culture samples were found outside the cells in duplicate cultures, indicating that the rate of peptidoglycan turnover is variable in *Bacillus subtilis*. Besides peptidoglycan, anionic wall polymers were detected in the culture supernatant, teichoic acid in magnesium-limited cultures, and teichuronic acid in phosphate-limited cultures. In several samples, the ratio between the peptidoglycan and the anionic-polymer concentrations was significantly lower in the extracellular fluid than in the walls. This divergency was attributed to the occurrence of direct secretion of anionic polymers after their synthesis.

An enzyme which catalyses the hydrolysis of acetamido groups of 2-acetamido-2-deoxy-D-glucopyranosyl residues in cell-wall peptidoglycan was found in the supernatant and 20,000g pellet fractions of Bacillus cereus.[28] Autolysis of the latter fraction resulted in solubilization and activation of the deacetylase. Among various bacteria, strains of Bacillus cereus which contain high proportions of 2-amino-2-deoxy-D-glucopyranosyl residues are particularly rich in the deacetylase. The peptidoglycan deacetylase is distinguishable from 2-acetamido-2-deoxy-D-glucose-6-phosphate deacetylase on the basis of their cellular distribution and chromatographic behaviour. The rate of reaction of the deacetylase with (D-GlcpNAc-MurNAc)$_3$ is less than 1/100 of that with peptidoglycan, while the enzyme is inactive towards (D-GlcpNAc-MurNAc)$_2$-D-GlcpNAc-MurNAc, and monomeric 2-acetamido-2-deoxy-D-glucose derivatives. The enzyme also deacetylates partially O-hydroxyethylated chitin, for half-maximum activities were found to be 0.29 and 6.9 mg per ml (0.17 and 20mM with respect to 2-acetamido-2-deoxy-D-glucopyranosyl residues), respectively. The occurrence of this enzyme accounts for the formation of cell-wall peptidoglycan containing non-acetylated 2-amino-2-deoxy-D-glucopyranosyl residues.

The cell walls isolated from axenically grown leprosy-derived corynebacteria were submitted to various chemical and enzymatic degradations. The glycan strands of the wall peptidoglycan were essentially composed of 2-acetamido-2-deoxy-D-glucopyranosyl-N-acetylmuramic acid disaccharide units.[29] Small amounts of 2-acetamido-2-deoxy-D-glucopyranosyl-N-glycolylmuramic acid (less than 10%) were also detected. The muramic acid residues of adjacent glycan strands are substituted by amidated tetrapeptide units which, in turn, are cross-linked through direct linkages extending between the C-terminal D-alanine residue of one tetrapeptide and the meso-diaminopimelic acid residue of another tetrapeptide. Such a structure is very similar to that of the wall peptidoglycan found in the taxonomically related micro-organisms of Corynebacterium, Mycobacterium, and Norcardia groups.

Two different cell-wall peptidoglycan synthetase systems are carried by penicillin-binding proteins 1A and 1B purified from

Escherichia coli. Both systems consist of two enzyme activities carrying out successive reactions of peptidoglycan synthesis from the lipid-linked precursor 2-acetamido-2-deoxy-D-glucopyranosyl-N-acetylmuramyl-(pentapeptide)-diphosphate-undecaprenol, namely those of peptidoglycan transglycosylase and β-lactam antibiotic-sensitive transpeptidase. The activities of the two enzyme systems differ in optimal conditions and sensitivities of β-lactam antibiotics. The properties of the purified penicillin-binding protein 1A have been reported.[30]

A purified preparation of N-acetylmuramoyl-L-alanine amidase, a murein hydrolase from Escherichia coli, was found to lose its activity during incubation in the presence of bacterial phospholipid suspensions.[31] Whether it was co-dispersed with the phospholipids or added to sonicated phospholipid suspension, the enzyme was inhibited (or inactivated) from the first minutes of incubation at 37°C. As the phosphatidylglycerol/cardiolipin ratio of the phospholipid suspension was increased (all other things being equal), a further decrease of amidase activity was observed. The highest losses of activity were found after co-dispersion of the enzyme and the substrate together with the phospholipids, the resulting suspension being formed of larger multilayered vesicles, as revealed by electron microscopy. In these conditions, the effect on enzyme activity was only partially accounted for by the proportion of the enzyme that was entrapped in the vesicles. The entrapment capacity of the enzyme (using a ^{35}S-labelled enzyme preparation) and of the substrate (^{3}H-labelled) by the multilamellar phospholipidic vesicles did not significantly change as a function of their relative content of phosphatidylglycerol and cardiolipin.

The mechanism by which amino acid starvation of Escherichia coli induces resistance against the lytic and bactericidal effects of penicillin has been studied.[32] Starvation of Escherichia coli strain W7 of the amino acids L-lysine or L-methionine resulted in the rapid development of resistance to autolytic cell-wall degradation, which may be effectively triggered in growing bacteria by a number of chemical or physical treatments. The mechanism of this effect in the amino acid-starved cells involved the production of a murein relatively resistant to the hydrolytic action of crude murein hydrolase extracts prepared from normally growing Escherichia coli. Resistance to the autolysins was not due to the covalently linked lipoprotein. Resistance to murein hydrolase developed most rapidy and most exclusively in the portion of cell wall synthesized after the onset of amino acid starvation. Lysozyme digests of the autolysin-resistant murein synthesized during the first 10 min of L-lysine starvation yielded (in addition to the characteristic degradation products) a high-molecular-weight material that was absent from the lysozyme digests of control cell-wall preparations. It has been proposed that inhibition of protein synthesis causes a rapid modification of murein structure at the cell-wall growth zone in such a manner that attachment of murein hydrolase molecules is inhibited. The mechanism may involve some aspects of the relaxed control system since protection against penicillin-induced lysis developed much slower in amino acid-starved relaxed controlled (relA) cells than in isogenic stringently controlled ($relA^+$) bacteria.

Physiological concentrations of guanosine 5'-diphosphate inhibited the synthesis of lipid intermediates and peptidoglycan catalysed by a particulate enzyme preparation from Escherichia coli.[33] The inhibition of these reactions was dependent on the concentration of guanosine 5'-diphosphate 3'-diphosphate and $MgCl_2$ in the assay. The degree of inhibition of lipid-intermediate synthesis decreased as the molar ratio of $MgCl_2$: guanosine 5'-diphosphate 3'-diphosphate was increased, and no inhibition was observed above a $MgCl_2$: guanosine 5'-disphosphate 3'-diphosphate ratio of 2.5. The synthesis of peptidoglycan was more sensitive to inhibition by guanosine 5'-disphosphate 3'-diphosphate, and significant inhibition occurred under conditions where lipid intermediate synthesis was unaffected (i.e. at $MgCl_2$: guanosine 5'-diphosphate 3'-diphosphate ratios of 2.5 or more). A variety of other nucleotides did not inhibit the synthesis of lipid intermediates and peptidoglycan.

Azureomycin B (10 µg/ml), a new antibiotic from Pseudonocardia azurea nov. sp., caused the accumulation of lipid intermediate and inhibition of peptidoglycan synthesis in an in vitro system using a particulate fraction from Bacillus megaterium KM with UDP-MurNAc-{^{3}H}pentapeptide and cold UDP-D-GlcpNAc or cold UDP-MurNAc-pentapeptide and UDP-{^{3}H}-D-GlcpNAc as substrates.[34] At higher concentrations of azureomycin B (over 100µg/ml), lipid-intermediate accumulation was also inhibited. When particulate fraction from Escherichia coli Y-10 and UDP-{^{14}C}-D-GlcpNAc and cold UDP-MurNAc-pentapeptide were used, accumulation of lipid intermediate and inhibition of peptidoglycan synthesis were also observed. This indicated that the primary target of azureomycin B is the transfer of the disaccharide peptide unit (D-GlcpNAc-MurNAc-pentapeptide) from lipid-bound precursor to acceptor.

meso-Lanthionine was identified in the hydrolysate of *Fusobacterium nucleatum* peptidoglycan by g.c.m.s. and the distribution of this unusual sulphur-containing dibasic amino acid among peptidoglycans of various species and strains of genus *Fusobacterium* investigated.[35] The peptidoglycans of *Fusobacterium* bacteria could be subdivided according to constituent dibasic amino acid into: a lanthionine type, a diaminopimelic acid type, and a mixed type.

The amino acids in the peptidoglycan of *Fusobacterium nucleatum* Fev 1 have been reported to be D-glutamic acid, *meso*-lanthionine, and D-(42%) and L-alanine (58%).[36] About 70% of the lanthionine residues were not susceptible to dinitrophenylation, evidently because they are involved in cross-linkages. Consequently, lysozyme digestion of the peptidoglycan yielded 20 to 25% uncross-linked disaccharide tri- and tetra-peptides. A chemical analysis of isolated glycopeptides indicated that the structure of the building block of this peptidoglycan is 2-acetamido-2-deoxy-D-glucose-*N*-acetylmuramic acid-L-alanine-D-glutamic acid-*meso*-lanthionine(-D-alanine). Evidence has been presented to support the classification of the *Fusobacterium nucleatum* Fev 1 peptidoglycan as a new A1δ, directly cross-linked, *meso*-lanthionine-containing peptidoglycan.

The density properties of bacterial cell walls, when centrifuged under conditions of low osmolality, support the evidence for an open structure for cell walls from *Micrococcus lysodeikticus*, *Bacillus licheniformis*, *Lactobacillus fermentum*, and *Proprionibacterium theonii*.[37]

The C-mycosidic glycopeptidolipid typing antigens from all serovars in the *Mycobacterium avium*/*Mycobacterium intracellulare*/*Mycobacterium scrofulaceum* complex have been examined to varying extents.[38] Detailed analysis of those from serovars 8, 9, 16, and 25 show that the antigens consist of short acetylated oligosaccharides linked to a common fatty acyl-peptidyl-*O*-(3,4-di-*O*-methyl-L-rhamnose) core (4). The oligosaccharide units, in a form suitable for chemical studies, were liberated as oligosaccharide alditols on treatment of the glycopeptidolipids with alkaline borohydride solution. The alditol in the reduced oligosaccharides from all sources was 6-deoxy-L-talitol. Moreover L-rhamnose was also always present, indicating that a basal disaccharide, L-rhamnosyl-6-deoxy-L-talosyl, is always linked to the *allo*-threonine in the acylpeptide. In addition the oligosaccharides from the glycopeptidolipids of each serovar were distinguished by their own individualistic sugars: 3-O-methyl-D-glucose in serovar 8 2,3-di-O-methyl-L-fucose in serovar 9, 2-O-methyl-L-fucose in serovar 25, 4-O-methyl-L-rhamnose in the oligosaccharide from one of the two glycopeptidolipids in serovar 16, and apparently another L-rhamnose substituent in the other oligosaccharide. The glycopeptidolipid antigens in their structural principals, cellular location, and physiological role bear a striking miniscular resemblance to cell wall components of other bacteria such as O-antigenic and R-antigenic lipopolysaccharides.

```
{Fatty Acyl}-D-Phe-D-αThr-D-Ala-L-Alaninol-O-L-Rhap3,4Me2
                      |
                      O
                      |
          {Oligosaccharide}-O-Ac
```

(4)

The specific oligosaccharide units of the C-mycisidic glycopeptidolipid antigens from serovarieties in the *Mycobacterium avium*/*Mycobacterium intracellulare*/*Mycobacterium scrofulaceum* complex were liberated as oligosaccharide alditols by treatment of the glycopeptidolipids with alkaline borohydride. The complete structures of the oligosaccharide alditols have been derived from the ^{1}H n.m.r. spectra or those of their permethylated (or pertrideuteriomethylated) derivatives, the mass spectra of the methylated derivatives, and from methylation fragmentation analysis.[39] Periodate oxidation was also used to confirm the position of the link between the ultimate and penultimate sugars. Structures (5)-(7) were proposed (with some extrapolation of enantiomeric configurations where evidence for assignment is not yet complete) for the oligosaccharide alditols from serovars 8, 9, and 25, respectively. 6-Deoxy-L-talitol (6-deoxy-L-talose in the original glycopeptidolipid) invariably occupies the reducing terminus. L-Rhamnose is invariably the penultimate sugar, and the link between L-rhamnose and 6-deoxy-L-talose is invariably (1→2). Moreover, the results point to the outer one or two appendages for the provision of individually distinctive features required for antigen specificity.

The surface properties of cells of *Mycobacterium* BCG, *Mycobacterium phlei*, *Mycobacterium smegmatis*, and *Mycobacterium microti* have been shown to be identical, irrespective of the growth medium, the age of the cells, and the colonial morphology, and after various vigorous treatments.[40] The negative surface charge for cells of all these species arises from the phosphate groups of phosphodiester linkages between the peptidoglycan and the arabinogalactan of the basic cell-wall structure, which is common to all species of *Mycobacteria*.

The peptidoglycan of *Neisseria gonorrhoeae* has been found to contain *O*-acetyl groups, by use of the hydroxamate formation reaction, and found to be only partly sensitive to lysozyme.[41]

β-*D*-Glc*p*3Me-(1→3)-α-*L*-Rha*p*-(1→2)-6-deoxy-*L*-Talitol

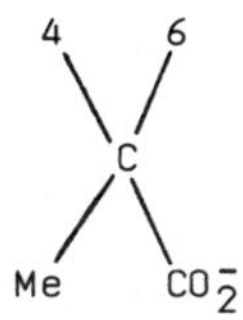

(5)

α-*L*-Fuc*p*2,3Me_2-(1→4)-α-*L*-Fuc*p*2,3Me_2-(1→3)-α-*L*-Rha*p*-(1→2)-
6-deoxy-*L*-Talitol

(6)

α-*L*-Fuc*p*2Me-(1→4)-α-*L*-Fuc*p*2Me-(1→3)-α-*L*-Rha*p*(1→2)-6-deoxy-*L*-Talitol

(7)

Low concentrations of β-lactam antibiotics caused an increased uptake of radioactive 2-amino-2-deoxy-*D*-glucose into the sodium dodecyl sulphate-insoluble peptidoglycan of growing *Neisseria gonorrhoeae*.[42] There was no appreciable change in the (small) amount of sodium dodecyl sulphate-soluble polymer present in the cultures. The sodium dodecyl sulphate-insoluble product in control cells was only partially dissolved by egg-white lysozyme (about 40%), but could all be released by the *Chalaropsis* B muramidase. In cells exposed to β-lactams the proportion of labelled peptidoglycan susceptible to lysozyme increased to 60%. Examination of the *Chalaropsis* B digests by thin-layer chromatography showed that they contained disaccharide-peptide monomers with and without *O*-acetylation and bis-disaccharide-peptide dimers with one or two *O*-acetyl groups, or with none. β-Lactam antibiotics caused a decrease in the degree of *O*-acetylation but did not greatly affect the amount of peptidoglycan cross-linking. They also enlarged the bacteria and conserved and thickened the septa that could be observed in thin sections by electron microscopy. The relationship between these results and effects of β-lactams on *in vitro* synthesis of peptidoglycan by ether-treated *Neisseria gonorrhoeae* has been discussed.

The murein of *Proteus mirabilis* is reported to be unique with respect to the *O*-acetylation of part of its *N*-acetylmuramic acid residues.[43] Working with synchronized cells and providing radioactive 2-acetamido-2-deoxy-*D*-glucose with the medium, the murein was pulse-labelled *in vivo* for 10 min periods at different times of the cell cycle. After isolation of murein and enzymatic cleavage with *endo*-*N*,*O*-diacetylmuramidase from *Chalaropsis*, the distribution of radio-activity between uncrosslinked murein subunits (monomers) and the peptide-crosslinked dimers was analysed. It was found that the murein synthesized at different times during the cell cycle did not show significant variation regarding the degree of crosslinkage and the degree of *O*-acetylation. However, the degree of *O*-acetylation of pulse-labelled murein was found to be about sevenfold lower than that of murein from continuously labelled asynchronous cultures. A pulse-chase experiment revealed that only non-*O*-acetylated murein subunits were incorporated into the growing murein sacculus, part of which became *O*-acetylated subsequently. Since a certain amount of newly incorporated subunits was lost during the chase, which reappeared in the sacculus about one generation later, limited murein turnover during the cell cycle became evident. The degree of crosslinkage of both the *O*-acetylated as well as the non-*O*-acetylated newly synthesized murein increased during subsequent growth. The labelled peptide-crosslinked dimers were cleaved by an endopeptidase isolated *Proteus mirabilis*. The distribution of radioactivity between the resulting non-*O*-acetylated and *O*-acetylated monomers indicated that most of the mono-*O*-acetylated dimer fraction carried radioactivity exclusively in the non-*O*-acetylated subunit which was non-*O*-acetylated during subsequent biosynthesis. Since only a small portion of the labelled mono-*O*-acetylated dimer fraction carried radioactivity in both subunits, and only at specific times of murein biosynthesis, it was concluded that this portion of the labelled mono-*O*-acetylated dimer fraction represented an intermediate in the *O*-acetylation process. A defined sequence of steps of murein biosynthesis has been proposed as a result of the results described.

A novel murein building block, 2-acetamido-2-deoxy-*D*-glycosyl-

N-acetylmuramoyl-dipeptide, formed during the cell-division cycle of Proteus mirabilis, has been reported.[44] The 2-acetamido-2-deoxy-D-glucose residue was found to contain, in some instances, O-acetyl groups.

The Pseudomonas aeruginosa outer membrane was isolated with attached peptidoglycan and fractioned with Triton X-100, ethylenediaminetetra-acetate, and lysozyme. The data suggested that major outer membrane proteins F, H2, and I are noncovalently associated with the peptidoglycan.[45]

The unique presence of a polyamine covalently linked to a peptidoglycan has been reported.[46,47] The polyamine cadaverine was found to exist as a component of cell-wall peptidoglycan of Selenomonas ruminantium, a strictly anaerobic bacterium. {^{14}C}Cadaverine added to the growth medium was incorporated into the cells, and about 70% of the total radioactivity incorporated was found in the peptidoglycan fraction. When the {^{14}C}cadaverine-labelled peptidoglycan preparation was acid hydrolysed, all of the ^{14}C counts were recovered as cadaverine. The {^{14}C}cadaverine-labelled peptidoglycan preparation was digested with lysozyme into three small fragments which were radioactive and were positive in ninhydrin reaction. One major spot, a compound of the fragments, was composed of DL-alanine, D-glutamic acid, diaminopimelic acid, cadaverine, muramic acid, and 2-amino-2-deoxy-D-glucose. One of the two amino groups of cadaverine was covalently linked to the peptidoglycan, and the other was free. The chemical composition of the peptidoglycan preparation of this strain was determined to be as follows: L-alanine-D-alanine-D-glutamic acid-meso-diaminopimelic acid-cadaverine-muramic acid-2-amino-2-deoxy-D-glucose (1.0:1.0:1.0:1.0:1.1:0.9:1.0).

The {^{14}C}cadaverine-labelled peptidoglycan was degraded with the lytic enzymes prepared from Streptomyces albus G into three small fragments including a major fragment (band A compound).[48] Band A compound was composed of L-alanine, D-glutamic acid, meso-diaminopimelic acid, D-alanine, and cadaverine in the molar ratio 0.98:1.0:1.0:0.98:0.97. Diaminopimelic acid, L-alanine, and cadaverine were N-terminal residues in band A compound. When the {^{14}C}cadaverine-labelled band A compound was subjected to partial acid hydrolysis, two peptide fragments were obtained. One of them consisted of diaminopimelic acid and D-alanine, diaminopimelic acid was the N-terminal amino acid, and the other fragment was composed of L-alanine, D-glutamic acid, and cadaverine, of which L-alanine and cadaverine were N-terminal. It was concluded that the primary peptide structure of band A compound is L-alanyl-D-glutamoyl-meso-diaminopimelyl-D-alanine and that cadaverine links covalently to the D-glutamic acid residue.

Streptococcus mutans BHT was grown in Todd-Hewitt dialysate medium containing 2-acetamido-2-deoxy-D-{^{14}C}glucose for 6 to 11 generations. After treatment with cold and hot trichloroacetic acid and trypsin, 52 to 65% of the radioactivity remained present in insoluble peptidoglycan-containing residues. Hen egg-white lysozyme or mutanolysin treatment of the peptidoglycan residues resulted in the release of 80 and 97%, respectively, of the ^{14}C label to the supernatant fraction.[49] Hydrochloric acid hydrolysates of such supernatants showed that essentially all of the radioactivity present in insoluble peptidoglycan fractions was present in compounds that comigrated on paper chromatography with 2-amino-2-deoxy-D-glucose (60%) or muramic acid (30%). Treatment of whole cells with low and high concentrations of lysozyme alone resulted in losses of 45 and 70% of the insoluble peptidoglycan, respectively, yet release of deoxyribonucleic acid from cells was not detected. Sequential addition of appropriate concentrations of selected inorganic salts after lysozyme treatment did result in the liberation of deoxyribonucleic acid. Deoxyribonucleic acid release was correlated with a further release of peptidoglycan from the insoluble fraction. However, the total amount of peptidoglycan lost effected by the low concentration of lysozyme and NaSCN (lysis) was significantly less than the amount of peptidoglycan hydrolysed by high concentrations of lysozyme alone (no lysis), which suggested that the overall amount of peptidoglycan lost did not correlate well with cellular lysis. The total amount of insoluble peptidoglycan lost at the highest salt concentrations tested was found to be greater than could be accounted for by lysozyme-sensitive linkages of the peptidoglycan, possibly implicating autolysins. The results obtained suggested that hydrolysis of peptidoglycan bonds in topologically localized, but strategically important sites was a more significant factor in the sequence that results in loss of cellular integrity (lysis).

A peptidoglycan layer of Treponema pallidum kazan was isolated by solubilization of whole cells with 1% warm sodium dodecyl sulphate and subsequent digestion of an insoluble residue with proteases. Electron microscopy revealed that the peptidoglycan was isolated as a single-layered sacculus of less than 5 nm thickness, freed from axial filaments and an envelope sheath.[50] An isolated peptidoglycan fraction was mainly composed of 2-amino-2-deoxy-D-glucose, muramic acid, alanine, D-glutamic acid, L-orthithine, and glycine in molar ratios of 0.65:0.68:1.63:1.00:0.75:1.03. Amino- and carboxyl-terminal amino acid analyses suggested the involvement of at least a part of the glycine residue in cross-linking between the amino group of L-ornithine residue at one strand of the stem peptide subunit and the carboxyl group of alanine of the neighbouring strand. The treponemal peptidoglycan lacked the immunoadjuvant activity both to stimulate antibody production and to induce delayed-type hypersensitivity against ovalbumin, as well as the properties necessary to stimulate guinea-pig and mouse splenocytes and guinea-pig peritoneal macrophages, unlike the cell walls or peptidoglycans (group A type of Schleifer and Kendlers' classification) isolated from many bacterial species parasitic to the mammal. However, the peptidoglycan activated the human complement system through the alternative pathway, as well as the classical one, and caused a liberation of 5-hydroxytryptamine in rabbit blood platelets in a similar manner to the cell-wall peptidoglycans of both group A and B types.

3 Lipopolysaccharides

The chemistry and biological significance of 3-deoxy-D-*manno*-2-octulosonic acid have been reviewed,[51] including its structure elucidation and its location and linkages in bacterial polysaccharides, chemical synthesis and monosaccharide chemistry, and enzymology of an inhibitor to metabolism. Bacterial lipopolysaccharide and its lipid A component has been reviewed briefly.[52] Confirmation of the structure of 3-deoxy-D-*manno*-2-octulosonic acid by X-ray crystallography has been brought nearer by the report of the X-ray structure of methyl (methyl 4,5,7,6-tetra-O-acetyl-3-deoxy-α-D-*manno*-2-octulopyranosid)onate (8).[53]

The lipopolysaccharide from *Actinobacillus actinomycetemcomitans* strains Y4 and N27 was isolated by the phenol-water procedure. Morphologically, the molecule consisted of ribbon and branched filaments which comprised 3% of the cellular dry weight.[54] Chemical analysis of the isolated and purified lipopolysaccharide of both strains showed them to consist of carbohydrate, lipid, 3-deoxy-D-*manno*-2-octulosonic acid, heptose, amino-deoxyhexose, and phosphate. The major fatty acids of the lipid A moiety were myristic and β-hydroxymyristic acids. L-Rhamnose, L-fucose, D-galactose, D-glucose, heptose, 2-amino-2-deoxy-D-glucose, and 2-amino-2-deoxy-D-galactose comprised the monosaccharide portion of the lipopolysaccharide. Biological-activity studies revealed both lipopolysaccharide molecules to be active in the Schwartzman reaction and *in vitro* ^{45}Ca bone resorption, as well as in macrophage activation and lethality and in platelet aggregation.

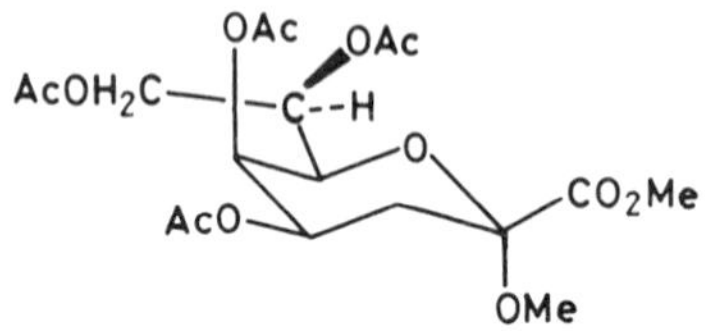

(8)

The core oligosaccharide of *Aeromonas hydrophila* (Chemotype III) lipopolysaccharide has been investigated.[55] The studies involved the use of methylation analysis, oxidation and chromium trioxide, partial hydrolysis with acid, periodate oxidation, Smith degradation, and tagging of the reducing end group. The core is unusual in having 3-acetamido-3,6-dideoxy-L-glucose as a constituent. Structure (9) was proposed for the core oligosaccharide.

```
                                             α-D-Glcp
                                                 1
                                                 |
                                                 6
dd-β-L-GlcpNAc-(1→3)-α-D-Galp-(1→3)-α-LD-Hepp-(1→2)-α-LD-Hepp
                                                 4
                                                 |
                                                 1
                                             α-D-Glcp
```

LD-Hepp = L-*glycero*-D-*manno*-heptopyranose

(9)

The sugar composition of the O-antigenic lipopolysaccharides isolated from Group F (once classified as *Aeromonas*) vibrios was analysed.[56] 3-Deoxy-D-manno-2-octulosonic acid was totally absent from the lipopolysaccharides. As common component sugars, D-glucose, D-galactose, L-glycero-D-manno-heptose, and 2-amino-2-deoxy-D-glucose were present. The Group F vibrios examined were found to be divided into two groups, designated tentatively as groups I and II, on the basis of the pattern of the sugar composition of their lipopolysaccharides. As additional sugar components, 2-amino-2-deoxy-D-mannose, 2-amino-2,6-dideoxy-D-glucose, and two unidentified amino sugars were present in group I, while L-rhamnose, 2-amino-2-deoxy-D-galactose, an unidentified amino sugar, and a relatively high content of D-glycero-D-manno-heptose were found in group II.

The main fatty acids present in lipopolysaccharides from *Bacteroides fragilis* NCTC 9343 were identified as 13-methyl-tetradecanoic, D-3-hydroxypentadecanoic, D-3-hydroxyhexadecanoic, D-3-hydroxy-15-methyl-hexadecanoic, and D-3-hydroxyheptadecanoic acids.[57] Of these 13-methyl-tetradecanoic acid is exclusively ester bound, and 3-hydroxy-15-methyl-hexadecanoic acid is exclusively involved in amide linkage. The other 3-hydroxy fatty acids are both ester and amide bound. All 3-hydroxy fatty acids possess the D-configuration, and the 3-hydroxyl group of ester-linked 3-hydroxy fatty acids is not substituted. Lipopolysaccharides of related *Bacteroides* species (*Bacteroides thetaiotaomicron*, *Bacteroides ovatus*, *Bacteroides distasonis*, and *Bacteroides vulgatus*) showed a fatty-acid spectrum with similar and distinct features compared to that of *Bacteroides fragilis* lipopolysaccharides.

Purified lipopolysaccharide extracted with phenol-water from smooth *Brucella abortus* was hydrolysed with 1% acetic acid at 100°C.[58] The degraded polysaccharide released gave reactions of identity with the native polysaccharide hapten in phenol-water- or trichloroacetic acid-extracted endotoxin preparations of *Brucella abortus* with the polysaccharide extracted by trichloroacetic acid from rough *Brucella melitensis* strain B115. The latter polysaccharide was present in the soluble cytoplasmic fraction, but not in the membrane fraction, of disrupted B115 cells. It could not be extracted from three rough mutants of *Brucella abortus* or from *Brucella canis* or *Brucella ovis* cells. Both the degraded polysaccharide and native polysaccharide hapten shared determinants present on smooth lipopolysaccharide and missing from the rough *Brucella melitensis* polysaccharide. Sugars found in purified lipopolysaccharide, the native hapten, and degraded polysaccharide

```
α-D-GlcpNAc-(1→2)-α-D-Glcp-(1→2)-α-D-Galp-(1→3)-α-D-Glcp-(1→
                                                          6
                                                          |
                                                          1
                                                      α-D-Galp
```

(10)

```
α-D-Galp-(1→2)-α-D-Galp-(1→2)-α-D-Glcp-(1→3)-α-D-Glcp-(1→
                               3
                               |
                               1
                           β-D-Glcp
```

(11)

```
α-D-GlcpNAc-(1→2)-α-D-Glcp-(1→2)-α-D-Glcp-(1→3)-α-D-Glcp-(1→
                                                          6
                                                          |
                                                          1
                                                      α-D-Galp
```

(12)

```
α-D-Glcp-(1→2)-α-D-Glcp-(1→2)-α-D-Galp-(1→3)-α-D-Glcp(1→
                               3
                               |
                               1
                          α-D-GlcpNAc
```

(13)

```
α-D-Galp-(1→2)-α-D-Galp-(1→2)-α-D-Glcp-(1→3)-α-D-Glcp-(1→
                               4
                               |
                               1
                           β-D-Galp
```

(14)

```
β-D-GlcpNAc-(1→6)-α-D-Glcp-(1→2)-α-D-Glcp-(1→3)-α-D-Glcp-(1→
                                                          6
                                                          |
                                                          1
                                                      α-D-Galp
```

(15)

included D-mannose, D-glucose, 2-amino-2,6-dideoxy-D-glucose, 2-amino-2-deoxy-D-glucose, and 3-deoxy-D-*manno*-2-octulosonic acid. The *Brucella melitensis* polysaccharide contained only a trace amount of 2-amino-2,6-dideoxy-D-glucose and no 3-deoxy-D-*manno*-octulosonic acid detectable by the thiobarbiturate assay. Sera from some rabbits immunized with pure smooth lipopolysaccharide and some, but not all, cows infected with field strain of *Brucella abortus* recognized the determinants missing from the *Brucella melitensis* polysaccharide. A subclass-specific enzyme-linked immunoassay showed that most of the antibody in sera from infected cows which binds to smooth lipopolysaccharide and to the native hapten is of the immunoglobulin G1 subclass.

The chemical structure of the O-specific polysaccharide of *Citrobacter* O36 lipopolysaccharide has been established by methylation analysis using gas-liquid chromatography and mass spectrometry.[59] This polysaccharide is a linear homopolymer composed of (1→2)-linked 4-deoxy-β-D-*arabino*-hexopyranosyl residues.

The structures for the hexose regions of cores from Enterobacteraceae lipopolysaccharides have been investigated, using specific degradation and ^{1}H n.m.r. studies as the principal methods.[60] Complete structures for these regions in the *Salmonella*, the *Escherichia coli* R1, R2, R3, R4, and *Eschericia coli* K12 and *Escherichia coli* B cores are proposed ((10)-(16), respectively). The apparent similarities between these structures indicate a close relationship.

α-D-Glc*p*(1→3)-α-D-Glc*p*(1→

(16)

They are all composed of five hexose residues. The *Salmonella* and the *Escherichia coli* R2 cores differ only in the nature of one hexose residue and the *Escherichia coli* RI, R2, R4 and *Escherichia coli* K12 cores all contain the disaccharide unit of the *Escherichia coli* core. There are only minor structural differences between R1 and the R4 cores. The *Salmonella* and the R3 core, finally, contain the same D-Glc-D-Gal-D-Glc trisaccharide unit. Some complementary information on the structure of the heptose region has also been obtained. This region is more uniform, and it would not be surprising if all complete heptose regions in *Salmonella*, *Arizona*, *Citrobacter*, *Escherichia coli*, *Shigella*, *Klebsiella*, *Enterobacter*, and *Serratia* proved to have essentially the same structure (17). One result of

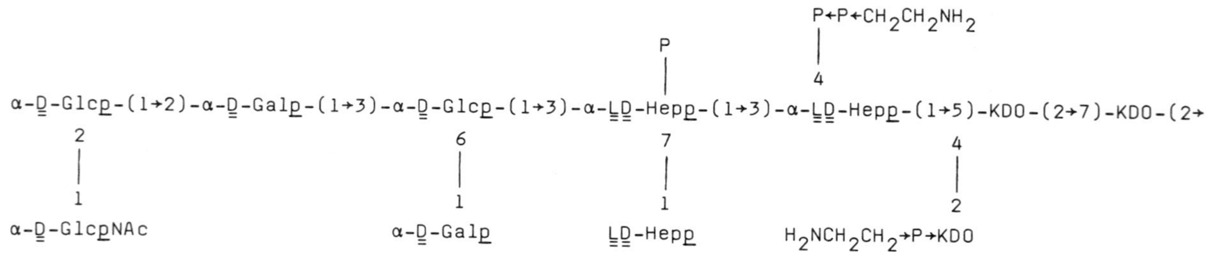

LD-Hepp = L-*glycero*-D-*manno*-heptopyranose

KDO = 3-deoxy-D-*manno*-octulosonic acid

(17)

this study[60] was the finding that all L-glycero-D-manno-heptopyranosyl residues seem to be α-linked.

The heptose region of the lipopolysaccharide of Escherichia coli K12 CR34 has been studied.[61] The D-glucosyl residue linked to the heptose II was found to be substituted by a D-galactosyl group and the linear chain of the core polysaccharide has two (1→3) linked heptoses. The heptose II is substituted by a lateral (1→7) linked heptose III and the heptose I is linked in (1→5) to 2-deoxy-D-manno-octulosonic acid. The three sugars of the linear chain, heptose I, heptose II, and D-glucose, are substituted by phosphate, pyrophosphate, or pyrophosphorylethanolamine groups linked to C-4 hydroxyl groups. However, in some polysaccharide chains, one or two substituting groups may be absent, which may explain the heterogeneity in the length of the core polysaccharide chains.

Structural studies of lipopolysaccharides of Escherichia coli K12 have demonstrated that smooth strains produce lipopolysaccharide with considerable heterogeneity with respect to the length of O-antigenic chains when ^{31}P n.m.r. was used, although traditional methods only provided an average composition over a range of heterogeneity.[62] Standard isolation procedures were shown to fail to extract some 30% of the total lipopolysaccharide present in the cells, the significance of which was discussed in relation to outer membrane structure.

When lipopolysaccharides from Escherichia coli B were sonicated together with pure spin-labelled phospholipids without the addition of unlabelled phospholipids, extensive line broadening was observed due to the close proximity of spin-labelled molecules to each other, a result suggesting that spin-labelled phospholipids existed in segregated domains containing few lipopolysaccharide molecules.[63] Such mixed bilayers were incubated under various conditions, including the addition of NaCl and $MgCl_2$ to the medium and the incorporation of the major outer-membrane protein, porin, into the bilayer, and the intermixing of the domains was followed by the decrease in line width. The diffusion of the labelled phospholipids into lipopolysaccharide domains was hardly detectable when the mixed bilayer contained spin-labelled phospholipids and lipopolysaccharide in approximately equimolar ratios. Although progressive diffusion was observed when a 17-fold molar excess of lipopolysaccharide was present, it is very slow even under the optimal conditions, usually requiring several days for a nearly complete mixing. In another series of experiments, spin-labelled phospholipids were diluted with a 100-fold excess of unlabelled phospholipids and then mixed with lipopolysaccharides. In these experiments, the fluidity of the domains containing spin-labelled phospholipids was shown to be identical, even after 3 days of incubation, with the fluidity of bilayers containing only phospholipids, in contrast to the expectation of the diminished fluidity if phospholipid molecules became finely interspersed with lipopolysaccharide molecules. These two different lines of approach therefore supported the idea that phospholipid (and most probably lipopolysaccharide) domains in mixed bilayers tend to be rather stable and persist for long periods of time.

The natural affinity of various bacterial glycopeptides and lipopolysaccharides for mammalian cell membranes has been estimated quantitatively by comparison with the adsorption of lipopolysaccharide from Escherichia coli NCTC 8623 to erythrocytes, thymocytes, bone-marrow cells, spleen cells, peritoneal lymphocytes, and macrophages.[64] Immunopotentiating activity was estimated by measuring the ability of the bacterial fractions to stimulate a humoral response to ovalbumin in HAM/1CR mice. When the affinity for mammalian cell membranes was compared with the stimulation of the antibody response, it was found that a negative correlation for peritoneal macrophages (r_s = -0.94, P 0.0005) and a positive correlation for peritoneal lymphocytes (r_s = +0.97, P 0.0005) and spleen cells (r_s = +0.76, 0.005) existed.

A heptose-deficient lipopolysaccharide strain of Escherichia coli O8, strain F151, was found to release portions of its outer membrane when cells were exposed to 10mM citrate buffer (pH 2.75) for 30 min and subsequently exposed to 100mM tris(hydroxymethyl)-aminomethane buffer (pH 8.0).[65] The outer-membrane component released was found to be composed of protein, lipopolysaccharide, phospholipid (cardiolipin, phosphatidylethanolamine, and phosphatidylglycerol), and alkaline phosphatase. The outer-membrane component was released from the cell envelope in the absence of cell lysis, as no D-glucose-6-phosphate dehydrogenase activity or succinate dehydrogenase activity was detected. Morphologically, the outer-membrane component appeared to consist of laminar fragments and vesicles which had an associated alkaline phosphatase activity.

In the chromatin of spleen cells of mice and rats immunized with lipopolysaccharides from Escherichia coli, a new species, and antigen non-specific fraction, of non-histone chromatin proteins has been described.[66] The possible role of this fraction in the regulatory process of gene activation during the immune response as expressed by the synthesis of the IgM class of antibodies was discussed.

Experimental evidence has been presented which indicates that 1,2-dihydro-1-hydroxy-6-methyl-2-(propanesulphonyl)-thieno(3,2-D-)(1,2,3)-diazaborine (18), a heterocyclic, boron-containing substance, prevents bacterial proliferation by inhibiting lipopolysaccharide biosynthesis.[67] There was a reduction in D-galactose incorporation into the lipopolysaccharide of whole cells, and no newly formed lipopolysaccharide chains appear in whole bacteria in the presence of the diazaborine compound, as demonstrated by electron microscopy. The 3-deoxy-D-manno-2-octulosonic acid metabolism seemed to be the target of the diazaborine compound as L-arabinose 5-phosphate incorporation into macromolecular material was affected.

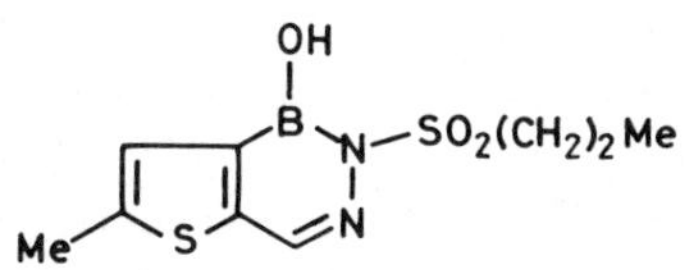

(18)

The purified lipopolysaccharide endoxin-like component from Listeria monocytogenes has been shown to act as both an in vitro mitogen for lymphocytes and in in vivo adjuvant in antibody production.[68]

A novel peptidoglycan-associated lipoprotein was found in the cell envelope of Proteus mirabilis.[69] This protein was shown to have an apparent molecular weight, in sodium dodecyl sulphate polyacrylamide gel, of 18,000. The protein was present in the cell envelope in a form very closely, but not covalently, associated with the peptidoglycan layer and was recovered predominantly from the outer membrane after separation of the cell envelopes. 1-^{14}C}-Palmitic acid and {2-^{3}H}glycerol were incorporated into the protein. A fatty-acid-containing polypeptide (lipopolypeptide) was isolated by digestion of the purified peptidoglycan-associated lipoprotein with trypsin in the presence of 0.05% sodium dodecyl sulphate. It was composed of 31 amino acid residues, an unidentified compound {X}, and ca. 3 fatty-acid residues.[70] A lipo-oligopeptide was also isolated after further digestion of lipopolypeptide with trypsin in the absence of sodium dodecyl sulphate. Lipooligopeptide was composed of 4 amino acid residues (L-Asx, 2 L-Ser, L-Lys), a compound {X}, and ca. 3 fatty-acid residues. The C-terminal amino acids of lipopolypeptide and lipooligopeptide were determined as L-arginine and L-lysine, respectively. The N-terminus of peptidoglycan-associated lipoprotein, lipopolypeptide, or lipooligopeptide could not be identified by conventional N-terminal analysis, indicating that the N-terminus is probably masked. The amino acid sequence of the lipooligopeptide derived from the N-terminal region of the peptidoglycan-associated lipoprotein of *Proteus mirabilis* was determined to be {X}-L-Ser-L-Ser-L-Asn-L-Lys.[71] The unidentified compound {X} present at the N-terminus was identified as glyceryl-L-cysteine {S-(propane-2',3'-diol)-3-thio-2-amino-propanoic acid}. The partial amino acid sequence of the lipopolypeptide, which contained the lipooligopeptide at its N-terminal part, was also determined, mainly by Edman degradation. The structure of the N-terminal part of peptidoglycan-associated lipoprotein was determined to be {3 fatty acids glyceryl-L-Cys-L-Ser-L-Ser-L-Asn-L-Lys-L-Asn-L-Asp-L-Asp-L-Glu-L-Thr-L-Asp-L-Thr-L-Ser.....}. These properties, except for molecular weight and non-covalent association with the peptidoglycan, showed resemblance to those of Braun's lipoprotein. However, the protein was distinct from Braun's lipoprotein in regard to amino acid composition. A similar peptidoglycan-associated lipoprotein was present widely in the cell envelopes of various Gram-negative bacteria. *Proteus mirabilis* contains about twelve times as much peptidoglycan-associated lipoprotein as *Escherichia coli*. Antiserum against peptidoglycan-associated lipoprotein of *Proteus mirabilis* was cross-reactive against peptidoglycan-associated lipoprotein of *Escherichia coli*, but not against Braun's lipoprotein of *Escherichia coli*.

The stimulation of incorporation of {^{3}H}D-galactose into membrane glycoconjugates, measured in a precipitation test, was used as a criterion for activation of bone-marrow cells.[72] In this assay, purified bacterial lipopolysaccharide, lipoprotein, and murein monomer and dimer fragments all activated rat bone-marrow cells *in vitro*. The response was dose dependent, followed a defined time course, and was not serum dependent. O-Acetylated murein dimer fragments from *Proteus mirabilis* were much less active than their unsubstituted counterparts, indicating a structural specificity for murein activation. Removal of adherent and phagocytizing cells from the marrow suspensions did not alter these results. The labelled, activated cells constituted a distinct population of buoyant density 1.064 to 1.069 g/cm^3 when centrifuged on a continuous gradient of Percoll. Enrichment of the target-cell population was achieved by a combination of adherent-cell removal and discontinuous density gradient centrifugation to remove granulocytes and erythropoietic cells. It was concluded that a population of myelopoietic precursors could be activated by direct contact with bacterial cell wall constituents. The stimulation of D-galactose incorporation was not coupled to active deoxyribonucleic acid synthesis in the marrow cells. Thus, the activation was interpreted as an induction of differentiation rather than a mitotic event.

To obtain information about the nature of the immunogens in the ribosomal vaccine (fraction II) of Pseudomonas aeruginosa, the specificity of rabbit antibodies to fraction II has been studied.[73] Crossed immunoelectrophoresis demonstrated the presence of antibodies which precipitated with ribosomal antigens, but not with lipopolysaccharide. By means of an enzyme-linked immunosorbent assay, antibodies to lipopolysaccharide were detected in antibodies to fraction II. Antibodies to fraction II could protect mice against a lethal challenge with Pseudomonas aeruginosa. Absorption experiments demonstrated that the protective ability of antibodies to fraction II was due to cell-envelope antigens, whereas antibodies to ribosomal antigens did not contribute to the protection. Antibodies to lipopolysaccharide could be detected in mice 1 week after a single vaccination with fraction II. It was concluded that the protective activity of fraction II was due, at least in part, to the presence of lipopolysaccharide in the ribosomal vaccine. Treatment of fraction II with ribonuclease decreased the protective activity of the ribosomal vaccine. Addition of synthetic polyadenylic acid-polyuridylic acid restored the protective activity of ribonuclease-treated fraction II, indicating that RNA in the ribosomal vaccine might act as an adjuvant or a carrier in the presentation of the contaminating cell-envelope antigens. The protective activity and the toxicity of fraction II were compared with the protective activity and the toxicity of fraction I, which contained cell-envelope components, including lipopolysaccharide, and of purified lipopolysaccharide. The results indicated that protection by the ribosomal vaccine was associated with a slightly higher toxicity than was protection by fraction I, whereas purified lipopolysaccharide was the most toxic vaccine.

Akaline treatment of Pseudomonas aeruginosa type 5 lipopolysaccharide resulted in reduced toxicity as measured by both the Limulus amoebocyte assay and the rabbit pyrogenicity test.[74] Chemical analysis of the deacylated lipopolysaccharide revealed that ester-linked fatty acids were removed while the amide-linked fatty acids remained intact. The neutral and amino sugar compositions for native lipopolysaccharide and deacylated lipopolysaccharide were identical within experimental error. Antigenic determinants for complement-dependent human opsonic antibody were retained under these deacylation conditions. To enhance its immunogenicity, deacylated lipopolysaccharide was covalently coupled to Pseudomonas pili and the 1,4-diaminobutyl derivatives of Pseudomonas exotoxin A and tetanus toxoid. Quantitative amino sugar analyses revealed that 2.6 and 3.2 mol of deacylated lipopolysaccharides were covalently bound to aminobutyl Pseudomonas exotoxin A and aminobutyl tetanus toxoid, respectively. Gel-electrophoresis data indicated at least 1 mol of deacylated lipopolysaccharide covalently bound per pilus subunit protein. Initial immunologic data indicated that antibody against deacylated lipopolysaccharide could be induced when the deacylated lipopolysaccharide is covalently attached to protein carriers.

The polysaccharide moiety was isolated by mild acid hydrolysis from the slime glycoprotein of Pseudomonas aeruginosa strain BI. After gel filtration, the polysaccharide obtained from the carbohydrate peak fractions was found to be lipid- and protein-free.[75] Analyses indicated that the polysaccharide contained the carbohydrate components of the parent glycolipoprotein. Molecular size of the polysaccharide was estimated by gel filtration as 70,000 to 10,000. The polysaccharide showed no indications of toxicity in mice at doses far in excess of the lethal dose for the patent glycolipoprotein, nor did the mice develop the leukopenia that characteristically follows intraperitoneal injection of glycolipoprotein. The polysaccharide acted as an inhibitor of indirect haemagglutination of glycolipoprotein-coated erthrocytes in the presence of anti-glycolipoprotein serum. However, it was not antigenic itself in rabbits. Coupled with methylated bovine serum albumin, the polysaccharide continued to lack the leukopenic and toxic properties of the parent glycolipoprotein, but the coupled polysaccharide was capable of stimulating indirect haemagglutinating antibody against both the polysaccharide and the glycolipoprotein coating erythrocytes. Moreover, the antibody to the coupled polysaccharide protected mice against challenge with lethal doses of viable Pseudomonas aeruginosa with the same effectiveness as anti-glycolipoprotein serum.

The identification of the new amino sugar 2,3-diacetamido-2,3-dideoxy-D-glucopranuronic acid (19) as a constituent of the O-specific polysaccharides of Pseudomonas aeruginosa strain 170014 has been brought about by use of ^{13}C n.m.r. and chemical methods.[76]

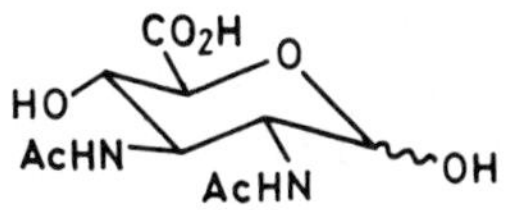

(19)

The procedure used for the isolation of the lipopolysaccharide from strains of Pseudomonas aeruginosa appears to lead to conflicting results in the interpretation of ^{31}P n.m.r. spectra. The use of the Westphal extraction procedure[77] gives a lipopolysaccharide with a higher phosphorus content than that obtained by used of the Bovin[78] procedure. The higher levels of phosphorus content have been attributed[77] by use of ^{31}P n.m.r. to the presence of triphosphate residues, probably in the lipid A and inner-core regions. The presence of triphosphate residues in Pseudomonas aeruginosa is in contrast to results obtained for Salmonella species and Escherichia coli.

Among the aerobic pseudomonads, the closely related species Pseudomonas diminuta and Pseudomonas vesicularis form an isolated subgroup, the inclusion of which in the genus Pseudomonas is becoming increasingly suspect. Chemotaxonomic grounds for the separate status of these species include the possession by their type strains of a unique range of polar lipids and lipopolysaccharides in which lipid A is based on 2,3-diamino-2,3-dideoxy-D-glucose rather than 2-amino-2-deoxy-D-glucose. The lipopolysaccharide from Pseudomonas diminuta NCTC 8545 has a number of other unusual features including the presence of D-threo-2-pentulose(D-xylulose).[79]

Bacteriophage 12P caused lysis of Pseudomonas phaseolicola but not other plant pathogenic or saprophytic pseudonomads.[80] Specificity of the bacteriophage may depend on recognition of some unique feature of the Pseudomonas phaseolicola cell-surface polysaccharides. Exopolysaccharide and lipopolysaccharide from Pseudomonas phaseolicola were more efficient at inhibiting bacteriophage attachment than were identically prepared extracts from five other pseudomonad species. Mutants of Pseudomonas phaseolicola with bacteriophage resistance produced exopolysaccharide and lipopolysaccharide different in amount and composition from those of the parental strain. Thus in some manner the exopolysaccharide and lipopolysaccharide structures of Pseudomonas phaseolicola are distinct from those of other plant pathogenic and saprophytic species. The difference may prevent recognition and initiation of resistance events when Pseudomonas phaseolicola challenges its susceptible host plant, bean.

The lipopolysaccharide of Rhodomicrobium vannielii ATCC 17100 has been studied and found to contain D-glucose and D-mannose (molar ratio 2:1) with trace amounts of D-glucose and D-xylose and 3-O-methyl-D-xylose.[81] 2-Acetamido-2-deoxy-D-glucose, uronic acids, and 3-deoxy-D-manno-2-octulosonic acid were also found to be major components together with β-hydroxypalmitic acid (exclusively amide-bound), β-hydroxymyristic acid, and myristic acid (mainly ester-bound). Heptoses and phosphate were found to be absent. The lipid A fraction (ca. 40% of the material dry weight) liberated by mild acid hydrolysis contains D-mannose (in the nonreducing terminal positions) in addition to 2-amino-2-deoxy-D-glucose and an unknown ninhydrin-positive compound, but the presence of D-mannose as a true constituent of lipid A (as opposed to it belonging to an additional, contaminating lipid as may be the case in lipopolysaccharides from Chromatium vinosum and Thiocapsa roseopersicina) has yet to be determined. The main constituents of the degraded polysaccharide fraction are D-glucose, D-mannose, 3-deoxy-D-manno-2-octulosonic acid, and uronic acids. It should be noted that uronic acids contribute negative charges in a phosphate-free lipopolysaccharide. Interestingly, D-xylose and 3-O-methyl-D-xylose were enriched relative to the other sugars when Rhodomicrobium vannielii ATCC 17100 was grown for several passages in the same R8AH medium.

There is a common structure (core region) in the lipopolysaccharides of Rhodospirillum tenue.[82] It is composed of a branched trisaccharide of L-glycero-D-manno-heptose (and of 3-deoxy-D-manno-2-octulosonic acid), as revealed by methylation analyses of degraded polysaccharides of four different Rhodospirillum tenue strains. The structure is similar or might even be identical to the inner core of enterobacterial O antigens. In addition, each of the four Rhodospirillum tenue lipopolysaccharides contains a strain-specific region that consists of heptose(s) (L-glycero-D-manno-heptose or D-glycero-D-manno-heptose or both) or hexoses. There is a partial substitution of the core region and the strain-specific region by phosphorus, showing microheterogeneity.

The lipopolysaccharide from the polymixin-resistant pmrA mutants of Salmonella typhimurium has been characterized and found to be an unusual type of phospholipid (20) built up of a central disaccharide of two 2-amino-2-deoxy-D-glucose residues.[83] Phosphate esters are bound in the 4- and 1-positions of residues II and I, respectively, to give the phospholipid backbone. The lipid character is given by substitution of the backbone with 7 mol of long-chain fatty acids: 4 mol of 3-hydroxymyristic acid and 3 mol of nonhydroxylated fatty acids, i.e. 1 mol each of lauric, myristic, and palmitic acids, respectively, with small amounts of 2-hydroxymyritic acid. The exact positions of these groups have only partly been identified.[52,83] The backbone contains two amino groups substituted by 3-hydroxymyristic acid and, besides the two phosphate-substituted hydroxyl groups, another four hydroxyl groups of which position 3 of residue II forms the link to the specific polysaccharide via 3-deoxy-D-manno-2-octulosonic acid. There is evidence that the two ester-bound 3-hydroxymyristic acids are linked directly to the backbone, substituting two of the three available hydroxyl groups. The normal fatty acids, in contrast, appear to be linked at the 3-hydroxy groups of the 3-hydroxymyristic acid residues: myristic and 2-hydroxymyristic acids substitute the ester-bound 3-hydroxymyristic acid, while preliminary evidence indicates the lauric and myristic acids to be ester-linked to the two amide-bound 3-hydroxymyristic acid. In this way lipid A is characterized by a high content of a rather unusual structure of long-chain fatty-acid esters of 3-hydroxymyristate. Thus only the 3-hydroxylated acids would substitute the backbone directly, two in amide and two in ester linkage, while the non-hydroxylated acids (and 2-hydroxymyristic acid) would be ester-linked to the 3-hydroxymyristic acid groups.

Mutations in gene rfaH of Salmonella typhimurium at 84 units on the linkage map make lipopolysaccharide of chemotypes Ra, Rb_2, Rb_3, and Rc.[84] F-Factor expression in $RfaH^-$ strains was reduced in the following properties when compared with $RfaH^+$ strains: transfer of Flac, number of phage f2 infective centres, lysis by and propagation of phages f2 and M13, proportion of cells with visible F-pili, and

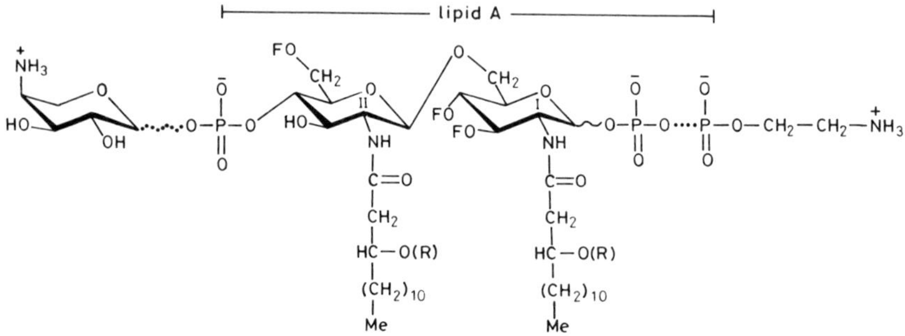

where F = -C=O (2 moles per 3 hydroxyl groups)

CH_2
HC-O(R)
$(CH_2)_{10}$
CH_3

and R = lauric acid (1 mole)
myristic acid (1 mole)
palmitic acid (1 mole)
2-hydroxymyristic acid (trace)

(20)

formation of mating aggregates with F^- cells. Inhibition of multiplication of Br60, a female-specific phage, was not reduced in Rfah$^-$ Flac strains. Plasmid transfer from RfaH$^-$ strains was reduced for Inc groups FI, FII, and T, unaffected for Inc groups β, Iα, L, N, P, and W, and increased for Inc group M when compared with plasmid transfer from RfaH$^+$ strains. Reduced F-factor function in RfaH$^-$ strains was not due to defective lipopolysaccharide since strains with mutations in other rfa genes were unaffected in plasmid transfer. Gene rfaH appears to be homologous with gene sfrB in Escherichia coli K-12, which maps at the same location, influences F-factor function, and affects synthesis of lipopolysaccharide. The gene product of sfrB has been proposed to be transcription antiterminator.

The mobility of membrane components, particularly lipopolysaccharide, is essential for biogenesis of the outer membrane and is a primary event in phage infection. To define the fluid dynamic properties of the outer membrane as related to function more accurately the capability to measure lateral diffusion coefficients in vivo pf rhodaminated G30 lipopolysaccharide fused into Salmonella typhimurium G30A filamentous bacteria has been developed.[85] The method used extends the FRAP procedure (fluorescence redistribution after photobleaching) to bacteria and the results demonstrate rapid diffusion of lipopolysaccharide ($D = (2.0 \pm 0.9) \times 10^{-10}\ cm^2 s^{-1}$) over micrometre distances.

A family of mutants of Salmonella tyhimurium with altered lipopolysaccharide core chain lengths were assessed for sensitivity to freeze-thaw and other stresses.[86] Deep rough strains with decreased chain length in the lipopolysaccharide core were more susceptible to novobiocin, polymyxin B, bacitracin, and sodium lauryl sulphate during growth, to ethylenediaminetetraacetic acid and sodium lauryl sulphate in resting suspension, and to slow and rapid freeze-thaw in water and saline, and these strains exhibited more outer-membrane damage than the wild type or less rough strains. Variations in the lipopolysaccharide chain length did not dramatically affect the sensitivity of the strains to tetracycline, neomycin, or NaCl in growth conditions or the degree of freeze-thaw-induced cytoplasmic-membrane damage. The deeper rough isogenic strains incorporated larger quantities of less-stable lipopolysaccharide and less protein into the outer membrane than did the wild type or less rough mutants, indicating that the mutations affected outer-membrane synthesis or organization or both.

Nikaido's model of the role of lipopolysaccharide and protein in determining the resistance of Gram-negative bacteria of low-molecular-weight hydrophobic antibiotics was discussed in relation to the stress of freeze-thaw.

The specific anti-lipopolysaccharide serum antibody response in BALB/c mice, before and after an oral booster with different non-pathogenic strains of Salmonella typhimurium, was measured by an ELISA technique.[87] Serum response to lipopolysaccharide was found to depend on the length of the lipopolysaccharide polysaccharide moiety in rough mutants. The M206 smooth mutant induced a less marked anti-lipopolysaccharide response. The immunogenicity of the polysaccharidic core appears to be modified by the O antigen. The specific response in the gut only becomes marked following after booster of Ra, Rc, and M206 mutants.

The O-haptenic polysaccharide of Salmonella montevideo has been reported to contain glyceraldehyde at its reducing terminus. However, O-hapten preparations from a pmi galE mutant contained products of partial hydrolysis of lipopolysaccharide, which in separate experiments gave {^{3}H}glycerol upon treatment with perchloric acid and {^{3}H}$NaBH_4$. Further study of the O-hapten reducing terminus suggested that it was actually D-mannose.[88]

The cell-wall polymers of the thermophilic cyanobacterium Synechococcus PCC6716 have been isolated and analysed, the major compounds of the thermophilic strain being similar but not identical to those of the mesophilic strains.[89] The lipopolysaccharide was found to contain β-hydroxypalmitic acid as the only hydroxylated fatty acid together with myristic, palmitic, and palmitoleic acids. The carbohydrates present were D-glucose, L-rhamnose, 2-amino-2-deoxy-D-glucose, and an unidentified ninhydrin-positive component with traces of D-mannose, D-galactose, L-fucose, and D-xylose. Heptoses and 3-deoxy-D-manno-2-octulosonic acid were found not to be present.

The ability of various bacterial lipopolysaccharides and mycoplasmal lipopolysaccharides (lipoglycans) to induce macrophage-mediated tumour-cell killing and Limulus amebocyte lysate clotting was determined.[90] Lipoglycans from the mycoplasma Acholeplasma axantum or Acholeplasma granularum had no activity or 10^4 to 10^5 less activity than lipopolysaccharides from Escherichia coli 0128:B12, Escherichia coli K235, or Salmonella minnesota R595 in causing Limulus lysate clotting and tumour-cell killing by peritoneal macrophages from normal or bacillus Calmette-Guerininfected mice. Previous studies have shown that the lipid A portion of bacterial lipopolysaccharide is responsible for the effects on macrophage-mediated tumour-cell killing and Limulus lysate clotting. The known differences in the lipid structures of bacterial lipopolysaccharides and mycoplasmal lipopolysaccharides (lipoglycans) may account for the noted differences in the biologic potencies observed.

4 Capsular Polysaccharides

The interaction of lectins and lectin-like substances with bacteria has been reviewed[90] and the interaction discussed as a means of distinguishing between microbial species, identifying the presence of a particular sugar residue and fractionating the polysaccharides from bacteria.[91]

Soluble antigens were obtained by extracting five serotype strains of Clostridium perfringens type A with water at 100°C.[92] The type-specific polysaccharides were precipitated with ethanol, and the common antigens were recovered from the ethanol supernatants by concentration, dialysis, and lyophilization. Refluxing the water-extracted cell residues with 1% acetic acid followed by concentration, dialysis, and lypophilization gave additional common antigen fractions. A comprehensive, side-by-side comparison of the antigen fractions, the ethanol precipitate, the ethanol supernatant, and the acetic acid supernatant, revealed that common antigens were recovered in all three fractions and that three distinct entities were responsible for the formation of the observed common immunoprecipitin lines. Whereas many fractions possessed all three immunoprecipitin lines, others contained only one or two. The serological homology observed between the various antigen fractions was apparently a consequence of 2-acetamido-2-deoxy-D-glucose and 2-acetamido-2-deoxy-D-mannose containing polymers. The common antigens were presumably associated with the cell envelope and may be the type of markers sought previously by others for the serological identification of Clostridium perfringens.

The bacterium formerly named Fusobacterium polysaccharolyticum and now reclassified as a Clostridium species has been shown to produce spores and have a cell-wall structure of the Gram-positive type desite its appearance as Gram-negative when stained.[93]

The capsular polysaccharides from Crytococcus neoformans

serotype A and from a possible, mutant strain thereof have been structurally analysed.[94] The latter material is similar to the capsular antigen of *Crytococcus neoformans* serotype D.

Epidemiological and immunological evidence indicates that the K1 capsular polysaccharide confers the property of virulence on *Escherichia coli*. One approach to studying whether the K1 antigen is sufficient to confer virulence or if other *Escherichia coli* structures are necessary is to isolate the K1 genes for genetic and biochemical analysis. Recombinant DNA methodology provided a powerful tool for such an approach. The molecular cloning of the *Escherichia coli* K1 antigen genes has been reported.[95] The cloned K1 genes synthesized a capsule in *Escherichia coli* K12 indistinguishable chemically and immunologically from that of wild type K1 strains.

The capsular polysaccharide was isolated from *Escherichia coli* O10:K5:H4: it could not be obtained from an uncapsulated ($K5^-$) mutant. It contains 2-acetamido-2-deoxy-D-glucose and D-glucuronic acid in a molar ratio of 1:1.[96] Acid hydrolysis of the acidic polysaccharide as well as Smith degradation and degradation by deamination of the carboxyl-reduced polysaccharide suggested that the polysaccharide is composed of a disaccharide repeating unit. The data obtained by methylation analysis and n.m.r. spectroscopy indicated that the repeating sequence of the capsular polysaccharide is similar to that of desulpho-heparin with the structure (21).

→4)-β-D-Glc*p*A-(1→4)-α-D-Glc*p*NAc-(1→

(21)

The linkage pattern of the K6 antigen was investigated using material from the urinary pathogen *Escherichia coli* LP 1092.[97] The polysaccharide consists of D-ribose and 3-deoxy-D-*manno*-2-octulosonate in a ratio of 2:1. Colorimetric procedures, Smith degradation, methylation analysis, and n.m.r. spectroscopy were applied to the whole polysaccharide and to a trisaccharide repeating unit obtained by mild-acid-catalysed hydrolysis. Together, the data were compatible only with a branched-chain structure (22).

A method has been developed to separate the cell envelope of encapsulated (type b) *Haemophilus influenzae* into its outer- and inner-membrane components with procedures that avoided two problems encountered in fractionation of this envelope: (i) the tendency of the outer and inner membranes to hybridize and (ii) the tendency of the apparently fragile inner membrane to fragment into difficultly sedimentable units.[98] Log phase cells, whose lipids were radioactively labelled, were lysed by passage through a French press. The lysate was applied to a discontinuous sucrose gradient, and envelope-rich material was collected by centrifugation onto a cushion of dense sucrose under carefully controlled conditions. This material was then further fractionated by isopycnic centrifugation in a sucrose gradient to yield four membrane fractions which were partially characterized. On the basis of their radioactivity, bouyant density, ultrastructure, polypeptide composition, and content of phospholipid, protein, lipopolysaccharide, and succinic dehydrogenase, these fractions were identified as follows: fraction 1 - outer-membrane vesicles with very little inner-membrane contamination (<4%), fraction 2 - outer-membrane vesicles containing entrapped inner membrane, fraction 3 - a protein-rich fraction of inner membrane, and fraction 4 - a protein-poor fraction of inner membrane. Fractions 3 and 4 contained about 25% outer-membrane contamination.

```
→3)-β-D-Ribf-(1→7)-KDO-(2→
               2
               |
               1
          β-D-Ribf
```

where KDO = 3-deoxy-β-D-*manno*-2-octulopyranosonic acid

(22)

The structure of the *Haemophilus influenzae* type e capsular polysaccharide has been determined by a combination of chemical and spectroscopic methods.[99] The structure of the repeating unit of the polymer was found to be linear (23).

→3)-β-D-Glc*p*NAc-(1→4)-β-D-Man*p*ANAc-(1→

(23)

The structures of two capsular polysaccharides elaborated by *Haemophilus influenzae* type e, strains NCTC 8455 and 8472, respectively, have been investigated, methylation analysis and n.m.r. spectrometry being the principal methods used.[100] It is concluded that the polysaccharides are composed of repeating units having the structure (24). In the polysaccharide from strain NCTC

8472, all of the repeating units contain the β-D-fructofuranosyl group. The polysaccharide from strain NCTC 8455, however, contains only traces of D-fructose, corresponding to approximately one group per 100 repeating units.

→3)-β-D-GlcpNAc-(→4)-β-D-ManpANAc-(1→
3
|
2
β-D-Fruf

(24)

The capsular polysaccharide from Klebsiella K4 contains the tetrasaccharide repeating sequence (25). ^{1}H n.m.r. spectroscopy and the measurement of optical rotation were used to establish the anomeric linkages in the polysaccharide and in the oligosaccharides derived by partial hydrolysis.[101] The repeating unit also contains one O-acetyl group.

→3)-α-D-Glcp-(1→2)-α-D-GlcpA-(1→3)-α-D-Manp-(1→3)-β-D-Glcp-(1→

(25)

A glycanase activity associated with the particles of Klebsiella bacteriophage No.6 catalyses cleavage of O-β-D-glucopyranosyl-(1→3)-4,6-O-(1-carboxyethylidene)-β-D-mannopyranose linkages in Klebsiella serotype 6 capsular polysaccharide.[102] Of 74 heterologous Klebsiella polysaccharides and two derivatives of the type 6 glycan, only the type 1 and type 57 polymers were additionally degraded by the phage 6 enzyme. The repeating units in the three substrates have a 1ax→3eq,1eq→eq-linked chain D-gluco- or D-galacto-pyranosyl residue in common (which constitutes the reducing end after glycanase action), and a carboxyl group on the next hexopyranosyl residue. Of the 72 polysaccharides not affected by the viral enzyme, at least the type 11 and type 21 glycans also contain the same homology of primary structure. This indicates that the conformation at the glycanase recognition site also constitutes an important feature of the substrates.

A β-D-galactosidase associated with bacteriophage ϕ18 has been used to depolymerize the capsular polysaccharide of Klebsiella serotype K18 into the hexasaccharide corresponding to one repeating unit of the polymer.[103] The n.m.r. spectra of the polymer and of this oligosaccharide are comparable, leading to the conclusion that the conformations in solution of the repeating unit of these substances are similar.

The structure of the capsular polysaccharide of Klebsiella K26 has been determined by using the techniques of methylation, periodate oxidation, partial hydrolysis, and β-elimination.[104] N.m.r. spectroscopy (^1H and ^{13}C) was used to establish the nature of the anomeric linkages and to identify oligosaccharides obtained by the different degradative techniques employed. The polysaccharide is comprised of repeating units of the heptasaccharide (26).

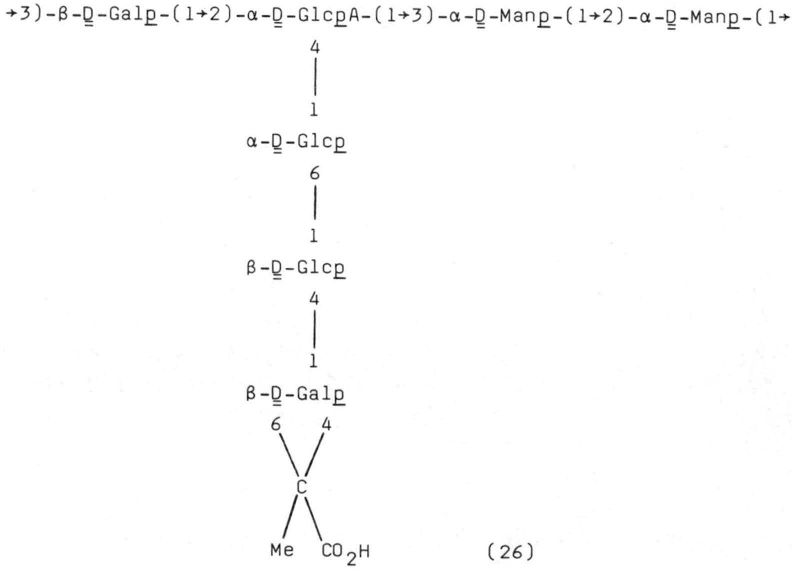

(26)

The capsular polysaccharide from Klebsiella type 73 (Enterobacter aerogenes) was found to contain equimolar amounts of D-galactose, D-glucose, L-rhamnose, and D-glucuronic acid.[105] Acid hydrolysis of the polysaccharide gave one aldobio- and aldotrio-uronic acid, whose structures were established by acid hydrolysis and by methylation analysis. The anomeric configurations of the different sugar residues were determined from specific rotations of the polysaccharide and the aldobio-and aldotrio-uronic acids, and also by oxidation of the native and the carboxyl-reduced polysaccharide with chromium trioxide. Methylation analysis of the polysaccharide provided information about the linkages of the different sugar residues. Based on all of these results, the structure (27) was assigned to the repeating unit of the polysaccharide.

The vaccinating power of a polysaccharide preparation from *Klebsiella pneumoniae* type I has been reported.[106] The essential part played by this extract in the vaccinating capacities of the ribosomal preparations from the same bacterium is demonstrated.

→3)-β-L-Rha*p*-(1→4)-β-D-Gal*p*-(1→4)-β-D-Glc*p*-(1→
3
|
1
β-D-Glc*p*A
(27)

Two peaks were obtained by caesium chloride density-gradient ultracentrifugation of *Klebsiella pneumoniae* ribosomal preparations.[107] Peak I contained capsular polysaccharide, lipopolysaccharide, protein, and less than 0.5% ribonucleic acid. Peak II consisted mainly of ribonucleic acid, with low amounts of protein and capsular polysaccharide. Expressed as capsular polysaccharide content, the 50% protective dose of peak I and of non-fractionated ribosomal preparations was nearly constant (2.6 and 1.2 ng, respectively). Since peak I contained less than 0.5% ribonucleic acid, these results provide evidence that ribosomal ribonucleic acid is not required for protection of mice by *Klebsiella pneumoniae* capsular polysaccharide, which contaminates ribosomal preparations.

An antitumour carboxymethylated (1→3)-β-D-glucan was found to have a potent ability to cause gelation of the amoebocyte lysate of horseshoe crab (*Limulus*) at concentrations as low as 10^{-6} mg/ml.[108] The gelation of the lysate and the activation of the proclotting enzyme in the lysate caused by this D-glucan were unique in that these reactions occurred at concentrations ranging from 10^{-6} to 10^{-3} mg/ml. The optimum concentration was 10^{-5} to 10^{-4} mg/ml, and at concentrations above 10^{-2} mg/ml no gelation occurred. This gelation pattern was also observed in common with other antitumour polysaccharides. The mechanism of the gelation caused by D-glucan was revealed to be distinctly different from that working in the gelation caused by endotoxins.

The proclotting enzyme, or factor B, is not sensitive to the carboxymethylated (1→3)-β-D-glucan and the coagulation pathway in Limulus amoebocytes is activated by sensitization of a third component (tentatively named factor G) by the D-glucan, which mediates the activation of the proclotting enzyme to its active form.[109]

Serum samples were collected from 120 healthy adult volunteers (105 Caucasians and 15 Negros) before and after immunization with meningococcal polysaccharide group B vaccine. Antibodies to meningococcal group B were measured and sera were typed for several Gm and Km(1) allotypes.[110] A significant association was found between the Km(1) allotype and immune response to meningococcal group B in Caucasians.

Colominic acid and meningococcal group B polysaccharide, both (2→8)-linked homopolymers of 5-acetamido-3,5-deoxy-D-glycero-α-D-galacto-2-nonulosonic acid, are made water-insoluble either by reaction with a carbodi-imide in aqueous solution at pH 4.75 or by treatment with 48% aqueous hydrofluoric acid at 0°C for 48 h. I.r. spectra of the products show a major band near 1750 cm^{-1}, consistent with ester formation. This band virtually disappeared after mild, alkali treatment.[111] Esterification also occurred by incubating the native polysaccharides below pH 6.0, and their i.r. spectra showed that the degree of esterification increases as the pH is lowered. The relatively low molecular weight of these partially esterified, water-soluble polymers is consistent with intra- rather than inter-molecular ester formation. Counter-current immunoelectrophoresis shows that esterification of ca. 9% is sufficient to abolish immunoprecipitation. ^{13}C N.m.r. spectroscopy of fully esterified colominic acid provides strong evidence for cross-linking between the carboxyl group of one residue and HO-9 of an adjoining residue. Meningococcal serogroup C polysaccharide, a (2→9)-linked homopolymer of 5-acetamido-3,5-dideoxy-D-glycero-α-D-galacto-2-nonulosonic acid containing O-acetyl groups at C-7 and/or C-8, does not undergo intramolecular esterification after carbodi-imide treatment. However, upon O-deacetylation of the polysaccharide, esterification occurs.

Hydrolysis of the meningococcal groups A, B, and C and Escherichia coli K92 polysaccharides by 60% aqueous hydrofluoric acid liberated various 1,2-diacylglycerols.[112] These were extracted with chloroform, trimethylsilylated, and analysed by g.c.m.s. Two 1,2-diacylglycerols were the major components isolated. In each polysaccharide, 80 to 90% dipalmitoyl glycerol and 10 to 20%

distearoyl glycerol were identified. No monoacylglycerols or mixed diacylglycerols were noted. The presence of the hydrophobic end causes the polysaccharides to aggregate in a micellar form and may be the entity by which the polysaccharide remains attached to the outer membrane of the bacterium, giving rise to the structure recognized as a capsule.

The currently United States-licensed group C *Neisseria meningitidis* vaccine, composed of the *O*-acetyl-positive capsular polysaccharide, is poorly immunogenic and does not afford protection from disease to infants and young children. Group C *Neisseria meningitidis* *O*-acetyl-negative polysaccharide vaccine induces higher titers in adults than does the *O*-acetyl-positive vaccine. The immunogenicity of these vaccines was compared in 2-year-old children.[113] Reactions were minimal and did not differ between the two vaccines. The postvaccination geometric mean titer was twofold greater in the *O*-acetyl-negative group (1.58 *versus* 0.73 μg of antibody per ml). The rates of decline in titer were similar in both groups. Further study regarding immunogenicity of and the anamnestic response to the *O*-acetyl-negative vaccine is warranted in the age group (<18 months) at highest risk for invasive meningococcal disease.

Phase-contrast and fluorescence microscopy observations showed that pea symbiont *Rhizobium leguminosarum* adsorbed to pea-root hairs, but non-symbiont rhizobial strains only adsorbed to a small extent.[114] ^{14}C-Labelled cells were used to assay the number of rhizobial cells adsorbed to a pea root. Capsular polysaccharides or lipopolysaccharides obtained from *Rhizobium leguminosarum* specifically inhibited the adsorption of ^{14}C-*Rhizobium leguminosarum* cells to a pea root and specifically adsorbed to pea-root hairs. Also, they reacted specifically with pea-seed lectins. These results suggest that capsular polysaccharides or lipopolysaccharides play an important role in host-specific adsorption. The interaction between the polysaccharides and pea lectins could be the key to determining host specificity in the infection process of *Rhizobium*-pea symbiosis.

The ultrastructural locations of the group-specific polysaccharide and the type-specific polysaccharides Ia, Ib, II, and III of group B streptococci (*Streptococcus agalactiae*) were studied on isolated walls by direct immunoferritin technique.[115] The types of polysaccharides were located exclusively on the outer side of the wall on which they formed a distinct capsule. Except for strain

58/59 (type Ia) the thickness of the capsule was characteristic of each strain investigated. In all strains the type-specific ferritin labelling was confined to the outer surface of the capsule. The group-specific polysaccharide could be demonstrated on the outer surface in strains 59/59 (type Ib) and 8/66 (group B variant) and on most of the walls of strain 58/59 (type Ia). The failure to detect this antigen on the outer side of the walls of strains 60/59 (type II) and 13/63 (type III) and some walls of strain 58/59 was probably due to the thickness of the type of polysaccharide capsule.

The release of serotype III group B streptococcal polysaccharides into the supernatant fluid was examined under a variety of physiological conditions.[116] Release of both high- and low-molecular-weight type III antigens was fairly constant throughout exponential growth, but increased markedly upon entering the stationary phase of growth. Increased D-glucose and decreased phosphate concentrations both caused a large increase in release of antigens. Inhibition of protein synthesis in exponentially growing cells by chloramphenicol (10 μg/ml) caused a condition of unbalanced growth in which antigen release was increased greatly over control values. Strain variability in antigen release was also observed. Strains which are known to be high neuraminidase producers released elevated levels of both low- and high-molecular-weight type III antigens. Non-neuraminidase-producing strains released considerably less high-molecular-weight antigen but similar levels of the low-molecular-weight antigen compared with the high neuraminidase producers. Strain D136C, a type III non-neuraminidase producer, released negligible quantities of the high-molecular-weight antigen in the supernatant fluid. These results indicate that both the physiological environment and the type III strain are important in determining the quantity of type-specific antigen released into the culture fluid.

Immunization of rabbits with group B type III *Streptococcus* organisms induces two distinct populations of antibodies with a specificity for determinants of the native capsular polysaccharide antigen of these organisms.[117] Some of the structural and conformational features of the two determinants responsible for the formation of these antibodies were elucidated by ^{13}C n.m.r. and serological studies on the native type III polysaccharide and some of its structurally modified analogues. The specificity of the determinant corresponding to the major population of antibodies is dependent on the presence of 5-acetamido-3,5-dideoxy-D-*glycero*-D-

galacto-2-nonulosonic acid residues on the native type III antigen, and, although these residues are not an integral part of the determinant, they exert conformational control over it. The carboxyl groups of the 5-acetamido-3,5-dideoxy-D-*glycero*-D-*galacto*-2-nonulosonic acid residues are an important factor in this control mechanism, which could possibly involve intramolecular hydrogen bonding. The terminal 5-acetamido-3,5-dideoxy-D-*glycero*-D-*galacto*-2-nonulosonic acid residues control the orientation of the penultimate β-D-galactopyranose residues with respect to the backbone of the native antigen. The orientation of these residues is critical to the determinant because the determinant is probably small and is located precisely at the junction of the same β-D-galactopyranose residues with the backbone of the native type III antigen. The determinant corresponding to the other population of antibodies is not 5-acetamido-3,5-dideoxy-D-*glycero*-D-*galacto*-2-nonulsonic acid dependent. This determinant is located on the backbone of the native antigen in the vicinity of the other determinant but on the opposite side to the oligosaccharide branches. In this position, its conformation is unaffected even by the removal of the oligosaccharide branches from the native antigen.

It has been demonstrated that mouse IgG3 antibodies are highly protective against infection with *Streptococcus pneumoniae*.[118]

The structure of the capsular polysaccharide from *Streptococcus pneumoniae* type 4 has been investigated.[119] Using n.m.r. spectroscopy, methylation analysis, and Smith degradation as the principal methods of structural investigation, it is concluded that the polysaccharide is composed of tetrasaccharide repeating units having the structure (28).

→4)-β-D-Man*p*NAc-(1→3)-α-L-Fuc*p*NAc-(1→3)-
α-D-Gal*p*NAc-(1→4)-α-D-Gal*p*-(1→
2 3
C
Me CO_2H

(28)

The chemical composition and immunochemical characterization of the four cross-reaction pneumococcal capsular polysaccharides within group 9 (types 9N, 9A, 9L, and 9V) have been investigated.[120] Their serological reactions were studied by using unabsorbed antisera prepared by immunizing rabbits with pneumococci of each of the four group 9 capsular polysaccharide types. Type 9A antiserum showed the most extensive cross-reactions with the four group 9 polysaccharides. Absorption with type 9N, 9L, or 9V polysaccharide removed 63, 96, or 87%, respectively, of the heterologous antibodies from the type 9A antiserum. All four of the group 9 polysaccharides contained D-glucose, 2-acetamido-2-deoxy-D-mannose, and D-glucuronic acid. In addition, types 9N and 9L had 2-acetamido-2-deoxy-D-glucose, and types 9A, 9L, and 9V contained D-galactose. Reduction of the uronic acid residues of the type 9 polysaccharides removed most of their homologous and much of their heterologous reactivities, indicating an important role for the uronic acid component of their antigenicity. The four group 9 polysaccharide preparations had comparable molecular sizes and only traces of protein and nucleic acid. Further studies to evaluate the most protective type among the group 9 strains to be included in the current pneumococcal vaccine have been discussed.

Mild acid hydrolysis of the capsular polysaccharide of Pneumococcus type IX (S IX) liberated a number of oligosaccharides which were separated into basic, acidic, and neutral fractions using ion-exchange column chromatography.[121] Each fraction was further separated into its constituent oligosaccharides using high-voltage electrophoresis. The individual oligosaccharides were subjected to methylation studies. When uronic acid was present the methylated oligomer was reduced and the methyl sugars in it were characterized. Based on these studies structures were assigned to the oligomers. The five oligosaccharides characterized support the revised structure assigned to the nonasaccharide repeating unit of type IX polysaccharide (29) obtained by periodate oxidation and Smith degradation linked to methylation analysis.[122] Chromium trioxide oxidation indicated that all glycosidic linkages were of the α-D-type.

→4)-α-D-GlcpA-(1→3)-α-D-Glcp-(1→4)-α-D-GlcpA-(1→3)-
α-D-GlcpNAc-(1→3)-α-D-GlcpNAc-(1→3)-α-D-Glcp-(1→3)-
α-D-ManpNAc-(1→4)-α-D-Glcp-(1→3)-α-D-ManpNAc-(1→

(29)

The capsular polysaccharide from Streptococcus pneumoniae type 12F is composed of D-glucosyl, D-galactosyl, 2-acetamido-2-deoxy-D-galactosyl, 2-acetamido-2,6-dideoxy-L-galactosyl, and 2-acetamido-2-

deoxy-D-mannopyranosyluronic acid residues in the proportions 2:1:1:1.[123] The main structural evidence was adduced from n.m.r. spectroscopy, methylation analysis, and specific degradations whereby it could be concluded that the polysaccharide is composed of hexasaccharide repeating units having the structure (30).

The type XXXIII *Streptococcus pneumoniae* (pneumococcal) capsular polysaccharide contains D-glucose (2 mol), D-galactose (1 mol), D-glucuronic acid (1 mol), and 2-acetamido-2-deoxy-D-mannose (1 mol).[124] Structural investigations by methylation analysis, specific degradations, and n.m.r. spectroscopy showed that the repeating unit was a linear pentasaccharide (31).

```
→4)-α-L-FucpNAc-(1→3)-β-D-GalpNAc-(1→4)-β-D-ManpANAc-(→
          3                                         3
          |                                         |
          1                                         1
      α-D-Galp                       α-D-Glcp-(1→2)-α-D-Glcp
```

(30)

→3)-α-D-Galp-(1→3)-β-D-ManpNAc-(1→4)-α-D-Glcp-(1→4)-
β-D-Glcp-(1→4)-α-D-GlcpA-(1→

(31)

The specific capsular polysaccharide of type 45 *Streptococcus pneumoniae* was isolated in pure form by chemical and chromatographic methods and found to be a high-molecular-weight, glycosidically linked polymer of a hexasaccharide repeating unit composed of D-galactose (2 mol), L-rhamnose (1 mol), 2-acetamido-2-deoxy-D-glucose (1 mol), 2-acetamido-2-deoxy-D-galactose (1 mol), 2-acetamido-2-deoxy-L-fucose (1 mol), and phosphate (1 mol).[125]

The specific capsular polysaccharide of type 46 *Streptococcus pneumoniae* (American type 73) was isolated in pure form and shown to be a high-molecular-weight, glycosidically linked polymer composed of D-galactose (2 mol), 2-acetamido-2-deoxy-D-glucose (1 mol), 2-acetamido-2-deoxy-D-galactose (1 mol), and 2-acetamido-2-deoxy-L-fucose (1 mol).[126]

The specific capsular polysaccharide produced by *Streptococcus pneumoniae* type 9V (American type 68) is composed of D-glucuronic acid (1 part), D-galactose (1 part), 2-acetamido-2-deoxy-D-mannose (1 part), D-glucose (2 parts), and O-acetyl (1.6 parts).[127] Methylation analysis, periodate oxidation, optical rotation, n.m.r.

spectroscopy, and partial hydrolysis showed that the polysaccharide is an unbranched high-molecular-weight linear polymer of a partially O-acetylated pentasaccharide repeating unit having the structure (32).

→4)-α-D-GlcpA?Ac-(1→3)-α-D-Galp-(1→3)-α-D-ManpNAc-(1→4)-
β-D-Glcp?Ac-(1→4)-α-D-Glcp-(1→
(32)

5 Extracellular and Intracellular Polysaccharides

An extracellular, acidic polysaccharide has been isolated from the culture medium of a spontaneous cellulose-negative strain of Acetobacter xylium.[128] Chemical analysis shows that the polymer is composed of D-glucose, D-mannose, L-rhamnose, and D-glururonic acid in a molar ratio approximating 3:1:1:1. No evidence for the presence of cellobiose units as structural parts in the polysaccharide has been found.

An organism producing extracellular polysaccharide was isolated from soil and identified as Aeromonas hydrophila (Chester) Stanier.[129] The effects of medium components and cultural conditions on production of the polysaccharide were studied. The optimal concentrations of carbon and nitrogen sources were 5% and 0.3%, respectively, for production of the polysaccharide. The optimal initial pH was 7 to 9. The maximum polysaccharide yield was obtained at 4 to 8 days fermentation. From sucrose and raffinose as carbon source, the organism produced levan and acidic polysaccharide in the ratio of 7:3 and 4:6, respectively. From D-glucose, D-galactose, D-fructose, D-mannose, maltose, and lactose mainly acidic polysaccharide was produced. The acidic polysaccharide was found to contain D-galactose, D-mannose, and D-glucuronic acid in a ratio of 5:4:2. The acidic polysaccharides obtained from sucrose and lactose seemed to be the same polysaccharide.

Curdlan gel shrinks even in water and causes syneresis. The syneresis is repressed almost completely by the addition of starch to curdlan before heating, but not repressed by the addition of different sugar compounds even of 10% concentration. The role of starch in repressing syneresis has been discussed.[130]

During growth in liquid culture containing a single cellulosic or non-cellulosic carbon source, a newly isolated strain of Aspergillus fumigatus released cellulases into the medium. The amounts produced depended on the nitrogen source, the type and concentration of the carbon source, pH, and temperature.[131] Extracellular cellulolytic activity was still increasing after incubation for 60 d with 1% (w/v) CF11 cellulose, $(NH_4)_2SO_4$ as nitrogen source, and a starting pH of 7. The activities of the new isolate were compared with those of Aspergillus fumigatus IMI 143864 and Trichoderma reesi QM6a (ATCC 13631), and it was shown to be a good producer of β-D-glucosidase.

The ability of Bacteroides thetaiotaomicron, an obligate anaerobe from human colonic microflora, to grow in a carbohydrate-limited continuous culture at generation times ranging from 3.5 to 28 h per division was investigated.[132] Four carbohydrates were tested: D-glucose, 2-acetamido-2-deoxy-D-glucose, D-glucuronic acid, and 2-amino-2-deoxy-D-glucose. At a generation time of 3.5 h per division, the growth yields for bacteria growing on D-glucose, 2-acetamido-2-deoxy-D-glucose, and D-glucuronic acid were 77, 68, and 50 g of cells (dry weight) per mol of substrate, respectively. Growth yields at 28 h per division were 61, 52, and 37 g/mol of substrate, respectively. When 2-amino-2-deoxy-D-glucose was the carbohydrate source, a stable population of bacteria was attainable only at generation times longer than 12 h per division. Growth yields at 15 and 32 h per division were 11 and 33 g/mol of substrate, respectively. There was no significant variation with increasing generation times in the specific activities of selected glycolipid enzymes, of disaccharides such as α- and β-D-glucosidases and -D-galactosidases, or of the polysaccharidase chondroitin AC lyase. By contrast, the pattern of fermentation products varied with both the generation time and the carbon source. At a generation time of 3.5 h per division, the main products from the fermentation of D-glucose were acetate and succinate, with a trace of propionate. At 28 h per division, propionate concentrations were higher and succinate concentrations were lower than at 3.5 h per division. The products from the fermentation of 2-acetamido-2-deoxy-D-glucose were the same as those from D-glucose fermentation. However, when 2-acetamido-2-deoxy-D-glucose was fermented, the concentration of acetate was much higher at all generation times than when D-glucose was the carbon source. When D-glucuronic acid was the carbon source, acetate was the main fermentation product, and only traces of propionate and succinate were detected. Another characteristic of the type that varied with the growth rate was the ability of *Bacteroides thetaiotaomicron* to produce the inducible enzyme α-D-glucosidase when exposed to maltose. The ability of the organism to produce this enzyme declined with increasing generation times.

The crude extracellular cellulase of *Clostridium thermocellum* LQRI (virgin strain) was very active and solubilized microcrystalline cellulose at one-half the rate observed for the extracellular cellulase of *Trichoderma reesi* QM9414 (mutant strain).[133] *Clostridium thermocellum* cellulase activity differed considerably from that of *Trichoderma reesi* as follows: higher *endo*-β-D-glucanase/*exo*-β-D-glucanase activity ratio, absence of extracellular cellobiase or β-D-xylosidase activity, long-chain oligosaccharides instead of short-chain oligosaccharides as initial (15 min) hydrolytic products on microcrystalline cellulose, mainly cellobiose or D-xylobiose as long-term (24 h) hydrolysis products of Avicel and MN300 or D-xylan, and high activity and stability at 60 to 70°C. Under optimized reaction conditions, the kinetic properties (V_{max} 0.4 μmol/min per mg of protein, energy of activation 33 kJ, temperature coefficient 1.8) of *Clostridium thermocellum* cellulose-solubilizing activity were comparable to those reported for *Trichoderma reesei*, except that the dyed Avicel concentration at half-maximal velocity was twofold higher (182 μM). The cellulose-solubilizing activity of the two crude cellulases differed considerably in response to various enzyme inhibitors. Most notably, Ag^{2+} and Hg^{2+} effectively inhibited *Clostridium thermocellum* but not *Trichoderma reesei* cellulase at <20 μM, whereas Ca^{2+}, Mg^{2+}, and Mn^{2+} inhibited *Trichoderma reesi* but not *Clostridium thermocellum* cellulase at >10 mM. Both enzymes were inhibited by Cu^{2+} (>29mM), Zn^{2+} (>1.0mM), and ethylene glycol-bis(β-aminoethyl ether)-*N*,*N*-tetraacetic acid (>10 mM). *Trichoderma reesei* but not *Clostridium thermocellum* cellulose-solubilizing activity was 20% inhibited by D-glucose (73 mM) and cellobiose (29 mM). Both cellulases preferentially cleaved the internal glycosidic bonds of cello-oligosaccharides. The overall rates of cello-oligosaccharide degradation were higher for *Trichoderma reesei* than for *Clostridium thermocellum* cellulase, except that the rates of conversion of cellohexaose to cellotriose were equivalent.

The formation and excretion of acetylmaltose and acetylated maltodextrins after accumulation of maltose in *Escherichia coli* have been studied.[134] As acetylmaltose was not a substrate for the maltose transport system it was proposed that the acetylation process was an effective detoxification mechanism.

When Escherichia coli strain CP78 (rel$^+$) was starved for L-isoleucine by the addition of L-valine, the amount of D-glucose in polymeric form in the cells increased markedly compared to that of the control cells.[135] In contrast, this phenomenon was not seen in strain CP79 (rel$^-$). The increase in CP78 was shown to be due to the increase of glycogen. These results indicate that glycogen synthesis was augmented under stringent control. This was confirmed using other isogenic pairs of rel$^+$ and rel$^-$ strains starved for other amino acids. When the cultivation temperature of strains 10B601 (rel$^+$) and 10B602 (rel$^-$) possessing temperature-sensitive L-valyl-tRNA synthetase was shifted from 30°C to 40°C, no difference was observed in the response of glycogen synthesis between the two strains. These results indicate that protein synthesis was necessary for the augmentation of glycogen synthesis and that guanosine 5'-diphosphate 3'-diphosphate did not exert its effect through stimulation of the activity of pre-existing enzyme(s) involved in glycogen synthesis. These conclusions were supported by the results of experiments using chloramphenicol and rifampicin. The rates of D-glucose utilization of CP78 and CP79 were decreased to nearly the same extent by L-valine addition. This suggests that the regulation site of glycogen synthesis under stringent control resides in a step after the transport of D-glucose by the phophotransferase system.

ADP-D-glucose pyrophosphorylase from Escherichia coli B mutant CL1136-504 has been purified to homogeneity. The native enzyme is a tetramer of 200,000 daltons composed of four identical 50,000 M_r subunits.[136] The structural genes for the glycogen biosynthetic enzymes of Escherichia coli, glycogen synthase, ADP-D-glucose pyrophosphorylase, and branching enzyme were cloned from chromosomal DNA bacterial plasmid, pBR322.[137]

A bacterial agglutinin was extracted from ground corn (WI hybrid 64A x W117) seed with phosphate-buffered saline (pH 6.0) and precipitated with $(NH_4)_2SO_4$ at 70% saturation. The activities of this agglutinin against 22 strains of Erwinia stewartii (agent of bacterial wilt of corn) that varied in virulence were determined.[138] Specific agglutination (agglutination titer per milligram of protein per milliliter) values were correlated negatively with virulence ratings. Strains with high specific agglutination values (15 or higher) were avirulent or weakly virulent, strains with low specific agglutination values (10 or lower) were highly virulent, with two exceptions. A virulent strain produced butyrous colonies and released only small amounts of extracellular polysaccharide into the medium, and the cells lacked capsules. Virulent strains produced fluidal colonies and released large amounts of extracellular polysaccharide produced by each strain (as determined by increase in viscosity of the medium) and the specific agglutination value. In contrast lipopolysaccharide compositions were similar in all strains. When cells of six fluidal strains were washed by repeatedly centifuging and resuspending them in buffer, they were agglutinated more strongly by corn agglutinin than were unwashed cells. When avirulent cells were washed, their specific agglutination values did not increase significantly. Eight extracellular polysaccharide-deficient mutants of *Erwinia stewartii*, selected for resistance to the capsule-dependent bacteriophage K9, had lower virulence by higher specific agglutination than did their corresponding wild-type parents. Production of extracellular polysaccharide appears to be essential for virulence. Extracellular may prevent agglutination of bacteria in the host, thus allowing their multiplication.

The dextran produced by a new strain of *Leuconostoc mesenteroides* obtained from contaminated case juice has been described.[139] Branch points at O-3 and O-2 (ratio 2:1) were determined by methylation whilst the oxidized dextran was found to liberate, on partial acid hydrolysis, 2-*O*-(α-*D*-glucopyranosyluronic acid)-*D*-glucose, *O*-(α-*D*-glucopyranosyluronic acid)-(1→6)-*O*-(*D*-glucopyranosyl)-(1→3)-*D*-glucose, and 6-*O*-(α-*D*-glucopyranosyluronic acid)-*D*-glucose (which may have been derived from the triouronic acid).

Rabbit anti-dextran B1355 sera prepared by injecting rabbits with *Leuconostoc mesenteroides* NRRL B1355 were separated on a Sephadex G75 column into two fractions, one binding and the other not binding to the column.[140] Oligosaccharide inhibition of precipitation of the two fractions with dextran B1355 indicated that both fractions had (1→3)-α-*D*-specificity. However, antibodies in the non-binding fraction were shown to be directed against *O*-α-*D*-glucopyranosyl-(1→3)-α-*D*-glucopyranosyl-(1→6)-glucose, whilst those in the binding fraction were directed against *O*-α-*D*-glucopyranosyl-(1→6)-α-*D*-glucopyranosyl-(1→3)-*D*-glucose.

(1→6)-α-*D*-Specific anti-dextran antibody was raised in rabbits by injecting N4 dextran-concanavalin A conjugate, and the interactions of five synthetic linear dextrans with rabbit anti-N4 dextran were studied.[141] The ability of D-glucans to precipitate antibody depended on their average molecular weight, samples with higher molecular weight precipitating more antibody, nitrogen under the same conditions. This phenomenon was shown to be due to solubility of the antigen-antibody complex. Oligosaccharide inhibition assay indicated that the maximum size of the (1→6)-α-D-specific antibody combining site corresponded to isomaltose. The precipitated antibody class was shown to be IgG immunoglobulin, and it was mostly directed to linear non-terminal α-D-glucosidic linkages. Determination of antibody nitrogen and D-glucan in the precipitates indicated that the ratio of antibody molecule to numbers of D-glucose residues was 1:16 in the extreme antibody excess region.

The concentration (c) and shear-rate (γ) dependence of viscosity (η) has been studied for a wide range of random-coil polysaccharide solutions (including dextran), and the following striking generalities are observed:[142] (i) the transition from dilute- to concentrated-solution behaviour occurs at a critical concentration $c^* \approx 4/\{\eta\}$, when zero-shear specific viscosity (η_{sp}) $\approx$ 10 (η_{sp} varies at $c^{1.4}$ for dilute solutions, and at $c^{3.3}$ for concentrated solutions), (ii) the shear-rate dependence of viscosity and frequency dependence of dynamic (oscillatory) viscosity are closely superimposable, (iii) double logarithmic plots of η/η_0 against $\gamma/\gamma_{0.1}$ (where η_0 is zero-shear viscosity, and $\gamma_{0.1}$ is the shear rate at which $\gamma = \gamma_0/10$) are essentially identical for all concentrated solutions studied, and thus the two parameters η_0 and $\gamma_{0.1}$ completely define the viscosity at all shear rates of practical importance.

Five strains of *Micrococcus*, representing three species, accumulated α,α-trehalose when grown in the presence of D-glucose.[143] α,α-Trehalose was not found in two *Staphylococcus* species or in *Planococcus citreus*.

Sudanophilic and iodine-staining cytoplasmic granules were isolated at various developmental stages during growth of *Nocardia asteroides* (strain 55) and identified as poly-beta-hydroxybutyrate and glycogen granules, respectively.[144] During growth in nutrient broth containing 1% D-glucose, maximal accumulation of poly-beta-hydroxybutyrate and glycogen granules, up to 10 and 20%, respectively, of its dry weight, was obtained in the filamentous cells at 16 h just prior to the onset of cell fragmentation during the stationary phase. The decrease of the cytoplasmic granules was concomitant with fragmentation of the cells to rod-like and spherical cells, suggesting that the polymers may serve as carbon and energy source during macrophogenesis. Both granules were detected in the three cell forms. At higher concentrations of D-glucose (4%) more glycogen granules (9 times) accumulated than poly-beta-hydroxybutyrate (4 times), and glycogen hydrolysis was also faster than that of poly-beta-hydroxybutyrate, suggesting preferences of glycogen over poly-beta-hydroxybutyrate as energy and carbon source under this growth condition. Growth and biosynthesis of granules were significantly reduced by very high D-glucose concentration (10%), and the usual pleomorphic developmental stages were reduced to a dimorphic life cycle. Thus, biosynthesis of both granules and morphogenesis are under catabolite repression. Both polymers were also found in Nocardia brasiliensis and Nocardia otitidis-caviarum, indicating that cytoplasmic accumulation of multiple granules is common in the genus.

The method used to isolate a high-molecular-weight immunogenic polysaccharide from Pseudomonas aeruginosa immunotype 1 (IT-1) was modified to permit the isolation of a similar polysaccharide from Pseudomonas aeruginosa IT-2.[145] This antigen was composed primarily of carbohydrate, had a complex monosaccharide composition, including sugars not found in the lipopolysaccharide, and was nonpyrogenic in rabbits and nontoxic in mice at high doses. This material protected mice from challenges with live homologous cells. Pseudomonas aeruginosa IT-2 polysaccharide gave a line of identity with the O side chain of the lipopolysaccharide, but differed from this polysaccharide in molecular weight, chemical composition, and ability to immunize mice actively. Lipopolysaccharide from Pseudomonas aeruginosa IT-2 contained an immunological determinant not found on Pseudomonas aeruginosa IT-2 polysaccharide, which was detected due to its stability during treatment with dilute alkali. Thus, a high-molecular-weight polysaccharide antigen from Pseudomonas aeruginosa IT-2 polysaccharide, which was serologically identical to the lipopolysaccharide O side chain but was chemically and physically distinct, was recovered. Also, like Pseudomonas aeruginosa IT-1 strains, Pseudomonas aeruginosa IT-2 contains an alkali-stable immunodeterminant on the lipopolysaccharide that may represent a core-like antigen.

Spontaneous alginate-producing (muc) variants were isolated from strains of Pseudomonas fluoresences, Pseudomonas putida, and Pseudomonas mendocina at a frequency of 1 in 10^8 by selecting for carbenicillin resistance.[146] The infrared spectrum of the bacterial exopolysaccharide was typical of an acetylated alginate similar to that previously described in Azotobacter vinelandii and in mucoid variants of Pseudomonas aeruginosa. Mucoid variants were not isolated from Pseudomonas stutzeri, Pseudomonas pseudoalcaligenes, Pseudomonas testosteroni, Pseudomonas diminuta, Pseudomonas acidovorans, Pseudomonas cepacia, or Pseudomonas maltophilia.

Gram-negative hydrogen bacterium Pseudomonas hydrogenovora was found to excrete an anthrone-H_2SO_4 positive viscous polysaccharide.[147] About 12 g/litre of the polysaccharide was produced autotrophically on gaseous hydrogen at the stationary phase of growth. Biosynthesis of the polysaccharide occurred under nitrogen-deficient condition. Its elementary composition was C 39.29%, H 6.23%, O 49.67%, N 0.21%, and ash 4.6%. The polysaccharide contained D-galactose, D-glucose, D-mannose, and L-rhamnose as its main components. The polysaccharide had anti-tobacco mosaic virus and anti-tumour activities.

The possibility of separating pyruvulated polysaccharides into pyruvate-rich and pyruvate-poor fractions has been demonstrated using an affinity matrix.[148] Thus in some polymers, a mixture of strand types exists. The affinity matrix was prepared by coupling antibodies to a Rhizobium polysaccharide to Sepharose gel. Elution was accomplished by the addition of pyruvate to the eluting buffer. Non-pyruvylated polysaccharides were not adsorbed.

Rhizobia are Gram-negative bacteria normally capable of nodulating the roots of leguminous plants, and the failure of five extracellular polysaccharide-deficient mutants to nodulate suggested that the polysaccharide is required for this nodulation. It has been reported that among a larger sample of mutants with altered polysaccharide production, isolated from two species of Rhizobium (including one of the original set plus 34 new ones), production of the extracellular polysaccharide does not correlate with ability to nodulate appropriate hosts.[149] Therefore, although involvement of a minor component cannot be ruled out, there is no evidence that the whole polysaccharide is required for nodulation.

Immunoelectron microscopy was combined with partial characterization of isolated exopolysaccharide to study binding of soybean lectin by Rhizobium japonicum strain USDA 138.[150] Lectin-binding activity resided in two forms of exopolysaccharide produced during growth: an apparently very high-molecular-weight capsular form and a lower-molecular-weight diffusible form. At low-speed centrifugation, the capsular form cosedimented with cells to form a viscous, white, cell-gel complex which was not diffusible in 1% agar, and the diffusible form remained in the cell-free supernatant. Electron-microscopic observation of the cell-gel complex after labelling with soybean lectin-ferritin conjugate revealed that capsular polysaccharides, frequently attached to one end of the cells, were receptors for lectin. The outer membrane of the cell bound no lectin. Various preparations of exopolysaccharide isolated from the culture supernatant were tested for lectin binding, interaction with homologous somatic antigen, and the presence of 3-deoxy-D-*manno*-2-octulosonic acid and were chromatographed in Sepharose 4B and 6B gel beds. Lectin binding was restricted to a polysaccharide component designated as lectin-binding polysaccharide. This polysaccharide, as present in the cell-free culture supernatant, was a diffusible acidic polysaccharide devoid of 3-deoxy-D-*manno*-2-octulosonic acid, with molecular weight of 2×10^6 to 5×10^6. It was concluded that soybean lectin-binding component of *Rhizobium japonicum* is an extracellular polysaccharide and not a lipopolysaccharide and that the diffusible lectin-binding polysaccharide probably differs from the very high-molecular-weight lectin-binding polysaccharide of the loose capsule (slime) only in the degree of polymerization.

The extracellular polysaccharide of *Rhizobium meliloti* 201 consists of two acidic polysaccharides, APS-I and APS-II.[151] APS-I is composed of D-glucose, D-mannose, and D-glucuronic acid in a molar ratio of 3:3:2, whereas APS-II is composed of D-glucose, D-galactose, D-mannose, and pyruvic acid in a molar ratio of 4:3:2:1. APS-II was separated from the extracellular polysaccharide preparation by hydrolysing APS-I to its octasaccharide repeating unit with a specific enzyme. APS-I and APS-II were also separated by treatment with cetylpyridinium chloride and by paper electrophoresis of the depyruvylated polysaccharide.

An improved, glycosyl-sequencing method has been used for elucidating the structure of the acidic polysaccharide secreted by *Rhizobium meliloti* strain 1021.[152] The polysaccharide was methylated, the ether partially hydrolysed, the products were reduced, and the alditols ethylated. The resulting mixture of peralkylated oligosaccharide-alditols was analysed by h.p.l.c. m.s. Chemical-ionization mass spectra were obtained at 2 s intervals, as the various, peralkylated oligosaccharide-alditols were eluted from the h.p.l.c. column. Peralkylated mono-, di-, tri-, and tetrasaccharide-alditols could be readily detected, and at least partially identified, by their M + 1 ions, in conjunction with other characteristic ions present in their chemical-ionization mass spectra. The peralkylated di- and tri-saccharide-alditols were further analysed by g.c. m.s. in the electron-impact mode. The structure of the acidic polysaccharide secreted by *Rhizobium meliloti* strain 1021 is the same as that of the previously characterized polysaccharide secreted by *Rhizobium meliloti* strain U-27.

The structure of an extracellular, acidic polysaccharide from *Rhizobium meliloti* IFO 13336 was studied by a method involving successive fragmentation with specific β-D-glycanases of *Flavobacterium* M64.[153] The polysaccharide is composed of repeating units of the octasaccharide (33). An acidic component was identified as D-riburonic acid.

Five cultures of *Rhizobium meliloti* (J7017, 202, 204, 207, 209) and one of *Rhizobium trifolii* (560) produced water-soluble polysaccharides containing D-glucose, D-galactose, and pyruvic acid in a molar ratio of 7:1:1 and some succinic and acetic acids.[154] These were identified as succinoglycan-like polysaccharides on the basis of their components, methylation analysis, and fragmentation with two specific β-D-glycanases. One culture of *Rhizobium meliloti* (IFO 13336) produced water-soluble polysaccharide containing D-glucose, D-galactose, D-glucuronic acid, and acetic acid in a molar ratio of 5:1:1:2 and an unidentified component. Two cultures of *Rhizobium meliloti* (201, 206) produced water-soluble polysaccharides containing D-glucose, D-galactose, D-mannose, and D-glucuronic acid in a molar ratio of 4:2:3:1 and 4:1:2:1, respectively, and some pyruvic acid. *Rhizobium trifolii* IFO 13337 and *Rhizobium japonicum* IFO 13338 produced water-soluble polysaccharides containing D-glucose, D-glactactose, D-glucuronic acid, pyruvic acid, and acetic acid in a molar ratio of 6:1:1:2:1. Two isolates from the stock culture of *Rhizobium trifolii* J60 produced large amounts of the water-insoluble polysaccharide curdlan. This is the first report in *Rhizobium* of the occurrence of curdlan and of spontaneous mutations in ability to produce succinoglycan-like polysaccharide and curdlan.

Carbohydrate mutants were isolated from *Rhizobium meliloti* L5-30 using the translocatable drug-resistance element Tn5.[155] Enzyme assays with cell-free extracts of four mutants showed that they lacked D-mannitol dehydrogenase, D-ribose kinase, D-xylose isomerase, and D-fructose kinase, respectively. An L-arabinose mutant was also

→4)-β-D-Glcp-(1→4)-β-D-Glcp-(1→4)-β-D-Glcp-(1→3)-β-D-Galp-(1→

6
|
1

α-D-RibfA-(1→4)-α-D-GlcpA-(1→4)-β-D-Glcp-(1→6)-β-D-Glcp

(33)

isolated. Uptake studies showed that the D-ribose, D-xylose, and D-fructose mutants still utilized the sugars on which they were unable to grow, possibly indicating that Rhizobium meliloti possesses alternative metabolic routes which do not result in growth. All the mutants were able to nodulate alfalfa plants, but the D-fructose kinase mutant was unable to fix nitrogen and the L-arabinose mutant showed either late or no nitrogen-fixing ability.

The ADP-D-glucose pyrophosphorylases from Rhodospirillum fulvum, Rhodospirillum molischianum, and Rhodospirillum tenue were partially purified, and their kinetic properties were studied.[156] The enzyme from the three organisms was found to be activated by pyruvate and thus was similar to the Rhodospirillum rubrum enzyme that had been previously studied. The enzymes from Rhodospirillum fulvum, Rhodospirillum molischianum, and Rhodospirillum tenue were also activated by oxamate, an analog of pyruvate. Other α-keto acids, α-ketobutyrate and hydroxypyruvate, activated to a smaller extent. The presence of pyruvate increased the apparent affinity for adenosine 5'-triphosphate and $MgCl_2$ for all three enzymes. The Rhodospirillum molischianum enzyme has very little sensitivity to inhibition by adenosine 5'-monophosphate, ADP, or inorganic phosphate. However, Rhodospirillum tenue ADP-D-glucose pyrophosphorylase is very sensitive to inhibition by adenosine 5'-monophosphate, and the Rhodospirillum fulvum enzyme is inhibited by ADP. Increasing pyruvate concentration reversed the inhibition caused by adenosine 5'-monophosphate or ADP. Since ADP-D-glucose is the glycosyl donor for synthesis of glycogen, it is possible that in vivo glycogen synthesis is regulated by the concentration of pyruvate and, in the case of Rhodospirillum fulvum and Rhodospirillum tenue, by the ratio of pyruvate concentration to inhibitor concentration.

ADP-D-glucose synthetase from the photosynthetic bacterium Rhodospirillum tenue has been purified greater than 95%. The molecular weight of the enzyme is approximately 215,000, with a subunit molecular weight of about 51,000. The enzyme appears to be composed of four similar if not identical subunits.[157] Although the amino acid composition of the enzyme is similar to that of Escherichia coli and Salmonella typhimurium, no apparent homology has been observed between their N-terminal amino acid sequences. Antisera prepared against the Rhodospirillum tenue enzyme can partially inhibit the activities of ADP-D-glucose synthetases from other photosynthetic bacteria.

D-Mannitol, not previously reported as an intracellular component of bacteria, although it has been found as an extracellular end product of anaerobic carbohydrate metabolism, accumulated within strains of all 10 staphylococcal species tested after aerobic incubation of washed cell suspensions in phosphate-buffered 1% D-glucose for 2 h.[158] Phenol extracts of the cells, before and after incubation, were analysed for D-mannitol content by periodate utilization and paper chromatography and for D-mannitol-phosphate content, with D-mannitol 1-phosphate dehydrogenase. In Staphylococcus aureus Towler, the content of D-mannitol increased from a 0 h value of <2.3 to 16 μmol/g (dry weight) after incubation, and the level of D-mannitol-phosphate increased from a 0 h value of 1 to 8 μmol/g. The identification of D-mannitol was confirmed as the per-O-acetyl ester by gas-liquid chromatography and as the per-O-methyl ether by mass spectrometry. Also tested were 5 additional Staphylococcus aureus strains and 32 coagulase-negative staphylococcal strains. All strains accumulated D-mannitol, even those strains that could not utilize exogenous D-mannitol during aerobic growth, usually in the range 4 to 25 μmol/g. Furthermore, three strains accumulated very high D-mannitol levels. Bacteria from several other genera were tested, and some were found to accumulate low to moderate levels of D-mannitol under similar incubation conditions. The metabolic conversion of D-glucose to intracellular D-mannitol, probably via D-mannitol 1-phosphate, is a common feature of staphylococci and also occurs in some other bacteria.

Soluble extracellular antigens were prepared from the culture supernatant of exponential growing cells of Streptococcus sanguis OMZ 9 by a combination of ammonium sulphate precipitation and chromatography on a Bio-Gel P6 column.[159] Soluble cell-wall antigens were obtained from the bacterial pellet by extraction with 1 M phosphate buffer (pH 6). Antisera against whole cells of Streptococcus sanguis and Streptococcus mutans of different serotypes, 10% trichloracetic extracts of bacterial cell walls, dextran, extracellular antigens, and wall antigens were prepared by injecting the different antigens several times in rabbits. Extracellular antigens and wall antigens were prepared from a representative strain of Bratthall's seven serological groups, Lactobacillus salivarius, and Actinomyces viscosus. All sera showed various agglutinin titers against heat-killed cells, and titers were generally higher with homologous cells. The comparison of the different antigens, using agar-gel diffusion and immunoelectrophoresis, showed the presence of extracellular common antigens in both extracellular antigens and wall antigens between the different strains. Absorption of anti-extracellular antigens sera with wall antigens and anti-wall antigens sera with extracellular antigens showed the existence of a specific antigen common to all bacteria in each fraction. Enzymatic treatment of the antigen before immunodiffusion demonstrated the protein nature of the two antigens present in extracellular antigens and wall antigens.

The covalently bound carbohydrate antigen extracted from the cell wall of group B Streptococcus type 1b is common to type 1 serotypes.[160]

Two sialic acid-containing type III group B streptococcal antigens were obtained from a supernatant growth medium, purified by anion exchange or gel filtration, and found to be free of group B reactivity.[161] Quantitation of the high-molecular-weight extracellular type III antigen indicated that approximately 20-fold more antigen was recoverable from the grown medium than could be obtained by neutral buffer extraction of whole cells.

Starved cells of Streptococcus lactis ML_3 (grown previously on D-galactose, lactose, or maltose) accumulated methyl 1-thio-β-D-galactopyranoside by the lactose:phosphotransferase system. More than 98% of accumulated sugar was present as a phosphorylate derivative methyl 1-thio-β-D-galactopyranoside-6-phosphate.[162] When a phosphotransferase-system sugar (D-glucose, D-mannose, 2-deoxy-D-glucose, or lactose) was added to the medium simultaneously with methyl 1-thio-β-D-galactopyranoside, the β-D-galactoside was excluded from the cells. D-Galactose enhanced the accumulation of methyl 1-thio-β-D-galactopyranoside-6-phosphate. D-Glucose, D-mannose, lactose, or maltose plus L-arginine, when added to a suspension of methyl 1-thio-β-D-galactopyranoside-6-phosphate-loaded cells of Streptococcus lactis ML_3, elicited rapid expulsion of intracellular solute. The material recovered in the medium was exclusively free methyl 1-thio-β-D-galactopyranoside. Expulsion of D-galactoside required both entry and metabolism of an appropriate sugar, and intracellular dephosphorylation of methyl 1-thio-β-D-galactopyranoside-6-phosphate preceded efflux of methyl 1-thio-β-D-galactopyranoside. The rate of dephosphorylation of methyl 1-thio-β-D-galactopyranoside-6-phosphate by permeabilized cells was increased two to threefold by adenosine 5'-triphosphate but was strongly inhibited by fluoride. Streptococcus lactis ML_3 (DG^r) was derived from Streptococcus lactis ML_3 by positive selection for resistance to 2-deoxy-D-glucose and was defective in the enzyme II^{Man} component of the D-glucose:phosphotransferace system. Neither D-glucose nor D-mannose excluded methyl 1-thio-β-D-galactopyranoside from cells of Streptococcus lactis ML_3 (DG^r), and these two sugars failed to elicit methyl 1-thio-β-D-galactopyranoside expulsion from preloaded cells of the mutant strain. Accumulation of methyl 1-thio-β-D-galactopyranoside-6-phosphate by Streptococcus lactis ML_3 can be regulated by two independent mechanisms whose activities promote exclusion or expulsion of D-galactoside from the cell.

Antibodies directed against Streptococcus mutans GS-5 intracellular invertase and D-glucosyltransferase fractions capable of synthesizing primarily water-soluble or insoluble D-glucans were used to localize ultrastructurally the enzymes by means of the unlabelled antibody peroxidase-antiperoxidase method.[163] This immunocytochemical procedure revealed that the intracellular invertase was associated primarily with the cytoplasmic membrane of the cariogenic organism. The D-glucosyl-transferase complex responsible for insoluble D-glucan synthesis was localized as aggregates attached to the cell surface or extracellular polysaccharides of strain GS-5. In contrast, the D-glucosyl-transferase activity synthesizing primarily water-soluble D-glucans was distributed uniformly over the cell surface or in association with extracellular polysaccharides. The results were discussed relative to the sucrose-metabolizing ability of Streptococcus mutans.

The isolation of the disaggregated water-insoluble D-glucan-synthesizing D-glucosyltransferases from the culture fluid of Streptococcus mutans strain B-13 (serotype d) has been described.[164] Adhesive water-insoluble D-glucans were shown to be synthesized through an overall reaction by two protein components.

The D-glucosyltransferase from Streptococcus mutans 6715 has been separated into three enzymic fractions that differ in their binding to dextran and in their synthesis of dextran from sucrose.[165] One enzymic fraction (AFF-I) does not bind to insoluble dextran, and it produces an insoluble D-glucan. Fraction AFF-IIU was eluted from a dextran affinity column by either dextran or urea, whereas fraction AFF-IID was eluted only by dextran. Both of these fractions produce insoluble D-glucans from sucrose.

Specific inhibition by periodate-oxidized dextrans of the synthesis of α-D-glucan by Streptococcus mutans D-glucosyltransferase prompted a search for structurally related inhibitors that might be effective as anticaries agent.[166] Clinical dextran derivatives in which from 5 to 50% of the D-glucose units were oxidized acted as potent and specific enzyme-inhibitors, as did 10%-oxidized derivatives of dextran fractions ranging in molecular weight from 10^4 to 2×10^6. Within these limits, differences in oxidation or molecular weight did not significantly affect the high inhibitory potency of the derivatives. In contrast, periodate oxidation of (1→6)-, (1→3)-, and (1→4)-linked oligosaccharides containing <~15 α-D-glucose units and of sucrose and structurally related trisaccharides yielded derivatives that were poor inhibitors. Enzymic hydrolysis of oxidized dextrans caused a loss of their inhibitory power and indicated that, to act as specific inhibitors, oxidized molecules must contain at least 16 to 20 D-glucosyl residues. The similar, minimum size required in order that unoxidized oligosaccharides may act as efficient acceptors in the D-glucosyltransferase reaction suggests that the inhibitory potencies of oxidized derivatives may reflect their relative abilities to bind at the acceptor site of the enzyme.

Manganese stimulated the utilization of D-glucose by *Streptococcus mutans*, *Streptococcus sanguis*, *Streptococcus mitior*, and *Streptococcus milleri*.[167] *Streptococcus mutans* serotype C strains formed larger amounts of intracellular–polysaccharide lactic acid and produced a lower terminal pH in D-glucose broth when grown in the presence of 0.5 mM manganese ion. Manganese ion stimulated the utilization of D-glucose by resting cell suspensions of each of the four oral streptococcal species examined. Fluoride inhibited the utilization of D-glucose by the streptococci, and manganese ion, but not calcium or magnesium ions, was found to counteract the inhibitory effect of fluoride.

Streptococcus sanguis isolated from human dental plaque has been shown to adhere to smooth surfaces (glass), and this interaction is affected by *Streptococcus mutans* D-glucosyltransferase.[168]

The cell-wall component of *Streptococcus salivarius* HB, which mediates in the coaggregation with *Veillonella alcalescens* V1, has been identified.[169]

A comparison of the use of the quaternary ammonium salts cetyltrimethylammonium bromide (CTAB) and the commercial mixture Cetavlon for the isolation of xanthan gum from fermentations of *Xanthomonas campestris* indicated that the former was the more efficient complexating agent.[170] Although in both cases more than the stoichiometric requirement was necessary to achieve quantitative recovery of the polysaccharide, CTAB left only 1.7% material in the supernatant from the precipitation of xanthan gum compared to 15% left by Cetavlon, which was in the agreement with the view that the efficiency of quaternary ammonium salts increases with increased paraffin chain length. An assessment of the use of Cetavlon for the isolation of xanthan gum in a recycle procedure showed that an 11.5% loss of precipitant per cycle occurred. In the procedure, the xanthan gum was precipitated as the purified K^+ salt from a dispersion of its quaternary ammonium complex in 2-propanol. Concentration of the 2-propanol wash permitted recovery of the quaternary ammonium salt.

The complexation reaction of CTAB with xanthan has been studied using ^{14}C-labelled CTAB.[171] It was found that large proportions of precipitant were present in solution during the reaction but at the end-point only 3.2% of the residual precipitant remained in solution.

Mutagenesis of Xanthomonas campestris yielded two major classes of mutant, both having cell-surface polysaccharides fundamentally different from the wild type.[172] The wild-type bacterium produced copious amounts of extracellular slime polysaccharide containing D-glucose, D-mannose, and D-glucuronic acid in a ratio of 2:2:1. Non-mucoid mutants produced trace amounts of exopolysaccharide identical to the wild-type product. Crenated mutants produced material with an unusual composition containing sugars normally found in the lipopolysaccharide. Analysis of lipopolysaccharide fractions from these strains showed that the wild-type polysaccharide fraction contained predominantly D-glucose. Polysaccharides from the two classes of mutant bacteria were similar and contained L-rhamnose, D-galactose, and smaller amounts of D-glucose.

The role of lipid-linked sugars in the biosynthesis of an exopolysaccharide freely liberated into the culture medium has been demonstrated for the first time using Xanthomonas campestris strain NRRLB-1459.[173]

It has been shown, using EDTA-treated cells and different combinations of the adequate {^{14}C}labelled donors, that phosphoenol pyruvate provides the acetal residues and that the transfer occurs on the terminal D-mannose residue of the pentasaccharide-phosphate-phosphate-lipid, as judged by the solubility properties and DEAE-cellulose column chromatography of the compounds formed as well as by the behaviour of the substances liberated by mild acid and alkaline treatments and Smith degradation.[174] The subsequent polymerization process leads to pyruvulated xanthan gum.

6 Miscellaneous Bacterial Polysaccharides

Calcofluor White ST, a stilbene derivative used commercially as an optical brightener for cellulose, increased the rate of D-glucose polymerization into cellulose by resting cells of the Gram-negative bacterium Acetobacter xylinum.[175] This bacterium normally produced a ribbon of cellulose that is a composite of crystalline microfibrils. In concentrations above 0.1 mM, Calcofluor disrupts the assembly of crystalline cellulose I microfibrils and their integration into a composite ribbon by stoichiometric binding to D-glucose residues of newly polymerized D-glucan chains. Under these conditions, the rate of D-glucose polymerization increases up to 4 times the control rate, whereas oxygen uptake increases only 10 to 15%. These observed effects are readily reversible. If free Calcofluor is washed away or depleted below the threshold value of binding to cellulose as polymerization continues, ribbon production and the normal rate of polymerization resume. It was concluded that polymerization and crystallization are cell-directed, coupled processes and that the rate of crystallization determines the rate of polymerization. It was suggested that coupling must be maintained for biogenesis of crystalline cellulose I.

During growth of two strains of human colonic Bacteroides, Bacteroides ovatus (VPI 0038-1) and Bacteroides eggerthii (VPI B8-51), on L-arabino-D-xylans several oligomers, the smallest of which was D-xylobiose, were released into the medium.[176]

The low-temperature-dependent cell-division-defective mutant Escherichia coli $15T^-R_1$ displays a wide range of cell lengths (2 to 500μm) in late logarithmic phase cultures at 24°C. When these cells were stained with a periodic acid-Schiff technique and viewed under light microscopy, polysaccharide accumulations appeared as discretely stained areas at the poles, and at sites throughout the cytoplasm of elongated cells.[177] A statistical analysis of the relationship between the size of stained areas and cell size indicated that the total amount of polysaccharide increased with increasing cell length. This occurred both by an increase in size of existing polysaccharide-stained areas and by the creation of new areas. Interestingly, over a wide range of cell size, the ratio of cell volume occupied by polysaccharide to total cell volume remained nearly constant with a mean of about 0.16. These data suggest the existence of a homeostatic mechanism for regulating polysaccharide concentration during elongation. Electron microscopy of preparations specifically stained for polysaccharide revealed clusters of granules with a similar distribution pattern to that of the dense areas seen by phase-contrast microscopy.[178] The granules were susceptible to α-amylase digestion, and chemical analysis of the extracted and purified polysaccharide showed that it consisted of poly-D-glucose, including glycogen. At 24°C the R_1 cells contained about twice as much poly-D-glucose and four times as much glycogen as at 37°C.

Studies were carried out to elucidate the nature of phosphate and sugar linkages to F-like conjugative pili encoded by the ColB2Fdr (compatibility group FII) and EDP208 (compatibility group FV) plasmids.[179]

Both types of pili, when in the intact undissociated state, were found to contain approximately 3 mol of phosphate and 3 mol of sugar per mol of pilin. However, further purification of the two types of pilin by gel-filtration chromatography in the presence of sodium dodecyl sulphate removed all of the carbohydrate from EDP208 pilin and approximately 65% of the carbohydrate from ColB2 pilin. Approximately 0.8 to 1.0 mol of D-glucose per mol of protein remained associated with ColB2 pilin after gel-filtration chromatography, but it was not possible to determine whether this was covalently linked to the pilin or tightly associated in a sodium dodecyl sulphate-resistant manner. Sodium dodecyl sulphate-gel chromatography did not remove phosphate from either ColB2 or EDP208 pilins. ^{31}P N.m.r. spectroscopy indicated that the pilin-associated phosphate is involved in a phosphodiester linkage. Acetone precipitation of chloroform-methanol extraction of the purified pilin material reduced the phosphate associated with EDP208 pilin to less than 0.04 molecule per pilin monomer. ColB2 pilin, under the same conditions, retained approximately 0.5 phosphate per pilin monomer. The extracted phosphate-containing moieties were identified as phosphatidylglycerol and phosphatidylethanolamine by thin-layer chromatography. Since the ^{31}P n.m.r. spectra for both ColB2 and EDP208 were identical and no signal other than that of a phosphodiester was detected in the ColB2 spectrum, the phosphate remaining with the ColB2 pilin after chloroform-methanol extraction is most likely due to a tightly bound noncovalent residue.

When D-glucose is the carbon source, glycogen was found to accumulate in partial (NH_4^+) or total (NH_4^+ and required amino acids) nitrogen starvation in Escherichia coli strains that are relA$^+$ or relA$^-$.[180] However, glycogen accumulated only in the relA$^+$ strain when required amino acids were depleted or L-isoleucine starvation was induced by L-valine addition. When glycerol is the carbon source, glycogen accumulates in both relA$^+$ and relA$^-$ strains after L-valine addition. It was concluded that the relA gene is not required for glycogen accumulation when synthesis of all nitrogen-containing compounds of the cell is limited or abolished and that, although the relA gene is needed for glycogen to accumulate in amino acid starvation, this requirement can be replaced by a high cellular concentration of 3',5'-cyclic AMP.

Maltose and lactose transport systems have been used to investigate the action of procaine on insertion and activity of membrane proteins and translocation of exported proteins in Escherichia coli.[181] Procaine mildly inhibited growth on lactose. The level of inhibition was consistent with the small reduction observed in active and facilitated transport functions of the lac permease. However, procaine caused a severe reduction of growth rate on maltose, as well as an inhibition of induction of maltose regulon activities. In both constitutive and inducible strains, the synthesis of both maltose transport activity (malB operon) and amyloglycosidase activity (malA operon) was inhibited. Coordinate inhibition of soluble and membrane products was not observed with the lac operon. β-D-Galactosidase synthesis proceeded normally during growth on procaine, whereas the appearance of new transport activity was reduced. Regardless of carbon source, procaine specifically inhibited the appearance of ompF protein in the membrane fraction.

Escherichia coli is capable of growing on L-fucose or L-rhamnose as a sole source of carbon and energy. When grown under anaerobic conditions on either sugar, a nicotinamide adenine dinucleotide-linked L-lactaldehyde reductase activity is induced. The functioning of this enzyme results in the regeneration of oxidized nicotinamide adenine dinucleotide. Conditions of induction of the enzyme activity were studied and were found to display different characteristics on each sugar.[182] In the L-rhamnose-grown cells, the increase in enzyme activity detected under inducing conditions was accompanied by the synthesis of the reductase, as measured by the appearance in the extracts of a protein that reacts with propanediol oxidoreductase antibodies. In contrast, in L-fucose-grown cells, the level of the reductase as measured by enzyme antibody-reacting material was high under noninducing and inducing conditions. Thus, the increase in enzyme activity detected in going from noninducing to inducing conditions in L-fucose-grown cells did not depend on the appearance of the specific protein but on the activation of the reductase already present in the cells in an inactive form. The reductase of both homologous systems should consequently be regulated by different control mechanisms.

A high-D-xylulose mixture (D-xylose:D-xylulose = 33:67) was prepared from the cold ethanol extract of preisomerized D-xylose solution (D-xylose:D-xylulose = 77:23). Fusarium oxysporum f. sp. lini and Aspergillus niger were demonstrated to preferentially utilize D-xylose in the mixture of D-xylose and D-xylulose.[183] Chromatographically pure D-xylulose was thus obtained in 90% yield. A high-D-xylulose mixture was also incubated with Rhodotorula toruloides, Klebsiella pneumoniae, Candida utilis, or Mucor rouxi. D-Xylose and D-xylulose were simultaneously consumed. When borate was added to the mixture D-xylulose-borate complex was formed, and it could be used to protect D-xylulose from being utilized.

The sulphated heterosaccharide from the cell-wall glycoprotein of Halobacterium halobium has been shown to contain a repetitive structure which arises from a sulphated lipid-linked precursor via a biosynthetic pathway different from that known for glycosaminoglycans.[184]

A regular array of exhibiting hexagonal periodicity with a centre-to-centre spacing of about 6 nm was found on the wall surface of freeze-etched Lactobacillus buchneri. The regular array was composed of a subunit protein with a molecular weight of about 55000 as determined by sodium dodecyl sulphate-polyacrylamide gel electrophoresis.[185] Subunits isolated from the wall by extraction with guanidine hydrochloride reassembled into the original regular array after dialysis against distilled water. The subunits could reattach to wall fragments from which most of the teichoic acid had been extracted with cold trichloroacetic acid, but not to wall fragments from which most of the neutral polysaccharide had been removed by treatment with hot formamide. Heterologous reattachment of the subunits took place on to Lactobacillus casei subsp. casei wall fragments which had been partially hydrolysed under mild acid conditions. These observations suggest that the subunit protein binds to a neutral polysaccharide moiety in the underlying wall layer but not to peptidoglycan or teichoic acid.

The fermentation of cellulose by monocultures of Acetivibrio cellulolyticus and cocultures of Acetivibrio cellulolyticus-Methanosarcina barkeri, Acetivibrio cellulolyticus-Desulfovibrio sp., and Acetivibrio cellulolyticus-Methanosarcina barkeri-Desulfovibrio sp. was studied.[186] The monoculture produced ethanol, acetate, H_2, and CO_2. More acetate and less ethanol was formed by the cocultures than by the monoculture. Acetate was utilized by Methanosarcina barkeri in coculture with Acetivibrio cellulolyticus after a lag period, whereas ethanol was metabolized by the sulphate reducer only under conditions of low H_2 partial pressure, i.e. when cocultured with Acetivibrio cellulolyticus-Methanosarcina barkeri or when grown together with the methanogen. Only the three-component culture carried out the rapid conversion of cellulose to CO_2 and methane. Furthermore, this culture hydrolysed the most cellulose - 85% of that initially present. This amount was increased to 90% by increasing the population of Methanosarcina barkeri in the triculture. Methane production was also increased, and a quicker fermentation rate was achieved.

2-Amino-2-deoxy hexose metabolism in relation to the spherule-wall synthesis in Physarum polycephalum was studied by the incorporation of labelled sugar into the wall and intermediary compounds in the biosynthesis of wall polysaccharide.[187] The incorporation of 2-amino-2-deoxy-D-{^{14}C}galactose into the wall material occurred after a lag period of about 10 h in an induction medium. Nucleotides and sugar phosphates in the acid-soluble fraction of spherulating plasmodia were analysed by column chromatography of Dowex 1-X8 (formate). The primary labelled products formed in the spherulating plasmodia after incubation with 2-amino-2-deoxy-D-{^{14}C}galactose were 2-amino-2-deoxy-D-galactose 1-phosphate, UDP-2-amino-2-deoxy-D-galactose, 2-amino-2-deoxy-hexose 6-phosphate, and UDP-2-amino-2-deoxy-hexose. Spherulation was insensitive to polyoxin D, while it was completely blocked by cycloheximide. The activity of D-galactokinase and D-galactose-1-phosphate uridyltransferase increased 4.5-fold during the spherulation.

Four lysozyme-sensitive mutants were isolated after nitrosoguanidine mutagenesis of a lysozyme-insensitive strain of Staphylococcus aureus.[188] One mutant was sufficiently effective for the isolation of macromolecules, such as plasmid deoxy-ribonucleic acids, from a cell after lysozyme-induced cell lysis.

7 Fungal Polysaccharides

The genetic manipulation of industrial yeasts has been reviewed with particular emphasis on ethanol production and tolerance, the genetics of brewer's yeast strains, metabolism of wort carbohydrates, flocculation, transformation, and protoplast fusion.[189]

The interaction of lectins and lectin-like substances with fungi for identification of particular sugar residues or fungal species or the separation of polysaccharides from fungal cell walls has been reviewed.[91]

The inhibitory influence of the higher concentration of crotonaldehyde was followed during the batch and long-term continuous fermentation of Candida utilis growing on synthetic ethanol.[190] Most crotonaldehyde was removed from the medium by biotransformation. Crotonaldehyde inhibits the growth, lengthens the lag phase, and decreases the biomass yield and the content of crude proteins in the biomass. The yeast Candida utilis is capable of growing on media containing very high concentrations of inhibitor in the in-flow during continuous cultivation. Uncharacteristic transport oscillations of the content of crotonaldehyde were observed for which acidic groups on the cell membrane are probably responsible. A sensitive method which is suitable for measuring very low concentration of crotonaldehyde in aqueous solutions has been described. Crotonaldehyde acts as an uncompetitive inhibitor with slight mixed type of inhibition. An equation describing the kinetics of inhibition has been derived.

The adherence of Candida albicans to acrylic has been measured in vitro after growth of the yeast to stationary phase in defined medium containing D-glucose, sucrose, D-galactose, D-fructose, or maltose as the carbon source.[191] In each case, yeast adherence was proportional to the concentration of sugar in the growth medium, but equimolar concentrations of different sugars promoted adherence to different extents. In vitro adherence was further increased by the addition of divalent cations to assay mixtures but was inhibited when saliva-treated acrylic strips were used or when yeasts were suspended in mixed saliva during the assay. The rate of spheroplast formation of yeasts grown in media containing a 500 mM concentration of the different sugars correlated well with the relative adherence of the cells to acrylic. D-Galactose-grown yeasts were more resistant to spheroplast formation with Zymolyase-500 and most adherent to acrylic, whereas D-fructose-grown organisms were least resistant to spheroplast formation and least adherent to acrylic. These results indicate that when grown to stationary phase in media containing high concentrations of certain sugars, Candida albicans undergoes a change in cell-surface composition which facilitates its adherence to acrylic surfaces. Electron microscopy of yeasts harvested from such media revealed the presence of an additional surface layer which may be responsible for this enhanced adherence.

The new yeast species Hanseniaspora nodinigri has been described to accommodate members of the genus Hanseniaspora that are unable to assimilate D-glucono-1,4-lactone and isolated from stromatal tissue of black knots (Dibotryon morbosum) of chokecherry, Prunus virginiana.[192] The newly described taxon showed much resemblance, by other criteria, to Hanseniaspora vineae van der Walt et Tscheuschner and Hanseniaspora osmophila (Niehaus) Phaff, Miller, and Shifrine.

In strain IGC 4052 of the amylolytic yeast Lipomyces kononenkoae growing in starch-linked chemostat cultures the critical dilution rate was reduced to about half of its theoretical value due to severe catabolite repression of amylase formation while its value in a repression-resistant mutant was near its theoretical value.[193] The enzyme-yield coefficients and the specific production rates of α-amylase and glucoamylase passed through maxima at intermediate dilution rates. The shapes of the respective curves were partly determined by catabolite repression (parent strain) or its absence (mutant strain) while induction did not seem to play a role. An additional growth-linked regulatory mechanism seemed to be involved. The use of continuous culture as compared with batch culture, increased the maximum biomass productivity by a factor of 2.2 in the mutant strain and by a factor of 1.4 in the parent strain.

The ability of non-pathogenic Mycobacterium species to grow on D-galactose as the source of carbon has been studied.[194] The outline of the catabolic pathway has been presented.

Glycogen phosphorylase from macroplasmodia of Physarum polycephalum was purified 76-fold to homogeneity.[195] The native enzyme migrated as a single protein band on analytical disc gel electrophoresis coinciding with phosphorylase activity. After reduction in the presence of sodium dodecyl sulphate one protein band was detectable which corresponded to an M_r of 93000. The molecular weight of the native enzyme determined by gel sieving or gradient-polyacrylamide gel electrophoresis was 172000 and 186000, respectively. The enzyme contained about 1 mol pyridoxal 5'-phosphate and less than 0.1 mol covalently bound phosphate per mol subunit. The amino acid composition of the enzyme was determined. In the direction of phosphorolysis the kinetic data were determined by initial velocity studies, assuming a rapid equilibrium random mechanism. D-Glucose 1-phosphate and GDP-D-glucose were competitive inhibitors toward phosphate and non-competitive to glycogen. AMP, a weak activator of the enzyme, counteracted the D-glucose-1-phosphate inhibition completely. Physarum phosphorylase was compared with phosphorylases from other sources on the basis of chemical and kinetic properties. No evidence for the presence of phosphorylated forms was found.

A specimen of a rhinosporidial nasal polyp of Indian origin was used to study the chemistry and development of the wall of Rhinosporidium seeberi using fluorescein isothiocyanate-labelled lectins.[196]

The screening of twenty yeast strains for ethanol productivity at high osmotic pressure at temperatures ranging from 32°C to 45°C has been described.[197] Shake flask fermentations of 30°C, 40°C, and 50°C Bx cane molasses were performed. The effect of temperature on productivity at a non-inhibitory ethanol level is weakly pronounced. Most strains fermented poorly at 50°C Bx molasses but two Schizosaccharomyces pombe and one commercial baker's yeast, Saccharomyces cerevisiae, performed well at all concentrations of molasses. In an extended study with Schizosaccharomyces pombe (CBS 352) and Saccharomyces cerevisiae (SJAB, fresh yeast) simulating a continuous run, it was shown that Schizosaccharomyces pombe was less sensitive to high DS than Saccharomyces cerevisiae. At 25% DS the productivity of Schizosaccharomyces pombe is almost twice that of Saccharomyces cerevisiae.

A plasmid has been constructed that allows the selection in vivo of gene fusions between the Escherichia coli β-D-galactosidase gene and the yeast (Saccharomyces cerevisae) URA3 gene.[198] A large yeast DNA fragment containing the URA3 gene was placed upstream of an amino-terminally deleted version of the lacZ gene. The plasmid vehicle contains sequences that allow selection and maintenance of the plasmid in both years and Escherichia coli. Selection for Lac^+ in Escherichia coli yielded numerous deletions that fused the lacZ gene to the URA3 gene and flanking yeast sequences, to the bacterial tetracycline-resistance gene from the parent plasmid pBR322, and to the yeast 2-μm plasmid DNA. Some of these fusion plasmids produced β-D-galactosidase activity when introduced into yeast. One of the fusions to the URA3 gene itself has been shown to place the expression of β-D-galactosidase activity under uracil regulation in yeast.

The chemical modification of 18S ribosomal RNA in the 40S ribosomal subunit of yeast Saccharomyces cerevisiae with dimethyl sulphate has been reported.[199] The data were used to test the model of the secondary structure of eukaryotic 18S RNA.

Commercial baker's yeast (Saccharomyces cerevisiae) has been used to study the conversion of D-xylulose to ethanol in the presence of D-xylose. The rate of ethanol production increased with an increase in yeast-cell density. The rate of temperature for D-xylulose fermentation was 35°C, and the optimal pH range was 4 to 6. The fermentation of D-xylulose by yeast resulted in the production of ethanol as the major product. Small amounts of xylitol and glycerol were also produced.[200] The production of xylitol was influenced by pH as well as temperature. High pH values and low temperatures enhanced xylitol production. The rate of D-xylulose fermentation decreased when the production of ethanol yielded concentrations of 4% or more. The slow conversion rate of D-xylulose to ethanol was increased by increasing the yeast-cell density. The overall production of ethanol from D-xylulose by yeast cell under optimal conditions was 90% of the theoretical yield.

D-Glucose, D-galactose, and D-fructose uptake has been studied during the germination of Streptomyces antibioticus spores. Dormant spores had only the D-glucose uptake system, whatever the carbon source of the sporulation medium. Even when sporulation took place in the presence of D-galactose or D-fructose as the sole carbon and energy source, neither D-galactose nor D-fructose uptake was detected.[201] During germination the uptake systems for these two sugars were inducible by the appropriate sugar. The induction time was less when the germination process was more advanced. Synthesis of all uptake systems was immediately blocked by the addition of inhibitors of protein or RNA synthesis. Vegetative mycelium in the exponential phase of growth showed a similar pattern of sugar uptake. D-Glucose caused repression of the D-galactose and D-fructose uptake systems. The three sugar uptake systems were strongly inhibited by potassium cyanide, 2,4-dinitrophenol, and 4-chloromercuribenzoate, and showed optimum temperatures for uptake at 40°C (D-glucose) and 45°C (D-galactose and D-fructose).

Growth within the pH range 2 to 8 of a strain of the yeast

Torulopsis pintolopesii was tested in media containing various sugars as carbon and energy sources. Of the sugars tested, only D-glucose, D-fructose, and D-mannose supported growth of the yeast.[202] In media containing those sugars, the organism grew over the entire pH range tested.

Microscopic examination of horse caecum contents revealed vegetative growth of phycomycete fungi on particles of digesta, and uniflagellated cells similar to fungal zoospores in the liquid phase.[203] Three morphologically distinct isolates of strictly anaerobic phycomycete fungi were obtained from the caecum contents and cultures in vitro. Two of the isolates were able to utilize a wide range of plant carbohydrates for growth, including α-cellulose, D-xylan and particulate starch, and extensively digested water-insoluble plant tissues.

Fungi and yeasts have been cultivated on media prepared from carrageenan from Eucheuma striatum (Schmitz).[204] The cultures were found to compare well with those grown on agar media.

D-Glucans. - Chemical analyses of the hyphal wall of Achlya ambisexualis Raper have been performed in an attempt to provide a basis for understanding the molecular architecture of the wall.[205] Walls were isolated, purified, and fractionated by differential solubility. The wall and its fractions were then hydrolysed by acid or enzymes. The resulting hydrolysates were analysed by paper and gas-liquid chromatography for their neutral sugar content and type of glycosidic linkages. The degree of branching was estimated by periodate oxidation. The only monosaccharide detected was D-glucose. Three disaccharides, laminaribiose, gentiobiose, and cellobiose, were detected, thus indicating (1→3)-, (1→6)-, and (1→4)-linkages. Acid-soluble (1→3)-β-D-glucan with single (1→6)-linked residues represented 40% of the wall. Alkali-soluble D-glucan, a linear polymer of (1→3)- and (1→4)-linkages with occasional (1→6)-side chains, represented 7% of the wall. Cellulose, as determined by solubility and X-ray diffraction, represented 21% of the wall. The insoluble residuum (6%) which remained after these solubilizations had a linkage pattern similar to the alkali-soluble fraction. An insoluble component consisting of 2-amino-2-deoxy-D-glucose represented 3% of the wall. Protein comprised 10% of the wall and on hydrolysis gave the full spectrum of amino acids including hydroxy-L-proline. Small amounts of phosphorus were detected.

The surface configuration of the hyphal walls and various fractions of the wall of Achlya ambisexualis Raper were analysed by electron microscopy using carbon-platinum replicas.[206] Fractionation was carried out with both enzymes and chemical solvents. Laminarinase with or without protease and acid with alkali removed the acid- and alkali-soluble D-glucans, revealing an underlying pattern of microfibrils. The combination of cellulase, laminarinase, and protease essentially dissolved the hyphae. The cellulose solvent cadoxen removed the microfibrillar pattern which was exposed following acid and alkali treatment. The acid-soluble fraction is amorphous, the alkali-soluble and the insoluble residuum is faintly microfibrillar, and the cellulose II preparation is strongly microfibrillar. Both the cellulose I and the chitinlike fractions are uniformly microfibrillar. Morphologically, the wall consists of an outer matrix of (1→3)- and (1→6)-β-D-glucans covering an inner cellulosic protein core.

The polymorphic fungus Aureobasidium pullulans has been fractionated into mycelium and single cells at three points in the growth cycle approximately corresponding to early, middle, and late exponential phase. The ability of cells to divert assimilated D-glucose to form the extracellular polysaccharide pullulan varies throughout the cycle. The spores are the major source of polysaccharide, although the hyphae are also capable of producing it.[207] Thus, the commencement of pullulan excretion, which has been associated with blastospore production in the early phases of the growth cycle, probably arises through some stimulation concurrent with, though not necessarily identical to, the initiation of spore formation.

Lamellar single crystals of the α-D-glucan nigeran from Aspergillus niger have been subjected to controlled enzymic degradation as an approach to understanding polysaccharide organization in these crystals.[208] Analysis of both reactant and products by various methods argues that a situation exists where the crystals are inaccessible at 20°C but become increasingly accessible as the temperature approaches that of solution for the crystals in water. It was concluded that chain folds are inaccessible to enzymes at 20°C and become increasingly accessible as the melting temperature is approached. This could be due to increased mobility of the surface chains as temperature is raised, which produces large dynamic loops and makes enzyme degradation possible. The general conclusion was that nigeran single-crystal surfaces are disordered and increasingly mobile close to their dissolution temperature in water.

Antitumour activities of two (1→6)-branched (1→3)-β-D-glucans, isolated from the fruiting body of Auricularia auricula-judae (kikurage, an edible mushroom), and other branched polysaccharides containing a backbone chain of (1→3)-α-D-glucosidic or (1→3)-α-D-mannosidic linkages {and their corresponding (1→3)-D-glycans, derived by mild, Smith degradation} have been compared.[209] Among these polysaccharides, a water-soluble, branched (1→3)-β-D-glucan (D-glucan I) of Auricularia auricula-judae exhibited potent, inhibitory activity against implanted Sarcoma 180 solid tumour in mice. The alkali-insoluble branched (1→3)-β-D-glucan (D-glucan II), a major constituent of the fruiting body, showed essentially no inhibitory activity. When the latter D-glucan, having numerous branches attached, was modified by controlled, periodate oxidation, borohydride reduction, and mild, acid hydrolysis, the resulting, water-soluble, degraded D-glucan, having covalently linked polyhydroxy groups attached at O-6 of the (1→3)-linked D-glucosyl residues, exhibited potent antitumour activity. Further investigations using the D-glucan-polyalcohol indicated that the attachment of the polyhydroxy groups to the (1→3)-β-D-glucan backbone may enhance the antitumour potency of the D-glucan. On the other hand, partial introduction of carboxymethyl groups into D-glucan II (d.s. 0.47-0.86), which altered the insolubility property, failed to enhance the antitumour activity. The interrelation between the antitumour activity and the structure of the branched (1→3)-β-D-glucan has been discussed, on the basis of methylation and ^{13}C n.m.r. studies of the periodate-modified D-glucans.

Cultured blastospores of Candida albicans exhibit cytoplasmic granularity which obliterates the identification of its organelles. Abundant glycogen particles, identified by light and electron microscopy, contributed to this phenomenon to a large extent.[210]

The inhibitory effect of papulacandin B and aculeacin A on the (1→3)-β-D-glucan synthase from Geotrichum lactis has been shown to be specific and to occur both in vitro and in vivo, leading to the production of cell-free extracts with a partially inactive synthase.[211]

Studies of carbon requirements for sporulation by Nadsonia fulvescens showed an unexpected appearance of spores in the absence of an exogenous carbon source, when the cells were grown on ethanol prior to induction of sporulation.[212] D-Glucose-grown cells taken from either the logarithmic or stationary phase required a carbon source (D-glucose, ethanol, or glycerol) for sporulation. An inhibitory effect by acetate upon sporulation was demonstrated, and appeared to be due to a repression mechanism.

The effect of medium components and cultural conditions has been studied on the production of polysaccharide (pendulan) and growth of mycelia in Porodisculus pendulus JTS-3009.[213] D-Glucose, D-mannose, D-fructose, cellobiose, sucrose, D-mannitol, and various starches were found to be suitable carbon sources. Pendulan was produced and mycelia grew well on ammonium, nitrate, and organic nitrogen. The optimal concentration of nitrogen was 0.03% for production of pendulan and 0.04% for growth of mycelia. The addition of suitable amounts of yeast extract aided pendulan production and mycelium growth. The optimal temperature for pendulan production was about 29°C and for mycelium growth 25 to 30°C. The optimal pH for pendulan production was between pH 6 and 8 and for mycelium growth between 5 and 6.5. The seed culture was stable after storage at 5°C or at -20°C with the addition of 20 to 30% (v/v) glycerin. The maximum yield of pendulan and mycelia was obtained after 90 to 120 hr of fermentation.

Mouse macrophages were treated with 42 different glycans in vitro. Macrophages were stimulated - as judged by morphology, cell size, 5'-nucleotidase activity, and the incorporation of 2-amino-2-deoxy-D-{^{14}C}glucose by some insoluble glycans, e.g. yeast D-glucan.[214] Not all insoluble glycans were stimulatory (e.g. chitin). Laminaran, with a monosaccharide content and bonding pattern similar to those of yeast D-glucan, could be rendered stimulatory by cross-linking and insolubilization. There was a clear correlation between stimulatory effect and the ability to convert complement factor C3. Preincubation of sera with yeast, presumably producing complement cleavage products, potentiated the stimulatory effect of yeast D-glucan. It was suggested that endocytosis of glycans with subsequent intracellular triggering of a complement reaction is the underlying mechanism of glycan stimulation.

In Saccharomyces cerevisiae, Neurospora crassa, Aspergillus nidulans, and Coprinus cinereus most of the alkali-insoluble (1→3)-β-D-/(1→6)-β-D-glucan of the wall can be extracted with dimethyl sulphoxide.[215] The same fraction, and in Saccaromyces cerevisiae a small additional fraction, can be extracted by a destructive procedure involving 40% sodium hydroxide at 100°C. The small fraction of the D-glucan which resists this treatment becomes soluble after a subsequent treatment with nitrous acid, indicating that it is covalently linked to chitin in the wall. In contrast, in Schizophyllum commune and Agaricus bisporus, nearly all the (1→3)-β-D/(1→6)-β-D-glucan appears to be held insoluble by linkage to chitin.

When cells of Saccharomyces cerevisiae growing exponentially on D-glucose as sole carbon source were washed and transferred to buffered yeast nitrogen base containing 100 mM acetate, they were unable to resume growth for several days, whereas they adapted within a few hours to grow (slowly) on ethanol, and within 12 h to grow on pyruvate.[216] After the cell transfer, oxygen consumption and ATP concentration decreased rapidly but recovered within a few hours on ethanol, more slowly on pyruvate, and only after 70 h on acetate. When the acetate culture had lost all detectable ATP, the viable cell titre slowly decreased until after 70 h enough cell had adapted to resume growth. At lower acetate concentrations (optimally 5 to 15 mM), ATP decreased less, and growth resumed within 1 day. After transfer from D-glucose medium to buffer plus a carbon source, cells sporulated equally well at ethanol concentrations from 20 to 150 M and at pH 5.5 or 7.0. With dihydroxyacetone, another uncharged carbon source, sporulation was optimal at concentrations between 30 and 50 mM and about equal at pH 5.5 and 7.0. In contrast, after transfer from D-glucose medium to buffer plus acetate, cells sporulated at pH 5.5 optimally with 15 mM acetate but not with 50 mM acetate or more. At pH 7.0 sporulation showed a broader optimum of acetate concentration about 50 mM. The results indicated that, in cells not adapted to gluconeogenesis, high concentrations of neutral acetic acid molecules caused complete consumption of intracellular ATP; consequently the cells could not adapt to gluconeogenesis for a long time.

D-Glucan synthase activity in cell-free extracts of Saccharomyces cerevisiae was partially stabilized when cells were broken in the presence of sucrose.[217] Under these conditions a significant amount of enzyme activity remained in the supernatant, after high-speed centrifugation. When this supernatant fraction was incubated with UDP-D-glucose, microfibrils were synthesized. Microfibrils were insoluble in water, ethanol, and acid, and soluble in alkali. Under the electron microscope they appeared more or less uniform with an average length of about 0.5 μm. Alkali-insoluble residue appeared in the form of densely packed longer microfibrils. After acidification of alkali-solubilized D-glucan, shorter microfibrils were reprecipitated. Microfibrils were digested by both *endo*- and *exo*-1,3-β-D-glucanase. In the latter case D-glucose was the only product indicating that the microfibrils consist of 1,3-β-D-glucan with no detectable branches.

The production of amyloglucosidase by those strains of *Saccharomyces diastaticus* which are able to hydrolyse starch or dextrin has been exploited to devise a rapid method for the recognition of such strains.[218]

The triple helix of a polysaccharide *Schizophyllum commune* (schizophyllan) of viscosity-average molecular weight 4.8×10^5 g mol^{-1} (in water) was melted at denaturation temperatures between 5 and 60°C in water-dmso mixtures. The solutions for different denaturation temperature values were studied by viscometry and ultracentrifugation.[219] As denaturation temperature was increased, the intrinsic viscosity of the mixture containing 12.76% (by weight) of water decreased sharply and, at about 50°C, it approached the value expected for the single chain of schizophyllan in pure dmso. Schlieren patterns of the sample solvent denatured at temperatures between 25 and 45°C in the same mixed solvent showed the presence of two solute species. The fast-sedimenting species dominating at denaturation temperatures of 25°C almost disappeared at denaturation temperature of 45°C. From sedimentation-coefficient data for the mixture, pure water, and dmso the fast- and slow-sedimenting species could be identified with the triple helix and the single chain of schizophyllan, respectively. These findings led to the conclusion that, in water-dmso mixtures containing ~13% of water, the schizophyllan triple helix melts into single chains in all-or-none fashion with increasing temperature.

D-Mannans. - Cell-wall components of mycelia and conidia of *Aspergillus fumigatus* were found to contain alkali-soluble polysaccharides comprised of D-galacto-D-mannans which coprecipitated with small proportions of a D-glucan, tentatively identified as glycogen.[220] The fine structures of the D-galacto-D-mannans varied as a function of the cell type. In a 5-day-old mycelium, the polysaccharide consisted of a main chain of (1→6)-linked α-D-mannopyranosyl residues substituted at O-2 by 1 to 3 α-D-mannopyranosyl residues that are (1→2)-interlinked. β-D-Galactofuranosyl residues are (1→6)-linked to the D-mannan core, being components of side chains of average length of ca. 6 residues, which are (1→5)-interlinked. The 10-day-old mycelium had a similar D-galacto-D-mannan, but the proportion of glycogen was smaller. Conidia contain polysaccharides of different structure, as shown by the ^{13}C n.m.r. spectrum and by methylation analysis. Side chains composed of a single β-D-galactofuranosyl group linked (1→6) to adjacent D-mannopyranosyl residues were identified with a minor proportion of 6-O-substituted D-galactofuranosyl residues. Also present were nonreducing D-galactopyranosyl groups and 2-amino-2-deoxy-glycosyl residues. The D-glucan component was not glycogen. Conidial walls have much less protein than mycelial walls. Predominant amino acids in the latter were L-aspartic and L-glutamic acids, L-tyrosine, L-alanine, and glycine. Palmitic, stearic, oleic, and linoleic acids were present in the mycelial and condial walls; linoleic acid was present in minor amounts in the mycelial wall, but was a major component of the lipid fraction from whole cells.

The possibility that dissolved D-mannan polysaccharide from Candida may have an inhibitory influence on the host defence mechanisms associated with neutrophils and thereby contribute to the pathogenesis of candidoses has been investigated.[221] A model for this hypothesis was created by using D-mannan derived from baker's yeast (Saccharomyces cerevisiae) and D-mannan isolated from the serum of a patient with chronic mucocutaneous candidosis. It was observed that these D-mannans inhibited the respiratory burst and release of myeloperoxidase stimulated by phagocytosis of serum-opsonized zymosan in vitro. D-Mannan did not inhibit release of three other lysosomal enzymes. The selective inhibition of release of myeloperoxidase was attributable to a carbohydrate-binding property of the enzyme, with the D-mannan causing co-sedimentation of the enzyme with the cellular fraction.

Various ultrastructural methods have been applied to study the localization of chemical and/or antigenic components of Candida albicans blastospore cell wall.[222] They concerned: PATAg reaction for the detection of polysaccharides on ultrathin sections associated with enzymatic digestions or polysaccharide extraction; the indirect immunoferritin method on intact cells; the indirect immunoperoxidase method on ultrathin section of water-soluble embedding medium; the indirect immunofluorescence test, using patients and experimental sera. The cytochemical results confirmed the previously described eight-layer organization. The immunocytochemical methods, also in agreement with this stratification, showed in addition to external structures that the layer located near the plasmalemma must be considered as an important antigenic area. The D-mannans responsible for antigenic differences between strains of Candida albicans and namely those supporting the serotype A activity were shown to be, at least, distributed among two of the described peripheral layers.

The cell-wall mutant of a hydrocarbon-assimilating yeast, Candida sp. M-7002, and its wild type have shown a significant difference in D-content. Each D-mannan was isolated from the mutant and the wild-type cells by fractionation with Cetavlon and copper reagent.[223] Both D-mannans contain D-mannose, D-glucose, and phosphate. The mutant D-mannan has a relatively high content of protein (18% in weight bases) whereas the wild-type D-mannan has a low protein content (5.1%) with a high amount of carbohydrate (>90%). Structural analyses by enzymatic and chemical methods showed that both D-mannans had a (1→6)-linked α-D-mannosidic back bone substituted at O-2 by side chains of varying lenth. The side chains of the mutant D-mannan were shown to consist of single D-mannose units and disaccharide units whose linkages were predominantly (1→2)-linked, while the wild-type D-mannan had two additional side chains of disaccharides. These additional side chains had (1→3)-linkages which were scarcely found in the mutant D-mannan. β-Elimination reaction demonstrated that the D-mannans also contain D-mannosyl oligosaccharides linked to protein through O-glycosidic linkage. The chemical properties of D-mannan of the Candida mutant indicates that the mutation might occur not only in the side-chain structure but also in the (1→6)-linked α-D-mannan back bone.

The structures of the proteomannans from the cell-wall mutant Candida sp. and its parent strain were further studied by immunochemical methods. Alkaline titration showed that the phosphate is present in the form of diester, and mild-acid treatment released D-mannose from both of the D-mannans.[224] In the wild-type D-mannan, the phosphorylated side-chain oligosaccharides gave 85% inhibition and a D-mannopentaose side chain having (1→3)-α-D-linkage in the nonreducing terminal also gave 40% inhibition in the homologous precipitin reaction. This indicated that the immunodominant side chain of the Candida sp. wild-type D-mannan is the phosphorylated D-mannopentaose which has an (1→3)-α-D-linked nonreducing terminal. In the mutant system, a significant reactivity in the cross-reaction between the mutant antiserum and

Kloeckera brevis D-mannan indicates that the immunodeterminant of the mutant D-mannan may be a structural analogue of the side-chain oligosaccharide from Kloeckera brevis, which has phosphorylated (1→2)-linked α-D-mannotriose. However, no definite results were obtained in the inhibition studies. On the basis of the results from chemical, enzymatical, and immunological experiments, the side-chain structure and macromolecular models of Candida sp. wild-type and mutant mannans were proposed.

The D-galacto-D-mannan from Conidiobolus coronatus has been subjected to methylation studies and found to contain (1→2)-, (1→3)-, and (1→6)-linked D-mannosyl residues with evidence of branching at C-2, C-3, and C-6 of the D-mannosyl residues.[225] D-Mannose was found also to be a terminal nonreducing residue. The D-galactose was present as (1→3)-linked residues, whilst half the D-galactose represented terminal nonreducing residues. The polysaccharide appears to have similar structural features to those present in yeasts and filamentous fungi.

An alkali-soluble polysaccharide, designated as S-Iawe, has been isolated from the mycelia of Epidermophyton floccosum. Methylation, periodate oxidation, and acetolysis studies suggested that S-Iawe is composed of (1→6)-O-α-D-mannopyranosyl-(1→6)-O-{α-D-mannopyranosyl-(1→2)}-O-α-D-mannopyranosyl repeating units.[226] Condensation of 2,3,4,6-tetra-O-acetyl-α-D-mannopyranosyl bromide with methyl 3-O-benzyl-4,6-O-benzylidene-α-D-mannopyranoside in the presence of mercuric cyanide gave, in 70% yield, methyl 3-O-benzyl-4,6-O-benzylidene-2-O-(2,3,4,6-tetra-O-acetyl-α-D-mannopyranosyl)-α-D-mannopyranoside. Condensation of the debenzylidenated disaccharide with 2,3,4,6-tetra-O-acetyl-α-D-mannopyranosyl bromide afforded the corresponding trisaccharide repeating unit.

Field-desorption mass spectrometry has been used to analyse carbohydrate polymers with 5 to 14 hexose units without prior derivatization. In all examples, the molecular weight of the oligosaccharide could be determined by means of the abundant quasimolecular ions of the type MNa^+, MH^+, $MNa_2{}^{2+}$, and $MNa_3{}^{3+}$.[227] Fragmentation at glycosidic linkages was observed in various extents. The reduced oligosaccharide $(D\text{-mannose})_8$-2-acetamido-2-deoxy-D-glucitol, obtained from IgM, gave quasimolecular-ion signals MNa^+ at m/z 1544, MH^+ at m/z 1522, $MNa_2{}^{2+}$ at m/z 784, and $MNa_3{}^{3+}$ at m/z 530, all corresponding to its assumed molecular weight of 1519.5. Mycobacterial methyl-D-mannose polysaccharides with the general structure $(D\text{-mannose})_x$-(3-O-methyl-D-mannose)$_y$-3-O-methyl

methyl-D-mannoside were also successfully analysed. D-Mannose-(3-O-methyl-D-mannose)$_{12}$-3-O-methyl methyl-D-mannoside, the largest homologue, gave the expected signal of the quasimolecular ion MNa^+ at m/z 2506. The larger polysaccharides were analysed by using a KRATOS MS-50 mass spectrometer with a high-field magnet enabling full sensitivity to be maintained up to 3000 atomic mass units. Polysaccharides up to m/z 1978 were analysed by using a KRATOS MS-9 mass spectrometer operated at 4 kV. The signal-to-noise ratio, which becomes a serious problem in field-desorption mass spectrometry at low accelerating voltages, and the low instrument sensitivity were improved considerably by use of a method of adding scans with low total ion currents obtained over a longer desorption time. In this way, complete sequence information was obtained on 3-O-methyl-D-mannose polysaccharides up to D-mannose-(3-O-methyl-D-mannose)$_8$-3-O-methyl methyl-D-mannoside (MNa^+ at m/z 1802). Analysis of a presumed D-mannose (3-O-methyl-D-mannose)$_6$-3-O-methyl methyl-D-mannose gave a spectrum consistent only with the structure (D-mannose)$_2$-(3-O-methyl-D-mannose)$_3$-3-O-methyl methyl-D-mannoside, revealing the existence of a 3-O-methyl-D-mannose homologue with 2 unmethylated D-mannoses at the nonreducing end of the chain.

Saccharomyces cerevisiae wild-type and mutant cells affected in the structure of D-mannan outer chain were shown to possess *in vivo* one major dolichol-phosphate-phosphate-bound oligosaccharide. The size, monosaccharide composition, and pattern obtained upon acetolysis and paper chromatography of the oligosaccharide were the same for all strains and for the main corresponding compound isolated from animal tissues.[228] Evidence is presented indicating that the dolichol-diphosphate derivative occurring *in vivo*, and containing 2-acetamido-2-deoxy-D-glucose (2 mols), D-mannose (9 mols), and D-glucose (3 mols), is the intermediate involved in yeast protein glycosylation. The transfer of the oligosaccharide to protein was followed *in vivo* by the excision of the D-glucose and at the most one D-mannose residue. D-Mannoses were then added to the trimmed saccharide moiety. No difference between the first stages (*i.e.* excision of monosaccharides) of the processing of the protein-bound oligosaccharides by wild-type and mutant cells was found. However, mutants carrying the *mnn 1* mutation, which are known to be devoid of terminal (1→3)-linked-α-D-mannosyl residues in the D-mannan outer chain and inner core, were found not to add such D-mannose residues to the already D-glucose-free protein-bound oligosaccharide.

The D-mannan of baker's yeast (a wild-type strain of Saccharomyces cerevisiae), prepared by means of cetyltrimethylammonium bromide, was investigated for its immunochemical properties.[229] Upon treatment with 10 mM HCl at 100°C for 60 min, this D-mannan gave a modified D-mannan, D-mannobiose, and D-mannose. By the action of 100 mM NaOH at room temperature for 18 h, another modified D-mannan was obtained together with D-mannotetraose, D-mannotriose, D-mannobiose, and D-mannose. In both acid- and alkali-degradation products, D-mannobiose was a common major oligosaccharide component. The results of methylation analyses indicated that D-mannobiose in the former degradation product consisted solely of (1→3)-α-D-linked residues, whereas the latter one contained (1→2)- and (1→3)-α-D-linked residues in a molar ratio of 1:8:1.0. The findings of quantitative precipitin assay of the parent D-mannan and the two modified D-mannans against the homologous anti-Saccharomyces cerevisiae serum revealed that the amounts of precipitated antibody by each modified D-mannan were nearly identical, and were significantly smaller than that of the parent D-mannan. Because the antibody-precipitating activities of both modified D-mannans were exactly identical to that of the D-mannan prepared by the Fehling solution method, it is evident that the (1→3)-linked α-D-mannobiosyl residues dominate a large part of the serological activity of the parent intact D-mannan, and that the D-mannan prepared by the Fehling solution method corresponds to a degradation product lacking these immunodominant D-mannobiosyl residues.

Carboxypeptidase Y, a vacuolar enzyme from Saccharomyces cerevisiae, was digested with endo-β-D-2-acetamido-2-deoxy-D-glucosidase H to release the four oligosaccharide chains that are linked to L-asparagine in the glycoprotein. The oligosaccharides were fractionated into a neutral and acidic component, and the latter proved to be phosphorylated.[230] From its gel-filtration pattern, the neutral fraction was shown to be a mixture of at least four homologues, the smallest of which had a ^{1}H n.m.r. spectrum almost identical to that given by an IgM oligosaccharide with eight D-mannoses and one 2-acetamido-2-deoxy-D-glucose. The yeast oligosaccharide has one additional α-D-mannosyl residue in a (1→3)- or (1→6)-linkage, whereas the larger homologues appear to have two, three, and four more D-mannose residues. One phosphorylated oligosaccharide with D-mannose/phosphate ratio of 12.5 was reduced with NaB^3H_4 and then subjected to mild-acid hydrolysis. This released D-mannose and D-mannobiose that were glycosidically linked to the phosphate group, whereas complete acid hydrolysis yielded D-mannose 6-phosphate. The recovered oligosaccharide phosphomonoester, which contained 11 or 12 D-mannose residues, was digested exhaustively with α-D-mannosidase, and the product of this reaction was treated with alkaline phosphatase, which yielded radioactive (D-mannose)$_3$-2-acetamido-2-deoxy-D-glucitol. These results suggested that the α-D-mannosidase-resistant phosphorylated oligosaccharide has the structure (34), in which some of the phosphate groups are substituted with D-mannobiose instead of D-mannose. A second phosphorylated oligosaccharide with a D-mannose/phosphate ratio of 6.5 probably contains two phosphodiester groups, but its structure has not been investigated in detail.

```
      α-D-Manp-(1→?)-α-D-Manp-(1→6)-β-D-Manp-(1→4)-D-GlcNAcH2
          6
          |
 α-D-Manp→P


          |
where    →P = orthophosphate diester linkage
```

(34)

Cell walls of flocculent strains (0006) and non-flocculent strains (0019) of Saccharomyces uvarum (Carlsbergensis), grown in different media and taken in both growth and stationary phases, were treated with water and with 2% (w/v) potassium hydroxide. This treatment yielded four fractions (FI, FII, FIII, and FR).[231] The fractions FI isolated from the flocculent cell walls contained more D-mannose and less protein than the corresponding fractions FI isolated from the non-flocculent cell walls. The amino acid composition was also different between the two types of fractions. A radioactive-labelling technique revealed that the FI and the walls from flocculent cells bound on average two to three times as much ^{45}Ca as did the FI and the walls from non-flocculent yeast. The substitution of carboxyl groups in FI and walls with glycine methyl ester led to a great drop of the ^{45}Ca binding capacity, suggesting that carboxyl groups of the cell walls are involved in the flocculation process. However, flocculation seems to be a phenomenon more complex that the simple formation of Ca^{2+} bridge, the involvement of lectin-like components easily removed from the cell walls, should not be rejected.

Chitin. - The structure of poly(2-amino-2-deoxy-D-galactose) purified from the culture fluid of Aspergillus parasiticus AHU 7165 was studied. Partial acid hydrolysis of this polysaccharide, in which 55 to 65% of the monosaccharide residues were N-unsubstituted, gave a series of 2-amino-2-deoxy-D-galactose oligosaccharides (dimer to hexamer) in a good yield.[232] From the data on analyses of the polysaccharide and its oligosaccharides by gel filtration, periodate oxidation, methylation analysis, and ^{1}H n.m.r. measurement, the polysaccharide was characterized as a linear chain of (1→4)-linked 2-amino-2-deoxy-α-D-galactosyl residues. The N-unsubstituted 2-amino-2-deoxy-D-galactosyl residues are probably distributed in a random fashion over the polysaccharide chain.

The uptake of 2-amino-2-deoxy-D-glucose by the strongly halotolerant Debaryomyces hansenii and the non-tolerant Saccharomyces cerevisiae has been studied to determine respiration and fermentation rates for these yeasts.[233]

Antisera specific for purified cell walls of Fusarium solani f. sp. pisi and phaseoli and of shrimp-shell chitosan were utilized as immunochemical probes to determine the location of fungal components in the pea-Fusarium interaction.[234] Within 15 minutes after inoculation, fungal cell-wall components appeared to enter the plant cell and to accumulate inside the plant cell wall as fungal growth on the plant tissue was inhibited. The accumulation patterns of chitosan and all components containing 2-amino-2-deoxy-hexose polymers resembled those of the fungal wall components. Chitosan is present on, and released from, the outer surface of the fungal spore. Within 15 minutes after applying {^{3}H}chitosan to the surface of the plant tissue, the label was readily detectable within the plant cytoplasm and conspicuously detectable within the plant nucleus. It was proposed that the potential for transport of chitosan between the spores of Fusarium solani and pea cells, in addition to its potential to inhibit fungal growth and elicit disease-resistance responses, suggests chitosan has a major regulatory role in this host-parasite interaction.

The in vivo differential rates of chitin-plus-chitosan biosynthesis in Mucor racemosus were determined under a variety of conditions, leading to yeast cell or mycelial morphology. Chitin-chitosan was determined as hot sodium hydroxide-insoluble radioactivity derived from 2-acetamido-2-deoxy-D-{1-^{3}H}glucose in the medium. Control experiments demonstrated that the labelled material possessed the properties of chitin-plus-chitosan. The results indicated that Mucor yeasts have a relatively low differential rate of chitin-plus-chitosan synthesis and that mycelial cells have a threefold-elevated differential rate.[235] Treatment of aerobic cells with exogenous N^6,O^2-dibutyryl cyclic adenosine 3',5'-monophosphate, an agent which induces yeast-cell morphology, also resulted in a lowered rate of chitin-plus-chitosan synthesis. Control experiments eliminated the possibility that the observed rate changes were due to changes in endogenous pool size, uptake of exogenous 2-acetamido-2-deoxy-D-{1-^{3}H}glucose, or alterations in growth rate. Therefore, the changes were seemingly linked to morphogenesis. These results strengthen the idea that cyclic adenosine 3',5'-monophosphate plays an important role in dimorphism in Mucor. In addition, pulse-chase experiments suggested that considerable modification of newly synthesized chitin plus chitosan in both yeast cells and mycelia occurs in vivo.

The components and structure of the cell wall of Rhizopus delemar have been investigated using purified lytic enzymes, protease and chitosanase from Bacillus R-4, and chitinase II from Streptomyces orientalis. When these enzymes were used individually they only partially lysed the cell wall, but when allowed to react on the cell wall together a complete lysis was achieved by cooperative action.[236] These modes of action on the cell wall and the chemical and morphological data suggested that the cell-wall structure was different in Rhizopus delemar of Zygomycetes from filamentous fungi of Euascomycetes and that its wall structure might be composed mainly of chitin fibres cemented by chitosan and protein or peptides scattered in a mosaic manner.

During culture of Rhodotorula glutinis the fall in pH from 4.5 to 2 stimulated the synthesis of chitin. Alternatively keeping the pH at 4.5 caused the chitin content to fall largely.[237] However, there was no proportional value between the chitin-content fall and that of total 2-amino-2-deoxy-hexoses.

Miscellaneous Cell-wall Polysaccharides. - During the development of a cell aggregate of Dictyostelium discoideum into a fruiting body, an antigenic acid mucopolysaccharide is synthesized only in the prespore cells of a cell mass. The subcellular distributions of UDP-D-galactose:polysaccharide transferase and UDP-D-glucose pyrophosphorylase involved in biosynthesis of mucopolysaccharide have been determined.[238] The transferase was specifically localized in the smaller vesicles with lighter density than the prespore-specific vacuoles identifiable by electron microscopy. In contrast to the enzyme, the antigenic mucopolysaccharide was exclusively localized in the prespore-specific vacuoles. Unlike the transferase, UDP-D-glucose pyrophosphorylase was confined to the soluble fraction. The sucrose gradient profiles of the transferase activity in the 5000 x g supernatant gave two main peaks. When the profiles were compared among standing and migrating slugs and culminating cell mass, the difference in the profiles closely reflected the state of biosynthesis of the acid mucopolysaccharide in each developmental stage.

An L-fuco-D-galactan and a D-manno-L-fuco-D-galactan were isolated from the fruit bodies of *Ganoderma applanatum*, by employing a column of concanavalin A-Sepharose 4B.[239] The former consists predominantly of a main chain of (1→6)-linked α-D-galactopyranose residues, *ca*. 30% of which are substituted at O-2 by single-unit α-L-fucopyranosyl side chains. The latter consists mainly of a backbone of (1→6)-linked α-D-galactopyranosyl residues, *ca*. 50% of which are substituted with either a 3-O-α-D-mannopyranosyl-α-L-fucopyranosyl or α-L-fucopyranosyl side chains. Both are partially substituted by O-acetyl groups. These structural attributions were supported by results of methylation analysis, g.l.c. analysis of polyhydric compounds, and by ^{1}H and ^{13}C n.m.r. spectroscopy.

The structures of four oligosaccharides, isolated from a partial acid-hydrolysate of the glycuronan protuberic acid isolated from *Kobayasia nipponica*, have been examined by chemical analysis, and by ^{1}H and ^{13}C n.m.r. spectroscopy. They were identified as O-(β-D-glucopyranosyluronic acid)-(1→4)-D-glucuronic acid, O-(α-L-idopyranosyluronic acid)-(1→4)-D-glucuronic acid, O-(β-D-glucopyranosyluronic acid)-(1→4)-O-(α-L-idopyranosyluronic acid)-(1→4)-D-glucuronic acid, and O-(β-D-glucopyranosyluronic acid)-(1→4)-O-(β-D-glucopyranosyluronic acid)-(1→4)-O-(α-L-idopyranosyluronic acid)-(1→4)-D-glucuronic acid.[240] The acid-resistant portion of protuberic acid had properties similar to those of the original protuberic acid. These results suggested that the linear structural sequence of protuberic acid is derived from the trisaccharide repeating unit (35).

The major water-soluble polysaccharide isolated from the sclerotia of *Omphalia lapidescence*, and purified by chromatography on DEAE-cellulose and zone electrophoresis, is a heteroglycan

$\{\alpha\}_D^{20}$ + 56° (c 1, water), which is composed of D-glucose, 2-acetamido-2-deoxy-D-glucose, and D-glucuronic acid in the molar ratios 1.0:1.95:2.0.[241] The results of periodate oxidation, Smith degradation, and methylation analysis showed that the polysaccharide has a highly branched structure involving 1,3-, 1,4-, 1,6-, and 1,3,6-linked D-glucopyranosyl, 1,3,4-linked 2-acetamido-2-deoxy-D-glucopyranosyl, and 1,3?- and 1,4-linked D-glucopyranosyluronic acid residues. The nonreducing terminal positions are occupied by the three component sugars. This branched glycosaminoglycan is the first to be found in the fungi.

→4)-β-D-GlcpA-(1→4)-α-L-IdopA-(1→4)-β-D-GlcpA-(→

(35)

Antigens soluble in hot trichloracetic acid from eighteen strains of *Pityrosporum ovale* were studied using rabbit antiserum to American Type Culture Collection strains 24027, 12078, and 14521. These strains were found to share two antigens, both of which were present in cells of *Pityrosporum orbiculare* and one of which was present in cells of *Pityrosporum pachydermatis*, *Pityrosporum canus*, *Candida albicans*, and *Rhodotorula rubrum*.[242]

The chemical properties of the hot water-soluble and dilute alkali-soluble polysaccharides prepared from the fruit bodies of the mushroom, *Psalliota campestris* (Linn) Fr., have been studied.[243] Five purified polysaccharides were obtained, three of which contained 2-amino-2-deoxy-D-glucose and D-glucose (with D-mannose also present in one polysaccharide) whilst the remaining two were D-glucans. From specific rotation studies all the polysaccharides appeared to be composed of β-D-glycosidic linkages expect the D-mannose-containing polysaccharide, which was composed of α-D-glycosidic linkages.

L-Fuco-D-galacto-D-mannan was obtained from the walls of *Rodotorula glutinis* K-24 by alkaline extraction, pronase digestion, and alcohol precipitation. The purified polysaccharide was composed of L-fucose, D-galactose, and D-mannose in the molar ratio of 1:2.5:2, and occupied about 20% of the cell wall.[244] In the successive Smith degradation, D-mannose residues were resistant to periodate oxidation, whereas most of D-galactose residues were degraded, releasing D-threitol and L-arabinose. It appeared that D-mannose residues consisted of a backbone of the polysaccharide, and all or part of the D-galactose residues might be of a furanose type.

By partial acid hydrolysis of the L-fuco-D-galacto-D-mannan, three kinds of oligosaccharide containing L-fucose and L-galactose were obtained (36-38). The methylation analysis revealed that the polysaccharide consisted of (1→3)-linked D-mannose residues, all of which were branched at the C-2 position with L-fucose and D-galactose residues. Most of the L-fucose residues existed in the nonreducing end of the side chain. D-Galactose residues were mostly contained as (1→2)- and (1→6)-linkages in the side chain. With the aid of electron microscopy the localization of the D-gluco-D-mannan and L-fuco-D-galacto-D-mannan in the cell wall of Rhodutorula glutinis and formation of spheroplast of living yeast cells has been described.[245]

D-Gal*f*-(1→6)-D-Gal*f*
(36)

L-Fuc*p*-(1→2)-D-Gal*f*-(1→2)-D-Gal*f*-(1→6)-D-Gal*f*
(37)

L-Fuc*p*-(1→2)-D-Gal*f*-(1→6)-D-Gal*f*-(1→2)-D-Gal*f*-(1→6)-D-Gal*f*
(38)

Evidence of the presence of polysaccharide polymers has been shown in *Streptomyces* sp. for the first time.[246] Cytochemical methods revealed the occurrence of polysaccharide granules in sporulating hyphae of *Streptomyces viridochromogenes*. Onset of the sporogenesis coincided with the appearance of the granules, which reached a maximum number during the early stages of maturation. The later stages of maturation showed a decrease of these granules, and in mature spores no granules were observed.

The locations of O-acetyl groups in the acidic polysaccharides AC and BC isolated from the fruit bodies of *Tremella fuciformis* BERK have been determined by employing an improved procedure. The results indicated that the O-acetyl groups in AC and BC are located at positions 4,6 and both 4 and 6 of a part of the D-mannose residues, and at positions 2,4 and both 2 and 4 of a part of the D-glucuronic acid residues.[247] Polysaccharides AC and BC differ significantly in that the amount of the O-acetyl groups linked to the D-mannose residues in BC was higher than that in AC. The improved procedure should be widely applicable for determining the locations of O-acetyl groups on acidic polysaccharides.

(References begin overleaf)

References

1 J.B. Ward, Microbiol. Rev., 1981, 45, 211.
2 I.C. Hancock, G. Wiseman and J. Baddiley, J. Bacteriol., 1981, 147, 698.
3 S.C. Cheah, H. Hussey and J. Baddiley, Eur. J. Biochem., 1981, 118, 497.
4 K.C. Bertram, I.C. Hancock and J. Baddiley, J. Bacteriol., 1981, 148, 406.
5 I.C. Hancock, Eur. J. Biochem., 1981, 119, 85.
6 M.T. Perez Urena, P. Lopez, M. Espinosa and A. Portoles, FEMS Microbiol. Lett., 1980, 9, 315.
7 J.J. Boon, W.R. De Boer, F.J. Kruyssen and J.T.M. Wouters, J. Gen. Microbiol., 1981, 122, 119.
8 N. Murazumi, Y. Sasaki, J. Okada, Y. Araki and E. Ito, Biochem. Biophys. Res. Commun., 1981, 99, 504.
9 A.J. Anderson and A.R. Archibald, FEMS Microbiol. Lett., 1981, 10, 7.
10 V.M. Brautigan, W.C. Childs III and F.C. Neuhaus, J. Bacteriol., 1981, 146, 239.
11 S.D. Johnson, K.P. Lacher and J.S. Anderson, Biochemistry, 1981, 20, 4781.
12 J. Leaver, I.C. Hancock and J. Baddiley, J. Bacteriol., 1981, 146, 847.
13 H.A.I. McArthur, I.C. Hancock and J. Baddiley, J. Bacteriol., 1981, 145, 1222.
14 A. Ndulue, J.-P. Flandrois and D. Marmet, J. Chromatogr., 1981, 209, 323.
15 W. Fischer, P. Rösel and H.U. Koch, J. Bacteriol., 1981, 146, 467.
16 L. Hardy, N.A. Jacques, H. Forester, L.K. Campbell, K.W. Knox and A.J. Wicken, Infect. Immun., 1981, 31, 78.
17 R.R.B. Russell, FEMS Microbiol. Lett., 1981, 11, 279.
18 H. Courtney, I. Ofek, W.A. Simpson and E.H. Beachey, Infect. Immun., 1981, 32, 625.
19 M. Batley, K. Knox, J. Redmond and A. Wicken, Proc. Aust. Biochem. Soc., 1981, 14, 110.
20 D.E.S. Stewart-Tull, Ann. Rev. Microbiol., 1980, 34, 311.
21 B. Ladešić, J. Tomašić, S. Kveder and I. Hršak, Biochem. Biophys. Acta, 1981, 678, 12.
22 M. Humphries and J.S. Thompson, Anal. Biochem., 1981, 113, 398.
23 M.R. Pincus and H.A. Scheraga, Biochemistry, 1981, 20, 3960.
24 B. Flouret, D. Mengin-Lecreulx and J. Van Heijenoort, Anal. Biochem., 1981, 114, 59.
25 E. Benedetti, B. Di Blasio, V. Pavone, C. Pedone, C. Toniolo and G.M. Bonora, J. Biol. Chem., 1981, 257, 9229.
26 D.J. Waxman, W. Yu and J.L. Strominger, J. Biol. Chem., 1980, 255, 11577.
27 W.R. De Boer, F.J. Kruyssen and T.M. Wouters, J. Bacteriol., 1981, 146, 877.
28 Y. Araki, S. Oba, S. Araki and E. Ito, J. Biochem. (Jpn.), 1980, 88, 469.
29 E. Janczura, M. Leyh-Bouille, C. Cocito and J.M. Ghuysen, J. Bacteriol., 1981, 145, 775.
30 F. Ishino, K. Mitsui, S. Tamaki and M. Matsuhashi, Biochem. Biophys. Res. Commun., 1980, 97, 287.
31 E. Vanderwinkel, M. De Vlieghere and L. De Tanhoffer De Volcsey, Biochem. Biophys. Acta, 1981, 663, 46.
32 W. Goodell and A. Tomasz, J. Bacteriol., 1980, 144, 1009.
33 E.E. Ishiguro and W.D. Ramey, Can. J. Microbiol., 1980, 26, 1514.
34 P. Spiri-Nakagawa, R. Oiwa, Y. Tanaka, H. Tanaka and S. Omura, J. Biochem. (Jpn.), 1980, 88, 565.
35 K. Kato, T. Umemoto, H. Fukuhara, H. Sagawa and S. Kotani, FEMS Microbiol. Lett., 1981, 10, 81.
36 E.N. Vasstrans, Infect. Immun., 1981, 33, 75.
37 M. Humphries, A.E. Wilkinson, B. Edwards and J.S. Thompson, Biochem. Soc. Trans., 1981, 9, 436.
38 P.J. Brennan, H. Mayer, G.O. Aspinall and J.E. Nam Shin, Eur. J. Biochem., 1981, 115, 7.

39 P.J. Brennan, G.O. Aspinall and J.E. Nam Shin, J. Biol. Chem., 1981, 256, 6817.
40 L.E. Hardham and A.M. James, Microbios, 1981, 30, 87.
41 J.K. Blundell, G.J. Smith and H.R. Perkins, FEMS Microbiol. Lett., 1980, 9, 259.
42 J.K. Blundell and H.R. Perkins, J. Bacteriol., 1981, 147, 633.
43 J. Gmeiner and H.-P. Kroll, Eur. J. Biochem., 1981, 117, 171.
44 J. Gmeiner and H.-P. Kroll, FEBS Lett., 1981, 129, 142.
45 R.E.W. Hancock, R.T. Irvin, J.W. Costerton and A.M. Carey, J. Bacteriol., 1981, 145, 628.
46 Y. Kamio, Y. Itoh, Y. Terawaki and T. Kusano, Agric. Biol. Chem. (Jpn.), 1980, 44, 2523.
47 Y. Kamio, Y. Itoh, Y. Terawaki and T. Kusano, J. Bacteriol., 1981, 145, 122.
48 Y. Kamio, Y. Itoh and Y. Terawaki, J. Bacteriol., 1981, 146, 49.
49 H. Goodman, J.J. Pollock, V.J. Iacono, W. Wong and G.D. Shockman, J. Bacteriol., 1981, 146, 755.
50 T. Umemoto, T. Ota, H. Sagawa, K. Kato, H. Takada, M. Tsujimoto, A. Kawasaki, T. Ogawa, K. Harada and S. Kotani, Infect. Immun., 1981, 31, 767.
51 F.M. Unger, Adv. Carbohydr. Chem. Biochem., 1981, 38, 323.
52 O. Westphal, O. Lüderitz, E. Th. Rietschel and C. Galanos, Biochem. Soc. Trans., 1981, 9, 191.
53 C. Kratky, D. Stix and F.M. Unger, Carbohydr. Res., 1981, 92, 299.
54 P. Kiley and S.C. Holt, Infect. Immun., 1980, 30, 862.
55 J.H. Banoub and D.H. Shaw, Carbohydr. Res., 1981, 98, 93.
56 S. Kondo, T. Iguchi and K. Hisatsune, Biochem. Biophys. Res. Commun., 1980, 97, 437.
57 H.-W. Wollenweber, E.T. Reitschel, T. Hofstad, A. Weintraub and A.A. Lindberg, J. Bacteriol., 1980, 144, 898.
58 E. Moreno, S.L. Speth, L.M. Jones and D.T. Berman, Infect. Immun., 1981, 31, 214.
59 E. Romanowska, A. Romanowska, C. Lugowski and E. Katzenellenbogen, Eur. J. Biochem., 1981, 121, 119.
60 P.E. Jansson, A.A. Lindberg, B. Lindberg and R. Wollin, Eur. J. Biochem., 1981, 115, 571.
61 D. Blache, M. Bruneteau and G. Michel, Eur. J. Biochem., 1981, 113, 563.
62 M. Batley, N. Packer, J. Redmond and P. Reeves, Proc. Aust. Biochem. Soc., 1981, 14, 109.
63 Y. Takeuchi and H. Nikaido, Biochemistry, 1981, 20, 523.
64 M. Davies and D.E.S. Stewart-Tull, Biochim. Biophys. Acta, 1981, 643, 17.
65 R.T. Irvin, T.J. MacAlister, R. Chan and J.W. Costerton, J. Bacteriol., 1981, 145, 1386.
66 A. Horst and E.M. Rakowicz-Szulczyńska, Mol. Cell. Biochem., 1981, 37, 21.
67 G. Hogenauer and M. Woisetschlager, Nature (London), 1981, 293, 662.
68 J.D. Oppenheim, M.S. Nachbar, N. Bhardwaj and H. Wexler, FEMS Microbiol. Lett., 1981, 10, 25.
69 T. Mizuno, J. Biochem. (Jpn.), 1981, 89, 1039.
70 T. Mizuno, J. Biochem. (Jpn.), 1981, 89, 1051.
71 T. Mizuno, J. Biochem. (Jpn.), 1981, 89, 1059.
72 D.A. Monner, J. Gmeiner and P.F. Mühlradt, Infect. Immun., 1981, 31, 957.
73 R. Gonggrijp, M.P.W. Volleberg, P.J.M.R. Lemmens and C.P.A. Van Boven, Infect. Immun., 1981, 31, 896.
74 R.C. Seid and J.C. Sadoff, J. Biol. Chem., 1981, 256, 7305.
75 L.H. Koepp, T. Orr and P.F. Bartell, Infect. Immun., 1981, 33, 788.
76 Y.A. Knirel, N.A. Kocharova, A.S. Shashkov, B.A. Dmitriev and N.K. Kochetkov, Carbohydr. Res., 1981, 93, C12.
77 S.G. Wilkinson, Biochem. J., 1981, 199, 833.
78 D. Horton and D.A. Riley, Biochem. J., 1981, 199, 835.
79 S.G. Wilkinson, Carbohydr. Res., 1981, 98, 247.
80 A.J. Anderson, Can. J. Microbiol., 1980, 26, 1422.
81 O. Holst, U. Hunger, E. Gerstner and J. Weckesser, FEMS Microbiol. Lett., 1981, 10, 165.

82 J. Radziejewska-Lebrecht, U. Feige, H. Mayer and J. Weckesser, J. Bacteriol., 1981, 145, 138.
83 M. Vaara, T. Vaara, M. Jensen, I. Helander, M. Nurminen, E. Th. Rietschel and P.H. Mäkelä, FEBS Lett., 1981, 129, 145.
84 K.E. Sanderson and B.A.D. Stocker, J. Bacteriol., 1981, 146, 535.
85 M. Schindler, M.J. Osborn and D.E. Koppel, Nature (London), 285, 1980, 261.
86 G.M. Bennett, A. Seaver and P.H. Calcott, Appl. Envir. Microbiol., 1981, 41, 843.
87 S. Lafont, J.M. Dumont, B. Ivanoff and R. Fontanges, Ann. Microbiol., 1981, 132A, 59.
88 F.A. Heasley, J. Bacteriol., 1981, 145, 624.
89 M. Schrader, G. Drews and J. Weckesser, FEMS Microbiol. Lett., 1981, 11, 37.
90 J.B. Weinberg, P.F. Smith and I. Kahane, Biochem. Biophys. Res. Commun., 1980, 97, 493.
91 T.G. Pistole, Ann. Rev. Microbiol., 1981, 35, 85.
92 K.I. Dayalu, R. Cherniak and C.L. Hatheway, Infect. Immun., 1981, 31, 608.
93 N.O. Van Gylswyk, E.J. Morris and H.J. Els, J. Gen. Microbiol., 1980, 121, 491.
94 A.K. Bhattacharjee, K.J. Kwon-Chung and C.P.J. Glaudemans, Carbohydr. Res., 1981, 95, 237.
95 R.P. Silver, C.W. Finn, W.F. Vann, W. Aaronson, R. Schneerson, P.J. Kretschmer and C.F. Garon, Nature (London), 1981, 289, 696.
96 W.F. Vann, M.A. Schmidt, B.Jann and K. Jann, Eur. J. Biochem., 1981, 116, 359.
97 P. Messner and F.M. Unger, Biochem. Biophys. Res. Commun., 1980, 96, 1003.
98 M.R. Loeb, A.L. Zacary and D.H. Smith, J. Bacteriol., 1981, 145, 596.
99 F.-P. Tsui, R. Schneerson and W. Egan, Carbohydr. Res., 1981, 88, 85.
100 P. Branefors-Helander, L. Kenne, B. Lindberg, K. Petersson and P. Unger, Carbohydr. Res., 1981, 88, 77.
101 E.H. Merrifield and A.M. Stephen, Carbohydr. Res., 1981, 96, 113.
102 U. Elsässer-Beile and S. Stirm, Carbohydr. Res., 1981, 88, 315.
103 G.G.S. Dutton, A.V. Savage and M. Vignon, Can. J. Chem., 1980, 58, 2588.
104 J. Di. Fabio and G.G.S. Dutton, Carbohydr. Res., 1981, 92, 287.
105 L. Batavyal and N. Roy, Carbohydr. Res., 1981, 98, 105
106 D.Robert, S. Lafont, P. Bienvenu, G. Normier, L.D. D'Hinterland and R. Fontanges, C. R. Acad. Sc. Paris, Ser. III, 1981, 292, 281.
107 M.-M. Riottot and J.-M. Fournier, Infect. Immun., 1981, 34, 126.
108 A. Kakinuma, T. Asano, H. Torii and Y. Sugino, Biochem. Biophys. Res. Commun., 1981, 101, 434.
109 T. Morita, S. Tanaka, T. Nakamura and S. Iwanage, FEBS Lett., 1981, 129, 318.
110 J.P. Pandey, W.D. Zollinger, H.H. Fudenberg and C.B. Loadholt, J. Clin. Invest., 1981, 68, 1378.
111 M.R. Lifely, A.S. Gilbert and C. Moreno, Carbohydr. Res., 1981, 94, 193.
112 E.C. Gotschlich, B.A. Fraser, O. Nishimura, J.B. Robbins and T.-Y. Liu, J. Biol. Chem., 1981, 256, 8915.
113 M.C. Steinhoff, E.B. Lewin, E.C. Gotschlich, J.B. Robbins and Panorama Pediatric Group, Infect. Immun., 1981, 34, 144.
114 G. Kato, Y. Maruyama and M. Nakamura, Agric. Biol. Chem. (Jpn.), 1981, 44, 2843.
115 M. Wagner, B. Wagner and V. Kubin, J. Gen. Microbiol., 1980, 120, 369.
116 T.I. Doran, D.C. Straus and S.J. Mattingly, Infect. Immun., 1981, 31, 615.
117 H.J. Jennings, C. Lugowski and D.L. Kasper, Biochemistry, 1981, 20, 4511.
118 D.E. Briles, J.L. Claflin, K. Schroer and C. Forman, Nature (London)., 1981, 294, 88.
119 P.-E. Jansson, B. Lindberg and U. Lindquist, Carbohydr. Res., 1981, 95, 73.
120 S. Szu, C.-J. Lee, D. Carlo and J. Henrichsen, Infect. Immun., 1981, 31, 371.
121 J. Pal, S.B. Bhattacharya and C.V.N. Rao, J. Chem. Chem. Soc., Perkin Trans. 1, 1981, 1393.
122 S.B.Bhattacharya and C.V.N. Rao, J. Chem. Soc., Perkin Trans. 1, 1981, 278.

123 K. Leontein, B. Lindberg and J. Lönngren, Can. J. Chem., 1981, 59, 2081.
124 L.G. Bennett and C.T. Bishop, Can. J. Chem., 1980, 58, 2724.
125 V. Daoust, D.J. Carlo, J.Y. Zeltner and M.B. Perry, Infect. Immun., 1981, 32, 1028.
126 L.Benzing, D.J. Carlo and M.B. Perry, Infect. Immun., 1981, 32, 1024.
127 M.B. Perry, V. Daoust and D.J. Carlo, Can. J. Biochem., 1981, 59, 524.
128 S. Valla and J. Kjosbakken, Can. J. Microbiol., 1981, 27, 599.
129 S. Ueda, F. Momii, K. Osajima and K. Ito, Agric. Biol. Chem. (Jpn.), 1981, 45, 1977.
130 K. Ishida and T. Takeuchi, Agric. Biol. Chem. (Jpn.), 1981, 45, 1409.
131 J.C. Stewart and J.B. Parry, J. Gen. Microbiol., 1981, 125, 33.
132 S.F. Kotarski and A.A. Salyers, J. Bacteriol., 1981, 146, 853.
133 T.K. Ng and J.G. Zeikus, Appl. Environ. Microbiol., 1981, 42, 231.
134 W. Boos, T. Ferenci and H.A. Shuman, J. Bacteriol., 1981, 146, 725.
135 M. Taguchi, K. Izui and H. Katsuki, J. Biochem. (Jpn.), 1980, 88, 379.
136 W.K. Kappel and J. Preiss, Arch. Biochem. Biophys., 1981, 209, 15.
137 T.W. Okita, R.L. Rodriguez and J. Preiss, J. Biol. Chem., 1981, 256, 6944.
138 J.J. Bradshaw-Rouse, M.H. Whatley, D.L. Coplin, A. Woods, L. Sequeira and A. Kelman, Appl. Environ. Microbiol., 1981, 42, 344.
139 R. Bhatnagar, N. Banerjee, H.C. Srivastava and K.A. Prabhu, Carbohydr. Res., 1981, 89, 346.
140 M. Torii, B.P. Alberto and S. Tanaka, J. Biochem. (Jpn.), 1980, 88, 1855.
141 M. Torii, S. Tanaka, T.Uryu and K. Matsuzaki, J. Biochem. (Jpn.), 1981, 89, 823.
142 E.R. Morris, A.N. Cutler, S.B. Ross-Murphy, D.A. Rees and J. Price, Carbohydr. Polymers, 1981, 1, 5.
143 Z.I. Ahmad, J.R. Alden and M.D. Montague, J. Gen. Microbiol., 1980, 121, 483.
144 A.C. Emeruwa, Ann. Microbiol., 1981, 132B, 13.
145 G.B. Pier, H.F. Sidberry and J.C. Sadoff, Infect. Immun., 1981, 34, 461.
146 J.R.W. Govan, J.A.M. Fyfe and T.R. Jarman, J. Gen. Microbiol., 1981, 125, 217.
147 B.T. Nguyen, T. Kodama and Y. Minoda, Agric. Biol. Chem. (Jpn.)., 1980, 44, 2925.
148 I.W. Sutherland, J. Chromatogr., 1981, 213, 301.
149 R. Sanders, E. Raleigh and E. Signer, Nature (London), 1981, 292, 148.
150 H.C. Tsien and E.L. Schmidt, J. Bacteriol., 1981, 145, 1063.
151 N.-X. Yu, M. Hisamatsu, A. Amemura and T. Harada, Carbohydr. Polymers, 1981, 1, 23.
152 P. Åman, M. McNeil, L.-E. Franzén, A.G. Darvill and P. Albersheim, Carbohydr. Res., 1981, 95, 263.
153 A. Amemura, M. Hisamatsu, S.K. Ghai and T. Harada, Carbohydr. Res., 1981, 91, 59.
154 S.K. Ghai, M. Hisamatsu, A. Amemura and T. Harada, J. Gen. Microbiol., 1981, 122, 33.
155 M.J. Duncan, J. Gen. Microbiol., 1981, 122, 61.
156 J. Preiss and E. Greenberg, J. Bacteriol., 1981, 147, 711.
157 S.-G. Yung and J. Preiss, J. Bacteriol., 1981, 147, 101.
158 K.G. Edwards, H.J. Blumenthal, M. Khan and M.E. Slodki, J. Bacteriol., 1981, 146, 1020.
159 M. Schöller, J.P. Klein and R.M. Frank, Infect. Immun., 1981, 31, 52.
160 C.G. Cumming, P.W. Ross and I.R. Poxton, J. Microbiol., 1981, 47, 93.
161 T.I. Doran, D.C. Straus and S.J. Mattingly, Infect. Immun., 1980, 30, 890.
162 J. Thompson and M.H. Saier, Jr., J. Bacteriol., 1981, 146, 885.
163 J.J. Bozzola, H.K. Kuramitsu and M.T. Maynard, Infect. Immun., 1981, 32, 830.
164 K. Fukushima, R. Motoda, K. Takada and T. Ikeda, FEBS Lett., 1981, 128, 213.
165 W.R. Figures and J.R. Edwards, Carbohydr. Res., 1981, 88, 107.
166 K. Ono, D.W. Nuessle and E.E. Smith, Carbohydr. Res., 1981, 88, 119.
167 D. Beighton, Microbios, 1980, 28, 149.
168 S. Hamada, M. Torii, S. Kotani and Y. Tsuchitani, Infect. Immun., 1981, 32,

364.
169 A.H. Weerkamp and B.C. McBride, Infect. Immun., 1981, 32, 723.
170 J.F. Kennedy, S.A. Barker, I.J. Bradshaw and P. Jones, Carbohydr. Polymers, 1981, 1, 55.
171 J.F. Kennedy and I.J. Bradshaw, Carbohydr. Res., 1981, 95, 132.
172 C. Whitfield, I.W. Sutherland and R.E. Cripps, J. Gen. Microbiol., 1981, 124, 385.
173 L. Ielpi, R.O. Couso and M.A. Dankert, FEBS Lett., 1981, 130, 253.
174 L. Ielpi, R.O. Couso and M.A. Dankert, Biochem. Biophys. Res. Commun., 1981, 102, 1400.
175 M. Benziman, C.H. Haigler, R.M. Brown, A.R. White and K.M. Cooper, Proc. Natl. Acad. Sci. USA, 1980, 77, 6678.
176 A.A. Salyers, F. Gherardini and M. O'Brien, Appl. Environ. Microbiol., 1981, 41, 1065.
177 J.M. Schoemaker, P.J.H. Schoemaker and J.J. Saukkonen, Microbios, 1980, 29, 149.
178 J.M. Schoemaker, J.M. Clark and J.J. Saukkonen, J. Gen. Microbiol., 1981, 123, 323.
179 G.D. Armstrong, L.S. Frost, H.J. Vogel and W. Paranchych, J. Bacteriol., 1981, 145, 1167.
180 M.P. Leckie, V.L. Tieber, S.E. Porter and D.N. Dietzler, Biochem. Biophys. Res. Commun., 1980, 95, 924.
181 S. Granett and M. Villaerejo, J. Bacteriol., 1981, 147, 289.
182 A. Boronat and J. Aguilar, J. Bacteriol., 1981, 147, 181.
183 L.-C. Chiang, H.-Y. Hsiao, P.P. Ueng and G.T. Tsao, Appl. Environ. Microbiol., 1981, 42., 66.
184 F. Wieland, J. Lechner, G. Bernhardt and M. Sumper, FEBS Lett., 1981, 132, 319.
185 K. Masuda and T. Kawata, J. Gen. Microbiol., 1981, 124, 81.
186 V.M. Laube and S.M. Martin, Appl. Environ. Microbiol., 1981, 42, 413.
187 A. Nishi, K. Masumoto, T. Asamizu, T. Katoh and H. Kinpara, Biochim. Biophys. Acta, 1981, 673, 243.
188 M. Inoue, T. Okubo, H. Oshima, T. Saito, M. Kato and S. Mitsuhashi, J. Bacteriol., 1980, 144, 1186.
189 G.G. Stewart, Can. J. Microbiol., 1981, 27, 973.
190 M. Rychtera, V. Kren and V. Grégr, Eur. J. Appl. Microbiol. Biotechnol., 1981, 13, 39.
191 J. McCourtie and L.J. Douglas, Infect. Immun., 1981, 32, 1234.
192 M.-A. Lachance, Can. Microbiol., 1981, 27, 651.
193 I. Sã-Correia and N. van Uden, Eur. J. Appl. Microbiol. Biotechnol., 1981, 13, 24.
194 T. Szumilo, FEMS Microbiol. Lett., 1981, 11, 171.
195 W. Nader and J.-U. Becker, Eur. J. Biochem., 1979, 102, 345.
196 R.W. Stoddard and D. Sanyal, Biochem. Soc. Trans., 1981, 9, 458.
197 Å. Haraldson and T. Björling, Eur. J. Appl. Microbiol. Biotechnol., 1981, 13, 34.
198 M. Rose, M.J. Casadaban and D. Botstein, Proc. Natl. Acad. Sci. USA, 1981, 78, 2460.
199 A.S. Mankin, A.M. Kopylov and A.A. Bogdanov, FEBS Lett., 1981, 134, 11.
200 L.-C. Chiang, C.-S. Gong, L.-F. Chen and G.T. Tsao, Appl. Environ. Microbiol., 1981, 42, 284.
201 J.A. Salas and C. Hardisson, J. Gen. Microbiol., 1981, 125, 25.
202 C. Huelsmann and D.C. Savage, Appl. Environ. Microbiol., 1981, 42, 554.
203 C.G. Orpin, J. Gen. Microbiol., 1981, 123, 287.
204 E.C. Laserna, F. Uyenco, E. Epifanio, R.L. Veroy and G.J.B. Cajipe, Appl. Environ. Microbiol., 1981, 42, 174.
205 J.B. Reiskind and J.T. Mullins, Can. J. Microbiol., 1981, 27, 1092.
206 J.B. Reiskind and J.T. Mullins, Can. J. Microbiol., 1981, 27, 1100.
207 B.J. Catley, J. Gen. Microbiol., 1980, 120, 265.
208 R.H. Marchessault, J.-F. Revol, T.F. Bobbitt and J.H. Nordin, Biopolymers, 1980, 19, 1069.

209 A. Misaki, M. Kakuta, T. Sasaki, M. Tanaka and H. Miyaji, Carbohydr. Res., 1981, 92, 115.
210 K.C. Rajasingham and R.A. Cawson, Microbios, 1980, 27, 163.
211 P. Pérez, R. Varona, I. Garcia-Acha and A. Durán, FEBS Lett., 1981, 129, 249.
212 G.E. Novak, Can. J. Microbiol., 1981, 27, 967.
213 Y. Iwamuro, M. Murata, K. Kanamaru, Y. Mikami and T. Kisaki, Agric. Biol. Chem., (Jpn.), 1981, 45, 653.
214 R. Seljelid, J. Bogwald and A. Lundwall, Exp. Cell Res., 1981, 131, 121.
215 J.H. Sietsma and J.G.H. Wessels, J. Gen. Microbiol., 1981, 125, 209.
216 M.I. Chu, A. Hartig, E.B. Freese and E. Freese, J. Gen. Microbiol., 1981, 125, 421.
217 G. Larriba, M. Morales and J. Ruiz-herrera, J. Gen. Microbiol., 1981, 124, 375.
218 B.A. Searle and R.S. Tubb, FEMS Microbiol. Lett., 1981, 11, 211.
219 T. Sato, T. Norisuye and H. Fujita, Carbohydr. Res., 1981, 95, 195.
220 R. Barreto-Bergter, P.A.J. Gorin and L.R. Travassos, Carbohydr. Res., 1981, 95, 205.
221 C.D. Wright, M.J. Herron, G.R. Gray, B. Holmes and R.D. Nelson, Infect. Immun., 1981, 32, 731.
222 D. Poulain, G. Tronchin, S. Jouvert, J. Herbaut and J. Biguet, Ann. Microbiol., 1981, 132A, 219.
223 T. Hamada, T. Nakajima, K. Izaki and K. Matsuda, Eur. J. Biochem., 1981, 119, 365.
224 T. Hamada, T. Nakajima and K. Matsuda, Eur. J. Biochem., 1981, 119, 373.
225 C. De Bièvre and F. Mariat, Carbohydr. Res., 1981, 95, 313.
226 T. Takeda, I. Kawarasaki and Y. Ogihara, Carbohydr. Res., 1981, 89, 301.
227 M. Linscheid, J. D'Angona, A.L. Burlingame, A. Dell and C.E. Ballou, Proc. Nat. Acad. Sci. USA, 1981, 78, 1471.
228 A.J. Parodi, Arch. Biochem. Biophys., 1981, 210, 372.
229 Y. Okubo, N. Shibata, T. Ichikawa, S. Chaki and S. Suzuki, Arch. Biochem. Biophys., 1981, 212, 204.
230 C. Hashimoto, R.E. Cohen, W.-J. Zhang and C.E. Ballou, Proc. Natl. Acad. Sci. USA, 1981, 78, 2244.
231 M.A. Amri, R. Bonaly, B. Duteurtre and M. Moll, Biochimie, 1981, 63, 575.
232 H. Takada, Y. Araki and E. Ito, J. Biochem. (Jpn.), 1981, 89, 1265.
233 B. Lindman, FEMS Microbiol. Lett., 1981, 10, 379.
234 L.A. Hadwiger, J.M. Beckman and M.J. Adams, Plant Physiol., 1981, 67, 170.
235 D.B. Domek and P.T. Borgia, J. Bacterol., 1981, 146, 945.
236 Y. Tominaga and Y. Tsujisaka, Agric. Biol. Chem. (Jpn.), 1981, 45, 1569.
237 M.C. Berthe, C. Charpentier, J. Lematre and R. Bonaly, Biochem. Biophys. Res. Commun., 1981, 100, 1504.
238 T. Ikeda, Biochim. Biophys. Acta, 1981, 657, 69.
239 T. Usui, Y. Iwasaki and T. Mizuno, Carbohydr. Res., 1981, 92, 103.
240 H. Tsuchihasi, T. Yadomae and T. Miyazaki, Carbohydr. Res., 1981, 98, 65.
241 T. Miyazaki and M. Nishijima, Carbohydr. Res., 1981, 96, 105.
242 G.F. Webster and K.J. McGinley, Microbios, 1980, 28, 41.
243 H. Kusunose, H. Matsumura and M. Sawamure, Agric. Biol. Chem. (Jpn.), 1980, 44, 743.
244 T.H. Lee, M. Arai and S. Murao, Agric. Biol. Chem. (Jpn.), 1981, 45, 1301.
245 T.H. Lee, M. Arai and S. Murao, Agric. Biol. Chem. (Jpn.), 1981, 45, 2343.
246 A.F. Brana, M.B. Manzanal and C. Hardisson, J. Bacteriol., 1980, 144, 1139.
247 T.Kiho, C. Hara and S. Ukai, Chem. Pharm. Bull., 1981, 29, 225.

5
Glycoproteins, Glycopeptides, Proteoglycans, and Animal Polysaccharides

BY R. J. STURGEON

1 Microbial Glycoproteins

Fungal Glycoproteins. — The effects of glycosylation on the thermal and proteolytic stability of catalase from Aspergillus niger have been reported.[1] Deglycosylation of the enzyme results in an increased susceptibility to proteolysis and a small decrease in thermostability.

Virulent and nonvirulent species of Cryptococcus when grown in culture with maximum capsule production elaborate an extracellular glycoprotein.[2] The glycoprotein from C. neoformans contains O-acetyl groups and more D-mannosyl- and -uronic acid residues than does the corresponding glycoprotein from the non-pathogenic species, which does not contain O-acetyl groups.

Changes in the plasma membrane glycoprotein pattern during cell development in Dictyostelium discoideum have been described.[3] Regulation of these glycoprotein patterns involve starvation pulses of cAMP and cell-cell contact. Three glycoproteins (mol. wts. 3.3×10^4, 3.1×10^4, and 2.8×10^4) have been isolated from D. discoideum by affinity chromatography on immobilized discoidin-1, the lectin of D. discoideum.[4] These endogenous glycoproteins are potent inhibitors of the haemagglutination activity of the lectin, and they appear to have an aggregation-enhancing effect on differentiated cells.

A thermolabile antigen on the surface of baker's yeast (Saccharomyces cerevisiae) has been purified by immunoadsorption chromatography.[5] The sugar moiety is composed on residues of D-galactose, D-mannose, 2-acetamido-2-deoxy-D-glucose, and L-fucose, whilst the protein moiety is composed of β-amino acids, four of which (Gly, L-Glu, L-Ser, and L-Asp) comprise 70% of the amino acid content. The antigenic determinant of this antigen is distinct from that of cell-wall D-mannan. Cell-wall D-mannoprotein of

Saccharomyces kluyveri cells has short oligosaccharide chains attached to *L*-serone or *L*-threonine, and long polysaccharide chains probably connected to *L*-asparagine residues through *NN*'-diacetylchitobiosyl residues.[6] The immunodominant structure of the *D*-mannoprotein is contained in the unit (1), which is found in both the *O*- and *N*-glycosidic chains of the molecule.[7]

```
β-D-Manp-(1→2)-α-D-Manp-(1→3)-α-D-Manp(1→
                                      6
                                      |
                                      1
                                  α-D-Manp
```

(1)

Viral Glycoproteins. — The following topics related to the chemistry and biochemistry of viral carbohydrates have recently been reviewed: carbohydrates of viral envelopes,[8] glycosylation of viral envelopes,[9] combinations of haemagglutinins and neuraminidase sub-types of influenza A viruses,[10] and the synthesis and assembly of vesicular stomatitis virus and Sindbis virus glycoproteins.[11]

The binding of human immunoglobulin to retroviruses has been reported.[12] Inhibition studies indicate that this binding is directed to the oligosaccharide chains of the viral glycoproteins and is non-specific in character.

In animals infected by bovine leukaemia virus, cytotoxic antibodies to the envelope glycoprotein are directed against the carbohydrate moiety.[13] Immune sera prepared in rabbits using purified envelope glycoprotein as immunogen are directed against determinants of the polypeptide backbone of the antigen. The carbohydrate moiety of two bovine leukaemia virus glycoproteins is an important antigenic determinant, since a rapid loss of antigenic activity occurs after treatment with glycosidases.[14]

The major virus-specific protein released from cells infected with feline leukaemia virus is a glycoprotein (mol. wt. 4.0×10^4) which carries *gag* gene product antigenic determinants.[15]

Replication-defective Friend murine leukaemia virus particles containing decreased levels of envelope glycoprotein have been described.[16] The G_{1X} antigen, which is expressed on the surface of thymocytes of certain mouse strains, is an antigenic determinant of the major envelope glycoprotein of murine leukaemia virus.[17] Comparison of the glycoprotein precursors from two closely related murine leukaemia viruses, one of which carries the antigenic determinant for G_{IX}, suggests that the G_{IX}^- virus codes for an extra glycosylation site in the envelope region of the genome, relative to the protein from the G_{IX} + virus. The gag glycoprotein (mol. wt. 8.0 x 10^4) of Rauscher murine leukaemia virus contains two N-glycosylated carbohydrate chains.[18] One occurs in the C-terminal half of the protein at L-Asn-177 of the p30 sequence, and the other is found in a fragment (mol. wt. 2.3 x 10^4) located in the N-terminal region.

Glycopeptides obtained after extensive proteolytic digestion of the env precursors of ecotropic, xenotropic, and dual-tropic murine leukaemia viruses have been compared.[19] The env precursor glycoproteins of dual-tropic viruses showed the unusual property of migrating to the cell surface without undergoing the normal oligosaccharide processing and proteolytic cleavage events that are observed for ecotropic and xenotropic murine leukaemia viral glycoproteins. The subcellular distribution of env-related glycoproteins following the transformation of murine erythroleukaemia cells by Friend leukaemia virus has been studied by subcellular fractionation techniques and pulse-chase experiments.[20] A rapid association of a glycoprotein (mol. wt. 5.5 x 10^4) with the nuclear membrane and endoplasmic reticulum is proposed with little redistribution after this initial association. Antibodies which precipitate purified virus envelope glycoprotein antigens of Simian sarcoma-associated virus and Friend leukaemia virus have been detected in the majority of normal human sera.[21] Immunoprevention of Friend virus-induced erythroleukaemia by vaccination with purified viral envelope glycoprotein has been reported.[22]

The structures of the oligosaccharides of the glycoprotein encoded by early region E3 of adenovirus 2 have been established.[23] Infection by this virus does not affect the usual cellular high D-mannose glycosylation pathway, and, despite being virus-coded, the glycoprotein is glycosylated in the same manner as a typical mammalian cell-coded glycoprotein. m-RNAs encoded in early region 3 of the adenovirus type 2 genome have been purified, and proteins encoded by different mRNAs have been identified in a cell-free protein-synthesizing system.[24] The primary amino acid structure of a glycoprotein contains a hydrophobic signal sequence, two presumptive glycosylation sites, and a hydrophobic region close to the C-terminus. The early region E3 of adenovirus type 7 encodes a glycoprotein (mol. wt. 2.8 x 10^4) and a protein that are nonessential for growth of the virus in culture.[25] These molecules may be analogues of the glycoprotein (mol. wt. 2.5×10^4) and protein coded by E3 of adenovirus type 2.

Hepatitis B surface antigen contains a protein (mol. wt. 2.5×10^4) and a glycoprotein (mol. wt. 3.0×10^4) of identical amino acid composition.[26] It is proposed that the two antigen components may differ only in the presence of carbohydrate in the glycoprotein and that they are in fact the same gene product. Some physicochemical and immunological properties of a glycopeptide obtained after protelytic cleavage of hepatitis B surface antigen have been reported.[27]

A method for determining specific reactivity to herpes simplex virus types 1 and 2 has been developed, using human sera to immunoprecipitate specific glycoprotein antigens from virus-infected cell extracts.[28] The viral glycoproteins precipitated from these extracts were then analysed by SDS-polyacrylamide gel electrophoresis.

Monoclonal antibodies directed against herpes simplex virus glycoproteins have been used to protect mice against acute virus-induced neurological disease.[29] Evidence that the virus-induced glycoprotein gC expresses type-specific antigenic determinants, whereas glycoprotein gD expresses type-common determinants, is reported.

By using a temperature-sensitive mutant of herpes simplex virus type 1, and inhibitors of glycosylation, specific glycoproteins of the virus have been shown to be involved in T cell-mediated cytolysis of virus-infected cells.[30] Two-dimensional gel electrophoresis has been used to identify polypeptides and glycoproteins of herpes simplex virus type 1.[31] Pulse chase experiments and treatment of infected cells with neuraminidase suggests that three glycoproteins contain neuraminic acid and that synthesis of two of the three occurs by at least ten discrete steps. The herpes simplex virus type 1-specific glycoprotein has been purified by affinity chromatography on immobilized soya bean agglutinin and *Helix pomatia* lectin.[32] The binding of the glycoprotein to these lectins indicates the presence of a terminal 2-acetamido-2-deoxy-α-D-galactosyl residue in an oligosaccharide, a finding not previously reported for glycoproteins of enveloped viruses.

Two monoclonal antibody preparations which are herpes simplex virus type 2-specific have been isolated.[33] One of the antibodies precipitated a glycoprotein with characteristics similar to glycoprotein E of the virus whereas the other antibody preparation precipitated a glycoprotein not previously described. This glycoprotein is tentatively designated glycoprotein F.

Energy depletion of viral cells caused by the addition to the culture medium of carbonyl cyanide 3-chlorophenylhydrazone (CCCP), an uncoupler of oxidative phosphorylation, does not affect the formation of N,N'-diacetylchitobiosyl pyrophosphoryl dolichol, or the pool size of D-glucosyl phosphoryl dolichol, and only slightly affects the pool size of GDP-D-mannose.[34] Using this system, influenza virus haemagglutinin can be glycosylated in chick embryo fibroblasts *via* either a penta-D-mannosyl or a nona-D-mannosyl derivative. The construction of a recombinant viral genome consisting of an SV40 vector and a cloned full-length DNA coding for the haemagglutinin surface glycoprotein of influenza virus has been reported.[35] Infection of primate cells with the hybrid virus produced a glycoprotein similar in molecular size to the haemagglutinin of influenza virus.

Some biochemical and biological properties of influenza C virus, as well as some morphological aspects of the organization of the glycoproteins on the viral envelope and the plasma membranes of infected cells, have been reported.[36] The carbohydrate portion on the haemagglutinin of influenza virus does not appear to be of major importance in defining the antigenicity of the haemagglutinin.[37] Although the ability of untreated and glycosidase-treated virus to inhibit the binding of antibodies directed against the haemagglutinin is almost indistinguishable, release of 50% of the carbohydrate from intact virus particles significantly affected haemagglutinating activity. The haemagglutinin membrane glycoprotein of influenza virus is composed of a triple-stranded coiled coil of α-helices extending 76Å from the membrane, and a globulin region of antiparallel β-sheet, which contains the receptor-binding site and the variable antigenic determinants, is positioned on top of this stem.[38] Each subunit has an unusual loop-like topology, starting at the membrane, extending 135Å distally, and folding back to enter the membrane. Four antigenic sites on the three-dimensional structure have been identified.[39] At least one amino acid substitution in each site seems to be required for the production of new epidemic strains.

Glycosylated tryptic peptides of the haemagglutinin of influenza A virus have been separated using a combination of ion-exchange chromatography and gel filtration.[40] Seven different glycosylated tryptic peptide classes were obtained from the HA1 polypeptide, which is derived from the N-terminal segment of the haemagglutinin, and only one glycosylated peptide from the C-terminal portion (HA2). Three types of carbohydrate chains, complex, high D-mannose, and hybrid type, are present in the HA1 polypeptide whereas HA2 contains the complex-type oligosaccharide chains. The oligosaccharide side chains of the major glycoprotein HA_1 of influenza virus haemagglutinin have been examined for antigenic activity using a solid-phase radioimmunoassay.[41] Although qualitative and quantitative differences between these oligosaccharide side chains were detected, the antigenically active side chains are cross-reactive. The amino acid sequence of a Hong Kong influenza haemagglutinin light (HA_2) chain has been elucidated, with the exception of a highly aggregated region near the C-terminus.[42] A single glycosylated L-asparagine residue bearing a complex oligosaccharide structure was noted. The amino acid sequence and oligosaccharide distribution for the haemagglutinin from the early Hong Kong influenza virus A/Aichi/2/68(X-31) have been reported.[43] Although the oligosaccharide distribution in this variant was found to be identical to that found in the haemagglutinin from A/Memphis/72, with sugar units attached to six specific L-asparaginyl residues, the monosaccharide compositions of the individual carbohydrate units were different. Amino acid sequence analysis shows the existence of structural relationships between the Hav 7 and the Hong Kong (H3) haemagglutinins of influenza virus.[44] Two of the six oligosaccharide units on the heavy chain of the haemagglutinin from the Hong Kong variant A/Memphis/102/72 have been shown to be antigenically related to the host antigen.[45] They contain the N-acetyl-lactosamine-type residues attached to L-Asn-8 and L-Asn-22. The antigenic structure of influenza virus haemagglutinin has been defined using hybridoma antibodies.[46]

The uncleaved (HN_0) and the cleaved (HN) forms of the haemagglutinin neuraminidase of Newcastle disease virus, strain Ulster, have been subjected to electrophoresis under reducing and non-reducing conditions.[47] The conversion of one form to the other involves the release by proteolysis of a glycopeptide (mol. wt. 8.0 x 10^3). The isoelectric points have been determined and the c.d. spectra analysed for both post-translational proteolytically cleaved and uncleaved Newcastle disease virus glycoproteins.[48] It is demonstrated that cleavage is paralleled by a conformational change of the glycoproteins.

The dissociation and reconstitution of the Semliki Forest virus membrane using non-ionic detergent have been studied.[49] Different mechanisms of reconstitution give rise to symmetric and asymmetric vesicles. Glycoprotein E1 of the virus in ts-3-infected cells is glycosylated and inserted into the endoplasmic reticulum membrane without simultaneous synthesis of the envelope protein (mol. wt. 6.2 x 10^4), which is a precursor of two other envelope glycoproteins.[50] A reversible defect in the glycosylation of the membrane proteins of Semliki Forest virus ts-1 mutant has been described.[51] At a restrictive temperature the viral membrane glycoproteins are arrested in the rough endoplasmic reticulum, but are transported to the Golgi complex once the cultures are shifted to a permissive temperature. An efficient method for the isolation of large quantities of Sendai virus F-glycoprotein by reduction of the intact virus particles with dithiothreitol and solubilization of the reduced virus with non-ionic detergent has been reported.[52]

Two membrane glycoproteins (HANA protein and F-protein) of HVJ (Sendai)virus contain residues of L-fucose, D-galactose, D-mannose, and 2-acetamido-2-deoxy-D-glucose, but no 2-acetamido-2-deoxy-D-galactose or neuraminic acid.[53] Both glycoproteins contain oligosaccharide chains of the high D-mannose type and of the complex type. The complex-type oligosaccharides contain 2-acetamido-2-deoxy-3-O-α-L-fucosyl-4-O-β-D-galactosyl-D-glucosyl groups in their outer-chain moieties.

Paramyxovirus fusion (F) proteins are actively involved in the induction of membrane fusion.[54] The F protein of Sendai virus is activated by proteolytic cleavage to yield two disulphide-linked polypeptides (F_1 and F_2). Conformations of the inactive uncleaved precursor glycoprotein and the active cleaved form have been compared. Direct evidence was obtained for the presence of hydrophobic binding sites on the active form of the F protein, but not on the inactive form.

Growth-dependent changes in D-mannosyl oligosaccharides derived from a single species of membrane glycoprotein have been studied in a viral system.[55] E1 and E2 glycoproteins of Sindbis virus express different relative quantities of four oligosaccharides. Metabolically labelled Sindbis virion and cell-associated viral glycopeptides have been analysed by a combination of gel filtration and specific glycosidase digestion.[56] The general model of L-asparaginyl oligosaccharide processing for the neutral and acidic-type structures in normal and lectin-resistant Chinese hamster ovary cells was confirmed. The presence of unusual small neutral structures in the mature viral glycoprotein from the cell lines was also noted.

The role of carbohydrate in the morphogenesis of vesicular stomatitis virus has been studied using tunicamycin to inhibit glycosylation.[57] It is confirmed that different glycoproteins have different carbohydrate requirements for plasma-membrane insertion. Simple mutational changes in a protein can also affect this requirement. Micelles formed by glycoproteins of this virus spontaneously partition into sonicated phosphatidyl-choline vesicles.[58] The hydrophobic tail fragment of the glycoprotein is resistant to proteolysis when the glycoprotein is inserted into the vesicle bilayer. The intracellular location of an integral membrane glycoprotein of vesicular stomatitis virus has been examined.[59] Its passage to the cell surface has been synchronized and thus stages in the intracellular pathway that it follows were mapped. Glycoprotein of vesicular stomatitis virus has been isolated and its binding to cell surfaces studied.[60] The structures of the oligosaccharides of the virus grown in different teratocarcinoma cell lines are not identical for all cell lines, as evidenced by differences in sensitivity to digestion by a mixture of endo- and exo-glycosidases.[61] Baby-hamster kidney cells infected with vesicular stomatitis virus and starved of D-glucose are capable of synthesizing D-mannosylated lipid-linked oligosaccharides but cannot efficiently complete the assembly of D-Glcp_3-D-Manp_9-D-Glc$p$NAc$_2$-pyrophosphoryl dolichol, resulting in an abnormally glycosylated G protein.[62]

Polyoma virus requires specific cell-surface sialo-oligosaccharide receptors for infection of 3T6 cells.[63] A trisaccharide sequence (2) can serve as a specific cell-surface receptor both for polyma virus-mediated haemagglutination and for polyoma virus infection of host cells.

α-Neu$p$5Ac-(2→3)-β-D-Gal$p$-(1→4)-D-Gal$p$NAc

(2)

The action of toyocamycin on the biosynthesis of viral glycoproteins in cells chronically infected with a murine retrovirus is in reducing the content of envelope glycoprotein (mol. wt. 7.0 x

10^4).[64] However, the precursor of this glycoprotein, having a molecular weight of 8.5×10^4, is synthesized normally and is processed into its final products, which accumulate in the cells.

The purified major envelope glycoprotein (mol. wt. 8.5×10^4) from avian myeloblastosis virus contains 45%carbohydrate including 25% 2-acetamido-2-deoxy-D-glucose.[65] Oligosaccharide chains are distributed between seven to nine points on the polypeptide backbone. One structural model, which reconciles the hydrodynamic properties of the glycoprotein with nearly spherical architecture observed by electron microscopy, requires the organization of the polypeptide chain and approximately half of the carbohydrate into a globular form. The remaining covalently linked oligosaccharides would extend outwardly from the globular structure as randomly oriented chains.

In order to characterize Belmont virus as a possible bunyavirus, evidence has been presented of similar translational control of the proteins and glycoproteins specified by the two viruses in mammalian cells.[66]

Cauliflower mosaic virus capsid polypeptide binds to fluorescein-derivatized concanavalin A, thus demonstrating the glycoproteinaceous nature of the molecule.[67]

The synthesis, glycosylation, and identification of early proteins induced by Epstein-Barr virus have been reported.[68] The spontaneous and induced syntheses of the virus antigens in Raji cells immobilized on surfaces coated with anti-lymphocyte globulin have been studied.[69] A major glycoprotein (gp 350/220) of Epstein-Barr virus has been detected by immunofluorescence with monoclonal antibodies on both the plasma membrane and in the cytoplasm.[70]

Four structural polypeptides of Hazara virus have been resolved by sodium dodecyl sulphate polyacrylamide gel electrophoresis.[71] Three glycoproteins (mol. wts. 8.4×10^3, 4.5×10^3, and 3.0×10^3) are associated with the virion envelope, and a nonglycosylated polypeptide is associated with the nucleocapsid.

The G_{1X} antigen of murine retrovirus has been analysed using monoclonal antibodies,[72] reactivity with which appears to require a stable configurational change in the env precursor protein coinciding with glycosylation, rather than direct participation of carbohydrate in the antigenic site. Thus later enzymic removal of carbohydrate chains does not alter reactivity with the antibodies.

Two measles virus glycoproteins, a haemagglutinin and a fusion protein, have been incorporated into artificial lipid bilayers.[73]

The resultant 'virosomes' had visible 'spikes' and possessed haemagglutinating activity, and should provide a reagent for studying immune responses to measles virus, independent of the immunosuppressive effects of the whole virus.

Monospecific antibodies have been prepared to each of two envelope paramyxovirus glycoproteins (HN and F).[74] Antibodies to the HN glycoprotein inhibited haemagglutinating and neuraminidase activities. Anti-F antibodies inhibited haemolysis.

The complete nucleotide sequence of the glycoprotein gene in rabies virus has been determined.[75] This has allowed a prediction to be made of several features of the glycoprotein, from the deduced amino acid sequence.

The effects of cell transformation on the glycosylation and processing of Prague C Rous sarcoma virus envelope polypeptides have been reported.[76] The major detectable difference is an increase in the relative amount of larger acidic-type oligosaccharides containing residues of neuraminic acid, D-galactose, and 2-acetamido-2-deoxy-D-glucose, compared with the neutral-type oligosaccharides. Glycopeptides derived from a glycoprotein (mol. wt. 8.5×10^4) have been examined after sequential glycosidase digestions followed by gel filtration.[77] The major hybrid species has an oligo-D-mannosyl core of five D-mannosyl- and two 2-acetamido-2-deoxy-D-glucosyl residues together with substitution of one of the D-mannosyl residues with an acidic side chain composed of residues of N-acetylneuraminic acid, D-galactopyranose, and 2-acetamido-2-deoxy-D-glucopyranose.

Immunological studies on the relationship of the major envelope glycoprotein of HEL-12 virus and the analogous glycoproteins of Simian sarcoma-Simian associated virus and baboon endogenous virus have been reported.[78] A glycopeptide (mol. wt. 2.0×10^5) associated with the transformation of Simian sarcoma virus has been identified.[79]

A glycoprotein specified by spleen focus-forming virus in three cell lines varies in molecular size and peptide composition but retains both xenotropic and ectotropic murine leukaemia virus antigenicity.[80] Both qualitative and quantitative differences in the post-translational processing of the envelope glycoproteins of polycythaemia- and anaemia-inducing strains of spleen focus-forming virus have been reported.[81]

The *in vitro* translation of Unkuniemi virus-specific RNA has led to the identification of a non-structural protein and a precursor to membrane glycoproteins.[82]

Vaccinia virus-induced haemagglutinin is a single glycoprotein (mol. wt. 8.5×10^4) which appears at the plasma membrane as a virus-induced function.[83] Although N-linked oligosaccharides predominate in the molecule, about 25% of the oligosaccharide chains which are O-glycosidically linked to the protein are responsible for the biological activity of the molecule. Glycoproteins and phosphoproteins have been identified as structural elements of this virus.[84]

Immunogenic glycoproteins (mol. wt. 6.5×10^4, 9.8×10^4, and 1.18×10^5) isolated from laboratory and vaccine strains of varicella-zoster virus have been identified.[85]

In reconstitution experiments using the envelope glycoproteins E1 and E2 of western equine encephalitis virus, the haemolytic activity is associated with the E1 glycoprotein which acts as a haemagglutinin.[86]

2 Plant Glycoproteins

A large number of legume-seed polypeptides have been identified as glycoproteins.[87] A high degree of binding between seed glycoproteins and their corresponding lectins might have some functional significance in maintaining large aggregates of protein in compact insoluble form.

Twelve N-glycosylated polypeptides have been identified in glyoxysomal membranes of castor beans.[88] Major bands on SDS-polyacrylamide electrophoresis correspond to glycoproteins of mol. wts. 9.0×10^4, 7.1×10^4, 5.6×10^4, 4.7×10^4, and 4.3×10^4.

A glycoprotein, involved with the control of intercellular recognition and of pollen incompatibility has been isolated from stigmas of S-allele genotype S_2S_2 of *Brassica oleracea*.[89] The glycoprotein, which is not precipitated by concanavalin A, contains residues of L-arabinose, D-galactose, D-glucose, and D-mannose.

Evidence that lipid-linked oligosaccharides can be assembled in a sequential manner from UDP-2-acetamido-2-deoxy-D-glucose and GDP-D-mannose by membrane preparations from developing soybean coyledons has been presented.[90] Tissue slices synthesize products which resemble the completed 7S storage glycoproteins.

Concanavalin A has been shown to bind to the endoplasmic reticulum and the starch grain surface of root statocytes of cress (*Lepidium sativum* L9).[91] The receptor for the lectin appears to be a high-molecular-weight glycoprotein which is resistant to solubilization after glutaraldehyde fixation of the cells.

Two soluble glycoprotein components from rye grass (Lolium perenne) have been isolated and partially characterized.[92] They are chemically and immunologically distinct. One of the glycoproteins is the major allergen of rye-grass pollen and binds specifically to immunoglobulin E of sensitized human subjects.[93] Immunological cross-reactivity that is due, at least in part, to common antigenic carbohydrate components is reported.

Two types of protein glycosylation in cell-free preparations of developing cotyledons of Phaseolus vulgaris have been postulated.[94] In an in vitro system, the attachment of 2-acetamido-2-deoxy-D-glucosyl residues to phytohaemagglutinin and phaseolin does not involve lipid-linked intermediates whilst the glycosylation of a class of proteins of heterogeneous molecular weights involves the lipid pathway.

Aqueous extracts of leaves, roots, and stems of green bean, and leaves of mung bean, soybean, and lima bean, cause agglutination of cells of a saprophytic bacterium Pseudomonas putida.[95] Purified preparations of the agglutinin are composed of glycoproteins, containing L-arabinose, D-galactose, and D-galacturonic acid as the predominant carbohydrate components. The properties of laccase, isolated from Schinus molle trees, including molecular weight, amino acid and carbohydrate composition, have been described.[96] The presence or absence of an inducing hormonal glycoprotein in the growth medium of Volvox larteri determines whether the algae grow sexually or asexually.[97] The site and time of formation of this glycoprotein have been established.

3 Lectins

Review articles dealing with lectins include a study of lectins in higher plants.[98] The endogenous biological functions of lectins have been reviewed.[99] Examples from plants, cellular slime moulds, and animal tissues are used in the analysis of this problem. The carbohydrate binding specificity and cell-membrane interaction of Vicia sativa lectin have been reported.[100] Affinity chromatography for the purification of lectins has been described.[101] Stable, high capacity affinity adsorbents used in the isolation of lectins have been prepared by derivatization of epoxy-agarose with various carbohydrates.[102]

Isoelectric focusing in polyacrylamide gels containing immobilized sugars results in a shift in the position of protein bands capable of interaction with these ligands.[103] These affinity gels have been used to demonstrate interactions with lectins. A sensitive assay has been developed for measuring the binding of lectins to glycoconjugates in wells of polyvinyl chloride plates.[104]

A chromophore-labelled D-galacto-D-mannan has been prepared by coupling a monochlorotriazinyl dye to guaran.[105] The product has been used for the quantitative estimation of lectins having affinity for D-galactosyl or 2-acetamido-2-deoxy-D-galactosyl residues.

The following lectins have been purified and some of their physical and chemical characteristics recorded (Table I).[106-114]

Iodinated lectins have been used for the determination of the degree of deglycosylation of high D-mannosyl glycoproteins following digestion with *endo*-β-2-acetamido-2-deoxy-D-glucanase H.[115]

Intermolecular forces in lectin-carbohydrate interaction have been analysed on the basis of their structure and chemical features.[116] The role of water, as well as that of other physico-chemical parameters influencing the formation of such complexes in aqueous media, is also discussed. A model depicting the importance of hydrogen bonding and charge-transfer interactions as the main sources of complex stability in the association between lectins and carbohydrates is proposed.

By performing electrophoresis perpendicular to a stationary pH gradient in polyacrylamide gels containing a specific ligand either covalently fixed or entrapped in the gel matrix, it is possible to measure dissociation constants of lectins and their pH dependence in a pH range.[117] The method has been used to study the lectins from *Ricinus communis* and *Lens culinaris*.

Aqueous extracts from leaves, roots, and stems of green bean and leaves of mung bean, soybean, and lima bean agglutinate cells of a saprophytic bacteria, *Pseudomonas putida*.[118] Purified preparations contain both carbohydrate and protein, require Mg^{2+} for activity, and are heat stable. No agglutination of other plant pathogenic bacteria or nonpseudomonad saprophytes was observed.

A survey of twenty-two commonly ingested fruits, vegetables, and seeds has indicated the presence, in many, of lectins which can interact with components of human saliva and *Streptococcus mutans* cells.[119] Interactions of soybean and peanut lectins with rhizobia that nodulate soybean, peanut, or both plants have been studied.[120]

Table 1 Composition and properties of some purified lectins

Source	Mol. wt.	Sub-units	Carbohydrate % Neutral	Carbohydrate % Amino sugar	Human blood type	Sugar specificity	Ref.
Plant							
Datura stramonium	8.6×10^4	2	37			Chitooligosaccharides	106
Erythrina arborescens	2.8×10^4		5.7		ABO	α-D-Galp: α-D-GalpNAc	107
Erythrina indica	6.8×10^4	2	9.0	1.5	ABO	β-D-Galp: D-GalpNAc	107
Erythrina lithosperma	5.4×10^4	2	4.1		ABO	α-D-Galp: D-GalpNAc	107
Erythrina suberosa	2.8×10^4		6.8		ABO	D-GalpNAc	107
Erythrina variegata			9-12		ABO	D-Galp, D-GalpNAc	108
Momordica charantia	1.2×10^5	4	10	2.5		β-D-Galp	109
Psophocarpus tetragonolobus	4.1×10^4				ABO	D-GalpNAc	110
Ulex europaeus II	1.1×10^5	4			H(O)	Chitotriose	111
Viscum album	1.2×10^5	4			ABO	D-Galp	112
	6.0×10^4	2			ABO	D-Galp, D-GalpNAc	112
	5.0×10^4	2			ABO	D-GalpNAc	112
Animal							
Electrophorus electricus	3.5×10^4	2	2.2			β-D-Galp	113
Other							
Azolla-Anabaena azollae (symbiosis)	1.3×10^5	6			B	α-D-Galp	114

No relationship between the ability of peanut or soybean rhizobia to nodulate the reciprocal host plant and their ability to bind to the lectin of that plant could be demonstrated.

The seeds of Pisum sativum, Canavalia ensiformis, Vicia faba, V. sativa, and Ricinus communis contain proteins, termed lectin binders, which are associated with their respective lectins.[121] The pea lectin binder is a tetrameric glycoprotein composed of identical subunits (mol. wt. 5.1×10^4), and its interaction with the lectin is abolished by low pH or by D-glucose. The concanavalin A binder is a protein (mol. wt. 3.5×10^4) containing no covalently bound carbohydrate. A study of the tissues of soybean and jack bean seeds demonstrates that the lectin of a particular species may exhibit a very narrow range of specificity towards endogenous glycoconjugates.[122] This suggests that not only is there one major lectin in the cotyledons of a given plant but also one soluble glycoprotein in the cotyledons that the lectin can recognize. The striking specificity shown between homologous lectin and receptor is not maintained when the lectin is used to isolate glycoproteins from the same tissue in other plant species.

Glycopeptides and oligosaccharides derived from N-glycosyl proteins have been used to define the specificity of twelve different lectins.[123] Lectins considered identical in terms of monosaccharide specificity possess an ability to recognize fine differences in more complex structures. The observation that glycoasparagines, glycopeptides, and glycoproteins possess a higher affinity for lectins than the related oligosaccharides has been explained by the fact that the glycan-amino acid linkage leads to structures more rigid than those of the oligosaccharides themselves. Many of the lectins used in this study seem to possess in or near their carbohydrate-binding site a hydrophobic area such as that described for concanavalin A, and this could be the cause of non-specific hydrophobic interactions between the lectins and residues of glycopeptides or glycoproteins, as described for concanavalin A.

Fluorescence polarization has been used to investigate the binding of 4-methylumbelliferyl-β-D-galactopyranoside to Abrus precatorius agglutinin.[124]

Lectin activity has been detected in seeds of twenty Acacia species of Australian origin.[125] Fourteen of the species had been reported by other workers to contain no lectin.

Peanut lectin has been purified by adsorption on glutaraldehyde cross-linked hematic desialylated and reticulated human stroma.[126]

The lectin (mol. wt. 1.25×10^5), although homogeneous on electrophoresis, yields eight major bands on isoelectric focusing.

The screening of peanut (Arachis) and its wild relatives for multiple molecular forms of peanut lectin and for genotypes devoid of the lectin has been reported.[127] A peanut lectin-negative phenotype was detected in four genotypes (<0.1% of those examined). Lectin preparations from lectin-containing genotypes were resolved into a small number of defined and clearly related isolectin profiles. Association constants and thermodynamic parameters for the binding of sugars to peanut agglutin by u.v. difference spectroscopy have been reported.[128] The binding kinetics of methyl α- and methyl β-D-galactopyranosides to the lectin have been studied by ^{13}C n.m.r. spectroscopy employing methyl D-galactopyranosides specially enriched in ^{13}C at C-1.[129] A two-step binding model is presented. Peanut agglutinin exhibits cryoinsolubility, which is partially reversible and totally inhibited in the presence of D-galactosides.[130] The efficacy of the sugars as inhibitors of cryoinsolubility is related to their affinity for the lectin. Charge-charge interactions are of little importance, but hydrogen bonds and/or van der Waals interactions are most probably responsible for the formation of cryoprecipitates. The effects of chemical modification of the conformation and biological activity of peanut agglutin have been recorded.[131] When free amino groups are modified with succinic anhydride and 1-isothiocyanate-4-benzenesulphonic acid, the derivatives retain their sugar-binding capacity, although the agglutinating activity with neuraminidase-treated erythrocytes and various tumour cells is reduced. When L-tyrosine residues are modified firstly with tetranitromethane and then with 4-aminophenyl-α-D-glucopyranoside and with 2-(4-aminobenzyl)-α-D-neuraminic acid, the agglutinating and mitogenic activities are not markedly altered. The influence of these modifications on the conformation of the lectin was examined by c.d. spectral studies.

The distribution of the five tetrameric Bandeiraea simplicifolia I isolectins (A_4, A_3B, A_2B_2, AB_3, B_4) isolated from seeds of different trees, has been analysed.[132] The amount of each isolectin varies from seed to seed, each seed displaying a unique distribution of the five isolectins. It is postulated that the intact B. simplicifolia I B_4 isolectin undergoes proteolytic cleavage by an endo- or exo-protease to form first the AB_3 isolectin, which in turn is then processed further to yield the A_2B_2

isolectin, and so on until A_4 is produced. 4-Methylumbelliferyl glycosides have been used in binding studies with the lectins BSI-A_4, BSI-B_4, and BS II from Bandeiraea (Griffonia) simplicifolia.[133] Equilibrium dialysis, fluorescent enhancement, quantitative precipitation, and hapten inhibition techniques have been used to investigate the binding characteristics of four of the B. simplicifolia I isolectins.[134] One of the most unusual features is that, although different, both the A and B subunits have the same affinity for α-D-galactosyl residues. Although both subunits also bind 2-acetamido-2-deoxy-D-galactosyl residues, the A subunit has an association constant for the amino sugar more than 1000-fold greater than the B subunit.

Affinity chromatography of divalent concanavalin A on immobilized 4-aminophenyl β-D-glucopyranoside has been analysed by means of equations based on the simplest model, in which the lectin has equivalent and independent binding sites and one lectin molecule binds only one immobilized ligand at a time.[135]

Polyacrylamide gradient electrophoresis of concanavalin A reveals the presence of various minor bands which are not detected by polyacrylamide disc electrophoresis.[136] The minor components are suggested to be different molecular species separating by virtue of the combined effect of electrophoretic mobility and molecular sieving.

Over wide ranges of temperature and pH, concanavalin A consists of only dimers and tetramers.[137] The large fraction of hydrolysed subunits in commercial preparations causes significant populations of dimeric species that associate only weakly or not at all. The effect of the binding of saccharide ligands on the reversible dimer-tetramer equilibrium of the lectin has been studied by high-speed sedimentation equilibrium.[138] Contrary to earlier published work, saccharide binding does not appear to affect the dimer-tetramer equilibrium of the native concanavalin A, although it does introduce irreversible behaviour in preparations containing significant proportions of hydrolysed subunits. The dimer-tetramer equilibrium is linked to the weak binding of calcium at a site probably in the dimer-dimer interface. Studies have been reported of the interaction of concanavalin A with 4-nitrophenyl 2-O-α-D-glucopyranosyl-α-D-mannopyranoside, 4-nitrophenyl-2-O-α-D-galactopyranosyl-α-D-mannopyranoside, and 4-nitrophenyl maltoside.[139] The existence of two binding modes for the disaccharides is proposed. Some of the specificity requirements for the interaction of concanavalin A with the non-reducing glycosyl group are characterized, and are summarized in Scheme 1.

Temperature-jump relaxation studies using 4-methylumbelliferyl α-D-mannopyranoside as a fluorescence indicator ligand have been used in an examination of the binding kinetics of methyl α-D-mannopyranoside to concanavalin A.[140]

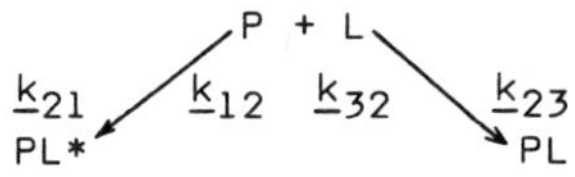

Scheme 1

An i.r.-attenuated total-reflectance spectroscopic method has been used to study the interaction of monolayers of concanavalin A with mono- and poly-saccharides.[141] The effects of film pressure, pH, urea, Mn^{2+}, and Ca^{2+} on the binding of dextran, methyl α-D-mannopyranoside, and D-galactose to the lectin were studied. The thermodynamic parameters of the Mn^{2+}-binding reaction with concanavalin A in the dimer state have been determined by a calorimetric technique.[142] The results indicate that the free energy change for the process is dominated by a positive entropy change and that the two identical S1 sites of the dimer are independent. Calorimetric studies have also been used to examine the binding of saccharides to concanavalin A.[143] The binding process is induced by both favourable enthalpic and entropic effects. Kinetic and equilibrium studies of concanavalin A activation of Ca^{2+} indicate that ion binding at S2 produces a general ordering of the ligands at this site which in turn orders those side-chain residues involved in saccharide recognition.[144] The carbohydrate-binding activity of concanavalin A containing various levels of Ca^{2+} and Mn^{2+} has been reported.[145] The precipitation activity of the lectin with glycogen as well as the binding of lectin to 4-nitrophenyl α-D-mannopyranoside are strongly affected by the number of bound Ca^{2+} ions. The kinetics of the interactions of Ca^{2+} with concanavalin A in the presence of Co^{2+}, Mn^{2+}, and/or Zn^{2+} have been followed by measurement of the quenching of fluorescence of 4-methylumbelliferyl α-D-mannopyranoside when bound to protein.[146] Some observations on metal-ion requirements and sugar binding by *Bandeiraea simplicifolia* I lectin are also reported.

The co-operative binding properties of concanavalin A with

glycoconjugates are dependent on the presence of hydrophobic binding sites.[147] The degree of co-operativity can be substantially altered by conformational changes of the ligand.

Irradiation of tetrameric concanavalin A with a high-pressure mercury lamp in the presence of chloracetamide results in the lectin showing a monomeric molecular weight with retention of binding capacity.[148]

Fluorescein-concanavalin A conjugates are capable of distinguishing between normal and malignant human cells.[149]

The four lectins concanavalin A, *Lens culinaris* agglutinin, *Vicia faba* agglutinin, and *Pisum sativum* agglutinin, considered to be identical in terms of specificity for α-D-mannose or α-D-glucose, possess the ability to recognize fine differences in more complex carbohydrate structures.[150]

A lectin from the seeds of *Croton tiglium* is capable of agglutinating sheep and other erythrocytes as well as inducing haemolysis of rabbit erythrocytes.[151] These effects are inhibited by sheep erythrocyte glycopeptides.

Datura stramonium lectin has been purified on columns of *N*,*N'*-diacetylchitobiose immobilized on agarose.[106] The lectin contains 37% carbohydrate, of which L-arabinose is the most predominant sugar, and a high content of L-cysteine, glycine, L-serine, and hydroxy-L-proline residues. This lectin is similar to, but not identical with, the lectin from potato.[152] Although they have similar specificities for chito-oligosaccharides, the *Datura* lectin shows its greatest affinity for glycopeptides and the potato lectin for chito-oligosaccharides. In both lectins, the hydroxy-L-proline residues are substituted with β-L-arabinofuranosides, and there are L-serine residues in the glycosylated region that are substituted with α-D-galactopyranosyl residues. The structure of the glycosylated region of the lectins is very similar to that of the hydroxy-L-proline-rich glycopeptides of plant cell walls, and these lectins could be precursors of such material.

Metal-chelate affinity chromatography has been used in the purification of the *Dolichos biflorus* seed lectin.[153] Equilibrium dialysis studies show that the lectin has two combining sites per molecule for methyl 2-acetamido-2-deoxy-D-galactoside.[154] Subunit I is primarily responsible for the carbohydrate-binding properties of the lectin. Monoclonal antibodies specific for subunit I of the *Dolichos biflorus* lectin combine with the C-terminal portion of the subunit and may be interacting with the active site.[155] This antibody does not react with another lectin-like protein from the stems and leaves of the plant.[156] The monoclonal antibody is of importance in that it can distinguish subunit I from subunit II of the seed lectin. Although these two subunits appear to differ from one another only at their C-terminal ends, subunit I is active whereas subunit II is not.[157] Atomic absorption spectrophotometry has established the presence of Ca^{2+}, Mg^{2+}, Mn^{2+}, Zn^{2+}, and Cu^{2+} in native Dolichos biflorus lectin.[158] The effects of removal and subsequent addition of different metal ions on the association constant of the lectin for 2-acetamido-2-deoxy-D-galactose are reported.

The lectin of Erythrina corallodendron seeds is similar to soybean lectin, being a glycoprotein (mol. wt. 1.1×10^5) and binding to 2-amino-2-deoxy-D-galactose, 2-acetamido-2-deoxy-D-galactose, α- and β-D-galactosides, and D-galactose.[159]

A lectin isolated from the roots of soybean (Glycine max) seedlings is similar but not identical to the corresponding seed lectin.[160] The lectin is associated with the outer surfaces of the root and is concentrated in the segments of the root at which root hair and early secondary roots are observed. From examination of Le and lele genotypes of Glycine max for membrane-bound and buffer-soluble soybean lectin, it seems unlikely that the membrane lectin is of cytoplasmic origin.[161] In contrast to other reports soybean (Glycine soja) lectin has been shown to retain its erythroagglutinating activity after complete removal of Mn^{2+} from the whole molecule.[162]

The subunit structure and complete amino acid sequence of lentil (Lens culinaris) lectin have been determined.[163] A comparison between the secondary structure of concanavalin A and the probable secondary structure of lentil lectin as predicted by two different methods indicates that the folding of these two polypeptides has been well conserved during evolution.

In a re-examination of the high-affinity carbohydrate-binding specifities of pea and lentil lectins, the ability of various glycopeptides to bind to columns of the immobilized lectins was studied.[164] L-Fucose residues are indeed an important determinant in tight binding to pea and lentil lectin but not to concanavalin A.

The lectin from Maclura pomifera, after purification by ion-exchange and affinity chromatography, has been resolved into five structurally related protein components of similar haemagglutinating and carbohydrate-binding activity.[165] Purification of the lectin has also been achieved by adsorption onto polyleucyl hog A+H blood-group substances, followed by elution with melibiose or 2-acetamido-2-deoxy-D-galactose.[166] Three subunits (mol. wt. 1.45×10^4, 1.35×10^4, and 1.2×10^4) are present in the intact lectin. Hydrophobic interaction is important for binding of the lectin to sugars. The combining site of the lectin appears to be as large as a disaccharide.

The interaction of mistletoe lectin I with carbohydrate has been studied.[167] Inhibition data suggest that unmodified hydroxyl groups at the C-2, C-3, and C-4 positions of the D-galactopyranosyl ring are essential for binding to the active site of the lectin.

The association constants for the binding of a series of ligands with the D-galactose-specific lectin from Momordica charantia have been determined through the light-induced quenching of protein fluorescence.[168] Analysis of the iodide quenching suggests that there is a slight increase in the accessibility of L-tryptophan residues of the lectin on binding lactose. Studies of the binding characteristics of M. charantia lectin to its specific sugar by fluorescence spectroscopy using 4-methylumbelliferyl β-D-galactopyranoside show the quenching of the sugar upon binding to the lectin.[169] The binding is carbohydrate-specific and is inhibited by lactose.

Considerable homology in structures and biological activities has been reported for 62 cultivars of Phaseolus vulgaris, on the basis of isolectin patterns.[170] A number of cultivars are sufficiently different to warrant subclassification and further characterization.

Inclusion of pure lectins from the seeds of P. vulgaris in diets of rats increases both faecal and urinary losses of nitrogen.[171] The animals develop antibodies of low avidity for the dietary lectins. Lectin toxicity is tentatively suggested to result from the combined effects of interference with normal intestinal digestion and/or absorption of protein through damaged enterocytes and of systemic responses of the rat to the internalised lectin.

The subunit compositions of individual tetrameric phytohaemagglutinin isolectins from Phaseolus vulgaris have been examined by isoelectric focusing and sodium dodecylsulphate electrophoresis.[172] The proportion of subunits resulting from each isolectin was determined for a direct physical conformation of the isolectin subunit structures.

Early events in the synthesis of polypeptides of Ricinis

communis agglutinin type 1 have been reported.[173] Co-translational translocation across the endoplasmic reticulum membrane is accompanied by proteolysis and, where appropriate, core glycosylation.

The agglutinins from Solanum tuberosum and wheat-germ agglutinin interact with keratan sulphate and chitin sulphate.[174] Differences in the reactivity between the polysaccharides and the two lectins are reported. Chemical modification of amino or carboxyl groups, L-arginine, L-methionine, and L-histidine residues does not alter significantly the haemagglutinating activity of potato lectin.[175] L-Tyrosine and L-tryptophan residues are closely involved in the activity, findings which are similar to those reported for other proteins that bind oligomers of 2-acetamido-2-deoxy-D-glucose.

The dependence of the Sophora japonica lectin on bivalent metal ions as well as on the pH range for optimal haemagglutination and precipitation in the presence of bivalent cations has been established.[176] Further characterization of the combining site of this lectin is reported.

Ulex europaeus haemagglutinin II (Cytisus-type anti-H(O)) is inhibited by di-N-acetylchitobiose, and has been purified by affinity chromatography.[111] The lectin is a glycoprotein in which D-mannose is the predominant sugar with smaller amounts of D-glucose, D-xylose, D-galactose, L-fucose, and 2-amino-2-deoxy-D-glucose.

Ribitol teichoic acid of Staphylococcus aureus, which contains terminal 2-acetamido-2-deoxy-β-D-glucosyl residues, possesses an abnormal affinity for immobilized wheat-germ agglutinin in binding at high ionic strength.[177]

Interactions of monomeric amino sugars with wheat-germ agglutinin have been studied by ^{1}H and ^{19}F n.m.r. spectroscopy.[178] N-Acetylneuraminic acid and 2-acetamido-2-deoxy-D-glucose have a common binding site on the lectin. Deuterium n.m.r. resonance has been used to delineate the molecular dynamics of sugars bound to lectins.[179] ^{2}H relaxation times and rotational correlation times for 2-{^{2}H}-acetamido-2-deoxy-D-3-{^{2}H}-glucose in the presence and absence of wheat-germ agglutinin were studied. The pyranose ring appears to have negligible motional freedom relative to the protein on binding. The acetamido side chain is also immobilized in the binding site, the only motion available being rotation of the $C^{2}H_3$ group about its threefold axis.

The elution profiles of various glycopeptides modified by glycosidase treatment, Smith degradation, acetolysis, and hydrazinolysis show that the structure (3) is essential for the binding of glycopeptides to immobilized wheat-germ agglutinin.[180] Both the N,N'-diacetylchitobiose moiety and the β-2-acetamido-2-deoxy-D-glucosyl residue linked to C-4 of the β-linked D-mannosyl residue contribute to the interaction of the glycopeptide with the lectin. The substitution at C-6 of the innermost β-2-acetamido-2-deoxy-D-glucosyl residue by an α-L-fucosyl residue or at C-6 of the β-linked D-mannosyl residue by another D-mannose residue reduces the affinity of glycopeptides for the column. In contrast to results from other workers, no interaction of glycopeptides containing N-acetylneuraminic acid residues was observed with the lectin.

Some properties of *Vicia graminea* lectin purified by affinity chromatography have been described, and the binding of the lectin to various erythrocytes has been characterized.[181]

The complete amino acid sequence of the α-subunit of *Vicia sativa* lectin has been established and compared with the known sequences of lectins from *Pisum sativum*, *Lens culinaris*, *V. faba*, and *Canavalia ensiformis* (concanavalin A).[182]

A partially purified haemagglutinin associated with cell walls from the hypocotyls of *Vigna radiata* exhibits both haemagglutinating activity and α-D-galactosidase activity.[183] It is suggested that, similar to the seed lectin of the same plant, both activities might be present in the same molecule. The native molecule could dissociate into smaller units exhibiting either of the two activities.

β-D-GlcpNAc-(1→4)-β-D-Manp-(1→4)-β-D-GlcpNAc-(1→4)-
β-D-GlcpNAc-1-L-Asn

(3)

A tetrameric D-galactose-binding protein of mung bean (*Vigna radiata*) that displays both haemagglutinin activity and α-D-galactosidase activity can be reversibly dissociated into a low-molecular-weight monomeric form which possesses only enzymic activity.[184] The possibility of *in vivo* changes in subunit equilibria, when combined with the accompanying alterations in activity, is suggested as a possible physiological role of phytohaemagglutinins.

Abrin A, purified from the seeds of Abrus precatorius, has physical and biological properties similar to the better known abrin C.[185] Possible structural relationships between the two glycoproteins and a related agglutinin from the same source were examined. Although covalently attached carbohydrate is a relatively minor part of the overall structure of ricin, oxidation with sodium periodate results in the destruction of D-mannosyl residues and loss of toxicity on the HeLa cell, while A-chain and B-chain (cell binding) activities are not significantly altered.[186]

A hybrid molecule consisting of the B chain of ricin linked to insulin via a disulphide bond binds to target cells and maintains several insulin-like biological activities.[187]

Several human and murine tumour cell lines undergo extensive aggregation in the presence of fetuin and asialofetuin.[188] This activity can be inhibited by lactose and to a lesser extent by D-galactose, 2-amino-2-deoxy-D-galactose, and 2-acetamido-2-deoxy-D-galactose. The implications of the existence of a carbohydrate-binding protein(s) on the surface of malignant cells on their in vivo behaviour are discussed.

Serum amyloid P component (9.5S α_1-glycoprotein), Clq (subcomponent of the Cl component of complement), and C-reactive protein have lectin-like combining sites for certain D-galactans.[189] Two new additional combining sites of the C-reactive protein have been described.

Foetal-calf skeletal muscle contains two lectins having binding activities inhibited by lactose and a separate haemagglutinating activity which is not inhibited by lactose.[190]

The binding of asialoorosomucoid to the isolated rabbit hepatic lectin appears to be a saturable and reversible process as shown by steady-state and kinetic analysis.[191] The minimal molecular weight (1.04×10^5) of the intact rat hepatic receptor for asialoglycoproteins has been measured directly in plasma membranes.[192] The value obtained is shown to be significantly affected by the presence of detergent employed in solubilization and isolation procedures.

Protein haemagglutinins have been detected in homogenates of each of the six coxal depressor muscles in the leg of the cockroach.[193] The quantitative ability of some glycosaminoglycans to inhibit these haemagglutinins correlates with the muscles' innervation by identified motor neurons.

The complete amino acid sequence of chicken hepatic lectin has been documented.[194] The β-D-galactoside-specific lectin present in embryonic-chick skeletal muscle does not appear to be involved in myotube formation in vitro.[195] A developmentally regulated lectin from chick-embryo kidney binds strongly to asialoglycoconjugates.[196] Its affinity towards 2-amino-2-deoxy-D-glucosyl, 2-amino-2-deoxy-D-galactosyl, and 2-amino-2-deoxy-D-mannosyl residues is low. The inhibitory effect of the amino sugars in haemagglutination assays is not due to simple electrostatic interaction.[197] A neuraminic acid-binding lectin isolated from the horseshoe crab Carcinoscorpius rotunda cauda has been purified by affinity chromatography on immobilized fetuin.[198] The lectin, which is a glycoprotein (mol. wt. 4.2×10^5) and contains subunits (mol. wts. 2.7×10^4 and 2.8×10^4), is antigenically unrelated to the other neuraminic acid-binding lectin, limulin, from the horseshoe crab Limulus polyphemus. Limulin binds specifically to gangliosides bearing an N-glycolylneuraminic acid residue.[199] The binding requires a free carboxyl group and a free hydroxyl group at C-4 of the neuraminic acid but the side chain of the N-glycolylneuraminic acid is not required. Wheat-germ agglutinin binds preferentially to gangliosides bearing an N-acetylneuraminic acid residue. The acetamido group but not the carboxyl group is involved in the binding.

Lectins of differing specificities have been detected in the crop, midgut, and haemolymph of the insect Rhodinus prolixus.[200] These are specific for 2-acetamido-2-deoxy-D-mannose, 2-acetamido-2-deoxy-D-galactose, and α- and β-D-galactose. Receptors for the lectins were detected in epimastigote forms of Trypanosoma cruzi, a protozoan parasite of the insect and of humans.

A possible mechanism of induction of the insect lectin of Sarcophaga peregrina on injury of the body wall has been described.[201] The larger subunit of the lectin is converted to the smaller subunit by proteolysis, in a process essential for constructing active lectin having affinity for D-galactose.

The sponge Halichondria panicea contains a lectin that has been isolated and purified.[202] From the same sponge species, bacteria have been isolated and identified as Pseudomonas insolita. The lectin, detected on the surface of mucoid cells from H. panicea, may be involved in a symbiotic relationship between the sponge and the bacterium.

A mycobacterial haemagglutinin, a development-specific lectin of Myxococcus xanthus, has been purified.[203,204] While simple sugars fail to inhibit the haemagglutinating activity of the lectin, a glycopeptide containing the trisaccharide (4), common to many glycoproteins, is a potent inhibitor.

Two lectins have been purified from Agaricus edulis.[205] Both are capable of agglutinating erythrocytes and immunoglobulins A, G, and M, but no monosaccharide was capable of inhibiting the interaction.

Neu*p*5Ac-(2→3)-β-D-Gal*p*-(1→3)-D-GalNAc

(4)

A chitin-binding haemagglutinin has been isolated from culture filtrates of Conidiobolus lampranges following affinity chromatography on formalinized erythrocytes.[206] Haemagglutination is strongly inhibited by chito-oligosaccharides. A pronase-like protease, produced by C. lampranges, acts, in company with the lectin, on erythrocytes to accelerate the activity of the lectin.[207] The lectin has chitin-binding properties similar to those of wheat germ agglutinin.

Purpurin, the lectin from Dictyostelium purpureum, is made up of seven tetrameric forms, assembled from four distinct subunits.[208] The isolectins, although related, can be functionally discriminated on the basis of their relative affinities for columns derivatized with complementary saccharides.

Oleic acid and dioleolyl phosphatidic acid at low concentrations agglutinate rabbit and rat erythrocytes, while dioleolyl phosphatidyl choline is not haemagglutinating.[209] Erythrocyte agglutination by lipids is indistinguishable in some respects from that caused by lectins, in particular both haemagglutination reactions exhibit cell specificity and are inhibited by certain glycoproteins.

4 Fibronectin

A review dealing with current concepts concerning the structure and function of fibronectin has been published.[210]

Rat plasma fibronectin has been isolated and characterized and monospecific antibodies prepared to it.[211] The antibodies only weakly cross-react with plasma fibronectins of chicken, horse, and human. A second glycoprotein (mol. wt. 7.0 x 10^4), also having collagen-binding activity, was isolated from rat plasma. Immunologically pure fibronectin cannot be prepared from human plasma by one-stage affinity chromatography on immobilized gelatin or immobilized fibronectin due to non-specific absorption of other proteins on the support.[212] Purified fibronectin can be obtained using a second separation by gel permeation chromatography.

Although human amniotic fluid fibronectin and plasma fibronectin are immunologically indistinguishable and are equally active promoters of cell attachment, they differ in carbohydrate composition.[213]

A solid-phase radioimmunoassay has been developed for the determination of fibronectin levels in plasma.[214] An immunological assay has been used to measure the uptake of endogenous and exogenous fibronectin by cells in culture.[215] The cellular uptake of fibronectin from the medium is influenced by cell shape and by the presence of collagen.

Cultured human-mammary epithelial cells derived from both normal tissues and primary carcinomas produce large quantities of fibronectin.[216] The presence of cell-associated fibronectin as well as the synthesis and secretion of fibronectin into the medium by these cells and by fibroblasts derived from these tissues were studied.

A structural comparison of fibronectins from normal and transformed hamster cell lines has been reported.[217] Although the major sites of phosphorylation in fibronectin are the same in both types of cells, fibronectin from transformed cells appears to be phosphorylated to a much higher extent than that from normal cells. Human germ-cell tumours produce fibronectin resembling its amniotic fluid variant, but differ from plasma fibronectin.[218] Such fibronectins have been suggested to be potential oncodevelopmental markers.

The specific calcium-dependent binding reactivity for human serum amyloid P component for fibronectin and a C4-binding protein is reported.[219]

Human peripheral blood monocytes possess a trypsin-sensitive plasma membrane receptor for surface-bound fibronectin.[220] Binding of monocytes to fibronectin leads to enhanced functional expression of their plasma-membrane receptors for the Fc portion of immunoglobulin G and for the third component of complement. It is proposed that plasma fibronectin acts as a circulating probe for fibrin or for exposed or altered collagen at sites of tissue damage.

Treatment of fibronectin with either methyl mercuric bromide or iodoacetamide as sulphydryl blocking agents protects the glycoprotein against precipitational losses during preparation.[221]

A combination of electron microscopy, c.d. and i.r. spectroscopy, and scanning microcalorimetry has been used in a study of the structural organization of human plasma fibronectin.[222]

Four functionally distinct domains in hamster plasma fibronectin have been isolated and characterized following proteolytic digestion of the glycoprotein.[223] Intact fibronectin is considered to contain four carbohydrate chains per subunit, three being attached to a domain of mol. wt. 4.0×10^4 and one to a domain of mol. wt. 1.4×10^5.

Heat denaturation of human plasma fibronectin proceeds through at least three stages, with endothermal denaturing transitions observed at 68, 82, and 119°C.[224] A three-domain structure for the glycoprotein is proposed in which the domain which unfolds at 68° is associated with gelatin binding and cell binding, while that unfolding at 82° appears to be associated with much of the immunological activity. The 119° domain has heparin binding activity as well as some immunological activity. The gelatin-binding domain of fibronectin is immunogenic, and antisera against this domain recognize cellular fibronectin gelatin-binding sites.[225] Inhibition of gelatin binding but not cell spreading by anti-gelatin binding domain Fab' fragments confirms the hypothesis that fibronectin has separate sites mediating these activities.

Plasma fibrinogen, on digestion with cathepsin D, releases two pools of long-chain polypeptides of different molecular weights and with graded affinity for immobilized heparin.[226] The fraction having weaker affinity contains the N-terminal region of fibronectin; that having the stronger affinity consists of two disulphide-linked peptide chains of different length and sharing a related structure. The higher-molecular-weight polypeptide contains an additional transamidase-sensitive site. Human plasma fibronectin has been treated with chymotrypsin.[227] Characterization of the isolated fragments has shown that the order of the functional domains from the N-terminus is staphylococcal binding, fibrin cross-linking and binding sites, gelatin-binding site, cell attachment area, and heparin-binding site.

Horse serum fibronectin has been reduced and alkylated to produce a modified glycoprotein which no longer binds to gelatin, but still provides a substrate for myoblast attachment.[228] Evidence was obtained to demonstrate that a chymotryptic fragment of fibronectin (mol. wt. 6.0×10^4) was unable to promote cell attachment unless fibronectin was present in the medium, suggesting that the gelatin-binding fragment also contains a site involved in fibronectin-fibronectin binding. A specific actin-binding site in fibronectin from chicken fibroblasts has been identified.[229] After affinity chromatography of fragments released from fibronectin by limited proteolysis, a fragment (mol. wt. 2.7×10^4) was isolated. The actin-binding site is close to, but not identical with, the reported collagen-binding site. The binding of human plasma fibronectin to actin has been studied using actin-coated microtitre wells in enzyme-linked immunosorbent assays.[230] The binding is inhibited by both actin and gelatin, but not by heparin or bovine serum albumin. The binding of collagen and actin to fibronectin may not take place simultaneously. Since the binding of collagen does not interfere with cell attachment, the cell binding site of fibronectin is probably different from the actin-binding site.

Structural differences between plasmas and fibroblast cellular fibronectins have been identified in at least three different regions.[231] One structural difference occurs near the gelatin-binding site and the others are near heparin-binding sites.

A number of cell types (HeLa, Ehrlich ascites, and viral transformed rat kidney) show a low rate of Ca^{2+} uptake, which is stimulated after incubation with fibronectin.[232]

The interaction of isolated plasma fibronectin with stimulated and nonstimulated human platelets has been characterized.[233] The specific binding of a large number of fibronectin molecules to thrombin-stimulated platelets suggests that fibronectin may play a role in modulating platelet adhesion and/or aggregation induced by this stimulus.

The incorporation of L-proline and glycine into fibronectin and collagen by cultures of skin fibroblasts obtained from patients with insulin-dependent diabetes mellitus has been reported.[234] Changes in the ratio of fibronectin to collagen between skin fibroblasts of normal and diabetic subjects during *in vitro* aging were noted.

The covalent binding of various structural proteins by factor XIII_a has been examined in a system in which proteins, immobilized on macroporous agarose, are incubated with labelled proteins in the presence of factor XIII_a, and the radioactivity of the urea-washed beads is determined.[235] Coupling occurred between fibrin and fibronectin and between myosin and fibronectin. The amino acid sequence in bovine fibronectin which contains L-glutamine residues labelled with 1,4-{^{14}C}-putrescine by factor XIII has been determined.[236] This L-glutamine residue is located at position 3 from the N-terminus of fibronectin.

Fibronectin and procollagen are released from the matrix of human fibroblasts by thrombin.[237] Collagenase alone is unable to bring about the release of fibronectin. The release of these two glycoproteins by thrombin may be involved in wound healing *in vivo*.

Human plasma fibronectin at physiological concentrations stimulates the level of total platelet adhesion and their spreading on surfaces coated with fibrillar collagen.[238] The stimulating effect of fibronectin is due to its interaction with collagen since free fibronectin does not influence the adhesion or spreading of platelets.

Plasma fibronectin mediates the binding of soluble native collagen type III to trypsinized macrophages.[239] The association is augmented by heparin or heparan sulphate, but hyaluronic acid reverses the effect.

Fibronectin has been shown to bind to aggregating collagen fibres.[240] Since fibronectin inhibits the rate of collagen fibrillogenesis it may regulate the size of collagen fibres. Cartilage proteoglycans inhibit fibronectin-mediated adhesion to collagen.[241] The presence of fibronectin in collagen fibrils of human fibroblasts has been shown, using immunological techniques.[242]

A gelatin-binding protein (mol. wt. 7.0×10^4), which is quite distinct from fibronectin, is synthesized by normal and malignant adherent cells.[243] The glycoprotein seems not to be a fragment of fibronectin and shows no immunological crossreactivity with it.

5 Collagen

The biochemical properties of basement membrane components, including collagen, have been reviewed.[244] The proceedings of a connective-tissue-research symposium include chapters dealing with recent advances in methods for investigating the molecular structure of collagens, chromatographic and electrophoretic methods for the purification of collagen, and tissue-specific differences in collagen.[245] A symposium has been devoted to the biology of collagen.[246]

In a novel structural model for collagen, it is proposed that there are no inter-chain hydrogen bonds and van der Waals' contacts.[247] Instead, a triple-helical structure is stabilized by water molecules, which form interchain bridges through hydrogen bonds with carbonyl groups.

Collagen has been selectively solubilized from mixtures of collagen and elastin using phenol, acetic acid, and water mixtures,[248] and also using cycles involving pepsin, pancreatic elastase, and dithiothreitol.[249] Preparations from normal hamster and baboon lungs contain principally types I and III collagen.

A chromatographic procedure consisting of gel permeation and anion-exchange chromatography for purifying native types I, II, and III collagens from a variety of sources has been reported.[250] The native collagen prepared by this method is more stable than that prepared by salt fractionation, as reflected by the higher and sharper thermal transition temperature.

Radioimmunoassay procedures for 7S collagen and laminin have been used to analyse these proteins in serum, cell cultures, and tissues.[251]

The relative amounts of types I and III collagens have been estimated after extraction of non-collagenous protein from rabbit lung tissue in sodium dodecyl sulphate.[252] The method has been applied to small tissue samples such as would be obtained by lung biopsy.

The skin collagens from the lamprey, *Entosphenus japonicus*, and the great blue shark, *Prionace glauca*, contain two distinct α-components and are classified as type I-like collagen.[253]

The heterogeneity of cyanogen bromide-cleavage peptides of human collagens, types I, II, III, and V, have been demonstrated by two-dimensional polyacrylamide gel electrophoresis.[254]

The amino acid sequence of the oligosaccharide attachment site within the proα-2 chain of chick type I collagen has been identified.[255] In addition D-mannose-rich glycopeptides obtained from both proα-1(I) and proα-2 carboxyl propeptides have been isolated and characterized.

Evidence for a structural mutation of procollagen type I in a patient with Ehlers-Danlos syndrome, type VII, has been reported.[256] The fibroblasts of the patient synthesize both an abnormal proα-2 chain and a normal proα chain.

Human type II collagen differs from type I collagen in molecular packing, as demonstrated by X-ray diffraction studies.[257]

Under physiological conditions the type II collagen is shown to contain 50-100% more water than the type I collagen. The content of glycosylated hydroxy-L-lysine residues is thought to be the variable most likely to be modulating fibrillar hydration.

Collagens synthesized by cultured bovine corneal endothelial cells have been characterized.[258] Type III collagen is the major component both deposited in the extracellular matrix and secreted into the media. Basement membrane collagens, type IV and V, are also found in each compartment.

The α2(v) chain of human collagen when cleaved with cyanogen bromide contains ten peptides, accounting for 956 amino acid residues.[259] Possible homologies between the α2(v) peptides and peptides derived from other collagen chains were noted.

Antibodies have been prepared to collagenase-resistant terminal regions of pro-type IV collagen.[260] Competition radioimmunoassay experiments show recognition of the antibodies for reduced and alkylated 7S collagen, supporting the possibility that this collagen is a crosslinked domain of type IV collagen.

A high-molecular-weight collagenous protein has been isolated, after limited pepsin digestion, from the endometrial villi of bovine placenta.[261] It resembles the protein isolated from human aortas in containing three polypeptide chains with similar amino acid composition.[262] The polypeptide chains show characteristics typical of basement membrane collagens, having a high content of hydroxy-L-lysine, which is almost all glycosylated. The occurrence of similar amounts of L-cysteine, L-tyrosine, and 2-amino-2-deoxy-D-glucose is not observed in other collagen types.

An unusual collagen fraction has been isolated and characterized from bovine hyaline cartilage and intervertebral disc.[263] Evidence supports a model containing three identical chains (mol. wt. 3.3×10^4) linked by interchain disulphide bonds to form a short triple-helical pseudoα-sized component.

The collagenous domain of bovine glomerular basement membrane has been isolated in soluble form and shown to contain the previously designated polypeptide XIV.[264] The amino acid and carbohydrate compositions and cyanogen bromide patterns indicate that polypeptide XIV has a structure similar to that of C chain isolated from other vascular tissues and that it may represent a major structural segment of the entire collagenous domain. The collagenous domain of rabbit renal tubular basement membrane is composed of a multiplicity of components ranging from mol. wts. 3.4

$\times 10^4$ to 10^6.[265]

As a self-assembly system, guinea-pig collagen chains tend to renature as triple-stranded molecules whose chain compositions are determined by the ratios of the α-chains in the mixture.[266]

The collagens from the cuticle of the nematode *Caenorhabditis elegans* have been separated by molecular-sieve chromatography and shown to contain no hydroxy-L-lysine residues.[267]

The collagen isolated from the platyhelminth *Cysticerus cellulosae* contains an amino acid composition characteristic of vertebrate collagen except that it contains no hydroxy-L-proline.[268]

An extra-embronic base membrane, Reichert's membrane, of mouse and rat contains collagen together with four non-collagenous glycoproteins.[269] One of the glycoproteins (mol. wt. 5.0×10^4) is unrelated to laminin, whilst the other three (mol. wts. 4.15×10^5, 2.45×10^5, and 1.70×10^5, respectively) share some immunochemical characteristics of laminin.

A 7S collagen has been identified as a crosslinking domain of mouse tumour matrix type IV collagen.[270] Transformed epithelial bovine lens cells synthesize and secrete in the culture medium only type IV collagen.[271] Like the interstitial procollagen, the molecule is a disulphide-bonded glycoprotein containing both a collagenous and a non-collagenous domain.

The absence of maturation of collagen cross-links in fish skin has been reported.[272]

A close similarity exists between octopus skin collagen and calf skin type I collagen.[273] Both collagens resemble each other, not only in chain composition but also in chemical composition.

The molecular organization of the collagenous components of the intestinal basement membrane of *Ascaris suum* has been investigated by characterization of their physical properties in both the native and denatured states.[274] The collagenous domain consists mainly of a component composed of two essentially identical triple-helical subunits joined end to end by disulphide bonds with each subunit being cross-linked by disulphide bonds between all constituent chains.

A study of collagen-sulphated glycosaminoglycuronan interactions in developing rat tail tendon has been reported in terms of fibrillogenesis and fibre maturation.[275]

Bovine corneal endothelial cells synthesize in culture predominantly type III collagen, with lesser amounts of types I and V and apparently little if any of type IV.[276,277]

M3A, a derivative of the clonal skeletal muscle myoblast line L6, secretes into the culture medium an abnormal form of collagen (mol. wt. 5.6×10^4) and a greatly reduced amount of high-molecular-weight collagen α-chains.[278] In addition M3A does not secrete unsulphated chondroitin while L6 does. Comparison of the behaviour of leucine aminopeptidase and carboxypeptidase-modified collagens suggests that the carboxyl telopeptide has a major role in the growth stages of self-assembly.[279]

Matrix-free cells from chick embryo tendons synthesize pro-γ collagen chains which are either not triple-helical or are in an unstable triple-helical conformation.[280] These chains contain less hydroxy-L-proline than the triple-helical procollagen secreted by tye cells. Their content of glycosylated hydroxy-L-lysine is, however, greater than that of the procollagen secreted by the cells.

The biosynthesis and processing of type V procollagens in several chick tissues produce procollagens $(\text{pro}\alpha1V)_2$,$(\text{pro}\alpha2V)_2$, and $(\text{pro}\alpha1V)_3$.[281] A comparative biosynthetic study of these glycoproteins by chick tendon fibroblasts and by hamster lung cell cultures has been reported.[282]

The N-glycosylation of L-lysine and hydroxy-L-lysine residues in type I collagen from streptozotocin-diabetic rats has been confirmed, and the stability of the complex has been shown to be due to an Amadori rearrangement.[283] This type of glycosylation does not represent an acceleration of the normal maturation process in collagen involving the reducible cross-links.

Post-translational modifications in the biosynthesis of type IV collagen by a human tumour cell line have been studied.[284] The differing extents of modifications shown by this cell line and normal human skin fibroblasts are explained by differences in the activities of L-lysylhydroxylase and hydroxy-L-lysyl-glucosyl transferases between the two types.

Collagen hydroxylases and glycosyltransferases of cultured human–foetal lung fibroblasts might not be co-ordinately regulated.[285] Regardless of hydroxylation events, glycosylation of the peptide might be limited to a specific fraction of the hydroxy-L-lysine residues during the post-translational modification of collagen.

6 Glycogen

A review dealing with the comparative biochemistry of starch and glycogen has been published.[286]

A highly branched, polydisperse, high-molecular-weight glycogen from rat liver has been isolated using centrifugation, gentle heating, and gel chromatography.[287] Using adult fasted rats, this glycogen was shown to be better than high-molecular-weight cold water-ethanol extracted glycogen for the binding of glycogen metabolizing enzymes.

The simultaneous demonstration of glycogen and protein in cardiac tissue glycosomes has been achieved histochemically.[288]

An abnormal urinary oligosaccharide pattern has been observed in patients with glycogen storage disease type III.[289] No characterization of the oligosaccharides was attempted.

Structural aspects of the catalytic and regulatory function of glycogen phosphorylase have been reviewed.[290] The regulation of glycogen phosphorylase and glycogen synthase by adrenalin in Soleus muscle in phosphorylase kinase-deficient mice has been reported.[291]

Data have been produced which are not consistent with the view that insulin activates glycogen synthase by producing an inhibitor of 3'5' cyclic adenosine monophosphate-dependent protein kinase in rat skeletal muscle.[292] Nor do they support the hypothesis that insulin acts by decreasing the activity of an inhibitor of a multi-substrate phosphoprotein kinase.

Bovine heart glycogen synthase has been specifically phosphorylated with a glycogen synthase kinase isolated from heart.[293] The combination of the enzyme and 3'5' cyclic AMP-dependent protein kinase leads to specific phosphorylation in three enzyme sites. Each form of the glycogen synthase, phosphorylated in specific enzyme sites, has been isolated and studied kinetically.

A protein kinase from rabbit reticulocytes, able to phosphorylate the β-subunit of eukaryotic initiation factor 2 (IF2), and glycogen synthase kinase from rabbit skeletal muscle are, if not identical, closely related proteins.[294]

Yeast 1,4-α-D-glucosidase, homologous albumin, and IgG raised against isolated rat hepatocytes have been cross-linked using glutaraldehyde to produce a stable, active enzyme-polymer complex (mol. wt. 1.0×10^6).[295] After intravenous injection into rats, the complex is preferentially associated with the Kupffer cells of the liver. The procedure has been used as a model for enzyme therapy for the possible lowering of hepatocyte glycogen content in type II glycogenesis (Pompe's disease).

7 Glycosaminoglycans and Proteoglycans

Analysis. — A semi-quantitative micromethod for the determination of free glycosaminoglycans in serum has been developed as a screening method for mucopolysaccharidosis.[296] Glycosaminoglycans are adsorbed in DEAE-cellusose paper before staining with Alcian Blue and measurement of the optical density of the resulting dye complex.

Glycosaminoglycans have been identified and measured after a selective and stepwise sequence of treatments with hyaluronidase, chondroitin sulphate lyase AC and ABC, endo-β-D-galactanase, and nitrous acid.[297] To determine the result of each degradative procedure the remaining glycosaminoglycans are subjected to cellulose acetate electrophoresis.

Hyaluronic acid and the isomeric chondroitin sulphates have been simultaneously determined after hydrolysis with chondroitinase ABC (EC 4.2.2.4.).[298] The method is reproducible and sensitive (detection limit 0.02 μmol disaccharide).

The specificity of a radioimmunoassay method for measuring human articular cartilaginous proteoglycans has been reported.[299] The antibodies used show a species specificity since there is no cross-reaction with the corresponding rat or dog proteoglycans. Antigenic sites appear to be located in the protein region and not in the glycosaminoglycan region of the molecule.

Complexes formed between serum poly-anions and cetyl pyridinium chloride have been quantified by laser nephelometry.[300] Measurement of the complexes in serum before and after digestion with specific enzymes enables the measurement of hyaluronic acid and chondroitin sulphate in small volumes of serum. The uronic acid levels in chondroitin 4- and 6-sulphates and dermatan sulphate have been determined after electrophoresis of the liberated uronic acids on Titan III cellulose acetate plates.[301]

An h.p.l.c. method has been described for the analysis of reduced unsaturated disaccharides derived from enzymic digestion and sodium borohydride reduction of chondroitin sulphates, dermatan sulphate, heparan sulphate, and heparin.[302] The method avoids the possibility of obtaining anomeric forms of unsaturated disaccharides.

A modification of an amino sugar analysis has been used to measure the sulphaminohexose content of heparin and heparan sulphate.[303] Polymer-bound N-sulphated 2-amino-2-deoxyhexoses are specifically deaminated to yield free 2,5-anhydro-D-mannose, which is then measured using the 3-methyl-2-benzothiazolinone reagent.

The degradation of proteoglycan and collagen by cells in culture may be measured simultaneously in a plate assay by monitoring the release of soluble radioactive degradation products from a dish coated with a gel containing ^{3}H-labelled proteoglycan and ^{14}C-labelled collagen.[304] Synovial cells are able to degrade both substrates, owing to the release of a proteoglycan-degrading neutral proteinase and of collagenase.

A monodimensional cellulose acetate electrophoresis system is capable of resolving keratan sulphate, heparan sulphate, dermatan sulphate, heparin, chondroitin 4-sulphate, and hyaluronic acid, on the basis of their differential migration in H_4-e.d.t.a. and lithium chloride solutions.[305] The electrophoretic behaviour of hyaluronic acid, desulphated chondroitin, heparan sulphate, and chondroitin sulphate with two different buffers under varying conditions of buffer concentration, electrophoresis time, and voltage has been examined.[306] A two-buffer monodirectional electrophoresis technique, which offers some of the advantages of resolution seen with two-dimensional methods yet retains the advantages of band comparison and quantitative analysis characteristic of one-dimensional electrophoresis, has been proposed.

Intact chick limb bud proteoglycan monomer is highly flexible.[308] Restrictions on chain flexibility occur in the neighbourhood of the linkage between protein and polysaccharide chains where high local chain concentrations cause collisions and entanglements. Proteoglycans from bovine nasal, bovine articular, and rat chondrosarcoma cartilage have been analysed by h.p.l.c. or a silica-based material bonded with an amide phase.[307]

Cation binding to multi-chain and single-chain glycosaminoglycan peptides has been studied by $^{23}Na^{+}$ n.m.r. spectroscopy.[309]

Hyperlipidemic rabbit serum impairs the precipitation of glycosaminoglycuronans from tissue-culture medium by cetylpyridinium chloride.[310] The interfering compounds can be removed by extraction of samples in lipid solvents. Various glycosaminoglycans and derivatives thereof have been subjected to affinity chromatography on low-density lipoprotein-substituted agarose.[311] By use of a variety of modification and degradation procedures, the chemical characteristics that maximize the affinity of a glycan for the lipoprotein have been assessed. The interaction between the two classes of macromolecules is dependent on the presence of sulphate groups and increases with charge density. The interaction of serum lipoproteins and a proteoglycan from bovine aorta has been reported.[312] Positive charges on the protein moiety of low-density lipoproteins are required for the interaction, as are presence of the protein core and the glycosaminoglycan chains of the proteoglycan.

Purified proteoglycans from human articular cartilage have been used in the development of a radioimmunoassay procedure.[313] Conditions for labelling and subsequently purifying the proteoglycan are described, and optimal conditions for the assay are detailed.

Occurrence, Isolation, and Structure. — The fractionation and characterization of proteoglycans isolated from chondrocyte cell cultures have been described.[314] The proteoglycans were compared with the proteoglycans extracted from bovine nasal or tracheal cartilage. Proteoglycans have been prepared from human femoral-head articular cartilage.[315] As a result of gel chromatography, light scattering, and ultracentrifugal studies, a physical model of the proteoglycan molecule in cartilage has been proposed. Three proteoglycan fractions produced by human-embryo lung fibroblasts have been isolated and characterized.[316] The fractions contain mainly heparan sulphate chains (mol. wt. 4.0×10^4), chondroitin sulphate chains (mol. wt. 4.0×10^4), and dermatan sulphate chains (mol. wt. 2.5×10^4), respectively. Glycosaminoglycans isolated from tryptic digests of atria of the human heart have been identified as containing hyaluronic acid, heparan sulphate, chondroitin 4- and 6-sulphates, and dermatan sulphate.[317] The proteoglycans of adult human meniscus appear to be of two types.[318] Preparations isolated at high buoyant density on caesium chloride density-gradient centrifugation contain molecules of large hydrodynamic size that are composed of protein, chondroitin sulphate, keratan sulphate, and neuraminic acid and possess subunit and aggregate structures comparable with those of hyaline-cartilage proteoglycans. Preparations from low buoyant density, which are enriched in dermatan sulphate, contain smaller-molecular-weight proteoglycans which do not interact with hyaluronic acid. The glycosaminoglycan compositions of canine menisci have been reported.[319] Although the composition is the same in different regions of the menisci, the total amounts vary considerably. Articular cartilage of young baboons (*Papio papio*) contains a proteoglycan of high electrophoretic mobility on large-porosity gels and is characterized as having a high protein content.[320] Proteoglycans from pig aorta have been extracted sequentially with inorganic salt solutions under associative and dissociative conditions, followed by fractionation by gel permeation chromatography in order to characterize and compare their chemical properties and hydrodynamic sizes.[321] Low-buoyant-density proteoglycans found in the link fraction from avian cartilage contain some of the antigenic determinants found on the proteoglycan monomer.[322] However, there is at least one monomer determinant which is not present in the link fraction.

Proteoglycan subunits of bovine nasal cartilage contain phosphate ester groups that are sensitive to acid phosphatase and are considered to be a regular constituent of the proteoglycan molecule.[323] Structural studies on the D-xylosyl-L-serine linkage in proteoglycans of human, porcine, and shark cartilages have been reported.[324] The finding that human and porcine proteoglycans have the same amino acid sequence in the linkage region as that reported previously for bovine proteoglycan suggests that there is a common structural feature in the core proteins of these mammalian cartilage proteoglycans.

Zonal-rate centrifugation in sucrose gradients has been used to study interactions in proteoglycan aggregation.[325] The link protein was shown to interact with the isolated hyaluronic acid-binding region of the proteoglycan monomer. Since it also interacts with isolated hyaluronic acid, the binding of proteoglycans to hyaluronate in the presence of link proteins is trifunctional. The binding properties of cartilage link protein and the hyaluronate-binding region of cartilage proteoglycan to hyaluronic acid immobilized on agarose have been compared and shown to have similar properties.[326] A gross physical characterization of the hyaluronic acid-binding region of proteoglycan from pig laryngeal cartilage has been achieved using densitometric and small-angle neutron studies.[327]

Platelet factor 4 is a basic tetrameric protein which interacts with sulphated polymers.[328] The transfer of this factor from its proteoglycan carrier to natural and synthetic polymers has been reported. The results of electric birefringence studies show that chondroitin 4-sulphate molecules have a very high degree of mobility within the proteoglycan structure.[329]

A novel low-molecular-weight chondroitin sulphate proteoglycan (mol. wt. 7.6×10^4) has been isolated from bovine nasal cartilage, and in different layers of bovine hip articular cartilage.[330] The proteoglycan contains two or three chondroitin sulphate chains and some oligosaccharides, and high contents of L-leucine and L-cysteine as compared with other proteoglycans. The inhibitor of Clq, a subunit of the first component of complement, has been identified as a chondroitin 4-sulphate proteoglycan.[331] Unlike the major species of cartilage proteoglycan, the serum proteoglycan does not form a complex with hyaluronic acid.

Rat brain contains a single polydisperse macromolecule in which chondroitin sulphate chains and oligosaccharides are both covalently linked to a common protein core.[332] The N- and O-glycosidically linked oligosaccharides of the glycoprotein appear to be randomly distributed in the proteoglycan. The brain proteoglycan appears to be capable of at least a limited degree of interaction with hyaluronic acid to produce larger aggregates. Some bovine aortic chondroitin sulphate- and dermatan sulphate-containing proteoglycans have been isolated and chemically characterized.[333]

Disaccharides liberated from chondroitin sulphates and dermatan sulphate by digestion with chondroitin lyase AC or chondroitin lyase ABC are completely separated on cellulose acetate plates by electrophoresis in barium acetate or calcium acetate.[334] Glycosaminoglycans have been isolated from the femurs of oestrogen-treated male Japanese quail.[335] Chondroitin 4-sulphate and keratan sulphate were found in the femurs, but only the keratan sulphate correlates with medullary bone production. A proteoglycan isolated from ascites fluid produced by a rat yolk-sac tumour has been partially characterized as a chondroitin sulphate proteoglycan.[336] Dermatan sulphate proteoglycans have been isolated and characterized from bovine sclera.[337] Two proteoglycans, differing with respect to the nature of their protein cores and the co-polymeric structure of their side chains, were isolated. Isopycnic centrifugation studies using caesium chloride and caesium sulphate have been reported on these proteoglycans.[338] Bovine sclera contains two proteodermatan sulphates, the larger exhibiting self-association in both gel chromatography and light-scattering experiments, and the smaller showing aggregation only in light-scattering experiments.[339] It is proposed that the aggregation is solely or partly mediated by interaction between dermatan sulphate side chains, although other forms of self-association, e.g. via polysaccharide-protein and protein-protein interactions, have not been excluded.

Calf skin proteodermatan sulphate is composed of three or four chains of dermatan sulphate (mol. wt. 1.7×10^4) covalently linked to a core protein (mol. wt. 5.6×10^4) via O-glycosidic linkages.[340] A proteodermatan sulphate from rat skin has been isolated and characterized.[341] The proteoglycan (mol. wt. 3.6×10^4) contains 46% protein which is bound to dermatan sulphate via an O-glycosidic linkage. Rat tail tendon has been stained with a cationic phthalocyanin dye, Cupromeronic Blue, to demonstrate the location of proteoglycan by electron microscopy.[342] A dermatan sulphate-rich proteoglycan is distributed about the collagen fibrils in an orthogonal array, the transverse elements of which are located almost exclusively at the d band in the gap zone.

Several different proteoglycan species, a D-glucuronic acid-rich and an L-iduronic acid-rich proteodermatan sulphate, together with two different proteoheparan sulphates, have been isolated from human skin fibroblast cultures.[343] Three different types of dermatan polysulphate peptides I, II, and III, isolated from hagfish notochord, hagfish skin, and shark skin, all contain L-serine, D-xylose, and D-galactose.[344] The D-xylosyl-O-seryl linkage was detected in polypeptide III. Polypeptide I contains an O-glycosyl linkage between 2-acetamido-2-deoxy-D-galactose and L-serine.

Chromatography on Sephadex^R G-50 results in optimal isolation of a heparin component that is highly active as measured by several assay methods.[345] The binding of Methylene Blue to heparin is highly co-operative, exothermic, and stabilized by entropic factors.[346] An examination of fundamental conditions for hydrophobic-interaction chromatography of heparin on hydrophobic gels has been reported.[347] Column dimensions, amount of heparin applied, flow rate, electrolyte and acidity of the elution medium, and temperature, influence the distribution of heparin among the fractions separated on the gel.

A direct method has been developed for the determination of the dissociation constant between heparin and bovine serum albumin, by using fluorescently labelled heparin and monitoring the fluorescence intensity change induced by the protein.[348]

Gel formation of thymus lymphocytes, spleen or bone marrow cells with heparin has been observed.[349] Since the major, if not only, source of heparin is the mast cell, and since heparin is released along with histamine during certain types of allergic and inflammatory reactions, it is suggested that heparin promotes the accumulation of leucocytes in the extravascular space.

Heparin and chondroitin 6-sulphate are the only glycosaminoglycans which at pH 6.0 precipitate low-density lipoprotein from aqueous solutions and in the presence of Ca^{2+}.[350] However, chondroitin 6-sulphate is the only one that binds to immobilized low-density lipoprotein in the presence of Ca^{2+} at the same pH. The binding of Ca^{2+} and Mg^{2+} ions to heparin in the presence of added univalent salts has been studied using a dye spectroscopic method.[351] The interaction of Mg^{2+} with heparin appears to be independent of the nature of the charged groups on the polyanion, but Ca^{2+} binding is considerably stronger than Mg^{2+} binding and is in excess of theoretical predictions, suggesting a localized or specific interaction with heparin. A selective binding of Zn^{2+} ions to heparin rather than to other glycosaminoglycans has been observed.[352] This observation suggests that, contrary to a previously proposed hypothesis, simple electrostatic interactions between the negatively charged carboxy groups of the glycosaminoglycans and Zn^{2+}cations cannot explain the observed binding.

Some chemical, enzymic, and physical properties of disaccharides derived from beef lung heparin by nitrous acid degradation followed by sodium borotritide reduction have been reported.[353] Partial N-desulphation of the heparin, prior to degradative deamination, yields larger-molecular-weight fragments, from which a tetrasaccharide has been partially characterized.[354] Heparin, after carboxyl reduction with sodium borotritide, has been degraded under conditions of acid hydrolysis that selectively cleaves the 2-O-sulpho-L-idopyranosidic linkages.[355] The resulting purified radio-labelled di-saccharides (5-13) were characterized and evaluated before and after N-acetylation or N-sulphation, as substrates for sulphamidase, acetylcoenzyme A:2-amino-2-deoxy-α-D-glucoside N-acetyltransferase, 2-acetamido-2-deoxy-α-D-glucosidase, or 2-acetamido-2-deoxy-D-glucose 6-sulphate sulphatase in human skin fibroblasts.

Tritiated di- and tetra-saccharides, representative of the various repeat sequences of heparin, have been prepared and evaluated as substrates for the estimation of sulphamidase activity present in skin fibroblasts, and for the detection of San-fillipo type A syndrome.[356]

A simple method for fractionating heparin uses antithrombin III which is noncovalently bound to immobilized concanavalin A.[357] Active antithrombin is eluted from the immobilized lectin using methyl α-D-glucoside. An electrophoretic method using agarose beads has been developed for the determination of heparin fractions having a high affinity for antithrombin III.[358]

A high-molecular-weight heparin fraction that includes components with as much as five to ten times the activities of the original heterogeneous preparations has been isolated from beef lung.[359] The fraction was characterized with respect to its molecular weight (3.5 x 10^4), composition, and binding to antithrombin

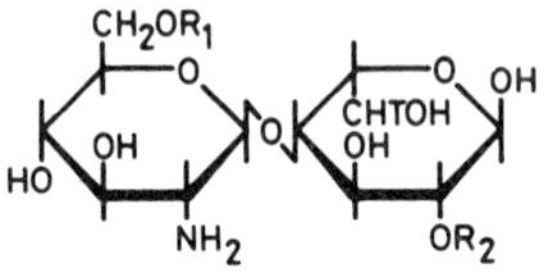

(5) $R_1 = R_2 = H$
(6) $R_1 = SO_3^-$, $R_2 = H$
(7) $R_1 = H$, $R_2 = SO_3^-$
(8) $R_1 = R_2 = SO_3^-$

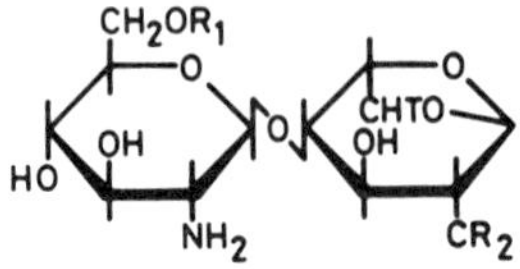

(9) $R_1 = R_2 = H$
(10) $R_1 = SO_3^-$, $R_2 = H$
(11) $R_1 = H$, $R_2 = SO_3^-$
(12) $R_1 = R_2 = SO_3^-$

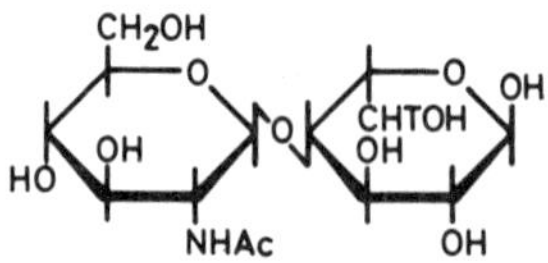

(13)

III. Hog mucosal heparin, after partial N-desulphation, has been conjugated to fluoresceinylthiocarbamoyl groups.[360] The fluorescent heparin was separated into low-affinity and high-affinity fractions with respect to antithrombin III. Substitution of the heparin molecules had no effect on the biological interaction with antithrombin III. Evidence for a 3-O-substituted 2-amino-2-deoxy-D-glucosyl residue in the antithrombin-binding sequence of heparin has been reported.[361] This sulphated amino sugar is a unique component of high-affinity heparin, located at a specific position in the antithrombin-binding sequence of the molecule. Confirmation of the existence of this component has been obtained from ^{13}C n.m.r. spectroscopic studies.[362] A highly active and a relatively inactive form of whale heparin have been subjected separately to enzymic degradation.[363] The disaccharide 2-acetamido-2-deoxy-3-O-(4-deoxy-L-*threo*-

pyranosyluronic acid)-D-glucose was formed exclusively from the highly active form and may arise from an essential binding section for antithrombin III.

Binding constants for the binding of high-affinity heparin to antithrombin III at different ionic strengths have been determined by fluorescence titrations and estimated from dissociation curves of the heparin-antithrombin III complex.[364] A thermodynamic evaluation of the dissociation of the complex suggests that maximally five to six charged groups are involved in the interaction. The binding of antithrombin III to heparin fractions with different molecular weights has been studied.[365] Binding constants and stoichiometries for the binding of these fractions to antithrombin III are presented and correlated with the anticoagulant activities of the fractions.

Porcine intestinal heparin has been partially degraded to produce an octasaccharide with high affinity for antithrombin III.[366] The octasaccharide exhibited more potent inactivation activities for thrombin, factor X_a, and plasma coagulation than the native heparin. The high-affinity binding of heparin to antithrombin III requires the presence of two consecutive N-sulphated 2-amino-2-deoxy-D-glucosyl residues in specific positions of the antithrombin-binding sequence.[367] Loss of either one of these N-sulphate groups, with or without N-acetylation, results in a distinct and appreciable decrease in binding affinity and in anticoagulant activity.

Porcine heparin has been cleaved randomly by chemical techniques to produce hexasaccharides, octasaccharides, decasaccharides, and fragments containing 14 and 16 residues that are able to complex with protease inhibitors.[368] As a result of studies on the avidity of these fractions for protease inhibitors, their binding to antithrombin, and the catalysis of the Factor Xa-antithrombin interaction, it is proposed that heparin possesses multiple discrete structural domains that modulate different functions of antithrombin III. The chemical composition and ^{13}C n.m.r. spectra of heparin octasaccharides having high affinity for antithrombin III and high anti-(factor Xa) activity, and prepared by three independent approaches, have been studied and compared with those of the corresponding inactive species.[369] Combined with chemical data, the spectra of the active oligosaccharides and of their fragmentation products afforded information on composition and sequence.

The effects of high-activity heparin fractions of differing molecular weight on antithrombin III inhibition of each of the serine proteases known to occur in the intrinsic pathway of the coagulation cascade have been reported.[370] Inhibition of thrombin, Factor IXa, and Factor XIa showed similarities in the dependence on molecular weight of heparin, and was found to decrease with decreasing molecular weight of the polysaccharide. Inactivation of Factor Xa, Factor XIIa, and kallikrein was less dependent on the size of the polysaccharide.

Low-molecular-weight heparin of low anticoagulant activity and high-molecular-weight heparin of corresponding high activity have been subjected to N-desulphation and the resulting amino groups re-acylated with dansyl chloride or rhodamine B isothiocyanate.[371] The results of binding studies of the modified heparins to antithrombin III are consistent with the proposal that a single high-molecular-weight high-activity heparin occupies two sites when it binds to antithrombin III whereas low-molecular-weight low-activity heparin binds to the two sites separately.

Proteoheparan sulphate prepared from rat liver plasma membranes inhibits the growth of AH-130 ascites hepatoma cells.[372] Heparan sulphate is less active.

A non-collagen protein, in association with heparan sulphate, has been identified as a normal constitutent of a naturally occurring rat basement membrane.[373] The material resembles proteoglycans recently isolated from the EHS sarcoma and from bovine glomerular basement membrane.

Heparan sulphate has been identified as the major glycosaminoglycan in bovine retinal capillary basement membrane.[374]

The binding of glycosaminoglycans to fibronectin at adhesion sites of murine fibroblasts has been reported.[375] Only heparan sulphate proteoglycans, containing highly N-sulphated sequences on the heparan sulphate, mediate substratum adhesion to the cells by co-ordinate binding to fibronectin on the cell surface and serum fibronectin.

Self-association between various heparan sulphate species and oligosaccharide fragments thereof have been studied by affinity chromatography.[376,377] Segments comprising both L-iduronate and D-glucuronate may serve as contact zones for the self-association, and the strength of binding is dependent on co-operative interactions between a number of such zones. However, various heparan sulphate subfractions from transformed cells have no affinity for agarose gels substituted with the corresponding heparan sulphate species.[378] The lack of association of the heparan sulphate may be caused by minor changes in the sequential arrangements of L-iduronate- and D-glucuronate-bearing repeating segments of the polysaccharide.

Heparan sulphate-cleaving enzyme has been used to treat cells of an ascites hepatoma AH66, for which the occurrence of heparan sulphate at the cell surface of the plasma membrane has been established.[379]

Two pools of heparan sulphate proteoglycans with different buoyant densities have been selectively solubilized from rat liver plasma membranes by successive incubations with heparin and detergent.[380] The detergent-extracted heparan sulphate represents a proteoglycan species that has its core protein in the lipid bilayer of the plasma membrane. The presence of heparan sulphate having a high metabolic activity has been demonstrated in isolated rat intestinal epithelial cells.[381] A high degree of sulphation of its chains was detected. Heparan sulphate is a potent inhibitor of DNA synthesis.[382] The inhibition is not simply related to gross charge density.

Bovine vitreous body hyaluronic acid has been purified, without the use of proteolytic enzymes, by ion-exchange chromatography.[383] ^{125}I-Labelled hyaluronate-binding proteins from cartilage have been prepared.[384] Their adsorption on an immobilized hyaluronate gel in the presence of free competing hyaluronic acid has been used as an assay for free hyaluronic acid.

^{1}H n.m.r. spectra in {$^{2}H_{6}$}-dimethylsulphoxide of dodecyltrimethylammonium salts of chondroitin sulphates and hyaluronate or sodium salts of oligomers from hyaluronate show unambiguous NH signals.[385] A secondary structure in hyaluronate and some chondroitin sulphates, involving a hydrogen-bonded acetamido NH, was observed. Sodium hyaluronate has been cleaved into an homologous series of oligosaccharides by the action of bovine testicular hyaluronidase.[386] C.d. analysis of the oligosaccharides showed that the c.d. spectrum of hyaluronate in aqueous solution at neutral pH does not reflect to any substantial degree a polymer conformation which requires cooperative interaction between several repeating residues for stabilization.

The sequential hydrolysis of rooster comb hyaluronic acid by β-D-glucuronidase and β-D-2-acetamido-2-deoxy-glycosidase has failed to demonstrate the presence of any branching in the polysaccharide chain.[387] No evidence was found for the presence of neutral sugars. The temperature dependence of binding hyaluronate oligomers to bovine nasal proteoglycan has been reported.[388]

Hyaluronectin, a human brain glycoprotein capable of binding to hyaluronic acid, has been isolated.[389] The binding is inhibited by hyaluronic acid and by the products of hydrolysis by hyaluronidase but not by either glycosaminoglycans or monosaccharides.

The interactions of sulphated polysaccharides such as keratan sulphate and chitin sulphate with *Solanum tuberosum* agglutinin and wheat germ agglutinin have been reported.[390] The presence of the sulphate groups in these polysaccharides did not interfere with their specific interactions with the lectins.

The degradation of keratan sulphate has been achieved by β-2-acetamido-2-deoxy-D-glucosidases A and B (EC 3.2.1.52).[391] For enzymic activity a substrate with exposed non-sulphated 2-acetamido-2-deoxy-D-glucosyl residues at the non-reducing end of the polymer is required.

Keratan sulphate has been degraded by an extract of a *Pseudomonas sp*. to give 2-acetamido-2-deoxy-3-O-D-galactosyl-β-D-glucose 6-sulphate.[392] This disaccharide after reduction with sodium borotritide can be used as a substrate for the measurement of 2-acetamido-2-deoxy-D-glucose 6-sulphate sulphatase, whilst the desulphated reduced disaccharide can be used as a substrate for the measurement of (1→3)-2-acetamido-2-deoxy-β-D-glucosidase. Peptido-keratan sulphate from bovine cornea, degraded by desulphation, hydrazinolysis, deamination and sodium borohydride reduction, releases two tetrasaccharide fractions, the structure of one of which (14) has been established.[393] Both terminal D-mannosyl residues bear keratan sulphate chains. A structure in which an L-

```
α-D-Manp-(1→6)-β-D-Manp-(1→?)-D-GlcpNAc-L-Asn
                3
                |
                1
            α-D-Manp
```

(14)

fucosyl residue and an NN'-diacetylchitobiosyl sequence replace the 2-acetamido-2-deoxy-D-glucosyl residue attached to the protein chain has also been proposed. Studies on the polydispersity and heterogeneity of proteokeratan sulphate from calf and porcine cornea have been reported.[394] The linkage region of bovine corneal proteokeratan sulphate is branced at a D-mannose residue (15).[395] The structural characteristics show that there is a great similarity

```
 (SO3)0-2
   |
(D-Galp-D-GlcpNAc)n-(D-Galp-D-GlcpNAc)2-α-D-Manp-(1→2)-β-D-Manp-D-GlcpNAc-D-GlcpNAc-L-Asn
                                                            2
                                                            |
                                                            1
           (D-Galp-D-GlcpNAc)m-(D-Galp-DGlcpNAc)2-α-D-Manp
              |
             (SO3⁻)0-2
```

(15)

between the linkage region of the proteokeratan sulphate and the N-glycosidically linked L-asparagine moiety of many glycoproteins.

Biosynthesis. -- Hyaluronic acid synthesis by the apical ectodermal ridge of chick limb bud has been reported.[396] The effects of phase of growth, cell density and also serum concentration are important contributors for the incorporation of 2-amino-2-deoxy-D-{^{3}H}-glucose into glycosaminoglycans by rabbit aortic smooth muscle cells.[397] The synthesis of sulphated glycosaminoglycans in rat liver explants prepared from regenerating tissue, in which the factors of cell necrosis and consecutive inflammation associated with most models of experimental hepatic fibrosis are excluded, shows a depression by about 50% of the incorporation of 2-amino-2-deoxy-D-{^{14}C}-glucose into total glycosaminoglycan.[398]

Rat liver parenchymal cells have been evaluated for their ability to synthesize and accumulate heparan sulphate as the major component and low-sulphated chondroitin sulphate, dermatan sulphate, chondroitin sulphate and hyaluronic acid as the minor ones.[399] Glycosaminoglycans are synthesized by rat glomeruli and transported to glomerular basement membranes. Heparan sulphate is the major product synthesized, although smaller amounts of hyaluronic acid and chondroitin sulphates are also produced.[400] The transfer of platelet factor 4, a basic tetrameric protein complexed with a series of glycosaminoglycans, to heparin has been detected and studied by measuring changes in the anisotropy of the dansyl derivative of platelet factor 4.[401] The release of platelet factor4-proteoglycan complex from the platelet in vivo results in the transfer of the protein to heparin. Moderate quantities of dermatan sulphate or heparan sulphate do not prevent the transfer.

The effect of cycloheximide on the synthesis of proteoglycans by cultured chondrocytes from the Swarm rat chondrosarcoma has been reported.[402] The cultured chondrocytes contain a large pool of core protein available for the addition of chondroitin sulphate chains. In the absence of entry of newly synthesized core protein into this pool following cycloheximide treatment, the remaining core protein molecules are processed to completed proteoglycans with nearly first order exponential kinetics. The biochemical pathways for the synthesis of chondroitin sulphate were not drastically affected by cycloheximide. Inhibition of protein synthesis by cycloheximide reduces the incorporation of {^{35}S}-sulphate into heparan sulphate to about 5% of untreated hepatocytes. 4-Nitrophenyl β-D-xyloside partially reverses the inhibitory effect.[403] Free heparan sulphate chains not bound to core protein were synthesized under these conditions.

The pattern of glycosaminoglycan synthesis by smooth-muscle cells from pig aortic media has been studied.[404] Proteoglycans synthesized by media cells exhibit different glycosaminoglycan distribution patterns according to their localization.

Cultured cells from red dermal tissue, containing a large proportion of cells from the hair follicles, produce hyaluronic acid, dermatan sulphate, chondroitin 4-sulphate, heparin and heparan sulphate.[405] The patterns of synthesis and sulphation of the glycosaminoglycans changed with continued time in culture.

Proteoglycan synthesis in normal and chronically hydrocortisone-suppressed rabbit articular cartilage has been reported.[406] The report confirms that hydrocortisone can effect UDP-2-acetamido-2-deoxy-D-hexose pool sizes, and also the level of D-galactosyltransferases, but not D-xylosyltransferase involved in the synthesis of the chondroitin sulphate carbohydrate-protein linkage region. The major effect is the depression of proteoglycan core-protein synthesis. No net change in the proteoglycan structure was observed.

Insulin and rat-derived, somatomedin-like polypeptide multiplication-stimulating activity stimulates proteoglycan synthesis in rat chondrosarcoma chondrocytes.[407] The structures of the proteoglycans synthesized in response to the hormones compared with those synthesized in the medium alone are similar, but the chondroitin sulphate chains, synthesized in the presence of hormones, increase in molecular weight by 25-30%.[408] The distribution of glycosaminoglycan synthesis by rat glomeruli *in vivo* and *in vitro* has been reported.[409]

Administration of a single dose of 2-amino-2-deoxy-D-galactose to rats causes time-dependent biphasic changes of total glycosaminoglycan synthesis in liver.[410] A rapidly occurring inhibition is followed by a significantly enhanced production of {^{35}S}-labelled glycosaminoglycans in later stages. Undersulphated β-D-xyloside glycosaminoglycans are secreted from chick-embryo chondrocytes in the presence of the ionophore monensin.[411] Stimulated peripheral blood monocytes contain a factor(s) which affects synovial cells in culture, depressing markedly collagen synthesis while enhancing glycosaminoglycan production.[412] The involvement of such biologically active monocyte-derived substance(s) in connective-tissue metabolism of inflammatory lesions is discussed.

The assembly of proteoglycan aggregates in cultures of chondrocytes from bovine tracheal cartilage has been studied.[413] Proteoglycan monomers and hyaluronic acid are exported separately and combined extracellularly. The size of the aggregates increases gradually with time as the proportion of monomers bound to hyaluronic acid increases. Immunochemical analysis of the proteoglycans has identified cross-reactivity of molecules isolated from different species.[414] The intracellular precursor pool of core protein in this chondrocyte system has been identified and partially characterized.[415]

A study of acceptors and primers for chondroitin sulphate-chain biosynthesis by microsomal preparations from chick-embryo epiphyseal cartilage has been reported.[416] Incorporation of sugars into glycosaminoglycans has been used to obtain information on the entire proteoglycan structure at the site of glycosaminoglycan synthesis.

A soluble enzyme from quail oviduct incorporates sulphate at O-6 of the non-reducing 2-acetamido-2-deoxy-D-galactosyl 4-sulphate residue of chondroitin sulphate.[417] The reaction catalysed by this enzyme is different from the sulphation mediated by other known sulphotransferases.

The formation of L-idopyranosyluronic acid during the synthesis of dermatan sulphate has been studied in human fibroblast cultures and microsomes of the same cells.[418] Epimerization of D-glucopyranosyluronic acid to L-idopyranosyluronic acid residues during the biosynthesis of dermatan sulphate involves an abstraction of the C-5 hydrogen of the uronosyl residue.

Substrates for 2-acetamido-2-deoxy-D-glucosyltransferase, which acts in conjunction with D-glucuronosyl transferase in the biosynthesis of heparin, have been prepared from a low sulphated, D-glucuronic acid-rich heparan sulphate fraction.[419] Using these oligosaccharides, the products of the reaction have been characterized and some of the kinetic parameters of the enzyme have been determined. Some of the oligosaccharides were shown to substitute for the endogenous substrates and serve as primers in polysaccharide chain elongation.

The secretion of heparan sulphate by cultured rat hepatocytes is increased in the presence of the flavenoid (+) catechin.[420] This increase is due to a new species of heparan sulphate lacking the carbohydrate-protein linkage between D-xylose and L-serine in the normal heparan sulphate proteoglycan.

Extracellular hyaluronic acid is the major glycosaminoglycan synthesized by the epidermis when intact skin slices are maintained in organ culture.[421] In the absence of the dermis this synthesis is considerably reduced, and the synthesis of sulphated glycosaminoglycans, mainly heparan sulphate, is unaffected.

The formation of both N- and O-sulphate groups (sulphamido and ester sulphate groups, respectively) of the 2-amino-2-deoxy-D-glucosyl residues in porcine aorta has been demonstrated by the study of sulphate incorporation from labelled 3'-phosphoadenyl sulphate to heparan sulphate of microsomal fractions.[422] The preferential N-sulphation is obtained for incorporation of labelled precursor over a short period, with O-sulphation occurring on previously N-sulphated heparan sulphate. The stimulation of hyaluronic acid synthesis in mouse skin in response to estrogen treatment is mediated through estrogen receptors and involves the induction of hyaluronic acid synthetase.[423] Glycosaminoglycan secretion from perfused monolayer cultures of rabbit ear chondrocytes has been reported.[424] Colchicine has been shown to affect the level of glycosaminoglycan synthesis and transport. A short-term incubation system, developed for the study of glycosaminoglycan synthesis during the early stages of medullary bone formation in Japanese quail in the presence of estrogen, has identified keratan sulphate and chondroitin sulphate.[425]

Pathology. — The polydispersity present in aggregating proteoglycans newly synthesized by chondrocytes from rat chondrosarcoma has been attributed to the amount of chondroitin sulphate/core protein and not to any detectable differences in the size of the core protein.[426,427] Glycosaminoglycans in urine from patients representing the major different mucopolysaccharidoses have been separated and measured by use of a procedure requiring small volumes of urine.[428] Approximately half the patients excreted small amounts of heparin. The keratan sulphate in samples from Morquio's disease patients migrated differently from authentic keratan sulphate unless digested with chondroitin sulphate lyase ABC.

Total sulphated glycosaminoglycan levels in commercially available preparations of hyaluronic acid and in the urine of normal individuals and the synovial fluid of individuals with rheumatoid arthritis and osteoarthritis have been measured spectrophotometrically as their Alcian blue complexes.[429] Lowe's syndrome results in an undersulphation of chondroitin sulphates.[430] This is now shown to be a consequence of a lower level of active sulphate (adenosine 3'-phosphate 5'-phosphosulphate) in Lowe's syndrome fibroblasts, caused by an elevation of nucleotide pyrophosphatase activity degrading adenosine 3'-phosphate 5'-phosphosulphate.

A low urinary excretion of heparan sulphate in three patients with Lowe's syndrome has been observed, although urinary excretion of total glycosaminoglycan and chondroitin sulphate isomers shows no significant differences between Lowe's syndrome subjects and age-matched controls.[431]

A new disease, tentatively designated Sanfilippo disease type D, has been recorded.[432] It is characterized by the clinical features of the Sanfilippo syndrome, excessive excretion of heparan sulphate, and the inability to release inorganic sulphate from 2-acetamido-2-deoxy-D-glucosyl 6-sulphate residues in heparan sulphate-derived oligosaccharides. However, the requirements for enzymatic hydrolysis of 2-acetamido-2-deoxy-D-glucosyl 6-sulphate linkages are different for heparan sulphate and keratan sulphate, when compared with those polysaccharides from ordinary Sanfilippo syndrome patients.

Glycosaminoglycan excretion in urine from patients with nephrotic syndrome has been investigated.[433] A protein-associated low-charge form of chondroitin sulphate which is not excreted by normal subjects was identified.

Chondroitin 6-sulphate with low sulphate content is excreted in the urine of patients with an unusual form of spondyloepiphyseal dysplasia.[434] The sera of these patients show a low activity of 3'-phosphoadenosine 5'-phosphosulphate chondroitin sulphate sulphotransferase, while the urinary undersulphated chondroitin sulphate is a much better acceptor of sulphate than standard chondroitin sulphate on incubation with 3'-phosphoadenosine 5'-phosphosulphate and normal sulphotransferases.

Precipitation with cetylpyridinium chloride followed by uronic acid assay has been used to study urinary glycosaminoglycan excretion in normal subjects and calcium-stone formers.[435] No significant difference in daily levels of glycosaminoglycans was detected.

Proteoglycans present in normal and atherosclerotic human aortas have been isolated and partially characterized.[436]

The glycosaminoglycan composition of samples of liver from patients with alcoholic cirrhosis has been compared with that of normal liver.[437] Increased levels of hydroxy-L-proline, hyaluronic acid, heparan sulphate, chondroitin 4- and 6-sulphates, and dermatan sulphate were detected in the cirrhotic liver samples.

Age-related changes in hydrodynamic size of the proteoglycan monomers of human cartilage and of the specific {^{35}S}-labelling of their chondroitin sulphate and keratan sulphate chains have been described.[438] In order to exclude degeneration (osteoarthrotic) processes that interfere with age-specific changes, cartilage of the processus xyphoid was investigated.

Autologous human knee and shoulder cartilage show age-related changes: the proteoglycan content decreases in content and proteoglycan subunits decrease in size.[439] In contrast, the proportions of keratan sulphate to chondroitin sulphate, protein to glycosaminoglycan, and chondroitin 6-sulphate to chondroitin 4-sulphate all increase with age.

The structures of the proteoglycans of articular cartilages of bovine foetuses at various stages of development have been compared with those of the calf and adult animal.[440] During foetal life, articular cartilage proteoglycans are rich in chondroitin sulphate and contain at all ages some keratan sulphate and O-glycosidically linked oligosaccharides in both a keratan-rich region and in the remainder of the polysaccharide attachment region.

An analysis of changes of sulphated glycosaminoglycans and related glycosidases during foetal development shows that chondroitin 6-sulphate increases in concentration up to fifty days of foetal development, and then decreases progressively until its complete disappearance in adult tissues.[441] The levels of heparatin sulphate and dermatan sulphate do not change during this period.

The major proteoglycans in the articular cartilage of foetal, calf, and adult animals differ in the content, types of structure of the chondroitin sulphate, keratan sulphate, and oligosaccharide constituents.[442] These changes in structure reflect the gross age-related changes in the chemical composition of the tissue.

Proteoglycans of the nucleus pulposus and annulus fibrosus of normal and degenerate invertebral discs have been examined.[443] A large proportion of the proteoglycans found in disc undergo a degree of degradation in the hyaluronate-binding region in degenerate tissue.

Glycosaminoglycans of cultured human-foetal melanocytes have been compared with those produced by cultured human melanoma cells.[444] By manipulating cell surfaces with enzymes specific for cleaving glycosaminoglycans from cells, it is possible to change the functional relationships such as adhesion and adhesion impairment among bone-marrow fibroblast-like cells cultured in vitro, mature leucocytes and blasts from normal marrow and peripheral blood of leukaemia patients.[445]

The absence of proteoglycan core protein in cartilage from omd/cmd (cartilage matrix deficiency) mouse[446] and in cartilage mutant nanom-elia[447] has been reported.

8 Cell and Tissue Glycoproteins

Glycoproteins from adult-rat brain synaptic vesicles have been fractionated by sequential affinity chromatography in the presence of sodium dodecyl sulphate on four different immobilized lectins.[448] Nine fractions containing only glycoproteins and differing in their terminal sugars were separated and analysed for their carbohydrate composition and electrophoretic profiles. A considerable heterogeneity of the glycoprotein population was observed which cannot be explained solely by the microheterogeneity of the glycans of the synaptic vesicle glycoproteins.

A group of structurally related glycoproteins from human brain and fibroblasts, which account for all the Thy-1 species cross-reactive antigenic activity present in these tissues, has been purified.[449] The isolated products are closely related to each other and could be the product of a single gene. They show many properties in common with the purified forms of rodent Thy-1 antigen. Human brain glycoproteins depleted of Thy-1 antigen have been used for monoclonal antibody production.[450] The antibody interacts with a determinant present on all human brain subregions (cerebral, cortical grey matter, white matter, caudate, thalmus, dentate nucleus, putamen, and cerebellar cortex) but is absent from a wide range of other tissues (heart, liver, spleen). The determinant is conserved in mammalian evolution, as the brains of rat and dog have amounts equal to that found in human brain.

Some biochemical characteristics, including the carbohydrate compositions, of brain Thy-1 glycoprotein of dog and man have been examined and compared with similar characteristics in rat and mouse.[451]

A new caprine neurocervical storage disease in related Nubian goats has been characterized by the presence in brain tissue of the trisaccharide O-β-D-mannopyranosyl-(1→4)-2-acetamido-2-deoxy-D-glucopyranosyl-(1→4)-2-acetamido-2-deoxy-D-glucose (β-D-mannosylchitobiose).[452] The accumulation of the trisaccharide suggests the possibility of a genetic defect in β-D-mannosidase in the catabolic pathway for N-linked complex glycopeptides. A deficiency of β-D-mannosidase activity was demonstrated in a number of tissues from the affected goats.[453]

The levels of N-acetylneuraminic acid in water-soluble and membrane fractions of various rat brain cortex tissues have been measured.[454]

Developmental changes in the biosynthesis and concentration of individual L-fucosyl- and N-acetylneuraminosyl-containing glycoproteins associated with synaptic membranes and synaptic junctions have been reported.[455]

The incorporation of L-fucose into synaptic mitochondrial and plasma-membrane glycoproteins has been recorded.[456] Synaptosomal mitochondria in situ have a greater capacity for the incorporation of L-fucose into protein than free mitochondria.

Synaptic membranes from sheep brain cortex incorporate inorganic sulphate into a range of glycoproteins (mol. wts. $1.6 \times 10^4 - 1.6 \times 10^5$).[457]

Evidence has been presented for the incorporation of L-fucose and the identification of L-fucosyl glycoproteins in sheep cortex synaptosomes.[458]

The carbohydrate structure of glycoproteins associated with rat central-nervous-system myelin is developmentally regulated.[459] Although the precise association of these membrane components with the axolemma, the oligodendroglial plasma membrane, or myelin has not been determined, the observed changes that occur during development support a probable functional role for many of them in the mechanism of myelinogenesis.

Fluorescent-labelled lectins have been used to examine the glycoprotein composition of rabbit peripheral-nervous-system myelin subfractions.[460] The glycoprotein composition of two major fractions is more complex than previously reported.

The sequence of events in the destruction of myelin proteins and glycoproteins in experimental demyelination of the rabbit optic nerve has been reported.[461]

The presence of myelin-associated glycoprotein in the

peripheral nervous system of rats has been demonstrated.[462] The presence of this glycoprotein in the periaxonal region of both peripheral and central myelin sheaths is consistent with a similar involvement of the glycoprotein in axon-sheath cell interactions in the peripheral nervous system and the central nervous system. Rat brain myelin contains several major glycoproteins in addition to those (mol. wt. 1.2×10^5 and 2.7×10^4) already characterized.[463] Four additional major glycoproteins and many minor glycoproteins are revealed after treatment of isolated myelin membrane with neuraminidase before oxidation with D-galactose oxidase and subsequent reduction with sodium borotritide. A glycoprotein soluble in organic solvents isolated from rat myelin has been partially purified and characterized as a sulphated L-fucosyl glycoprotein.[464]

The major glycoprotein (PO) of chick sciatic-nerve myelin has been purified and partially characterized.[465] The amino acid composition indicates a definite species difference when compared with mammalian PO glycoproteins.

Dopamine β-hydroxylase (EC 1.14.2.1) is a tetrameric glycoprotein with subunits of mol. wt. 7.5×10^4.[466] It is present in catecholamine storage vesicles of both adrenal medulla and sympathetic nerve terminals. The presence of neuraminic acid residues on oligosaccharide chains greatly increases the resistance of the enzyme to tryptic proteolysis.

An age-related increase of non-enzymatic glycosylation in normal bovine lens crystallins has been observed.[467] It appears to be a general characteristic of the ageing process.

Bovine thyroglobulin has been subjected to sequential treatment with glycohydrolases in order to define the components of the oligosaccharide chain which are important in binding the glycoprotein to bovine thyroid membranes.[468] Preparations of asialogalactothyroglobulin exhibit the best binding, suggesting that exposed 2-acetamido-2-deoxy-D-glucose residues on the B carbohydrate chain of the hormone are important in the interaction with the thyroid membranes. L-Tyrosine residues play an important role in thyroglobulin interactions with thyroid membranes.[469] The attachment of carbohydrate to thyroglobulin involves an intermediate (D-Glcp_3-D-Manp_9-D-Glc$p$NAc$_2$-pyrophosphoryldolichol) and is followed rapidly by the initiation of modifying reactions including the removal of D-glucosyl- and some D-mannosyl residues.[470] The inhibition of incorporation of carbohydrate into thyroglobulin by tunicamycin has been reported.[471]

Parathyroid secretory protein (mol. wt. 6.7 x 10^4 and containing 18% carboydrate) has been purified from extracts of bovine parathyroid glands.[472]

Secretory protein I, a glycoprotein (mol. wt. 7.0 x 10^4) of unknown biological activity, has been isolated and partially characterized from bovine parathyroid glands.[473] It contains 2.6% carbohydrate.

The N-terminal fragment of porcine pro-opiomelanocortin, the ACTH/LPH precursor, is a major product of maturation and is glycosylated.[474] The amino acid sequence is homologous to similar peptides in ox and man, with two sites of glycosylation probably at L-Thr-45 and L-Asn-65.

Some of the relationships between protein synthesis, glycosylation, cleavage, and secretion of pro-ACTH-endorphin in mouse pituitary tumour cells have been described.[475]

β-D-Glucuronidase has been isolated from rat preputial gland.[476] Two oligosaccharide alditols (16-17), released by reductive alkaline hydrolysis, have been characterized.

D-Man$_3$-α-{(α-D-Manp-D-GlcpNAc)-D-Manp}-β-D-GlcpNAc-β-D-GlcNAc-ol
(16)

D-Manp$_{3-4}$-α{D-Manp$_2$-α-D-Manp}-β-D-GlcpNAc-β-D-GlcNAc-ol
(17)

Glycoproteins and pepsinogen are associated in the secretory granules of fundic epithelial cells isolated from guinea-pig stomach.[477]

Incorporation of monosaccharides into rat liver ferritin can occur non-enzymically under physiological conditions of pH and ionic strength, in a process which is dependent on the concentration of monosaccharides, the concentration of ferritin, time, and temperature.[478] Varying degrees of glycosylation might account for the occurrence of isoferritins.

The structure and dynamic behaviour of the oligosaccharide side chain of bovine pancreatic ribonuclease B have been studied by ^{13}C n.m.r. spectroscopy.[479] The conclusions about the primary structure of the carbohydrate side chains are consistent with results based on chemical methods.

Chemical analyses, together with histochemical assessments,

have been obtained for specimens of adenocarcinoma of the colon and histologically normal colonic epithelium.[480] In epithelial glycoproteins of normal tissue, the majority of the neuraminic acid residues contain a side-chain O-acyl substituent at C-8, whereas side-chain substitution of the neuraminic acids of tumour glycoproteins is markedly reduced. Differences in the susceptibility to neuraminidase treatment were also noted between the two sources of glycoproteins.

Human colon tumour antigen which is recognized by leucocyte adherence inhibition assay is unrelated to carcinoembryonic antigen (CEA).[481] The former antigen, but not CEA, binds specifically to immobilized human β_2-microglobulin antibodies.

Blood-group determinants have been found on five CEA preparations.[482] One of the preparations has an A_1 determinant, another has a B determinant, and all have H, Le^a, Le^b, and MN blood-group determinants. The findings are consistent with the idea that incomplete or unexpected glycosylation patterns occur in glycoproteins produced by tumour cells. Since antibodies directed against blood-group substances cross-react with carbohydrate determinants on CEA, clinical determinations of CEA or anti-CEA levels in serum may be adversely affected. Concurrent measurements of carcinoembryonic antigen, D-glucose phosphate isomerase, γ-glutamyltransferase, and lactate dehydrogenase in malignant, normal adult, and fetal colon tissues give results that are consistent with earlier observations that the activities of these markers are increased significantly in the serum of patients with metastatic colon cancer.[483] Carcinoembryonic antigen-related antigens have been purified and characterized in normal adult faeces.[484]

Glycoproteins metabolically labelled with 2-amino-2-deoxy-D-{^{3}H}-glucose and isolated from human renal cancer and normal kidney epithelial cell cultures have been compared by two-dimensional polyacrylamide gel electrophoresis.[485] Although overall profiles of glycoproteins from tumours and normal cells were extremely similar, several significant differences were observed that were consistent even among allogeneic comparisons.

A glycoprotein inhibitor (mol. wt. 1.33×10^4) of calcium oxalate monohydrate crystal growth has been isolated from human kidney tissue culture medium.[486] This glycoprotein contains 39.4% carbohydrate, consisting of residues of L-fucose, D-glucose, D-galactose, and neuraminic acid.

Studies at the ultrastructural level indicate that the

glycoprotein GP-2 is both a cell membrane and a basement membrane protein in mouse kidney cells.[487]

L-Fucose can be incorporated into a series of low-molecular-weight amino acid L-fucosides in normal rat kidney cells.[488] The monosaccharide sequence and position of the linkage of one of the L-fucosides (18) and the anomeric configurations of the carbohydrate linkages of three other L-fucosides (19-21) have been identified.[489] Tentative evidence is given to suggest that these unusual glycosides are derived from an L-fucoprotein.

β-D-Glcp-(1→3)-α-L-Fucp-1-L-Ser
(18)

α-L-Fucp-1→L-Thr
(19)

α-L-Fucp-1→L-Ser
(20)

β-D-Glcp(1→)-α-L-Fucp-1→L-Thr
(21)

A particulate folate-binding protein from human placenta has been isolated and characterized.[490] Immunofluorescent studies using antibody directed against this glycoprotein demonstrate that an immunologically similar protein is present on the plasma membrane of human erythrocytes, suggesting that it may function as a folate receptor.

Some physicochemical properties of an acidic glycoprotein secreted by rat epididymal cells have been characterized.[491] The carbohydrate moiety contains residues of D-mannose, D-galactose, 2-acetamido-2-deoxy-D-glucose, neuraminic acid, and, unexpectedly, D-glucose.

From amino acid sequence studies of glycosylated component C_3 of rat prostatic binding protein, a glycopeptide containing 77 amino acids has been characterized.[492] An oligosaccharide chain is attached to the peptide by an N-glycosidic bond to L-Asn-17.

Bovine cartilage matrix glycoprotein (mol. wt. 1.48 x 10^5) contains three subunits.[493] Glycopeptides isolated after proteolytic digestion are heterogeneous on gel chromatography and contain D-mannosyl and 2-acetamido-2-deoxy-D-glucosyl residues, probably derived from oligosaccharides of the high-D-mannose type bound to protein by N-glycosidic linkages.

Two non-collagenous insoluble structural glycoprotein fractions from uncalcified canine-puppy rib cartilage are both associated with lipid.[494] The complexes formed are stable enough to be of possible significance in the metabolism of connective tissues.

An intrinsic glycoprotein (mol. wt. 5.3×10^4) of the sarcoplasmic reticulum of rabbit skeletal muscle has been separated from the Ca^{2+} + Mg^{2+} adenosine triphosphatase.[495] The glycoprotein contains 48% nonpolar amino acids in addition to two glycan chains, each containing nine D-mannosyl and two 2-acetamido-2-deoxy-D-glucosyl residues.

A sulphated glycoprotein, entactin (mol. wt. 1.58×10^5), has been isolated from an extracellular basement membrane-like matrix elaborated in cell culture by a mouse endodermal cell line.[496] Immunoelectron microscopic studies show that the mouse kidney antigen is predominantly localized at the surface of epithelial cells of tubules and glomeruli adjacent to the basement membrane.

Laminin, present in the lamina lucida of the basement membrane, is cleaved by thrombin and plasmin.[497] Fragments cleaved by the enzymes have been assessed for their ability to mediate binding of epithelial cells to type IV collagen. Proteolysis of laminin from human placental and renal basement membranes has been achieved with pepsin.[498] The binding of laminin to glycosaminoglycans (aggregating and non-aggregating subsets of heparan sulphates and dermatan sulphates as well as heparin, chondroitin sulphates, and hyaluronic acid) has been studied by affinity chromatography.[499] The binding is specific and occurs on a single binding site of laminin. Studies on the biosynthesis of laminin by murine parietal endoderm cells have been reported.[500] Evidence is given for the presence of N-linked oligosaccharide side chains on all four polypeptide components of this glycoprotein.

Two extracellular matrix glycoproteins, similar to laminin, have been isolated from cultures of a mouse endodermal cell line.[501]

A structural glycoprotein from bovine ligamentum muchae exhibits a dual amine oxidase activity.[502] If exposed to copper ions the glycoprotein is capable of forming fibrils about 11 nm in diameter and of the same size as the microfibrils associated mainly with elastic fibres in both embryonic and mature tissues.

Modifications in the levels of glycosylamines and crosslinks observed in glomerular basement membranes in streptozotocin diabetic rats have been reported.[503] Enhanced O-glycosylation of intermolecular crosslinks and increased formation of glycosylamine could affect the physiocochemical properties of the membrane.

A glycoprotein (mol. wt. 1.9 x 10^5) is secreted by aortic endothelial cells in culture.[504] Immunological and structural studies indicate that the glycoprotein is either identical to or closely related to thrombospondin, which is contained in platelet granules and released in response to thrombin-induced aggregation.

Antibodies to human serum amyloid P bind immunospecifically to the peripheral microfibrillar mantle of elastic fibres in skin and blood vessels of normal adults.[505]

A sialoglycoprotein fraction from herring eggs contains both alkali-labile and alkali-stable carbohydrate units.[506] Each peptide chain has an average of four oligosaccharide chains O-glycosidically linked *via* 2-acetamido-2-deoxy-D-galactosyl and L-threonine residues.

A probable structure of the glycan portion of a sialoglycopeptide fraction isolated from the skin of the fish *Labeo rohita* has been suggested from methylation analysis and enzymic analysis data.[507]

The protein composition of isolated human platelet α-granules has been characterized.[508] Crossed affinity immunoelectrophoresis using lectins identified at least seven glycoproteins, six of which appeared to be sialoglycoproteins. The α-granule component is the origin of their enhanced haemagglutination activity.[509] An insufficiency of the platelet-bound lectin may be the cause of the inability of gray platelets to aggregate normally in response to thrombin.

Thrombospondin is one of the glycoproteins released from human blood platelets in response to thrombin.[510] A hydrodynamic model of thrombospondin is proposed from physical solution measurements of molecular weight, partial specific volume, sedimentation coefficient, and intrinsic viscosity.

Platelet glycoproteins have been shown to be useful phenotypic markers for cells of megakaryocyte lineage.[511] Immunofluorescent staining with an antiserum to highly purified platelet glycoproteins exclusively labels human megakaryocytes, platelets, and a population of small human bone-marrow mononuclear cells.

Differences between human leucocyte and fibroblast interferons have been reviewed.[512] A book on interferon includes a chapter dealing with the purification and characterization of interferons.[513]

Human fibroblast interferon has been purified by zinc-chelate chromatography,[514] as well as by chromatography on Blue Sepharose^{R}.[515] Evidence for the existence of this interferon as a dimer (mol.wt. 4.0 x 10^4) as well as a monomer (2.0 x 10^4) is reported. A combination of hydrophobic chromatography, copper chelate chromatography, and affinity chromatography on Blue Sepharose^{R} has been used in the purification of human lymphoblastoid interferon.[516]

Size characteristics of human leucocyte interferon under reducing and non-reducing conditions have been determined.[517]

Metabolic carbohydrate labelling in the mouse system shows that not only mitogen-activated T and B lymphocytes, but some of their subpopulations can be distinguished by their specific glycoprotein labelling patterns.[518,519] Human lymphocyte subpopulations also display specific sets of glycoproteins, which are easily detectable by metabolic carbohydrate labelling and electrophoretic techniques.[520]

The binding pattern of horseradish peroxidase-conjugated peanut lectin on sections of lymphoid tissue from different sources has been investigated.[521] The binding, which is highly species-dependent, is greatest to lymphocytes in thymic cortex and germinal centres in man, mouse, and sheep. Lymphoid tissue from hamster, guinea pig, and rabbit did not react, whilst in rat only weak binding to cells in the thymic cortex and germinal centres was detected.

The oxidation of human peripheral mononuclear cells with sodium periodate results in lymphocyte activation.[522] The mitogenic effect of the oxidant is accompanied by the oxidation of membrane-associated neuraminosyl, D-galactosyl, and L-fucosyl residues.

The receptor-mediated pinocytosis of glycoproteins terminated by D-mannosyl and 2-acetamido-2-deoxy-D-glucosyl residues by macrophages has been reviewed.[523] Glycoconjugates bearing terminal L-fucosyl or D-mannosyl residues are recognized and internalized by macrophages via the same receptor that mediates plasma clearance and uptake of lysosomal glycosidases.[524] The macrophage receptor has a broad specificity and is able to bind various glycoconjugates.

In human fibroblasts, the recognition of lysosomal enzymes by cell-surface receptors is mediated by D-mannosyl 6-phosphate residues located on oligosaccharide chains on these enzymes.[525] Both cathepsin D and β-hexosaminidase contain the phosphate groups which are diester-linked to 2-acetamido-2-deoxy-D-glucosyl residues.

Developmental changes in the glycoproteins synthesized by normal and dystrophic satellite mouse cells grown in culture have been observed.[526] Similar, but not identical, L-fucosylated glycoproteins appear to be synthesized exclusively by normal multinucleated myolubes, but not by duplicating mononucleated satellite cells or by other non-myogenic cells from both sources.

Changes in glycoproteins (mol. wts. 1.0 x 10^5 and 9.0 x 10^4, respectively) in malignant transformation have been shown to be directly related to the uptake of D-glucose by these cells.[527]

The purification and partial structures of the major glycopeptides isolated from cultured human melanoma cells and from the culture media have been reported.[528] The glycopeptides obtained from the highly tumorigenic human melanoma cells are markedly different from those obtained from nontumorigenic human fetal uveal melanocytes.

The plasminogen activator secreted by human melanoma cells in culture is very similar to, or identical with, the plasminogen activator found in normal tissue, but is different from urokinase.[529] The glycoprotein (mol. wt. 7.2 x 10^4), purified using immobilized concanavalin A, exists as a dimer. Receptors for Dolichos biflorus agglutinin on embryonal carcinoma cells have been isolated and shown to be glycoproteins (mol. wt. >7.0 x 10^4).[530] The lectin does not bind to differentiated cells. Differences in the content of neuraminic acid of low cancer cells, high cancer cells, and normal mouse lung counterparts have been observed.[531] Normal lung cells contain higher neuraminic acid levels per cell in comparison with the transformed ones.

Three types of mononucleosomes isolated from Ehrlich ascites nuclei have been characterized.[532] Specific glycoproteins are associated with two of the mononucleosome types with the principal glycoprotein having an apparent mol. wt. of 1.3 x 10^5.

A carbohydrate-rich, water-soluble glycoprotein (mol. wt. 4.5 x 10^4) has been isolated from delipidated lung lavage fluid from a patient with pulmonary alveolar proteinosis.[533] The very high carbohydrate content is made up of residues of neuraminic acid (24), 2-acetamido-2-deoxy-D-glucose (23), 2-acetamido-2-deoxy-D-galactose (6), D-galactose (19), D-mannose (4), L-fucose (1), and D-glucose (1), per molecule of glycoprotein. Unlike a number of collagen-related glycoproteins which have been isolated from insoluble lung contents in pulmonary proteinosis, this glycoprotein does not contain hydroxy-L-proline or hydroxy-L-lysine.

```
α-Neup5Ac-(2→6)-β-D-Galp-(1→4)-β-D-GlcpNAc-(1→2)-α-D-Manp-(1→3)-β-D-Manp-(1→4)-β-D-GlcpNAc-(1→4)-
                                                            6              β-D-GlcpNAc-1-L-Asn
                                                            |                   |
                                                            1                   R2
                   R1-β-D-Galp-(1→4)-β-D-GlcpNAc-(1→2)-α-D-Manp
                                              (22)  R1 = H, R2 = H
                                              (23)  R1 = α-Neup5Ac, R2 = α-LFucp-(1→6)
```

A glycoprotein (mol. wt. 6.2 x 10^4), isolated from the lung lavage material of patients with alveolar proteinosis, contains 72 residues of glycine and 5 residues of hydroxy-L-proline per molecule, in addition to D-mannose, D-galactose, L-fucose, 2-acetamido-2-deoxy-D-glucose, and neuraminic acid in the molar ratio of 5:4:1:7:3.[534] Partial amino acid sequence analysis of peptides derived from the digestion of the glycoprotein indicates the presence of alternating collagenous and non-collagenous regions in the same polypeptide chain.

A human alveolar glycoprotein releases two major glycopeptide fractions on proteolysis.[535] From structural analytical studies the tentative structures (22-23) have been proposed.

Human α_1-antichymotrypsin has been purified from human pleural fluid and from human serum.[536] The glycoprotein (mol. wt. 5.8 x 10^3) isolated from the two sources has the same chemical composition.

A glycoprotein (mol. wt. 3.6 x 10^3) with similar properties to collagen has been isolated from the lung lavage of normal rabbit.[537]

Blood-group-specific glycoproteins obtained from ovarian-cyst fluids of A_1 and A_2 persons have been submitted to alkaline borohydride elimination, de-N-acetylation, partial acid hydrolysis, and subsequently re-N-acetylation.[538] The trisaccharides (24-26) were characterized from both the A_1 and A_2 materials. Oligosaccharide (26) has not previously been detected in human glycoproteins.

α-D-GalpNAc-(1→3)-β-D-Galp-(1→3)-D-GlcNAc
(24)

α-D-GalpNAc-(1→3)-β-D-Galp-(1→4)-D-GlcNAc
(25)

α-D-GalpNAc-(1→3)-β-D-Galp-(1→3)-GalNAc-ol
(26)

Lubricating glycoprotein 1 has been prepared from bovine synovial fluid.[539] Electron-microscopic studies show that the glycoprotein is an asymmetric molecule. The molecular model that best fits the molecular dimensions (222 nm long and 1-2 nm diameter) is that of a partially extended flexible rod.

Human amniotic fluid contains an irreversible tissue inhibitor of collagenase.[540] Although this glycoprotein is closely associated with fibronectin, it does not cross-react with antibodies to fibronectin.

The microheterogeneity of the different carbohydrate chains of purified rat α-fetoprotein variants has been studied.[541] The methodology used includes the release of carbohydrate chains from the glycoprotein by hydrazinolysis, the labelling of glycans by re-N-acetylation with {^{14}C}-acetic anhydride, the fractionation of the labelled oligosaccharides on immobilized concanavalin A, and characterization of the labelled glycans by t.l.c. α-Fetoprotein has been isolated from human ascites fluid.[542] Oligosaccharides (27-29) were released from the polypeptide chain by hydrazinolysis.

Human enterokinase (enteropeptidase EC 3.4.21.9), purified from accumulated duodenal fluid, contains three glycosylated subunits (mol. wts. 5.4×10^4, 1.02×10^5, and 1.4×10^5).[543] The smallest subunit contains the active-site L-serine residue, and the oligosaccharide chains of this unit appear to be N-glycosidically linked.

The NAD glycohydrolase from Bungarus fasciatus (banded krait) venom has been purified to electrophoretic homogeneity.[544] The enzyme is a glycoprotein (mol. wt. 1.2×10^5) containing two subunits.

9 Cell-surface Glycoproteins

Two recently published treatises have included chapters dealing with carbohydrates and cell membranes,[545] and with the structures and functions of the carbohydrate moieties of membrane glycoproteins and glycolipids.[546] Reviews dealing with glycosyltransferases and T-cell recognition,[547] and with glycoconjugates of surface membranes of normal and cancerous cells,[548] have been published. The importance of glycoconjugates in the organization of the cell membrane and the multiple roles played by these membrane constituents in malignant transformations have been reviewed.[549] A detailed account of histochemical methods for characterizing complex cell-surface glycoconjugates by light microscopy has been published.[550]

Electron microscopy has been used to map the loci of immunochemical active sites on individual glycoprotein molecules.[551] This involves the complexing of a group-specific macromolecule, such

```
R1-β-D-Galp-(1→4)-β-D-GlcpNAc-(1→2)-α-D-Manp-(1→6)-β-D-Manp-(1→4)-β-D-GlcpNAc-(1→4)-D-GlcNAc-ol
                                                        3                                         6
                                                        |                                         |
                                                        1                                         1
        R2-β-D-Galp-(1→4)-β-D-GlcpNAc-(1→2)-α-D-Manp                                      (α-L-Fucp)n
```

n = 0 or 1

(27) $R_1 = R_2 = H$

(28) $R_1 = H$, R_2 = Neu5Ac-(2→6)

(29) $R_1 = R_2$ = Neu5Ac-(2→6)

as a lectin or an antibody, with an asymmetric glycoprotein followed by the direct visualization of the complex in the electron microscope.

A procedure for determining membrane purity of isolated cells has been reported.[552] Intact cells are labelled with peroxidase-conjugated wheat-germ agglutinin. After disruption of the intact labelled cells, plasma membrane purification can be monitored by ultrastructural histochemical examination of the peroxidase-conjugated lectin.

Histochemical fixation of cell glycoproteins using periodate oxidation, followed by complexing with L-lysine and paraformaldehyde, has previously been assumed to crosslink specifically plasma membrane glycoproteins.[553] The results of an evaluation using Novikoff rat ascites hepatocellular carcinoma cells suggest a complex interaction between periodate-oxidized glycoproteins and polymeric complexes of formaldehyde and L-lysine.

Cell-surface glycoproteins, but not glycolipids, from erythrocytes, colonic tumour cells, and skin fibroblasts, bind to immobilized *Ricinus communis* agglutinin.[554] The lack of binding of the solubilized glycolipids to the immobilized lectin could be due to the fact that their monovalency results in too low an affinity for the lectin to cause retention on the column.

Apparent minor changes in cell-preparation methods have been found to result in major alterations in cell behaviour, reflecting differences in cell-surface properties of rat hepatocytes.[555]

Plasma membrane glycoproteins from rat hepatocytes that had been exposed to subendothelial spaces of Disse have been selectively labelled with D-galactose oxidase-sodium borotritiide. Approximately forty glycoproteins were resolved by two-dimensional electrophoresis.

Different cell-surface glycoproteins seem to be involved in the initial reactions of cell-cell and cell-collagen adhesion of rat hepatocytes.[557]

Glycopeptides obtained by pronase digestion of rat liver plasma membranes have been fractionated by affinity chromatography on immobilized concanavalin A.[558] The primary structure (29) of a biantennary glycan from one of the fractions has been identified by high-resolution ^{1}H n.m.r. spectroscopy and methylation analysis.

The kinetic behaviour of trypsin in non-covalent electrostatic interaction with a sialoglycopeptide fraction isolated from major hepatoma cell surfaces has been reported.[559]

Immunological evidence for the transmembrane nature of the rat liver receptor for asialoglycoproteins has been presented.[560] The receptor is exposed on the cytoplasmic, as well as the external, surfaces of the hepatocyte plasma membrane, and determinants on each surface are antigenically distinguishable. The biosynthesis and insertion of the receptor into endoplasmic reticulum membranes have been studied using antibody monospecific for the binding protein.[561] The binding protein is exclusively synthesized on the membrane-bound ribosomes and spans the microsomal membrane, probably exposing the C-terminal segment on the cytoplasmic surface and the N-terminal segment containing carbohydrate moieties on the luminal surface.

Demonstrating a new route of clearance of glycoproteins from rat plasma, a binding protein specific for D-mannosyl and/or 2-acetamido-2-deoxy-D-glucosyl residues has been isolated from rat liver.[562] This protein is analogous to the binding protein specific for asialoglycoproteins, which mediates the original route of clearance, in its predominant localization in the liver and in its requirement of calcium binding.

The asialoglycoprotein receptor, requiring Ca^{2+} for ligand binding, has been identified on a continuous human hepatoma cell line, Hep G2.[563] There are approximately 150,000 ligand molecules bound per cell at 4°C.

Differential scanning calorimetry has been used to examine the thermal denaturation of rabbit hepatic D-galactosyl-binding protein.[564] Ca^{2+} and ligand binding influence the denaturation process.

Human asialotransferrins, types 1, 2, and 3, and rabbit asialotransferrin exhibit unequal binding with purified rat liver plasma membranes.[565] The asialoglycoprotein receptor on cultured rat hepatocytes mediates the toxic effects of a conjugate of asialofetuin and fragment A of diphtheria toxin.[566] Asialoglycoproteins but not the corresponding sialoglycoproteins are able to block the action of the conjugate.

Evidence for the *in vivo* hepatic uptake by the parenchymal cells of a D-galactosyl-terminated glycoprotein (asialofetuin) in fish (*Salmo alpinus*) has been reported.[567] The lectin-dependent recognition of desialylated erythrocytes by Kupfer cells has been discussed.[568]

The specificity and kinetics of endocytosis of glycopeptides and glycoproteins by isolated rat reticuloendothelial cell preparations have been examined.[569] The cell receptor is highly selective in its recognition of oligosaccharides which are subsequently internalized, and the minimum structure (30) required for recognition and endocytosis is proposed.

```
R-(1→6)-α-D-Manp-(1→6)-β-D-Manp-(1→4)-β-D-GlcpNAc-(1→4)-
                               3               β-D-GlcpNAc-L-Asn
                               |
                               1
                          α-D-Manp
```

(30) R = α-D-Manp or β-D-GlcpNAc

The cell-surface glycoconjugates of baby-hamster kidney cells and of four ricin-resistant variants have been investigated by the use of ^{125}I-substituted ricin which binds to D-galactosyl residues, and by the use of lectins which bind to neuraminosyl and 2-acetamido-2-deoxy-D-glucosyl residues.[570] Major differences in the number of lectin and/or ricin receptors were found between the cell surface of wild-type cells and that of ricin-resistant variants. The ricin resistance of variant cells is concomitant to modifications of the concentration of certain glycoconjugate structures which are accessible to the sugar-binding proteins.

Monkey kidney cells infected with Yaba tumour pox virus exhibit plasma membrane alterations when treated with both fluorescein-labelled and unlabelled concanavalin A.[571]

A combination of immunological and biochemical methods have been used to identify surface membrane components involved in cell-substratum adhesion.[572] Highly purified adhesion-related materials consisting of integral membrane glycoproteins (mol. wt. 1.4×10^5) that are exposed to the extracellular environment were isolated. Their role in the process of cell-substratum adhesion is not yet understood.

Mutant Chinese-hamster ovary cells selected for resistance to a conjugate of ricin and ω-(6-phospho)-D-mannopentaose have altered D-mannose 6-phosphate receptors.[573] These mutants exhibit reduced uptake and altered binding of exogenously added acid hydrolase. The mutants secrete two to six times more α-D-mannosidase, β-D-glucuronidase, and L-fucosidase than the parent cells.[574] This elevated secretion is not due to alteration of the D-mannose 6-phosphate recognition marker on the enzymes, but appears to result from alterations in the D-mannose 6-phosphate receptor.

A review has been presented on the cell-surface properties of the glycoconjugate patterns of normal, differentiating, and neoplastic pancreatic acinar cells.[575]

Solubilization with sodium dodecyl sulphate and subsequent electrophoretic fractionation has demonstrated the heterogeneity of proteins and glycoproteins in the intestinal brush border membrane isolated from tadpole, young and adult Rana catesbeiana.[576] Glycoprotein bands that co-migrate with maltase, glucoamylase, and alkaline phosphatase activity have been identified.

Sucrase-isomaltase, an integral membrane glycoprotein that appears in intestinal microvillus membranes during differentiation of intestinal epithelial cells, has been purified on immobilized monoclonal antibody-protein A columns.[577] The major part of the carbohydrate associated with the glycoprotein is in the form of complex oligosaccharides linked to L-asparagine.

Lectin binding to sealed and disrupted disc membranes of vertebrate retinal rod cells supports the view that rhodopsin is a transmembrane protein with its oligosaccharide components oriented toward the interior of the discs.[578] Indirect evidence favouring the transmembrane disposition of a rod outer–segment disc glycoprotein is also considered.

A human spermatozoal antigen, containing residues of neuraminic acid, amino sugars, and hexose, has been purified.[579] The antigen reacts with immunoglobulin G and immunoglobulin M antibodies and with both types of antisera.

The major glycoprotein on the plasma membrane of rat testicular spermatozoa (mol. wt. 1.1×10^5) has been isolated.[580] As spermatozoa pass through the epididymis, this glycoprotein disappears and is replaced by another glycoprotein (mol. wt. 3.2×10^4), which is a major component of epididymal secretions.

A glycoprotein (mol. wt. 1.6×10^5) containing residues of D-galactose, D-mannose, 2-acetamido-2-deoxy-D-glucose, and 2-acetamido-2-deoxy-D-galactose has been identified as a cell-surface receptor of boar spermatozoa for concanavalin A.[581]

Enzymic removal of cell-surface neuraminic acid leads to an unmasking and stimulation of gonadotrophin-receptor activity in plasma membranes of bovine corpus luteum.[582] A relationship between gonadotrophin-receptor interaction and glycoprotein-bound neuraminic acid has been demonstrated.

The sodium channel saxitoxin-binding component of rat hind-limb sarcolemma contains residues of D-mannose, D-galactose, 2-acetamido-2-deoxy-D-galactose, 2-acetamido-2-deoxy-D-glucose, and neuraminic acid.[583] The toxin-binding site is spatially separated from binding sites for lectins such as concanavalin A, wheat-germ agglutinin, and *Ricinus communis*.

The binding sites for wheat-germ agglutinin, *Ricinus communis* agglutinin, and concanavalin A on mouse neuroblastoma cell membranes have been identified.[584] Concanavalin A binds to over twenty different glycopeptides, most with molecular weights greater than 5.0×10^4, but four of these glycopeptides were identified as the major cell-surface receptors for the lectin.

Detergents have been used to solubilize transferrin-transferrin receptor complexes from rabbit reticulocyte membrane.[585] Estimates of the molecular weight of the transferrin receptor depend on the conditions of electrophoresis. It is suggested that the receptor is partially modified, perhaps by glycosylation.

A soluble haemagglutination activity has been obtained from teratocarcinoma stem cells.[586] This haemagglutinin is sensitive to trypsin and is inhibited by specific carbohydrates such as D-mannans and L-fucans, indicating that it is a lectin. The carbohydrate- and erythrocyte-binding specificities of the solubilized lectin are very similar to that of the cell-surface carbohydrate-binding component on intact stem cells, suggesting identity of the two components.

The effect of dexamethazone on cell-surface glycoprotein accumulation in HeLa cells is related to increased sugar-linked dolichol synthesis.[587]

The major plasma membrane sialoglycoprotein of 13762 rat ascites mammary adenocarcinoma cells is released from non-xenotransplantable and xenotransplantable sublines by proteolytic cleavage.[588] The sialoglycoprotein probably binds to the membrane *via* a hydrophobic portion of the polypeptide which remains with the membrane on release.

Evidence has been presented for the *in vivo* association of two cell glycoproteins of mammary ascites tumour cells, based upon studies on the redistribution of fluorescent lectins.[589] Redistribution of concanavalin A receptors was observed in the presence of peanut agglutinin, but redistribution of peanut agglutinin receptors was observed only upon treatment of the cells with concanavalin A.

A pea lectin receptor from 6C3HED murine ascites tumour cells has been degraded by trypsin to release a glycopeptide (mol. wt. 2.0×10^5).[590] This high-molecular-weight glycopeptide is absent from non-specific protease digests of viable cells. Sequential lectin affinity chromatography of detergent extracts of cells has permitted the probable identification of the native glycoprotein.

Cell-free matrix material from human fibroblast cultures contains a cell-surface glycoprotein (mol. wt. 1.4×10^5) which seems to be closely associated with the pericellular fibronectin.[591] Like fibronectin, this glycoprotein can form an extensive disulphide crosslinked structure, possibly with a second glycoprotein component (mol. wt. 2.5×10^5).[592] The possibility that covalent and/or non-covalent bonds bind all three glycoproteins in a matrix was discussed.

Rat Thy-1 antigen, a glycoprotein of 111 amino acid residues, is a major membrane molecule of thymocytes and brain.[593] The full amino acid sequence of rat brain Thy-1 glycoprotein has been reported. An apparent sequence and structural homology with immunoglobulin was observed. Evidence for an evolutionary relationship between Thy-1 antigen and immunoglobulins has been presented and a hypothesis for the functional significance of the Thy-1 antigen proposed.[594] The antigen exists in two allelic forms, either Thy-1.1 or Thy-1.2.[595] The Thy-1.2 antigen has been identified by monoclonal-antibody techniques and shown to be located on the Thy-1 glycoprotein.

Metabolic carbohydrate labelling for the analysis of glycoproteins from activated human lymphocytes has been used in a study of the responding cell population.[596] Numerous glycoproteins, isolated from lymphocyte subsets after stimulation with mitogen and alloantigen, have been detected. Differences in the carbohydrate moieties of high-molecular-weight glycoproteins from human lymphocytes of T and B origins have been demonstrated using monoclonal antibodies with anti-I and anti-i specificities.[597] The binding of chicken anti-(bovine α_2-macroglobulin) antibodies to Namalva lymphoblastoid cells has been demonstrated.[598] The possible role of the glycoprotein on the cells is discussed.

Treatment of cultured human lymphocytes with tunicamycin results in a decrease of incorporation of carbohydrate into protein, together with a decrease in the binding of insulin and growth hormone to the cells.[599] For the hormones, the decreases are mostly in receptor-binding capacity and not affinity. Antibodies directed at surface glycoproteins inhibit cytolytic but not suppressor functions of human T lymphocytes.[600] A monoclonal antibody, specific for the 2.2×10^5 molecular weight form of T200 (the major cell-surface glycoprotein on lymphoid cells), has been used to show that the glycoprotein is expressed only on B cells and a subset of bone-marrow cells which includes B cell precursors.[601]

A comparison of binding sites for wheat-germ agglutinin on Raji lymphoblastoid cells and their isolated nuclei and plasma membranes indicates similarities in the lectin receptors on the outer surface of lymphoblastoid cells and the cell nuclei.[602] Differences were obtained with isolated membranes, and these may be due to inversion of the membrane vesicles or to their decreased rigidity as compared with the intact cell.

Periodate oxidation, borohydride reduction, and acid hydrolysis of human peripheral lymphocytes, but not mouse splenocytes or calf lymph node cells, release glycerol, propan-1,2-diol, and the seven carbon analogue of neuraminic acid.[603] These fragments probably arise from D-galactosyl, L-fucosyl, and N-acetylneuraminosyl residues. A subpopulation of chicken B lymphocytes that reacts with the L-fucose-specific lectin from *Lotus tetragonolobus* has been identified.[604] The distribution of the lectin-reactive cells varies with the age of the chicken. Differential glycosylation of murine B cell and spleen adherent cell tA antigens has been reported.[605]

Three predominant glycoproteins have been isolated from rat thymocyte plasma membranes.[606] Two of the glycoproteins have a carbohydrate composition characteristic of N-glycosylated proteins. The other glycoprotein, containing about twenty O-glycosylated carbohydrate units per 100 amino acids, shows structural similarities to glycophorin, the major sialoglycoprotein of human erythrocytes. The monoclonal antibodies anti-T1 and anti-T3 both react with all human peripheral thymus-derived lymphocytes and with 10% of thymocytes.[607] Each, however, recognizes different cell-surface structures, with the target antigen of anti-T1 being a glycoprotein (mol. wt. 6.9×10^4) and that of the anti-T3 being a glycoprotein (mol. wt. 1.9×10^4). A human cell-surface glycoprotein (mol. wt. 1.0×10^5) that is selectively expressed by proliferating human leukaemic thymus-derived cells has been identified as a receptor for serum transferrin.[608] This glycoprotein contains both complex and high D-*manno*-oligosaccharides linked to L-asparagine.[609] Glycosylation is not an absolute requirement for the receptor to act as an acceptor for fatty acid, nor for transport to the cell surface.

Two Qa-1 antigens from mice splenocytes have been isolated from both biosynthetically labelled cells and from surface iodinated cells.[610] These glycoproteins have been shown to be distinct from

TL antigens.

Primary structures of the transplantation antigen of the murine H-2 major histocompatibility complex have been reviewed.[611] Partially purified, papain-solubilized mouse liver $H-2^a$ histocompatibility antigens have been purified further by reverse-phase h.p.l.c. under acid conditions, yielding β_2-microglobulin and $H-2^a$ heavy chains.[612] An amino acid sequence of the first 98 amino acid residues at the N-terminus of the murine major histocompatibility antigen [613] and of 98 residues of the murine histocompatibility antigen $H-2K^d$ has been determined.[614] Comparison of the sequences with $H-2K^b$ and $H-2D^b$ antigens reveals extensive localized differences. Comparison of the primary structures of murine H-2 antigens and human HLA molecules suggests an evolutionary origin of major histocompatibility products.[615] Data presented on the amino acid sequence of an intramembranous hydrophobic segment of the $H-2K^b$ murine histocompatibility antigen[616] and of the C-terminal hydrophobic region [617] have provided the complete primary structure of the glycoprotein.

Evidence for the existence of three carbohydrate prosthetic groups on mouse histocompatibility antigens $H-2k^d$ and $H-2D^b$ has been reported.[618]

Purified HLA-A2 antigen from lymphoblastoid cell membranes which has been fluorescently labelled specifically in its C-terminal region interacts with lymphoblastoid cytoskeletal proteins when recombined *in vitro*.[619] This system allows study of the interactions of a membrane protein with cytoskeletal elements, and may be used to explore the structural basis and regulation of such interactions.

A number of unusual properties of the invariant (Ii) chain of the murine Ia antigens has been described.[620] While all of the Ia polypeptide chains possess N-linked oligosaccharide units, they also show that the invariant chain differs from the polymorphic chains in both the number and type of carbohydrate units. Cell-surface HLA-DR, HLA-ABC antigens, and glycophorin are all expressed at different stages during erythroid differentiation.[621]

In the biosynthesis of HLA antigens in two lymphoblastoid cell lines, Daudi and Raji, the antigen heavy chains in Daudi cells are synthesized normally, but, although core glycosylation takes place, terminal glycosylation does not.[622] The biosynthesis of the heavy chains of mouse lymphoblastoid histocompatibility antigens and of immunoglobulin M involves the same intracellular pathway as secretory proteins.[623] These polypeptides pass through the Golgi subsite, defined by monensin, and acquire terminal neuraminic acid residues distal to this site. _In vitro_ translation studies on the biosynthesis of HLR-DR antigens indicate that the α- and β-chains are synthesized as precursors with signal sequences, in common with many other membranes and secreted glycoproteins.[624] In the presence of a heterologous microsomal system, the signal peptide is cleaved and 'core' oligosaccharide units are added.[625] The events involved in the biosynthesis and maturation of the antigens _in vivo_ have also been described.

The histocompatibility Y (H-Y) antigen is a minor histocompatibility antigen detected on the cell surface from the heterogametic sexes of mammals, birds, amphibian, and invertebrate species.[626] The serological determinant of the antigen resides in the carbohydrate portion of the glycoprotein.

Three cell-surface antigens, two glycoproteins and a protein, structurally related to the major human histocompatibility antigens have been characterized from the leukaemic T cell line MOLT-4.[627] One antigen is a glycoprotein (mol. wt. 4.9×10^4) and is associated with β_2-microglobulin. The other glycoprotein (mol. wt. 4.3×10^4) is also associated with β_2-microglobulin but is serologically distinct from the higher-weight glycoprotein. The present state of knowledge of the primary structure of the murine H-2 alloantigens has been reviewed.[628] A speculative model of the antigen on the membrane is presented, and interpretations of the results of primary structural comparisons are used to probe possible answers to the questions of the generation of polymorphism and genetic relationships.

Wax-bean agglutinin mimics the activities of insulin by mediating antilipolysis and the oxidation of D-glucose.[629] It does not appear to compete with the binding of the hormone to its receptor, but exerts its insulin-like action _via_ membrane glycoproteins which are distinct from insulin-binding sites. The possibility that a glycoprotein effector molecule is an integral part of the hormonal mediated system and that the lectin-mediated oxidation of D-glucose and antilipolysis may be triggered _via_ different glycoprotein receptors is discussed.

The previously observed inhibition by concanavalin A of insulin binding to intact rat fat cells may be due to the interaction of the lectin with a carbohydrate-containing moiety on the cell membrane distinct from, but capable of modifying, the insulin receptor as previously postulated.[630] Insulin receptors from human placenta and cultured IM-9 lymphocytes display similar specificites for twelve immobilized lectins.[631] Based on binding studies with lectins the carbohydrate moiety of the receptor contains 2-acetamido-2-deoxy-D-glucosyl, α-D-mannosyl, and β-D-galactosyl residues. The increase in affinity of the insulin receptor for insulin after its desorption from lectins may be due to the removal of an associated inhibitor.

Tunicamycin, which inhibits N-linked oligosaccharide chain addition to nascent polypeptides, interrupts glycosylation of the insulin receptor in 3T3-L1 adipocytes giving rise to inactive receptors.[632] The inactive aglycoinsulin receptor accumulates post-translationally during chronic treatment with tunicamycin and then re-enters the glycosylation pathway when the inhibitor is removed, giving rise to a functional insulin receptor.

A guinea-pig peritoneal macrophage surface glycoprotein (mol. wt. 1.6×10^5) is unique among major surface components in its sensitivity to mild trypsin treatment.[633] Neuraminic acid residue(s) were located solely in one of the fragments (mol. wt. 8.5×10^4).

Heterogeneity in surface glycoproteins of mouse peritoneal macrophage populations has been observed.[634] Upon activation of the macrophages with the culture medium of cell-free tumour ascites fluid, a new surface glycoprotein (mol. wt. 1.35×10^5) is expressed. The possibility that this molecule is merely a variation of a pre-existing protein differing in its extent of glycosylation, or even only a higher rate of synthesis of a normally minor surface glycoprotein, was not excluded.

Two previously unknown macrophage-specific antigens (mol. wts. 3.2×10^4 and 1.1×10^5) have been isolated by use of a combination of cell hybridization and removal of previously recognized antigens by immunoadsorbents.[635]

On the basis of agglutination studies with a number of lectins, guinea-pig phagocytic vesicles have been shown to have lectin receptors on their cytoplasmic side.[636] D-Galactosyl, 2-acetamido-2-deoxy-D-galactosyl, D-mannosyl, D-glucosyl, 2-acetamido-2-deoxy-D-glucosyl, and N,N'-diacetylchitobiosyl residues are probably present on the receptors.

Lectin-like receptors capable of binding *Staphylococcus albus* have been demonstrated in the membranes of phagocytes including macrophages, neutrophils, and eosinophils from various sources and species.[637] These receptors are likely to contribute to bacterial adherence and phagocytosis in the non-immune animal.

Four independent monoclonal antibodies precipitate a murine cell-surface glycoprotein which has been identified as a polymorphic differentiation antigen of murine mesenchymal cells.[638]

Treatment of melanoma cells with tunicamycin selectively inhibits the shedding of melanoma-associated glycoproteins without affecting their cell-surface expression.[639]

K-562 cells, originating from the pleural effusion of a chronic myelogenous leukaemia patient, expresses an N-linked oligosaccharide (erythroglycan, mol. wt. 7.0×10^3 - 1.1×10^4), specific cell-surface gycoprotein.[640] The early stages of erythroglycan biosynthesis are the same as those of the transferrin-type oligosaccharide.[641] However, maturation of the oligosaccharide is influenced by protein structure so that erythroglycan is only expressed on specific glycoproteins. The K-562 cells express the foetal type (L) antigen on different glycoproteins from those of erythrocytes.[642]

Carbohydrate moieties of receptors for immunoglobulin E on rat basophilic leukaemia cells and rat mast cells are not directly located in the binding site of the receptor.[643] However, the receptor carbohydrate may play a part in transport of the receptor to the plasma membrane or in its orientation thereafter. Cell-surface receptors for immunoglobulin E have been isolated from rat basophilic leukaemia cells.[644] The effect of tunicamycin on the molecular properties of the receptor have been evaluated. The usual diffuse receptor band (mol. wt. 4.5×10^4 - 6.2×10^4) observed on sodium dodecylsulphate polyacrylamide gel electrophoresis is replaced by one (mol. wt. 3.8×10^4). A number of lines of evidence indicate that this lower-molecular-weight protein represents carbohydrate-deficient receptor.

Characteristic glycoprotein profiles and carbohydrate structures are expressed on different leukaemic cell lines blocked at different stages of myeloid cell differentiation.[645] Different stages of granulocyte differentiation can be identified by specific cell-surface structures.

A murine cell-surface glycoprotein (mol. wt. 8.0×10^4) has been identified and studied by use of monoclonal antibodies.[646] The determinants of this major plasma-membrane constituent, recognized by the antibodies, are allospecific, and the expression of the glycoprotein is specific to certain types of cells. During differentiation receptors for concanavalin A have been shown to be distributed in patches over the entire cell-surface neuroblastoma cells.[647]

The importance of the platelet-isolation technique to platelet recovery and platelet-membrane integrity has been stressed.[648] The loss of platelet-membrane glycoproteins during washing procedures is probably the result of membrane loss from platelets and is not due to the isolation of a selected platelet subpopulation. Membrane glycoproteins have been isolated from human platelets after affinity chromatography on immobilized wheat-germ agglutinin,[649] and by high-voltage free-flow electrophoresis.[650]

Monoclonal antibodies have been used in isolation and quantitation of the platelet-membrane glycoprotein deficient in Glanzmann's thrombasthenia.[651] The membrane glycoprotein is a complex of two platelet glycoproteins designated IIb and IIIa. Membrane glycoproteins of platelets from normal and Glanzmann's thrombasthenic subjects are identical in terms of isoelectric points and molecular weights, and differ only in the amount of certain glycoproteins.[652] The molecular defect in platelets from patients with this disease is due to a deficiency of two membrane glycoproteins.[653] The possibility of additional membrane defects in thrombasthenic platelets may be related to another surface glycoprotein (mol. wt. 9.3×10^4). The glycoproteins of platelet membranes from normal patients and those with Glanzmann's thrombasthenia have been compared using carbohydrate-specific or protein-specific labelling techniques and high-resolution two-dimensional gel electrophoresis.[654] In Glanzmann's thrombasthenia the absence or reduction of two major membrane glycoproteins is observed, while a number of other glycoproteins contain increased levels of neuraminic acid.

The existence in human platelet plasma membranes of a macromolecular component (mol. wt. $>4.0 \times 10^6$) with the chemical composition of a phosphoglycoprotein has been confirmed.[655] The phosphoglycoprotein inhibits the ADP, epinephrine, collagen, and thrombin-induced platelet reaction with concomitant changes of the plasma-membrane structure.

Analysis of the glycoproteins and proteins of platelets of four Bernard-Soulier patients has established that these platelets possess a specific and characteristic membrane-glycoprotein lesion.[656] A comparison with the results obtained for control platelets after treatment with neuraminidase has shown that a decreased neuraminic acid content cannot alone account for the observed Bernard-Soulier surface alterations.

Receptors for bovine von Willebrand factor have been identified on whole human platelets, as well as on platelet membranes.[657]

The kinetics of formation of factor IX_a-factor VIII complex on the surface of human platelets have been reported.[658]

Thrombin-induced platelet aggregation has been inhibited by a derivative of wheat-germ agglutinin.[659] The immobilized lectin was used for the affinity-chromatographic isolation from platelet membranes of a glycoprotein (mol. wt. 7.4×10^4) capable of inhibiting platelet aggregation by thrombin. This glycoprotein may be a physiologic receptor of thrombin in human platelets and differs from the ristocetin or ristocetin—von Willebrand factor receptor. Platelet plasma membranes retain functional aggregation sites following cell lysis and membrane isolation.[660] Isolated platelet plasma-membrane glycoproteins have a greater affinity for thrombin-activated platelets than control platelets. An immunological procedure has been used to explore the concept that a redistribution of membrane receptors may be involved in human platelet secretion.[661] A non-agglutinating derivative of wheat-germ agglutinin has been prepared that binds to platelets and precipitates an antibody to the lectin. Platelets treated with this inactive derivative release 5-hydroxytryptamine when exposed to bivalent (but not monovalent) $F(ab')_2$ fragments of the lectin antibody.

Asialoglycoprotein Ib and asialoglycocalicin have been isolated from the membranes and from the supernatant, respectively, of neuraminidase-treated human platelets.[662] The two glycoproteins have been shown to be closely related, and evidence supporting the idea that glycocalicin is derived from the membrane-bound glycoprotein is reported.

The presence of ectosialyltransferase activity has been demonstrated at the external surface of human platelets.[663] The major endogenous acceptor is the plasma-membrane glycoprotein GPIIb. Platelet-membrane glycoproteins IIb and IIIa have been isolated and purified from human platelet membrane.[664] The glycoproteins are antigenically different and structurally distinct glycoproteins, which in the native state may form a macromolecular complex with a role in mediating platelet-platelet interactions. The major constituents of human platelet membranes GPIIb and GPIIIa have been purified, and the generation of monospecific antisera to these antigenically different and structurally distinct glycoproteins has been described.[665,666] Morphological evidence demonstrating macromolecular complex formation of these glycoproteins in the platelet membrane after thrombin stimulation is also reported.

Platelet glycoprotein III and α-actinin have been differentiated on the basis of the ability of the former to bind to immobilized wheat-germ agglutinin.[667]

Further evidence has been shown to support the hypothesis that young rabbit platelets are hemostatically most effective when they are released from bone marrow and that they undergo significant changes in the circulation as they age.[668] This is shown by decreased size, symmetrical loss of membrane glycoproteins, and decreased hemostatic effectiveness of the platelets.

A major glycoprotein (mol. wt. 1.1×10^5) carrying binding sites for concanavalin A and soybean agglutinin has been purified from mosquito plasma membranes.[669] The presence of N-glycosidically linked oligosaccharide chains composed of residues of D-mannose and 2-acetamido-2-deoxy-D-glucose (in the molar ratio 9:2) and O-glycosidically linked 2-acetamido-2-deoxy-D-galactosyl residues was established.

Concanavalin A binds to a surface-membrane receptor of the insect trypanosomatid *Crithidia fasciculata*.[670] The receptor (mol. wt. 1.4×10^4) may be identical to a previously reported D-mannan from that source.

The *in vivo* synthesis, membrane insertion, and glycosylation of a multisubunit integral membrane protein of the *Torpedo electroplax* acetylcholine receptor have been studied.[671] Each of the four subunits is synthesized as an individual polypeptide, and all four polypeptide chains are independently integrated into dog pancreas microsomes as transmembrane proteins. The mechanism of integration appears to involve a cotranslational process analogous to that described for viral membrane glycoproteins.

10 Glycoprotein Hormones

The structure and function of glycoprotein hormones,[672] and the regulation of glycopeptide hormone synthesis in cell culture,[673] have been reviewed. A book dealing with the evolution of protein structure and function includes a chapter on glycoprotein hormones.[674]

Four gonadotrophin components have been purified from the basic protein fraction of bullfrog pituitary glands.[675] The amino acid composition of the four components is appreciably different from that of mammalian luteinizing hormone.

The binding sites of human gonadotrophin in the intracellular organelles of bovine corpora lutea have been characterized and compared with plasma-membrane sites.[676] The synthesis of glycoprotein hormone α-subunit and placental alkaline phosphatase has been followed in the presence of tunicamycin, 2-deoxy-D-*arabino*-hexose, and sodium butyrate.[677] The glycosylation inhibitors reduced the butyrate-stimulated synthesis of the subunit and alkaline phosphatase to a greater extent than their basal levels of synthesis.

Pregnant-mare serum gonadotrophin dissociates at acid pH with a parallel loss of follicle-stimulating hormone and luteinizing-hormone activities, supporting the hypothesis that the same molecular entity bears the two binding sites for the receptors of follicle-stimulating hormone and luteinizing hormone.[678] Molecular weight studies of the gonadotrophin indicate a value of 4.5×10^4 as opposed to previously reported values of $(5.3 — 6.4) \times 10^4$.

A book dealing with the chemistry of and homology between human chorionic gonadotrophin and pituitary luteinizing hormone has been published.[679] The current understanding of immunological specificity of the β-subunit of HCG is elucidated.

The human serum hormones lutrophin and the β-subunit of choriogonadotrophin are adsorbed by immobilized concanavalin A and by immobilized wheat-germ agglutinin, which also adsorbs follitrophin and thyrotrophin.[680] The hormone-free serum thus produced is suitable as a matrix for standards in the radioimmunoassay of these hormones. A highly specific enzyme immunoassay for human chorionic gonadotrophin has been established.[681]

Cultured human choriocarcinoma cells synthesize high D-mannosyl oligosaccharide-containing forms of the α-subunit (mol. wt. 1.5×10^4 and 1.8×10^4) and β-subunit forms (mol. wt. 1.8×10^4 and 2.4×10^4) of human chorionic gonadotrophin.[682] The secretion of the α- and β-subunits of the hormone involves the formation and accumulation of incompletely processed forms of these subunits intracellularly. The free α-subunit produced by the cells is a larger polypeptide and has a different carbohydrate composition than the corresponding α-subunit of the placental human chorionic gonadotrophin.

A one-step procedure for the isolation of human chorionic gonadotrophin in milligram amounts, using selective steady-state stacking on polyacrylamide gel, has been reported.[683]

The production of human chorionic gonadotrophin subunits by hormone-producing human cancer cells growing in nude mice is similar but not identical to that observed in culture.[684] There appears to be a shift toward more complete hormone production and a decrease in free subunit production as compared to the pattern in cultured cells. Aerobic tumour-associated bacteria have been isolated and identified from malignant tissues from patients.[685] All but one of the isolates were capable of producing the β-subunit of human chorionic gonadotrophin.

Differences in the molecular weights of the α-subunit of human chorionic gonadotrophin excreted in the urine and secreted by HeLa cells have been reported.[686] The subunit secreted by the tumour cells has a higher molecular weight and may be due to an increase in the neuraminic acid content of the protein.

Human chorionic gonadotrophin has been covalently crosslinked to its receptor in rat testes.[687] Analysis of the crosslinked complexes indicates that the membrane receptor may consist of a dimer of two binding subunits which binds to the α-subunit of the hormone.

An assessment has been made of the effects of various modifications of the terminal D-galactosyl residues of human asialo choriogonadotrophin.[688] The preparation of the derivatives and the influence of derivatization on their rates of plasma clearance, uptake by the liver, immuno-reactivity, and target tissue binding are described. The role of carbohydrate in the subunit interactions and in the binding to the receptor has been assessed by deglycosylation of individual α- and β-subunits of human chorionic gonadotrophin.[689] Removal of about 90% and 80% of the carbohydrate from the α- and β-subunits, respectively, had no effect on the immunological activities, on the apparent molecular weights, or on the ability of the subunits to undergo reassociation. Fluorescent-labelled α-subunits of human chorionic gonadotrophin recombine normally with native β-subunits. Labelling procedures appear to compromise the ability of the β-subunits to recombine.[690] Since the method of introducing the fluorescent label involves prior periodate oxidation of the intact hormone, it is possible that primary amino groups of the protein compete with added fluorescent amines for the generated aldehyde groups to produce covalent crosslinking between subunits.

The presence of the β-subunit of human chorionic gonadotrophin on various tumours is reported to be rare on the basis of immunohistochemical staining of various tumours with anti-β-HCG antibodies.[691]

L-Methionine residues of the α- and β-subunits of bovine lutrophin and bovine thyrotrophin have been specifically alkylated with iodoacetic acid.[692] The modified residues were identified and the effects of modification on recombination with the unmodified counterpart subunit, receptor-binding activity, and conformation were assessed. Terminal O-sulphated residues of 2-acetamido-2-deoxy-D-glucose have been identified in bovine lutrophin.[693] Human pituitary lutrophin is at least partially sulphated, in contrast to its placental counterpart, LCG.

Evidence for the amino acid sequence of the α-subunits of ovine follicle-stimulating hormone and luteinizing hormone has been presented.[694]

Similar studies of the amino acid sequence of the β-subunit of ovine follicle-stimulating hormone have shown a remarkable degree of preservation of structural information in the β-subunit of this and other hormones.[695] Two forms of equine follicle-stimulating hormone have been isolated.[696] A highly specific homologous radioreceptor assay for the hormone is described. An enzyme immunoassay for human follicle-stimulating hormone has been developed using the hormone covalently attached to D-glucose oxidase and antiserum to the hormone.[697] A radioimmunoassay procedure for human thyroid-stimulating hormone has been improved by reducing the heterogeneity of the ^{125}I-human thyroid-stimulating hormone tracer.[698]

The cellular processing, assembly, and release of subunits in thyroid-stimulating-hormone biosynthesis have been studied.[699] The α- and β-subunits are initially synthesized in equivalent amounts. A subsequent excess of α-subunits is observed, resulting from the intracellular degradation of β-subunits. It is proposed that this imbalance in the net production and secretion of the subunits originates at the level of post-translational degradation rather than at the translational process level.

The biosynthesis of calcitonin, a 32-amino acid hormone (mol. wt. 3.5×10^3), involves the glycosylation and proteolytic cleavage of a newly synthesized precursor.[700] The precursor undergoes extensive co- and post-translational processing to the smaller non-glycosylated hormone.

Similar D-glucosyl-containing oligosaccharide lipids are synthesized by thyroid rough microsomes and by thyroid cells.[701] The transfer of D-glucose to these oligosaccharide lipids is depressed by the addition of GDP-D-mannose. From the results of kinetic experiments, it seems unlikely that the sole participation of newly synthesized oligosaccharide lipids results in the extensive transfer of D-glucose from UDP-D-glucose to thyroid endogenous proteins.[702]

Two large glycoprotein fragments related to the N-terminal part of the adrenocorticotrophin-β-lipoprotein precursor are the end products of the maturation process in the rat pars intermedia.[703]

A novel human pituitary glycopeptide composed of 39 amino acid residues and one oligosaccharide chain has been isolated from whole pituitaries.[704] This glycopeptide exhibits masked sequence homology to pig posterior pituitary glycopeptide, and to whole pituitary glycopeptides isolated from ox, sheep, and pig.

The isolation and purification of a human glycopeptide representing the major immunoreactive form of the pituitary N-terminal segment of pro-opiomelanocortin have been reported.[705] It bears an O-glycosylation site at L-Thr-45 and an N-glycosidic site at L-Asn-65.

Human and rhesus-monkey pituitaries have been stained immunocytochemically with antibodies against the β-subunits of pituitary glycoprotein hormones.[706]

11 Milk Glycoproteins

The structures (31) of a homologous series of oligosaccharides from the milk of the tamar wallaby (Macropus eugenii) have been elucidated mainly by ^{13}C n.m.r. spectroscopy.[707]

$$\{\beta\text{-D-Gal}p\text{-}(1\rightarrow 3)\}_{2-5}\text{-}\beta\text{-D-Gal}p\text{-}(1\rightarrow 4)\text{-D-Glc}$$

(31)

Interactions of substrates and α-lactalbumin with D-galactosyltransferase have been measured by difference spectroscopy.[708] The conversion of native α-lactalbumin to a conformation I or D and also the conversion of I to D have been studied by u.v. difference spectroscopy and c.d. spectroscopy.[709] The data have been used to calculate standard free-energy changes for each of the transitions. Laser photo-chemically induced dynamic nuclear polarization n.m.r. studies for five α-lactalbumins from different animal species confirm a high degree of homology from species to species with only minor differences in chemical-shift environment between them.[710] A tightly bound impurity from milk has a pronounced effect on the L-tryptophan fluorescence of goat α-lactalbumin.[711]

The binding of one Ca^{2+} ion to bovine α-lactalbumin causes a conformational change reflected in a decrease of the L-tryptophan fluorescence and a spectral shift towards shorter wavelengths.[712]

The site of crosslinking on α-lactalbumin for D-galactosyltransferase in the lactose synthetase complex has been studied using the mutual exclusivity of acetylation and amidation with bis(imidoesters).[713] The results indicate that L-lysine-108 is situated 6.1 to 7.3 Å from an amino group on galactosyltransferase in the crosslinked complex.

Electrophoretic examination of milk from Bali cattle Bos (Bibos) javanicus has led to the identification of three new electrophoretically distinct β-lactoglobulins, designated E, F, and G.[714] Deletions or substitutions of specific amino acids in the protein are responsible for the different electrophoretic properties of these variants. A new α-lactalbumin variant was also identified.[715]

Glycopeptides have been prepared from bovine colostral K-casein after treatment with cyanogen bromide and proteases.[716] A sequence of thirteen amino acid residues on one of the glycopeptides was established. Three of four L-threonine residues in this sequence are thought to be the sites of attachment of three polysaccharide chains in the casein. Glycopeptides released from bovine colostrum K-casein glycopeptide obtained immediately after calving contain one additional prosthetic sugar group linked to L-threonine instead of only one, and up to ten in the case of bovine (normal) and human caseinoglycopeptides, respectively.[717] The tetrasaccharide (32) has been isolated from bovine colostrum K-casein after alkaline borohydride treatment of caseinoglycopeptide.[718] One neutral oligosaccharide alditol (33) and three acidic oligosacchsaccharide alditols (34-36) have been identified from bovine colostrum K-casein taken six hours after parturition.[719] Oligosaccharides (33-34) are characteristic oligosaccharides having 2-acetamido-2-deoxy-3-O-β-D-galactosyl-D-glucosyl groups at the non-reducing end and have not been reported to occur in normal K-casein. The other two oligosaccharides (35-36) have already been identified in K-casein from normal bovine milk. Oligosaccharide (34) has also been characterized in a 500 MHz ^{1}H n.m.r. study.[720]

The localization of glycosylated K-casein in isolated bovine casein micelles has been studied at the ultra-structural level using gold granules labelled with *Ricinus communis* lectin, which specifically interacts with β-D-galactosyl residues.[721] No evidence was obtained for the presence of glycosylated K-casein on the surface of glutaraldehyde-fixed micelles whether or not they had been treated with neuraminidase. The glycoprotein was mainly located in the bridging network interconnecting the micelles, and appeared to be loosely associated with the micelles.

β-D-GlcpNAc-(1→3)-D-Galp-(1→3)-D-GalNAc-ol
6
|
2
α-Neup5Ac

(32)

β-D-Galp-(1→4)-β-D-GlcpNAc-(1→6)-D-GalNAc-ol
3
|
1
β-D-Galp
3
|
2
R

(33) R = H
(34) R = α-Neup5Ac

α-Neup5Ac-(2→3)-β-D-Galp-(1→3)-D-GalNAc-ol
6
|
2
R

(35) R = H
(36) R = α-Neup5Ac

The interaction between bovine K-casein and β-lactoglobulin has been examined.[722] The oligosaccharide portion of the glycoprotein influences complex formation, and Ca^{2+} ions have an inhibitory effect on the interaction.

Subcellular fractionation of ovine mammary-gland membranes indicates that 2-acetamido-2-deoxy-D-galactosyl : polypeptide transferase and D-galactosyl : 2-acetamido-2-deoxy-α-D-galactosyl transferase, which are involved in the assembly of disaccharide units of x-casein, are localized chiefly in Golgi membranes.[723] A neuraminosyl transferase also occurs in the same fraction, probably suggesting a post-translational glycosylation process of x-casein occurring simultaneously with the transport of these molecules to the Golgi region. A casein component (C-2 casein, mol. wt. 3.4×10^4) from rat milk has been purified by ion-exchange and gel filtration chromatography.[724] 2-Acetamido-2-deoxy-D-glucose was the only sugar detected in this glycoprotein. Alkali-labile oligosaccharides liberated from the glycopeptides of human milk fat globule membrane have been characterized and compared with the corresponding oligosaccharides isolated from bovine milk fat globule membranes.[725] The formation of lipid-linked sugars into explants of developing rabbit mammary gland, stimulated to differentiate in culture with hormones, and the appearance of novel glycoproteins, whose synthesis is dependent on this pathway, have been described.[726] Incubation of a membrane preparation from lactating bovine mammary tissue with UDP-2-acetamido-2-deoxy-D-glucose and GDP-D-mannose results in the synthesis of an entire range of lipid-linked saccharides that differ from one another by one monosaccharide unit.[727] They appear to be precursor-product type of intermediates for the glycosylation of L-asparagine-linked glycoproteins. Two isomeric structures for the lipid-linked hexa- and hepta-saccharides are proposed.[728]

The elucidation of the structures of two carbohydrate units (37-38), N-glycosidically linked to an L-asparagine residue of lactotransferrin, using 500 MHz ^{1}H n.m.r. spectroscopy has been described.[729]

Structural relatedness between human lactotransferrin and human ceruloplasmin has been established.[730] The comparison of amino acid sequences of two ceruloplasmin fragments corresponding to a total of 564 amino acid residues and 70% of the lactotransferrin molecule (445 residues) indicates ten homologous areas including 133 amino acids.

```
R-(1→2)-α-D-Manp-(1→2)-α-D-Manp-(1→3)-β-D-Manp-(1→4)-β-D-GlcpNAc-(1→4)-β-D-GlcpNAc-1→L-Asn
                                           6
                                           |
                                           1
          α-D-Manp-(1→2)-α-D-Manp-(1→3)-α-D-Manp
                                           6
                                           |
                                           1
                                     α-D-Manp
                                           2
                                           |
                                           1
                                     α-D-Manp
```

(37) R = H

(38) R = α-D-Manp

12 Serum Glycoproteins

Rat α_2-macroglobulin has been purified from sera by sequential use of dextran sulphate, ion-exchange chromatography, and gel-permeation chromatography.[731] An antiserum prepared against the glycoprotein was used to assay the concentration of α_2-macroglobulin in the sera of normal rats and in the sera of rats undergoing an inflammatory response.

Evidence for the existence of two types of complexes, one of which is covalent and the other of which is not between α_2-macroglobulin and proteases, has been reported.[732] The conversion of the non-covalent to the covalent form requires the presence of unblocked amino groups.

Soybean trypsin inhibitor has been used to assess the degree of inaccessibility of porcine trypsin within the α_2-macroglobulin-trypsin complex.[733] Proteinase-binding sites have been identified on the α_2-macroglobulin surface. The two subunits of this protein do not appear to be identical, or they are not symmetrically arranged. Evidence for the cleavage of a thiol ester in α_2-macroglobulin during proteinase-complex formation has been discussed.[734] The native human glycoprotein contains one reactive labile thiol ester in each of its four identical subunits.[735] The cystine residue constituting one part of the thiol ester structure is located in an identical sequence to that found in human complement component C3. Trypsin-induced activation of thiol esters of α_2-macroglobulin generates a short-lived intermediate that can react rapidly to incorporate not only methylamine or putrescine but also proteins lacking protease activity.[736] A model for the interaction of trypsin with this protease inhibitor has been proposed in which the inhibitor can bind to the enzyme in three distinct modes.[737]

Human α_2-macroglobulin, after derivatization with monodansylcadaverine by transglutaminase-catalysed incorporation, contains two γ-glutamyl residues per conjugated molecule, and shows no differences in trypsin binding or methylamine inactivation when compared with the native α_2-macroglobulin.[738] After trypsin or methylamine treatment, the α_2-macroglobulin conjugate is not effectively recognized by its cellular receptor. The covalent binding of trypsin to α_2-macroglobulin occurs by a primary amino group on the protease, and is inhibited by primary amines when present during complex formation.[739] Covalent incorporation of the same primary amines during trypsin binding to α_2-macroglobulin indicates transient activation of the methylamine-reactive site during complexation. The covalent linkage of non-proteolytic proteins to α_2-macroglobulin during its reaction with proteases has been further characterized.[740]

The amino acid sequence location of a putative transglutaminase crosslinking site in human α_2-macroglobulin has been identified.[741] The amino acid sequence of a 35-residue internal stretch of human α_2-macroglobulin, which contains two sites cleaved by elastase and one site cleaved by trypsin, plasmin, and thrombin, has been established.[742] Trypsin cleaves a L-Lys-L-Leu bond in α_2-macroglobulin, while *Staphylococcus aureus* V8 proteinase cleaves an L-Glu-Gly bond.[743]

α_1-Acid glycoprotein has been measured by laser nephelometry in a quick and precise method for detection and follow-up of bacterial infections in newborn infants.[744] The complete nucleotide sequence of rat α_1-acid glycoprotein messenger RNA has been described.[745] The inferred amino acid sequence has been compared to the previously reported sequence of human α_1-acid glycoprotein. Isoelectric focusing has been used to reveal the two α_1-acid glycoprotein populations separated by affinity chromatography on immobilized concanavalin A and derived from whole normal serum and sera from patients undergoing an acute inflammatory response.[746] A significant difference is observed between patterns obtained from normal and inflammatory sera.

α_1-Acid glycoprotein produces three peaks on crossed immunoaffinoelectrophoresis with concanavalin A in the first dimension, thus implying a degree of heterogeneity in the carbohydrate portion of this glycoprotein.[747] These variations have been studied under conditions associated with alterations in serum concentration of female sex hormones. A new type of microheterogeneity with regard to the position of attachment of an L-fucosyl residue on α_1-acid glycoprotein has been observed from the improved resolving power and enhanced sensitivity of 500 MHz ^{1}H n.m.r. spectroscopy.[748]

α_1-Acid glycoprotein and its deglycosylated derivatives, produced by enzymatic cleavage of neuraminosyl, D-galactosyl, D-mannosyl, and 2-acetamido-2-deoxy-D-glucosyl residues, may function to regulate immune responses in various physiological and pathological conditions.[749] The precursor forms of plasma proteins synthesized in human liver have been investigated.[750] By comparing the chromatographic behaviour of thirteen different plasma proteins isolated from either liver or blood, five intracellular precursor proteins can be identified. Two of them, transferrin and α_1-acid glycoprotein, were purified and characterized by comparing their structures with those of the corresponding plasma forms. The liver forms of both glycoproteins lacked neuraminic acid, and in addition two liver forms of α_1-acid glycoprotein are described as having lower neutral and amino sugar content.

Rat α-1-antitrypsin has been characterized with respect to its amino acid and carbohydrate composition.[751] Some of its chemical properties have been compared to those of human α-1-antitrypsin. Antibodies prepared to the rat glycoprotein have been shown to be monospecific. α-1-Antitrypsin has been fractionated on immobilized concanavalin A into two components which differ in the degree of branching of their oligosaccharide side chains.[752] One form contains three biantennary side chains and the other contains two biantennary and one triantennary side chain. These alternative structures have been termed isoforms, a term used to differentiate them from variants of genetic or post-secretory origin.

A latent trypsin inhibitor is released from denatured human serum proteins by digestion with thermolysin.[753] Immunological cross-reation identified this inhibitor as a complex between the inhibitory active part of the glycoprotein and immunoglobulin G.

α_1-Protease inhibitor from rat serum exists in five forms.[754] These multiple forms are derived from three original forms by sequential addition of neuraminic acid to each one. A modified end-point laser nephelometric procedure[755] and a sandwich enzyme immunoassay[756] have been developed for measuring antithrombin III. The thermal denaturation of antithrombin III is stabilized by heparin and also by anions such as phosphate, sulphate, and e.d.t.a.[757] The effects of dithiothreitol on the modulation of antithrombin III activity by sulphated polysaccharides have been reported.[758] Kinetic analysis of the heparin-enhanced plasmin-antithrombin III reaction supports the theory that heparin must bind to the enzyme to accelerate the plasmin-antithrombin III reaction.[759] The binding of high-affinity heparin to antithrombin III has been investigated following characterization of the protein fluorescence enhancement of L-tryptophan residues in the antithrombin in the presence and absence of heparin.[760] Stopped-flow kinetic studies of the binding interaction have also been reported.[761]

In a study of the plasma of patients congenitally deficient in antithrombin III, only two of four electrophoretically separated forms identified in normal patients' plasma could be detected.[762] Isoelectric-focusing profiles of human antithrombin III have demonstrated heterogeneity in lyophilized salt-free products and in aged products.[763]

A single proteolytic cleavage site in human antithrombin has been identified between L-Arg-385 and L-Ser-386 following digestion with human thrombin.[764] This is at a position homologous to that observed for the bovine glycoprotein.

The structure and function of factor VIII complex have been reviewed.[765]

Factor VIII-related antigen has been measured using a fluoroimmunoassay.[766] The assay, which uses a fluorescent-labelled anti-factor VIII antibody as the single major reagent, correlates with an existing method. The molecular forms of factor VIII coagulant antigen in clinical concentrates of factor IX have been measured using ^{125}I-labelled factor VIII:C antibodies.[767] Factor VIII clotting antigen has been measured with two different immunoradiometric assays, one utilizing two acquired antibodies and one a haemophilic antibody.[768] The results confirm the division of both moderate and mild haemophilia A into three sub-groups. In haemophilia A, there is a quantitative reduction rather than a qualitative defect in the protein responsible for the factor VIII clotting activity.

A radioreceptor assay for quantitating plasma levels of human factor VIII/von Willebrand's factor has been developed using receptor sites on human platelets.[769] The plasma glycoprotein concentrations correlate with levels measured by the ristocetin-induced platelet-aggregation method.

Aluminium hydroxide but not barium chloride absorption of factor VIII procoagulant antigen is associated with variable recoveries of the antigen.[770]

Immobilized phospholipid vesicles have been used in the purification of a number of coagulation factors, including factors II, VIII, IX, and X.[771]

Fibril suspensions of types I, II, and III collagen are all capable of adsorbing ristocetin cofactor activity and factor VIII related antigen activity from normal plasma.[772] The binding of factor VIII/von Willebrand factor has been detected by following the adsorption of the labelled antigen to collagen films.[773] In contrast to reports from other workers, binding to bovine tendon fibres was also detected.

The application of affinity-chromatographic techniques for separations and studies on molecular interactions of the components of the blood-coagulation system has been reviewed.[774]

An electrophoretic system, employing large-pore polyacrylamide gels in neutral buffer in the absence of sodium dodecylsulphate, has been used to fractionate single species of factor VIII multimers under conditions where biological properties of factor VIII are preserved.[775] Factor VIII:C, factor VIII:RCof, and factor VIII:Ag activities are extractable from the gels after electrophoresis and can be ascribed to individual factor VIII multimers.

Human anti-haemophilic factor (mol. wt. 1.16×10^5), containing no detectable von Willebrand factor, has been prepared from normal human plasma.[776] Partial reduction with dithiothreitol exposes critical sulphydryl groups, which when alkylated maintain the anti-haemophilic factor without inactivating procoagulant activity. Further purification of the anti-haemophilic factor was achieved after conversion of protein-2-pyridyl mixed disulphide conjugates followed by affinity chromatography on thiopropyl-agarose.[777]

Fractionation of partially purified bovine factor VIII on calcium citrate columns produces a material which contains high levels of factor VIII procoagulant activity and factor VIII-related antigen but does not aggregate human platelets.[778] The apparent molecular weight of the procoagulant factor is significantly lower than that of the forms of factor VIII which contain platelet-aggregating activity, and it exhibits a higher mobility on electrophoresis.

Bovine factor VIII has been prepared free from platelet-aggregating activity.[779] The preparation migrated as a triplet on sodium dodecylsulphate-polyacrylamide gel electrophoresis with apparent mol. wts. 9.3×10^4, 8.8×10^4, and 8.5×10^4.

Both plasmin and thrombin destroy a protein (mol. wt. 8.5×10^4), present in highly purified factor VIII/von Willebrand factor, at a rate that parallels the loss of factor VIIIC.[780] This protein may be necessary for full expression of factor VIIIC and may be defective in cases of haemophilia.

The levels of factor VIII coagulant antigen in normal, thrombin-treated, and haemophilic plasma have been analysed.[781] The antigen detected in normal plasma is not covalently linked to factor VIII/von Willebrand factor multimers and is absent in plasma from a number of haemophilia patients. Thrombin-induced proteolysis of the antigen has been observed in normal plasma.

Thrombin is more effective in potentiating factor VIII procoagulant activity when the amount of thrombin relative to the concentration of the factor VIII/von Willebrand factor is stoichiometric rather than catalytic.[782] In contrast to its activation, the inactivation of thrombin-activated factor VIII/von Willebrand factor is not a proteolytic process, and activation of this factor is a prerequisite for its inactivation.

Further evidence of the formation of stable immune complexes between human factor VIII and its homologous antibodies has been published.[783] Some evidence is also provided which supports the view that factor VIII consists of two dissimilar subunits which are noncovalently linked.

Heterogeneity of molecular size of factor VIII/von Willebrand factor in von Willebrand's disease has been demonstrated by a combination of SDS polyacrylamide gel electrophoresis and crossed-immunoelectrophoresis.[784] The factor VIII/von Willebrand glycoprotein consists of a series of multimeric structures ranging in molecular weight from 2.4×10^6 to greater than 10×10^6.[785] Differences in the specific activity of the polymeric forms are a function not only of their size but also of quaternary conformation that may be dictated by disulphide bond arrangements.

A technique has been developed in order to study the relationship between factor VIII-related antigen (VIIIRAg) and factor VIII clotting antigen (VIIICAg).[786] The VIIIC part of the complex is pre-incubated with ^{125}I-anti-VIIICAg prior to immunoelectrophoresis. Autoradiography of the immunoprecipitin line produced with anti-VIIIRAg indicated the presence or absence of VIIICAg (VIIIC) associated with VIIIRAg.

Human factor VIII procoagulant activity interacts with certain phospholipids.[787] Further evidence has shown that this activity and the factor VIII/von Willebrand factor are separate entities. Variant forms of procoagulant-like factor VIII in haemophiliacs have been reported.[788]

The activation of factor X <u>via</u> the intrinsic pathway in the blood-coagulation system involves phospholipid and factor VIII.[789] The stimulating effect of phospholipid in factor X activation is mainly due to an effect on the K_m for factor X, while the stimulatory effect of factor VIII is explained by its 200,000-fold increase in the V_{max} of factor Xa formation.

A metal-ion-catalysed conformational change affecting factor X_1 is a prerequisite for glycoprotein-lectin precipitin interaction.[790] When Ca^{2+} or Mn^{2+}, but not Mg^{2+}, is present, specific sugars on the carbohydrate chains of factor X_1 become accessible to wheat-germ agglutinin. Whether this cation-glycoprotein interaction takes place through a direct interaction between carbohydrate moieties and cation or by a direct interaction between protein moiety and cation, or both, is not known.

A three-step method has been developed for the determination of factor VII using a chromogenic substrate.[791]

A preparative scheme for the isolation of multiple components of complement from one large pool of human plasma in which the purified products are characterized functionally, physicochemically, and immunochemically has been described.[792]

The purified subcomponent Clq of the first component of mouse complement contains hydroxy-L-proline, hydroxy-L-lysine, glycine, and carbohydrate.[793]

The interaction of isolated human complement components Clr and Cls with cell-bound and fluid-phase Clq has been reported.[794] The binding of Clq to the Fc portion of an antibody influences its activity twoards Clr and Cls and its ability to activate these Cl components. Binding of Clr and Cls to Clq in the fluid phase appears to change or fix the Clq conformation so that Clq is then able to bind only to those binding sites which can trigger the internal activation.

The initial reaction of binding of human complement C3 and C4 to surfaces involves similar mechanisms for both proteins, with the binding being by means of a covalent bond (ester or amide) of which the carboxyl group is donated by C3 or C4.[795] Haemolytically active C3 and C4 undergo peptide-bond cleavage in the α-chain when incubated under denaturing conditions. In each case only a single peptide bond is split.

The noncovalently associated α- and β-subunits of the eighth component of human complement (C8) recombine in solution to form native C8 with concomitant appearance of C8 haemolytic activity.[796] Structural domains which mediate specific interaction of C8 with its native cytolytic complex are located in the β-subunit.

Direct chemical evidence shows that non-enzymatic glycosylation of haemoglobin involves the initial formation of a reversible aldimine (Schiff base) precursor which slowly rearranges to a stable ketoamine by the Amadori rearrangement.[797] An estimate of the distribution of labile aldimine and stable ketoamine in erythrocytes from normal and diabetic sources is obtained from the rate constants for the production of the two forms.

Unless strictly standardized conditions are applied to the measurement of non-enzymically glycosylated serum proteins using the thiobarbituric acid assay, erroneous results may be obtained.[798] The accuracy of the method has been improved by a number of modifications to assay conditions. The conditions for the thiobarbituric acid assay for non-enzymic glycosylation of human serum albumin have been examined, and those of optimal sensitivity and reproducibility are reported.[799]

A colorimetric method for the estimation of glycosylated haemoglobin, based on the phenol-sulphuric acid reaction of carbohydrate, is claimed to be more reliable and more sensitive than the thiobarbituric acid method.[800] In a different approach, the carbohydrate content of glycosylated haemoglobin is determined by measuring the release of formaldehyde following periodate oxidation of the carbohydrate moieties.[801] The formaldehyde is estimated as the fluorescent condensation product formed with acetylacetone and ammonia. An improved colorimetric assay for glycosylated haemoglobin depends on the conversion of the D-glucose moiety to 5-hydroxymethylfurfuraldehyde by oxalic acid, followed by colorimetric estimation of the 5-hydroxymethylfurfuraldehyde with 2-thiobarbituric acid.[802] The utility of glycosylated-haemoglobin measurement as an index of chronic control in diabetes can be adversely affected by interference from a labile glycosylated fraction that changes rapidly with blood D-glucose concentration.[803] This fraction may be measured by h.p.l.c. or by electrophoresis.

Interference by D-glucose in the colorimetric estimation of glycosylated plasma protein has been eliminated by precipitation of the proteins with trichloracetic acid.[804]

Glycosylated minor components of human fetal haemoglobin have been separated by ion-exchange chromatography, identified, and functionally characterized.[805] Sources of variation in the ion-exchange column chromatographic determination of glycosylated haemoglobin (HbA) have been described.[806]

A simple, commercially available method for separating glycosylated haemoglobin electrophoretically on agar-gel films has been assessed.[807] The method correlated well with a column chromatographic technique.

D-Glucose also binds covalently to ε-amino groups of L-lysine of human apolipoproteins.[808] The level of glycosylated apoprotein B of low-density lipoproteins is increased in serum from diabetic patients.

Nonenzymatic glycosylation of proteins in diabetics extends beyond haemoglobin to the proteins of the erythrocyte membrane, and probably affects other proteins that have slow turnover and are exposed to high concentration of D-glucose.[809]

Human plasma contains a factor, terminal autorosette inhibition factor, which can inhibit the ability of murine lymphocytes to bind autologous erythrocytes.[810] The purified factor has been identified as a glycoprotein (mol. wt. 8.0×10^4), being a larger-molecular-weight precursor from which histidine-rich glycoprotein and, possibly, plasminogen-binding protein are derived by proteolysis during purification.

Histidine-rich glycoprotein from rabbit serum binds Cu^{2+}, Zn^{2+}, Hg^{2+}, Cd^{2+}, Ni^{2+}, and Co^{2+}, with high affinity, in vitro, which is consistent with this protein having a role in transport or homeostasis of metals in serum.[811]

Human β_2-microglobulin forms complexes with rat serum glycoprotein.[812] After injection in vivo the serum contains a complex (mol. wt. 5.5×10^4 - 6.7×10^4), but with in vitro incubation the β_2-microglobulin forms additional complexes (mol. wt. 2.0×10^5).

A comparison of serum pregnancy-specific β_1-glycoprotein analysis by enzyme-linked immunosorbent assay and radioimmunoassay has given good correlation between the two methods.[813]

A glycoprotein (mol. wt. 5.0×10^4) associated with malignancy has been isolated from human sera from cancer patients.[814] Reaction of this glycoprotein with various antisera directed against normal serum glycoproteins has established that it is not one of the major acute-phase reactant glycoproteins associated with malignancy and can be distinguished from carcinoembryonic antigen and α-fetoprotein by its molecular weight and chromatographic behaviour.

The influence of temperature, pH, and various enzymes on the intermediate structures formed in the transformation of fibrinogen to fibrin has been reported.[815] Light-scattering studies show that rod-like intermediates are formed at the beginning of the aggregation process when fibrinogen is treated with thrombin, Babroxobin (an enzyme preparation for the venom of the snake Bothrops atrox majoensis), and Contortrix (an enzyme preparation from the venom of the copperhead snake Ancistrodon contortrix

contortrix).

Extensive inactivation of the iron-binding capacity of human transferrin occurs when approximately four L-tyrosine residues are oxidized with sodium periodate.[816] The correlation of destruction of this amino acid with the loss of iron-binding activity suggests that the L-tyrosine residues are those that function as ligands to metal ions bound to the protein.

Human serum thyroxine-binding globulin has been affinity labelled with N-bromoacetyl-L-thyroxine.[817] L-Methionine was identified as the major residue labelled.

Microheterogeneity of α-fetoprotein in human serum has been demonstrated by lectin affinoelectrophoresis.[818] A practical and economic enzyme immunoassay for human α-fetoprotein has been developed and its performance evaluated.[819] Some studies on binding and internalization of asialoglycoproteins by isolated rat hepatocytes have been reported.[820]

The conformations of the N-glycosylated oligosaccharide chains of a glycopeptide derived from fetuin have been examined by hydrogen-exchange techniques.[821] Eight acetamido hydrogens are present, originating from five 2-acetamido-2-deoxy-D-glucosyl residues and from three N-acetylneuraminic acid residues, together with one hydrogen from the oligosaccharide-peptide linkage. Seven of the hydrogens exchange at a rate indicating free contact with solvent, but the remaining two hydrogens exchange at a significantly slower rate, implying their involvement in intramolecular hydrogen bonds.

The total carbohydrate content and the monosaccharide composition of plasma apolipoproteins isolated from normal and 2-amino-2-deoxy-D-galactose-treated rats have been recorded.[822]

Radioimmunoassay techniques have been used to evaluate the contribution of the carbohydrate moiety to the immunological reactivity of human serum low-density lipoprotein.[823] Neuraminic acid and D-mannose, the terminal residues of low-density lipoprotein glycopeptides I and II, are implicated in the antigenic site(s).

The interactions of fibrinogen and asialofibrinogen with concanavalin A and the blood-clotting properties of the complexes have been described.[824] Two of the four possible binding sites on the fibrinogen molecule are not accessible to concanavalin A tetramers. Asialofibrinogen and its concanavalin A complexes coagulate twice as fast as those of fibrinogen. The translated products of rat liver messenger RNA, total polysomes, and rough microsomes include glycosylated subunits of fibrinogen.[825] Both the Bβ- and the γ-chains are glycosylated <u>via</u> L-asparagine moieties and the Aα-chains are not glycosylated. The γ-chain is glycosylated as an early cotranslational event, while the Bβ-chain is glycosylated at or slightly after the termination of polypeptide synthesis of that chain.

The presence of oligosaccharide chains on L-asparagine-288 of human plasminogen decreased the binding of fragments containing the high-affinity L-lysine-binding site (residues 79-337 or 79-353) to both α_2-antiplasmin and fibrin.[826]

A technique for the identification of glycoprotein antigens in immune complexes isolated from sera of patients with Burkitt's lymphoma and nasopharyngeal carcinoma has been reported.[827] Two closely related glycoproteins were identified in 80% of the sera examined from patients with those two diseases, but were not present in patients with a variety of unrelated tumours or in sera of healthy individuals.

The comparative analysis of the lectin-binding properties of plasma glycoproteins from control and cystic-fibrosis patients has been reported.[828] The results of a comparison of plasma glycoproteins from normal controls, cystic-fibrosis patients, and cystic-fibrosis heterozygotes do not support the hypothesis of a generalized defect in glycoprotein metabolism as the basic defect in cystic fibrosis.[829] However, the possibility of a more limited defect in the metabolism of certain other types of glycoproteins, possibly those involved in the processes of exocrine excretion, is considered.

Transcortin, the major human plasma glucocorticoid, has been isolated and partially characterized.[830] Some of its physicochemical parameters, particularly its microheterogeneity and its propensity for aggregation, are described.

Isopiestic data indicate that fish antifreeze glycoproteins bind significantly more water than haemoglobin or cytochrome <u>c</u>, observations which are consistent with n.m.r. spectroscopic studies.[831]

An improved procedure for the determination of antifreeze glycoprotein activity using a freezing-point osmometer has been described.[832] The sensitivity of the method has allowed the identification of a hitherto overlooked antifreeze activity in the Atlantic cod (<u>Gadus morhua</u>).

13 Immunoglobulins

A rapid method for simultaneously screening the immune complexes of numerous sera for the presence of an unsuspected antigen has been developed.[833] An antibody-enzyme reagent can identify free x-casein when no antibody is present and also x-casein if it is bound in an immune complex absorbed onto Raji cells. The level of sensitivity allowed the demonstration of the development and subsequent disappearance of x-casein immune complexes in the sera of either immunoglobulin A-deficient or milk-allergic patients who have ingested a small amount of milk.

Isoelectric focusing of antibodies in agarose gels has been reported using inexpensive laboratory-synthesized ampholytes.[834] The resolving power of these and commercially available ampholytes are compared, and the advantages of substituting agarose gels bonded to plastic films for polyacrylamide gels are discussed. Soluble immune complexes have been studied using a procedure of binding to immobilized protein A followed by isoelectric focusing to desorb and separate the immune complexes.[835] By this approach IgG-type immunoglobulins and putative antigens can be detected in biological fluid from human specimens.

Differences in binding of rat immunoglobulin G sub-classes (IgG1, IgG2, IgG2b, and IgG2c) to staphylococcal protein A have been reported.[836] A procedure for the isolation of pure IgG1 and IgG2c from normal rat serum was developed. Immunoglobulins associated with human platelets bind to immobilized protein A, with retention of the platelets.[837] This technique may be of value in the study of thrombocytopenic patients where greatly increased amounts of associated IgG are found.

The covalent attachment of immunoglobulin G and $F(ab')_2$ fragments to liposomes is achieved by the prior periodate oxidation of glycosphingolipids in liposomes to produce aldehyde groups on the vesicle surface, followed by protein coupling by reductive amination.[838] Under the conditions used, a large proportion of the external glycolipid, but not internal glycolipid, is oxidized. Encapsulated glycerol 1-phosphate is not oxidized within the vesicles and encapsulated carboxyfluorescein is substantially retained.

The $F(ab')_2$ fragment of goat immunoglobulin G has been covalently coupled to phosphatidylethanolamine with subsequent association of this complex with liposomes.[839] Binding assays with antigen-coated Staphylococcus aureus cells are used to evaluate the immunologically specific recognition and the membrane stability of the liposomes. Comparative kinetic studies of ligand dissociation and D_2O-induced fluorescent enhancement have been performed with both heterogeneous and homogeneous antifluoresceinyl immunoglobulin G antibodies.[840] The anti-fluorescein antibody active site contains both solvent-accessible and relatively-inaccessible components, with the binding of ligand involving both exchangeable hydrogen atoms and other as yet unresolved interactions. The mechanism of D_2O fluorescence enhancement is discussed in terms of its complexity involving heterogeneous rate mechanisms.

Several biological effector functions mediated by sites on the Fc region of human immunoglobulin G1 have been studied in two variant immunoglobulin G1K monoclonal proteins which contain deletions corresponding to the entire hinge region of the heavy chains.[841] The functional lack of potency of these immunoglobulins is related to the close association between Fab and Fc regions in these molecules and the limited degree of segmental flexibility permitted in the absence of the hinge region. A major role for the C-γ2 domain in mediating effector functions in normal immunoglobulin G1 is proposed.

A model system, based on the desialylation of immunoglobulin G and generating one of the serological changes normally associated with the pathological condition of rheumatoid arthritis, has been developed.[842] Rabbit asialo-immunoglobulin G tends to aggregate and is immunogenic in autologous hosts. The neuraminic acid content of rheumatoid factor isolated from the serum of a rheumatoid patient and identified as immunoglobulin G and immunoglobulin M is also lower than that of the corresponding normal immunoglobulins.

Immobilized pepsin has been used for the limited cleavage of human immunoglobulin G.[843] The $F(ab')_2$-rich fraction was isolated for intravenous use.

Proximity relationships within the Fc segment of rabbit immunoglobulin G have been analysed by resonance-energy transfer.[844] The principle forces maintaining the integrity of the native functional Fc fragment are the strong noncovalent interactions of the C_H3 domains and the single inter-heavy-chain disulphide bond.

The carbohydrate structures (39, 40) and the complete amino acid sequence of a human λ-type immunoglobulin light chain, protein Sm, have been determined.[845] One O-glycosidically linked chain (40)

```
α-Neup5Ac-(2→6)-β-D-Galp-(1→4)-β-D-GlcpNAc-(1→2)-α-D-Manp-(1→6)-β-D-Manp-(1→4)-β-D-GlcpNAc-(1→4)-
                                                        3                  β-D-GlcpNAc-L-Asn
                                                        |                         6
                                                        1                         |
          ?-Neup5Ac-(2→6)-β-D-Galp-(1→4)-β-D-GlcpNAc-(1→2)-α-D-Manp               1
                                                                           α-L-Fucp
```

(39)

is attached to L-Ser-21. This appears to be the first report of a tetrasaccharide containing two neuraminic acid residues in the light chain of an immunoglobulin.

Oligosaccharides of similar structure occur in human λ-type monoclonal immunoglobulin light chains.[846]

```
α-Neup5Ac-(2→3)-β-D-Galp-(1→3)-D-GalpNAc-L-Ser
                                   6
                                   |
                                   2
                              α-Neup5Ac
```

(40)

A human monoclonal immunoglobulin G1 possesses two species of light chain (mol. wts. 2.3×10^5 and 2.8×10^5, respectively).[847] The difference in mass is due to the presence of a carbohydrate moiety on the 2.8×10^4 species, attached within the J region of the K-chains. Not all of the K-chains are glycosylated, even though no differences in the primary structure between the two L-chains species could be found.

A mouse myeloma cell line produces an immunoglobulin $G2_b$ with two carbohydrate attachment sites.[848] Alterations in glycosylation also occur in two variant cell lines. The complete primary structure of the constant region of L-chains derived from a homogeneous Balilea rabbit antibody specific for type II pneumococcal polysaccharide has been established,[849] as has the amino acid sequence of an euglobulin-like λ Bence-Jones protein NIG-58.[850]

Antibodies have been raised to numerous glycoproteins, glycolipids, and polysaccharides. Those antibodies known to interact with carbohydrate determinants on the antigen molecule are listed in Table II.

Further studies of the inhibition of the monoclonal anti-Ma(group 1) antibody in quantitative precipitin assays have been reported.[852] It is confirmed that the antibody binds to a portion of a β-D-galactopyranosyl-(1→4)-O-β-D-2-acetamido-2-deoxy-glucopyranosyl-(1→6)-O-β-D-galactopyranosyl unit of an oligosaccharide structure in a specific conformation. The conformational requirements indicate that the determinant is accepted into the antibody-combining site along a hydrophobic portion of the structure which includes the regions about C6-C1'-

Table II Antibodies interacting with glycoprotein, glycolipid or polysaccharide antigens

Antigen	Determinant(s) on antigen	Antibody type	Ref.
Glycoproteins			
Histocompatibility antigen (H-2K)	α-D-Man*p*, Neu*p*5Ac, α,β-D-Gal*p*, D-Man*p*	Monoclonal	851
Ma (group 1)	β-D-Gal*p*-(1→4)-β-D-Glc*p*NAc-(1→6)-β-D-Gal*p*	Monoclonal	852
Glucoamylase	α-D-Man*p*	Polyclonal	853
Bovine leukaemia envelope viral glycoprotein	Intact oligosaccharide chain	Polyclonal	13
Retrovirus envelope glycoprotein	D-GlcNAc, D-Gal, methyl α-D-Man*p*	Polyclonal	12, 21
Polysaccharides			
Dextran B512		Monoclonal	854
Dextran B1355	α-D-Glc*p*-(1→3)-α-D-Glc*p*-(1→6)-D-Glc, α-D-Glc*p*-(1→6)-α-D-Glc*p*-(1→3)-D-Glc	Polyclonal	855
D-Galactan	β-(1→6)-D-galactopyranan	Monoclonal	856

(Continued overleaf)

Table II (Continued)

Antigen	Determinant(s) on antigen	Antibody type	Ref.
Glycolipids			
Lewis[b] blood group	See a below	Monoclonal	857
Lewis[x] blood group	α-L-Fucp-(1→4)-D-GlcpNAc	Polyclonal	858
Glycosphingolipids	β-D-Galp-(1→4)-D-GlcpNAc, α-L-Fucp-(1→2)-β-D-Galp-(1→4)-D-GlcpNAc	Monoclonal	859
SSEA-1	β-D-Galp-(1→4)-β-D-GlcpNAc 3 | 1 α-L-Fucp	Monoclonal	860

a α-L-Fucp-(1→2)-β-D-Galp-(1→3)-β-D-GlcpNAc-β-D-Galp-(1→4)-D-Glc
4
|
1
α-L-Fucp

C3'-O5''-C6'' as represented in Scheme 2. It is assumed that O3' is intramolecularly hydrogen bonded to O5'' and that the other hydroxyl groups of the determinant remain engaged in bonding with solvent water.

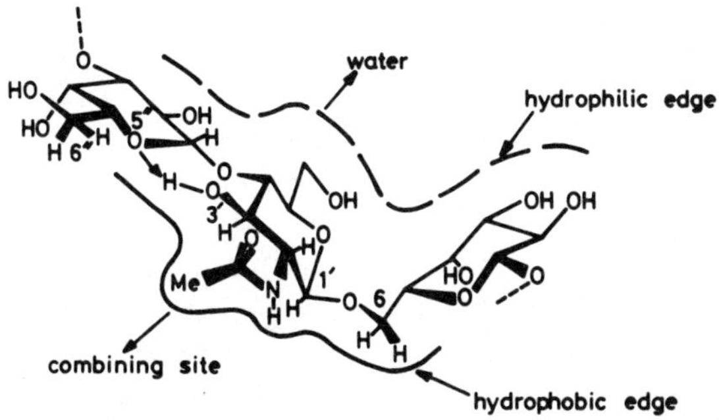

Display showing that the conformation of the I Ma antigenic determinant possesses a region along C6-C1'-C3'-O5''-C6'' which can be expected to be compatible with a hydrophobic region for the combining site

Scheme 2

Rabbit anti-dextran B1355 has been separated into two fractions.[855] Antibodies of a non-binding fraction to Sephadex^{R} G-75 are directed against O-α-D-glucopyranosyl-(1→6)-O-α-D-glucopyranosyl-(1→3)-D-glucosyl residues.

The affinities of two monoclonal anti-D-galactan immunoglobulin Fab' fragments have been measured with a linear and a highly branched D-galactan.[856] Affinity data show that the immunoglobulins can bind intercatenarily to a (1→6)-β-D-galactopyranan but not to a (1→3)-β-D-galactopyranan. Four monoclonal antibodies produced by hybridomas obtained from a mouse immunized with a human adenocarcinoma cell line are directed against the Le^{b} antigen of the human Lewis blood-group system.[857] Their specificities were established by binding studies using purified Le^{b}-active glycolipids and by hapten inhibition studies with oligosaccharides obtained from human milk.

The antigen specificities of different anti-Lewis^{x} sera have been examined by immunoadsorption studies using adsorbents with well defined carbohydrate units covalently bound to an inorganic matrix.[858] In contrast to those of normal anti-Le$^{\underline{a}}$ and anti-Le$^{\underline{b}}$ sera, the antibody-binding site of Le$^{\underline{x}}$ antibodies was found to be considerably smaller, comprising the structure α-L-Fucp-(1→4)-D-GlcpNAc.

Two monoclonal anticarbohydrate antibodies directed toward the carbohydrate moiety of glycosphingolipids containing a lacto-N-glycosyl Type II chain have been prepared.[859] One is specific for the Type II chain H determinant (41) and the other is directed to the non-reducing terminal 2-acetamido-2-deoxy-4-O-β-D-galactosyl-D-glucosyl disaccharide.

α-L-Fucp-(1→2)-β-D-Galp-(1→4)-β-D-GlcpNAc-1→R

(41) R = rest of oligosaccharide chain

A monoclonal antibody reacting with early mouse embryos and murine embryonal carcinoma cells defines the stage-specific embryonic antigen (S SEA-1) and is specific for type 2 blood-group antigens.[860] The combining site of the antibody is complementary to the sequence (42). The antigen has been identified as a complex glycolipid.[861,862]

A cold agglutinin resembling anti-I antibodies has been isolated from a patient suffering from immunoblastoma.[863]

High-titre antisera specific for Lewis$^{\underline{a}}$, $^{\underline{b}}$, and $^{\underline{d}}$ determinants have been obtained by the immunization of rabbits with artificial antigens prepared by coupling chemically synthesized haptens to bovine serum albumin.[864] These reagents have been used in the immunohistochemical analysis of these antigens in the mucosa of the human stomach and duodenum of H group individuals.

```
β-D-Galp-(1→4)-β-D-GlcpNAc-(1→6)R
                  3
                  |
                  1
              α-L-Fucp
```

(42) R = remaining oligosaccharide chain

Monoclonal antibodies, which bind to an antigen present in highly purified carcinoembryonic antigen (CEA) from various tumours, have been isolated.[865] They are consistent with the concept that different immunological forms of CEA exist which can be distinguished by appropriate antisera. However, in view of the immunlogical heterogeneity of CEA and the polyspecific nature of many conventional anti-CEA sera, the possibility that these monoclonal sera are not recognizing a unique antigen not related to CEA cannot be excluded.

Evaluation of a baboon antiserum to CEA has given similar results to a commercially available goat anti-CEA.[866]

Specific antibody to α-fetoprotein has been prepared and found to bind specifically to α-fetoprotein-producing tumour cells.[867]

Human antibodies to factor VIII have been isolated from nine haemophilic plasmas.[868] Subsequent immunological characterizations revealing restricted heterogeneity in both light-chain type and heavy-chain sub-class have been described. Factor VIII clotting antigen has been measured with two different immunoradiometric assays, one utilizing two acquired antibodies and one a haemophilic antibody.[768]

Monoclonal antibodies to von Willebrand's factor have been isolated and their reactivity with porcine and human antigens reported.[869]

Monoclonal antibodies against human acid α-glucosidase have been prepared.[870] The antibodies were used to show that two major components of the purified enzyme contain at least one antigenic component in common.

An enzyme immunoassay for autoantibodies to thyroglobulin in human serum has been developed.[871] Evaluation of the technique by comparison with an indirect haemagglutination test for anti-thyroglobulin antibodies gave a correlation coefficient of 0.85.

Sera from patients with Graves' disease contain antibodies which inhibit the binding of thyrotrophin to thyroid membranes.[872] Immunoglobulin G, isolated from the sera, inhibits the binding of the hormone by binding directly to the hormone receptor.

Antibodies used in an enzyme immunoassay for human chorionic gonadotrophin (hCG) have been isolated in a highly purified form after affinity chromatography on colums of immobilized tricosapeptide of the hCG β-chains.[681]

Rabbit transferrin has been modified by introduction of methionine sulphone hydrazide either into its neuraminic acid residues or into both neuraminic acid and D-galactosyl residues.[873] Rabbits immunized with the modified transferrin produced antibodies which reacted with the native antigen.

Mouse immunoglobulin G3 antibodies to the phosphocholine determinant of pneumococcal C carbohydrate and to type 3 pneumococcal carbohydrate are highly protective against experimental type 3 pneumococcal infection.[874] This is one of the first reports that antibodies of the immunoglobulin G3 sub-class can protect against bacterial infections.

Antibodies against D-galactosyl-hydroxy-L-lysyl:D-glucosyltransferase from chick embryos have been tested for their ability to inhibit the enzyme reaction with various substrates.[875] The cross-reactivity of the glucosyltransferase between different species is low.

Monoclonal rat immunoglobulin G_{2b} antibody, which is directed against the Thy 1.2 antigen, has been covalently bound to ricin.[876] Addition of lactose saturates the D-galactosyl-binding sites on ricin and inhibits the toxin from binding and killing cells *via* the D-galactosyl-containing receptors. Ricin-monoclonal antibody hybrids of this type combine a high degree of cell-type selectivity and may have pharmacological use as antitumour reagents.

Highly specific antibodies directed against cell-surface immunoglobulins on normal or neoplastic murine B lymphocytes have been covalently coupled to the A chain of ricin.[877] The hybrid molecules maintain both their antigen-binding capacity and their toxic properties. Minute amounts of the conjugates are effective in killing target cells *in vitro*.

The 8aB5-9 cloned hybridoma cell line produces a monoclonal antibody *Tab*, directed against a major platelet-surface glycoprotein deficient in platelets of patients with thrombasthenia.[878] This monoclonal antibody, which is directed against a platelet-membrane complex consisting of glycoproteins IIb and IIIa, has been used to enumerate the number of these complexes on normal platelets and to indicate the virtual absence of the complex on the surface of thrombasthenic platelets.

Four independent monoclonal antibodies that recognize an alloantigenic structure on a murine cell-surface glycoprotein have been used to characterize the complexity of protein polymorphism.[879] Each of the antibodies appears to react with the same or a closely related epitope on the glycoprotein.

The localization of chitosan and other fungal cell-wall components in the pea-*Fusarium* interaction have been studied immunochemically with anti-chitosan and anti-fungal cell-wall antisera.[880]

Monoclonal antibodies specific for subunit 1 of the lectin from Dolichos biflorus combine with the C-terminal portion of the subunit and may be interacting with the active site or with a determinant conformationally independent of the active site.[155] This antibody, which does not react with another lectin-like protein from the stems and leaves of the plant,[156] can distinguish subunit 1 from subunit 2 of the seed lectin.[157]

The binding of human immunoglobulin to type C retroviruses has been analysed by radioimmunoassay.[12] Binding is directed to the oligosaccharide chains of the viral glycoproteins and is non-specific in character.

Cytotoxic antibodies present in bovine leukaemia virus-infected animals are mostly directed against the carbohydrate part of the viral envelope glycoprotein.[13] Immune sera prepared from purified envelope glycoproteins are directed against determinants of the polypeptide backbone of the antigen. The majority of, if not all, normal human sera contain naturally occurring heterophil antibodies that react with the carbohydrate moieties of retrovirus envelope glycoproteins.[21] These antiviral antibodies are not detectable in cord sera but show an age-dependent peak in early childhood and subsequently persist life-long.

Monoclonal antibodies directed against herpes simplex virus glycoproteins have been used to protect mice against acute virus-induced neurological disease.[29] Virus glycoprotein gC expresses type-specific antigenic determinants, whereas glycoprotein gD expresses type-common determinants. Monoclonal antibodies to herpes simplex virus type 2 have been prepared and found to precipitate two different viral glycoproteins.[33]

The major glycoprotein (Gp 350/220) of Epstein-Barr virus has been detected both on the plasma membrane and in the cytoplasm by immunofluorescence with antibodies.[70]

Two monoclonal antibodies reacting with AKR leukaemic cells recognize closely related, if not identical, antigenic determinants on the G_{1x} glycoprotein.[72]

Monospecific antibodies to each of two envelope glycoproteins of paramyxovirus have been prepared and characterized.[74] The effects of these antibodies on the infectious process and the other biological activities of the virions and of the isolated glycoproteins are discussed.

Amino acid sequence differences in the V region of murine antibodies to α(1→3)-D-glucans (dextrans) have been correlated with the expression of cross-reactive and individual idiotypes, through an analysis of twelve different dextran antibodies.[881] Idiotype reagents can recognize two amino acid differences with V and D segments of variable regions. In anti-dextran antibodies, cross-reactive idiotypes involve V-region determinants, whereas individual idotype determinants correlate with D-segment variation. Pure secretory immunoglobulin A has been isolated from human milk.[882] The hinge region containing four *O*-glycosidically linked oligosaccharides is released *en bloc* after proteolytic digestion. The structures (43-46) of these oligosaccharides, which are linked to *L*-serine or *L*-threonine residues, have been identified. These oligosaccharides are more complex and heterogeneous than the oligosaccharides linked to *L*-serine residues of the hinge region from myeloma sera immunoglobulin A.

A partially characterized glycopeptide from human milk immunoglobulin A has been shown to be a mixture of four glycopeptides (47-50) following hydrazinolysis and nitrous acid deamination of the glycopeptide.[883]

The membrane receptor for immunoglobulin A, isolated from rabbit liver and mammary gland, is structurally related to secretory component.[884] The membrane secretory component is composed of two groups of amphiphilic molecules which are synthesized as transmembrane precursors and inserted in the cell-surface membrane where they act as receptors for polymeric immunoglobulin A.

R_1-(2→3)-β-*D*-Gal*p*-(1→3)-*D*-GalNAc
6
|
1
R_2

(43) R_1 = R_2 = H
(44) R_1 = H, R_2 = β-*D*-Glc*p*NAc
(45) R_1 = H, R_2 = β-*D*-Gal*p*-(1→4)-β-*D*-Glc*p*NAc
(46) R_1 = α-Neu*p*5Ac, R_2 = H

Spectroscopic properties of light-chain derivatives of murine MOPC-315 immunoglobulin A have been reported.[885] A comparison between c.d. spectra of the three free and hapten-bound light-chain derivatives demonstrates the existence of conformational transitions in these proteins induced by hapten binding.

Biosynthetic and *in vitro* translation studies have demonstrated

```
α-Neup5Ac-(2→6)-β-D-Galp-(1→4)-β-D-GlcpNAc-(1→2)-α-D-Manp-(1→3)-β-D-Manp(1→4)-β-D-GlcpNAc-(1→4)-
                                                                 6       β-D-GlcpNAc-1-L-Asn-peptide
                                                                 |             6
                                                                 1             |
                    R1-β-D-Galp-(1→4)-β-D-GlcpNAc-(1→2)-α-D-Manp               1
                                          |                             α-L-Fucp
                                          R2
```

(47) $R_1 = R_2 = H$

(48) R_1 = β-D-Galp-(1→3), $R_2 = H$

(49) $R_1 = H$, R_2 = α-L-Fucp-(1→3)

(50) R_1 = α-L-Fucp-(1→6), $R_2 = H$

that membrane-bound and secreted immunoglobulin A molecules contain different α-polypeptide chains that are encoded by different α-mRNA's.[886]

Variants have been isolated from the J558 Balb/c mouse myeloma (immunoglobulin A,λ: anti-α-(1→3)-dextran) with altered glycosylation of its heavy chain and decreased reactivity with dextran.[887] The carbohydrate mutant immunoglobulin contains more neuraminic acid than does the carbohydrate of the wild-type immunoglobulin. However, the mutant immunoglobulin appears to have the same reactivity pattern with oligosaccharides as does the wild-type immunoglobulin, and thus the size of the antibody-combining sites may be similar.

By measuring the affinities at various temperatures between anti-D-galactan immunoglobulin A J539 and antidextran immunoglobulin A W3129, and their respective ligands, the enthalpies and entropies of binding have been determined.[888] The interpretation of the results is in agreement with previous work showing that J539 has a groove-type combining region and W3129 has a cavity-type region.

The interaction of saccharide haptens with three homogeneous immunoglobulin A mouse myeloma proteins which have specificity for saccharides has been investigated in a 270 MHz n.m.r. study.[889] Difference spectra show that comparatively few protein resonances are perturbed on binding hapten, suggesting that accompanying conformational changes are limited to the combining site.

The primary structure of the L-asparagine-563-linked carbohydrate chain of an immunoglobulin M from a patient with Waldenstrom's macroglobulinemia has been reinvestigated using 500 MHz n.m.r. spectroscopy, and a new structure (51) has been proposed.[890] No evidence was obtained to support an earlier claim[891] that only one amino sugar residue is present in an unusual core structure. In the present work the n.m.r. spectrum shows all the spectral features characteristic of an NN'-diacetylchitobiosyl unit with heterogeneity existing in both the carbohydrate and peptide moieties.

Glycosylation is not required for membrane localization or secretion of immunoglobulin M in a mouse B cell lymphoma.[892] However, glycosylation does confer increased intracellular stability and resistance to proteolytic digestion. The molecular weight of the high D-mannose core oligosaccharide of immunoglobulin M has been measured by field-desorption m.s.[893]

High-affinity monoclonal immunoglobulin M antibodies directed

```
                                                                                              L-Tyr(NH2)
                                                                                                  |
R-(1→2)-α-D-Manp-(1→2)-α-D-Manp-(1→3)-β-D-Manp-(1→4)-β-D-GlcpNAc-(1→4)-α-D-GlcpNAc-1-L-Asn
                                             6                                                    |
                                             |                                                L-Val
                                             1                                                    |
                   D-Manp-(1→3)-α-D-Manp                                                    L-Ser(COOH)
                                             6
                                             |
                                             1
                                   α-D-Manp
                                             2
                                             |
                                             1
                                             R
```

(51) R = D-Manp or H

towards epitopes on hepatitis B surface antigen have been used in the immunodiagnosis of hepatitis B.[894]

A D-mannose-rich glycopeptide (52) from a human pathological immunoglobulin M contains only D-mannosyl and 2-acetamido-2-deoxy-D-glucosyl residues (molar ratio 10:2).[895] The recovery of one molecule of this glycopeptide per molecule of heavy chain and the determination of the amino acid sequence led to the location of the glycopeptide on L-Asn-402 of the Fc portion of the heavy chain μ of the immunoglobulin.

The amino acid sequence and location of three glycopeptides in the Fc region of the γ chain of human immunoglobulin IgD have been identified.[896] These glycopeptides have oligosaccharide chains linked N-glycosidically to L-Asn-68, L-Asn-159, and L-Asn-210.

A tentative amino acid sequence of the J_H region, the Cγ1 domain, and the hinge region of the γ heavy chain of the immunoglobulin has been published.[897] The hinge region of the γ chain has an unusual structure, in which four or five oligosaccharides containing 2-amino-2-deoxy-D-galactose residues are attached at the N-terminal half of the chain, whereas the C-terminal half lacks carbohydrate, is dissimilar in sequence, and has a high charge.

A hybrid mouse cell line (B1-8) secretes immunoglobulin D which is heavily glycosylated, as indicated by the fact that the unglycosylated ∂ chain has a molecular weight of 4.4×10^4 as compared to the value of 6.1×10^4 for the major species of glycosylated γ chain.[898]

Alkali-extracted human erythrocyte ghost membranes from Rh_o(D)-positive donors complex with human immune anti-Rh_o(D) immunoglobulin D.[899] The antibody was shown to bind to band 3 glycoproteins of the erythrocyte membrane.

14 Erythrocyte Glycoproteins

A review dealing with the red-cell membrane and its cytoskeleton has been published[900]

A new approach for the isolation of lectin receptors without the use of detergents involves the mechanical shearing of cells bound to lectin-coated macroporous agarose beads, whereupon the receptors remain attached to the beads and are released specifically by inhibitory sugars.[901] In this way a glycoprotein released from α-D-Manp-(1→2)-α-D-Manp-(1→2)-α-D-Manp-(1→3/6)-β-D-Manp-(1→4)-β-D-GlcpNAc-(1→4)-β-GlcpNAc-1-L-Asn

6/3
|
1

α-D-Manp-(1→2)-α-D-Manp-(1→3/6)-α-D-Manp

6/3
|
1

α-D-Manp-(1→2)-α-D-Manp

L-Tyr
|
L-Asn
|
L-Ala
|
L-Thr
|
L-Ser
|
Gly

(52)

neuraminidase-treated human erythrocytes after binding to immobilized peanut agglutinin or to immobilized soybean agglutinin was characterized as asialoglycophorin.

Blood-group H sites of human O cells have been converted in vitro into group A sites by transfer of 2-acetamido-2-deoxy-D-{^{14}C}-galactosyl residues using the blood-group A gene-dependent α-D-2-acetamido-2-deoxy-galactosyl transferase from human A_1 plasma.[902] The majority of the radioactivity is tightly bound to the membranes and is only released by proteolysis, implying that the bulk of blood-group H determinants are bound to glycoprotein material.

The human erythrocyte receptor for the L-fucose-specific lectin produced by Streptomyces species binds to human erythrocytes bearing ABO determinants.[903] In a haemagglutination-inhibition assay poly(glycosyl)-ceramide was the best inhibitor with a glycoprotein fraction showing only slight inhibitory activity. The distribution of A and B determinants in the polyglycosyl peptides of human red cells of blood group AB has been studied by isolating blood-group ABH-active components from delipidated human blood–group AB erythrocyte membranes.[904] The A and B antigenic sites are located in different polyglycosyl chains. Approximately half of the blood-group ABH-active glycopeptides carry blood-group A determinants and half carry blood-group B determinants.

Human erythrocytes have been labelled at the cell-surface D-galactosyl and 2-acetamido-2-deoxy-D-galactosyl residues with sodium borotritide.[905] A significant decrease in the number of residues during ageing of the erythrocytes was detected.

Two hereditary platelet disorders, Glanzmann's thrombasthenia and Bernard-Soulier syndrome, are characterized by defects involving adhesion and aggregation and are associated with different membrane glycoprotein deficiencies.[906] No glycoprotein defects have been observed on analysis of the erythrocyte membrane glycoproteins of patients of both disorders.

Immunochemical evidence for the existence of hybrid sialoglycoproteins of human erythrocytes has been reported.[907]

Using human erythrocyte membranes as a model system, three related procedures, based on the specific interaction of lectins with membrane carbohydrate, have been developed for the isolation of inside-out vesicles from heterogeneous populations of membrane fragments.[908] The techniques should prove applicable to other biological membranes where it is not possible to induce controlled sealing of membrane fragments in a defined orientation as is the

```
β-D-Galp-(1→4)-β-D-GlcpNAc-(1→2)-α-D-Manp
                                     1
                                     |
                                     6
         β-D-GlcpNAc-(1→4)-β-D-Manp-(1→4)-β-D-GlcpNAc-(1→4)-D-GlcpNAc
                                     3                            6
                                     |                            |
                                     1                            1
β-D-Galp-(1→4)-β-D-GlcpNAc-(1→2)-α-D-Manp                   α-L-Fucp
```

(53)

case for erythrocyte membranes.

Glycophorin A has been reconstituted in small lipid vesicles (250 - 300 Å in diameter) by using cholate detergent solubilization followed by rapid removal of cholate on a molecular-sieve column.[909] Incorporation is in a transbilayer fashion, with a high concentration of the molecules orientated with the carbohydrate-containing N-terminus to the vesicle exterior. Interaction of the protein with the hydrophobic portion of the bilayer can be shown by ^1H n.m.r. spectroscopy.

The translational mobility of glycophorin in bilayer membranes of dimyristoylphosphatidylcholine has been reported.[910]

Evidence for multiple components of blood-group A and B antigens in human erythrocyte membranes - suggestive of different carbohydrate structures - has been reported.[911]

An L-asparagine-linked oligosaccharide chain (53) has been isolated from a tryptic fragment of human glycophorin A.[912] Structural similarities between this oligosaccharide and the sugar chains of immunoglobulin A are evident. The blood-group M and N specific determinants of human erythrocyte glycophorin A have three identical carbohydrate moieties (54) per peptide.[913] The M or N determinant is the N-terminal amino acid and a neuraminic acid residue(s). The authors argue that there is no chemical basis for assertions in the literature that M and N antigens differ in their oligosaccharide structure or that the N antigen is biosynthetically transformed to the M antigen by sialylation.

α-Neup5Ac-(2→3)-β-D-Galp-(1→3)-α-D-GalpNAc-1-O-L-Ser/L-Thr
6
|
2
α-Neup5Ac

(54)

The O-glycosidically linked oligosaccharide chains of glycophorin from bovine erythrocyte membranes have been isolated as reduced oligosaccharides by alkaline borohydride treatment, and purified by ion-exchange chromatography followed by gel filtration.[914] The structures of three oligosaccharides (55-57), a

D-Galp-(1→3)-D-Galp-(1→4)-D-GlcpNAc-(1→3)-
D-Galp-(1→3)-D-GalNAc-ol

(55)

D-Galp-(1→3)-D-Galp-(1→4)-D-GlcpNAc-(1→3)-D-Galp-(1→4)-D-GlcpNAc-(1→3)-D-Galp-(1→3)-D-GalNAc-ol

(56)

D-Galp-(1→3)-D-Galp-(1→4)-D-GlcpNAc-(1→6)-D-Galp-(1→4)-D-GlcpNAc-(1→3)-D-Galp-(1→3)-GalNAc-ol

3
|
1

D-Galp-(1→3)-D-Galp-(1→4)-D-GlcpNAc

(57)

penta-, a hepta-, and a deca-saccharide obtained as major components of a neutral fraction, have been established. The complete amino acid sequence of porcine erythrocyte glycophorin has been established.[915] The twelve oligosaccharide chains are located at the N-terminal end of the molecule. Ten of these oligosaccharides are linked O-glycosidically to L-threonine or L-serine residues and the remaining two are attached N-glycosidically to L-asparagine residues.

Horse erythrocyte membranes contain two distinct glycophorins.[916] The major component has a blocked N-terminus and consists of 70% protein and 30% carbohydrate, while the minor component has L-threonine.

The chemistry and biochemistry of glycophorin A as a model membrane glycoprotein have been reviewed.[917]

Amino acid and carbohydrate structural variants of major erythrocyte glycoproteins have been reported.[918] The MM(Miltenberger III) glycoprotein appears to be a product of the M gene and contains O-glycosidically linked oligosaccharide chains partially identified by the structure (58). The M^{g} is an allomorph of the M-N locus, probably evolved from a single base substitution in the N gene. The resulting single amino acid substitution affects the post-translational glycosylation of neighbouring L-serine or L-threonine resides.

The amino acid sequence of a N-terminal tryptic glycopeptide of the blood-group M^{g}-specific major human erythrocyte membrane sialoglycoprotein has been reported.[919] The amino acid sequence and the sites of glycosylation of the twelve amino acids at the N-terminus of blood-group M^{c}-specific major erythrocyte sialoglycoprotein have been established.[920] The variant M^{c} represents an evolutionary link between blood-group M- and N-specific glycoproteins.

```
Neup5Ac-D-Galp-(1→3)-D-GalpNAc-1→L-Thr
                            |
                       D-GlcpNAc
```

(58)

The human leukaemia cell line K562 synthesizes glycophorin A.[921] An isolated messenger RNA fraction from the cells has been used in a rabbit reticulocyte cell-free system for the synthesis of the glycoprotein. The presence of membranes is required for N-

```
                                                   R2
                                                   |
R1→β-D-GlcpNAc-(1→2)-α-D-Manp-(1→6)-βD-Manp-(1→4)-β-D-GlcpNAc-(1→4)-β-D-GlcpNAc
                                                   3                          |
                                                   |                          R4
                                                   1
                                               α-D-Manp
                                                   2
                                                   |
                                                   1
                                               β-D-GlcpNAc
                                                   |
                                                   R3
```

(59) $R_1 = R_3$ = β-D-Galp-(1→4), R_2 = H, R_4 = α-L-Fucp-(1→6)

(60) $R_1 = R_3$ = H, β-D-GlcpNAc-(1→4), R_4 = α-L-Fucp$_n$-(1→4) where n = 0 or 1

glycosylation and O-glycosylation of the apoprotein.

Despite their heterogeneous appearance after sodium dodecyl sulphate gel electrophoresis, band 3 (the major transmembrane glycoprotein of human erythrocytes) and its proteolytic fragments are homogeneous polypeptides.[922] The apparent heterogeneity of band 3 and its C-terminal region may reflect variability of glycosylation on sodium dodecyl sulphate binding. The characteristic diffusion of band 3 glycoprotein as seen by SDS-polyacrylamide gel elctrophoresis has been attributed to heterogeneity of the molecular size of oligosaccharide chains.[923] The structures (53, 59, 60) of three further oligosaccharides originating from band 3 glycoprotein have been established.[924] The mechanism of biogenesis of band 3 glycoprotein is similar to that of cotranslationally inserted glycoproteins like vesicular stomatitis virus G, HLA-A antigen, and glycophorin.[925] The major sialoglycoproteins of rat erythrocyte membrane have been purified by hot phenol partitioning followed by cation-exchange chromatography.[926] Further chromatographic resolution has yielded a single purified glycoprotein (mol. wt. 1.9 x 10^4) and a mixture of higher-molecular-weight glycoproteins accounting for at least 23% of the rat-erythrocyte-membrane neuraminic acid.

15 Salivary and Mucous Glycoproteins

The following aspects of this area have been reviewed: gastric mucosal protection and gastric glycoproteins,[927] neuraminic acid and mucous rheology,[928] and the isolation and characterization of neutral oligosaccharides from human bronchial glycoproteins of cystic-fibrosis patients.[929]

A gel-filtration method has been developed to isolate simultaneously and quantitate specifically radiolabelled mucous glycoprotein from multiple samples of culture media.[930]

Mucin, isolated from a patient with cystic fibrosis, consists of an extensively glycosylated core protein which is linked to at least two other proteins.[931] Direct evidence for the role of the mucins in the rheological behaviour of sputum has been summarized.

Human parotid L-proline-rich glycoprotein (mol. wt. 3.6 x 10^4) contains a biantennary oligosaccharide structure similar to that described for several serum glycoproteins.[932]

Mild acid hydrolysis has been used to cleave a sulphated

monosaccharide from salivary mucin of the stump tail monkey (*Macaca arctoides*).[933] The monosaccharide was purified by ion-exchange chromatography and identified as 2-amino-2-deoxy-D-glucose 4-sulphate.

The incorporation of L-{^{3}H}-leucine and 2-acetamido-2-deoxy-D-{^{14}C}-mannose into proteins and glycoproteins of the parotid, submandibular, and sublingual glands of the mouse has been compared.[934]

A novel neuraminic acid, 4-O-acetyl-9-O-lactyl-N-acetylneuraminic acid, has been identified as a constituent of horse submandibular-gland glycoproteins.[935] The structure has been established by g.l.c.-m.s.

A fluorimetric method for measuring α-L-fucose released from porcine submaxillary mucine (A^{+}) by α-L-fucosidase has been developed.[936]

Alkaline borohydride reductive cleavage of hog submaxillary glycoproteins from three immunologically determined phenotypes results in the release of a series of neutral and acidic oligosaccharide-alditols.[937] All have been identified as partial structures representing the possible complete and biosynthetically incomplete stages of the chain of the pentasaccharide (61), present in the glycoprotein with blood-group A activity. The N-glycolylneuraminic acid residues, when present, are only linked to 2-acetamido-2-deoxy-D-galactitol residues and not to D-galactosyl residues, as had been previously reported.

```
α-D-GalpNAc-(1→3)-β-D-Galp-(1→3)-D-GalNAc-ol
                      2            6
                      |            |
                      1            2
                  α-L-Fucp      α-Neup5Ac
```

(61)

A 2-acetamido-2-deoxy-D-glucosyltransferase which acts on mucin substrates has been detected in canine submaxillary glands.[938] Using mucin or benzyl 2-acetamido-2-deoxy-3-O-β-D-galactosyl-α-D-galactoside as acceptors, the transfer of a 2-acetamido-2-deoxy-D-glucosyl residue to either O-4 or O-6 of the 2-acetamido-2-deoxy-D-galactosyl residue of the acceptor occurs. Detailed substrate specificity of the 2-acetamido-2-deoxy-D-glucosyltransferase when tested with a variety of potential glycoprotein and synthetic disaccharide acceptors has been reported.[939] Porcine liver microsomes can introduce residues of neuraminic acid from CMP-neuraminic acid into porcine submaxillary asialo-afuco-mucin, a D-galactosyl ovine submaxillary asialomucin and ganglioside G_{M1}.[940] No evidence was obtained to indicate any transfer of neuraminic acid to 2-acetamido-2-deoxy-D-galactosyl residues, and the enzyme transfers neuraminic acid solely by formation of an α(2→3)-glycosidic linkage to D-galactosyl residues. Rat submaxillary mucous glycoprotein has been purified after aqueous extraction of the glands followed by recycling gel-filtration chromatography.[941] The chemical composition of this mucin resembles that of the human counterpart and is dissimilar to the composition of rat sublingual glycoprotein.

Chemical properties of and affinity of lectins for human bronchial mucous glycoproteins have been compared.[942]

The actions of mercaptoethanol on soluble mucous glycoproteins obtained from both cystic-fibrotic and chronic-bronchitic sputum have been compared.[943,944] The effect of the reducing agent under non-reducing conditions on the glycoproteins cannot be explained solely by the disulphide bond-breaking action, but appears to involve the existence of a mercaptoethanol-inducible mucolytic system which varies from one sputum to another.

Oligosaccharides obtained by alkaline degradation of human bronchial mucous glycoproteins have been fractionated by ion exchange, gel-permeation chromatography, and h.p.l.c.[945]

Sulphated glycoproteins from human gastric juice have been isolated after affinity chromatography on columns of immobilized L-lysine.[946] Two major glycoprotein fractions differing in their O-sulphate and neuraminic acid content were isolated.

The nature of the non-covalent interactions between glycoprotein molecules of human and pig gastric mucous has been studied by mechanical spectroscopy, using storage and loss moduli to characterize, respectively, solid-like and liquid-like behaviour.[947]

Oligosaccharides having blood-group Ii activity have been isolated from sheep gastric glycoproteins which had been enriched for these blood-group activities by affinity chromatography on an anti-I adsorbent column.[948] The fractions of smallest molecular weight with both I and i activities are mixtures of hexa- and octa-saccharides. From studies on the hexa- to octa-saccharide fractions a structural model (62) is proposed which consists of (a) a core region, (b) a backbone region having (1→3)- and (1→6)-linked 2-

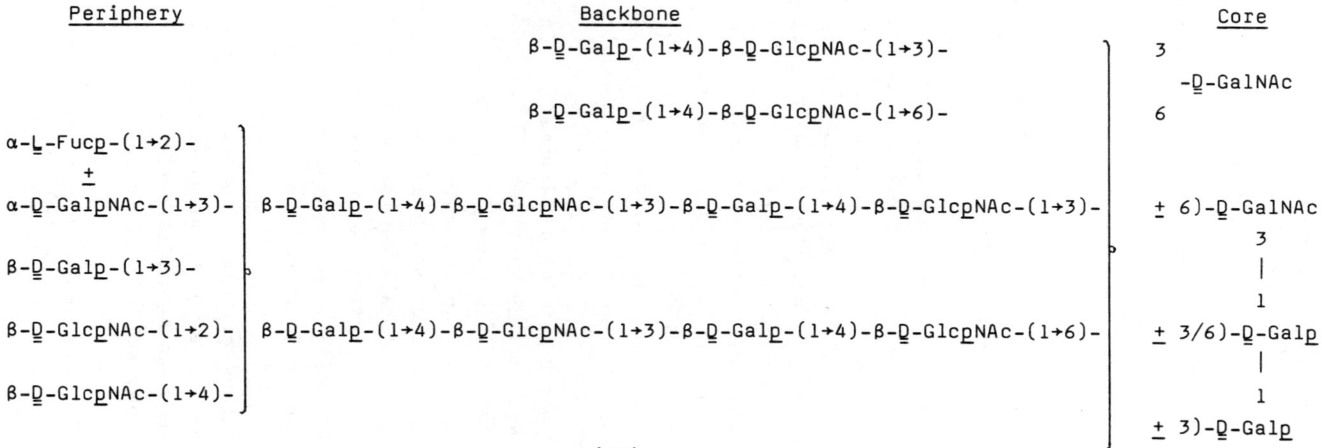
Periphery
Backbone
Core
β-D-Galp-(1→4)-β-D-GlcpNAc-(1→3)-
β-D-Galp-(1→4)-β-D-GlcpNAc-(1→6)-
3
-D-GalNAc
6
α-L-Fucp-(1→2)-
±
α-D-GalpNAc-(1→3)-
β-D-Galp-(1→3)-
β-D-GlcpNAc-(1→2)-
β-D-GlcpNAc-(1→4)-
β-D-Galp-(1→4)-β-D-GlcpNAc-(1→3)-β-D-Galp-(1→4)-β-D-GlcpNAc-(1→3)-
β-D-Galp-(1→4)-β-D-GlcpNAc-(1→3)-β-D-Galp-(1→4)-β-D-GlcpNAc-(1→6)-
± 6)-D-GalNAc
3
1
± 3/6)-D-Galp
1
± 3)-D-Galp

(62)

acetamido-2-deoxy-4-O-β-D-galactosyl-D-glucosyl branches with I activities and linear repeating (1→3)-linked 2-acetamido-2-deoxy-4-O-β-D-galactosyl-D-glucose units with i activities, and (c) a peripheral region with blood-group isotype activities.[949]

A protein (mol. wt. 7.0 x 10^4) is linked to pig gastric mucous glycoprotein by disulphide bridges.[950] On dissociation of the glycoprotein into subunits by reduction or proteolysis, the protein is either released in the reduced form or lost on proteolytic digestion.[951] Pig gastric and small-intestinal mucous glycoproteins are cleaved into four subunits by proteolytic enzymes.[952] A model for the gastric mucous glycoprotein (mol. wt. 2 x 10^6), where on average four subunits are each joined to a protein (mol. wt. 7.0 x 10^4) by disulphide binding, is proposed. The intestinal glycoprotein probably has a similar overall structure.

Rat gastric mucosal cells are capable of the synthesis and secretion of at least two families of mucous glycoproteins of widely different molecular weights and rheological properties.[953]

Oligosaccharide chains bearing the Forssman antigenic determinant are associated with glycoproteins of the dog gastric mucous.[954] Enzymic and immunological evidence indicates that these glycoproteins, like Forssman glycosphingolipids, contain the 2-acetamido-2-deoxy-3-O-(2-acetamido-2-deoxy-α-D-galactosyl)-D-galactose structure, which determines the immunological specificity of the Forssman antigen.

The viscosity and gel-forming potential of pig small-intestinal mucous increase with progressive removal of non-covalently bound protein during isolation.[955] This glycoprotein (mol. wt. 1.7 x 10^6) exhibits A and H blood-group activity and contains residues of L-fucose, D-mannose, D-galactose, 2-acetamido-2-deoxy-D-glucose, 2-acetamido-2-deoxy-D-galactose, and neuraminic acid. The undegraded glycoprotein contains interchain disulphide bridges and is dissociated by proteolysis or reduction into subunits that are different in size and structure from pig gastric mucous glycoprotein.[956]

Rat colonic mucous glycoproteins have been solubilized in the absence of proteolytic enzymes or denaturing solvents.[957] The presence of several high-molecular-weight components, varying in their carbohydrate and sulphate contents, blood-group antigenicity and net charge, was observed.

Evidence for the existence of two immunochemically distinct mucins isolated after proteolytic digestion of human colonic mucin

```
D-Galp-(1→3)-D-Galp-(1→4)-D-GlcpNAc-(1→3)-D-Galp-(1→3)-D-GalpNAc-(1→3)-L-Ser/L-Thr
                                                          6
                                                          |
                                                          2
                                                        Neup5Ac
```

(63)

```
D-Galp-(1→3)-D-Galp-(1→3)-D-Galp-(1→4)-D-GlcpNAc-(1→3)-D-GalpNAc-(1→3)-L-Ser/L-Thr
                                                          6
                                                          |
                                                          2
                                                        Neup5Ac
```

(64)

has been reported.[958] It is not clear whether these two mucin fragments originate from a single mucin molecule or from two different mucins.

Glycoproteins from the periovulatory phase mucins of the bonnet monkey, Macaca radiata, have been degraded with proteolytic enzymes.[959] The partial structures (63-64) of two oligosaccharide chains were identified.

16 Urinary Glycoproteins, Glycopeptides, and Oligosaccharides

Some biochemical findings concerning four different types of sialidoses, on the basis of the specific α-neuraminidase deficiency and the nature of the storage material, have been reviewed.[960]

From the knowledge of the numerous structures of glycoprotein glycans, it is possible to demonstrate that the oligosaccharides and glycoasparagines accumulating in tissues and urines of patients with diseases characterized by a lack in lysosomal glycosidases, originate from glycoprotein glycans incompletely catabolized.[961] A scheme for the normal and pathological catabolism of glycoproteins has been proposed.

Three sialoglycoproteins (mol. wts. 7.79×10^4, 3.7×10^4, and 5.3×10^3, respectively) have been isolated from normal human urine.[962] Alkaline borohydride degradation of the sialoglycoprotein of lowest molecular weight yielded two sialylated oligosaccharides which were partially characterized.

Patients with tubular or mixed glomerular-tubular proteinuria excrete an α_1-microglobulin which has been shown to have three N-glycosidically linked carbohydrate chains of structural similarity to those derived from many other serum glycoproteins.[963]

A glycoprotein (mol. wt. 2.0×10^4) which completely inhibits trypsin at a 1:1 molar ratio has been isolated from human urine.[964] It is generated from a precursor molecule which in turn is derived from plasma inter-α-trypsin inhibitor. An acid-resistant trypsin inhibitor (mol. wt. 3.0×10^4) from human urine is composed of two Kunitz-type domains.[965] The C-terminal domain is responsible for antitryptic activity but no inhibitory activity has been designated to the other domain. Both the N-terminal extension peptide and the latter domain are linked O-glycosidically and N-glycosidically, respectively. The O-glycosyl residues are attached to L-serine-10 while the N-glycosyl residues are to L-asparagine-24 on the first Kunitz-type domain.[966]

Glycopeptides have been isolated from proteolytic digests of human Tamm-Horsfall glycoprotein and its asialo derivative.[967] The glycoprotein contains at least five L-asparagine residues substituted by complex carbohydrate moieties, three being of one type, relatively rich in D-galactosyl residues, and two containing less D-galactose but more neuraminic acid.

The structures (65-67) of three oligo-D-mannoside-type L-asparagines isolated from the urine of a patient with Gaucher's disease have been elucidated using 500 MHz ^{1}H n.m.r. spectroscopy.[968]

A reduction or absence of activity of the glycoprotein degrading lysosomal amidase 1-aspartamido-β-2-acetamido-2-deoxy-D-glucose amidohydrolase (EC 3.5.1.26) is the basic enzymic defect in aspartylglycosaminuria.[969] Urinary sialyglycoconjugates have been studied in a number of patients with this inherited deficiency. The levels of the main metabolite, 2-acetamido-2-deoxy-1-β-L-asparaginyl-D-glucose, in various neural and extraneural tissues of a patient suffering from aspartylglycosaminuria have been measured by g.l.c.[970] Correlation of the results with the clinical manifestations of the disease did not reveal any direct relationship between the amount of 2-acetamido-2-deoxy-1-β-L-asparaginyl-D-glucose stored and the degree of organ dysfunction.

Six mono- and di-saccharides have been determined simultaneously by h.p.l.c. of deionized human urine.[971] Improvements in a detector based on optical activity of the components allow the detection of 100 ng of sugar.

R_1-(1→3)-β-D-Manp-(1→4)-β-D-GlcpNAc-(1→4)-β-D-GlcpNAc-1→L-Asn
6
|
1
R_2-(1→3)-α-D-Manp
6
|
1
R_3

(65) $R_1 = R_2 = R_3 = H$
(66) R_1 = α-D-Manp, $R_2 = R_3 = H$
(67) $R_1 = R_2 = R_3$ = α-D-Manp

The structure of an oligosaccharide (68) containing nine D-mannosyl residues and isolated from patients with mannosidosis has been similarly assigned.[972] Excessive gingival hyperplasia with storage of D-mannose-rich oligosaccharides appears to be a unique feature present in a patient exhibiting mannosidosis.[973] Eight characteristic D-manno-oligosaccharides each terminated at the reducing end with a 2-acetamido-2-deoxy-D-glucosyl residue were isolated from the patients' urines. In contrast to the urine in which a trisaccharide is predominant, tetrasaccharides and pentasaccharides are more abundant in gingiva.

The unique structure (69) of a trisaccharide from a patient with mannosidosis has been deduced by ^{13}C n.m.r. spectroscopy in terms of sugar composition, ring forms, anomeric configurations, sequence, and inter-residue linkage positions.[974]

In a case of feline mannosidosis, a marked deficiency of α-D-mannosidase in brain, kidney, and liver tissue is accompanied by high concentrations of neutral oligosaccharides in the urine.[975] A D-mannosyl-rich hexasaccharide was found to be the predominant oligosaccharide.

A set of procedures for screening of patients' urine to detect oligosaccharide-storage diseases has been described.[976] Urines from patients with mucolipidosis I, mannosidosis, fucosidosis, aspartylglycosaminuria, and type I glycogen-storage disease can be distinguished by t.l.c. β-D-Galactosidase deficiency in patients can be detected by examination of the urine using a combination of

```
α-D-Manp-(1→2)-α-D-Manp-(1→2)-α-D-Manp
                                  1
                                  |
                                  3
                              β-D-Manp(1→4)-D-GlcNAc
                                  6
                                  |
                                  1
α-D-Manp-(1→2)-α-D-Manp-(1→3)-α-D-Manp
                                  6
                                  |
                                  1
               α-D-Manp-(1→2)-α-D-Manp
```

(68)

ion-exchange chromatography and t.l.c. Excess sialyloligosaccharide excretion is detected by gel filtration followed by estimation of neuraminic acid.

Oligosaccharide patterns obtained by gel filtration of urine of G_{M1} gangliosidosis (Type 1) patients differ from those of G_{M1} gangliosidosis (Type 2) patients.[977] The amount of total oligosaccharide excreted in the urine of Type 2 patients is 10-20% of that in the urine of Type 1 patients. Oligosaccharides having the structures (70-84) occur in the urine of Type 1 patients. The same oligosaccharides with the exception of (76, 77, 79, 84) are excreted in the urines of Type 2 patients.

α-D-Manp-(1→3)-β-D-Manp-(1→4)-D-GlcNAc

(69)

Four positional isomers of neuraminosyl oligosaccharides have been isolated from the urine of a patient with sialidosis with partial deficiency of β-D-galactosidase.[978] Their structures are identical to oligosaccharides (70-71) bearing either (2→3)-α-neuraminosyl or (2→6)-α-neuraminosyl terminal residues.

Two siblings with neuronal ceroid lipofucosinosis excreted the blood-group A trisaccharide, 3-O-(2-acetamido-2-deoxy-α-D-galactopyranosyl)-2-O-(α-L-fucopyranosyl)-D-galactose, in the urine.[979]

17 Avian Glycoproteins

Absorption spectra, c.d. spectra, and sedimentation-equilibrium values have been reported for complexes of lutein with egg ovalbumin.[980]

Ovalbumin has been subfractionated on immobilized concanavalin A.[981] A distinctive carbohydrate composition in each of four fractions was noted. High-field ^{1}H n.m.r. spectroscopy has been used to demonstrate heterogeneity of chick ovalbumin glycopeptides.[982] An unambiguous assignment of all C1-H and D-mannose C2-H resonances was made, enabling the structures of the glycopeptides to be assigned.[983]

The complete amino acid sequence of hen ovalbumin, comprising 385 residues, has been determined.[984] Ovalbumin has four sites of postsynthetic modification: in addition to the acetylated N-terminus, the carbohydrate moiety is located at L-Asn-292, and the

```
β-D-Galp-(1→4)-β-D-GlcpNAc-(1→2)-α-D-Manp-(1→3)-β-D-Manp-(1→4)-D-GlcNAc
                              (70)

β-D-Galp-(1→4)-β-D-GlcpNAc-(1→2)-α-D-Manp-(1→6)-β-D-Manp-(1→4)-D-GlcNAc
                              (71)

                                                  α-D-Manp
                                                     1
                                                     |
                                                     6
β-D-Galp-(1→4)-β-D-GlcpNAc-(1→2)-α-D-Manp-(1→3)-β-D-Manp-(1→4)-D-GlcNAc
                              (72)

β-D-Galp-(1→4)-β-D-GlcpNAc-(1→2)-α-D-Manp-(1→6)-β-D-Manp-(1→4)-D-GlcNAc
                                                     3
                                                     |
                                                     1
          β-D-Galp-(1→4)-β-D-GlcpNAc-(1→2)-α-D-Manp
                              (73)
```

```
β-D-Galp-(1→4)-β-D-GlcpNAc-(1→2)-α-D-Manp-(1→6)-β-D-Manp-(1→4)-D-GlcNAc
                                                        3
                                                        |
                                                        1
            β-D-Galp-(1→4)-β-D-GlcpNAc-(1→2)-α-D-Manp
                                                        2
                                                        |
                                                        1
                              β-D-Galp-(1→4)-β-D-GlcpNAc
                               (74)

β-D-Galp-(1→4)-β-D-GlcpNAc-(1→6)-α-D-Manp-(1→6)-β-D-Manp-(1→4)-D-GlcNAc
                                        2               3
                                        |               |
                                        1               1
                                 β-D-GlcpNAc      α-D-Manp
                                        4               2
                                        |               |
                                        1               1
                                 β-D-Galp         β-D-GlcpNAc
                                                        4
                                                        |
                                                        1
                                                  β-D-Galp
                                  (75)
```

```
β-D-Galp-(1→4)-β-D-GlcpNAc-(1→2)-α-D-Manp-(1→6)-β-D-Manp-(1→4)-D-GlcNAc
                                                      3
                                                      |
                                                      1
β-D-Galp-(1→4)-β-D-GlcpNAc-(1→3)-β-D-Galp-(1→4)-β-D-GlcpNAc-(1→2)-α-D-Manp
```

(76)

```
β-D-Galp-(1→4)-β-D-GlcpNAc-(1→3)-β-D-Galp-(1→4)-β-D-GlcpNAc-(1→2)-α-D-Manp-(1→6)-β-D-Manp-(1→4)-D-GlcNAc
                                                                                         3
                                                                                         |
                                                                                         1
                                             β-D-Galp-(1→4)-β-D-GlcpNAc-(1→2)-α-D-Manp
```

(77)

```
                          β-D-Galp-(1→4)-β-D-GlcpNAc
                                               1
                                               |
                                               2
β-D-Galp-(1→4)-β-D-GlcpNAc-(1→6)-α-D-Manp-(1→6)-β-D-Manp-(1→4)-D-GlcNAc
                                                          3
                                                          |
                                                          1
                β-D-Galp-(1→4)-β-D-GlcpNAc-(1→4)-α-D-Manp
                                                          2
                                                          |
                                                          1
                               β-D-Galp-(1→4)-β-D-GlcpNAc
                                (78)

R1-β-D-Galp-(1→4)-β-D-GlcpNAc-(1→2)-α-D-Manp-(1→6)-β-D-Manp-(1→4)-D-GlcNAc
                                                            3
                                                            |
                                                            1
             R2-β-D-Galp-(1→4)-β-D-GlcpNAc-(1→4)-α-D-Manp
                                                            2
                                                            |
                                                            1
                              R3-β-D-Galp-(1→4)-β-D-GlcpNAc
                             (79)  R1 = β-D-Galp-(1→4)-β-D-GlcpNAc-(1→3), R2 = R3 = H
                             (80)  R2 = β-D-Galp-(1→4)-β-D-GlcpNAc-(1→3), R1 = R3 = H
                             (81)  R3 = β-D-Galp-(1→4)-β-D-GlcpNAc-(1→3), R1 = R2 = H
```

```
                      R2-β-D-Galp-(1→4)-β-D-GlcpNAc
                                              1
                                              |
                                              2
R1-β-D-Galp-(1→4)-β-D-GlcpNAc-(1→6)-α-D-Manp-(1→6)-β-D-Manp-(1→4)-β-D-GlcNAc
                                                             3
                                                             |
                                                             1
                    R3-β-D-Galp-(1→4)-β-D-GlcpNAc(1→2)-α-D-Manp
```

(82) R_1 = β-D-Galp-(1→4)-β-D-GlcpNAc-(1→3)-
R_2 = R_3 = H

(83) R_2 = β-D-Galp-(1→4)-β-D-GlcpNAc-(1→3)-
R_1 = R_3 = H

(84) R_3 = β-D-Galp-(1→4)-β-D-GlcpNAc-(1→3)-
R_1 = R_2 = H

two phosphorylated L-serines are at residues 68 and 344.

Two glycopeptides have been isolated from pepsin-treated chicken ovalbumin.[985] Both high D-mannose-type and hybrid-type oligosaccharides were released by an almond emulsin glycopeptidase.

Natural-abundance ^{13}C n.m.r. spectroscopy has been used to show that some α-D-mannopyranosyl-(1→3)-D-mannopyranosyl linkages in hen ovalbumin glycopeptides are cleaved much faster than the corresponding (1→6)-linkages by jack-bean α-D-mannosidase.[986] In addition the technique was used to examine the structures of partially cleaved species produced during the action of glycosidases on oligosaccharides, glycopeptides, and glycoproteins.

Ovotransferrin, in addition to human serum transferrin, is oxidized by sodium periodate with a resulting destruction of L-tyrosine residues and a loss of iron-binding activity.

The structural determination of one of the glycopeptides obtained by proteolysis of turtle-dove ovomucoid has revealed the presence of a terminal trisaccharide (85), a specific blood-group P_1 antigenic determinant previously described for human erythrocyte P_1 antigen.[987]

The secondary structures of chicken ovomucoid contain α-helix (20%), β-structure (46%), and random coil (18%).[988] Secondary-structure predictions for chicken egg-white ovomucoid have been published.[989] Some of the earlier predictions on structure are incorporated in this model, and discussion includes conformational characteristics of the glycoprotein, the secondary structure at the reactive site, and the structural homology among three domains.

18 Miscellaneous Glycoproteins and Chitin

The structures of fifteen oligosaccharides (86-100) isolated from

α-D-Galp-(1→4)-β-D-Galp-(1→4)-D-GlcNAc

(85)

β-D-Galp-(1→3)-D-GlcNAc

(86)

β-D-Galp-(1→3)-D-GalNAc

(87)

```
β-D-Galp-(1→4)-D-GlcNAc
                  3
                  |
                  1
            α-L-Fucp
          (88)
```

```
R-(1→2)-β-D-Galp-(1→4)-D-Glc
                           3
                           |
                           1
                     α-L-Fucp
            (89)   R = H
            (90)   R = α-L-Fucp
```

```
β-D-Galp(1→3)-β-D-GlcpNAc-(1→3)-D-Glc
                   (91)
```

```
β-D-Galp-(1→-4)-β-D-GlcpNAc-(1→3)-D-Gal
                   (92)
```

```
                R1-(1→2)-α-D-Manp
                               1
                               |
                               6
R2-(1→2)-α-D-Manp-(1→3)-β-D-Manp-(1→4)-β-D-GlcNAc
                               4
                               |
                               1
                        β-D-GlcpNAc
A = α-Neup5Ac-(2→6)-β-D-Galp-(1→-4)-β-D-GlcpNAc
B = β-D-GlcpNAc
C = β-D-Galp-(1→4)-β-D-GlcpNAc

          (93)   R1 = H, R2 = A
          (94)   R1 = A, R2 = H
          (95)   R1 = A, R2 = B
          (96)   R1 = B, R2 = A
          (97)   R1 = C, R2 = A
          (98)   R1 = A, R2 = C
          (99)   R1 = R2 = A
         (100)   R1 = B, R2 = C
                 or R1 = C, R2 = B
```

newborn meconium have been identified.[990] It is proposed that endoglycanases are involved in the production of these compounds from glycoproteins.

D-Galactans have been isolated from the albumen glands from snails of the genus *Biomphalaria*.[991] The results of methylation analysis and periodate oxidation data indicate the presence of a multibranched structure containing (1→3)- and (1→6)-linkages. Polysaccharides containing both D- and L-galactosyl residues have been isolated from the albumen glands of four species of snail.[992] Antisera to the four galactans showed various degrees of cross-reactivity, indicating structural differences, ascribable in part to determinants involving galactose phosphate, and probably also to the linkage and position of L-galactosyl residues in the molecule. A β-D-galactopyranan, isolated from the albumen gland of the snail *Pomacea lineata*, contains 3,4-O-(1-carboxyethylidene) groups, as shown by the liberation of pyruvic acid on acid hydrolysis and by methylation data.[993] The structure of the 5-membered acetal groups of pyruvic acid was studied by ^{13}C n.m.r. spectroscopy.

The role of glycosylation in the secretion of various avian and mammalian proteins synthesized in *Xenopus laevis* oocytes under the direction of heterologous messenger RNA has been reported.[994] The presence of N-glycosyl chains on the proteins did not appear to facilitate secretion. Oligosaccharides with a new type of core structure (101) have been isolated from trout egg glycoprotein.[995] The occurrence of N-glycolylneuraminic acid linked (2→3) to the internal 2-acetamido-2-deoxy-D-galactosyl residue has not been described in other glycoproteins or glycolipids.

```
                                          Neup5G
                                            2
                                            |
                                            3
β-D-GalpNAc-(1→4)-β-D-GalpNAc-(1→4)-β-D-GalpNAc-(1→3)-
                                          β-D-Galp-(1→3)-GalNAc-ol
```

(101)

Bovine hypothalmic mRNA translated in a reticulocyte lysate system yields a common precursor (mol. wt. 2.1 x 10^4) by microsomal membranes from dog pancreas.[996] This precursor binds to immobilized concanavalin A and is sensitive to treatment with α-D-mannosidase.

Purified variant-specific glycoprotein antigens of *Trypanosoma brucei* exist in solution as dimers and occasionally as higher oligomers.[997] Two variant surface glycoproteins of *T. brucei* have been shown to have a conserved C-terminal amino acid sequence.[998] The glycosylation of the major variable surface coat glycoprotein of *T. brucei* is inhibited by tunicamycin.[999] *N*-Linked glycosylation occurs subsequent to synthesis of the protein.

Variant antigens of *Trypanosoma congolense* have been purified by lectin affinity chromatography.[1000]

A molecular-weight analysis of the major polypeptides and glycopeptides of the protozoan *Toxoplasma gondii* has been reported.[1001]

A sulphated glycoprotein (mol. wt. 1.85×10^5) is synthesized in the multicellular organism *Velvox carteri* only during the limited period of embryogenesis.[1002] Evidence for an embryonic control function of this cell-surface glycoprotein is provided.

Small amounts of chitin in arthropod cuticles can be measured by estimating the derivatized soluble chitosan.[1003] The method is not affected by the presence of non-chitinous cuticular components.

Covalently bound chitin-protein complexes of several species of marine invertebrates have been isolated.[1004] After deproteinization, all the samples contained small amounts of residual amino acids with marked species variation in their identity and content.

A comparable series of chitins derived from blue, red, stone, and horseshoe crabs by treatment with mild acid, H_4 e.d.t.a., and alkali have been isolated.[1005] The polysaccharides comprise a family of closely related products with variable solubility, molecular weight, optical rotation, and acetyl values, which are a function of both the species from which they originated and their method of preparation.

Increased rates of hydrolysis by chitinase and lysozyme of chitin and chitosan after oxidation with sodium metaperiodate have been observed.[1006]

19 Analysis of Glycoproteins

The structure, elucidation of structures, and functions of carbohydrate residues on glycoproteins have been reviewed.[1007] Methods for analysing complex mixtures of antigens and glycoproteins have been reviewed.[1008]

Sialoglycoproteins have been fractionated on an immobilized

neuraminic acid-binding lectin from the horseshoe crab <u>Carcinoscorpius</u> <u>rotunda</u> <u>cauda</u>.[1009] The resolution of the isoenzymes of alkaline phosphatase was achieved.

Polyacrylamide gel electrophoresis has been used to resolve fluorescein-derivatized asialoglycopeptides.[1010] Specific exoglycosidase digestion of the glycopeptides generates a series of degradation products that are resolved.

A method for monitoring the removal of poly-D-mannosyl chains from glycoproteins by <u>endo</u>-2-acetamido-2-deoxy-D-glycanase H has been described.[115] After polyacrylamide gel electrophoresis of the glycoproteins and their enzymically degraded products, the gels are overlaid with iodinated lectins. The degree to which these lectins bind is then measured by autoradiography.

A peptide:N-glycanase has been purified from almond emulsin and its specificity examined with glycopeptides prepared from ovalbumin and immunoglobulin M.[1011] Both high D-mannose and complex glycopeptides are hydrolysed at the β-L-aspartylglycosylamine linkage.

A method for the fluorescent labelling of sialoglycoproteins involves oxidation with sodium periodate followed by condensation with fluoresceinamine or monodansyl-1,2-diaminoethane followed by stabilization of the Schiff's bases by reductive amination.[690]

Lectins have been used for the detection of glycoproteins transferred to nitrocellulose sheets, after their electrophoretic separation on polyacrylamide gels.[1012]

Oligosaccharides derived from glycoproteins by enzymic release with endoglycosidases or chemical release by hydrazinolysis have been reduced by sodium borotritide and separated by h.p.l.c.[1013]

Treatment of glycoproteins with trifluoromethanesulphonic acid at room temperature results in the rapid cleavage of peripheral sugars, a slow loss of L-serine and L-threonine-linked 2-acetamido-2-deoxy-D-galactosyl residues, and retention of N-glycosidically linked 2-acetamido-2-deoxy-D-glucosyl residues.[1014] The reagent is safer to use than either hydrogen fluoride or its pyridine complex.

Glycoproteins have been iodinated using the solid-phase oxidizing agent 1,3,4,6-tetrachloro-3α,6α-diphenylglycouril (iodogen).[1015] The method compares favourably with the solid-phase lactoperoxidase and chloramine T methods.

A method for determination of μl volumes of glycoprotein using spot analysis on cellulose acetate layers is described.[1016] With D-glucose oxidase and fluorescein-labelled concanavalin A sensitivity of 10 ng is reached. This level of detection is lowered to 1 ng of D-glucose oxidase when the reagents are used in combination with horse-radish peroxidase.

^{13}C n.m.r. spectroscopic data for 3-*O*-(2-acetamido-2-deoxy-α-D-galactopyranosyl)-*N*-acetyl-L-serine, 3-*O*-(2-acetamido-2-deoxy-α-D-galactopyranosyl)-*N*-acetyl-L-threonine, 3-*O*-α,β-D-galactopyranosyl-L-serine, and 3-*O*-α,β-D-galactopyranosyl-L-threonine have been recorded.[1017] These model compounds represent common carbohydrate-protein linkage regions of many glycoproteins.

The structures of the oligosaccharide components of a number of glycopeptides have been established following hydrazinolysis-nitrous acid deamination, which leads to the specific cleavage of 2-amino-2-deoxy-D-glucosyl linkages.[883] The derived 2,5-anhydro-D-mannose-containing oligosaccharides are reduced with sodium borohydride and analysed by g.l.c.-m.s. of their methylated ethers.

The flexibility of bi- and tri-antennary glycans of the *N*-acetyl-lactosamine type has been studied by spin-labelling the neuraminic acid residues in glycopeptides of known structure.[1018]

A detailed analysis of the 360 MHz ^{1}H n.m.r. spectral parameters for the anomeric and C2-H resonances of a large number of glycopeptides and oligosaccharides of known structure reveals a general method for the determination of the primary structure of glycopeptides for most currently known classes of structures.[1019] A two-dimensional display formed by plotting D-mannosyl C1-H against C2-H chemical shifts demonstrates that these pairs of values are sensitive to long-range perturbation by remote substitution by hexoses as well as to direct substitution effects. A total of forty-one of these chemical-shift clusters have been defined which characterize unique structural microenvironments. On this basis the sequence and branching pattern for most structures can be derived.

The hydrodynamic behaviour of reduced glycopolypeptides has been studied by gel filtration in guanidine hydrochloride in conjunction with h.p.l.c.[1020] Although carbohydrate-rich glycopolypeptides may occasionally yield underestimated values of molecular weights, the method is suitable for the rapid estimation of molecular weights of simple polypeptides and glycopolypeptides.

Neuraminic acid residues are involved in the electro-immunodiffusion but not the radial immunodiffusion of α_1-acid glycoprotein.[1021] Differential determination of the glycoprotein by the two immunological methods offers a method for the estimation of the degree of sialylation of this and other serum glycoproteins.

The structures of sialo-oligosaccharides as determined by ^1H n.m.r. spectroscopy together with methylation analysis have been shown to vary with those derived from periodate oxidation data alone.[1022] Examination of acidic tetrasaccharides obtained from A^+, H^+, and A^-H^- hog submaxillary mucin indicates that neuraminic acid residues are not released from oligosaccharides when periodate oxidation is carried out at pH 4.5. The anionic properties of the neuraminic acid residues were used to separate the periodate oxidation products, thereby giving information on the location of neuraminic acid on the oligosaccharide chain.

The monosaccharide components of glycoproteins have been reacted with dansyl hydrazine and the resulting derivatives separated by h.p.l.c.[1023] Fluorescent detection provides a sensitivity of 10 pmol per injection.

Partially methylated O-methyl O-acetyl methyl glycosides obtained by methanolysis and acetylation of glycoproteins have been identified by g.l.c.-m.s. analysis.[1024] A scheme for the rapid determination of the position of the methyl and acetyl residues is proposed. The method was applied to sialoglycopeptides isolated from α_1-acid glycoprotein.

Methylation techniques used in the structural analysis of glycoproteins and glycolipids have been reviewed.[1025] A modification of a g.l.c. method allowing simultaneous separation of neutral and amino sugar alditol acetates has been described.[1026] The method has been applied to the monitoring of the fractionation of complex mixtures of glycoproteins and glycosaminoglycans. A rapid isocratic h.p.l.c. method for the estimation of neuraminic acid in serum has been described.[1027] Separation is achieved on a cation-exchange resin using a 0.006N sulphuric acid mobile phase.

Reviews dealing with glycosyltransferases and their use in assessing oligosaccharide structure and structure-function relationships have been published.[1028,1029]

Lysosomal enzymes contain D-mannosyl 6-phosphate moieties which mediate their translocation to lysosomes.[1030] A new assay for α-D-2-acetamido-2-deoxy-glucosyl phosphotransferase using α-methyl-D-mannoside as acceptor has been reported.

A coupled enzyme assay for 2-acetamido-2-deoxy-D-glucose : UDP-D-galactose D-galactosyl transferase has been developed that allows the enzyme to be assayed spectrophotometrically and in denaturing polyacrylamide gels.[1031] In a sequence of linked enzyme reactions the liberated UDP is successively converted to UTP, UDP-D-glucose, and finally UDP-D-glucuronic acid with the subsequent production of NADH. NADH may then be estimated spectrophotometrically or detected in polyacrylamide gels fluorimetrically.

20 Biosynthesis of Glycoproteins

A review of the transmembrane assembly of membrane and secretory glycoproteins has been published.[1032] Blood-group antigens and the enzymes involved in their synthesis have been reviewed.[1033]

Exposure of rat hepatoma tissue culture cells to dexamethasone results in the appearance of a new glycoprotein (gp35-50, mol. wt. 3.5×10^4 - 5.0×10^4) and increased synthesis of another glycoprotein (mol. wt. 5.0×10^4).[1034] The former glycoprotein is expressed in normal liver, whereas the latter is expressed in hepatoma cells and is regulated differently by steroid hormones.

Treatment of human kidney tumour cells with butyrate increases markedly the synthesis and sulphation of tumour-cell glycoproteins.[1035] Whereas untreated cultured cells release most of their sulphated glycoproteins into the medium, butyrate-treated cells preferentially accumulate these products into the cell layer.

The synthesis and accumulation of D-mannose-containing glycopeptides in human fibroblast cells have been studied.[1036] Results are consistent with the hypothesis that multiple pathways for L-asparagine-linked glycoprotein biosynthesis are possible. The incorporation of D-mannose, D-glucose, and 2-acetamido-2-deoxy-D-glucose into endogenous membrane proteins in calf pituitary has been reported.[1037]

Glycosylation of proteins in the protozoan *Crithidia fasciculata* does not involve glucosylated lipid-bound oligosaccharides as intermediates.[1038] An oligosaccharide containing seven D-mannosyl and two 2-acetamido-2-deoxy-D-glucosyl residues is transferred to protein before undergoing processing.

Carbohydrate-depleted or chemically denatured glycoproteins or synthetic L-Asn-X-L-Ser/L-Thr-containing peptides have been glycosylated using hen oviduct membranes as the enzyme source.[1039] Tripeptides may be glycosylated but the level of glycosylation increases as the length of the peptide chain increases. The peptides were examined by c.d. in aqueous lipid mixtures, when it was observed that the presence of secondary structure in the lipid state promotes the level of glycosylation. The synthesis and processing of L-asparagine-linked oligosaccharides have been reviewed.[1040] The assembly of the lipid-linked precursor oligosaccharide through its transfer to protein and its conversion to the diverse array of final products is discussed. Glycoprotein biosynthesis in mouse L cells involves lipid-linked saccharides in the synthesis of L-asparagine-linked glycoproteins.[1041] The formation of high D-mannosyl oligosaccharide bound to protein can occur independently of higher lipid-linked oligosaccharide synthesis.

Dolichol phosphate appears to control the rate of protein glycosylation during the development of sea-urchin embryos.[1042] Compactin, an inhibitor of polyisoprenoid biosynthesis, inhibits dolichol phosphate synthesis and thus glycosylation of protein. A second effect of the inhibitor is that a significant fraction of the oligosaccharide chains synthesized in the presence of the inhibitor and transferred to protein appear to be altered so that they are more negatively charged. Both de novo synthesis of dolichol as well as its phosphorylation may play an important role in the increase in glycoprotein synthesis during embryonic development of the sea urchin.[1043] The possible role of dolichol kinase in the activation of stored dolichol, and perhaps the pathway for its de novo biosynthesis in this and other biological systems, is discussed.

Turpentine-induced inflammation in rats causes increased synthesis of glycosylated dolichol phosphate derivatives, which are intermediates in the biosynthesis of L-asparagine-linked glycoproteins, and also increased levels of a CTP-dependent dolichol kinase.[1044]

Immature chick oviduct membranes contain a transferase responsible for transfer of oligosaccharide pyrophosphoryl dolichol to protein acceptors.[1045] The enzyme is stimulated by estrogen treatment.

The metabolism of lipid-linked D-mannose intermediates in glycoprotein synthesis during liver regeneration has been studied in rat liver microsomes.[1046] Following partial hepatectomy, the incorporation of D-{^{14}C}-mannose from GDP-D-{^{14}C}-mannose into dolichyl phosphoryl D-mannose and protein decreases as compared with controls. Evidence is given suggesting that the incorporation of D-mannose into glycoprotein in regenerating rat liver may be regulated at the step of oligosaccharide transfer.

Using the Thy-1$^-$ mutant mouse lymphoma cells of the class E complementation group which lack GDP-D-mannose : dolichol phosphoryl mannosyltransferase and which are therefore unable to interconvert GDP-D-mannose and D-mannosyl phosphoryl dolichol, further evidence has been provided that five of the nine D-mannosyl residues which are added to the growing lipid-linked oligosaccharides are donated directly by GDP-D-mannose.[1047] The remaining four residues arise from D-mannosyl phosphoryl dolichol.

The reactions involved in the addition of D-glucosyl residues to the oligosaccharide-lipid intermediates by thyroid microsomes have been studied.[1048] The properties of the enzymes involved in the transfer of D-glucose from its phosphoryl dolichol derivative to endogenous acceptors are described and the products are characterized in terms of their structure and capacity to serve as donors in the glycosylation of proteins. The biosynthesis and partial characterization of a D-glucose-containing insect lipid-linked oligosaccharide have been reported.[1049] Its properties are closely related to the D-glucosylated dolichyl oligosaccharide obtained from mammalian systems.

Amphomycin inhibits the transfer of D-mannose, D-glucose, and 2-acetamido-2-deoxy-D-glucose 1-phosphate from their respective nucleotide derivatives to dolichol phosphate by membrane preparations from calf brain membranes.[1050] Other D-mannosyl-, D-glucosyl-, 2-acetamido-2-deoxy-D-glucosyl-transferases associated with the same preparation are not affected by the antibiotic.

UDP-2-Deoxy-D-*arabino*-hexose (UDP-2-deoxy-D-glucose) inhibits the formation of D-glucosylphosphoryl dolichol in chick-embryo cell membranes, but has no effect on the formation of *NN*'-diacetylchitobiosyl pyrophosphoryl dolichol.[1051] GDP-2-Deoxy-D-*arabino*-hexose inhibits the formation of both the lipid intermediates by competition with physiological nucleotide sugars for dolichol phosphate.

The influence of various aryl phosphates and phosphonates on the synthesis of glycosylphosphoryl dolichol, oligosaccharides, and glycoproteins by rat liver microsomal fractions has been investigated.[1052] Phenyl phosphate directly effects the synthesis of D-mannosyl phosphoryl dolichol, and the inhibition of lipid-linked oligosaccharides and glycoproteins may be a consequence of the effect.

The antibiotic tsushimycin inhibits the formation of D-glucosyl-, D-mannosyl-, and 2-acetamido-2-deoxy-D-glucosyl phosphoryl dolichol, but allows the incorporation of sugars into lipid-linked oligosaccharides in a particulate enzyme preparation from pig aorta.[1053] The major oligosaccharide formed from GDP-D-mannose in the presence of the antibiotic is (D-Man*p*)$_5$-(D-Glc*p*NAc)$_2$.

Tunicamycin has little effect on the proteins in the epidermis when pig skin slices are maintained in organ culture, although the synthesis of specific epidermal glycoconjugates is affected.[1054] The glycosylation of particulate glycoproteins is decreased, whereas that of soluble glycoproteins is hardly affected. Inhibition of sulphated glycosaminoglycans but not of hyaluronic acid was also noted.

The peptide antibiotic tridecaptin stimulates the incorporation of D-mannose from GDP-D-{^{14}C}-mannose and D-glucose from UDP-D-{^{3}H}-glucose into lipid-linked monosaccharides by enzyme fractions from pig aorta.[1055] The antibiotic also stimulates the incorporation of sugars into lipid-linked oligosaccharides.

Epithelial cells of the rat small intestine synthesize D-mannosyl phosphoryl dolichol and oligosaccharyl pyrophosphoryl dolichol, which are similar to those found in a variety of other tissues.[1056] The importance of monitoring and controlling the degradation of nucleotide sugars when glycosyltransferase activities are being compared in differentiating cells is demonstrated.

Incubation of cell-free extracts of the brine shrimp (*Artemia salina*) with GDP-D-mannose or UDP-D-glucose results in the formation of the corresponding dolichyl derivatives of the sugars.[1057] The enzymic activities were detected early in the development of the encysted *Artemia* embryos.

Starvation of D-glucose alters lipid-linked oligosaccharide biosynthesis in Chinese hamster ovary cells.[1058] The cells quickly cease synthesizing the (D-Glc*p*)$_3$(D-Man*p*)$_9$-(D-Glc*p*NAc)$_2$ species and instead make predominantly the (D-Man*p*)$_5$(D-Glc*p*NAc)$_2$ species, which is glucosylated and transferred to protein where it is subsequently processed, thus demonstrating the use of an alternate glycosylation pathway.

An oligosaccharyl pyrophosphoryl lipid from porcine liver is composed of a tetrasaccharide with equal quantities of D-mannose and 2-acetamido-2-deoxy-D-glucose.[1059]

D-Mannosyltransferase II has been purified from rabbit liver microsomes.[1060] The enzyme catalyses direct transfer from GDP-D-mannose to oligosaccharyl-pyrophosphoryl lipids with the resulting formation of α(1→3)-D-mannosyl-D-mannose linkages.

A number of cell lines of ricin-resistant baby-hamster kidney cells have been assayed for gross carbohydrate composition of cellular glycoproteins, for glycosidase and glycosyltransferase activity.[1061] None of the changes in glycosyl-transfer reactions in these lines is due to enhanced glycosidase or sugar nucleotidase activities in the mutant cells.

Lactose synthetase A protein from human serum has been purified and characterized.[1062] The remarkable immunological dissimilarity between bovine serum galactosyltransferase and this enzyme may be related to major differences in the carbohydrate moieties of the two enzymes.

The biosynthesis of erythroglycan-like products by incubation of a microsomal fraction derived from chronic myelogenous leukaemia-derived cells with UDP-2-acetamido-2-deoxy-D-{6-^{3}H}-glucose and UDP-D-{6-^{3}H}-galactose has been described.[1063] The high-molecular-weight products of the transferase-catalysed reactions were partially characterized after enzymic degradation as a di-, tri-, and unidentified oligo-saccharide.

The presence of glycosyltransferases on chromatin acceptors in monkey liver nuclear membranes has been demonstrated.[1064] The eventual role of these enzymes in the glycosylation of non-histone proteins is discussed.

An α-D-mannose : β1,2-(2-acetamido-2-deoxy-D-glucosyl) transferase has been isolated and purified from porcine tracheal mucosa.[1065] The enzyme forms β1,2 bonds between 2-acetamido-2-deoxy-D-glucose and terminal branched D-mannosyl residues of glycoproteins and glycopeptides.

Iodine is readily incorporated from iodine chloride into D-galactosyltransferase from bovine milk with total loss of enzymatic activity.[1066] Substrates and α-lactalbumin protect against inactivation and the only amino acid modified is L-tyrosine.

An α-D-galactosyltransferase activity in Ehrlich ascites tumour cells has been characterized.[1067] The enzyme has been shown to transfer D-galactosyl residues from UDP-D-galactose to N-acetyl-lactosamine in the synthesis of the unique trisaccharide (99). The endogenous localization of UDP-galactose : asialomucin galactosyltransferase activity in rat liver endoplasmic reticulum and Golgi apparatus has been reported.[1068]

Partially purified UDP-galactosyltransferase from bovine milk has been used to synthesize millimolar amounts of 4-O-β-D-galactopyranosyl-D-glucose, 2-acetamido-2-deoxy-4-O-β-D-galactosyl-D-glucosyl-β-hexanolamine, and O-β-D-galactosyl-(1→4)-O-β-2-

acetamido-2-deoxy-D-glucosyl-(1→4)-2-acetamido-2-deoxy-β-D-glucose.[1069] ^{13}C-Enriched D-galactopyranosyl moieties of the same oligosaccharide were also synthesized.

Peptides containing a tri-L-prolyl sequence carboxyl to an L-threonine residue can be O-glycosylated by crude extracts of porcine submaxillary glands containing UDP-2-acetamido-2-deoxy-D-galactose : polypeptide 2-acetamido-2-deoxy-D-galactose transferase.[1070] Using eleven synthetic peptide substrates, it is revealed that L-threonine cannot be glycosylated without a carboxyl tri-L-prolyl sequence and without blocking of the N-terminal group of the L-threonine.

One form of 2-α-L-fucosyltransferase and one form of 3-α-L-fucosyltransferase have been detected in human serum.[1071] Human milk contains three forms of 3-α-L-fucosyltransferase and one form of 4-α-L-fucosyltransferase.

Three α-L-fucosyltransferase activities have been purified from human milk.[1072] The three activities appear to be contained in a single enzyme. Evidence for two distinct 3-α-L-fucosyltransferase activities in human saliva has been reported.[1073] Both enzymes appear to catalyse the transfer of L-fucose to O-3 of 2-acetamido-2-deoxy-β-D-glucosyl residues of blood-group glycoproteins, but only the transferase dependent on the expression of the Le gene has the capacity to transfer L-fucose to O-3 of D-glucosyl residues.

Evidence for the existence of the postulated UDP-2-acetamido-2-deoxy-D-glucose : glycoprotein 2-acetamido-2-deoxy-D-glucose 1-phosphotransferase has been established.[1074] This enzyme is involved in the phosphorylation of the high D-mannose oligosaccharide units of lysosomal enzymes.

The D-mannosyl 6-phosphate moieties act as recognition markers on lysosomal enzymes.[1030] The marker is synthesized by the sequential action of UDP-2-acetamido-2-deoxy-D-glucose : lysosomal enzyme 2-acetamido-2-deoxy-D-glucose 1-phosphotransferase and an α-D-2-acetamido-2-deoxyglucosyl phosphodiesterase. It is proposed that the transferase catalyses the initial determining step by which newly synthesized acid hydrolases are distinguished from other newly synthesized glycoproteins and thus are eventually targeted to lysosomes.

Glycosylated derivatives of lysozyme have been used as substrates for the determination of the specificity of sialyltransferases.[1075] Characteristic differences between various sialyltransferases were demonstrated.

Fetal-calf liver, embryonic-chicken brain, human placenta, and several other tissues contain CMP-N-acetylneuraminyl : β-D-galactosyl-(1→4)-2-acetamido-2-deoxy-D-glucosyl-α-(2→3)-sialyltransferase.[1076] This sialyltransferase activity appears to be due to an enzyme which is distinct from all other known sialyltransferases.

Elevated sialyltransferase activities have been detected in the intestinal lymph of colchicine-treated rats.[1077] Enterectomy does not prevent the rise of sèrum sialyltransferase, suggesting that the intestine is not the major source of the serum enzyme.

Colchicine has an effect on glycoprotein synthesis in rat liver Golgi membranes by affecting glycosyltransferases.[1078] Although both sialyl- and galactosyl-transferases are inhibited, their sensitivity to the drug differs.

(References begin opposite)

References

1 B.P. Wasserman and H.P. Hultin, Arch. Biochem. Biophys., 1981, 212, 385.
2 A. Ross and I.E.P. Taylor, Infection and Immunity, 1981, 31, 911.
3 K. Toda, K. Ono and H. Ochiai, J. Biochem. (Tokyo), 1981, 90, 1429.
4 W. Breuer and C.H. Sui, Proc. Natl. Acad. Sci. USA, 1981, 78, 2115.
5 Y. Tamai, H. Shinmoto and M. Takakowa, Agric. Biol. Chem. (Jpn), 1981, 45, 2713.
6 W.J. Zhang and C.E. Ballou, J. Biol. Chem., 1981, 256, 10073.
7 D. Pang, L. Ballou, W.J. Zhang and C.E. Ballou, J. Biol. Chem., 1981, 256, 10080.
8 H.D. Klenk, Cell Membranes and Viral Envelopes, 1981, 2, 519.
9 W.J. Grimes, G.N. Irwin and L.M. Patt, Cell Membranes and Viral Envelopes, 1980, 2, 541.
10 V.S. Hinshaw, R.G. Webster and R.J. Rodriguez, Arch. Virol., 1981, 67, 191.
11 H.F. Lodish, A. Zilberstein and M. Porter, Perspectives in Virology XI, ed. M. Pollard, A.R. Liss (New York) p31.
12 E.A. Cardoso, Brit. J. Cancer, 1981, 43, 517.
13 D. Portetelle, C. Bruck, M. Mammerickx and A. Burny, Virology, 1980, 105, 223.
14 M.J.F. Schmerr, J.M. Miller and M.J. van der Maaten, Virology, 1981, 109, 431.
15 J.C. Neil, J.E. Smart, M.J. Hayman and O. Jarrett, Virology, 1980, 105, 250.
16 J.K. Collins and B. Chesebro, J. Virol., 1981, 37, 161.
17 M.R. Rosner, J.S. Tung, N. Hopkins and P.W. Robbins, Proc. Natl. Acad. Sci. USA, 1980, 77, 6420.
18 A.M. Schultz, S.M. Lockhart, E.M. Rabin and S. Oroszlan, J. Virol., 1981, 38, 518.
19 M.C. Kemp, N.G. Famulari and R.W. Compans, J. Virol., 1981, 39, 463.
20 D.S. Lyles and K.A. McConnell, J. Virol., 1981, 39, 263.
21 J. Löwer, E.A. Davidson, N.M. Teich, R.A. Weiss, A.P. Joseph and R. Kurth, Virology, 1981, 109, 409.
22 G. Hunsmann, J. Schneider and A. Schulz, Virology, 1981, 113, 602.
23 R. Kornfeld and W.S.M. Wold, J. Virol., 1981, 40, 440.
24 H. Persson, H. Jörnvall and J. Zabielski, Proc. Natl. Acad. Sci. USA, 1980, 77, 6349.
25 Q.S. Kapoor, W.S.M. Wold and G. Chinnadurai, Virology, 1981, 112, 780.
26 D.I. Peterson, J. Biol. Chem., 1981, 256, 6975.
27 V.P. Karelin and V.M. Zhdanov, Mol. Immunol., 1981, 18, 237.
28 R. Eberle and R.J. Courtney, Infection and Immunity, 1981, 31, 1062.
29 R.D. Dix, L. Pereira and J.R. Baringer, Infection and Immunity, 1981, 34, 192.
30 V.C. Carter, P.A. Schaffer and S.S. Tevethia, J. Immunol., 1981, 126, 1655.
31 L. Haar and H.S. Marsden, J. Gen. Virol., 1981, 52, 77.
32 S. Olofsson, S. Jeansson and E. Lycke, J. Virol., 1981, 38, 564.
33 N. Balachandran, D. Harnish, R.A. Killington, S. Bcchetti and W.E. Rawls, J. Virol., 1981, 39, 438.
34 R. Datema and R.T. Schwarz, J. Biol. Chem., 1981, 256, 11191.
35 M.M. Sveda and C.J. Lai, Proc. Natl. Acad. Sci. USA, 1981, 78, 5488.
36 G. Herrler, A. Nagele, H. Meier-Ewert, A.S. Bhown and R.W. Compans, Virology, 1981, 113, 439.
37 A.K. Gitelman, V.A. Berezin and I.G. Karitonenkov, Arch. Virol., 1981, 62, 253.
38 I.A. Wilson, J.J. Skehel and D.C. Wiley, Nature, 1981, 289, 366.
39 D.C. Wiley, I.A. Wilson and J.J. Skehel, Nature, 1981, 289, 373.
40 S. Basak, D.G. Pritchard, A.S. Bhown and R.W. Compans, J. Virol., 1981, 37, 549.
41 D.C. Jackson, L.E. Brown and D.O. White, J. Gen. Virol., 1981, 52, 163.

42 T.A. Dopheide and C.W. Ward, J. Gen. Virol., 1980, 50, 329.
43 C.W. Ward and T.A. Dopheide, Biochem. J., 1981, 193, 953.
44 C.W. Ward and T.A. Dopheide, Biochem. J., 1981, 195, 337.
45 C.W. Ward, L.E. Brown, J.C. Downie and D.C. Jackson, Virology, 1981, 108, 71.
46 W. Gerhard, J. Yewdell, M.E. Frankel and R. Webster, Nature, 1981, 290, 713.
47 W. Garten, T. Kohama and H.D. Klenk, J. Gen. Virol., 1980, 51, 207.
48 T. Kohama, W. Garten and H.D. Klenk, Virology, 1981, 111, 364.
49 A. Helenius, M. Sarvas and K. Simons, Eur. J. Biochem., 1981, 115, 27.
50 K. Hashimoto, S. Erdei, S. Keränen, J. Saraste and L. Kääriänen, J. Virol., 1981, 38, 34.
51 M. Personen, J. Saraste, K. Hashimoto and L. Kääriänen, Virology, 1981, 1981, 109, 165.
52 M. Tomasi and A. Loyter, FEBS Lett., 1981, 131, 381.
53 H. Yoshima, M. Nakanishi, Y. Okada and A. Kobata, J. Biol. Chem., 1956, 256, 5355.
54 M. Hsu, A. Scheid and P.W. Choppin, J. Biol. Chem., 1981, 256, 2557.
55 J. Hakimi and P.H. Atkinson, Biochemistry, 1980, 19, 5619.
56 L.A. Hunt, Virology, 1981, 113, 534.
57 P.A. Chatis and T.G. Morrison, J. Virol., 1981, 37, 307.
58 W.A. Petri and R.R. Wagner, Virology, 1980, 107, 543.
59 J.E. Bergman, K.T. Tokuyasu and S.J. Singer, Proc. Nat. Acad. Sci. USA, 1981, 78, 1746.
60 R.L. Thimmig, J.V. Hughes, R.J. Kinders, A.G. Milenkovic and T.C. Johnson, J. Gen. Virol., 1980, 50, 279.
61 J.R. Etchison, D.F. Summers and C. Georgopoulos, J. Biol. Chem., 1981, 256, 3366.
62 S. Turco, Arch. Biochem. Biophys., 1981, 205, 330.
63 H. Fried, L.D. Cahan and J.C. Paulson, Virology, 1981, 109, 188.
64 D. Mathieu-Mahul, J.P. Barque, M. Mauchauffe, L. Peraudeau and C.J. Larsen, Biochimie, 1981, 63, 403.
65 R.W. Green and J.H. Shaper, Biochim. Biophys. Acta, 1981, 668, 439.
66 D.A. McPhee and E.G. Westaway, J. Gen. Virol., 1981, 54, 149.
67 D.H. du Plessis and P. Smith, Virology, 1981, 109, 403.
68 R.J. Feighny, B.E. Henry and J.S. Pagano, J. Virol., 1981, 39, 651.
69 G.J. Bayliss and H. Wolf, J. Gen. Virol., 1981, 54, 397.
70 D.A. Thorleylawson and K. Geilinger, Proc. Nat. Acad. Sci. USA, 1980, 77, 5307.
71 R.S. Foulke, R.R. Rosato and G.R. French, J. Gen. Virol., 1981, 53, 169.
72 M. Pierotti, A.B. de Leo, A. Pinter, P.V. O'Donnell, U. Hämmerling and E. Fleissner, Virology, 1981, 112, 450.
73 P. Casali, J.G.P. Sissons, R.S. Fujinami and M.B.S. Oldstone, J. Gen. Virol., 1981, 54, 161.
74 D.C. Merz, A. Scheid and P.W. Choppin, Virology, 1981, 109, 94.
75 A. Anilionis, W.J. Wunner and P.J. Curtis, Nature, 1981, 294, 275.
76 L.A. Hunt, W. Lamph and S.E. Wright, J. Virol, 1981, 37, 207.
77 L.A. Hunt and S.E. Wright, J. Virol., 1981, 39, 646.
78 J.T. Reynolds and S. Panem, Virology, 1981, 113, 214.
79 H.J. Thiel, T.J. Matthews, E.M. Broughton, A.W. Butchko and D.P. Bolognesi, Virology, 1981, 112, 642.
80 M. Yoshida and H. Yoshikura, J. Gen. Virol., 1981, 52, 183.
81 S.K. Ruscetti, J.A. Feild and E.M. Scolnick, Nature, 1981, 294, 663.
82 I. Ulmanen, P. Seppala and R.F. Pettersson, J. Virol., 1981, 37, 72.
83 H. Shida and S. Dales, Virology, 1981, 111, 56.
84 M. Oie and Y. Ichihashi, Virology, 1981, 113, 263.
85 C. Grose, B.J. Edmond and W.E. Friedrichs, Infection and Immunity, 1981, 31, 1044.
86 K. Yamamoto, K. Suzuki and B. Simizu, Virology, 1981, 109, 452.
87 S.M.M. Basha and R.M. Roberts, Plant Physiol., 1981, 67, 936.
88 U. Bergner and W. Tanner, FEBS Lett., 1981, 131, 68.
89 T.E. Ferrari, D. Bruns and D.H. Wallace, Plant Physiol., 1981, 67, 270.
90 D.S. Bailey, V. Deluca, M. Durr, D.P.S. Verma and G.A. MacLachlan, Plant

Physiol., 1980, 66, 1113.
91 E.M. Schneider and A. Sievers, Planta, 1981, 152, 177.
92 B.J. Howlett and A.E. Clarke, Biochem. J., 1981, 197, 707.
93 B.J. Howlett and A.E. Clarke, Biochem. J., 1981, 197, 695.
94 H.M. Davies and D.P. Delmer, Plant Physiol., 1981, 68, 284.
95 C.A. Jasalavich and A.J. Andersen, Can. J. Bot., 1981, 59, 264.
96 N. Bar-Nun, A.M. Mayer and N. Sharon, Phytochemistry, 1981, 20, 407.
97 R. Gilles, C. Bittner and L. Jaenicke, FEBS Lett., 1981, 124, 57.
98 H. Lis and N. Sharon, in 'Lectins in Higher Plants; Biochemistry of Plants, ed. A. Marcus, Acad. Press (New York), 1981, Vol 6.
99 S.H. Barondes, Annu. Rev. Biochem., 1981, 50, 207.
100 C.A. Rossi, A. Falasca, C. Franceschi and F. Stirpe, Bull. Mol. Biol. Med., 1979, 4, 59.
101 H. Lis and N. Sharon, J. Chromatogr., 1981, 215, 361.
102 I. Matsumoto, H.. Kitagaki, Y. Akai, Y. Ito and N. Seno, Anal. Biochem., 1981, 116, 103.
103 V. Horejsí and M. Tichá, Anal. Biochem., 1981, 116, 22.
104 B.J. Howlett and A.E. Clarke, Biochem. Int., 1981, 2, 553.
105 B.S. Rathaur, G.S. Khatri, K.C. Gupta, C.K. Narang and N.K. Nathur, Anal. Biochem., 1981, 112, 55.
106 J.F. Crowley and I.J. Goldstein, FEBS Lett., 1981, 130, 149.
107 L. Bhattacharyya, P.K. Das and A. Sen, Arch. Biochem. Biophys., 1981, 211, 459.
108 T.K. Datta and P.S. Basu, Biochem. J., 1981, 197, 751.
109 T. Mazumber, N. Gaur and A. Surolia, Eur. J. Biochem., 1981, 113, 463.
110 P.S. Appokattan and D. Basu, Anal. Biochem., 1981, 113, 253.
111 Y. Konami, T. Tsuji, I. Matsumoto and T. Osawa, Hoppe Seyler's Z. Physiol. Chem., 1981, 362, 983.
112 H. Franz, P. Ziska and A. Kindt, Biochem. J., 1981, 195, 481.
113 G. Levi and V.L. Teichberg, J. Biol. Chem., 1981, 156, 5735.
114 R.B. Mellor, G.M. Gadd, P. Rowell and W.D.P. Stewart, Biochem. Biophys. Res. Comm., 1981, 99, 1348.
115 F.K. Chu, F. Maley and A.L. Tarentino, Anal. Biochem., 1981, 116, 152.
116 J.L. Ochoa, J. Chromatogr., 1981, 215, 251.
117 K. Ek, E. Gianazza and P.G. Righetti, Biochem. Biophys. Acta, 1980, 626, 356.
118 C.A. Jasalavich and A.J. Anderson, Can. J. Bot., 1981, 59, 264.
119 R.J. Gibbons and I. Dankers, Appl. Environ. Microbiol., 1981, 41, 880.
120 S.G. Pueppke, T.G. Freund, B.C. Schulz and H.P. Friedman, Can. J. Microbiol., 1980, 26, 1489.
121 R. Gansera, H. Schurz and H. Rüdiger, Hoppe Seyler's Z. Physiol. Chem., 1979, 360, 1579.
122 D.J. Bowles and S. Marcus, FEBS Lett., 1981, 129, 135.
123 H. Debray, D. Decout, G. Strecker, G. Spik and J. Montreuil, Eur. J. Biochem., 1981, 117, 41.
124 M.I. Khan, N. Surolia, M.K. Mathew, P. Balaram and A. Surolia, Eur. J. Biochem., 1981, 115, 149.
125 K.D. Hapner and M.A. Jermyn, Ann. Bot. (London), 1981, 48, 89.
126 M. Caron, J. Ohanessian, A. Faure and M. Felon, C.R. Hebd. Seances Acad. Sci. Ser. III, 1981, 292, 183.
127 S.G. Pueppke, Arch. Biochem. Biophys., 1981, 12, 254.
128 K.J. Neurohr, N.M. Young and H.H. Mantsch, J. Biol. Chem., 1980, 255, 9205.
129 K.J. Neurohr, N.M. Young, I.C.P. Smith and H.H. Mantsch, Biochemistry, 1981, 20, 3499.
130 M. Decastel, R. Rourrillon and J.P. Frénoy, J. Biol. Chem., 1981, 256, 9003.
131 D. Nonnenmacher and R. Brossmer, Biochim. Biophys. Acta, 1981, 668, 149.
132 J.E. Lamb, F.L. Bookstein, I.J. Goldstein and L.E. Newton, J. Biol. Chem., 1981, 256, 5874.
133 H. De Boeck, F.G. Loontiens, F.M. Delmotte and C.K. De Bruyne, FEBS Lett., 1981, 126, 227.
134 I.J. Goldstein, D.A. Blake, S. Ebisu, T.J. Williams and L.A. Murphy,

J. Biol. Chem., 1981, 256, 3890.
135 Y. Oda, K.I. Kasai and S.I. Ishii, J. Biochem. (Tokyo), 1981, 89, 285.
136 B. Karkstam, J. Chromatogr., 1981, 211, 233.
137 D.F. Senear and D.C. Teller, Biochemistry, 1981, 20, 3076.
138 D.F. Senear and D.C. Teller, Biochemistry, 1981, 20, 3083.
139 T.J. Williams, L.D. Homer, J.A. Shafer, I.J. Goldstein, P.J. Garegg, H. Hultberg, T. Iverson and R. Johansson, Arch. Biochem. Biophys., 1981, 209, 555.
140 R.M. Clegg, F.G. Loontiens, A. van Landschoot and T.M. Jovin, Biochemistry, 1981, 20, 4687.
141 N. Ockman, Biochim. Biophys. Acta, 1981, 643, 220.
142 M. Toselli, E. Battistel, F. Manca and G. Rialdi, Biochim. Biophys. Acta, 1981, 667, 99.
143 M. Dani, F. Manca and G. Rialdi, Biochim. Biophys. Acta, 1981, 667, 108.
144 A.D. Sherry, M.A. Lindorfer, P. Adams-Stemler and J.H. Oppenheimer, Biochemistry, 1981, 20, 3492.
145 F. Obata, R. Sakai and H. Shiokawa, J. Biochem. (Tokyo), 1981, 89, 1475.
146 P.C. Harrington, R. Moreno and R.G. Wilkins, Isr. J. Chem., 1981, 21, 48.
147 E. Köttgen, C. Bauer, U. Ehlerding and W. Gerok, Biochem. J., 1981, 193, 659.
148 I. Tanaka, Y. Abe, T. Hamada, O. Yonemitsu and S. Ishii, J. Biochem.(Tokyo), 1981, 89, 1643.
149 C.J. Louis and R.G. Wyllie, Experientia, 1981, 37, 508.
150 H. Debray and J. Montreuil, in 'Lectins - Biology, Biochemistry, Clinical Biochemistry, ed. T.C. Bog-Hansen, Walter de Gruyter, Berlin, 1981, 1, 221.
151 K.K. Banerjee and A. Sen, Arch. Biochem. Biophys., 1981, 212, 740.
152 N.N. Desai, A.K. Allen and A. Neuberger, Biochem. J., 1981, 197, 345.
153 C.A.K. Borrebaeck, B. Lönnerdal and M.E. Etzler, FEBS Lett., 1981, 130, 194.
154 M.E. Etzler, S. Gupta and C.A.K. Borrebaeck, J. Biol. Chem., 1981, 256, 2367.
155 C.A.K. Borrebaeck and M.E. Etzler, J. Biol. Chem., 1981, 256, 4723.
156 M.E. Etzler and C.A.K. Borrebaeck, Biochem. Biophys. Res. Commun., 1980, 96, 92.
157 M.E. Etzler, S. Gupta and C.A.K. Borrebaeck, J. Biol. Chem., 1981, 256, 2367.
158 C.A.K. Borrebaeck, B. Lönnerdal and M.E. Etzler, Biochemistry., 1981, 20, 4119.
159 N. Gilboa-Barber and L. Mizrahi, Can. J. Biochem., 1981, 59, 315.
160 W. Gade, M.A. Jack, J.B. Dahl, E.L. Schmidt and F. Wold, J. Biol. Chem., 1981, 256, 12905.
161 S.G. Pueppke, H.P. Friedman and L.-C. Su, Plant Physiol., 1981, 68, 905.
162 J. Nováková, M. Tichá and J. Kocourek, Biochem. Biophys. Acta, 1981, 670, 401.
163 A. Foriers, E. Lebrun, R. van Rapenbusch, R. de Neve and A.D. Strosberg, J. Biol. Chem., 1981, 256, 5550.
164 K. Kornfeld, M.L. Reitman and R. Kornfeld, J. Biol. Chem., 1981, 256, 6633.
165 J.N. Bausch, J. Richey and R.D. Poretz, Biochemistry, 1981, 20, 2618.
166 M. Sarkar, A.M. Wu and E.A. Kabat, Arch. Biochem. Biophys., 1981, 209, 204.
167 P. Ziska and H. Franz, Experientia, 1981 37, 219.
168 M.K. Das, M.I. Khan and A. Surolia, Biochem. J., 1981, 195, 341.
169 M.I. Khan, T. Mazumber, D. Pain, N. Gaur and A. Surolia, Eur. J. Biochem., 1981, 113, 471.
170 R.L. Felsted, J. Li, G. Pokrywka, M.J. Egorin, J. Spiegel and R.M.K. Dale, Int. J. Biochem., 1981, 13, 549.
171 A. Pusztai, E.M.W. Clarke, G. Grant and T.P. King, J. Sc. Food Agric., 1981, 32, 1037.
172 R.L. Felsted, R.D. Leavitt, C. Chen, N.R. Bachur and R.M.K. Dale, Biochim. Biophys. Acta, 1981, 668, 132.
173 L.M. Roberts and J.M. Lord, Eur. J. Biochem., 1981, 119, 31.
174 N. Toda, A. Doi, A. Jimbo, I. Matsumoto and N. Seno, J. Biol. Chem., 1981, 256, 5345.
175 D. Ashford, R. Menon, A.K. Allen and A. Neuberger, Biochem. J., 1981, 199,

299.
176 A.M. Wu, E.A. Kabat, F.G. Gruezo and R.D. Poretz, Arch. Biochem. Biophys., 1981, 209, 191.
177 A. Ndulue, J.P. Flandrois and D. Marmet, J. Chromatogr., 1981, 209, 323.
178 F. Jordan, H. Bahr, J. Patrick and P.W.K. Woo, Arch. Biochem. Biophys., 1981, 207, 81.
179 K.J. Neurohr, N. Lacelle, H.H. Mantsch and I.C.P. Smith, Biophys. J., 1980, 32, 931.
180 K. Yamamoto, T. Tsuji, I. Matsumoto and T. Osawa, Biochemistry, 1981, 20, 5894.
181 M. Duk and E. Lisowska, Eur. J. Biochem., 1981, 118, 131.
182 G. Gebauer, E. Schiltz and H. Rüdiger, Eur. J. Biochem., 1981, 113, 319.
183 D. Haass, R. Frey, M. Thiessen and H. Dauss, Planta, 1981, 151, 490.
184 E. del Campillo, L.M. Shannon and C.N. Hankins, J. Biol. Chem., 1981, 256, 7177.
185 M.S. Herrmann and W.D. Behnke, Biochim. Biophys. Acta, 1981, 667, 397.
186 L.S. Simeral, W. Kapmeyer, W.P. MacConnell and N.O. Kaplan, J. Biol. Chem., 1980, 255, 11098.
187 R.A. Roth, B.A. Maddux, K.Y. Wong, Y. Iwamoto and I.D. Goldfine, J. Biol. Chem., 1981, 256, 5350.
188 A. Raz nd R. Lotan, Cancer Res., 1981, 41, 3642.
189 G. Uhlenbruck, J. Sölter, E. Hanssen and H. Haupt, Hoppe-Seyler's Z. Physiol. Chem., 1981, 362, 1167.
190 G.J. Montelione, S. Callahan and T.R. Podleski, Biochim. Biophys. ACTA, 1981, 670, 110.
191 D.T. Connolly, C.A. Hoppe, M.K. Hobish and Y.C. Lee, J. Biol. Chem., 1981, 256, 12940.
192 C.J. Steer, E.S. Kempner and G. Ashwell, J. Biol. Chem., 1981, 256, 5851.
193 J.L. Denburg, Biochem. Biophys. Res. Commun., 1980, 97, 33.
194 K. Drickamer, J. Biol. Chem., 1981, 256, 5827.
195 H. Den and J.H. Chin, J. Biol. Chem., 1981, 256, 8069.
196 M.J. Pitts and D.C.H. Yang, Biochem. Biophys Res. Commun., 1980, 95, 750.
197 M.J. Pitts and D.C.H. Yang, Biochem. J.., 1981, 195, 435.
198 D.T. Dorai, B.K. Bachhawat, S. Bishayee, K. Kannan and D.R. Ras, Arch. Biochem. Biophys., 1981, 209, 325.
199 R. Maget-Dana, R.W. Vem, M. Sander, A.C. Roche, R. Schauer and M. Monsigny, Eur. J. Biochem., 1981, 114, 11.
200 M.E.A. Pereira, A.F.B. Andrade and J.M.C. Ribeiro, Science, 1981, 211, 597.
201 H. Komano, D. Mizuno and S. Natori, J. Biol. Chem., 1981, 256, 7087.
202 W.E.G. Muller, R.K. Zahn, B. Kurelec, C. Lucu, I. Müller and G. Uhlenbruck, J. Bact., 1981, 145, 548.
203 M.G. Cumsky and D.R. Zusman, J. Biol. Chem., 1981, 256, 12581.
204 D.R. Nelson, M.G. Cumsky and D.R. Zusman, J. Biol. Chem., 1981, 256, 12589.
205 R. Eifler and P. Ziska, Experientia, 1980, 36, 1285.
206 F. Ishikawa, K. Oishi and K. Aida, Agric. Biol. Chem. (Jpn), 1981, 45, 557.
207 F. Ishikawa, T. Kameyama, A. Takenaka, K. Oishi and K. Aida, Agric. Biol. Chem. (Jpn), 1981, 45, 2105.
208 D.N. Cooper and S.H. Barondes, J. Biol. Chem., 1981, 256, 5046.
209 T. Tsivion and N. Sharon, Biochim. Biophys. Acta, 1981, 642, 336.
210 E. Ruoslahti and E. Engwall, Collagen and Related Res., 1981, 1, 95.
211 R.E. Weiss and A.H. Reddi, Biochem. J, 1981, 197, 529.
212 D.L. Scott, P.A. Bedford and K.W. Walton, J. Immunol. Methods, 1981, 43, 29.
213 E. Ruoslahti, E. Engvall, E.G. Hayman and R.G. Spiro, Biochem. J., 1981, 193, 295.
214 E. Pearlstein and L. Baez, Anal. Biochem., 1981, 116, 292.
215 S.I. Rennard, M.L. Wind, A.T. Hewitt and H.K. Kleinman, Arch. Biochem. Biophys., 1981, 206, 205.
216 M.R. Stampfer, I. Vlodavsky, H.S. Smith, R. Ford, F.F. Becker and J. Riggs, J. Nat. Cancer Inst., 1981, 67, 253.
217 I.U. Ali and T. Hunter, J. Biol. Chem., 1981, 256, 7671.
218 E. Ruoslahti, H. Jalanico, D.E. Comings, A.M. Neville and D. Raghavan,

Int. J. Cancer, 1981, 27, 763.
219 F.C. de Beer, M.L. Baltz, S. Holford, A. Feinstein and M.B. Pepys, J. Exp. Med., 1981, 154, 1134.
220 M.P. Bevilacqua, D. Amrani, M.W. Mosesson and C. Biano, J. Exp. Med., 1981, 153, 42.
221 S.R. Gonda and J.R. Shainoff, Thrombosis Res., 1980, 20, 581.
222 V.E. Koteliansky, M.A. Gludhova, M.V. Bejanian, V.N. Smirnov, V.V. Filimonov, O.M. Zalite and S.Y. Venyaminov, Eur. J. Biochem., 1981, 19, 619.
223 K. Sekiguchi, M. Fukuda and S.I. Hakomori, J. Biol. Chem., 1981, 256, 6452.
224 D.G. Wallace, J.W. Donovan, P.M. Schneider, A.M. Meunier and J.L. Lundblad, Arch. Biochem. Biophys., 1981, 212, 515.
225 J.A. McDonald, T.J. Broedelmann, D.G. Kelley and B. Villiger, J. Biol. Chem., 1981, 256, 5583.
226 H. Richter, M. Seidl and H. Hormann, Hoppe-Seyler's Z. Physiol. Chem., 1981, 362, 399.
227 E. Ruoslahti, E.G. Hayman, E. Engvall, W.C. Cothran and W.T. Butler, J. Biol. Chem., 1981, 256, 7277.
228 R. Ehrismann, M. Chiquet and D.C. Turner, J. Biol. Chem., 1981, 255, 4056.
229 J. Keski-Oja and K.M. Yamada, Biochem. J., 1981, 193, 615.
230 J. Keski-Oja, E. Ruoslahti and E. Engvall, Biochem. Biophys. Res. Commun., 1981, 100, 1515.
231 M. Hayashi and K.M. Yamada, J. Biol. Chem., 1981, 256, 11293.
232 A. Rephaeli, M. Spector and E. Racker, J. Biol. Chem., 1981, 256, 6069.
233 E.F. Plow and M.H. Ginsberg, J. Biol. Chem., 1981, 256, 9477.
234 B.D. Smith and C.K. Silbert, Biochem. Biophys. Res. Commun., 1981, 100, 275.
235 D.R. Kahn and I. Cohen, Biochim. Biophys. Acta., 1981, 668, 490.
236 R.P. McDonagh, J. McDonagh, T.E. Petersen, H.C. Thogersen, K. Skorstengaard, L. Sottrup-Jensen and S. Magnusson, FEBS Lett., 1981, 127, 174.
237 J. Keski-Oja, G.J. Todaro and A. Vameri, Biochim. Biophys. Acta, 1981, 673, 323.
238 V.E. Koteliasky, V.L. Leytin, D.D. Sviridov and V.N. Smirnov, FEBS Lett., 1981, 123, 59.
239 H. Hörmann and V. Jelinić, Hoppe-Seyler's Z. Physiol. Chem., 1981, 362, 87.
240 H. Kleinman, C.M. Wilkes and G.R. Martin, Biochemistry, 1981, 20, 2325.
241 A.M. Rich, E. Pearlstein, G. Weissmann and S.T. Hoffstein, Nature, 1981, 293, 224.
242 L.T. Furcht, D. Smith, G. Wendelschafer-Crabb, D.F. Mosher and J.M. Foidart, J. Histochem. Cytochem., 1980, 28, 1319.
243 T. Vartio and A. Vaheri, J. Biol. Chem., 1981,256, 13085.
244 L.D. Johnson, Clin. Biochem., 1980, 13, 204.
245 Prog. Clin. Biol. Res., 1981, 54, (1).
246 Biology of Collagen, ed. A. Viidik and J. Vuust, Acad. Press (London), 1980.
247 V.I. Lim, FEBS Lett., 1981, 132, 1.
248 P.F. Davis and Z.M. Mackle, Anal. Biochem., 1981, 115, 11.
249 J.W. Stock, M.C. Henning and J.F. Collins, Anal. Biochem., 1981, 110, 323.
250 S.C. Meredith and F.J. Kezdy, Biochim. Biophys. Acta, 1981, 668, 357.
251 J. Risteli, H. Rohde and R. Timpl, Anal. Biochem., 1981, 113, 372.
252 G.J. Laurent, P. Cockerill, R.J. McAnulty and J.R.B. Hastings, Anal. Biochem., 1981, 113, 301.
253 S. Kimura, T. Kamimura, Y. Takema and M. Kubota, Biochim. Biophys. Acta, 1981, 669, 251.
254 W.G. Cole and D. Cham, Biochem. J., 1981, 197, 377.
255 D.M. Pesciotta, L.A. Dickson, A.M. Showalter, Eikenberry, B. de Crombrugghe, P.P. Fietzek and B.R. Olsen, FEBS Lett., 1981, 125, 170.
256 B. Steinmann, L. Tuderman, L. Peltonen, G.R. Martin, V.A. McKusik and D.J. Prockop, J. Biol. Chem., 1980, 255, 8887.
257 M.D. Grynpas, D.R. Eyre and D.A. Kirschner, Biochim. Biophys. Acta, 1980, 626, 346.
258 S.C.G. Tseng, N. Savion, D. Gospodarowicz and R. Stern, J. Biol. Chem., 1981, 56, 3361.
259 R.K. Rhodes, K.D. Gibson and E.J. Miller, Biochemistry, 1981, 20, 3117.

260 S. Garbisa, L.A. Liotta, K. Tryggvason and G.P. Siegal, FEBS Lett., 1981, 127, 257.
261 R. Jander, J. Rauterberg, B. Voss and D.B. von Bassewitz, Eur. J. Biochem., 1981, 14, 17.
262 D.K. Furuto and E.J. Miller, J. Biol. Chem., 1980, 255, 290.
263 S. Ayad, M.Z. Abedin, S.M. Grundy and J.B. Weiss, FEBS Lett., 1981, 123, 195.
264 D.C. Dean, B.D. Peczon, M.E. Noelken and B.G. Hudson, J. Biol. Chem., 1981, 256, 7543.
265 R.J. Butkowski, G.S. Brungardt, J.J. Grantham and B.G. Hudson, J. Biol. Chem., 1981, 256, 7603.
266 J. Glowacki and J. Cross, Biochim. Biophys. Acta, 1981, 668, 216.
267 R. Ouazana and D. Herbage, Biochim. Biophys. Acta, 1981, 669, 236.
268 A. Torre-Blanco and I. Toledo, J. Biol. Chem., 1981, 256, 5926.
269 K.K. Smith and S. Strickland, J. Biol. Chem., 1981, 256, 4654.
270 K. Kuhn, H. Wiedemann, R. Timpl, J. Risteli, H. Dieringer, T. Voss and R.W. Glanville, FEBS Lett., 1981, 125, 123.
271 M. Laurent, P. Kern and F. Regnault, Int. J. Biochem., 1981, 13, 63.
272 L. Cohen-Solal, M. le Lous, J.C. Allain and F. Meunier, FEBS Lett., 1981, 123, 282.
273 S. Kimura, Y. Takema and M. Kubota, J. Biol. Chem., 1981, 256, 13235.
274 C.H. Hung, M.E. Noelken and B.G. Hudson, J. Biol. Chem., 1981, 256, 3822.
275 J.E. Scott, C.R. Orford and E.W. Hughes, Biochem. Soc. Trans., 1981, 9, 226.
276 E.A. Sankey, F.E. Brown, L.F. Morton, D.M. Scott and M.J. Barnes, Biochem. J., 1981, 198, 707.
277 S.C.G. Tseng, N. Savion, D. Gospodarowicz and R. Stern, J. Biol. Chem., 1981, 256, 3361.
278 D. Schubert and M. La Corbiere, J. Biol. Chem., 1980, 255, 11557.
279 D.L. Helseth and A. Veis, J. Biol. Chem., 1981, 256, 7118.
280 V.J. Uitto, J. Uitto and D.J. Prockop, Arch. Biochem. Biophys., 1981, 210, 445.
281 L.I. Fessler, C.A. Kumamoto, M.E. Meis and J.G. Fessler, J. Biol. Chem., 1981, 56, 9640.
282 L.I. Fessler, W.J. Robinson and J.H. Fessler, J. Biol. Chem., 1981, 256, 9646.
283 A. le Pape, J.P. Muh and A.J. Bailey, Biochem. J., 1981, 197, 405.
284 T. Pihlajaniemi, R. Myllylä, K. Alitolo, A. Vaheri and K.I. Kivirikko, Biochemistry, 1981, 20, 7409.
285 H.H. Carnicero, A.M. Adamany and S. Englard, Arch. Biochem. Biophys., 1981, 210, 678.
286 J. Preiss and D.A. Walsh in 'Biology of Carbohydrates', ed. V. Ginsburg and P. Robbins, John Wiley, New York, Vol 1, p199.
287 D.M. Haverstick and A.H. Gold, Anal. Biochem., 1981, 111, 137.
288 K. Rybicka, J.Histochem. Cytochem., 1981, 29, 4.
289 J. McLaren, W.G. Ng and R. Toe, Clin. Clem., 1980, 26, 1924.
290 V. Dombrádi, Int. J. Biochem., 1981, 13, 125.
291 P.T.W. Cohen, Y. le Marchand Brustel and P. Cohen, Eur. J. Biochem., 1981, 115, 619.
292 H. Shikama, J.L. Chiasson and J.H. Exton, J. Biol. Chem., 1981, 256, 4450.
293 J.W. Mitchell and J.A. Thomas, J. Biol. Chem., 1981, 256, 6160.
294 A.A. de Paoli-Roach, P.J. Roach, K. Pham, G. Kramer and B. Hardesty, J. Biol. Chem., 1981, 256, 8871.
295 M.J. Poznansky and D. Bhardwaj, Biochem. J., 1981, 196, 89.
296 J. Melet, G.J.M. Hoogh Winkel, M.A.H. Giesberts and H.H. van Geldern, Clin. Chim. Acta, 1980, 108, 179.
297 M. Breen, P.A. Knepper, H.G. Weinstein, L.J. Blacik, D.G. Lewandowski and B.M. Baltros, Anal. Biochem., 1981, 113, 416.
298 T.O. Kleine and B. Merten, Analyt. Biochem., 1981, 118, 185.
299 P. Gysen, G. Heynen and P. Franchimont, C.R. Seances Soc. Biol. Ses Fil., 1981, 175, 538.
300 C. Warren and G. Manley, Clin. Chim. Acta, 1981, 116, 369.

301 I.Miyamoto and S.Nagase, Anal. Biochem., 1981, 115, 308.
302 G.J.L. Lee, D.W. Liu, J.W. Pav and H. Tieckelmann, J. Chromatogr., 1981, 212, 65.
303 R.E. Hurst and J.M. Settine, Anal. Biochem., 1981, 115, 88.
304 C. Peeters-Joris, X. Emonds-Alt and G. Vaes, Biochem. J., 1981, 196, 95.
305 E.H. Schuchman and R.J. Desnick, Anal. Biochem., 1981, 117, 419.
306 J.E. Morris, M.W. Canoy and L.S. Rynd, J. Chromatogr., 1981, 224, 407.
307 E.R. Schwartz, J. Stevens and D.D. Schmidt, Anal. Biochem., 1981, 112, 170.
308 D.A. Torchia, M.A. Hasson and N.C. Hascall, J. Biol. Chem., 1981, 256, 7129.
309 H. Gustavsson, G. Siegel, B. Lindman and L.A. Fransson, Biochem. Biophys. Acta., 1981, 677, 23.
310 K. Pietilä and T. Nikkari, Anal. Biochem., 1981, 116, 317.
311 L.A. Fransson and B.G. Johansson, Int. J. Biol. Macromol., 1981, 3, 25.
312 P. Vijayagopal, S.R. Srinivasan, B. Radhakrishnamurthy and G.S. Berenson, J. Biol. Chem., 1981, 256, 8234.
313 P. Gysen, G. Heynen and P. Franchimont, C.R. Seances Soc. Biol. Ses Fil., 1980, 174, 867.
314 S. Björnsson and D. Heinegard, Biochem. J., 1981, 197, 249.
315 D.W. Bullimore, C.F. Phelps and M.I.V. Jayson, Biochem. J., 1981, 199, 433.
316 K.G. Vogel and D.W. Peterson, J. Biol. Chem., 1981, 256, 13235.
317 H. Masuda and S. Schichijo, Int. J. Biochem., 1981, 13, 603.
318 P.J. Roughley, D. McNicol, V. Santer and J. Buckwalter, Biochem. J., 1981, 197, 77.
319 M.E. Adams and H. Muir, Biochem. J., 1981, 197, 385.
320 V. Stanescu and M.B.E. Sweet, Biochim. Biophys. Acta, 1981, 673, 101.
321 M. Breton, J. Picard and E. Berrou, Biochime, 1981, 63, 515.
322 P.J. McKeown-Longo, K.J. Sparks and P.F. Goetinck, Arch. Biochem. Biophys., 1981, 212, 216.
323 C. Peters, J. Schwermann, A. Schmidt and E. Buddecke, Biochim. Biophys. Acta, 1981, 673, 270.
324 M. Isemura, T. Hanyu, H. Kosaka, T. Ono and T. Ikenaka, J. Biochem. (Tokyo), 1981, 89, 1113.
325 A. Fransén, S. Björnsson and D. Heinegard, Biochem. J., 1981, 197, 669.
326 A. Tengblad, Biochem. J., 1981, 199, 297.
327 S.J. Perkins, A. Miller, T.E. Hardingham and H. Muir, J. Mol. Biol., 1981, 150, 69.
328 M. Luscombe, S.E. Marshall, D.S. Pepper and J.J. Holbrook, Biochim. Biophys. Acta, 1981, 678, 137.
329 K. Kikuchi, B.R. Jennings and M. Isles, Int. J. Biol., 1981, 3, 53.
330 D. Heinegard, M. Paulsson, S. Inerot and C. Carlström, Biochem. J., 1981, 197, 355.
331 L. Silvestri, J.R. Baker, L. Roden and R.M. Stroud, J. Biol. Chem., 1981, 256, 7383
332 W.L. Kiang, R.U. Margolis and R.K. Margolis, J. Biol. Chem., 1981, 256, 10529.
333 R. Kapoor, C.F. Phelps, L. Cöster and L.A. Fransson, Biochem. J., 1981, 197, 259.
334 I. Miyamoto, H. Watanabe and S. Nagase, Anal. Biochem., 1981, 110, 39.
335 L.W. Fisher and H. Schraer, Arch. Biochem. Biophys., 1980, 205, 396.
336 A. ldberg, E. Ghayman and E. Ruoslahti, J. Biol. Chem., 1981, 256, 10847.
337 L. Cöster and L.A. Fransson,Biochem. J., 1981, 193, 143.
338 J.K. Sheehan, I. Carlstedt, L. Cöster, A. Malmström and L.A. Fransson, Biochem. J., 1981, 199, 581.
339 L. Cöster, L.A. Fransson, J. Sheehan, I.A. Nieduszynski and C.F. Phelps, Biochem. J., 1981, 197, 483.
340 N. Fujii and Y. Nagai, J. Biochem. (Tokyo), 1981, 90, 1249.
341 I. Miyamoto and S. Nagase, J. Biochem. (Tokyo), 1980, 88, 1793.
342 J.E. Scott and C.R. Orford, Biochem. J., 1981, 197, 213.
343 I. Carlstedt, L. Cöster and A. Malmström, Biochem. J., 1981, 197, 217.
344 F. Akiyama and N. Seno, Biochim. Biophys. Acta, 1981, 674, 289.
345 R. Losito, H. Gattiker and G. Bilodeau, J. Chromatogr., 1981, 226, 61.

346 S. Kadokura, T. Miyamoto and H. Inagaki, Biopolymers, 1981, 20, 1113.
347 A. Ogamo, K. Matsuzaki, H. Uchiyama and K. Nagasawa, J. Chromatogr., 1981, 213, 439.
348 G. Oshima, H. Uchiyama and K. Nagasawa, Anal. Biochem., 1981, 111, 366.
349 A.P. Almeida and M.A. Beavan, Life Sci., 1980, 26, 549.
350 P.A.S. Mourão, S. Pillai, and N. Di Ferrante, Biochim. Biophys. Acta, 1981, 674, 178.
351 J. Mattai and J.C.T. Kwak, Biochim. Biophys. Acta, 1981, 677, 303.
352 R.F. Parrish and W.R. Fair, Biochem. J., 1981, 193, 407.
353 B. Weissmann, H. Chao and P. Chow, Carbohydr. Res., 1981, 92, 239.
354 B. Weissmann and H. Chao, Carbohydr. Res., 1981, 92, 255.
355 J.J. Hopwood and H. Elliott, Carbohydr. Res., 1981, 91, 165.
356 J.J. Hopwood and H. Elliott, Clin. Chim. Acta, 1981, 112, 55.
357 J. Denton, W.E. Leyis, I.A. Nieduszynski and C.F. Phelps, Anal. Biochem., 1981, 118, 388.
358 H.K. Takahashi, H.B. Nader and C.P. Dietrich, Anal. Biochem., 1981, 116, 456.
359 S. Radoff and I. Danishefsky, Thromb. Res., 1981, 22, 353.
360 H. Uchiyama and K. Nagasawa, J. Biochem. (Tokyo), 1981, 89, 185.
361 U. Lindshi, G. Bäckström, L. Thunberg and I.G. Leder, Proc. Natl. Acad. Sci. USA, 1980, 77, 6551.
362 B. Meyer, L. Thunberg, U. Lindahl, O. Larm and I.G. Leder, Carbohydr. Res., 1981, 88, Cl.
363 N. Ototani, M. Kikuchi and Z. Yosizawa, J. Biochem. (Tokyo), 1981, 90, 241.
364 B. Nordenman and I. Björk, Biochim. Biophys. Acta, 1981, 672, 227.
365 A. Danielsson and I. Björk, Biochem. J., 1981, 193, 427.
366 N. Ototani and Z. Yosizawa, J. Biochem. (Tokyo), 1981, 90, 1553.
367 J. Riesenfeld, L. Thunberg, M. Höök and U. Lindahl, J. Biol. Chem., 1981, 256, 2389.
368 G.M. Oosta, W.T. Gardner, D.L. Beeler and R.D. Rosenberg, Proc. Natl. Acad. Sci. USA, 1981, 78, 829.
369 B. Casu, P. Oreste, G. Torri, G. Zoppetti, J. Choay, J.C. Lormeau, M. Petitou and P. Sinay, Biochem. J., 1981, 197, 599.
370 E. Holmer, K. Kurachi and G. Söderstoöm, Biochem. J., 1981, 193, 395.
371 M.W. Piepkorn, D. Lagunoff and G. Schmer, Arch. Biochem. Biophys., 1980, 205, 315.
372 H. Kawakami and H. Terayama, Biochim. Biophys. Acta, 1981, 646, 161.
373 M.P. Cohen, V.Y. Wu and M.L. Surma, Biochim. Biophys. Acta, 1981, 678, 322.
374 M.P. Cohen and C.J. Ciborowski, Biochim. Biophys. Acta, 1981, 674, 400.
375 J. Laterra, R. Ansbacher and L.A. Culp, Proc. Natl. Acad. Sci. USA, 1980, 77, 6662.
376 L.A. Fransson, Eur. J. Biochem., 1981, 120, 251.
377 L.A. Fransson, B. Havsmark and J.K. Sheehan, J. Biol. Chem., 1981, 256, 13039.
378 L.A. Fransson, I. Sjöberg and V.P. Chiarug, J. Biol. Chem., 1981, 256, 13044.
379 Y. Ohkubo, I. Funakoshi and I. Yamashina, J.Biochem. (Tokyo), 1981, 89, 161.
380 L. Kjellén, I. Pettersson and M. Höök, Proc. Natl. Acad. Sci. USA., 1981, 78, 5371.
381 P. Levy, J. Picard and A. Bruel, FEBS Lett., 1981, 131, 257.
382 D.J. Winterbourne and J.G. Salisbury, Biochem. Biophys. Res. Commun., 1981, 101, 30.
383 O.Schmut and H.Hofmann, Biochim. Biophys. Acta, 1981, 673, 192.
384 A. Tengblad, Biochem. J., 1981, 185, 101.
385 J.E. Scott, F. Heatley, D. Moorcroft and A.H. Olavsen, Biochem. J., 1981, 199, 829
386 M.K. Cowman, E.A. Balazs, C.W. Bergmann and K. Meyer, Biochemistry, 1981, 20, 1379
387 M.O. Longas and K. Meyer, Biochem. J., 1981, 197, 275.
388 R.L. Cleland, Arch. Biochem. Biophys., 1981, 210, 565.
389 B. Delphech and C. Halavent, J. Neurochem., 1981, 36, 855.

390 N. Toda, A. Doi, A. Jimbo, I. Matsumoto and N. Seno, J. Biol. Chem., 1981, 256, 5345.
391 T. Ludolph, E. Paschke, J. Glössl and H. Kresse, Biochem. J., 1981, 193, 811.
392 S. Toma, D.T. di Ferrante, R. Tenni, N. di Ferrante and Y.M. Michelacci, Carbohydr. Res., 1981, 88, 93.
393 A. Brekle and G. Mersmann, Biochim. Biophys. Acta, 1981, 675, 322.
394 H.W. Stuhlsatz, F. Hirtzel, R. Keller, S. Cosma and H. Greiling, Hoppe-Seyler's Z. Physiol. Chem., 1981, 362, 841.
395 R. Keller, T. Stein, H.W. Stuhlsatz, H. Greiling, E. Ohst, E. Müller and H.D. Scharf, Hoppe-Seyler's Z. Physiol. Chem., 1981, 362, 327.
396 R.A. Kosher and M.P. Savage, Nature, 1981, 291, 231.
397 K. Pietilä, Biochim. Biophys. Acta, 1981, 677, 318.
398 A.M. Gressner, J.E. Cadenbach and H. Greiling, J. Clin. Chem. Clin. Biochem., 1981, 19, 465.
399 Y. Ninomiya, R.I. Hata and Y. Nagai, Biochim. Biophys. Acta, 1981, 675, 248.
400 M.C. Lemkin, M.G. Farquhar, Proc. Nat. Acad. Sci. USA, 1981, 78, 1726.
401 M. Luscombe, S.E. Marshall, D.S. Pepper and J.J. Holbrook, Biochim. Biophys. Acta, 1981, 678, 137.
402 J.H. Kimura, C.B. Caputo and V.C. Hascall, J. Biol. Chem., 1981, 256, 4368.
403 P.R. Sudhakaran, W. Sinn and K. von Figura, Hoppe-Seyler's Z. Physiol. Chem., 1981, 362, 39.
404 E. Deudon, M. Breton, E. Berrou and J. Picard, Biochimie, 1981, 62, 811.
405 R. Frater and D. Hewish, Aust. J. Biol. Sci., 1981, 34, 171.
406 T.R. Oegema and F. Behrens, Arch. Biochem. Biophys., 1981, 206, 277.
407 R.L. Stevens, S.P. Nissley, J.H. Kimura, M. Rechler, A.I. Kaplan and V.C. Hascall, J. Biol. Chem., 1981, 256, 2045.
408 R.L. Stevens and V.C. Hascall, J. Biol. Chem., 1981, 256, 2053.
409 D.M. Brown, A.F. Michael and T.R. Oegema, Biochim. Biophys. Acta, 1981, 674, 96.
410 A.M. Gressner and W. Kostereiserfunke, J. Clin. Chem. Clin. Biochem., 1981, 19, 363.
411 T. Kajiwara and M.L. Tanzer, FEBS Lett., 1981, 134, 43.
412 M. Daireaux, H. Penfornis, M. Langris, J. Bocquet, J.P. Pujol, R. Beliard and G. Loyau, FEBS Lett., 1981, 132, 93.
413 S. Björnsson and D. Heinegard, Biochem. J., 1981, 199, 17.
414 J. Wieslander and D. Heinegard, Biochem. J., 1981, 199, 81.
415 J.H. Kimura, E.J.M. Thonar, V.C. Hascall, A. Reiner and R. Poole, J. Biol. Chem., 1981, 256, 7890.
416 C.R. Faltynek and J.E. Silbert, J. Biol. Chem., 1981, 256, 7202.
417 Y. Nakanishi, M. Shimizu, K. Otsu, S. Kato, M. Tsuji and S. Suzuki, J. Biol. Chem., 1981, 256, 5443.
418 A. Malmström, Biochem. J., 1981, 198, 669.
419 W.T. Forsee and L. Roden, J. Biol. Chem., 1981, 256, 7240
420 W. Sinn, P.R. Sudhararan and K. von Figura, Biochem. J., 1981, 200, 51.
421 I.A. King, Biochim. Biophys. Acta, 1981, 674, 87.
422 P. Levy, J. Picard and A. Bruel, Eur. J. Biochem., 1981, 115, 397.
423 M. Uzuka, K. Nakajima, S. Ohta and Y. Mori, Biochim. Biophys. Acta, 1981, 673, 387.
424 K. Madsen, Arch. Biochem. Biphys., 1981, 211, 368.
425 S.J. Hunter and H. Schraer, Arch. Biochem. Biophys., 1981, 210, 647.
426 S.A. Fellini, J.H. Kimura and V.C. Hascall, J. Biol. Chem., 1981, 256, 7883.
427 J.H. Kimura, E.J.M. Thonar, V.C. Hascall, A. Reiner and A.R. Poole, J. Biol. Chem., 1981, 256, 7890.
428 R.W. Burlingame, G.H. Thomas, R.L. Stevens, K. Schmid and H.W. Moser, Clin. Chem., 1981, 27, 124.
429 E.W. Gold, Biochim. Biophys. Acta, 1981, 673, 408.
430 S. Fukui, H. Yoshida, T. Tanaka, T. Sakano, T. Usui and I. Yamashina, J. Biol Chem., 1981, 256, 10313.
431 T. Yokoi, T. Uozaki, M. Kasei, T. Sato and N. Taniguchi, Clin. Chim. Acta, 1981, 116, 153.

432 H. Kresse, E. Paschke, K.Von Figura, W. Gilberg and W. Fuchs, Proc. Nat. Acad. Sci. USA., 1980, 77, 6822.
433 I. Straprans, S.J. Garon, J. Hopper and J.M. Felts, Biochim. Biophys. Acta, 1981, 678, 414.
434 P.A.S. Mourao, S. Kato and P.V. Donnelly, Biochem. Biophys. Res. Commun., 1981, 98, 388.
435 C.T. Samuell, Clin. Chim. Acta, 1981, 117, 63.
436 B.G.J. Salisbury and W.D. Wagner, J. Biol. Chem., 1981, 266, 8050.
437 H.W. Stuhlsatz, S. Vierhaus and H. Greiling, Z. Analyt. Chem., 1980, 301, 102.
438 G.F. Triphaus, A. Schmidt and E. Buddecke, Hoppe-Seyler's Z. Physiol. Chem., 1980, 361, 1773.
439 P.J. Roughley, R.J. White and V. Santer, J. Biol. Chem., 1981, 256, 12699.
440 E.J.M.A. Thonar and M.B.E. Sweet, Arch. Biochem. Biophys., 1981, 208, 535.
441 L.O. Samaio and C.P. Dietrich, J. Biol. Chem., 1981, 256, 9205.
442 H.G. Garg and D.A. Swann, Biochem. J.., 1981, 193, 459.
443 G. Lyons, S.M. Eisenstein and M.B.E. Sweet, Biochim. Biophys. Acta, 1981, 673, 443.
444 V.P. Bhavanandan, Biochemistry, 1981, 20, 5595.
445 M. del Rosso, R. Cappalette, G. Dini, G. Fibbi, S. Vannuchi, V. Chiarugi and C. Guazzelli, Biochim. Biophys. Acta, 1981, 676, 129.
446 K. Kimata, H.J. Barrach, K.S. Brown and J.P. Pennypacker, J. Biol. Chem., 1981, 256, 6961.
447 W.S. Argraves, P.J. McKeown-Longo, and P.F. Goetinck, FEBS Lett., 1981, 131, 265.
448 J.P. Zanetta, A. Reeber and G. Vincendon, Biochim. Biophys. Acta, 1981, 670, 393.
449 S.F. Cotmore, S.A. Crowhurst, and M.D. Waterfield, Eur. J. Immunol., 1981, 11, 597.
450 K.H. Lakin, and J.W. Fabre, J. Neurochem., 1981, 37, 1170.
451 J.L. McKenzie, A.K. Allen and J.W. Fabre, Biochem. J., 1981, 197, 629.
452 M.Z. Jones and R.A. Laine, J. Biol. Chem., 1981, 256, 5181.
453 M.Z. Jones and G. Dawson, J. Biol. Chem., 1981, 256, 5185.
454 A.A. Neokesariiskii and V.S. Turovskii, Biochemistry (U.S.S.R.), 1980, 45, 1385.
455 S.C. Fu, T.F. Cruz and J.W. Gurd, J. Neurochem., 1981, 36, 1338.
456 C.J.B. White and D.J. Mullins, Biochem. Soc. Trans., 1980, 8, 619.
457 C.J.B. White, Biochem. J., 1981, 200, 373.
458 C.J.B. White, ed. J.M. Dawson, J. Neurochem., 1981, 37, 1155.
459 J.F. Poduslo, J. Neurochem., 1981, 36, 1924.
460 C. Linington and T.V. Waehneldt, J. Neurochem, 1981, 36, 1528.
461 J. Reigner, J.M. Matthieu, R. Kraus-Ruppert, H. Lassman and J.F. Poduslo, J. Neurochem., 1981, 36, 1986.
462 D.A. Figlewicz, R.A. Quarles, D. Johnson, G.R. Barbarash and N.H. Sternberger, J. Neurochem., 1981, 37, 749.
463 E.E. Mena, B.W. Moore, S. Hagen and H.C. Agrawal, Biochem. J., 1981, 195, 525.
464 C.J.B. White and M.E. Hounslow, Biochem. Soc. Trans., 1980, 8, 618.
465 C. Mezei and J.A. Verpoorte, J. Neurochem., 1981, 37, 550.
466 D.Auino, R. Wong, R.U. Margolis and R.K. Morgolis, FEBS Lett., 1980, 112, 195.
467 S.H. Chiou, L.T. Chylack, W.H. Tung and H.F. Bunn, J. Biol. Chem., 1981, 256, 5176.
468 E. Consiglio, S. Shifrin, Z. Yavin, F.S. Ambesi-Impiombato, J.E. Rall, G. Salvatore and L.D. Kohn, J. Biol. Chem., 1981, 256, 10592.
469 S. Shifrin and L.D. Kohn, J. Biol. Chem., 1981, 256, 10600.
470 D. Godelaine, M.J. Spiro and R.G. Spiro, J. Biol. Chem., 1981, 256, 10161.
471 F. Monaco, R. Dominici, F. Carlini, C. Carducci, R. Depirro, P. Felli and J. Roche, C.R. Seances Soc. Biol. Ses Fil., 1981, 175, 446.
472 K. Takatsuki, A.B. Schneider, K.Y. Shin and L.M. Sherwood, J. Biol. Chem., 1981, 256, 2342.

473 D.V. Cohn, J.J. Morrissey, J.W. Hamilton, R.E. Shofstall, F.L. Smardo and L.L.H. Chu, Biochemistry, 1981, 20, 4135.
474 N. Larivière, N.G. Seidah, G. De Serres, J. Rochemont and M. Chrétien, FEBS Lett., 1980, 122, 279.
475 M.A. Phillips, M.L. Budarf and E. Herbert, Biochemistry, 1981, 20, 1666.
476 J.C. Byrd and O. Touster, Biochim. Biophys. Acta, 1981, 677, 69.
477 D. Bouhours, J.F. Bouhours and P.A. Bryon, Biochim. Biophys. Acta, 1981, 672, 288.
478 Z. Zeman and R.L. Verwilghen, Biochim. Biophys. Acta, 1981, 669, 120.
479 E. Berman, D.E. Walters and A. Allerhand, J. Biol. Chem., 1981, 256, 3853.
480 P.E. Reid, C.F.A. Culling, W.L. Dunn, C.M. Ramey, A.B. Magil and M.G. Clay, J. Histochem. Cytochem., 1980, 28, 217.
481 D.M.P. Thompson, D.N. Tataryn, J.C. Weatherhead, P. Friedlander, J. Rauch, R. Schwartz, P. Gold and J. Schuster, Eur. J. Cancer, 1980, 16, 539.
482 R. Pompecki, J.E. Shively and C.W. Todd, Cancer Res., 1981, 41, 1905.
483 D.D. Munjal, Clin. Chem., 1980, 26, 1809.
484 M. Kuroki, Y. Koga and Y. Matsuoka, Cancer Res., 1981, 41, 713.
485 S.I. Ogata, R. Ueda, K.O. Lloyd, Proc. Natl. Acad. Sci. USA, 1981, 78, 770.
486 Y. Nakagawa, H.C. Margolis, S. Yokoyama, F.J. Kézdy, E. Kaiser and F.L. Coe, J. Biol. Chem., 1981, 256, 3936.
487 T.D. Oberley, A.E. Chung, J.E. Murphy-Ullrich and D.F. Mosher, J.Histochem. Cytochem., 1981, 29, 1237.
488 D.P. Via, S. Sramek, G. Larriba and S. Steiner, J. Cell. Biol., 1980, 84, 225.
489 M.M. Klinger, R.A. Laine and S.M. Steiner, J. Biol. Chem., 1981, 256, 7932.
490 A.C. Antony, C. Utley, K.C. van Horne and J.F. Kolhouse, J. Biol. Chem., 1981, 256, 9684.
491 O.A. Lea and F.S. French, Biochim. Biophys. Acta, 1981, 668, 370.
492 B. Peeters, W. Rombauts, J. Mous and W. Heyns, Eur. J. Biochem., 1981, 15, 115.
493 M. Paulsson and D. Heinegård, Biochem. J., 1981, 197, 367.
494 R.K. Chopra and J.M. Bowness, Can. J. Biochem., 1981, 59, 191.
495 K.P. Campbell and D.H. MacLennan, J. Biol. Chem., 1981, 256, 4626.
496 B. Carlin, R. Jaffe, B. Bender and A.E. Chung, J. Biol. Chem., 1981, 256, 5209.
497 L.A. Liotta, R.H. Goldfarb and V.P. Terranova, Thromb. Res., 1981, 21, 663.
498 L. Risteli and R. Timpl, Biochem. J., 1981, 193, 749.
499 M. Del Rosso, R. Cappelletti, M. Viti, S. Vannucchi and V. Chiarugi, Biochem. J., 1981 199, 699.
500 A.R. Cooper, M. Kurkinen, A. Taylor and B.L.M. Hogan, Eur. J. Biochem., 1981, 119, 189.
501 S. Sakashita and E. Ruoslahti, Arch. Biochem. Biophys., 1980, 205, 283.
502 A. Serafini-Fracassini, G. Ventrella, M.J. Field, J. Hinnie, N.I. Onyzili and R. Griffiths, Biochemistry, 1981, 20, 5424.
503 A. le Pape, J.D. Guitton and J.P. Muh, Biochem. Biophys. Res. Comm., 1981, 100, 1214.
504 J. McPherson, H. Sage and P. Bornstein, J. Biol. Chem., 1981, 256, 11330.
505 S.M. Breathnach, S.M. Melrose, B. Bhogal, F.C. de Beer, R.F. Dyck, G. Tennent, M.M. Black and M.B. Pepys, Nature, 1981, 293, 652.
506 M. Iwasaki and S. Inoue, J. Biochem. (Tokyo), 1981, 89, 1067.
507 S.K. Sikder and A. Das, Carbohydr. Res., 1981, 95, 249.
508 G.O. Gogstad, I. Hagen, R. Korsmo and N.O. Solum, Biochim. Biophys. Acta, 1981, 70, 150.
509 T.K. Gartner, J.M. Gerrard, J.G. White and D.C. Williams, Blood, 1981, 58, 153.
510 S.S. Margossian, J.W. Lawler and H.S. Slayter, J. Biol. Chem., 1981, 255, 7495.
511 E.M. Mazur, F. Hoffman, J. Chasis, S. Marchesi and E. Bruno, Blood, 1981, 57, 277.
512 T.G. Hayes, Arch. Virol., 1981, 67, 267.
513 E. Knight in "Interferon", ed. I. Gresser, Acad. Press, (London), 1980,

Vol.2, p.1.
514 J.W. Heine, J. Vandamme, M. Deley, A. Billiau and P. Desomer, J. Gen. Virol., 1981, 54, 47.
515 E. Knight and D. Fahey, J. Biol. Chem., 1981, 256, 3609.
516 S. Yonehara, Y. Yanase, T. Sano, M. Imai, S. Nakawawa and H. Mori, J. Biol. Chem., 1981, 256, 3770.
517 I.A. Braude, L.S. Lin and W.E. Stewart, Biochem. J., 1981, 193, 947.
518 R.V.W. van Eijk and P.F. Mühlradt, Eur. J. Immunol., 1979, 9, 506.
519 R.V.W. van Eijk, G. Rosenfelder and P.F. Mühlradt, Eur. J. Biochem, 1979, 101, 185.
520 R.V.W. van Eijk and P.F. Mühlradt, Eur. J. Biochem., 1981, 115, 23.
521 M.L. Rose and F. Malchiodi, Immunology, 1981, 42, 583.
522 J.F.J. Banchereau, D.M. Danois, M. Guenounou, G.M. Durand and J.C.L. Agneray, Biochim. Biophys. Acta, 1981, 678, 98.
523 P.D. Stahl and P.H. Schesinger, Trends Biochem. Sci., 1980, 5, 194.
524 V.L. Shepherd, Y.C. Lee, P.H. Schlesinger, P.D. Stahl, Proc. Natl. Acad. Sci. USA, 1981, 78, 1019.
525 A. Hasilik, U. Klein, A. Waheed, G. Strecker, K. Von Figura, Proc. Natl. Acad. Sci. USA, 1980, 77, 7074.
526 G. Cossu, M. Pacifici, M. Marino, B. Zani, M. Coletta and M. Molinaro, Exp. Cell Res., 1981, 132, 349.
527 M.E. Bramwell, Biochem. Soc. Trans., 1980, 8, 697.
528 V.P. Bhavanandan, A.W. Katlic, J. Banks, J.G. Kemper and E.A. Davidson, Biochemistry, 1981, 20, 5586.
529 D.C. Rijken and D. Collen, J. Biol. Chem., 1981, 256, 7035.
530 T. Muramatsu, H. Muramatsu and M. Ozawa, J. Biochem. (Tokyo), 1981, 89, 473.
531 M.J. Vilarem, J. Jouanneau and R. Bourrillon, Biochem. Biophys. Res. Commun., 1981, 98, 7.
532 B.L.A. Miki, J.W. Gurd and I.R. Brown, Can. J. Biochem., 1980, 58, 1261.
533 R.H. Glew, J.U. Balis, S. Shelley, T. Kuhlen Schmidt and R. Jaffe, Biochim. Biophys. Acta, 1981, 670, 101.
534 S.N. Bhattacharyya, Biochem. J., 1981, 193, 447.
535 S.C. Sahu and W.S. Lynn, Carbohydr. Res., 1981, 90, 251.
536 A. Laine and A. Hayem, Biochim. Biophys. Acta, 1981, 668, 429.
537 S.N. Bhattacharyya and W.S. Lynn, Biochim. Biophys. Acta, 1980, 626, 451.
538 A.S.R. Donald, Eur. J. Biochem., 1981, 120, 243.
539 D.A. Swann, H.S. Slayter and F.H. Silver, J. Biol. Chem., 1981, 256, 5291.
540 J. Aggeler, E. Engvall and Z. Werb, Biochem. Biophys. Res. Commun., 1981, 100, 1195.
541 B. Bayard and J.P. Kerckaert, Eur. J. Biochem., 1981, 113, 405.
542 H. Yoshima, T. Mizuochi, M. Ishii and A. Kobata, Cancer Res., 1980, 40, 4276.
543 A.G. Magee, D.A.W. Grant and J. Herman-Taylor, Clin. Chem. Acta, 1981, 115, 214.
544 A. Yost and B.M. Anderson, J. Biol. Chem., 1981, 256, 3647.
545 R.C. Hunt and N.F. Moore, in "Cell Membranes and Viral Envelopes, Vol.1, ed. H.A. Blough and J.M. Tiffany, Acad. Press (London), 1980, p.277.
546 C.G. Gahmberg in "Membrane Structure, New Comprehensive Biochemistry", ed. J.B. Finean and R.H. Michell, Elsevier, 1981, 1, 127.
547 C.R. Parish, H.C. O'Neil and T.J. Higgins, Immunology Today, 1981, 2, 98.
548 H. Debray, Bull. Cancer, 1979, 66, 353.
549 L. Warren and C.A. Buck, Clin. Biochem., 1980, 13, 191.
550 S.S. Spicer, D.A. Baron, A. Sato and B.A. Schulte, J. Histochem. Cytochem., 1981, 29, 994.
551 H.S. Slayter and J.F. Codington, Biochem. J., 1981, 193, 203.
552 J.P. Banga, P. Penfold and I.M. Riott, Biochem. J., 1981, 198, 421.
553 D.C. Hixson, J.M. Yep, J.R. Glenney, T. Hayes and G.F. Walborg, J. Histochem. Cytochem., 1981, 29, 561
554 D. Tsao and Y.S. Kim, J. Biol. Chem., 1981, 256, 4947.
555 S.A. Carlsen, E. Schmell, P.H. Weigel and S. Roseman, J. Biol. Chem., 1981, 256, 8058.

556 T. Sasaki and T. Sakagami, Cell Struc. Function, 1980, 5, 347.
557 C. Ocklind, K. Rubin, B. Obrink, FEBS Lett., 1980, 121, 47.
558 H. Debray, B. Fournet, J. Montreuil, L. Dorland and J.F.G. Vliegenthart, Eur. J. Biochem., 1981, 115, 559.
559 C. Bonfils, F. Nato, R. Bourrillon, C. Balny, FEBS Lett., 1981, 123, 222.
560 J. Harford and G.G. Ashwell, Proc. Natl. Acad. Sci. USA, 1981, 78, 1557.
561 H. Nakada, T. Sawamura and Y. Tashiro, J. Biochem. (Tokyo), 1981, 89, 135.
562 Y. Mizuno, Y. Kozutsumi, T. Kawasaki and I. Yamashina, J. Biol. Chem., 1981, 256, 4247.
563 A.L. Schwartz, S.E. Fridovich, B.B. Knowles and H.F. Lodish, J. Biol. Chem. 1981, 256 8878.
564 D.K. Strickland, T.T. Andersen, R.L. Hill and F.J. Castellino, Biochemistry, 1981, 20, 5294.
565 M.T. Debanne, P.A. Chindemi and E. Regoeczi, J. Biol. Chem., 1981, 256, 4928.
566 D.B. Cawley, D.L. Simpson and H.R. Herschman, Proc. Natl. Acad. Sci.USA., 1981, 78, 3383.
567 B.H. Dannevig, H. Tolleshaug and T. Berg, Biochim. Biophys. Acta, 1981, 667, 501.
568 J. Schlepper-Schäfer, V. Kolb-Bachofen and H. Kolb, Biochem. J., 1980, 186, 827.
569 Y. Maynard and J.U. Baenziger, J. Biol. Chem., 1981, 256, 8063.
570 A. Obrenovitch, C. Sené, A.C. Roche, M. Monsigny, P. Visher and R.C. Hughes, Biochimie, 1981, 63, 169
571 H. Rouhandeh and J.C. Richards, Biochim. Biophys. Acta, 1980, 600, 301.
572 K.A. Knudsen, P.E. Rao, C.H. Damsky, C.A. Buck, Proc. Natl. Acad. Sci. USA., 1981, 78, 6071.
573 A.R. Robbins, R. Myerowitz, R.J. Youle, G.J. Murray and D.M. Neville, J. Biol. Chem., 1981, 256, 10618.
574 A.R. Robbins and R. Myerowitz, J. Biol. Chem., 1981, 256, 10623.
575 J.D. Jamieson, D.E. Ingber, V. Muresan, B.E. Hull, M.P. Sarras, M.F. Maylie-Pfenniger and V. Iwanij, Cancer, 1981, 47, 1516.
576 M. Danca, J. Hourdry, J.S. Hugon, D. Menard, Comp. Biochem. Physiol. B., 1981, 69, 15.
577 A. Herscovics, A. Quaroni, B. Bugge and K. Kirsch, Biochem. J., 1981, 197, 511.
578 S.P. Clark and R.S. Molday, Biochemistry., 1979, 18, 5868.
579 A.B. Czuppon, L. Mettler, R. Schauer and V. Pawassarat, Hoppe-Seyler's Z. Physiol. Chem., 1981, 362, 963.
580 R. Jones, C. Pholpramool, B.P. Setchell and C.R. Brown, Biochem. J., 1981, 200, 457.
581 J. Herman and B. Keil, Biochim. Biophys. Acta, 1981, 643, 30.
582 S. Azhar and K.M.J. Menon, Eur. J. Biochem., 1981, 118, 25.
583 S. Cohen and R.L. Barchi, Biochim. Biphys. Acta, 1981, 645, 253.
584 P. Maher and R.S. Molday, Biochim. Biophys. Acta, 1981, 647, 259.
585 M.T. Nunez, J. Glass and E.S. Cole, Biochim. Biophys. Acta, 1981, 673, 137.
586 L.B. Grabel, C.G. Glabe, M.S. Singer, G.R. Martin and S.D. Rosen, Biochem. Biophys. Res. Commun., 1981, 102, 1165.
587 C.K. Ramachandran, C.E. Hignite, S.L. Gray and G. Melnykovych, Biochem. J., 1981, 198, 23.
588 S.C. Howard, A.P. Sherblom, J.W. Huggins, C.A.C. Carraway and K.L. Carraway, Arch. Biochem. Biophys., 1981, 207, 40.
589 R.M. Helm and K.L. Carraway, Exp. Cell Res., 1981, 135, 418.
590 H.J. Allen and E.A.Z. Johnson, Biochim. Biophys. Acta, 1980, 600, 320.
591 V.P. Lehto, T. Vartio and I. Virtanen, FEBS Lett., 1981, 124, 289.
592 W.G. Carter and S.I. Hakomori, J. Biol. Chem., 1981, 256, 6953.
593 D.G. Campbell, J. Gagnon, K.B.M. Reid and A.F. Williams, Biochem. J., 1981, 195, 15.
594 F.E. Cohen, J. Novotny, M.J.E. Sternberg, D.G. Campbell and A.F. Williams, Biochem. J., 1981, 195, 31.
595 M. Letarte, J. Addis, M.E. MacDonald, A. Bernstein and P. Lake, Can. J.

Biochem., 1980, 58, 1026.
596 R.V.W. van Eijk and P.F. Mühlradt, Eur. J. Biochem., 1981, 115, 23.
597 R.A. Childs and T. Feizi, Biochem. Biophys. Res. Commun., 1981, 102, 1158.
598 J. Ivanyi and L. Moyes, Mol. Immunol., 1981, 17, 1545.
599 L.M. Keefer and P. De Meyts, Biochem. Biophys. Res. Commun., 1981, 101, 22.
600 E.L. Reinherz, R.E. Hussey, F. Fitzgerald, P. Snow, C. Terhorst and S.F. Schlossman, Nature, 1981, 294, 168.
601 R.L. Coffman and I.L. Weissman, Nature, 1981, 289, 681.
602 M. Jett and G.A. Jamieson, Biochemistry, 1981, 20, 5221.
603 J. Banchereau, D. Danois, M. Guenounou, G. Durand and J. Agneray, C. R. Hebd. Seances Acad. Sci., 1980, 290, 191.
604 R.L.P. Flower ans G.E. Wilcox, J. Immunol. Methods, 1981, 46, 347.
605 S.E. Cullen, C.S. Kindle, D.C. Shreffler and C. Cowing, J. Immunol., 1981, 127, 1478.
606 W.R.A. Brown, A.N. Barclay, C.A. Sutherland and A.F. Williams, Nature, 1981, 289, 456.
607 A. Vanagthoven, C. Terhorst, E. Reinherz and S. Schlossman, Eur. J. Immunol., 1981, 11, 18.
608 I.S. Trowbridge and M.B. Omary, Proc. Natl. Acad. Sci. USA, 1981, 78, 3039.
609 M.B. Omary and I.S. Trowbridge, J. Biol. Chem., 1981, 256, 12888.
610 R.G. Cook, R.R. Rich and L. Flaherty, J. Immunol., 1981, 127, 1894.
611 S.G. Nathenson, H. Uehara, B.M. Ewenstein, T.J. Kindt and J.E. Coligan, Annu. Rev. Biochem., 1981, 50, 1025.
612 V.L. Alvarez, C.A. Roitsch and O. Henriksen, Anal. Biochem., 1981, 115, 353.
613 W.L. Maloy, S.G. Nathenson and J.E. Coligan, J. Biol. Chem., 1981, 256, 2863.
614 E.S. Kimball, S.G. Nathenson and J.E. Coligan, Biochemistry, 1981, 20, 3301.
615 J.E. Coligan, T.J. Kindt, H. Uehara, J. Martinko and S.G. Nathenson, Nature, 1981, 291, 35.
616 H. Uehara, J.E. Coligan and S.G. Nathenson, Biochemistry, 1981, 20, 5936.
617 H. Uehara, J.E. Coligan and S.G. Nathenson, Biochemistry, 1981, 20, 5940.
618 E.S. Kimball, W.L. Maloy and J.E. Coligan, Mol. Immunol., 1981, 18, 677.
619 J.S. Pober, B.C. Guild, J.L. Strominger and W.R. Veatch, Biochemistry., 1981, 20, 5625.
620 E. Sung and P.P. Jones, Mol. Immunol., 1981, 18, 899.
621 J. Robinson, C. Sieff, D. Delia, P.A.W. Edwards and M. Greaves, Nature, 1981, 289, 68.
622 K. Sege, L. Rask and P.S. Peterson, Biochemistry, 1981, 20, 4523.
623 A. Tartakoff, D. Hoessli and P. Vassalli, J. Mol. Biol., 1981, 150, 525.
624 A.J. Korman, L.L. Ploegh, J.F. Kaufmann, M.J. Owen and J.L. Strominger, J. Exp. Med., 1980, 152, 655.
625 M.J. Owen, A.M. Kissonerghis, H.F. Lodish and M.J. Crumpton, J. Biol. Chem., 1981, 256, 8987.
626 M. Shapiro and R. P. Erickson, Nature, 1981, 290, 503.
627 T. Cotner, H. Mashimo, P.C. Kung, G. Goldstein and J.L. Strominger, Proc. Natl. Acad. Sci. USA, 1981, 78, 3858.
628 S.G. Nathenson, H. Uehara, B.M. Ewenstein, T.J. Kindt and J.E. Coligan, Annu. Rev. Biochem., 1981, 50, 1025.
629 Y. Shechter and B.A. Sela, Biochem. Biophys. Res. Commun., 1981, 98, 367.
630 H.M. Katzen, D.D. Soderman and B.G. Green, Biochem. Biophys. Res. Commun., 1981, 98, 410.
631 J.A. Hedo, L.C. Harrison and J. Roth, Biochemistry, 1981, 20, 3385.
632 G.V. Ronnett and M.D. Laine, J. Biol. Chem., 1981, 256, 4704.
633 E. Remold-O'Donnell, J. Exp. Med., 1980, 152, 1699.
634 G. Kaplan and R. Olstad, Exp. Cell Res., 1981, 135, 379.
635 T.A. Springer, J. Biol. Chem., 1981, 256, 3833.
636 F. Amano and D. Mizuno, J. Biochem. (Tokyo), 1981, 89, 1149.
637 E. Glass, J. Stewart and D.M. Weir, Immunology, 1981, 44, 529.
638 G. Mengod, E.N. Hughes and J.T. August, Arch. Biochem. Biophys., 1981, 209, 718.

639 A.C. Morgan, D.R. Galloway, K. Imai and R.A. Reisfeld, J. Immunol., 1981, 126, 365.
640 S. Turco, J.S. Rush and R.A. Laine, J. Biol. Chem., 1980, 255, 3266.
641 J.S. Rush, S.J. Turco and R.A. Laine, Biochem. J., 1981, 193, 361.
642 M. Fukuda, Nature, 1980, 285, 405.
643 A.R. Pecoud, S. Ruddy and D.H. Conrad, J. Immunol., 1981, 126, 1624.
644 B.L. Hempstead, C.W. Parker and A. Kulczycki, J. Biol. Chem., 1981, 256, 10717.
645 M. Fukuda, H.P. Koeffler and J. Minowada, Proc. Natl. Acad. Sci. USA., 1981, 78, 6299.
646 E.N. Hughes, G. Mengod and J.T. August, J. Biol. Chem., 1981, 256, 7023.
647 M.C. Fishman, P.R. Dragsten and I. Spector, Nature, 1981, 290, 781.
648 J.N. George, L.L. Thoi and R.K. Morgan, Thromb. Res., 1981, 23, 69.
649 A. Rotman and S. Linder, Thromb. Res., 1981, 22, 227.
650 S. Menashi, H. Weintroub and N. Crawford, J. Biol. Chem., 1981, 256, 4095.
651 R.P. McEver, N.L. Baenziger and P.W. Majerus, J. Clin. Invest., 1980, 66, 1311.
652 D.M. Peterson and B. Wehring, Thromb. Res., 1981, 22, 53.
653 J.R. Holahan and G.C. White, Blood, 1981, 57, 174.
654 J.L. McGregor, K.J. Clemetson, E. James, A. Capitanio, T. Greenland, E.F. Lüscher and M. Dechavanne, Eur. J. Biochem., 1981, 116, 379.
655 R. Apitz-Castro, G. Fonesca, V. Michelena and M.R. Cruz, Thrombosis Haemostatis, 1981, 45, 130.
656 A.T. Nurden, D. Dupuis, T.J. Kunicki and J.P. Caen, J. Clin. Invest., 1981, 67, 1431.
657 L.P. Thuy, R.F. Baugh, J.E. Brown and C. Houghie, Biochim. Biophys. Acta, 1981, 678, 187.
658 S. Elodi, K. Varadi and E. Vörös, Thromb. Res., 1981, 21, 695.
659 P. Ganguly and N.G. Fossett, Blood, 1981, 57, 343.
660 H.R. Prasanna, H.H. Edwards and D.R. Phillips, Blood, 1981, 57, 305.
661 P. Ganguly and N.G. Fossett, Biochem. Biophya. Res. Commun., 1981, 99, 176.
662 K.J. Clemetson, H.Y. Naim, E.F. Lüscher, Proc. Natl. Acad. Sci. USA, 1981, 78, 2712.
663 B. Bauvois, R. Cacan, A.T. Nurden, J. Caen, J. Montreuil, A. Verbert, FEBS Lett., 1981, 125, 277.
664 L.L.K. Leung, T. Kinoshita and R.L. Nachman, J. Biol. Chem., 1981, 256, 1994.
665 M.J. Polley, L.L.K. Leung, F.Y. Clark and R.L. Nachman, J. Exp. Med., 1981, 154, 1058.
666 L.L.K. Leung, T. Kinoshita and R.L. Nachman, J. Biol. Chem., 1981, 256, 1994.
667 G.O. Gogstad, N.O. Solum and I. Hagen, Thromb. Res., 1981, 24, 157.
668 M.A. Blajchman, A.F. Senyi, J. Hirsh, E. Genton and J.N. George, J. Clin. Invest., 1981, 68, 1289.
669 T.D. Butters and R.C. Hughes, Biochim. Biophys. Acta, 1981, 640, 655.
670 N. Frantzis and C. Vakirtzi-Lemonias, Biochem. Soc. Trans., 1981, 9, 135.
671 D.J. Anderson and G. Blobel, Proc. Natl. Acad. Sci. USA, 1981, 78, 5598.
672 J.G. Pierce and T.F. Parsons, Annu. Rev. Biochem., 1981, 50, 465.
673 R.P. Cox and D.G. Day, Adv. Cell Culture, 1981, 1, 15.
674 J.G. Pierce and T.F. Parsons in "Evolution of protein structure and function", ed. D.S. Sigman and M.A.B. Brazier, Acad. Press (New York), 1980, p.99.
675 H. Takahashi and Y. Hanaoka, J. Biochem. (Tokyo), 1981, 90, 1333.
676 C.V. Rao, S. Mitra and F.R. Carmian, J. Biol. Chem., 1981, 256, 2628.
677 G.S. Cox, Biochemistry, 1981, 20, 4893.
678 Y. Combarnous, R. Salesse and J. Garnier, Biochim. Biphys. Acta, 1981, 667, 267.
679 Chorionic gonadotrophin, ed. S.J. Segal, Plenum Publishing Corp., New York, 1980.
680 M.C. Stuart, S. Ellis, L. Gowlland and S. Tuff, Clin. Chem., 1981, 27, 52.
681 S. Iwasa, C. Kitada, I. Yoshida, K. Kondo, M. Hori and M. Fujino,

J. Biochem. (Tokyo), 1981, 89, 1091.
682 R.W. Ruddon, A.H. Bryan, C.A. Hanson, F. Perini, L.M. Ceccorulli and B.P. Peters, J. Biol. Chem., 1981, 256, 5189.
683 G. Kapadia, J. Vaitukaitis and A. Ohrambach, Prep. Biochem., 1981, 11, 1.
684 R.W. Ruddon, A.H. Bryan, K.S. Meade-Cobun and V.A. Pollack, Cancer Res., 1980, 40, 4007.
685 B.T. Backus and L.F. Affronti, Infec. Immun., 1981, 32, 1211.
686 G.S. Cox, Biochem. Biophys. Res. Comm., 1981, 98, 942.
687 R.V. Rebois, F. Omedeo-Sale, R.O. Brady and P.H. Fishman, Proc. Natl. Acad. Sci. USA, 1981, 78, 2086.
688 H.M. Madnick, N.K. Kalyan, H.L. Segal and O.P. Baal, Arch. Biochem. Biophys., 1981, 212, 432.
689 N.K. Kalyan and O.P. Bahl, Biochem. Biophys. Res. Commun., 1981, 102, 1246.
690 K.C. Ingham and S.A. Brew, Biochim. Biophys. Acta, 1981, 670, 181.
691 D. Bellet, J.M. Arrang, G. Contesso, J.M. Caillaud and C. Bohaun, Eur. J. Cancer, 1980, 16, 433.
692 J.M. Governman and J.G. Pierce, J. Biol. Chem., 1981, 256, 9431.
693 T.F. Parsons and J.G. Pierce, Proc. Natl. Acad. Sci. USA, 1980, 77, 7089.
694 M.R. Sairam, Biochem. J., 1981, 197, 535.
695 M.R. Sairam, N.G. Seidah and M. Chrétien, Biochem. J., 1981, 197, 541.
696 Y. Cambarnous and M.H. Hengé, J. Biol. Chem., 1981, 256, 9567.
697 J.A. Dias, W.J. Driskell and L.E. Reichert, Anal. Biochem., 1981, 114, 268.
698 C.A. Spencer and J.T. Nicoloff, Clin. Chim. Acta, 1980, 108, 415.
699 W.W. Chin, F. Maloof and J.F. Habener, J. Biol. Chem., 1981, 256, 3059.
700 J.W. Jacobs, P.K. Lund, J.T. Potts, N.H. Bell and J.F. Habener, J. Biol. Chem., 1981, 256, 2803.
701 C. Ronin, C. Caseti and S. Bouchilloux, Biochim. Biophys. Acta, 1981, 674, 48.
702 C. Ronin and C. Caseti, Biochim. Biophys. Acta, 1981, 674, 58.
703 P. Crine, N.G. Seidah, L. Jeanotte and M. Chrétien, Can. J. Biochem., 1980, 58, 1318.
704 N.G. Seidah, S. Benjannet and M. Chrétien, Biochem. Biophys. Res. Commun., 1981, 100, 901.
705 N.G. Seidah, J. Rochemont, J. Hamelin, M. Lis and M. Chrétien, J. Biol. Chem., 1981, 256, 7977.
706 N.S. Halmi, J. Histochem. Cytochem., 1981, 29, 837.
707 J.G. Collins, J.H. Bradbury, E. Trifnoff and M. Messer, Carbohydr. Res., 1981, 92, 136.
708 K. Takase and K.E. Ebner, J. Biol. Chem., 1981, 256, 7269.
709 C.C. Contaxis and C.C. Bigelow, Biochemistry, 1981, 20, 1618.
710 L.J. Berliner and R. Kaptein, Biochemistry, 1981, 20, 799.
711 P.B. Sommers and M.J. Kronman, Biochim. Biophys. Acta, 1980, 626, 366.
712 E.A. Permyakov, V.V. Yarmolenko, L.P. Kalinichenko, L.A. Morozova and E.A. Burstein, Biochem. Biophys. Res. Commun., 1981, 100, 191.
713 S.K. Sinha and K. Brew, J. Biol. Chem., 1981, 256, 4193.
714 K. Bell, H.A. McKenzie and D.C. Shaw, Aust. J. Biol. Sci., 1981, 34, 133.
715 K. Bell, K.E. Hopper and H.A. McKenzie, Aust. J. Biol. Sci., 1981, 34, 149.
716 H. Doi, H. Kobatake, F. Ibuki and M. Kanamori, Agric. Biol. Chem., 1980, 44, 2605.
717 A.M. Fiat, J. Jollès, M.H. Loucheux-Lafebvre, C. Alais and P. Jollès, Hoppe-Seyler's Z. Physiol. Chem., 1981, 362, 1447.
718 T. Saito, T. Itoh and S. Adachi, Biochim. Biophys. Acta, 1981, 673, 487.
719 T. Saito, T. Itoh, S. Adachi, T. Suzuki and T. Usui, Biochim. Biophys. Acta, 1981, 678, 257.
720 H. van Halbeek, L. Dorland, J.F.G. Vliegenthart, A.M. Fiat, P. Jollès, FEBS Lett., 1981, 133, 45.
721 M. Horisberger and M. Vonlanthen, J. Dairy Res., 1980, 47, 185.
722 H. Doi, F. Ibuki and M. Kanamori, Agric. Biol. Chem., 1981, 45, 2351.
723 S. Soulier and P. Gaye, Biochimie, 1981, 63, 619.
724 M. Hirose, T. Kato, K. Omori, M. Maki, M. Yoshikawa, R. Sasaki and H. Chiba, Biochim. Biophys. Acta, 1981, 667, 309.

725 A. Safi and R. Harrison, Biochem. Soc. Trans., 1981, 9, 436.
726 J.P. Bradshaw and D.A. White, Biochem. J., 1981, 198, 683.
727 I.K. Vijay, G.H. Perdew and D.E. Lewis, J. Biol. Chem., 1980, 255, 11210.
728 I.K. Vijay and G.H. Perdew, J. Biol. Chem., 1980, 255, 11221.
729 H. van Halbeek, L. Dorland, J.F.G. Vliegenthart, G. Spik, A. Cheron and J. Montreuil, Biochim. Biophys. Acta, 1981, 675, 293.
730 M.H. Metz-Boutigue, J. Jollès, J. Mazurier, G. Spik, J. Montreuil and P. Jollès, FEBS Lett., 1981, 132, 239.
731 H. Okubo, O. Miyanaga, M. Nagano, H. Ishibashi, J. Kudo, T. Ikuta and K. Shibata, Biochim. Biophys. Acta, 1981, 668, 257.
732 K. Wu, D. Wang and R.D. Feinman, J. Biol. Chem., 1981, 256, 10409.
733 J.B. Bieth, M. Tourbez-Perrin and F. Pochon, J. Biol. Chem., 1981, 256, 7954.
734 L. Sottrup-Jensen, T.E. Petersen and S. Magnusson, FEBS Lett., 1980, 121, 275.
735 L. Sottrup-Jensen, H.F. Hansen, S.B. Mortensen, T.E. Petersen and S. Magnusson, FEBS Lett., 1981, 123, 145.
736 L. Sottrup-Jensen, T.E. Petersen and S. Magnusson, FEBS Lett., 1981, 128, 123.
737 L. Sottrup-Jensen, T.E. Petersen and S. Magnusson, FEBS Lett., 1981, 128, 127.
738 F. van Leuven, J.J. Cassiman and H. van den Berghe, J. Biol. Chem., 1981, 256, 9016
739 F. van Leuven, J.J. Cassiman and H. van den Berghe, J. Biol. Chem., 1981, 256, 9023.
740 G.S. Salvesen, C.A. Sayers and A.J. Barrett, Biochem. J., 1981, 195, 453.
741 S.B. Mortensen, L. Sottrup-Jensen, H.F. Hansen, D. Rider, T.E. Petersen, S. Magnusson, FEBS Lett., 1981, 129, 314.
742 L. Sottrup-Jensen, P.B. Lønblad, T.M. Stepanik, T.E. Petersen, S. Magnusson and H. Jörnvall, FEBS Lett., 1981, 127, 167.
743 P.K. Hall, L.P. Nelles, J. Travis and R.C. Roberts, Biochem. Biophys. Res. Commun., 1981, 100, 8.
744 J. Bienvenu, L. Sann, F. Bienvenu, C. Lahet, P. Divry, J. Cotte and M. Bethenod, Clin. Chem., 1981, 27, 721.
745 G.A. Ricca and J.M. Taylor, J. Biol. Chem., 1981, 256, 11199.
746 I. Nicollet, J.P. Lebreton, M. Fontaine and M. Hiron, Biochim. Biophys. Acta, 1981, 668, 235.
747 C. Wells, T.C. Bøg-Hansen, E.H. Cooper and M.R. Glass, Clin. Chim. Acta, 1981, 109, 59.
748 H. van Halbeek, L. Dorland, J.F. Vliegenthart, J. Montreuil, B. Fournet and K. Schmid, J. Biol. Chem., 1981, 256, 5588.
749 M. Bennett and K. Schmid, Proc. Natl. Acad. Sci. USA, 1980, 77, 6109.
750 G. Schreiber, H. Dryburgh, K. Weigand, M. Schreiber, I. Witt, H. Seydewitz and G. Howlett, Arch. Biochem. Biophys., 1981, 212, 319.
751 D.E. Roll and R.H. Glew, J. Biol. Chem., 1981, 256, 8190.
752 L. Vaughan and R. Carrell, Biochem. Int., 1981, 2, 461.
753 K. Hochstrasser, O.L. Schönberger, K. Lempart and M. Metzgar, Hoppe-Seyler's Z. Physiol. Chem., 1981, 362, 1363.
754 Y. Ikehara, M. Miyasoto, S. Ogata and K. Oda, Eur. J. Biochem., 1981, 115, 253.
755 Z. Parvez, J. Fareed, H.L. Messmore and R. Moncada, Thromb. Res., 1981, 24, 367.
756 M. Tanaka and K. Kato, Thromb. Res., 1981, 22, 67.
757 T.F. Busby, D.H. Atha and K.C. Ingham, J. Biol. Chem., 1981, 256, 12140.
758 W.F. Long and F.B. Williamson, Biochem. Soc. Trans., 1981, 9, 68.
759 R. Machovich, P.I. Bauer, P. Arányi, E. Kecskés, K.G. Büchi and I. Horváth, Biochem. J., 1981, 199, 521.
760 S.T. Olson and J.D. Shore, J. Biol. Chem., 1981, 256, 11065.
761 S.T. Olson, K.R. Srinivasan, I. Björk and J.D. Shore, J. Biol. Chem., 1981, 256, 11073.
762 L. Williams and G. Murano, Blood, 1981, 57, 229.

763 G. Murano, M. Miller-Anderson and L. Williams, Thromb. Res., 1981, 24, 489.
764 I. Björk, Å. Danielsson, J.W. Fenton and H. Jörnvall, FEBS Lett., 1981, 126, 257.
765 L.W. Hoyer, Blood, 1981, 58, 1.
766 J.M. Yoder, L.A. Schick and R.P. More, Thromb. Res., 1981, 24, 51.
767 M.J. Seghatchian and I.J. Mackie, Thromb. Res., 1981, 24, 473.
768 R. Ljung and L. Holmberg, Thromb. Res., 1981, 24, 45.
769 K.J. Kao, S.V. Pizzo and P.A. McKee, Blood, 1981, 57, 579.
770 G. Kemball-Cook, R.A. Furlong, I.R. Peake and T.W. Barrowcliffe, Thromb. Res., 1981, 23, 193.
771 L.O. Andersson, L.P. Thuy and J.E. Brown, Thromb. Res., 1981, 23, 481.
772 S.A. Santoro, Thromb. Res., 1981, 21, 689.
773 D.M. Scott, B. Griffin, D.S. Pepper and M.J. Barnes, Thromb. Res., 1981, 24, 467.
774 L.O. Andersson, J. Chromatogr., 1981, 215, 153.
775 B.A. Perret, M. Furlan, F. Kneubeuhl and E.A. Beck, Biochim. Biophys. Acta, 1981, 669, 98.
776 R.B. Harris, J. Newman and A.J. Johnson, Biochim. Biophys. Acta., 1981, 668, 456.
777 R.B. Harris, A.J. Johnson and L.T. Hodgins, Biochim. Biophys. Acta., 1981, 668, 471.
778 E.P. Kirby, Thromb. Haemostatis, 1981, 46, 479.
779 G.A. Vehar and E.W. Davie, Biochemistry, 1980, 19, 401.
780 C.G. Cockburn, R.J. Debeau Fre-Apps, J. Wilson and R.M. Hardisty, Thromb. Res., 1981, 21, 295.
781 M. Weinstein, L. Chute and D. Deykin, Proc. Natl. Acd. Sci. USA, 1981, 78, 5137.
782 M.E.P. Switzer and P.A. McKee, J. Biol. Chem., 1980, 255, 10606.
783 M.I. Kavanagh, C.N. Wood and J.F. Davidson, Thromb. Haemostasis, 1981, 45, 267.
784 S. Mikami, Y. Takahashi, M. Nishino, Y. Okubo and H. Fukui, Thromb. Haemostasis, 1981, 45, 272.
785 S.E. Martin, V.J. Marder, C.W. Francis and G.H. Barlow, Blood, 1981, 57, 313.
786 B.L. Davies, R.A. Furlong and I.R. Peake, Thromb. Res., 1981, 22, 87.
787 A. Lajmanovich, G. Hudry-Clergeon, J.M. Freyssinet and G. Marguerie, Biochim. Biophys. Acta, 1981, 678, 132.
788 G. Rock, W.H. Cruickshank and D.S. Palmer, Thromb. Res., 1981, 21, 53.
789 G. van Dieijen, G. Tans, J. Rosing and H.C. Hemker, J. Biol. Chem., 1981, 256, 3433.
790 J.M. Freyssinet, FEBS Lett., 1981, 124, 48.
791 P. Margaritella and E. Deutsch, Thromb. Res., 1981, 21, 585.
792 C.H. Hammer, G.H. Wirtz, L. Renfer, H.D. Gresham and B.F. Tack, J. Biol. Chem., 1981, 256, 3995.
793 K. Yonemasu and T. Sasaki, Biochem. J., 1981, 193, 621.
794 M.D. Golan, T. Hitschold and M. Loos, FEBS Lett., 1981, 128, 281.
795 R.B. Sim and E. Sim, Biochem. J., 1981, 193, 129.
796 J.B. Monahan and J.M. Sodetz, J. Biol. Chem., 1980, 255, 10579.
797 P.J. Higgins and H.F. Bunn, J. Biol. Chem., 1981, 256, 5204.
798 R. Dolhofer and O.H. Wieland, Clin. Chim. Acta, 1981, 112, 197.
799 K.A. Ney, K.J. Colley and S.V. Pizzo, Anal. Biochem., 1981, 118, 294.
800 S.S. Nayak and T.N. Pattabiraman, Clin. Chim. Acta, 1981, 109, 267.
801 P.M. Gallop, R. Flückiger, A. Hanneken, M.M. Mininsohn and K.H. Gabbay, Anal. Biochem., 1981, 117, 427.
802 K.M. Parker, J.D. England, J. da Costa, R.L. Hess and D.E. Goldstein, Clin. Chem., 1981, 27, 669.
803 D.M. Nathan, Clin. Chem., 1981, 27, 1261.
804 A. Ma, M.A. Naughton and D.P. Cameron, Clin. Chim. Acta, 1981, 115, 111.
805 E.C. Abraham, Biochim. Biophys. Acta, 1981, 667, 168.
806 A.P.M. Schellekens, G.T.B. Sanders, W. Thornton and T. van Groenestein, Clin. Chem., 1981, 27, 94.

807 A. Read, L. Tibi and A.F. Smith, Clin. Chim. Acta, 1981, 108, 487.
808 E. Schleicher, T. Deufel and O.H. Wieland, FEBS Lett., 1981, 129, 1.
809 J.A. Miller, E. Gravallese and H.F. Bunn, J. Clin. Invest., 1980, 65, 896.
810 D.B. Rylatt, D.Y. Sia, J.P. Mundy and C.R. Parish, Eur. J. Biochem., 1981, 119, 641.
811 W.T. Morgan, Biochemistry, 1981, 20, 1054.
812 C. Vincent, J.P. Revillard and B. Charpentier, C.R. Hebd. Seances, Acad. Sci. Ser. III, 1981, 292, 755.
813 D.R. Pledger, A. Belfield and P. Strömberg, J. Clin. Chem. Clin. Biochem., 1981, 19, 943.
814 S.D. Bolmer and E.A. Davidson, Biochemistry, 1981, 20, 1047.
815 M.Müller, H. Lasarcyk and W. Burchard, Int. J. Biol. Macromol., 1981, 3, 19.
816 K.F. Geoghegan, J.L. Dallas and R.E. Feeney, J. Biol. Chem., 1980, 255, 11429.
817 F. Erard, S.Y. Cheng and J. Robbins, Arch. Biochem. Biophys., 1981, 206, 15.
818 J. Breborowicz, A. Macklewicz and D. Breborowicz, Scand. J. Immunol., 1981, 14, 15.
819 M.I. Halliday and G.B. Wisdom, Biochem. Soc. Trans., 1981, 9, 330.
820 H. Tolleshaug, Int. J. Biochem., 1981, 13, 45.
821 J.C. Oberholtzer, S.W. Englander and A.F. Horwitz, Biochemistry, 1981, 20, 4785.
822 P. Kiss, F. Sharbert and R. Kattermann, Hoppe-Seyler's Z. Physiol. Chem., 1981, 362, 897.
823 S. Goldstein and M.J. Chapman, Biochemistry, 1981, 20, 1025.
824 T. Diaz-Mauriño and A. Albert, Biochim. Biophys. Acta, 1981, 677, 381.
825 J.M. Nickerson and G.M. Fuller, Biochemistry, 1981, 20, 2818.
826 H.R. Listen, B. van Hoef and D. Collen, Eur. J. Biochem., 1981, 120, 149.
827 Y. Gazitt, Z. Gilead, G. Klein and D. Sulitzeanu, J. Immunol. Methods, 1981, 43, 49.
828 F. Halliman and E. Tempany in Current Problems New Trends in Cystic Fibrosis: Monographs in Paediatrics, vol 14, ed. M. Schoni and R. Kraemer, S. Karger (Basel) 1981, 14, 40.
829 F. Hallinan and E. Tempany, Biochem. Soc. Trans., 1981, 9, 328.
830 W.W. Mueller and J.M. Potter, Biochem. J., 1981, 197, 645.
831 J.G. Duman, J.L. Patterson, J.J. Kozak and A.L. De Vries, Biochim. Biophys, Acta, 1980, 626, 332.
832 D. Slaughter and C.L. Hew, Anal. Biochem., 1981, 115, 212.
833 C. Cunningham-Rundles, Eur. J. Immunol., 1981, 11, 504.
834 S. Binion and L.S. Rodkey, Anal. Biochem., 1981, 112, 362.
835 B.W. Maidment, L.D. Papsidero, M. Gamarra, T. Nemoto and T.M. Chu, Anal. Biochem., 1981, 111, 336.
836 J. Rousseaux, M.T. Picque, H. Bazin and G. Biserte, Mol. Immunol., 1981, 18, 639.
837 D.J. McVeigh and C.N. Chesterman, Thromb. Res., 1981, 23, 559.
838 T.D. Heath, B.A. Macher and D. Papahadjopoulos, Biochim. Biophys. Acta, 1981, 640, 66.
839 V.K. Jansons and P.L. Mallett, Anal. Biochem., 1981, 111, 54.
840 D.M. Kranz, J.N. Herron, D.E. Giannis and E.W. Voss, J. Biol. Chem., 1981, 256, 4433.
841 M. Klein, N. Haeffnercavaillon, D.E. Isenman, C. Rivat, M.A. Navia, D.R. Davies and K.J. Dorrington, Proc. Natl. Acad. Sci., USA, 1981, 78, 524.
842 M.D. Dodon and G.A. Quash, Immunology, 1981, 42, 401.
843 T. Tomono, T. Suzuki and E. Tokunaga, Biochim. Biophys. Acta, 1981, 660, 186.
844 R. Luedtke, C.S. Owen, J.M. Vanderkooi and F. Karush, Biochemistry, 1981, 20, 2927.
845 F.A. Garver, L.S. Chang, C.R. Kiefer, J. Mendicino, E.V. Chandrasekaran, T. Isobe and E.F. Osserman, Eur. J. Biochem., 1981, 115, 643.
846 E.V. Chandrasekaran, A. Mendicino, F.A. Garver and J. Mendicino, J. Biol. Chem., 1981, 256, 1549.

847 G. Savvidou, M. Klein, C. Horne, T. Hofman and K.J. Dorrington, Mol. Immunol., 1981, 18, 793.
848 S. Weitzman and J. Protmore, J. Immunol., 1981, 127, 2095.
849 I. Garcia and J.C. Jaton, Biochem. J., 1981, 197, 177.
850 T. Takayasu, N. Takashi, T. Shinoda, T. Okuyama and H. Tomioka, J. Biochem. (Tokyo), 1981, 89, 421.
851 H.C. O'Neill, C.R. Parish and T.J. Higgins, Mol. Immunol., 1981, 18, 663.
852 E.A. Kabat, J. Liao, M.H. Burzynska, T.C. Wong, H.Thøgersen and R.U. Lemieux, Mol. Immunol., 1981, 18, 873.
853 J.H. Pazur, K.R. Forry, Y. Tominaga and E.M. Ball, Biochem. Biophys. Res. Commun., 1981, 100, 420.
854 J. Sharon, E.A. Kabat and S.L. Morrison, Mol. Immunol., 1981, 18, 831.
855 M. Torii, B.P. Alberto and S. Tanaka, J. Biochem. (Tokyo), 1980, 88, 1855.
856 A. Roy, B.N. Manjula and C.P.J. Glaudemans, Mol. Immunol., 1981, 18, 79.
857 M. Blockhaus, J.L. Magnani, M. Blaszczyk, Z. Steplewski, H. Koprowski, K.A. Karlsson, G. Larson and V. Ginsburg, J. Biol. Chem., 1981, 256, 13223.
858 H. Schenkel-Brunner and P. Hanfland, Vox Sanguinis, 1981, 40, 358.
859 W.W. Young, J. Portoukalian and S.I. Hakomori, J. Biol. Chem., 1981, 256, 10967.
860 E.F. Hounsell, H.C. Gooi and T. Feizi, FEBS Lett., 1981, 131, 279.
861 E. Nudelman, S.I. Hakomori, B.B. Knowles, D. Solter, R.C. Nowinski, M.R. Tam and W.W. Young, Biochem. Biophys. Res. Commun., 1980, 97, 443.
862 E. Nudelman, S.I. Hakomori, B.B. Knowles, D. Solter, R.C. Nowinski, M.R. Tam and W.W. Young, Biochem. Biophys. Res. Commun., 1981, 98, 348.
863 D. Roescke, Vox Sanguinis, 1981, 41, 98.
864 R.U. Lemieux, D.A. Baker, W.M. Weinstein and C.M. Switzer, Biochemistry, 1981, 20, 199.
865 G.T. Rogers, G.A. Rawlins and K.D. Bagshawe, Brit. J. Cancer, 1981, 43, 1.
866 D.E. Haagensen, C.E. Cox, W.G. Dilley, M. Hansley, J. Murdoch, E.S. Newman and S.A. Wells, Clin. Chem., 1980, 26, 1787.
867 H. Hirai, Y. Tsukada, A. Hara, N. Hibi, S. Nishi, H.T. Wepsic, T. Koji and N. Ishii, J. Chromatogr, 1981, 215, 195.
868 M.L. Kavanagh, C.N. Wood and J.F. Davidson, Thrombosis Haemostatis, 1981, 45, 60.
869 J.A. Katzmann, D.K. Mujwid, R.S. Miller and D.N. Fass, Blood, 1981, 58, 530.
870 J. Hilkens, J.M. Tager, F. Buijs, B. Brouwer-Kelder, G.M. van Thienen, F.P.W. Tegelaers and J. Hilgers, Biochim. Biophys. Acta, 1981, 678, 7.
871 J.M. Gallagher and G.B. Wisdom, Biochem. Soc. Trans, 1981, 9, 331.
872 C. Rickards, P. Buckland, B.R. Smith and R. Hall, FEBS Lett., 1981, 127, 17.
873 A.L. Barsoum and E.A. Davidson, Mol. Immunol., 1981, 18, 367.
874 D.E. Briles, J.L. Claflin, K. Schroer and C. Forman, Nature, 1981, 294, 88.
875 R. Myllylä, Biochim. Biophys. Acta, 1981, 658, 299.
876 R.J. Youle and D.M. Neville, Proc. Natl. Acad. Sci. USA, 1980, 77, 5483.
877 K.A. Krolick, C. Villemez, P. Isakson, J.W. Uhr and E.S. Vitetta, Proc. Natl. Acad. Sci. USA, 1980, 77, 5419.
878 R.P. McEver, N.L. Baenziger and P.W. Majerus, J. Clin. Invest., 1980, 66, 1311.
879 G. Mengod, E.N. Hughes and J.T. August, Arch. Biochem. Biophys., 1981, 209, 718.
880 L.A. Hadwiger, J.M. Beckman and M.J. Adams, Plant. Physiol., 1981, 67, 170.
881 B. Clevinger, J. Schilling, L. Hood and J.M. Davie, J. Exp. Med., 1980, 151, 1059.
882 A. Pierce-Cretel, M. Pamblanco, G. Strecker, J. Montreuil and G. Spik, Eur. J. Biochem., 1981, 114, 169.
883 G. Strecker, A. Pierce-Cretel, B. Fournet, G. Spik and J. Montreuil, Anal. Biochem., 1981, 111, 17.
884 L.C. Kühn and J.P. Kraehenbuhl, J. Biol. Chem., 1981, 256, 12490.
885 R. Zidovetzki, O. Farver and I. Pecht, Eur. J. Biochem., 1981, 114, 97.
886 H. Kikutani, R. Sitia, R.A. Good and J. Stavnezer, Proc. Natl. Acad. Sci., USA, 1981, 78, 6436.
887 L. Matsuuchi, L.A. Wims and S.L. Morrison, Biochemistry, 1981, 20, 4827.

888 A.K. Bhattacharjee, M.K. Das, A. Roy and C.P.J. Glaudemans, Mol. Immunol., 1981, 18, 277.
889 P. Gettins, J. Boyd, C.P.J. Glaudemans, H. Potter and R.A. Dwek, Biochemistry, 1981, 20, 7463.
890 H. van Haalbeek, L. Dorland, J.F.G. Vliegenthart, J. Jouanneau and R. Bourillon, Biochem. Biophys. Res. Commun., 1981, 99, 886.
891 J. Jouanneau and R. Bourillon, Biochem. Biophys. Res. Commun., 1979, 91, 1057.
892 C.H. Sibley and R.A. Wagner, J. Immunol., 1981, 126, 1868.
893 M. Lindsheid, J. D'Angona, A.L. Burlingame, A. Dell and C.E. Ballou, Proc. Natl. Acad. Sci. USA, 1981, 78, 1471.
894 J.R. Wands, R.I. Carlson, H. Schoemaker, K.J. Isselbacher and V.R. Zurawski, Proc. Natl. Acad. Sci. USA, 1981, 78, 1214.
895 J. Jouanneau, B. Fournet and R. Bourrillon, Biochim. Biophys. Acta, 1981, 667, 277.
896 T. Takayasu, N. Takashi and T. Shinoda, Biochem. Biophys. Res. Commun., 1980, 97, 635,
897 F.W. Putman, N. Takahashi, D. Tetaert, B. Debuire and L.C. Lin, Proc. Natl. Acad. Sci. USA, 1981, 78, 6168.
898 M.S. Neuberger and K. Rajewsky, Proc. Natl. Acad. Sci. USA, 1981, 78, 1138.
899 E.J. Victoria, L.C. Mahan and S.P. Masouredis, Proc. Natl. Acad. Sci. USA, 1981, 78, 2898.
900 W.B. Gratzer, Biochem. J., 1981, 198, 1.
901 A. Jakobovits, Y. Eshdat and N. Sharon, Biochem. Biphys. Res. Commun., 1981, 100, 1484.
902 H. Schenkel-Brunner, Eur. J. Biochem., 1980, 104, 529.
903 T. Kameyama, K. Oishi and K. Aida, Agric. Biol. Chem., 1981, 45, 975.
904 J. Viitala, K.K. Karhi, C.G. Gahmberg, J. Finne, J. Jarnefelt, G. Myllyla and T. Krusius, Eur. J. Biochem., 1981, 113, 259.
905 L. Gattegno, G. Perret, F. Fabia, D. Bladier and P. Cornillot, Carbohydr. Res., 1981, 95, 283.
906 A.T. Nurden, D. Dupuis, D. Pidard and T.J. Kunicki, Biochem. Soc. Trans., 1981, 8, 698.
907 W.J. Mawbry, D.J. Anstee and M.J.A. Tanner, Nature, 1981, 291, 161.
908 J.G. Lindsay, G.P. Reid and M.P. D'Souza, Biochim. Biophys. Acta, 1981, 640, 791.
909 R.L. Ong, V.T. Marchesi and J.H. Prestegard, Biochemistry, 1981, 20, 4283.
910 W.L.C. Vaz, H.G. Kapitza, J. Stumpel, E. Sackmann and T.M. Jovin, Biochemistry, 1981, 20, 1392.
911 H. Fujii and A. Yoshida, Proc. Natl. Acad. Sci. USA, 1980, 77, 2951.
912 T. Irimura, T. Tsuji, S. Tagami, K. Yamamoto and T. Osawa, Biochemistry, 1981, 20, 560.
913 R. Prohaska, T.A.W. Koerner, I.M. Armtage and H. Furthmayr, J. Biol. Chem., 1981, 256, 5781.
914 K. Fukuda, M. Tomita and A. Hamada, Biochim. Biophys. Acta, 1981, 677, 462.
915 K. Honma, M. Tomita and A. Hamada, J. Biochem. (Tokyo), 1980, 88, 1679.
916 J. Murayama, K. Takeshita, M. Tomita and A. Hamada, J. Biochem. (Tokyo), 1981, 89, 1593.
917 H. Furthmayr, in 'Biology of Carbohydrates' ed. V. Ginsburg and P. Robbins, John Wiley, New York, vol.1, p.123.
918 O.O. Blumenfeld, A.M. Adamany, K.V. Puglia, Proc. Natl. Acad. Sci. USA, 1981, 78, 747.
919 W. Dahr, K. Beyreuther, E. Gallasch, J. Krüger and P. Morel, Hoppe-Seyler's Z. Physiol. Chem., 1981, 362, 81.
920 W. Dahr, M. Dordowicz, K. Beyreuther and J. Krüger, Hoppe-Seyler's Z. Physiol. Chem., 1981, 362, 363.
921 M. Jokinen, I. Ulmanen, L.C. Andersson, L. Käariänen and C.G. Gahmberg, Eur. J. Biochem., 1981, 114, 393.
922 S. Markowitz and V.T. Marchesi, J. Biol. Chem., 1981, 256, 6463.
923 T. Tsuji, T. Irimura and T. Osawa, Carbohydr. Res., 1981, 92, 328.
924 T. Tsuji, T. Irimura and T. Osawa, J. Biol. Chem., 1981, 256, 10497.

925 W.A. Braell and H.F. Lodish, J. Biol. Chem., 1981, 256, 11337.
926 A.S.B. Edge and P. Weber, Arch. Biochem. Biophys., 1981, 209, 697.
927 K. Hotta, Seikagaku, 1981, 53, 101.
928 M. Guslandi, Clin. Chim. Acta, 1981, 117, 3.
929 G. Lamblin, M. Lhermitte, A. Boersma, P. Roussel, H. van Halbeek, L. Dorland and J.F.G. Vliegenthart, in 'Current problems and new trends in cystic fibrosis'. Monographs in Paediatrics, ed. M. Schoni and R. Kraemer, S. Karger (Basel), 1981, 14, 46.
930 P.W. Cheng, J.M. Sherman, T.F. Boat and M. Bruce, Anal. Biochem., 1981, 117, 301.
931 N.F. Tabachnk, P. Blackburn and A. Cerami, J. Biol. Chem., 1981, 256, 7161.
932 A. Boersma, M. Lhermitte, G. Lamblin and P. Degand, Carbohydr. Res., 1981, 95, 227.
933 L.A. Tabak and M.J. Levine, Arch. Oral Biol., 1981, 26, 315.
934 A.V. Nieuw Amerongen, M.E.G. Aarsman, A.P. Vreugdenhil and P.A. Roukema, Arch. Oral Biol., 1981, 26, 651.
935 G. Reuter, R. Pfeil, J.P. Kamerling, J.F.G. Vliegenthart and R. Schauer, Biochim. Biophys. Acta, 1980, 630, 306.
936 M.A. Cohenford, J.C. Urbanowski and J.A. Dain, Anal. Biochem., 1981, 112, 76.
937 H. van Halbeek, L. Dorland, J. Haverkamp, G.A. Veldink, J.F.G. Vliegenthart, B. Fournet, G. Ricart, J. Montreuil, W.D. Gathmann and D. Aminoff, Eur. J. Biochem, 1981, 118, 487.
938 D. Williams and H. Schachter, J. Biol. Chem., 1980, 255, 11247.
939 D. Williams, G. Longmore, K.L. Matta and H. Schachter, J. Biol. Chem., 1980, 255, 11253.
940 M.L.E. Bergh, G.J.M. Hooghwinkel and D.H. van den Eijnden, Biochim. Biophys. Acta, 1981, 660, 161.
941 C.E. Malinowski and A. Herp, Comp. Biochem. Physiol., 1981, 69 (B), 605.
942 M. Lehrmitte, G. Lamblin, J.J. Lafitte, P. Degand, P. Roussel and M. Mazzuca, Carbohydr. Res., 1981, 92, 333.
943 A. Le Treut, G. Lamblin, N. Houdret, P. Degand and P. Roussel, Biochimie, 1981, 63, 425.
944 H. Houdret, A. Le Treut, M. Lhermitte, G. Lamblin, P. Degand and P. Roussel, Biochim. Biophys. Acta, 1981, 668, 413.
945 A. Boersma, G. Lamblin, P. Degand and P. Roussel, Carbohydr. Res., 1981, 94, C7.
946 K. Hotta and K. Goso, Anal. Biochem., 1981, 110, 338.
947 A.E. Bell, A. Allen, E. Morris and D.A. Rees, Biochem. Soc. Trans., 1980, 8, 716.
948 E. Wood, E.F. Hounsell and T. Feizi, Carbohydr. Res., 1981, 90, 269.
949 E.F. Hounsell, E. Wood, T. Feizi, M. Fukuda, M.E. Powell and S. Hakomori, Carbohydr. Res., 1981, 90, 283.
950 J.P. Pearson and A. Allen, Biochem. Soc. Trans., 1980, 8, 388.
951 J.P. Pearson, A. Allen and S. Parry, Biochem. J., 1981, 197, 155.
952 M. Mantle, J. Pearson and A. Allen, Biochem. Soc. Trans., 1980, 8, 715.
953 C. Decaens, J. Bara, D. Waldron-Edward and J. Labat-Robert, Int. J. Biochem., 1981, 13, 261.
954 B.L. Slomiany, Z. Banas-Gruszka, K. Kojima, A. Herp and A. Slomiany, FEBS Lett., 1981, 130, 201.
955 M. Mantle and A. Allen, Biochem. J.., 1981, 195, 267.
956 M. Mantle, D. Mantle and A. Allen, Biochem. J., 1981, 195, 277.
957 J.T. LaMont and A.S. Ventola, Biochim. Biophys. Acta, 1980, 626, 234.
958 D.V. Gold, D. Schochat and F. Miller, J. Biol. Chem., 1981, 256, 6354.
959 Nasir-ud-Din, R.W. Jeanloz, J.R. Vercelotti and J.W. McArthur, Biochim. Biophys. Acta, 1981, 678, 483.
960 G. Strecker in 'Structure and Function of Gangliosides' ed. L. Svennerholm, H. Dreyfus and P.F. Urban, Plenum Publishing Corp., 1980, p.371.
961 J. Montreuil, C.R. Seances Soc. Biol. Ses Fil., 1981, 175, 694.
962 I. Berggård, B. Ekström, E. Lisowska and A. Lundblad, Eur. J. Biochem., 1981, 115, 183.

963 D. Ekström, A. Lundblad and S. Svensson, Eur. J. Biochem., 1981, 114, 663.
964 G. Fex, A. Grubb, C. Loeffler and J. Larsson, Biochim. Biophys. Acta, 1981, 667, 303.
965 E. Wachter and K. Hochstrasser, Hoppe-Seyler's Z. Physiol. Chem., 1981, 362, 1351.
966 H. Hochstrasser, O.L. Schönberger, I. Rossmanith and E. Wachter, Hoppe-Seyler's Z. Physiol. Chem., 1981, 362, 1357.
967 A.M.M. Afonso, P.A. Charlwood and R.D. Marshall, Carbohydr. Res., 1981, 89, 309.
968 H. van Halbeek, L. Dorland, G.A. Veldink, J.F.G. Vliengenthart, J.C. Michalski, J. Montreuil, G. Strecker and W.E. Hull, FEBS Lett., 1981, 121, 65.
969 C.P.J. Maury, Clin. Chim. Acta, 1981, 109, 219.
970 C.P.J. Maury and J. Paulo, Clin. Chim. Acta, 1980, 108, 293.
971 J.C. Kuo and E.S. Yeung, J. Chromatogr., 1981, 223, 321.
972 H. van Halbeek, L. Dorland, G.A. Veldink, J.F.G. Vliegenthart, G. Strecker, J.C. Michalski, J. Montreuil and W.E. Hull, FEBS Lett., 1980, 121, 71.
973 P.F. Daniel, D.F. Defeudis and I.T. Lott, Eur. J. Biochem., 1981, 114, 235.
974 H.A. Nunez, F. Matsura and C.C. Sweeley, Arch. Biochem. Biophys., 1981, 212, 638.
975 L.J. Burditt, K. Chotai, S. Hirani, P.G. Nugent, B.G. Winchester and W.F. Blakemore, Biochem. J., 1980, 189, 467.
976 A.C. Sewell, Clin. Chem., 1981, 27, 243.
977 K. Yamashita, T. Ohkura, S. Okada, H. Yabuuchi and A. Kobata, J. Biol. Chem., 1981, 256, 4789.
978 M. Kuriyama, T. Ariga, S. Ando, M. Suzuki, T. Yamada and T. Miyatake, J. Biol. Chem., 1981, 256, 12316.
979 M.H. Fischer, B.A. Lewis and W.N. Adkins, Clin. Chim. Acta, 1981, 114, 11.
980 S. Takagi, M. Shiroishi and T. Takagi, Agric. Biol. Chem., 1980, 44, 2111.
981 H. Iwase, Y. Kato, K. Hotta, J. Biol. Chem., 1981, 256, 5638.
982 P.H. Atkinson, A. Grey, J.P. Carver, J. Hakimi and C. Ceccarini, Biochemistry, 1981, 20, 3979.
983 J.P. Carver, A.A. Grey, F.M. Winnik, J. Hakimi, C. Ceccarini and P.H. Atkinson, Biochemistry, 1981, 20, 6600.
984 A.D. Nisbet, R.H. Saundry, A.G. Moir, L.A. Fothergill and J.E. Fothergill, Eur. J. Biochem., 1981, 115, 335.
985 H. Ishihara, N. Takahashi, J. Ito, E. Takeuchi and S. Tejima, Biochim. Biophys. Acta, 1981, 669, 216.
986 E. Berman and A. Allerhand, J. Biol. Chem., 1981, 256, 6657.
987 C. Francois-Gerard, J. Brocteur, A. Andre, C. Gerday, A. Pierce-Cretel, J. Montreuil and G. Spik, Blood Transfusion and Immunohaematology, 1980, 23, 579.
988 K. Watanabe, T. Matsuda and Y. Sato, Biochim. Biophys. Acta., 1981, 667, 242.
989 T. Matsuda, K. Watanabe and Y. Sato, Agric. Biol. Chem., 1981, 45, 417.
990 M.C. Halant-Peers, J. Montreuil, G. Strecker, L. Dorland, H. van Halbeek, G.A. Veldink and J.F.G. Vliegenthart, Eur. J. Biochem., 1981, 117, 291.
991 M. Iacomini, G.R. Duarte, E.R. Duarte, H.S. Duarte, J.D. Fontana and J.H. Duarte, Agric. Biol. Chem., 1981, 45, 1373.
992 H. Bretting, K. Königsmann-Lange, H.J. Thiem, N.F. Whittaker and E.A. Kabat, Carbohydr. Res., 1981, 98, 213.
993 P.A.J. Gorin, M. Mazurek, H.S. Duarte and J.H. Duarte, Carbohydr. Res., 1981, 92, Cl.
994 A. Colman, C.D. Lane, R. Craig, A. Boulton, T. Mohun and J. Morser, Eur. J. Biochem., 1981, 113, 339.
995 S. Inoue, M. Iwasaki and G. Matsumura, Biochem. Biophys. Res. Commun., 1981, 102, 1295.
996 R. Ivell, H. Schmale and D. Richter, Biochem. Biophys. Res. Commun., 1981, 102, 1230.
997 C.A. Auffret and M.J. Turner, Biochem. J., 1981, 193, 647.
998 G. Matthyssens, F. Michiels, R. Hamers, E. Pays and M. Steinert, Nature,

1981, 93, 230.
999 J.E. Strickler and C.L. Patton, Proc. Natl. Acad. Sci. USA, 1980, 77, 1529.
1000 E. Reinwald, P. Rautenberg and H.J. Risse, Biochim. Biophys. Acta, 1981, 668, 119.
1001 A.M. Johnson, P.J. McDonald and S.H. Neoh, Biochem. Biophys. Res. Commun., 1981, 100, 934.
1002 S. Wenzl and M. Sumper, Proc. Natl. Acad. Sci. USA, 1981, 1981, 78, 3716.
1003 R.H. Hackman and M. Goldberg, Anal. Biochem., 1981, 110, 277.
1004 C.J. Brine and P.R. Austin, Comp. Biochem. Physiol., 1981, 70B, 173.
1005 C.J. Brine and P.R. Austin, Comp. Biochem. Physiol., 1981, 69B, 283.
1006 S. Hirano and Y. Yagi, Carbohydr. Res., 1981, 92, 319.
1007 P.V. Wagh and O.P. Bahl, C.R.C. Critical Rev. Biochem., 1981, 10, 307.
1008 B.H. Anderton and R.C. Thorpe, Immunol. Today, 1980, 1, 122.
1009 D.T. Dorai, B.K. Bachhawat and S. Bishayee, Anal. Biochem., 1981, 115, 130.
1010 R.D. Poretz and G. Pieczenik, Arch. Biochem., 1981, 115, 170.
1011 T.H. Plummer and A.L. Tarentino, J. Biol. Chem., 1981, 256, 10243.
1012 W.F. Glass, R.C. Briggs and L.S. Hnilica, Anal. Biochem.,1981, 115, 219.
1013 S.J. Mellis and J.U. Baenziger, Anal. Biochem., 1981, 114, 276.
1014 A.S.B. Edge, C.R. Faltynek, L. Hof, L.E. Reichert and P. Weber, Anal. Biochem., 1981, 118, 131.
1015 P.R.P. Salacinski, C. McLean, J.E.C. Sykes, V.V. Clement-Jones and P.J. Lowry, Anal. Biochem., 1981, 117, 136.
1016 V. Neuhoff, E. Ewars and G. Huether, Hoppe-Seyler's Z. Physiol. Chem., 1981, 362, 1427.
1017 K. Dill, B. Ferrari, J.M. Lacombe and A.A. Pavia, Carbohydr. Res., 1981, 98, 132.
1018 J. Davoust, V. Michel, G. Spik, J. Montreuil and P.F. Devaux, FEBS Lett, 1981, 125, 271.
1019 J.P. Carver and A.A. Grey, Biochemistry, 1981, 20, 6607.
1020 N. Ui, J. Chromatogr., 1981, 215, 289.
1021 M.C. Bordas, D.R. Biou, J.M. Feger, G.M. Durand, D.H. Joziasse and D.H. van den Eijnden, Clin. Chim. Acta, 1981, 116, 17.
1022 W.D. Gathmann and D. Aminoff, Biochem. Biophys. Res. Commun., 1981, 100, 1453.
1023 W.F. Alpenfels, Anal. Biochem., 1981, 114, 153.
1024 B. Fournet, G. Strecker, Y. LeRoy and J. Montreuil, Anal. Biochem., 1981, 116, 489.
1025 H. Rauvala, J. Finne, T. Krusius, J. Kärkkäinen and J. Jarnefelt, Adv. Carb. Chem. Biochem., 1981, 38, 389.
1026 T. Anastassiades, R. Puzic and O. Puzic, J. Chromatogr., 1981, 225, 309.
1027 H.K.B. Silver, K.A. Karim, M.J. Gray and F.A. Salinas, J. Chromatogr., 1981, 224, 381.
1028 T.A. Beyer, J.E. Sadler, J.I. Rearick, J.C. Paulson and R.L. Hill, Adv. Enzymol, 1981, 52, 24.
1029 R.L. Hill in "Evolution of Protein Structure and Function", ed. D.S. Sigman and M.A.B. Brazier, Acad. Press, New York, 1980, p.63.
1030 M.L. Reitman and S. Kornfeld, J. Biol. Chem., 1981, 256, 11977.
1031 M. Pierce, R.D. Cummings and S. Roth, Anal. Biochem., 1980, 102, 441.
1032 J.A. Hanover and W.J. Lennarz, Arch. Biochem. Biophys., 1981, 211, 1.
1033 W.M. Watkins, A.D. Yates and P. Greenwell, Biochem. Soc. Trans., 1981, 9, 186.
1034 H. Baumann and W.A. Held, J. Biol. Chem., 1981, 256, 10145.
1035 A. Heifetz and M.D. Prager, J. Biol. Chem., 1981, 256, 6529.
1036 P. Cammarata, L. Chan and C. Ceccarini, Biochim. Biophys. Acta, 1981, 674, 128.
1037 J.A. Henner, M.J. Kessler and O.P. Bahl, J. Biol. Chem. 1981, 256, 5997.
1038 A.J. Parodi, L.A.Q. Allue and J.J. Cazzulo, Proc. Natl. Sci. USA, 1981, 78, 6201.
1039 J.P. Aubert, N. Helbecque and M.H. Loucheux-Lefebvre, Arch. Biochem. Biophys, 1981, 208, 20.
1040 S.C. Hubbard and R.J. Ivatt, Annu. Rev. Biochem., 1981, 50, 555.

1041 D.S. Bailey, J. Burke, R. Sinclair and B.B. Mukherjee, Biochem. J., 1981, 195, 139.
1042 D.D. Carson and W.J. Lennarz, J. Biol. Chem., 1981, 256, 4679.
1043 D.P. Rossignol, W.J. Lennarz and C.J. Waechter, J. Biol. Chem., 1981, 256, 10538.
1044 T. Coolbear and S. Mookerjea, J. Biol. Chem., 1981, 256, 4529.
1045 B.N. Singh and J.J. Lucas, J. Biol. Chem., 1981, 256, 12018.
1046 Y. Okamoto and N. Akamatsu, Int. J. Biochem., 1981, 13, 791.
1047 J.I. Rearick, K. Fujimoto and S. Kornfeld, J. Biol. Chem., 1981, 256, 3762.
1048 L.A. Murphy and R.G. Spiro, J. Biol. Chem., 1981, 256, 7487.
1049 L.A.Q. Allue, Mol. Cell. Biochem., 1980, 33, 149.
1050 D.K. Banerjee, M.G. Schaer and C.J. Waechter, Biochemistry, 1981, 20, 1561.
1051 R. Datema, R.P. Lezica, P.W. Robbins and R.T. Schwarz, Arch. Biochem. Biophys., 1981, 206, 65.
1052 S. Kato, M. Tsuji, Y. Nakanishi and S. Suzuki, Biochem. J., 1981, 196, 71.
1053 A.D. Elbein, Biochem. J., 1981, 193, 477.
1054 I.A. King and A. Tabiowo, Biochem. J., 1981, 198, 331.
1055 A.D. Elbein, Biochem. J., 1981, 195, 191.
1056 A. Herscovics and B. Bugge, Biochim. Biophys. Acta, 1980, 617, 122.
1057 L.A. Qallue, Biochem. J., 1981, 198, 429.
1058 J.I. Rearick, A. Chapman and S. Kornfeld, J. Biol. Chem., 1981, 256, 6255.
1059 J.W. Jensen, J.D. Springfield and J.S. Schutzbach, J. Biol. Chem., 1980, 255, 11268.
1060 J.W. Jensen and J.S. Schutzbach, J. Biol. Chem., 1981, 256, 12899.
1061 P. Vischer and R.C. Hughes, Eur. J. Biochem., 1981, 117, 275.
1062 Y. Fujita-Yamaguchi and A. Yoshida, J. Biol. Chem., 1981, 256, 2701.
1063 T.Z. Russin, R.A. Laine and S.J. Turco, Biochem. J., 1981, 197, 327.
1064 G. Berthillier, J.P. Benedetto and R. Got, Biochim. Biophys. Acta, 1980, 603, 245.
1065 J. Mendicino, E.V.Chandrasekaran, K.R. Anumula and M. Davilla, Biochemistry, 1981, 20, 967.
1066 J.S. Silvia and K.E. Ebner, J. Biol. Chem., 1980, 255, 11262.
1067 D.A. Blake and I.J. Goldstein, J. Biol. Chem., 1981, 256, 5387.
1068 G.N. Anderson and L.C. Eriksson, J. Biol. Chem., 1981, 256, 9633.
1069 H.A. Nunez and R. Barker, Biochemistry, 1980, 19, 489.
1070 J.P. Briand, S.P. Andrews, E. Cahill, N.A. Conway and J.D. Young, J. Biol. Chem., 1981, 256, 12205.
1071 C. Clamagirand-Mulet, J. Badet and J.P. Cartron, FEBS Lett., 1981, 126, 123.
1072 J.P. Prieels, D. Monnom, M. Dolmans, T.A. Beyer and R.L. Hill, J. Biol. Chem., 1981, 256, 10456.
1073 P.H. Johnson, A.D. Yates and W.M. Watkins, Biochem. Biophys. Res. Commun., 1981, 100, 1611.
1074 M.L. Reitman and S. Kornfeld, J. Biol. Chem., 1981, 256, 4275.
1075 R. Schauer, E. Moczar and M. Wember, Hoppe-Seyler's Z. Physiol. Chem., 1979, 360, 1587.
1076 D.H. van den Eijnden and W.E.C.M. Schiphorst, J. Biol. Chem., 1981, 256, 3159.
1077 S. Ratman, I.H. Fraser, J.M. Collins, J.A. Lawrence, J.A. Barrowman and S. Mookerjea, Biochim. Biophys. Acta, 1981, 673, 435.
1078 M.M. Mitranic, J.M. Boggs and M.A. Moscarello, Biochim. Biophys. Acta, 1981,672, 57.

6
Enzymes

BY J. F. KENNEDY AND D. P. ATKINS

1 Introduction

General Aspects and Nomenclature. -- Volume 1 of 'Enzymes' covers the development of enzymology from its beginnings in the eighteenth century, represented by the studies on digestion by Reaumur and Spallanzani, to the modern organic synthesis of ribonuclease by Merrifield.[1] The papers selected emphasize the chemical nature and mode of action of enzymes rather than the field of intermediary metabolism. More than twenty classic papers appear here in translation for the first time, including the work of Payen and Persoz on enzyme purification, Schwann on pepsin, Berzelius on catalysis, Liebig on the nature of fermentation, Buchner on cell-free fermentation, Fischer on stereospecific enzyme reactions, and Sorensen on pH.

The publication by the Nomenclature Committee of the International Union of Biochemistry (NC-IUB) 'Enzyme Nomenclature, Recommendations 1978' has seen a further supplement. 'Supplement 2: Corrections and Additions' is the second listing of amendments to 'Enzyme Nomenclature 1978' (Academic Press, New York, 1979); the first supplement was published in Eur.J.Biochem., 1980, 104, 1.[2]

Methods of Assay. -- 'Methods in Enzymology', Volume 73, Part B, continues the coverage of general immunochemical techniques presented in Volume 70, Part A.[3] The papers illustrate the ingenuity characteristic of workers who have adapted the antigen-antibody reaction to develop a variety of assays which are applicable to numerous biochemical and clinical problems.

The Proceedings of the 4th International Symposium, Veldhoven, The Netherlands, 22-26 June, 1981, give a survey of new developments and the status of affinity chromatography, illustrating the increasing importance of affinity techniques in industrial and biomedical/diagnostic applications.[4]

In Volume 74 of 'Methods in Enzymology' immunochemical techniques are described.[5] One section of this volume is particularly concerned with the use of antibodies to study enzymes.

The latest volume in the series 'Advances in Enzymology and Related Areas of Molecular Biology' contains a chapter which describes the use of glycosyltransferases in the assessment of oligosaccharide structure.[6]

Kinetics. -- A versatile BASIC program has been described for 'Analyzing Kinetic Data'.[7] Direct curve-fitting is uncontested as the method of choice for evaluating kinetic experiments. If carried out by computer, the numerical solution of such problems is superior to customary graphical procedures in any repsect. However, most programs available to date cannot be executed by small systems. The authors describe a BASIC program for desk-top microcomputers that performs linear and non-linear regression analysis of kinetic data. It may be directly applied to a number of problems common in enzyme kinetics, and is easily modified to handle further ones. The performance of the program is illustrated by some typical applications. In addition, fundamental statistical methods are discussed that allow a critical examination of the results obtained.

Mechanisms of Action. -- A book has been published on enzyme inhibitors which derives from the papers presented at the 1980 spring symposium of the Swiss Chemical Society.[8] The program consisted of 21 full papers and 3 abstracts from industrial and university laboratories reporting on theory, structures, syntheses and in vitro as well as in vivo effects of mechanism-based enzyme inhibitors. These include suicide substrates, transition-state analogues, paracatalytic self inactivators, and a number of compounds belonging to the class of substrate and coenzyme analogs. The fully indexed publication offers a wealth of valuable information to investigators seeking novel structures in the design of enzyme inhibitors.

Applications. -- Volume 5 of the series entitled 'Economic Microbiology' covers microbial enzymes and bioconversions.[9] This volume differs from the previous volumes in that it is concerned only with the development for industry of individual enzymes or short sequences of enzymes. Particularly relevant sections are amylases, amyloglucosidases and related glucanases, D-glucose oxidase, D-glucose dehydrogenase, D-glucose isomerase, β-D-galactosidase and invertase, pectic enzymes, cellulases, immobilized enzymes.

A book on cellulase and other natural polymer systems contains, in addition to chapters on the biosynthesis and structure of cellulose, chapters which deal with β-D-glucanases in higher plants and cellulose degradation.[10]

The proceedings of the 2nd Symposium on Enzyme Therapy in Genetic Diseases cover: human trials: cell and organ transplantation/other therapeutic approaches; human trials: direct enzyme replacement; enzyme manipulation: mechanisms and therapeutic trials; animal model studies: enzyme entrapment, neural delivery and transplantation, enzyme recognition and modification; enzyme availability: purification and characterization.[11]

Enzyme Immobilization. -- A book on the uses of immobilized enzymes for food processing covers the following areas: immobilized enzyme engineering; requirements unique to the food and beverage industry; manufacture of high-fructose corn syrup using immobilized glucose isomerase; potential and use of immobilized carbohydrases; application of lactose and immobilized lactase; immobilized proteases - potential application, application and potential of other enzymes in food processing: aminoacylase, aspartase, fumarase, glucose oxidase-catalase, sulphydryl oxidase, and controlled-release enzymes.[12]

In the proceedings of the 18th Prague IUPAC Micro Symposium on Macromolecules, published in a series of journal issues, there is a section which describes the immobilization of enzymes to different polysaccharides by graft copolymerization.[13]

The use of reversible enzyme immobilization has been reported to facilitate the recovery of peptide antibiotics.[14] Practical aspects of the facile immobilization of enzymes on hydrous metal oxides, a well established means of enzyme-movement restriction, are described. Various enzymes (e.g. glucoamylase, peroxidase, dextranase) have been immobilized by chelation of several hydrous metal oxides, those of titanium(IV) and zirconium(IV) proving to be the most satisfactory for practical purposes. Modification of the gel into a granular form could be achieved successfully with good enzyme-immobilization characteristics by using ion-exchange resin as an internal matrix. The immobilization process was highly efficient for the relative proportions of hydrous oxide to enzyme used, with usually > 90% of the available protein being insolubilized. Retention of enzyme activity was generally very good and the enzyme was stable to reuse and to conventional buffer conditions. Activities of the immobilized enzymes were partially stable to lyophilization or drying of the hydrous oxide gels. Modification of the hydrous metal oxide surface by drying or treatment with phosphate or carbonate led to a decrease in complexing ability. The effect of carbonate can be circumvented by lowering the pH of the solution to around 5 and removing any carbon dioxide formed, by aeration. Such treatment allowed compounds to chelate to hydrous zirconium oxide(IV) in the presence of carbonate, and therefore the hydrous oxide could be applied successfully to the concentration of peptide antibiotics from the fermentation medium in which they are being produced, including production at low concentrations.

A simple and novel method for the immobilization of enzymes by their coimmobilization in hen egg-white using glutaraldehyde has been described.[15]

Films of highly polymerized collagen prepared under industrial conditions by the Centre Technique du Cuir, Lyon, France, have been routinely used after acyl azide activation for the covalent immobilization of numerous enzymes from different classes.[16] The stability of the resulting membranes to operational and storage conditions and their excellent mechanical strength and resistance to bacterial degradation allowed their use for several purposes. Fundamental aspects of their heterogeneous enzymology, including diffusional effects and subunit interations, were examined. The enzyme-engineering aspect was developed with polymembrane bioreactors and enzyme electrodes. Besides the protein environment provided by the collagen matrix which prevents enzyme inactivation, the form of these membranes is advantageous for industrial purposes and for fundamental research.

2 β-D-2-Acetamido-2-deoxygalactosidases, β-D-2-Acetamido-2-deoxyglucosidases, and β-D-2-Acetamido-2-deoxyhexosidases

The bulk of rat brain neutral β-D-2-acetamido-2-deoxyhexosidases have been shown to be present in the cytosol fraction.[17] They were not bound by concanavalin A-Sepharose while the acid β-D-2-acetamido-2-deoxyhexosidases were all bound. The neutral β-D-2-acetamido-2-deoxyglucosidase had a pH optimum of 5.2 and K_m of 0.57 mM, while the neutral β-D-2-acetamido-2-deoxygalactosidase had the highest reaction rate at pH 6.0 with a K_m of 0.12 mM. No divalent ions activated either of the enzymes. The β-D-2-acetamido-2-deoxyglucosidase lost more than 90% of the activity in 30 min at 50°C. The β-D-2-acetamido-2-deoxygalactosidase was heat-stable and lost only 10-20% of its activity after 3 h at 50°C. The neutral β-D-2-acetamido-2-deoxyglucosidase was inhibited by free 2-acetamido-2-deoxy-β-D-glucose but not by 2-acetamido-2-deoxy-β-D-galactose. The reverse was found for the neutral β-D-2-acetamido-2-deoxygalactosidase. Two enzymes were separated almost completely by hydroxyapatite chromatography. Heat stability of the separated activity peaks suggested that the neutral β-D-2-acetamido-2-deoxygalactosidase, which was not bound to hydroxyapatite, may be specific to the D-galacto substrate. The neutral β-D-2-acetamido-2-deoxyglucosidase may, on the other hand, have some activity toward the D-galacto substrate. Both of the neutral enzyme activities were shown to be highest during the first postnatal week in rat brain, in contrast to the acidic enzyme, which showed peak activities during the second and third weeks.

Activity levels and cytochemical localization have been investigated in intact and regenerating planarian *Dugesia lugubris*.[18] In 24 hour regenerating planarians the β-D-2-acetamido-2-deoxyglucosidase was found to be localized in the lysosomes of dedifferentiating cells. This β-D-2-acetamido-2-deoxyglucosidase localization correlated with the increase in β-D-2-acetamido-2-deoxyglucosidase and β-D-2-acetamido-2-deoxygalactosidase activities during the first 24 hours of regeneration, confirming the involvement of the dedifferentiation in the origin of blastema. In non-regenerating animals the β-D-2-acetamido-2-deoxyglucosidase activity has been cytochemically demonstrated in the lysosomes of gastrodermal cells and in the mucous granules of basophilic secretory cells. This latter finding is interpreted as proof of a similarity between planarian and vertebrate mucous secretion.

High lysozyme and chitinase activities have been found in Leydig's organ of *Etmopterus spinax*, *Somniosus microcephalus*, and *Tonpedo nobiliana*, in Leydig's and epigonal organs and spleen of *Raj radiata*, and in the epigonal organ of *Rhinoptera bonasus*.[19] Strong chitinase activity with little or no lysozyme activity was noted in Leydig's and epigonal organs of Scyliorhinus canicula, and in the epigonal organ of Ginglymostom cirratum and Heterodontus francisci. Very high β-D-2-acetamido-2-deoxyglucosidase activity was found in lymphomyeloid organs of all species investigated and in the pancreas of Somniosus microcephalus. Thirteen other glycoside hydrolases were assayed in lymphomyeloid and digestive tissues of Ginglymostoma cirratum, Heterodontus francisci, and Etmopterus spinax. Activities of α-D-mannosidase, β-galactosidase, β-D-2-acetamido-2-deoxygalactosidase, β-D-glucuronidase, and α- and β-D-glucosidase usually were higher in lymphomyeloid tissues than in digestive tissues.

Nine glycoside hydrolases in bloodstream forms of Trypanosoma brucei brucei S42 have been partially characterized.[20] α-D-Glucosidase had similar physicochemical and enzymic properties to those of α-D-galactosidase, β-D-glucosidase, β-D-2-acetamido-2-deoxyglucosidase, and β-D-2-acetamido-2-deoxygalactosidase. It was suggested that the glycosidases studied may play a role in the turnover of trypanosomal glycoproteins, in particular the variant-specific surface antigen.

α-D-Galactosidase deficiency has been demonstrated in lymphoid cell lines established by Epstein-Barr virus transformation of B-lymphocytes from a Fabry patient, as in blood whole leukocytes and lymphocytes.[21] The residual activity was heat stable. α-D-2-Acetamido-2-deoxygalactosidase was present at normal level in leukocytes, lymphocytes, or lymphoid cells from normal and Fabry patients and was heat stable in all cases. α-D-2-Acetamido-2-deoxygalactosidase electrofocusing profiles presented one major peak (pI 4.5) and one minor peak (pI 5.1). Blood lymphocytes and lymphoid cell lines from Fabry disease showed similar defects of all the forms of α-D-galactosidase group A and the residual activity proceeded from the α-D-galactosidase form II, which is an α-D-2-acetamido-2-deoxygalactosidase. In the absence of animal models, those established lines seemed to be an accurate cellular system for in vitro experimental studies of Fabry disease.

The physiological activator protein for the degradation of ganglioside G_{M2} by β-D-2-acetamido-2-deoxygalactosidase has been employed to assess the capability of cultured human fibroblast extracts to catabolize this ganglioside.[22] This method permits a more reliable diagnosis of the different variants of G_{M2} gangliosides than the methods hitherto used. These either rely on artificial substrates or, when natural substrates are used, on detergents. The present method avoids a number of possible sources of error introduced by the unphysiological detergents, such as alteration of the isoenzymes' substrate specificity or inactivation of the enzymes. The range of application of the new method is discussed.

The patterns of β-D-acetamido-2-deoxyglucosidase excretion after renal transplantation have been investigated.[23] β-D-2-Acetamido-2-deoxyglucosidase activity was assayed in 750 early-morning urine samples from 25 renal-transplant patients during the post-operative period. Eighty-four per cent of all acute rejection episodes were preceded or accompanied by a greater than two-fold rise in β-D-2-acetamido-2-deoxyglucosidase activity. Similar increases were caused by dialysis, gentamicin therapy, and ureteric dehiscence. Only 9% of all significant increases in β-D-acetamido-2-deoxyglucosidase excretion could not be accounted for by any of these four processes. Analysis of the day-to-day pattern of β-D-2-acetamido-2-deoxyglucosidase activity as opposed to individual β-D-2-acetamido-2-deoxyglucosidase values provided a clue to the occurrence of rejection during immediate post-transplantation oliguria.

Urinary excretion of β-D-2-acetamido-2-deoxyglucosidase and L-alanine aminopeptidase has been measured in patients receiving amikacin or *cis*-platinum.[24] Urinary excretion of alanine aminopeptidase and β-D-2-acetamido-2-deoxyglucosidase was determined for 25-70 days in five patients receiving *cis*-platinum and for 8-53 days in six patients receiving amikacin. This study was performed to investigate if the excretion of urinary enzymes represents a sensitive parameter for the early detection of toxic kidney damage. In both patient groups an increase in the excretion of the two enzyme activities could be demonstrated. In patients receiving amikacin, the excretion of alanine aminopeptidase was always higher than that of β-D-2-acetamido-2-deoxyglucosidase, whereas in three patients receiving *cis*-platinum it was the opposite. In two *cis*-platinum patients the excretion of both enzymes was of the same size. The changes during amikacin therapy seem to be reversible, whereas in four *cis*-platinum patients these changes seemed to be partly irreversible. Serum creatinine concentration was less sensitive than the urinary enzyme excretion for detection of kidney damage.

Predictive values of urinary β-D-2-acetamido-2-deoxyglucosidase, L-alanine aminopeptidase, and β-2-microglobulin have been used in evaluating the nephrotoxicity of gentamicin.[25] Concentrations of β-D-2-acetamido-2-deoxyglucosidase, L-alanine aminopeptidase, and β-2-microglobulin were determined daily in the urine of 28 patients treated with gentamicin (2-3 mg kg^{-1} day^{-1}) for a mean of 15 days. All had normal renal function. Increased activity in β-D-2-acetamido-2-deoxyglucosidase and L-alanine aminopeptidase was observed for all patients, either immediately or after 2 or 3 days of treatment. The results were compared with serum creatinine concentrations and urinary β-2-microglobulin levels. This study indicates a relationship between the nephrotoxicity of gentamicin and initial urinary enzymic activity of β-D-2-acetamido-2-deoxyglucosidase prior to any treatment. The degree of β-D-2-acetamido-2-deoxyglucosidase response during the first ten days of treatment appeared as a second prognostic factor. Renal failure was observed for one out of the 12 patients with normal β-D-2-acetamido-2-deoxyglucosidase$_i$ (β-D-2-acetamido-2-deoxyglucosidase$_i$ < 200 μmol day^{-1}). Seven of them showed a marked enzyme activity response (> 1500 μmol day^{-1}) with an increase in β-2-microglobulin activity. Eleven out of the 16 patients with elevated β-D-2-acetamido-2-deoxyglucosidase$_i$ (β-D-acetamido-2-deoxyglucosidase$_i$ > 200 μmol day^{-1}) developed renal failure and showed an elevated maximal response. The concentration of L-alanine aminopeptidase appears to be of little prognostic value. The variation in individual maximal urinary enzyme responses observed among the 28 patients during the first ten days of treatment points to the existence of individual sensitivities to gentamicin, the exact mechanism of which remains unclear.

The conditions for maximal activity (pH, buffer, saturating substrate concentration, range of linear relationships between enzyme activity *versus* incubation time and *versus* enzyme concentration) in the fluorimetric assay of several glycoside hydrolases of lysosomal origin in human plasma and serum have been established.[26] The following enzymes were studied: α-D-galactosidase, β-D-2-acetamido-2-deoxyglucosidase, β-D-glucosidase, β-D-glucuronidase, α-D-mannosidase, α-L-fucosidase. All examined enzymes turned out to be more or less unstable upon storage at 37°C, 4°C, and -20°C in both serum and plasma. Generally the degree of instability was greater in serum than in plasma. The levels of some enzymes, including β-D-2-acetamido-2-deoxyglucosidase, were markedly higher in serum than in plasma. Conversely the levels of the same enzymes in platelet-free serum equalled those in plasma. This stresses the necessity to use freshly prepared plasma for lysosomal glycohydrolase assay. Under the procedural conditions recommended for the assay the methods for the determination of lysosomal glycohydrolases in plasma appeared to be simple, sensitive, and reproducible.

A spectrophotometric assay for urinary β-D-2-acetamido-2-deoxyglucosidase activity has been developed.[27] It involves (a) gel filtration to separate the enzyme from inhibitors in urine, (b) enzymic hydrolysis of 4-nitrophenyl 2-acetamido-2-deoxy-β-D-glucopyranoside at pH 4.4, and (c) spectrophotometry of the liberated 4-nitrophenate. Measurements of activity of the enzyme in 58 urine specimens correlated closely (r = 0.998) with results by an established procedure. The within-run coefficient of variation (CV) was 3.7%. The between-run CV averaged 6.8%. Reference values for the activity were established by assays or urine specimens from 135 healthy persons, aged two weeks to 52 years. Efficacy of the assay for detection of nephrotoxicity was demonstrated in rats after experimental induction of reversible renal insufficiency by intraperitoneal injection of nickel chloride. Clinical application of the assay in approximately 1000 patients corroborated its utility for detection and monitoring of renal disorders.

The digestion of L-asparagine-linked oligosaccharides by endo-β-D-2-acetamido-2-deoxyglucosidase in the human skin fibroblasts obtained from fucosidosis patients has been reported.[28] The substrate specificity of human endo-β-D-2-acetamido-2-deoxyglucosidase was studied by using the homogenate of cultured skin fibroblasts of fucosidosis patients as an enzyme source. The results indicate that biantennary-complex-type L-asparagine-linked sugar chains as well as high-D-mannose-type sugar chains are cleaved by the enzyme action. None of the sugar chains with a fucosyl residue on the proximal 2-acetamido-2-deoxy-D-glucose of their N,N'-diacetylchitobiose moieties was cleaved. These results proved enzymically the mechanism of production of oligosaccharides detected in the urine of various exo-glycosidase deficiencies.

Degradation of mucin oligosaccharides of human colon systems has been found to be associated with extracellular, but not with cell-bound, β-D-galactosidase, β-D-2-acetamido-2-deoxyglucosidase, and neuraminidase.[29] Among the seven subjects studied, the estimated most probable numbers (MPN) of faecal bacteria producing extracellular β-D-galactosidase, β-D-2-acetamido-2-deoxyglucosidase, and neuraminidase ranged from 10^6-10^{10} g^{-1} dry faecal wt, were comparable to the MPN of mucin-degrading bacteria, and were significantly smaller than the MPN of total faecal bacteria. These findings were interpreted as evidence for the existence of bacterial sub-populations in the normal faecal flora that produce extracellular glycosidases, and that these sub-populations have a major role in degrading the complex oligosaccharides of mucin in the gut lumen.

Adjuvant-induced arthritis in rats has been studied by the changes in serum and urinary protein-bound carbohydrate metabolites, changes in serum and tissue lysosomal glycohydrolases and lysosomal fragility.[30] The free activities of lysosomal glycohydrolases investigated, viz β-D-glucuronidase, β-D-acetamido-2-deoxyglucosidase, β-D-acetamido-2-deoxyglucosidase, β-D-galactosidase, α-D-mannosidase, and cathepsin D, are increased in liver and spleen in the acute phase. The free activities of β-D-glucuronidase, β-D-acetamido-2-deoxyglucosidase, and cathepsin D of kidney showed no change, whereas those of β-D-galactosidase and α-D-mannosidase are increased. In the chronic phase of the disease the free activities of all glycohydrolases are significantly increased in all tissues. Serum glycohydrolases are significantly increased in both acute and chronic phases. Studies in lysosomal preparations showed increased fragility of lysosomes derived from liver and kidney of arthritic rats in both phases of the disease.

The subunit and polypeptide structure of β-D-2-acetamido-2-deoxyglucosidases from human placenta has been reported.[31] Previous reports have indicated that the β-D-2-acetamido-2-deoxyglucosidases are tetrameric enzymes composed of either two dimers of β-chains (β-D-2-acetamido-2-deoxyglucosidase B) or a β-chain dimer and an α-chain dimer (β-D-2-acetamido-2-deoxyglucosidase A). β-D-2-Acetamido-2-deoxyglucosidase S contains only α-chains. The apparent molecular weight of both α- and β-chains is 25,000 dalton. β-D-2-Acetamido-2-deoxyglucosidase A and B were isolated and purified more than 6000-fold from β-D-2-acetamido-2-deoxyglucosidase A prepared and a more anodically migrating β-D-2-acetamido-2-deoxyglucosidase S, which contained only α-chains. After either extensive reduction and alkylation or performic acid oxidation, β-D-2-acetamido-2-deoxyglucosidase B gave only a single protein band in polyacrylamide gel electrophoresis in sodium dodecyl sulphate. β-D-2-Acetamido-2-deoxyglucosidase A consistently gave two bands, indicating it contained chains of two different molecular weights. Other reports have noted two bands in β-D-2-acetamido-2-deoxyglucosidase A but have dismissed the high-molecular-weight band as a dimer caused by either hydrophobic interaction or an unbroken disulphide bond. To rule out hydrophobic interaction, the molecular weight was determined by gel filtration in 6 M guanidinium chloride. To ensure complete disulphide-bond breakage, a harsher reduction and alkylation procedure than previously employed by others was used. The results were then confirmed by the use of two performic acid oxidation techniques, the stronger method resulting in extensive peptide-bond oxidation. Samples of β-D-2-acetamido-2-deoxyglucosidase A under either reduction and alkylation or the weaker performic acid oxidation procedure again chromatographed as two approximately equal peaks (50,000 and 25,000). β-D-2-Acetamido-2-deoxyglucosidase B and the β-D-2-acetamido-2-deoxyglucosidase B prepared from β-D-2-acetamido-2-deoxyglucosidase A resulted in one 25,000 dalton peak. β-D-2-Acetamido-2-deoxyglucosidase S chromatographed as a single 50,000 peak. The authors conclude that either the α-chain in β-D-2-acetamido-2-deoxyglucosidase has a molecular weight of 50,000 with one α-chain per α-subunit or that two 25,000 dalton chains are united by a non-disulphide crosslink (e.g. an isopeptide bond).

The disaccharide 2-acetamido-2-deoxy-β-D-glucopyranosyl-(1→3)-D-{1-^{3}H}-galactitol, prepared from keratan sulphate, has been found to be rapidly hydrolysed by the A and B isoenzymes of normal human liver β-D-2-acetamido-2-deoxyglucosidase and by the B isoenzyme prepared from the liver of a Tay-Sachs disease patient.[32] The disaccharide substrate was also hydrolysed by extracts of normal, cultured-skin fibroblasts, and fibroblasts of patients with Tay-Sachs disease, whereas it was not hydrolysed by fibroblast extracts of patients with Sandhoff disease. Thus, defective degradation of keratan sulphate, secondary to a defect of the subunits present in the A and B isoenzymes of β-D-2-acetamido-2-deoxyglucosidase, may contribute to the appearance of skeletal lesions in patients affected by Sandhoff disease.

The phospholipid acyl-chain dependence of the membrane-bound lysosomal β-D-2-acetamido-2-deoxyglucosidase has been examined on control membranes from rat brain primary cell cultures and on membranes modified by culturing the cells in media supplemented with polyunsaturated fatty acids.[33] The relationship between β-D-2-acetamido-2-deoxyglucosidase activity and the membrane phospholipid acyl-chain composition has been evaluated. An increase in the unsaturation level of phosphatidylethanolamines and phosphatidylcholines, the most abundant phospholipids in this membrane fraction, is related to the enzymic reaction. The Arrhenius plot of the enzyme activity in modified membranes showed break-temperatures, starting from approximately 15 °C. The apparent activation energy below and above the break-temperature was not correlated with phospholipid acyl-chain unsaturation.

D-Mannose, methyl-α-D-mannopyranoside, and 2-amino-2-deoxy-D-mannose have been found to inhibit β-D-2-acetamido-2-deoxy-glucosidase B purified from both porcine placenta and bovine epididymis.[34] The inhibition is competitive and the inhibitory constants are about 20 mM for D-mannose and methyl-α-D-mannopyrano-side and 2 mM for 2-amino-2-deoxy-D-mannose. The physiological role of this inhibitory effect is discussed.

The changes of sulphate-containing glycosaminoglycans and glycosaminoglycanases during bovine foetal development have been analysed.[35] It was shown that chondroitin 6-sulphate increases in concentration up to the 50th day of foetal development and then decreases progressively until its complete disappearance in most adult tissues. Likewise, hyaluronidase also reaches a peak on the 50th day and decreases in activity until its disappearance in adult tissues. On the other hand, heparan sulphate and dermatan sulphate as well as β-D-glucuronidase and β-D-2-acetamido-2-deoxyglucosidase remain without significant changes during the whole period. The foetal chondroitin 6-sulphate is tissue specific with different molecular weights depending on the tissue of origin. Some properties of foetal muscle- and brain-hyaluronidase are also described. The possible role of chondroitin 6-sulphate and hyaluronidase in the processes of differentiation and division is discussed in view of these findings.

The α-L-fucosidase, β-D-2-acetamido-2-deoxyglucosidase, α-D-mannosidase, β-D-glucuronidase, β-D-xylosidase, β-D-glucosidase, and α-L-arabinofuranosidase activities of normal and atrophic skeletal muscle of developing and adult rats have been collectively investigated.[36]

β-D-2-Acetamido-2-deoxyglucosidase has been characterized from mouse testes. Only one of the two isozymes of β-D-2-acetamido-2-deoxyglucosidase in testes was obtained in the final preparation, having a specific activity of 4.44 units mg^{-1} protein.[37] The purified enzyme showed a K_m of 0.24 mM and V_{max} of 0.165 μM min^{-1} mg^{-1} protein. The optimum temperature is 60°C and transition temperature ($T_{1/2}$) for heat denaturation is 63°C. The optimum pH of the enzyme is 4.5. The molecular weight, by Sephadex G-200 filtration, corresponds to 178,000 dalton. Ag^+, Hg^{2+}, and PCMB are potent inhibitors of β-D-2-acetamido-2-deoxyglucosidase, though N-ethylmaleimide and 2-acetamido-2-deoxy-D-glucose also inhibit the enzyme activity. Protective action by L-cysteine for the inhibition by $HgCl_2$ suggests the role of a thiol group at the catalytic site of the enzyme. The purified enzyme does not cross-react in immuno-diffusion plates with anti-mouse testicular hyaluronidase serum, suggesting that β-D-2-acetamido-2-deoxyglucosidase does not have common antigenic determinants in hyaluronidase of hyaluronidase-β-D-2-acetamido-2-deoxyglucosidase complex of acrosome or testes.

The activities of various glycosidases in homogenates of the small intestinal mucosa of two adults and 18 suckling tammar wallabies (M. eurgenii) aged from 6 to 50 weeks have been investigated.[38] β-D-Galactosidase, β-D-2-acetamido-2-deoxyglucosidase, α-L-fucosidase, and neuraminadase activities were high during the first 34 weeks post partum and then declined to very low levels. α-D-Glucosidase, limit dextrinase, α-D-glucosidase, and αα-trehalase activities were very low or absent during the first 34 weeks, and then increased. The β-D-galactosidase activity was unusual in being greater in the distal than the middle or proximal thirds of the intestine, and in its low pH optimum (pH 4.6), its inhibition by 4-chloromercuribenzene sulphonate but not by Tris, and the lack of β-D-glucosidase activity. These properties were those of a lysosomal acid β-D-galactosidase rather than of a brush border neutral β-D-galactosidase. The α-D-glucosidase activity had the characteristics of a lysosomal acid α-D-glucosidase early in lactation and of a brush border neutral α-D-galactosidase in adult animals. The significance of these findings was discussed in relation to changes in dietary carbohydrates during weaning and to the mode of digestion of milk carbohydrates by the pouch young.

The bulk of rat brain neutral β-D-2-acetamido-2-deoxyhexosidases have been shown to be present in the cytosol fraction.[17] They were not bound by concanavalin A-Sepharose while the acid β-D-2-acetamido-2-deoxyhexosidases were all bound. The neutral β-D-2-acetamido-2-deoxyglucosidase had a pH optimum of 5.2 and K_m of 0.57 mM, while the neutral β-D-2-acetamido-2-deoxygalatosidase had the highest reaction rate at pH 6.0 with a K_m of 0.12 mM. No divalent ions activated either of the enzymes. The β-D-2-acetamido-2-deoxyglucosidase lost more than 90% of the activity in 30 min at 50°C. The β-D-2-acetamido-2-

deoxygalactosidase was heat-stable and lost only 10-20% of its activity after 3 h at 50°C. The neutral β-D-2-acetamido-2-deoxyglucosidase was inhibited by free 2-acetamido 2-deoxy-β-D-glucose but not by 2-acetamido-2-deoxy-β-D-galactose. The reverse was found for the neutral β-D-2-acetamido-2-deoxygalactosidase. Two enzymes were separated almost completely by hydroxyapatite chromatography. Heat stability of the separated activity peaks suggested that the neutral β-D-2-acetamido-2-deoxygalactosidase, which was not bound to hydroxyapatite, may be specific to the D-galacto substrate. The neutral β-D-2-acetamido-2-deoxyglucosidase may, on the other hand, have some activity toward the D-galacto substrate. Both of the neutral enzyme activities were shown to be highest during the first postnatal week in rat brain, in contrast to the acid enzyme, which showed peak activities during the second and third weeks.

The separation and properties of β-D-2-acetamido-2-deoxyglucosidase A, B, and I from horse brain have been reported.[39] Three forms of β-D-2-acetamido-2-deoxyglucosidase (A, B, and I) have been separated for the first time from horse brain by ion-exchange chromatography on DEAE-cellulose. Form I has properties intermediate in the elution position from DEAE-cellulose pI and thermal stability, under the assay conditions, between those of forms A and B. After isoelectric focusing, especially forms A and I are transformed into more thermostable forms. D-Mannose, 2-amino-2-deoxy-D-mannose, 2-acetamido-2-deoxy-D-glucose, and methyl-α-D-mannopyranoside have been found to inhibit the three above-mentioned forms. D-Glucose and 2-amino-2-deoxy-D-galactose were weak inhibitors for β-D-2-acetamido-2-deoxyglucosidase B and I. D-Glucuronic acid was an inhibitor for forms A and I. The metabolic role of these inhibitions on specificities of β-D-2-acetamido-2-deoxyglucosidase towards natural substrates is discussed.

The induction of β-D-2-acetamido-2-deoxyglucosidase activity in the submaxillary glands of streptozotocin diabetic mice has been reported.[40] Streptozotocin diabetes strongly reduced the β-D-2-acetamido-2-deoxyglucosidase activity in the submaxillary glands of mice. A reciprocal change was observed between the enzyme activity and blood sugar in diabetes. Insulin treatment of the diabetic mice returned the reduced activity to the normal level. Isoelectric focusing analysis showed that the submaxillary-gland enzyme was composed of two isozymes having isoelectric points (pI) of 4.8 and 8.8. Only the pI 8.8 isozyme was affected by the diabetes.

Two major isoenzymes, A and B, of β-D-2-acetamido-2-deoxyglucosidase have been purified from adult-rat gastrocnemius muscle.[41] Studies were also done on adult-rat atrophic muscle after unilateral neurectomy of the sciatic nerve. In both normal and atrophic tissues three additional minor forms of the enzyme were identified.

Sucrose density gradients have been used to characterize lysosome-containing fractions from 0 h white prepupae of *Stomoxys calcitrans* L.[42] The acidic glycosidases (α-D-glucosidase, α-D-galactosidase, α-D-mannosidase, β-D-glucosidase, β-D-galactosidase, β-D-glucuronidase, and β-D-2-acetamido-2-deoxyglucosidase) equilibrate at the same density as does acid phosphatase. The mitochondrial marker enzyme cytochrome oxidase also equilibrates at the same density. Gomori staining showed the presence of acid phosphatase in the lysosomes.

β-D-2-Acetamido-2-deoxyglucosidase has been partially purified from the digestive gland of the three coastal New England bivalves *Mercenaria mercenaria*, *Spisula solidissima*, and *Mya arenaria* and their properties have been compared. Heat-inactivation studies on the β-D-2-acetamido-2-deoxyglucosidase preincubated at 45°C revealed that the preparation from *S. solidissima* was stable up to 60 min, while that from *M. arenaria* and *M. mercenaria* lost 47 and 91% of their original activities under these conditions, respectively. Inhibition studies indicated that D-glucurono-6,3-lactone is more inhibitory towards the *M. arenaria* enzyme, while $HgCl_2$ appears to be less inhibitory towards the *S. solidissima* enzyme. The V_{max} value for β-D-2-acetamido-2-deoxyglucosidase from *M. mercenaria* was approximately 2.5-fold greater than that from *S. solidissima* and *M. arenaria*. Other characteristics of the β-D-2-acetamido-2-deoxyglucosidase such as pH optimum, K_m, mol. wt., energy of activation, and effect of ionic strength on enzyme activity were found to be similar for all three species.

β-D-2-Acetamido-2-deoxyglucosidase and β-D-2-acetamido-2-deoxygalactosidase activity levels and cytochemical localization have been investigated in intact and regenerating planarian *Dugesia Lugubris*.[18] In 24 h regenerating planarians the β-D-2-acetamido-2-deoxyglucosidase was found to be localized in the lysosomes of dedifferentiating cells. The β-D-2-acetamido-2-deoxyglucosidase localization in dedifferentiating cells, correlated with the increase in β-D-2-acetamido-2-deoxyglucosidase and β-D-2-acetamido-2-deoxygalactosidase activities during the first 24 h of regeneration, confirms the involvement of the dedifferentiation in the origin of blastema. In non-regenerating animals the β-D-2-acetamido-2-deoxyglucosidase activity has been cytochemically demonstrated in the lysosomes of gastrodermal cells and in the mucous granules of basophilic secretory cells. This latter finding is interpreted as proof of a similarity between planarian and vertebrate mucous secretion.

Nine glycosidases in bloodstream forms of *Trypanosoma brucei brucei* S42 have been partially characterized. α-D-Glucosidase had similar physiochemical and enzymic properties to those of α-D-galactosidase, β-D-glucosidase, β-D-2-acetamido-2-deoxyglucosidase and β-D-2-acetamido-2-deoxygalactosidase.[20] α-D-Mannosidase was clearly distinct from the other glycosidases with respect to pH optimum, thermostability, specific activity during the course of parasitaemia, and subcellular location. It was suggested that the glycosidases studied may play a role in the turnover of trypanosomal glycoproteins, in particular the variant-specific surface antigen.

A comparative study of β-D-2-acetamido-2-deoxyglucosidase from *Mercenaria mercenaria*, *Mya arenaria*, and *Spisula solidissima* has been made.[43]

A marked increase in β-D-2-acetamido-2-deoxyglucosidase activity has been observed in the germinating cotyledon of cotton seeds.[44] The enzyme was isolated from cotton seedlings and purified to study its physiological function in the germination of cotton seeds. The purification procedure involves ammonium sulphate fractionation, ion-exchange chromatography, gel filtrations, and concanavalin A-Sepharose 4B chromatography, and the purified β-D-2-acetamido-2-deoxyglucosidase was shown to be homogeneous by disc electrophoresis. The molecular weight was estimated to be about 125,000 dalton by gel filtration. The enzyme hydrolysed both 4-nitrophenyl 2-acetamido-2-deoxy-β-D-glucopyranoside and 4-nitrophenyl 2-acetamido-2-deoxy-β-D-galactopyranoside. When 4-nitrophenyl 2-acetamido-2-deoxy-β-D-glucopyranoside was used as substrate, K_m and V_{max} were 0.625 nM and 228 moles min^{-1} mg^{-1}, respectively, and optimum activity was at pH 5.6. The enzyme liberated β-linked 2-acetamido-2-deoxy-D-glucopyranosyl residues from chitin, ovalbumin, and pronase-digested wheat-germ lectin.

Purified β-D-2-acetamido-2-deoxyglucosidase and purified Proteinase I, both isolated from the cellular slime mould *Dictyostelium discoideum*, have been shown to be immunologically cross-reactive with antiserum against the proteinase.[45] A peptide was isolated from a papain digest of Proteinase I that completely inhibited the immunoprecipitation of β-D-2-acetamido-2-deoxyglucosidase but only partially inhibited the immunoprecipitation of Proteinase I. This peptide inhibitor contained a high concentration of 2-acetamido-2-deoxy-D-glucose-1-phosphoryl-L-serine. It was proposed that these phosphoryl moieties might represent a common antigenic determinant in the two enzymes.

The secretion of lysosomal enzymes has been studied in *Dictyostelium discoideum*.[46] Using conditions which employ axonically grown amoebae in suspension in a simple starvation buffer, the lysosomal enzymes were secreted at rates that were at least as rapid as during normal development. Unlike normal growth or development, there was no appreciable lysosomal-enzyme synthesis under standard secretion conditions, so the enzyme activities acted as markers for the vesicles that contain them. One group of enzymes, including β-D-2-acetamido-2-deoxyglucosidase, β-D-mannosidase, β-D-glucosidase, β-D-galactosidase-I, and β-D-glucosidase-I were very efficiently secreted, with 10 to 80% of the total cellular activity becoming extracellular within a few hours. All of these enzymes have similar or identical secretion kinetics and may come from the same lysosomal vesicles. The kinetics of secretion from these vesicles are complex, with an initial lag period followed by a short period of rapid secretion and a final plateau with little or no secretion. The secretion of acid phosphatase is distinct from the secretion of the efficiently secreted glycosidases. It is secreted with linear kinetics for at least 6 h with only 13% becoming extracellular. This difference in secretion kinetics, along with differential effects of inhibitors, indicates that the secreted acid phosphatase activity may be localized within vesicles other than the above glycosidases. The authors have demonstrated that the egestion of indigestible material from residual or terminal phagosomes is kinetically distinct from the secretion of any of the lysosomal enzymes. Also β-D-galactosidase-II and perhaps other lysosomal enzymes are secreted either inefficiently or not at all under our standard secretion conditions. Thus, residual bodies, vesicles containing β-D-galactosidase-II, and perhaps other vesicles may constitute a third class of lysosomal vesicles in this organism which are essentially non-secretory under these defined conditions.

Lysosomal enzymes have been found to possess a common antigenic determinant in Dictyostelium discoideum.[47] Antisera have been prepared against two of the lysosomal enzymes. The two purified enzyme preparations used for immunization, β-D-2-acetamido-2-deoxyglucosidase and β-D-glucosidase-I, show no cross-contamination with each other and no significant contamination by other lysosomal enzymes. However, antisera raised against either enzyme bind equally well to seven different lysosomal enzymes and show no preference for the enzyme against which they were raised. A total of 10 different antisera have been examined and all show similar results. Preadsorption of antisera with either purified enzyme removes all antibody activity against the other enzyme. Evidence is presented which indicates that the same species of antibodies are responsible for the precipitation of seven lysosomal enzymes. These data are discussed in terms of the proposal that the antigen that is shared by the lysosomal enzymes is a post-translational modification of the enzyme proteins.

The relationship between carbohydrate moiety and thermostability of β-D-glucosidase from the thermophilic fungus Mucor meihei YH-10 has been investigated.[48] Two forms of thermostable β-D-glucosidase (a more thermostable G-a and less thermostable G-b) were purified to homogeneity from the culture filtrate. G-a and the G-b were glycoenzymes, and they contained 23.4% and 13.0% carbohydrate residues, respectively. The carbohydrate residues were D-mannose and 2-acetamido-2-deoxy-β-D-glucopyranose. The carbohydrate residues in G-a had a high content of 2-acetamido-2-deoxy-β-D-glucopyranose in the culture filtrate, and a high activity of β-D-2-acetamido-2-deoxyglucosidase was observed. The carbohydrate content of native G-a decreased to 13.0% when sufficiently digested with a commercial preparation of β-D-2-acetamido-2-deoxyglucosidase instead of β-D-2-acetamido-2-deoxyglucosidase produced by this fungus. The digested enzyme was identical with native G-b in carbohydrate content, molecular weight, amino acid composition, isoelectric point, hydrolysis curve for cellobiose, K_m value for β-D-glucosides, and in both thermo- and pH-dependency and stability. Therefore, the liberation of carbohydrates from more thermostable G-a by β-D-2-acetamido-2-deoxyglucosidase in culture filtrate led to the formation of less thermostable G-b.

Six glycoside hydrolases in the culture medium of Bacteroides fragilis - α-D-glucosidase, β-D-glucosidase, α-D-galactosidase, β-D-galactosidase, β-D-2-acetamido-2-

deoxyglucosidase, and α-L-fucosidase - were systematically purified by ammonium sulphate precipitation, gel-filtration chromatography, and density-gradient isoelectric focusing.[49] The isoelectric focusing resolved the glycosidases into distinct, well separated fractions and revealed three differently charged forms of β-D-2-acetamido-2-deoxyglucosidase and of α-L-fucosidase. Furthermore, α-D-glucosidase and β-D-2-acetamido-2-deoxyglucosidase were shown to possess dual affinities for the respective D-galactoside substrates, and β-D-galactosidase also hydrolysed β-D-fucoside. α-D-Glucosidase was purified to homogeneity, as indicated by a thin-layer isoelectric-focusing zymogram technique. The glycosidases, with exception of β-D-glucosidase and the acid α-L-fucosidase, were all separated from other glycosidic activities to 99%. The molecular weights varied between 58,000 and 125,000. The pH optima ranged from 4.8 to 6.9.

The effect of tunicamycin on secreted glycosidases of *Aspergillus niger* has been described.[50] The main glycosidases secreted into the culture medium during growth were β-D-glucosidase, α-D-galactosidase, and β-D-acetamido-2-deoxyglucosidase. In the presence of tunicamycin, an inhibitor of protein *N*-glycosylation, the activities of these enzymes in the culture medium were considerably decreased, whereas fungal growth and total mycelial protein content were not diminished. Intracellular glucosidase activity was also reduced in the presence of tunicamycin. The antibiotic also caused a decrease in the incorporation of D-mannose into total protein of the culture filtrate. Polyacrylamide gel electrophoresis of these proteins indicated that the majority of the D-mannosylated proteins were affected. The incorporation of L-leucine in total culture filtrate protein was not inhibited by tunicamycin. The amount of β-D-2-acetamido-2-deoxyglucosidase in the culture medium was estimated by immunotitration of enzyme activity with specific antiserum. The results suggest that the decreased enzyme activity observed during growth in the presence of tunicamycin is due to a reduced amount of enzyme.

In vitro cultures of *Entamoeba histolytica* strains HK-9 and NIH-200 have been found to contain considerable amounts of the acid hydrolase β-D-2-acetamido-2-deoxyglucosidase.[51] The results indicate that this enzyme is produced by the amoebae and secreted into the growth medium. The enzyme was purified 1000-2000 fold and preliminarily characterized. In cultures from NIH-200 two β-D-2-acetamido-2-deoxyglucosidases with distinct pIs, pH optima, and kinetics were demonstrated. From cultures HK-9 only one enzyme was isolated and it seems that it was identical with one of the NIH-200 enzymes. The molecular weight of all three enzymes was 125,000 ± 10,000. As it is known that hydrolases can be involved in inflammatory conditions it is suggested that released β-D-2-acetamido-2-deoxyglucosidase participates in the pathogenesis of amoebiases.

Sites of autolysis in Bacillus subtilis have been analysed by electron microscopy. Cell lysis in B. subtilis 168/s tpr thy and in an autolytic-deficient mutant, B. subtilis FJ6, was studied by following the release of wall material liberated by the two known autolysins, an N-acetylmuramoyl-L-alanine amidase and an endo-β-D-2-acetamido-2-deoxyglucanase.[52] Lysing organisms were examined in thin sections by electron microscopy, and measurements were made both of the location and size of holes produced by autolysin action and of peripheral wall thickness. In strain 168/s the initial site of lysis involved solubilization of the cross-wall and of one or both poles distal to the division site. Subsequently, the cylindrical wall was degraded through perforations at an increasing number of sites. Specific sites of autolysis, at the cross-wall and at one pole, were also observed in organisms prefixed with formaldehyde in a buffer of low ionic strength, although these effects were not observed in strain FJ6. In strain 168/s the bulk of the autolysin appeared to be located at the cross-wall and at one distal pole. The peripheral wall was reduced in thickness in autolysing cells, although it was uncertain whether the reduction was due to autolytic attack at the outer surface, to contraction of the wall in the suspending buffer or to shrinkage during preparation for electron microscopy. Lysis of strain FJ6 involved the initial breakdown of the cytoplasmic membrane and leakage of cellular contents through a limited number of sites along the wall. In contrast, lysis in B. subtilis MB21, the parental strain of FJ6, was similar to that described for strain 168/s.

The trypanocidal drug suramin has been found to be a potent inhibitor of β-D-2-acetamido-2-deoxyhexosidase A, with a K_i of about 4.5 μM, and to a lesser extent of β-D-2-acetamido-2-deoxyhexosidase (K_i 31.5 μM). β-D-2-Acetamido-2-deoxyhexosidase B remained active in the presence of 1.0 mM suramin whereas the activity of β-D-2-acetamido-2-deoxyhexosidase A was completely inhibited at 0.1 mM.[53]

A protein activator (G_{M2}-activator) specific for stimulating the hydrolysis of G_{M2} ganglioside by β-D-2-acetamido-2-

deoxyhexosidase A has been purified over 10^5-fold with a high yield from human liver.[54] The purification procedure includes the adjustment of the pH of liver extract to pH 4.3, followed by ammonium sulphate precipitation, Sephadex G-200 filtration, and column chromatographies on DEAE-Sephadex A-50, Matrex Gel Blue A, and octyl-Sepharose 4B. The physical properties of the G_{M2}-activator are: (a) moderately heat stable up to 50°C, (b) molecular weight about 23,500, and (c) isoelectric point about 4.8. Chemical analysis identifies this activator as a protein. G_{M2}-activator is very specific for stimulating the hydrolysis of G_{M2} ganglioside, but only slightly effective in stimulating the hydrolysis of asialo-G_{M2} or globotetraosylceramide catalysed by β-D-2-acetamido-2-deoxyhexosidase A. Unlike bile salts such as sodium taurodeoxycholate, this activator does not stimulate the hydrolysis of the above three glycosphingolipids catalysed by β-D-2-acetamido-2-deoxyhexosidase B. It does not stimulate the hydrolysis of synthetic substrates such as 4-methylumbelliferyl and 4-nitrophenyl 2-acetamido-2-deoxy-β-D-galactopyranosides. The oligosaccharide derived from G_{M2} ganglioside is not hydrolysed by β-D-2-acetamido-2-deoxyhexosidase A or B in the presence of either G_{M2}-activator or sodium taurodeoxycholate. The rate of the hydrolysis of G_{M2} ganglioside is affected by both the amount of G_{M2}-activator and β-D-2-acetamido-2-deoxyhexosidase A. The molar ratio of enzyme to G_{M2}-activator for obtaining the maximal hydrolysis of G_{M2} ganglioside is close to 1:1, while the molar ratio of G_{M2} ganglioside to G_{M2}-activator is about 300:1. These results may suggest that G_{M2}-activator interacts with β-D-2-acetamido-2-deoxyhexosidase A rather than the lipid substrate.

Pulse-chase experiments had shown that β-D-2-acetamido-2-deoxyhexosidase was synthesized in cultured human fibroblasts in precursor form and that during maturation of the enzyme the α-chains were converted from M_r = 67,000 to 54,000 and the β-chains from M_r = 63,000 to 29,000 plus smaller fragments, probably through an intermediate form of M_r = 52,000 (Hasilik, A. and Neufeld, E.F. (1980) J. Biol. Chem., 255, 4937).[55] The β-D-2-acetamido-2-deoxyhexosidase precursor which is present in NH_4^+-induced secretions has not been used as substrate to study these conversions in a cell-free system. A concentrate of such secretions, labelled biosynthetically, was incubated with cell fractions or purified proteinases. After the reaction, the β-D-2-acetamido-2-deoxyhexosidase was immunoprecipitated and its constituent polypeptides were denatured, reduced, and analysed by polyacrylamide gel electrophoresis and fluorography. A 3,000-12,000 g fraction of fibroblasts catalysed the conversion of the α-chain of the precursor β-D-2-acetamido-2-deoxyhexosidase from M_r = 67,000 to 56,000 and of the β-chain from M_r = 63,000 to 53,000. These products were similar to the mature α-chain and the intermediate form of the β-chain. The reaction occurred at acid pH, was stimulated by dithiothreitol, and was inhibited by iodoacetamide, N-ethylmaleimide, leupeptin, chymostatin, and antipain. Phenylmethylsulphonyl fluoride gave partial inhibition whereas H_4 edta, pepstatin A, and aprotinin had no effect. These properties indicate that the reaction was catalysed by a lysosomal thiol proteinase. However, the enzyme was probably distinct from cathepsin B, as purified cathepsin B itself had no effect on the β-D-2-acetamido-2-deoxyhexosidase precursor. A similar reaction was also catalysed by trypsin, chymotrypsin, and some related proteinases. Very high levels of trypsin released in addition polypeptides of M_r = 30,000 and lower molecular weight, which resembled the products of intracellular β-chain maturation. Formation of the 56,000 and 53,000 products by the lysosomal fraction, trypsin and chymotrypsin, proceeded equally well if the radioactive β-D-2-acetamido-2-deoxyhexosidase precursor was used as substrate in the form of an immunoprecipitate. The remainder of the precursor molecule appeared to be degraded. The limited proteolysis of the β-D-2-acetamido-2-deoxyhexosidase precursor did not affect its catalytic activity towards a synthetic substrate or its potential for uptake mediated by the D-mannose 6-phosphate recognition system.

The delivery of β-D-2-acetamido-2-deoxyhexosidase A to the cerebrum after osmotic modification of the blood-brain barrier has been investigated.[56] The studies were undertaken to evaluate the possibility that β-D-2-acetamido-2-deoxyhexosidase A, the enzyme deficient in Tay-Sachs disease, could be effectively delivered to the brain. Previous studies have shown that hypertonic mannitol can be used to osmotically produce reversible disruption of the blood-brain barrier in animals (rat and dog) and man without significant neurotoxicity, and that such barrier modification significantly increases the delivery of cytoreductive chemotherapy agents to selected areas of brain. By using the rat model of blood-brain barrier modification and radiolabelled enzyme, increased β-D-2-acetamido-2-deoxyhexosidase A delivery to brain has been demonstrated in more than 85 animals. The time of injection of β-D-

2-acetamido-2-deoxyhexosidase A after blood-brain barrier disruption is critical for maximum delivery. Rapid (over 30 sec) intra-arterial administration of β-D-2-acetamido-2-deoxyhexosidase A immediately after blood-brain barrier disruption resulted in a marked increase in enzyme delivery to the brain when the enzyme was administered 15-20 min after barrier disruption; 50% less β-D-2-acetamido-2-deoxyhexosidase A was delivered. When given 60-120 min after barrier modification, the amount delivered was the same as in control animals. This critical time course is very different from that seen in trials of low-molecular-weight chemotherapeutic agents (methotrexate and adriamycin). These preliminary studies suggest that β-D-2-acetamido-2-deoxyhexosidase A can be delivered to the brain by blood-brain barrier modification and may be indicative of the potential for enzyme replacement in patients who have Tay-Sachs disease.

The properties of β-D-2-acetamido-2-deoxyhexosidase have been investigated in isolated normal and I-cell lysosymes.[57] Using a combination of differential centrifugation and free-flow electrophoresis (Harms, E., Kern, H., and Schneider, J.A. (1980) Proc. Natl. Acad. Sci. U.S.A., 77, 6139), a single population of highly purified lysosomes was obtained from normal, I-cell disease type 1, and I-cell disease type 2 cultured fibroblasts. The authors' findings indicate that most of the residual acid hydrolase activities remaining within the I-cell fibroblasts are localized in the lysosomes, analogous to normal cells. Characterization of the carbohydrate-dependent properties of the lysosomal β-D-2-acetamido-2-deoxyhexosidase revealed that the I-cell and normal enzymes do not contain a significant proportion of neuraminidase-susceptible sialic acid residues, interact poorly with the β-D-galactose-specific lectin Ricinus communis, and are highly sensitive to endo-2-acetamido-2-deoxyhexosidase H treatment, indicating that the oligosaccharide units of both the I-cell and normal lysosomal β-D-2-acetamido-2-deoxyhexosidase are predominantly of the high D-mannose type. The I-cell and normal lysosomal β-D-2-acetamido-2-deoxyhexosidases, however, differed in their endocytotic properties. In contrast to the high rate of endocytosis of the normal lysosomal enzyme (7.8% mg^{-1} h^{-1}), the I-cell type-1 lysosomal enzyme failed to be endocytosed into Sandhoff cells, indicating an absent or altered phosphohexyl recognition marker on the I-cell enzyme. Examination of the normal extracellular β-D-2-acetamido-2-deoxyhexosidase revealed the presence of predominantly high D-mannose-type oligosaccharide units, similar to the corresponding lysosomal enzyme, although properties typical of complex-type oligosaccharide chains were also evident. In contrast, the secreted I-cell enzyme revealed the presence of oligosaccharide units predominantly of the complex type, indicating that the I-cell β-D-2-acetamido-2-deoxyhexosidase had high D-mannose-type oligosaccharide chains modified to complex type probably in the Golgi or GERL prior to secretion from the cell.

Human fibroblasts with a genetic deficiency of a single lysosomal enzyme and fibroblasts from a patient with I-cell disease with a multiple deficiency of lysosomal hydrolases have been used as recipient cells in studies on recognition and uptake of β-D-2-acetamido-2-deoxyhexosidase, β-D-glucuronidase, and β-D-galactosidase. Normal human fibroblasts and fibroblasts, hepatocytes, and hepatoma cells from the rat were used as donor cells.[58] The release of β-D-2-acetamido-2-deoxyhexosidase was found to be similar among these different cell types, but the extracellular activities of β-D-glucuronidase and β-D-galactosidase were much higher in rat cell cultures than in cultures of normal human fibroblasts. The enzymes released by rat fibroblasts were ingested by deficient human fibroblasts: enzyme from normal human fibroblasts was shown to be taken up by rat fibrosis by means of electrophoresis. This indicates that reciprocal transfer of lysosomal hydrolases occurs between human and rat fibroblasts. Rat hepatocytes released hydrolases that were poorly taken up by human recipient fibroblasts, and uptake of human fibroblast enzyme was not detected in the hepatocytes. Rat hepatoma cells, on the other hand, release lysosomal enzymes that were taken up by human deficient cells with a higher deficiency than those from fibroblasts. The uptake was subject to competitive inhibition by D-mannose 6-phosphate. The kinetics of this inhibition were comparable with those for high-uptake forms of lysosomal enzymes. Electrophoretic studies showed that rat hepatoma cells not only were capable of ingesting β-D-2-acetamido-2-deoxyhexosidase from normal human fibroblasts but also defectively processed enzyme release by I-cells. These findings make rat hepatoma cells a useful model for the study of recognition and uptake of lysosomal enzymes.

The distribution of the different types of oligosaccharides in cathepsin D and in β-D-2-acetamido-2-deoxyhexosidase synthesized in cultured human fibroblasts has been studied by using <u>endo</u>-β-D-2-acetamido-2-deoxyglucanase H as a probe for high D-mannose oligosaccharides.[59] The enzymes were specifically labelled in the protein or the carbohydrate moiety. In both enzymes, resistant and cleavable oligosaccharides prevailed in the secreted enzymes. Precursor molecules of cathepsin D contained two oligosaccharide side chains. Multiple forms of the precursor are synthesized with both, one, or none of two oligosaccharides sensitive to the action of the endo-β-D-2-acetamido-2-deoxyglucanase H. In fibroblasts unable to phosphorylate lysosomal enzymes (mucolipidoses II) the excessively secreted lysosomal enzymes contained predominantly oligosaccharides resistant to endo-β-D-2-acetamido-2-deoxyglucanase H.

A study has been made of lysosomal-enzyme activities in serum and leukocytes in chronic hepatic disease.[60] Five lysosomal-enzyme activites (arylsulphatase A, α-D-mannosidase, α-L-fucosidase, β-D-2-acetamido-2-deoxyhexosidase, and β-D-galactosidase) were determined in serum and leucocytes of 30 controls and 114 patients suffering from various liver diseases, including 30 with idiopathic hemochromatosis and 34 with alcoholic cirrhosis. The results show: (1) a decrease of the serum arylsulphatase A activity in diopathic haemochromatosis, (2) an increase of this activity in cholestatic jaundice, and (3) mainly a sharp rise of the leucocyte lysosomal-enzyme activities in the liver diseases studied (chiefly idiopathic haemochromatosis and alcoholic cirrhosis). The mechanism and the meaning of these disturbances are discussed.

The effect of tunicamycin and cycloheximide on the secretion of acid hydrolases from I-cell cultured fibroblasts has been investigated.[61] I-Cell cultured fibroblasts secrete excessive amounts of β-D-acetamido-2-deoxyhexosidase and α-L-fucosidase into the culture media as compared with normal fibroblasts. Addition of tunicamycin or cycloheximide at doses that inhibit the incorporation of {^{3}H}D-mannose (60-80%) and {^{14}C}L-leucine (40-50%) into trichloroacetic acid precipitate material decreased the secretion of these I-cell hydrolases to normal values within 24 hours but had no effect on the secretion of acid hydrolases from normal fibroblasts. These results indicated that I-cell cultured fibroblasts secreted at least two types of acid hydrolases: one of them was tunicamycin- and cycloheximide-sensitive and constituted the greater proportion of the secreted hydrolases, and a smaller proportion was insensitive to tunicamycin and cycloheximide, similar to the acid hydrolases secreted by normal cultured fibroblasts.

Enzymic cleavage of 2-acetamido-2-deoxy-β-D-glucosyl residues of keratan sulphate has been studied _in vitro_ by using as substrate 2-amino-2-deoxy-{^{3}H}glucose-labelled desulphated keratan sulphate containing 2-acetamido-2-deoxy-β-D-glucose residues at the non-reducing end.[62] Both lysosomal β-D-2-acetamido-2-deoxyhexosidases A and B were proposed to participate in the degradation of keratan sulphate on the basis of the following observations. Homogenates of fibroblasts from patients with Sandhoff disease, but not those from patients with Tay-Sachs disease, were unable to release significant amounts of 2-acetamido-2-deoxy-β-D-{^{3}H}glucose. On isoelectric focusing of β-D-2-acetamido-2-deoxyhexosidase from human liver the peaks of keratan sulphate-degrading activity coincided with the activity towards 4-nitrophenyl 2-acetamido-2-deoxy-β-D-glucopyranoside. A monospecific antibody against the human enzyme reacted with both enzyme forms and precipitated the keratan sulphate-degrading activity. Both isoenzymes had the same apparent K_m of 4 mM, but the B form was approximately twice as active as the A form when compared with the pH-activity profiles of both isoenzymes. Thermal inactivation of isoenzyme B was less pronounced towards the polymeric substrate than towards the 4-nitrophenyl derivative.

Changes in the kinetic properties of β-D-2-acetamido-2-deoxyhexosidase and arylsulphatase A have been reported upon immobilization on concanavalin A.[63] The immobilized β-D-2-acetamido-2-deoxyhexosidase showed a broad optimum at pH 3.5-4.5 compared with the free enzyme, which had a sharp optimum at pH 4.5. The optimum of free arylsulphatase A, pH 5.5, changed to pH 5.25 after immobilization. The apparent K_m values for the free and concanavalin A-immobilized β-D-2-acetamido-2-deoxyhexosidase and arylsulphatase A increased significantly (3-4 fold) on immobilization. The immobilization significantly increased the thermostability of these enzymes. The extent of inhibition by the metal ions was less for the immobilized β-D-2-acetamido-2-deoxyhexosidase and arylsulphatase A than for the free enzymes.

The _in vitro_ sialylation of purified human liver β-D-2-acetamido-2-deoxyhexosidase has been described.[64] In order to study structure-function relationships of lysosomal enzymes, human liver β-D-2-acetamido-2-deoxyhexosidase was purified by an extraction/affinity-chromatography/ion-exchange procedure. The isoenzymes A and B, native as well as neuraminidase-treated, were incubated with a partially purified preparation of bovine colostrum sialyltransferase (CMP-N-acetyl-neuraminate:D-galactosylglycoprotein

N-acetylneuraminyltransferase, EC 2.4.99.1). Native β-D-2-acetamido-2-deoxyhexosidases were found to be poor acceptors for the sialyltransferase used. However, incorporation of sialic acid into neuraminidase-treated β-D-2-acetamido-2-deoxyhexosidases A and B amounted to a 58 and 72% saturation of the theoretical acceptor sites, respectively. The acceptor specificity of the sialyltransferase suggests that D-Galp-(1 → 4)-D-GlcpNAc units may be present in at least part of the β-D-2-acetamido-2-deoxyhexosidase A and B molecules. However, oligomannosidic-type chains may also occur on the lysosomal enzyme, as shown by the sugar composition of the enzyme. The presence and/or amount of sialic acid residues does not appear to affect the kinetic properties of β-D-2-acetamido-2-deoxyhexosidases A and B towards 4-methylumbelliferyl glycoside substrate.

An unusual isoenzyme of β-D-2-acetamido-2-deoxyhexosidases has been characterized from a human colonic carcinoma cell line.[65] The separation and characterization of the isoenzymes of β-D-2-acetamido-2-deoxyhexosidases from each cell line were reported. The parental cell line contained A and B isoenzymes. The sub-line lacked the A-isoenzyme activity and contained an atypical B isoenzyme that was thermolabile, susceptible to alkylation, and of lower molecular weight.

A structural difference between the β-chains in β-D-2-acetamido-2-deoxyhexosidases B and A has been reported.[66] β-D-2-Acetamido-2-deoxyhexosidase B_A was prepared from purified placental A by treatment with merthiolate and rechromatography on DEAE Sepharose. The heat stability, isoelectric-focusing pattern, and peptide map of this isoenzyme were compared with those of naturally occuring purified placental β-D-2-acetamido-2-deoxyhexosidase B. B_A proved to be less heat stable at 60°C, have a slightly lower pI, and have one more peptide spot than did B. The lower pI of β-D-2-acetamido-2-deoxyhexosidase B_A is not caused by sialic acid residues since it was unaffected by neuraminidase treatment. The extra peptide spot in the β-D-2-acetamido-2-deoxyhexosidase B_A map cannot be explained by carbohydrate differences since no corresponding unpaired peptide spot was seen in the β-D-2-acetamido-2-deoxyhexosidase B map. Though β-D-2-acetamido-2-deoxyhexosidases B and B_A are products of the same gene, there is evidence that both are derived from a larger precursor polypeptide chain. The results indicated that enzymes B and B_A are processed differently, resulting in a small peptide present in enzyme B_A that is not found in β-D-2-

acetamido-2-deoxyhexosidase B.

Cultured rat peritoneal macrophages have been demonstrated to contain a receptor-mediated system for the uptake of the lysosomal enzyme {^{125}I}-β-D-2-acetamido-2-deoxyhexosidase A.[67] The uptake process was saturable (4.5 ng 10^{-6} cells) and had a K uptake of 5.9 X 10^{-8}M. The uptake was inhibited 68% by unlabelled β-D-2-acetamido-2-deoxyhexosidase A, 66% by D-mannan, and 24% by a hexosaminofetuin (D-mannose terminated). Methylamine, ammonium ion (both 20 mM), mepacrine, and chloroquine (both 0.1 mM) were found to inhibit strongly the specific uptake of {^{125}I}-β-D-2-acetamido-2-deoxyhexosidase A, while cytochalasin B (5 μg ml^{-1}) and colchicine (10^{-5}M) inhibited specific uptake by 35 and 28%, respectively. Mepacrine inhibited phagocytosis of latex beads to 76%. Methylamine, chloroquine, and cytochalasin B inhibited approximately to 50%. Colchicine and ammonium ion had a negligible effect. These results suggested that adsorptive pinocytosis and phagocytosis function independently of each other.

Purification and kinetic studies of β-D-2-acetamido-2-deoxyhexosidase A from porcine kidney have been reported.[68] Porcine kidney β-D-2-acetamido-2-deoxyhexosidase A was purified to electrophoretic homogeneity by procedures involving ammonium sulphate precipitation, gel filtration, and column chromatography on diethylaminoethylcellulose, octyl Sepharose CL-4B, and 6-aminohexanoyl derivatives of 2-acetamido-2-deoxy-β-D-glucopyranose coupled to Sepharose 4B. Analysis by sodium dodecyl sulphate gel electrophoresis suggested that the enzyme consists of three subunits with molecular weights of 61,000, 31,000, and 9,000. The molecular weight of the intact enzyme as determined by gel filtration was 129,000 at pH 8.0 and 302,000 at pH 5.0. Kinetic analysis with mixed substrates, as well as inhibition studies with various substrate analogues, indicated that the enzyme hydrolyses both 2-acetamido-2-deoxy-β-D-glucoside and -D-galactoside at the same active site. Among hexoses tested for inhibitory effect, only D-mannose was an effective competitive inhibitor. A comparison of D-mannosides with 2-acetamido-2-deoxy-D-glucosides with respect to binding ability to the enzyme in terms of K_i or K_m values implied the presence of a site for D-mannose other than the active site. On the other hand, kinetic studies with mixed inhibitors indicated that D-mannose competes with 2-acetamido-2-deoxy-D-glucose, 2-acetamido-2-deoxy-D-galactose, and acetamide.

The bulk of rat brain neutral β-D-2-acetamido-2-

deoxyhexosidases have been shown to be present in the cytosol fraction.[17] They were not bound by concanavalin A-Sepharose whereas the acid β-D-2-acetamido-2-deoxyhexosidases were all bound. The neutral β-D-2-acetamido-2-deoxyglucosidase had a pH optimum of 5.2 and K_m of 0.57 mM, while the neutral β-D-2-acetamido-2-deoxygalactosidase had the highest reaction rate at pH 6.0 with a K_m of 0.12 mM (see first citation of ref.17 for further details).

Acid hydrolases from the slime mould Dictyostelium discoideum have been reported to contain phosphomannosyl recognition markers.[69] Acid hydrolases from mammalian sources contain phosphorylated oligosaccharides which function as recognition markers for their receptor-mediated endocytosis by human fibroblasts. The discovery that glycopeptides derived from D. discoideum contain D-mannose 6-phosphate led to the suggestion that acid hydrolases from this source might also bear the marker. To test this hypothesis, the binding and endocytosis of purified β-D-glucosidase, β-D-2-acetamido-2-deoxyhexosidase, and β-D-mannosidase by human fibroblasts were investigated. These enzymes underwent endocytosis with efficiencies of 8.6 to 60% mg^{-1} h^{-1}, and 1 mM D-mannose 6-phosphate markedly inhibited their uptake (80 to 100%). The specificity of inhibition by sugar phosphates, the saturation kinetics of endocytosis, and the binding properties of D. discoideum acid hydrolases were similar to those reported for enzyme preparations derived from mammalian sources. In addition, 95 to 100% of the β-D-glucosidase or α-D-mannosidase molecules from D. discoideum preparations were competent in in vitro clearance. Furthermore, the three purified acid hydrolases contain 5 to 7 mol of D-mannose 6-phosphate per mol of enzyme. This indicates that, unlike many mammalian enzyme preparations, most if not all of these enzyme molecules from D. discoideum contain the phosphomannosyl recognition marker.

Several mollusc glycosidases have been studied for their activities towards natural substrates. α-L-Fucosidases from Chamelea gallina, Tapes rhomboideum, and Mytilus edulis hydrolyse oligosaccharides (di-, tri-, and penta-saccharides) with 1 → 2, 1 → 3, and 1 → 4 bonds, L-fucose-containing glycopeptides from bovine thyroglobulin, and porcine submandibular mucin (devoid of sialic acid). α-L-Fucosidase from Littorina littorea hydrolyses fucose-containing glycopeptides from bovine thyroglobulin.[70] β-D-Glucuronidase from L. littorea hydrolyses hyaluronic acid, chondroitin 4-sulphate, and heparin with a very low activity.

However, it is much more active on oligosaccharides (from the above-mentioned macromolecules) containing non-reducing terminal glucuronyl residues. β-D-2-Acetamido-2-deoxyhexosidase from *Helicella ericetorum* acts mainly with an *endo*-hydrolase activity on 2-acetamido-2-deoxy-β-D-hexopyranosyl-(1 → 4) linkages of ovalbumin, ovomucoid, chitin, hyaluronic acid, and chondroitin 4-sulphate. It also has a secondary *exo*-hydrolase activity on these substrates.

The substrate specificity of diplococcal β-D-2-acetamido-2 deoxyhexosidase has been studied in detail by using oligosaccharides of known structure. The enzyme cannot cleave D-GlcNAc-(1 → 4)-D-Man and D-GlcNAc-(1 → 6)-D-Man linkages, although it readily hydrolyses D-GlcNAc-(1 → 2)-D-Man, D-GlcNAc-(1 → 3)-D-Gal, and D-GlcNAc-(1 → 6)-D-Gal linkages.[71] The D-GlcNAc-(1 → 2)-D-Man linkage in D-GlcNAc-(1 → 4)-{D-GlcNAc-(1 → 2)}-D-Man group is cleaved by the enzyme but the linkage in D-GlcNAc-(1 → 6)-{D-GlcNAc-(1 → 2)}-D-Man group is not, probably because of the steric effect of D-GlcNAc-(1 → 6)-D-Man residue on D-GlcNAc-(1 → 2)-D-Man linkage. Similar steric effect is also observed in the cases of various other oligosaccharides. It is suggested that the substrate specificity of diplococcal β-D-2-acetamido-2-deoxyhexosidase can be used effectively for the structural studies of L-asparagine-linked sugar chains.

3 α-L-Arabinofuranosidases and α-L-Arabinopyranosidases

Active-site-directed irreversible inhibition of glycosidases by the corresponding glycosylmethyl-(4-nitrophenyl) triazines has been investigated.[72] β-D-Galactopyranosylmethyl-(4-nitrophenyl) triazine is an active-site-directed irreversible inhibitor (ASDIN) of the *ebg*^{0} and the *lacZ* β-D-galactosidases of *Escherichia coli*, of the β-D-galactosidase of human liver lysosomes, and, less effectively, of the α-D-galactosidase of green coffee beans. It has no effect on the *lac* repressor of *E. coli*. β-D-Glucopyranosylmethyl-(4-nitrophenyl) triazine is an ASDIN of both β-D-galactosidase activities of sweet-almond β-D-glucosidase B. It has no effect on yeast α-D-glucosidase, the *lacZ* β-D-galactosidase of *E. coli*, or glucoamylase from *Aspergillus niger*. β-D-Xylopyranosylmethyl-(4-nitrophenyl) triazine is an ASDIN of the β-D-xylosidase from *Pencillium wortmanni*, but has no effect on the β-D-xylosidase of *Bacillus pumilus*, except by virtue of its consumption of stabilizing dithiothreitol. α-L-Arabinofuranosylmethyl-(4-nitrophenyl) triazine is an ASDIN for the extracellular α-L-arabinofuranosidase of Monilinia fructigena of more alkaline pH optimum at pH 7. The other isoenzyme is inert at this pH, and more acidic pHs destroy the reagent. It is concluded that glycosylmethyl(aryl) triazines are ASDINs for glycosidases for which both substrate and product stereochemistries at the anomeric centre are the same as that of the reagent, that they can act, with lower effectiveness, on glycosidases for which both substrate and product stereochemistries are opposite to those of the reagents, and that they have no effect on glycosidases with opposite stereochemistries of substrate and product, or proteins of no catalytic function.

The α-L-fucosidase, β-D-2-acetamido-2-deoxyglucosidase, α-mannosidase, β-D-glucosidase, β-D-xylosidase, β-D-glucosidase, and α-L-arabinofuranosidase activities of normal and atrophic skeletal muscle of developing and adult rats have been collectively investigated.[73]

Substrate specificity of α-L-arabinofuranosidase from plant Scopolia japonica has been examined using three kinds of L-arabino-disaccharides prepared from natural sources or synthetically.[74] This enzyme hydrolysed α-L-arabinofuranosylarabinoses which had either a (1 → 3) or a (1 → 5) linkage, but hydrolysed α-L-arabinopyranosylarabinose having a (1 → 5) linkage to a lesser degree. α-L-Arabinofuranosidase, which was shown to be an exo-enzyme, degraded beet araban incompletely. Arabinose oligomers and galactose-containing fragments, isolated following acid hydrolysis of araban, were both incompletely degraded by the enzyme. The reasons for the incomplete degradation were explained by the novel finding of (1 → 2) linkages and L-arabinopyranosides and the inclusion of trace amounts of D-galactose into the carbohydrate chain of araban. This enzyme was practically non-reacting with the hydroxyprolyl-L-arabinose linkage of glycopeptides from plant cell walls.

An α-L-arabinofuranosidase produced by wild-type Streptomyces sp. No. 17-1 has been described.[75] One hundred and fifty strains of actinomycetes were isolated from soils on plate cultures containing beet arabinan as the sole carbon source. About one-third of the culture fluids were found to have α-L-arabinofuranosidase activity. A wild-type strain, Streptomyces sp. No. 17-1, was selected as the best producer of α-L-arabinofuranosidase. The highest enzymic activity was obtained in the culture fluid when the initial pH was adjusted to 9.0. An α-L-arabinofuranosidase was highly purified from the culture filtrate of No. 17-1 by combining column chromatography on DEAE-cellulose, gel filtration on Sephadex G-100, and isoelectric focusing. The molecular weight of the purified enzyme was estimated to be about 92,000, and its isoelectric point was pH 4.4. The enzymic activity was maximum at pH 6.0 and was completely inhibited by Hg^{2+}. The apparent K_m value of the enzyme for 4-nitrophenyl α-L-arabinofuranoside was determined to be 3.6mM.

4 β-D-Fructofuranosidases

A continuous photometric method has been described for the determination of β-D-fructofuranosidase from the small intestine.[76] The activity of β-D-fructofuranosidase is determined spectrophotometrically by the production of NADH during the conversion of β-D-glucose (a reaction product of β-D-fructofuranosidase) by D-glucose dehydrogenase to D-glucono-1,4-lactone.

A study of the distribution of some adsorbed and intrinsic enzymes between the mucosal cells of the rat small intestine and the apical glycocalyx separated from them has included work on β-D-fructofuranosidase.[77]

Solubilization of β-D-fructofuranosidase and alkaline phosphatase from guinea-pig intestinal brush-border membranes has been studied using various procedures such as treatments with papain, sodium dodecyl sulphate, sodium deoxycholate, and Triton X-100.[78] All the procedures tried were quite efficient in solubilizing the membrane proteins. Under optimum conditions, 65-90% β-D-fructofuranosidase activity was solubilized compared to 30-60% of the alkaline phosphatase. Differences in the extractability of these proteins are presumably due to their differential topological arrangements in the membrane matrix. Such differences in the localization of β-D-fructofuranosidase and alkaline phosphatase in brush borders may explain the observed disparities in the absorption rate of D-glucose released by the hydrolysis of sucrose and D-glucose 1-phosphate in the intestine {Miller, D. and Crane, R.K. (1961) Biochim. Biophys. Acta, 52, 281-293}.

The immobilization of β-D-fructofuranosidase on krill chitin has been investigated. The optimum pH value for adsorption was found to be pH 5.0.[79]

The separation and purification of β-D-fructofuranosidase and an inulinase from germinating garlic (*Allium sativum* L.) bulbs have been described.[80] Both the enzymes had apparent molecular weights of 76,000, as determined by gel chromatography on Sephadex G-200, and were inactivated by treatment with 4-chloromercuribenzoate, 5,5'-dithiobis(2-nitrobenzoic acid), and heat treatment at 55°C for 5 min at pH 7. The β-D-fructofuranosidase hydrolysed sucrose and raffinose but had no action on inulin. The apparent K_m value for sucrose was 26mM. The inulinase hydrolysed, apart from inulin, sucrose and raffinose. The K_m values for inulin and sucrose were 10mM and 25mM, respectively. While in the early stages of plant growth (up to 6 days) the inulinase and β-D-fructofuranosidase activities increased in a parallel fashion in the bulbs. At later stages the increase in β-D-fructofuranosidase activity was more than that of inulase activity.

β-D-Fructofuranosidases in sugar-cane leaf sheaths have been found to be firmly bound to the cell wall.[81] They consist of three different enzymes distinguished on the basis of optimum pH, K_m, and the response to inhibitors. This β-D-fructofuranosidase complex differs from the single enzyme described for both immature and mature stalk tissues.

Cytoplasmic and wall-bound β-D-fructofuranosidases have been prepared from the seedlings of sugar beet, and adsorption-release characteristics of the enzymes were compared by the use of a partially purified wall.[82] Adsorption of the bound enzyme occurred below 0.25 M in NaCl concentration and was unaffected by changes in pH, while the cytoplasmic enzyme was adsorbed only on acidic media of pH 3~4. Release of re-adsorbed enzymes with NaCl depended on its concentration, and the cytoplasmic enzyme was liberated more readily than the bound type, giving their maximum values at 0.2 M and 0.8 M NaCl, respectively. Thus, the two β-D-fructofuranosidases showed marked differences in adsorption to and release from the wall. Adsorption of the cytoplasmic enzyme to the wall was reversible with changes in pH of the medium. The results suggested that the wall-bound β-D-fructofuranosidase was essentially different from the cytoplasmic enzyme and consequently not an artifact.

Optimum conditions for β-D-fructofuranosidase detection in polyacrylamide and agarose gels have been defined from comparison of zymograms obtained with two staining methods including an original one.[83] Under all conditions tested in this study detection has been improved with the new staining procedure. The new staining medium developed here uses two intermediary enzymes, D-glucose oxidase and peroxidase, 3,3'-diaminobenzidine as final acceptor, and sucrose as substrate for β-D-fructofuranosidase. A β-D-fructofuranosidase zymogram is obtained either by gel immersion in this staining solution or, when highly crosslinked polyacrylamide separating gels are used, by overlaying the slab within a thin buffered agarose gel containing staining chemicals (sandwich technique). Examples are presented to illustrate the general principles involved and indicate the conditions necessary for optimal development of this precise and specific technique.

A soluble form of β-D-fructofuranosidase has been purified from radish (*Raphanus sativus*) seedlings.[84] Using Sephadex G-25 chromatography, ammonium sulphate precipitation, concanavalin A-Sepharose, Ultrogel AcA_{34}, AcA_{54}, and immunosorption chromatographies, this enzyme was purified 700 times. Crossed immunoelectrophoresis with a polyspecific antiserum showed an extremely low level of contaminants even when a high amount of final product was tested. The purified enzyme has an apparent molecular weight of 48,500 and a total sugar content of 7.7%. The carbohydrate moiety, with probably a poly-D-mannosidic structure, is not necessary for β-D-fructofuranosidase activity and has a limited role in enzyme stability or in protection against proteolysis. However, β-D-fructofuranosidase solubility is shown to depend on the carbohydrate moiety.

Radish β-D-fructofuranosidase has been shown to become inactivated by diluting the enzyme solution, and the activity can be restored by addition of bovine serum albumin or other proteins.[85] The use of detergent, high-molar salt solutions, or silicone-coated tubes showed that decrease of specific activity upon dilution was not linked to adsorption of the enzyme on to glass walls. Albumin neither protected the enzyme from denaturation by heat nor changed its stability during conservation at room temperature. The action of added proteins was not due to removal of an inhibitor from the enzyme solutions. Some polyanions or polycations had the same effect as albumin, but dialysis or chromatography showed that they did not act by reassociation of inactive products formed during dilution of the active enzyme. A molecular-weight heterogeneity was observed in the enzyme population when chromatography was performed without albumin. It was suggested that inactive forms, formed upon dilution, differed slightly in their molecular conformation from the active forms obtained at high protein concentration.

Soluble β-D-fructofuranosidase from radish cotyledons has been subjected to electrofocusing and electrophoresis.[86] The results suggested considerable heterogeneity of the enzyme.

A serological study has been made on a β-D-fructofuranosidase from Spacelia sorghi.[87] Antibodies raised against a purified β-D-fructofuranosidase from S. sorghi were cross-reacted with a purified β-D-fructofuranosidase from Claviceps purpurea. An immunodiffusion test indicated a high degree of serological identity between the S. sorghi and C. purpurea enzymes. A complement fixation test confirmed this result. The similarity in titres indicated a very close structural similarity between the enzymes.

The carbohydrate structure of yeast β-D-fructofuranosidase has been investigated.[88] β-D-Fructofuranosidase, extracted from broken cells of Saccharomyces cerevisiae X-2180 mnn^2 D-mannan mutant, was separated into a fraction insoluble in 75% ammonium sulphate (P75 β-D-fructofuranosidase, 36% carbohydrate) and a soluble fraction (S75 β-D-fructofuranosidase, 53% carbohydrate). The latter reacted with antibodies specific for the (1 → 6)-linked-α-D-mannose unit of the mannoprotein outer chain, whereas the P75 β-D-fructofuranosidase failed to react with this antiserum, although it did react with serum against terminal (1 → 3)-linked-α-D-mannose units that are characteristic mannoprotein core. A bacterial endo-(1 → 6)-α-D-mannanase removed the outer chains from the S75 invertase and converted it to a form that was similar in electrophoretic and immunochemical properties to the P75 β-D-acetamido-2-deoxyglucosidase, whereas the endo-D-mannanase had little effect on the latter β-D-acetamido-2-deoxyglucosidase. The results suggest that the P75 invertase is a form of the enzyme to which only the core oligosaccharide units had been added, and the S75 β-D-acetamido-2-deoxyglucosidase represents an enzyme fraction to which the polysaccharide outer chains were also attached. A strong anomeric ^{1}H n.m.r. signal for unsubstituted (1 → 6)-linked-α-D-mannose in the S75 β-D-acetamido-2-deoxyglucosidase and a much reduced signal in the P75 β-D-acetamido-2-deoxyglucosidase and endo-D-mannanase-digested S75 β-D-acetamido-2-deoxyglucosidase support these conclusions. endo-β-D-Acetamido-2-deoxyglucosidase digestion of the S75 and P75 β-D-2-deoxyglucosidase, as well as of a purified wild-type yeast β-D-acetamido-2-deoxyglucosidase, produced an apparently identical series of 3 to 4 carbohydrate-containing proteins that were separable by polyacrylamide gel electrophoresis in sodium dodecyl sulphate but that migrated as a single band on isoelectric focusing. The bands ranged from about 63,000 to 69,000 and differed by the size of one or more carbohydrate core units each of 15 D-mannose and 1 β-D-acetamido-2-deoxyglucopyranose units. The results suggested that the external β-D-acetamido-2-deoxyglucosidase molecules contain some core units without attached outer chains and that the cells contained a precursor form of the enzyme to which only the core units have been added. In support of this conclusion, ^1H n.m.r. spectra and chromatographic patterns showed that the core fragments from the P75, S75, and wild-type β-D-acetamido-2-deoxyglucosidase were essentially identical.

2-Deoxy-D-glucopyranose has been used in the selective isolation of mutants of *Trichoderma reesi* with enhanced β-D-glucosidase production.[89] By the use of 2-deoxy-D-glucose, a non-metabolizable catabolic repressor of certain yeast enzymes, mutants derepressed and hyperproductive with respect to β-D-fructofuranosidase, and α-amylase have been obtained in very high plate yields from the yeast *Saccharomyces cerevisiae* and *Lipomyces kononenkoae*, respectively. When 2-deoxy-D-glucose was applied to *T. reesai* QM 9414 in a similar way, it permitted the selective solution in very high plate yields of mutants derepressed and hyperproductive with respect to β-D-glycosidase.

Streptococcus salivarius strain 51 has been found to produce an extracellular β-D-fructofuranosidase when grown in a levan-containing medium.[90] The enzyme hydrolyses levan, and inulo-biose, -tetraose, -pentaose, and -hexaose, but has negligible effects on inulin, inulotriose, and inuloheptaose. The branching in *S. salivarius* strain 51 levan is through a single (2 → 1)-β-D linkage. Evidence has been presented for the selective breakdown of altered tomato β-D-fructofuranosidase molecules by a neutral protease from *Bacillus subtilis*.[91]

The recovery method and some industrial properties of non-specific β-D-fructofuranosidase from *Kluyveromyses fragilis* have been described.[92] The yeast was grown in continuous culture on a complex medium containing sucrose as the carbon source and limiting nutrient. The inulinase yield was 7,000 μg hexose min^{-1} mg^{-1} biomass, determined with 4% sucrose at 50°C and pH 5.0. Maximum amount of enzyme was formed at the lowest dilution rate (0.09 h^{-1}) tested. Any increase in dilution rate caused a severe reduction in activity per unit biomass due to carbon catabolite repression. β-D-Fructofuranosidase yields were constant in the pH range 3.5 to 6.0 and in the dissolved oxygen tension range 2.5 to 40% of saturation. Continuous cultivation on sucrose produced twice as much inulase as the previously used method of batch cultivation on the expensive substrate inulin. A high-quality industrial grade was prepared by autolysis of the yeast cells, ultrafiltration of the supernatant, and acetone precipitation. Under the assay conditions used the preparation showed an activity ratio toward sucrose and inulin of 10.5 as compared to 1,600 for baker's yeast invertase. Further comparison with invertase showed that it is at least as resistant to substrate inhibition, as thermostable, and has slightly less transferase activity. These results suggest that K. fragilis β-D-fructofuranosidase may represent an alternative to Saccharomyces cerevisiae invertase presently used in industry.

5 D-Fructose-1,6-Bisphosphatase

D-Fructose 2,6-bisphosphate, a known powerful stimulant of phosphofructokinase (Van Schaftingen, E., Hue, L., Other, H.G. (1980) Biochem J., 192, 897), has been found to inhibit, at micromolar concentrations, liver and muscle D-fructose-1,6-bisphosphatase.[93] The main characteristics of this inhibition are that (i) it is much stronger at low than at high substrate concentration, (ii) it changes the substrate saturation curve from almost hyperbolic to sigmoidal, and (iii) it is synergistic with the inhibition by AMP. This inhibition may play an important role in the stimulation of gluconeogenesis by glucagon, because this hormone is known to decrease the concentration of D-fructose-2,6-bisphosphatase in the liver.

The coupling between D-fructose-1,6-bisphosphatase and myo-inositol synthetase has been investigated.[94] Based mostly on biochemical and genetic studies with myo-inositol-deficient systems, a model is proposed by the authors which suggests a coupled operation of the gluconeogenic enzyme D-fructose-1,6-bisphosphatase and myo-inositol synthetase, as the probable regulatory event for cell survival under conditions of myo-inositol deficiency.

A Bacillus subtilis mutation, producing a deficiency of D-fructose-1,6-bisphosphatase has been isolated and genetically purified.[95] The mutant did not produce cross-reacting material. It grew on any carbon source that allowed growth of the standard strain except myo-inositol and D-gluconate. Because the mutant could grow on D-fructose, glycerol, or L-malate as the sole carbon source, B. subtilis can produce D-fructose 6-phosphate and the derived cell wall precursors from these carbon sources in the absence of D-fructose-1,6-bisphosphatase. In other words, during gluconeogenesis B. subtilis must be able to bypass this reaction.

Rat liver D-fructose-1,6-bisphosphatase and three other gluconeogenic D-fructose-bisphosphatases have been tested as substrates for the catalytic subunit of cyclic AMP-dependent protein kinase.[96] In contrast to the rat liver enzyme, homogeneous preparations of mouse liver, rabbit liver, and pig kidney D-fructose-bisphosphatase could not be phosphorylated by the kinase. Comparative sodium dodecyl sulphate-polyacrylamide gel electrophoresis of the above four D-fructose-bisphosphatases revealed that the subunit molecular weight of the isolated rat liver enzyme (ca. 40,000-42,000) was greater than that of mouse liver, rabbit liver, and pig kidney D-fructose-bisphosphatases (ca. 36,000-37,000). Treatment of ^{32}P-labelled rat liver D-fructose-bisphosphatase with trypsin resulted in the conversion of the rat liver enzyme to an active species with a subunit molecular weight identical to that of the three other enzymes, with complete loss of the ^{32}P-labelled site. Identical trypsin treatment of pig kidney D-fructose-bisphosphatase caused no change in the molecular weight of the enzyme.

6 β-D- and α-L-Fucosidases

Specificity patterns of different types of human fucosidase have been investigated through hydrolysis of fucosides, galactosides, and arabinosides by different types of fucosidase.[97] The comparative study of the splitting of α-L-fucoside and β-D-arabinoside with a similar structure of C1-C4 of the pyranose ring and the preparation of the enzyme by affinity chromatography showed that both substrates were hydrolysed by the same enzyme, α-L-fucosidase. The analogous investigation of the hydrolysis of β-L-fucoside, β-D-galactoside, and α-L-arabinoside with a similar structure at C1-C4 of the pyranose ring demonstrated that these glycosides were split by the same enzyme, β-D-fucosidase, the activity of which was decreased dramatically in G_{M1}-gangliosidosis. It is suggested that the specific action of different types of fucosidase is due to recognition by these enzymes of C1-C4 of the pyranose ring in the corresponding substrates. The problems of differential diagnosis of some glycosidoses (fucosidosis, G_{M1}-gangliosidosis, and Faby disease) are discussed on the basis of the data obtained.

It has been demonstrated that β-D-fucosidase, isolated from the digestive juice of Achatina balteata, was markedly inactivated by preincubation with three different probes, specific for L-tyrosine residues under the experimental conditions used.[98] An effective protection against inactivation of the enzyme was obtained in the presence of a substrate analogue. These data strongly suggested that β-D-fucosidase possesses essential L-tyrosine residues.

β-D-Fucosidases isolated from the digestive juice of Achatina balteata have been shown to catalyse the hydrolysis of β-D-fucosides, β-D-glucosides, and β-D-galactosides, and the catalytic efficiency is maximum towards β-D-fucosides.[99] The results of mixed-substrate incubation studies and inhibition by glycopyranoses indicate that there is at least one site at which all tested substrates are hydrolysed. In the absence of inhibitor, the reciprocal plots exhibit a significant downward curvature. If a substrate analogue is present, the plots can be straight lines. These results are consistent with the presence on the enzyme molecule of at least two distinct sites for the substrate molecules, one being an active site and the other being either a second active site with different kinetic parameters or a modifier site. Also, data are shown to fit quite well with the mechanism proposed for a mnemonic enzyme.

Crude liver α-L-fucosidase has been investigated in 10 vertebrate species ranging from fish to human using 4-methylumbelliferyl α-L-fucopyranoside as substrate.[100] The α-L-fucosidases were characterized by determinations of pH optima and isoelectric-focusing profiles, K_m values, thermostabilities after preincubation at 45°C and 60°C, and immunoprecipitation by anti-human liver α-L-fucosidase antibody. Results suggest differences in liver α-L-fucosidases with respect to pH optima, thermostabilities, and isoelectric-focusing profiles. However, the K_m values were largely similar, and 9 to 10 species were precipitated by the human IgG preparation, suggesting antigenic similarity.

In affinity chromatography of α-L-fucosidase from the human liver on 6-amino hexanoyl derivative of 2-amino-2-deoxy-β-L-fucopyranose coupled to Sepharose, the elution profile of the enzyme has been characterized by two components, arbitrarily denoted as α-L-fucosidases A and B.[101] One of the components (α-L-fucosidase A), in the absence of sodium azide in the buffer mixtures used, was retained on the affinity sorbent, while the other (α-L-fucosidase B) was adsorbed under the same conditions. This latter component was eluted by a solution containing the enzyme inhibitor L-fucose. On the basis of rechromatography data, it is suggested that there is an equilibrium in solution between α-L-fucosidases A and B, differing in their affinities for the affinity sorbent.

Purification and properties of two forms of human α-L-fucosidase have been described.[102] High-(100,000) and low-(50,000) molecular-weight forms of α-L-fucosidase (α-L-fucosidases I and II) have been purified to apparent homogeneity from human spleen, liver, brain, and kidney on the basis of differential affinity for 6-amino hexanoyl derivative of 2-amino-2-deoxy-L-fucose coupled to agarose bead columns. α-D-Fucosidase I (the bound form) consisted of two 50,000 molecular-weight monomers. However, both forms could aggregate to tetramer and hexamer forms. The bound (100,000) form was shown to be a sialoglycoprotein, whereas the unbound (50,000) form was a neutral D-mannose-rich glycoprotein. Other differences with respect to amino acid composition, pH optimum, electrophoretic mobility, K_m, thermal stability, and natural substrate specification were observed.

The conditions for maximal activity (pH, buffer, saturating substrate concentration, range of linear relationships between enzyme activity *versus* incubation time and *versus* enzyme concentration) in the fluorimetric assay of several glycohydrolases of lysosomal origin in human plasma and serum have been established.[26] The following enzymes were studied: α-D-galactosidase, β-D-galactosidase, β-D-glucuronidase, α-D-mannosidase, α-L-fucosidase. See initial citation of ref.26 for further details.

Alterations in the functions and structure of α-L-fucosidase have been described in cystic fibrosis.[103] In this communication data are presented which are relevant to other recently published reports on α-L-fucosidase activity in cystic-fibrosis plasma (Hosli, P. and Vogt, E. (1979) *Lancet ii*, 543) and serum (Alhadeff, J.A. and Watkins, P. (1979) *Clin. Chim. Acta*, **105**, 131). Hosli and Vogt (1979) reported an increased thermolability of several acid hydrolases in cystic-fibrosis plasma when compared to non-cystic-fibrosis plasma, and they predicted that α-L-fucosidase from cystic-fibrosis plasma would also have an increased thermolability. In the other report, Alhadeff and Watkins were unable to demonstrate a decreased activity of α-L-fucosidase in cystic-fibrosis serum. In addition, using conditions which differed from those of Hosli and

Vogt, they were unable to demonstrate an increased thermolability of the enzyme. The data reported here demonstrate that α-L-fucosidase in cystic-fibrosis plasma does not have an increased lability under the mild thermal conditions of Hosli and Vogt. Furthermore, it is emphasized that age-matched controls must be used to demonstrate the decreased activity of α-L-fucosidase in cystic-fibrosis serum or plasma.

It has been found that the activity of α-L-fucosidase was reduced 2-3 fold in cultured lymphoblasts derived from patients with cystic fibrosis relative to normals.[104] It was suggested that this change may be involved in the alterations in the L-fucose content of some of the abnormal glycoproteins which contribute to the respiratory and glandular obstructions in the disease.

The lysosomal enzyme α-L-fucosidase has been examined by thin layer gel and column isoelectric focusing in skin fibroblasts and liver from patients with cystic fibrosis and controls.[105] All three common phenotypes of the enzyme were observed in both control and cystic-fibrosis fibroblasts. When individuals of the same α-L-fucosidase phenotype were compared, no major differences between the isoenzyme profiles of cystic-fibrosis patients and controls were detected in either fibroblasts or liver tissue.

Human α-L-fucosidase in the normal and fucosidosis states has been purified and studied. An antiserum has been raised to purified α-L-fucosidase.[106] Levels of cross-reaction with serum of two unrelated fucosidosis patients and normal individuals with low activity are consistent with the presence of very low amounts of normal enzyme. Similarly no cross-reacting material could be found in cultured fibroblasts from fucosidosis patients. It was deduced that in these cases there is no production of mutant enzyme in quantities comparable to normal levels. Some observations on the interrelations of α-L-fucosidases I and II are reported.

Thermostability studies have been performed at different preincubation temperatures (37-65°C) on human α-L-fucosidases, purified serum and liver enzyme, the isoelectric forms of purified liver enzyme which were separated by preparative isoelectric focusing, crude adult and foetal liver supernatant enzyme, and neuraminidase-treated enzyme.[107] Very different thermostability curves were found for the various isoelectric forms of α-L-fucosidase. The most neutral form (I) was least thermostable and the most acidic form (VIII) most thermostable, with the intervening forms (II-VII) having intermediate thermostabilities. For the different isoelectric forms of liver α-L-fucosidase there appears to be a significant trend of increasing thermostability with increasing acidity (and, presumably, increasing amount of sialic acid). The purified sialic acid-rich serum enzyme was considerably more thermostable than the purified liver enzyme. The foetal liver enzyme (which is less acidic and may contain less sialic acid than the adult liver enzyme) was less thermostable than adult liver α-L-fucosidase. In contrast, neuraminidase treatment of human liver α-L-fucosidase did not change its thermostability, even when considerable desialylation had occurred as monitored by isoelectric focusing. Several possible interpretations of the data were presented.

The effect of tunicamycin and cycloheximide on the secretion of acid hydrolases from I-cell cultured fibroblasts has been investigated.[61] I-Cell cultured fibroblasts secrete excessive amounts of β-D-2-acetamido-2-deoxyhexosidase and α-L-fucosidase into the culture media as compared with normal fibroblasts. For further details see initial citation of ref.61.

Studies on human α-L-fucosidase have been performed on normal- and low-activity sera. After DEAE-Sephacel chromatography, two major forms of enzyme have been characterized in both types of sera. These enzymic forms were different with respect to their thermostability and electrofocusing behaviour.[108] The less acidic form (I) was thermolabile. In contrast, the more acidic one (II) was thermostable. But the forms I and II had respectively similar properties when isolated from either normal- or low-activity sera. These results suggest the presence in both types of sera and two major forms of α-L-fucosidase but in variable proportions. The low-activity sera contained a major proportion of the thermolabile less acidic form when compared with normal-activity sera.

The steric factors involved in the action of glycosidases and D-galactose oxidase have been investigated. α-(1 → 2)-L-Fucosidase, β-D-galactosidase, and D-galactose oxidase are sterically hindered by certain types of branching in the oligosaccharide chains.[109] β-D-Galactosidase will not cleave the D-galactosyl link when the penultimate sugar carries a sialic acid residue as is (1). D-Galactose oxidase will not oxidize the D-galactose residue in trisaccharide (1) but will in (2). Moreover, neither D-galactose nor 2-acetamido-2-deoxy-D-galactose, glycosidically bound as in (3), is susceptible to oxidation with D-galactose oxidase until the (1 → 3) linkage between them is cleaved by α-D-2-acetamido-2-

deoxygalactosidase. α-(1→2)-L-Fucosidase action is inhibited by a (1→3) 2-acetamido-2-deoxy-α-D-galactosyl or α-D-galactosyl residue, as in (3) and (4). Removal of the terminal sugars makes the α-L-fucosyl residue susceptible to α-L-fucosidase action.

α-NeupNGl
2
|
6
β-D-Gal-(1→3)-D-GalNAcol
(1)

α-L-Fucp
1
|
2
β-D-Gal-(1→3)-GalNAcol
(2)

α-D-GalNAc-(1→3)-β-D-Gal-(1→3)-D-GalNAcol
2
|
1
α-L-Fuc
(3)

α-D-Gal-(1→3)-β-D-Gal-(1→3)-D-GlcNAc
2
|
1
α-L-Fuc
(4)

The activities of various glycosidases in homogenates of the small intestinal mucosa of two adult and 18 suckling tammar wallabies (*M. eugenii*) aged from 6 to 50 weeks have been investigated.[38] β-D-Galactosidase, β-D-2-acetamido-2-deoxyglucosidase, α-L-fucosidase, and neuraminidase activities were high during the first 34 weeks *post partum* and then declined to very low levels. For further details see initial citation of ref.38.

The α-L-fucosidase, β-D-2-acetamido-2-deoxyglucosidase, α-D-mannosidase, β-D-glucosidase, β-D-xylosidase, β-D-glucosidase, and α-L-arabinofuranosidase activities of normal and atrophic skeletal muscle of developing and adult rats have been collectively investigated.[36]

Several mollusc glycosidases have been studied for their activities towards natural substrates. α-L-Fucosidases from Chamelea gallina, Tapes rhomnoideus, and Mytilus edulis hydrolyse oligosaccharides (di-, tri-, and penta-saccharides) with 1 → 2, 1 → 3, and 1 → 4 bonds, L-fucose-containing glycopeptides from bovine thyroglobulin and the porcine submandibular mucin (devoid of sialic acid). α-L-Fucosidase from Littorina littorea hydrolyses L-fucose-containing glycopeptides from bovine thyroglobulin.[70] For further details see initial citation of ref.70.

A novel synthetic chromogenic substrate, α-L-Fucp-(1 → 2)-β-D-Galp-(1 → $OC_6H_4NO_2$ - p), has been used for the rapid assay of α-L-fucosidases which hydrolyse the glycosidic linkage, α-L-Fucp-(1 → 2)-D-Gal.[110] The procedure is based on the sequential action of α-L-fucosidase and an exogenously added exo-β-D-galactosidase to release the easily measurable 4-nitrophenol moiety. This method detected α-L-fucosidases from Aspergillus niger, Clostridium perfringens, and almond, which are known to hydrolyse the (1 → 2) linkage specifically. The advantages of this procedure over procedures previously used for the detection of substrate-specific α-L-fucosidases are discussed.

Six glycoside hydrolases in the culture medium of Bacteroides fragilis - α-D-glucosidase, β-D-glucosidase, α-D-galactosidase, β-D-galactosidase, β-D-2-acetamido-2-deoxyglucosidase, and α-L-fucosidase - were systematically purified by ammonium sulphate precipitation, gel-filtration chromatography, and density-gradient isoelectric focusing.[49] The isoelectric focusing resolved the glycosidases into distinct, well separated fractions and revealed three differently charged forms of β-D-2-acetamido-2-deoxyglucosidase and α-L-fucosidase. For further details see initial citation of ref.49.

7 α- and β-D-Galactosidases and D-Galactolipid-oriented α- and β-D-Galactosidases

α-D-Galactosidase deficiency has been demonstrated in lymphoid cell lines established by Epstein-Barr virus transformation of B-lymphocytes from a Fabry patient, as in blood whole leukocytes and lymphocytes.[21] In normal-blood lymphocytes and lymphoid cell lines, α-D-galactosidase was separated by electrofocusing into three molecular forms: form I (pI 5.0), form II (pI 4.5), and form III (pI 4.3). For further details see initial citation of ref.21.

The conditions for maximal activity (pH buffer, saturating substrate concentration, and range of linear relationships between enzyme activity *versus* incubation time and *versus* enzyme concentration) in the fluorimetric assay of several glycohydrolases of lysosomal origin in human plasma and serum have been established.[26] The following enzymes were studied: α-D-galactosidase, β-D-galactosidase, β-D-glucosidase, β-D-glucuronidase, α-D-mannosidase, and α-L-fucosidase. All examined enzymes turned out to be more or less unstable upon storage at 37°C, 4°C, and -20°C in both serum and plasma. For further details see original citation of ref.26.

A simple and sensitive fluorimetric method has been described for the differential determination of the activity of lysosomal α-D-galactosidase A and α-D-galactosidase B.[111] The procedure employs 4-methylumbelliferyl α-D-galactopyranoside as substrate and 2-acetamido-2-deoxy-D-galactose as an inhibitor of α-D-galactosidase B, but not α-D-galactosidase A, to differentiate the two activities. This method was shown to be applicable in the differentiation of the two enzyme activities in human tissues and in the diagnosis of the heterozygous and hemizygous genotypes for Fabry's disease in cultured skin fibroblasts.

Two isoenzymic forms (I and II) of α-D-galactosidase have been isolated and purified for lysosomes of *Mytilus edulis* hepatopancreas.[112] The purification involved extraction, ammonium sulphate fractionation, ion-exchange chromatography, and gel filtration. The apparent molecular weights determined by gel filtration were estimated to be 70,000 and 50,000, respectively. Both forms showed different pH optimum, K_m, and V_{max} values with 4-nitrophenyl α-D-galactoside as substrate and differed in their thermal stabilities. The two forms were activated by Na^+ and K^+, while Hg^{2+}, Cu^{2+}, and Zn^{2+} all inhibited both forms. α-D-Galactosidase I and II exhibited activity on raffinose and melibiose.

Sucrose density gradients have been used to characterize lysosome-containing fractions from zero-hour white prepupae of *Stomoxys calcitrans*.[42] The acid glycosidases (α-D-glucosidase, α-D-galactosidase, α-D-mannosidase, β-D-glucosidase, β-D-galactosidase, β-D-glucuronidase, and β-D-2-acetamido-2-deoxyglucosidase) equilibrate at the same density as did acid phosphatase.

Nine glycosidases in bloodstream forms of *Trypanosoma brucei* have been partially characterized.[20] α-D-Glucosidase had similar physicochemical and enzymic properties to those of α-D-galactosidase, β-D-glucosidase, β-D-2-acetamido-2-deoxyglucosidase, and β-D-2-acetamido-2-deoxygalactosidase. For further details see initial citation of ref.20.

The α-D-galactosidase of *Poterioochromonas malhamensis* has been partially characterized.[113] In cell-free extracts of *P. malhamensis* α-D-galactosidase was stable at pH 8. The pH optimum of the enzyme was 7. The enzyme was purified 10-fold through chromatographic steps involving DEAE-cellulose, hydroxyapatite, and Sephadex G-200. The apparent molecular weight was 360,000 by sucrose density-gradient centrifugation. All activity was lost on subjecting the enzyme to polyacrylamide gel electrophoresis. The substrate specificity of the enzyme was examined and some kinetic values were determined. The enzyme displayed an unusual curve with respect to isofluoridoside.

Mung-bean seeds possess a tetrameric D-galactose-binding protein that displays two types of activities: (a) a haemagglutinin activity and (b) an α-D-galactosidase activity.[114] This protein can be reversibly converted by pH changes from a tetrameric form, which possesses both enzymic and phytohaemagglutinin activities, to a monomeric form, which possesses enzymic activity only. This suggested that the enzymic phytohaemagglutinin is an aggregated form of monomeric α-D-galactosidase. The tetrameric α-D-galactosidase has a pH optimum of about pH 7.0, while the monomeric form displays a pH optimum of 5.6. Circular-dichroism difference spectra and inhibition studies suggest that aggregation induces conformation changes in the subunits sufficient to alter their enzymatic properties. The possibility of *in vivo* changes in subunit equilibria, when combined with the accompanying alterations in activity, provides a new concept worthy of consideration with respect to the physiological role of phytohaemagglutinins.

The purification and properties of α-D-galactosidase from immature stalks of sugar cane (*Saccharum offinarum*) have been described.[115] The α-D-galactosidase was highly purified by ammonium sulphate fractionation, gel filtration on Sephadex G-100, ion-exchange chromatography on DEAE-cellulose and CM-cellulose, and heat treatment (60°C, 15 min) in the presence of 0.2 M D-galactose. In polyacrylamide gel electrophoresis, the purified enzyme was homogeneous, having a molecular weight of approximately 46,000. In gel filtration, it was approximately 47,000. The activity was optimum at pH 4.5 and at 60°C. The purified enzyme hydrolysed 4-nitrophenyl α-D-galactopyranoside (K_m 0.83 mM, V_{max} 25.0 μmol mg^{-1} min^{-1}), raffinose (K_m 25.9mM, V_{max} 15.4 μmol mg^{-1} min^{-1}), and stachyose (K_m, 13.0 mM, V_{max} 2.7 μmol mg^{-1} min^{-1}), in addition to melibiose, guar gum, and locust-bean gum. The hydrolysis of 4-nitrophenyl α-D-galactopyranoside was markedly inhibited by $HgCl_2$, $AgNO_3$, 4-chloromercuribenzoate L-ascorbic acid, melibiose, stachyose, and D-galactose. Also the purified enzyme showed a lectin activity with trypsinized erythrocytes.

α-D-Galactosidase has been isolated from sweet-chestnut (Castanae sativa) seeds.[116] The major sugars of fresh seeds were shown to be raffinose, stachyose, and sucrose. Drying seeds at 25°C for 14 weeks increased the ratio raffinose:stachyose from 1.1 to 3.5, reduced sucrose content by ca. 50%, and decreased total extractable α-D-galactosidase. The enzyme activity was resolved into two peaks, a high-molecular-weight form I (215,000) and a low-molecular-weight form II (53,000). The latter form was predominant in the extract of fresh seeds whereas the former was the main form in the 14-week dried seeds. An increase in the amount of enzyme I was also observed when a buffered extract (pH 5.5) of fresh seeds was stored at 4°C. Enzymes I and II had pH optima of 4.5 and 6, respectively. Both enzymes hydrolysed 4-nitrophenyl α-D-galactopyranoside at a much greater rate than the natural substrates raffinose, stachyose, locust-bean gum, and carob gum. However, enzyme I showed preferences for stachyose as compared to raffinose. The opposite order was observed for enzyme II.

Two forms of α-D-galactosidase which differed in molecular weight have been resolved from Cucumis sativus leaves.[117] The enzymes were partially purified using ammonium sulphate fractionation, Sephadex gel filtration, and diethylaminoethyl-Sephadex chromatography. The molecular weights of the two forms, by gel filtration, were 50,000 and 25,000. The 50,000 form comprised approximately 84% of the total α-D-galactosidase activity in crude extracts from mature leaves and was purified 132-fold. The partially purified 25,000 molecular-weight form rapidly lost activity unless stabilized with 0.2% albumin and accounted for 16% of the total α-D-galactosidase activity in the crude extract. The smaller-molecular-weight form was not found in older leaves. The two forms were similar in several ways including their pH optima, which were 5.2 and 5.5 for the 50,000 and 25,000 dalton forms, respectively, and activation energies, which were 15.4 and 18.9 kilocalories per mole for the larger and smaller forms, respectively. Both enzymes were inhibited by D-galactose as well as by excess concentrations of 4-nitrophenyl α-D-galactoside substrate - K_m values with this substrate and with raffinose and melibiose were different for each substrate, but similar for both forms of the enzyme. With stachyose, K_m values were 10 and 30 mM for the 50,000 and 25,000 molecular-weight forms, respectively.

Red-light- and gibberellic acid-enhanced α-D-galactosidase activity has been investigated in germinating lettuce seeds. Dry lettuce (*Lattuca sativa*) seeds contain α-D-galactosidase at a level which is maintained in the imbibed dormant state in darkness.[118] Both red light and gibberellic acid promote an increase in enzyme activity several hours prior to the completion of germination. Germination and enzyme activity are not essentially linked, however, for the latter can increase while the former is inhibited. α-D-Galactosidase activity increases within the cotyledons and the endosperm following red-light stimulation, but the axis is essential to perceive the stimulus and to promote and maintain the increase in enzyme activity. A diffusible factor (or factors) is produced by and/or released from irradiated axes, and it migrates to the cotyledons (and possibly endosperm) to promote and increase in α-D-galactosidase activity. Gibberellic acid, particularly in the presence of benzyladenine, can replace the requirement for irradiated axes.

The immediate phytochrome induction of α-D-galactosidase has been demonstrated in lettuce seeds.[119] The light-induced breaking of dormancy of Grand Rapids lettuce seeds imbibed in darkness at 25°C is a classical phytochrome-controlled phenomenon. Although it has been shown that irradiation of seeds 1-2 h after the start of imbibition produces the active form of phytochrome (Pfr) which acts rapidly after its formation, the completion of germination (observed as radicle protusion) is only evident several hours later. Clearly then radicle emergence is temporarily removed from the rapid (and presumably primary) action of phytochrome. Very little is known about the processes set in motion by Pfr that lead to the completion of germination. Here the author reports that red-light-induced stimulation of α-D-galactosidase activity occurs substantially before germination is completed, and shows that the action of phytochrome in promoting an increase in the activity of this enzyme is very rapid.

Six glycoside hydrolases in the culture medium of Bacteroides fragilis - α-D-glucosidase, β-D-glucosidase, α-D-galactosidase, β-D-galactosidase, β-D-2-acetamido-2-deoxyglucosidase, and α-L-fucosidase - have been systematically purified by ammonium sulphate precipitation, gel-filtration chromatography, and density-gradient isoelectric focusing.[49] The isoelectric focusing resolved the glycosides into distinct, well separated fractions. For further details see initial citation of ref.49.

The effect of tunicamycin on secreted glycosidases of Aspergillus niger has been described.[50] The main glycosidases secreted into the culture medium during growth were β-D-glucosidase, α-D-galactosidase, and β-D-2-acetamido-2-deoxyglucosidase. In the presence of tunicamycin the activities of these enzymes in the culture medium were considerably decreased. For further details see initial citation of ref.50.

The specificity of induction of the synthesis of α-D-galactosidase from Saccharomyces carlsbergensis has been investigated.[120] Besides D-galactose and the substrates of the enzyme (melibiose, raffinose, and stachyose), D-galaturonic acid, L-arabinose, D-tagatose, methyl α-D-galactopyranoside, lactose, and isopropyl β-D-thiogalactopyranoside were able to act as inducers. Of these, methyl α-D-galactopyranoside, lactose, isopropyl β-D-thiogalactopyranoside, and L-arabinose have been shown to be gratuitous inducers with which kinetic studies of induction have been carried out. Lactose was the most efficient inducer, giving a maximal differential rate of synthesis of the enzyme of 110 mU 10^{-7} cells at a concentration of 180 mM, followed by L-arabinose (60 mU 10^{-7} cells at 40 mM), isopropyl β-D-thiogalactopyranoside (43 mU 10^{-7} cells at 60 mM), and methyl α-D-galactopyranoside (25 mU 10^{-7} cells at 150 mM). The concentrations of inducer required to obtain half-maximal induction were similar to those for lactose, L-arabinose, and isopropyl β-D-thiogalactopyranoside and about 5-fold higher for methyl α-D-galactopyranoside. The property of the compounds to act as inducers was compared to their ability to interact with the enzyme, and the results were discussed in terms of the molecular structures which are recognized by the enzyme and by the induction machinery.

A method has been described for following continuously the

action of β-D-galactosidase on 4-methylumbelliferyl β-D-galactopyranoside at pH 4.5, in which 4-methylumbelliferone production was measured at fluorescence excitation and emission wavelengths of 324 and 444 nm, respectively.[121] Initial-rate studies have shown that the presence of salt activates β-D-galactosidase up to 100 mM but is inhibitory above that concentration. The enzyme was very unstable at 37°C and low ionic strength, but stability increased with ionic strength. The stability of the enzyme at 37°C decreased markedly with rising pH in the range 5.9-8.0. The gel-filtration pattern showed decreasing amounts of dimer with increasing pH. The correlation between activity, stability, and molecular form of β-D-galactosidase was discussed. It was suggested that the dimeric form of the enzyme is the most stable and active form. The implications of this finding for the assay of β-D-galactosidase under physiological conditions for prenatal diagnosis were discussed. Evidence for the possible occurrence of a 36,000 molecular-weight form of β-D-galactosidase was presented. A computer program for the calculation of initial rates has been deposited as Supplementary Publication SUP 50114 at the British Library Lending Division, Boston Spa.

It has been reported that the ageing of human liver cell lines is accompanied by a decrease in the lysosomal β-D-galactosidase activity.[122] In six cell lines, both catalytic assay using 4-methylumbelliferyl β-D-galactopyranoside and immunochemical titration by a monospecific anti-β-D-galactosidase anti-serum were performed. The reduced catalytic activity was not associated with the presence of cross-reacting material and probably did not result from a modification of the biosynthesis, nor from post-translational modification of the enzyme.

A rapid and sensitive method has been devised for determining β-D-galactosidase activity specific for galactocerebroside.[123] A fluorescent derivative for galactocerebroside, 1-O-β-D-galactosyl-2-N-1-dimethylaminonaphthalene-5-sulphonyl-sphingosine, was used as substrate, and the product, 2-N-1-dimethylaminonaphthalene-5-sulphonyl-sphingosine, was taken into organic solvent phase. Quantitative analysis of 2-N-dimethylaminonaphthalene-5-sulphonyl-sphingosine was carried out fluorimetrically by use of high-performance liquid chromatography on silica gel column.

The conditions for maximal activity (pH, buffer, saturating substrate concentration, range of linear relationships between enzyme activity *versus* incubation time and *versus* enzyme concentration) in the fluorimetric assay of several glycoside hydrolases of lysosmal origin in human plasma and serum have been established.[26] The following enzymes were studied: α-D-galactosidase, β-D-galactosidase, β-D-2-acetamido-2-deoxyglucosidase, β-D-glucosidase, β-D-glucuronidase, α-D-mannosidase, α-L-fucosidase. All examined enzymes turned out to be more or less unstable upon storage at 37°C, 4°C, and -20°C in both serum and plasma. For further details see initial citation of ref.26.

It has been reported that the combined deficiency of β-D-galactosidase and neuraminidase in human fibroblasts can be corrected to nearly normal values.[124] This can be accomplished by addition of concentrated culture medium obtained after NH_4Cl stimulation of different types of human fibroblasts, including those with an isolated β-D-galactosidase or neuraminidase deficiency. The corrective factor is a macromolecular glycoprotein, which is labile at 60°C. Its uptake by human fibroblasts is competitively inhibited by D-mannose-6-phosphate and its corrective action within β-gal$^-$/neur$^-$ fibroblasts continues during a chase of 72 hours.

It has been shown that β-D-galactosidase activity was partially restored by protease inhibitors leupeptin, chymostatin, and E-64 in cultured fibroblasts from three patients with β-D-galactosidase neuraminidase deficiency.[125] Pepstatin did not activate this enzyme. Neuraminidase was not affected by any of these compounds in the culture medium. It was concluded that the activating effect was produced by a specific inhibition of thiol proteases.

Human fibroblasts with a genetic deficiency of a single lysosomal enzyme and fibroblasts from a patient with I-cell disease with a multiple deficiency of lysosomal hydrolases have been used as recipient cells in studies on recognition and uptake of β-D-2-acetamido-2-deoxyhexosidase, β-D-glucuronidase, and β-D-galactosidase.[58] Normal human fibroblasts and fibroblast hepatocytes and hepatoma cells from the rat were used as donor cells. The extracellular activities of β-D-glucuronidase and β-D-galactosidase were much higher in the rat cell cultures than in cultures of normal human fibroblasts. For further details see initial citation of ref.58.

Isoelectric focusing of the acid β-D-galactosidases in normal crude liver supernatant fluids has demonstrated multiple isoelectric forms in the pH range 4.58-5.15, while corresponding I-cell disease samples have shown an absence of isoelectric forms in the pH range

4.99-5.15.[126] Concanavalin A-Sepharose 4B chromatography of the I-cell disease mutant C.A. demonstrated a 31% and 37% decrease in the binding of 4-methylumbelliferyl β-D-galactosidase and G_{M1} β-D-galactosidase activities, respectively, when compared with normal samples. Isoelectric-focusing profiles on the concanavalin A-Sepharose 4B methyl α-D-mannopyranoside effluents containing normal and I-cell disease acid β-D-galactosidase were generally similar, but the unadsorbed I-cell disease enzyme from concanavalin A-Sepharose 4B demonstrated more activity in the pH range 4.21-4.49 than normals. Normal and I-cell disease acid β-D-galactosidase A and B, separated by gel-column chromatography, were found to have similar properties with respect to apparent molecular weights, pH *vs*. activity profiles, and apparent K_m values for the 4-methylumbelliferyl β-D-galactosidase, G_{M1}-ganglioside, and asialofetuin substrates. However, the apparent V_{max} values for the I-cell disease samples were consistently reduced when compared to the results obtained with the corresponding normal fractions. The greatest decreases in apparent V_{max} were obtained for acid β-D-galactosidase activities in I-cell disease crude supernatant fluids and for the separated I-cell disease B enzyme. The differences in the isoelectric-focusing profiles, the altered binding to concanavalin A-Sepharose 4B, and the reduced V_{max} values with natural and synthetic substrates may be related to changes in carbohydrate composition of I-cell disease acid β-D-galactosidase.

The wheat-germ lectin binding of β-D-galactosidase has been studied in human normal and sialidosis type-II fibroblasts.[127] To assess whether this residual activity is due to decreased synthesis of normal enzyme or normal synthesis of a mutant enzyme, its behaviour on wheat-germ lectin-Sepharose 4B columns has been studied. The total applied activity passed through the column unbound, but on subsequent washing and elution with β-D-2-acetamido-2-deoxyglucose more activity, corresponding to 800% of the applied activity, was eluted. The sum of the unbound and bound activities was the same as that which might have been expected for normal cells. The explanation given for this finding was that there was an inhibitor present in sialidosis type-II cells that was strongly bound to the enzyme and only dissociable under the relatively harsh conditions of the lectin binding and elution. The authors suggest that neuraminidase deficiency may be the primary defect in sialidosis type-II cells.

A deficient activity of neutral β-D-galactosidase in liver and cultivated fibroblasts has been observed in patients with lactosyl ceramidosis.[128] Neutral β-D-galactosidase was partially purified from livers of normal controls and patients with Niemann-Pick disease type A or lactosyl ceramidosis, using concanavalin A-Sepharose adsorption and filtration. The partially purified fractions were essentially free of β-D-galactosylceramide β-D-galactosidase and G_{M1} β-D-galactosidase activities. The normal and Niemann-Pick fractions were found to hydrolyse lactosyl ceramide, in the presence of sodium taurodeoxycholate, at a pH optimum of 5.6 as well as aryl β-D-galactosides and aryl β-D-glucosides at pH 6.2. The corresponding fraction from the lactosyl ceramidosis liver contained only 1-4% of the normal activity towards artificial substrates and lactosyl ceramide. Cross-reacting material identical to the normal was demonstrated in this fraction with antiserum raised against purified neutral β-D-galactosidase, but no activity was observed in the precipitin line when stained with naphthol AS-LC-β-D-galactoside or naphthol AS-LC-β-D-glucoside. A similar deficiency of neutral β-D-galactosidase activity was demonstrated in cultivated fibroblasts of the patient with lactosyl ceramidosis. Following adsorption on concanavalin A-Sepharose and anti-G_{M1} β-D-galactosidase antibody-Sepharose conjugates and chromatography on DEAE cellulose, fibroblast lysates from the patient exhibited 3% of normal activity towards 4-methylumbelliferyl β-D-glucoside at pH 6.2 and 12% of normal activity towards lactosyl ceramide at pH 5.6. These data suggested that neutral β-D-galactosidase may have an in vivo role in the cleavage of lactosyl ceramide and that a deficiency of this activity may be related to the lactosyl ceramide accumulation observed in the patient with lactosyl ceramidosis.

The amounts of β-D-galactosidase, sucrose α-D-glucohydrolase, α-D-glucosidase, microvillus aminopeptidase, and dipeptidyl peptidase IV in tangentially sectioned biopsies from jejunum have been studied by quantitative immunoelectrophoresis and enzymic assays.[129] All enzymes had their maximum activities near the mid-region of the villi and their lowest activities at the bases of the crypts. The ratio between enzyme activity and immunoreactive protein was constant along the villus-crypt axis. This result was consistent with a continuous brush-border-enzyme synthesis as the enterocytes migrate up the villi.

A study has been made of lysosomal-enzyme activities in serum and leukocytes in chronic hepatic disease.[60] Five lysosomal-enzyme activities (arylsulphatase A, α-D-mannosidase, α-L-fucosidase, β-D-

2-acetamido-2-deoxyhexosidase, and β-D-galactosidase) were determined in serum and leucocytes of 30 controls and 114 patients suffering from various liver diseases, including 30 with idiopathic haemochromatosis and 34 with alcoholic cirrhosis. For further details see initial citation of ref.60.

Degradation of mucin oligosaccharides of human colon systems has been found to be associated with extracellular, but not with cell-bound, β-D-galactosidase, β-D-2-acetamido-2-deoxyglucosidase, and neuraminidase.[29] Most probable numbers (MPN) estimates of faecal bacteria producing extracellular β-D-galactosidase, β-D-2-acetamido-2-deoxyglucosidase, and neuraminidase ranged from 10^6-10^{10} g^{-1} dry faecal weight. For further details see initial citation of ref.29.

Adjuvant-induced arthritis in rats has been studied by the changes in serum and urinary protein-bound carbohydrate metabolites, changes in serum and tissue lysosomal glycohydrolases, and lysosomal fragility.[30] The free activities of lysosomal glycohydrolases investigated, viz. β-D-glucuronidase, β-D-2-acetamido-2-deoxyglucosidase, β-D-galactosidase, α-D-mannosidase, and cathepsin, D, are increased in liver and spleen in the acute phase. For further details see initial citation of ref.30.

The activities of various glycosidases in homogenates of the small intestinal mucosa of two adult and 18 suckling tammar wallabies (M. eugenii) aged from 6 to 50 weeks have been investigated.[38] β-D-Galactosidase, β-D-2-acetamido-2-deoxyglucosidase, α-L-fucosidase, and neuraminidase activities were high during the first 34 weeks post partum and then declined to low levels. For further details see initial citation of ref.38.

The steric factors involved in the action of glycosidases and D-galactose oxidase have been investigated.[109] α-(1 → 2)-L-Fucosidase, β-D-galactosidase, and D-galactose oxidase are sterically hindered by certain types of branching in the oligosaccharide chains. For further details see initial citation of ref.109.

Acid β-D-galactosidase of porcine adrenocortical lysosomes, assayed for its activity towards 4-nitrophenyl β-D-galactopyranoside, has been shown to possess two activity peaks on gel-filtration profile at pH 7.4, one corresponding to a molecular weight of approximately 270,000 (termed form A) and the other about 65,000 (termed form A1).[130] Another form of acid β-D-galactosidase with a molecular weight of about 130,000 (termed form A2) was found when the high-speed extract or partially purified frm A1 was chromatographed on Sephadex G-150 at pH 4.5. In the presence of 0.1 M NaCl or saturating amounts of substrate at pH 4.5 the high-speed extract showed the aggregation of form A2 yielding form A3. Dissociation of form A3 back to form A1 was observed on incubation at 37°C in 0.02 M sodium phosphate buffer, pH 7.4, and that was followed by irreversible enzyme inactivation. Dissociation of form A3 into form A1 and enzyme inactivation in phosphate buffer, pH 7.4, were prevented by addition of 0.1 M NaCl. The interconvertible enzymic forms showed the same pH-activity profiles and Michaelis constants. These results suggest that the lysosomal acid β-D-galactosidase in the porcine adrenal cortex exists *in vivo* as the dimer, and that the dimer may further aggregate into the tetramer.

Fifteen glycosidases have been assayed in lymphomyeloid and digestive tissues of *Ginglymostoma cirratum*, *Heterodontus francisci*, and *Etmopterus spinax*.[19] Activities of α-D-mannosidase β-D-galactosidase, β-D-2-acetamido-2-deoxygalactosidase, β-D-glucuronidase, and α- and β-D-glucosidases usually were higher in lymphomyeloid tissues than in digestive tissues.

Sucrose density gradients have been used to characterize lysosome-containing fractions from 0 h white prepupae of *Stomoxys calcitrans*.[42] The acidic glycosidases (α-D-glucosidase, α-D-galactosidase, α-D-mannosidase, β-D-glucosidase, β-D-galactosidase, β-D-glucuronidase, and β-D-2-acetamido-2-deoxyglucosidase) equilibrate at the same density as does acid phosphatase.

Purification and properties have been reported of the two major isoenzymic forms of β-D-galactosidase from lysosomes of *Mytilus edulis* hepatopancreas.[131] Four isoenzymic forms (I, II, III, and IV) of β-D-galactosidase were separated and purified from lysosomes of *M. edulis* hepatopancreas to homogeneity, and the physical and kinetic properties of isoenzymes III and IV were ascertained. Forms III and IV exhibit different molecular weight, pH optima, K_m, and V_{max} values with 4-nitrophenyl β-D-galactopyranoside as substrate. Form III hydrolyses lactose and G_{M1}-ganglioside; form IV hydrolyses lactose. Both forms hydrolyse aryl β-D-galactosidase. The two forms were activated by Cl^-, while Hg^{2+}, Cu^{2+}, and Zn^{2+} all inhibit both forms.

Multiple β-D-galactosidase activities have been identified, separated, and characterized in *Petunia hydrida*.[132] Using isoelectrofocusing, multiple forms of *Petunia* β-D-galactosidase activity could be detected. The β-D-galactosidase pattern showed only minor tissue-specific differences. There were, however, species-specific differences. _Zea mays_, for instance, showed two bands which differed from the zones obtained with _Petunia_ preparations. _Petunia_ and corn leaves were mixed and extracted commonly. The species-specific activity patterns remained unchanged. _Petunia_ preparations were inactivated by 8 M urea. Following dialysis, enzymatic activity and the _Petunia_-specific pattern were restored. The same holds true for a mixture of _Petunia_ and _E. coli_ β-D-galactosidase preparations. On refocusing isolated _Petunia_ zones, untreated or inactivated by 8 M urea and reactivated by dialysis, the original mobilites were shown. Therefore, it seems highly improbable that the β-D-galactosidase pattern was due to artefacts. Using a _Petunia_ line which was 'pure', also in respect to its β-D-galactosidase pattern, the four main bands were preparatively separated by isoelectrofocusing and characterized. They showed the same pH optimum (4.3), the same temperature optimum (55°C), the same inactivation kinetics by urea, the same sensitivity against Cl^-, and closely related K_m values. In sucrose gradient centrifugation they invariably showed _S_ values of 8 - 10. The multiple activities could not be separated by zone electrophoresis using various carrier systems or by gel filtration. It seems possible that they represent forms which differ only in isoelectric points, not in molecular weight.

β-D-Galactosidase activity has been found to occur in all of the organs of the sugar-cane plant, and is also of general occurrence among different cultivars and species.[133] Most of the activity was associated with the cell wall, and only _ca_. 12 - 16% was an intracellular form. Both activities possess similar optimum pH and K_m, both are activated by Mn^{2+} and ethanol and inhibited by Hg^{2+}, and both attack the same substrates.

A process for production of mould β-D-galactosidase has been developed.[134] Tests were carried out in pilot and industrial scale with an _Aspergillus niger_ strain selected after screening a number of moulds. A computer-coupled autoanalyser system was used for monitoring enzyme formation in the pilot fermentor. β-D-Galactosidase production was investigated using different pH and temperature profiles. _A. niger_ lactose has an acid pH optimum, a high temperature optimum, and good stability. It does not require any metal ions. It is suitable for immobilization for hydrolysis of lactose in acid whey. Three-fold enhancement of β-D-galactosidase production was obtained by mutagenizing _A. niger_ using NTG as mutagenic agent. The β-D-galactosidases produced by the mutants have the same pH and temperature optima and stability but the growth properties of the mutants were different from those of the original strain. Sufficient specific activity of the enzyme preparation for immobilization was obtained by purifying the enzyme by selective adsorption on Na-Ca-silicate.

The secretion of lysosomal enzymes has been studied in the cellular slime mould Dictyostelium discoideum.[46] Using conditions which employ axanically grown amoebae in suspension in a simple starvation buffer, the lysosomal enzymes are secreted at rates that are at least as rapid as during normal development. Unlike normal growth or development, there is no appreciable lysosomal-enzyme synthesis under standard secretion conditions, so the enzyme activities act as markers for the vesicles that contain them. One group of enzymes, including β-D-2-acetamido-2-deoxyglucosidase, β-D-mannosidase, β-D-glucosidase, β-D-galactosidase-I, and β-D-glucosidase-I, are very efficiently secreted, with up to 50% of the total cellular activity becoming extracellular within a few hours. All of these enzymes have similar or identical secretion kinetics and may come from the same lysosomal vesicles. For further details see initial citation of ref.46.

An enzyme from Rhizobium melitoti has been shown to be a β-D-glucosidase which has β-D-galactosidase activity.[135] An earlier study had shown the existence of two enzymes, A and B, with β-D-galactosidase activity in R. meliloti. A mutant, lacking β-D-galactosidase A, allowed a study to be made of the role of enzyme B. It was demonstrated that enzyme B, whose synthesis was inducible by cellobiose, is able to hydrolyse 4-nitrophenyl β-D-glucopyranoside and 4-nitrophenyl β-D-galactopyranoside (K_m 0.135 and 6.4 mM, respectively). Enzyme B is actually a β-D-glucosidase with β-D-galactosidase activity.

Extracellular β-D-galactosidase of Fusarium moniliforme grown in whey has been characterized.[136] The enzyme, prepared as an ethanol precipitate, functioned optimally at pH 3.8 to 5.0 at 60°C on both 2-nitrophenyl β-D-galactopyranoside and lactose. The activation energy of the enzymic hydrolysis of the glycoside and lactose in the range of 20 to 55°C was 8,500 and 7,200 cal (ca. 3.57 x 10^4 and 3.02 x 10^4 J mol^{-1}), respectively. The K_m values were 4.4 and 12.4 mM for the glycoside and lactose, respectively. At optimum pH, the enzyme lost half of its activity when it was heated at 50°C for 6 h. At the same pH, the loss was only 5% when the enzyme was heated at 37°C for 6 h. At optimum conditions, 50% of the lactose in whey was hydrolysed by 10 units of this enzyme in 50 h.

The β-D-galactosidase of Alternaria tenuis has been purified to a homogeneous state from a filtrate of the culture fluid by precipitation with acetone, ion-exchange chromatography on DEAE-cellulose, sorption on hydroxyapatite, and affinity chromatography on N-β-D-galactopyranosylthiocarbamoyl-ε-aminohexanoyl-AN-Sepharose 4B.[137] Homogeneity was confirmed by ultracentrifugation and electrophoresis in polyacrylamide gel with and without SDS. The specific activity of the homogeneous enzyme was 160 units per mg protein. The molecular weight, determined by various methods, was 142,000-176,000, pI 4.6, temperature optimum 60-65°C, pH optimum 3.8-4.4 for 2-nitrophenyl β-D-galactopyranoside and 3.6-4.8 for lactose, K_m = 0.21 x 10^{-3} M for 2-nitrophenyl β-D-galactopyranoside and 6.57 x 10^{-3} M for lactose. The enzyme is a glycoprotein and contains up to 30% carbohydrates. H_4 edta and 4-chloromercuribenzoate have no effect on the β-D-galactosidase activity. D-Galactose is a competitive inhibitor. D-Glucose has no inhibiting effect on the enzyme.

Aspergillus niger mutant strains producing elevated levels of β-D-galactosidase have been isolated and characterized.[138] An Aspergillus niger mutant strain, VTT-D-80144, with an improvement of three- to four-fold in the production of extracellular β-D-galactosidase was isolated after mutagenesis. The production of β-D-galactosidase by this mutant was unaffected by fermenter size, and the enzyme was also suitable for immobilization.

The use of brushite ($CaHPO_4.2H_2O$) for enzyme immobilization has been described and exemplified with β-D-galactosidase.[139] Immobilization was carried out at 5°C in a 0.01 M Tris buffer, pH 7.6, containing 0.01 M phosphate. The immobilized enzyme showed high activity over a long period of time (after 90 days the activity was still 44% of the initial value). The leakage of enzyme from these matrices is very low, amounting to no more than 4% upon washing a columnful saturated with enzyme with 300 column volumes of buffer. A method is described in which the enzyme leakage is virtually negligible. One advantage of brushite as a matrix for enzyme immobilization is that the immobilization can be done in almost any buffer, even at high salt concentration, provided that high concentrations of phosphate are avoided. The immobilized enzymes can be desorbed easily by increasing the phosphate concentration. Since the enzyme is adsorbed at the top of the column, no preconcentration of the enzyme solution is required prior to immobilization.

β-D-Galactosidase has been immobilized in a hollow-fibre ultrafiltration module.[140] The hydrolysis of 2-nitrophenyl β-D-galactopyranoside was significantly affected by enzyme loading, flow rate, and substrate concentration. Pretreatment of hollow fibres with a protein was necessary to minimize enzyme inactivation. Residence-time distribution studies indicated that the product of the reaction was significantly adsorbed by the fibres, which resulted in the reactor taking 10-30 h to achieve steady state. An equation based on Michaelis-Menten kinetics and a plug-flow model adequately described the performance of the reactor with regard to operating variables, even though some diffusion effects were observed.

Novobiocin (0.05 μg ml^{-1}) has been found to reduce the growth rate of cultures of Escherichia coli strain DK6 by about a factor of 2.[141] The lag in appearance of β-D-galactosidase-forming capacity was extended from 50 s to 85 s by the drug. This appeared to be the result of a reduced rate of nascent mRNA elongation.

The synthesis of β-D-galactosidase in E. coli has been shown to be repressed as a result of infection with single-stranded DNA phage OX174.[142] An amber mutant in OX174 cistron A, which codes for two proteins, does not inhibit the enzyme synthesis, while amber mutants in all other genes do cause repression. A mutant near the amino-terminal end of cistron A, which produces the small 35,000 molecular-weight cistron A polypeptide, also inhibits the synthesis of β-D-galactosidase. Inhibition is also observed in an E. coli rep mutant which does not support the replication of replicative-form DNA. Exogenous nucleotide bases and adenosine 3',5'-phosphate do not have any effect on the degree of repression.

The interaction between the transport systems of methyl α-D-glucopyranoside and β-D-galactosidases has been studied in E. coli K12.[143] It was shown that the addition of methyl α-D-glucopyranoside to the cells leads to suppression of the rate of accumulation of {^{14}C}lactose and the hydrolysis of 2-nitrophenyl β-D-galactopyranoside. Mutation damage to one of the components of the system of methyl α-D-glucopyranoside transport coupled with phosphoenolpyruvate-dependent phosphorylation is accompanied by elimination of the inhibiting action of glucoside. Intracellular phosphate esters of methyl α-D-glucopyranoside or D-glucose lower the effectiveness of the transmembrane transport of β-D-galactosides. The results obtained suggest that suppression of the β-D-galactoside permease activity during the transport of methyl β-D-glucopyranoside may be a consequence of two processes: translation of methyl β-D-glucopyranoside intrinsically coupled with phosphorylation and accumulation of methyl β-D-glucopyranoside-6-phosphate within the cell. An intramembrane interaction (direct or indirect) between the enzyme II^{Glc} and β-D-galactoside permease is suggested.

Oxygen-18 leaving-group kinetic isotope effects (KIEs) have been determined on both V_{max} (*V*) and V_{max}/K_m (*V*/*K*) for the β-D-galactoside-catalysed hydrolysis of 4-nitrophenyl β-D-galactoside (I) and 2,4-dinitrophenyl β-D-galactoside (II).[144] The former substrate exhibits KIEs of 1.022 ± 0.002 and 1.014 ± 0.003 on *V* and *V*/*K*, respectively, while corresponding KIEs for the latter are 1.002 ± 0.009 and 1.030 ± 0.003. These results indicate that bond scission is largely rate determining for I but not for II at substrate saturation. The first irreversible step for both substrates must involve cleavage of the bond to the nitrophenyl leaving group. The mechanism proposed for this reaction is characterized by two parallel pathways for substrate hydrolysis. The predominant route for all but the most reactive substrates includes an S_N2 nucleophilic displacement of aglycon by the enzyme to yield a covalent D-galactosyl enzyme, which in turn is hydrolysed *via* a nucleophilic attack by water. The most reactive substrates (*e.g.* II) form transiently an enzyme-bound D-galactosyl oxocarbonium ion which partitions between enzyme to give the covalent D-galactosyl enzyme and H_2O to yield D-galactose.

β-D-Galactosidase has been found to act on α-lactose slightly more than twice as rapidly as on β-lactose for both the hydrolysis and transgalactosylis reactions.[145] The effect was shown to be on the V_{max} values; the K_m values for the different anomeric forms were the same. The step of the reaction for which the enzyme has anomeric specificity was shown to be glycosidic bond breakage. The steps in D-glucose release or in the D-glucose acceptor reaction were not affected by anomeric composition. Neither allolactose hydrolysis nor transport of lactose into the cells by *lac* permease was sensitive to the anomeric composition of the substrate. The implications of these results for *lac* operon induction and for lactose metabolism are discussed.

The positions of the *lacZX90* mutation and hybridization between complete and incomplete β-D-galactosidase have been investigated.[146]

The position of the termination codon in lacZX90 was determined by isolation of a lac^{+} revertant L-lysine which was found to replace L-tyrosine at position 1,012 of β-D-galactosidase, indicating that X90 protein lacked the carboxy-terminal 10 residues. A heat- and urea-sensitive hybrid enzyme was formed in vivo when supC, which supplies L-tyrosine to the position in the polypeptide corresponding to the nonsense codon, was used to suppress lacZX90. This result shows that suppression that adds back the original amino acid may not lead to the production of the wild-type enzyme if the latter is multimeric, because incomplete chains can be incorporated into the oligomer.

Mechanism of the rel defect has been identified in β-D-galactosidase synthesis.[147] Relaxed (relA) mutants of Escherichia coli are defective in β-D-galactosidase synthesis during amino acid limitation. This defect was found to comprise both a transcriptional component and a translational component.

An in vitro coupled transcription-translation system has been asked to synthesize transaminase B and β-D-galactosidase in the presence of a deoxyribonucleic acid template containing lac deoxyribonucleic acid under normal lac-specific control and in the presence of several deoxyribonucleic acid templates containing lac deoxyribonucleic acid fused to the ilvD gene.[148] Time-course experiments revealed that transcription of the lacZ gene from the fusion template required a longer time than did that initiated at the lac promoter. With a phage template containing an intact ilvE gene but lacking the normal ilv-specific promoter, synthesis of ilvE message was completed before synthesis of lacZ message. A phage template that contained the normal ilv-specific promoter but from which part of ilvE had been deleted also allowed formation of β-D-galactosidase. Three plasmids containing the ilv-lac fusion were also used as templates. Two plasmids that contained both an intact ilvE gene and the normal ilv-specific promoter required longer times for lacZ transcription but were more efficient templates than was a plasmid in which the ilv-lac fusion, the ilvE gene, and the contiguous non-specific ilvE promoter were inverted with respect to the normal ilv-specific promoter. β-D-Galactosidase synthesis was stimulated by guanosine 3'-pyrophosphate-5'-pyrophosphate with all templates tested except that in which the ilv-lac fusion has been inverted.

A plasmid was constructed that allows the selection in vivo of gene fusions between the Escherichia coli β-D-galactosidase gene and the yeast (Saccharomyces cerevisiae) URA3 gene.[149] A large yeast DNA fragment containing URA3 gene was placed upstream of an amino-terminally depleted version of the lacZ gene. The plasmid vehicle contains sequences that allow selection and maintenance of the plasmid in both yeast and E. coli. Selection for lac^+ in E. coli yielded numerous deletions that fused the lacZ gene to the URA3 gene and flanking yeast sequences, to the bacterial tetracycline-resistance gene from the parent plasmid pBR322, and to the yeast 2 µm plasmid DNA. Some of these fusion plasmids produced β-D-galactosidase activity under uracil regulation in yeast.

Overlapping sequences have been detected in β-D-galactosidase α-complementation.[150] Enzyme activity is restored to two defective β-D-galactosidase molecules (M15 protein lacking amino acid residues 11-41 and M112 protein lacking residues 23-31) by incubation with peptide residues 3-92 of β-D-galactosidase. M15 and M112 proteins (α-acceptors) are dimers. Complemented enzyme, like wild type, has a tetrameric structure. Cleavage of the same peptide with L-glutamic acid-specific protease yielded a much smaller α-donor (3-41 peptide) which was also effective in complementation, indicating that the M15 protein can supply all of the residues from 42-92 for the structure of complemented enzyme. Treatment of M112 protein/3-41 peptide complemented enzyme with trypsin under very mild conditions followed by examination of the products demonstrated that α-donor peptide supplies the NH_2-terminal segment of complemented enzyme. Similar trypsin treatment of M15 protein/$CNBr_2$ indicated that in this complemented enzyme the polypeptide region beyond those residues missing in the α-acceptor can be provided by either the α-donor or the α-acceptor. Both M15 and M112 protein are more susceptible to mild tryptic proteolysis than complemented enzyme, indicating a more open structure. Several antipeptide antibodies that react with these two proteins do not react with β-D-galactosidase. M112 protein, like M15 protein, can be activated by anti-β-D-galactosidase but to a much higher level.

The effect of single amino acid substitutions in the β-D-galactosidase polypeptide chain has been studied.[151] Amino acid substitutions were made at four different positions, residues 17, 23, 36, and 41, in the β-D-galactosidase polypeptide chain. The effects on properties of the protein were studied. Substitutions of L-tyrosine for the normally occurring amino acids at these positions do not significantly affect enzyme activity, sensitivity in ureas, or ability to renature from urea. Enzymes, like the wild type, are tetrameric. β-D-Galactosidases containing L-tyrosine at 23, 36, or 41 are as heat stable as or more heat stable than wild type, but L-tyrosine, L-glutamine, or L-serine substitutions at position 17 make the enzyme less heat stable. The effects of amino acid substitutions on α-complementation were studied after isolating α-donor peptides (residues 3-92) from the substituted β-D-galactosidases. L-Tyrosine substitution at residues 23, 36, and 41 had little effect on α-donor activity or heat stability or complemented enzymes with other M15 protein or M112 protein as α-acceptor. In contrast, L-tyrosine substitution for the normally occurring L-glutamic acid at residue 17 leads to significantly decreased α-donor activity and decreased heat stability of complemented enzyme. Peptides with L-glutamine and L-serine substituted at residue 17 have nearly normal α-donor activities. The poor α-donor activities of the L-tyrosine 17 peptide can be attributed to its lower affinity for the α-acceptor proteins. Very little secondary structure was seen in $CNBr_2$ peptides by circular-dichroism measurements. Prediction of secondary structure by Chou and Fasman (Chou, P.Y. and Fasman, G.D. (1978) Adv. Enzymol., 47, 45-148) rules suggests that position 17 is between two β-bends. L-Tyrosine substitutions at residue 17 could cause a critical localized perturbation in structure before or after interaction with α-acceptor proteins.

Six hybridomas producing monoclonal antibodies Escherichia coli β-D-galactosidase have been derived from two separate somatic cell fusions.[152] Three of these antibodies can activate defective enzymes produced by strains of E. coli carrying Z-gene point mutations. In antigen excess, one monoclonal antibody shows similar enzyme-binding and mutant-activating capacity. Characteristically, the former reaction has a 200-fold higher equilibrium constant. These data provide direct evidence that the enzyme-activation reaction is a single-hit event in which one antibody site favours the correct conformation of one active centre of the enzyme. Because each activating hybridoma is able to activate several but not all point mutant enzymes tested, it appears that the correction of the genetic defect is produced by binding key sites of the protein three-dimensional structure rather than the sites affected by the mutation.

The mechanism by which large premature termination fragments of β-D-galactosidase are degraded by Escherichia coli has been studied using quantitative immunoprecipitation techniques.[153] Two different lacZ nonsense mutants which produced apparent primary translation products of 96,000 and 109,000 daltons, respectively, were shown to produce a second β-D-galactosidase-related polypeptide of M_r = 90,000. These 90,000 dalton polypeptides appeared to be the same in both strains since they co-migrated when analysed as a mixture of sodium dodecyl sulphate-polyacrylamide gels and were indistinguishable when analysed by one-dimensional peptide mapping. Pulse-chase experiments established a stoichiometric precursor-product relationship between the primary mutant gene products (called the A polypeptides) and the common 90,000 dalton polypeptide (called the B polypeptide). No intermediates were detected between the A and B polypeptides. The authors propose that there is a common pathway for the degradation of these different large fragments for β-D-galactosidase. According to this model, the first step would be a specific endoproteolytic cleavage of the primary translation product which produces the 90,000 dalton polypeptide as a common intermediate. The kinetic analysis demonstrated a first-order decay of both A and B polypeptides but, surprisingly, the first-order rate constant for the decay of A appeared dependent upon the induction regimen. This result suggested that degradation may possibly be autoregulated either by the intracellular level of A or by the other intermediates in the degradation pathway.

Active-site-directed irreversible inhibition of glycosidases by the corresponding glycosylmethyl-(4-nitrophenyl)triazenes has been investigated.[72] β-D-Galactopyranosylmethyl-(4-nitrophenyl)triazine is an active-site-directed irreversible inhibitor (ASDIN) of the ebg⁰ and the lacZ β-D-galactosidases of Escherichia coli of the β-D-galactosidase of human liver lysosomes, and, less effectively, of the β-D-galactosidase of green coffee beans. It has no effect on the lac repressor of E. coli. For further details see initial citation of ref.72.

The products of β-D-galactosidase action on lactose by intact E. coli cells have been found to appear as soon as lactose is added to the medium and the amount of product is equal to the lactose used.[154] No detectable levels of β-D-galactosidase were found in the medium, and lactose was not significantly broken down unless permease was present. The appearance did not depend upon the presence of any of the commonly known D-galactose or D-glucose permease systems. The K_m of product appearance from whole cells was equal to the K_t for lactose transport by lac permease. When the cells were broken the K_m became the normal β-D-galactosidase K_m.

The β-D-galactosidases of Medicago sativa cells have been investigated during suspension-cultured growth of lactose.[155] During the growth of suspension-cultured M. sativa cells, β-D-galactosidase activities were investigated, using the standard substrate 2-nitrophenyl-β-D-galactopyranoside, with respect to the lactose uptake and residual sugar levels in the cells and in the broth. The pH profiles showing two optima at pH 4 and pH 7 in crude extracts of cells strongly suggested the occurrence of two β-D-galactosidase activities. Only one optimum pH was detected in the broth at about pH 4.4. The β-D-galactosidase activity at pH 7 was found to be located inside the cells while that observed at pH 4 was bound to the cell wall. At the onset of the active growth phase an increase in the β-D-galactosidase activity occurred in the medium, probably due to a release of the cell-wall-linked enzyme. The separation of the intracellular β-D-galactosidases by gel filtration on a Sephadex G-200 column was achieved and the pH dependences of the enzymes were confirmed. The two enzymes were found to be efficient on lactose hydrolysis.

A new procedure for inducing and purifying endo-β-D-galactosidase from Escherichia freundii has been described.[156] The enzyme was found to be induced with high efficiency in culture medium containing Smith-degraded hog gastric mucin, which was prepared from a commercially available starting material. endo-β-D-Galactosidase was then purified by ammonium sulphate fractionation, DEAE-Sephadex chromatography, and affinity chromatography on Sepharose conjugated with the Smith-degraded mucin. The enzyme thus purified by only three steps showed no other glycosidase or protease activities and had higher specific activity compared to the previous method. This new method has a great advantage since the gastric mucin is abundantly available and the efficiency of enzyme production was high without significant induction of exoglycosidase. The hydrolysis of oligosaccharides, glycosphingolipid, and keratan sulphate was studied by using this newly purified enzyme. Kinetic data indicate that hydrolysability of these substrates is largely affected by substrate concentration, enzyme concentration, and the structure of substrates. Based on these results, the specificity of E. freundii endo-β-D-galactosidase was discussed.

β-D-Galactosidase activity in intact cells of 21 species of Streptomyces has been measured using 2-nitrophenyl β-D-galactopyranoside hydrolysis, without addition of a permeabilizing agent.[157] Differences in the induction efficiency of 2-nitrophenyl β-D-galactopyranoside hydrolytic activity by lactose or D-galactose, which could have taxomic implications, were observed among the species.

Carbohydrate fermentation of Streptococcus cremoris and S. lactis growing in agar gels has been investigated.[158] When lactic streptococci were embedded in agar gels and incubated at 30°C, the end products of carbohydrate fermentation depended on the initial cell density, which determined the subsequent distribution and size of colonies in the gel. With high initial cell densities, microcolonies formed close together and lactose and D-glucose were converted almost entirely to lactate. However, inoculation with a small number of cells, which then grew to form widely spaced and comparatively large colonies, resulted in up to 30% diversion of end-product, usually to formate, ethanol, and acetate. In these low-colony-density gel cultures, the initial rate of fermentation was exponential and only lactate was formed. However, this rate then became linear and fermentation became progressively more heterolactic. S. lactis ML_8 was the only strain among the 10 tested which remained homolactic. Incubation at temperatures either above or below the optimum for growth and metabolism decreased the diversion to end-products other than lactate. The change from homo- to hetero-lactic fermentation appears to be caused by carbohydrate depletion in the vicinity of the colony, so that fermentation is then limited by the diffusion of substrate. Growth of cells on gel surfaces exposed to air resulted in up to 40% diversion of end-product from lactate, mainly to CO_2, acetoin, 2,3-butanediol, and acetate. Six of the 12 S. cremoris strains tested remained homolactic under these aerobic conditions, whereas all 8 of the S. lactis strains tested, including ML_8, were heterolactic.

The general properties of β-D-galactosidase of Xanthomonas campestris have been reported.[159] The partially purified β-D-galactosidase required 32 to 37°C and pH 5.5 to 5.8 for optimum activity. The enzyme had low-affinity lactose hydrolysis (K_m = 22 mM) and was inhibited by thiol group reagents, H_4 edta, D-galactose, and D-galactal.

Six glycoside hydrolases in the culture medium of Bacteroides fragilis, α-D-glucosidase, β-D-glucosidase, α-D-galactosidase, β-D-galactosidase, β-D-2-acetamido-2-deoxyglucosidase, and α-L-fucosidase, were systematically purified by ammonium sulphate precipitation, gel-filtration chromatography, and density-gradient isoelectric focusing.[49] For further details see initial citation of ref.49.

An endo-β-D-galactosidase has been isolated in a highly purified form of a new organism, Flavobacterium keratolyticus.[160] The purification procedure included ammonium sulphate precipitation of the enzyme from the culture medium, Sephadex G-100 filtration, chromatography at the pI of the enzyme on a column which contained equal portions of CM-Sephadex C-50 and DEAE-Sephadex A-50, followed by Matrex gel blue A and DEAE-Sephadex A-50 chromatography. The final preparation showed one major but diffuse protein band on polyacrylamide gel electrophoresis. The molecular weight of this enzyme is about 30,000, and the optimal activity occurs at pH 6.0 using keratan sulphate as the substrate. Hg^{2+}, Ag^{+}, Cu^{2+}, and 4-chloromercuribenzoate are potent inhibitors for the enzyme, which is stabilized but not activated by Ca^{2+}. This enzyme hydrolyses endo-β-D-galactosyl linkages in keratan sulphates and glycoconjugates with N-acetyl-lactosamine repeating units and milk oligosaccharide. The specificity of this enzyme is similar to the one isolated from Escherichia freundii (Nakagawa, H., Yamada, T., Chien, J., Gardas, A., Kitamikado, M., Li, S.-C., and Li, Y.-T. (1980) J. Biol. Chem., 255, 5955-5959). The only difference is that the enzyme isolated from F. keratolyticus hydrolyses milk oligosaccharides faster than the one isolated from E. freundii and other organisms that produce endo-β-D-galactosidase; F. keratolyticus produces endo-β-D-galactosidase without induction by keratan sulphate. Thus this organism is the best source so far for the preparation of the endo-β-D-galactosidase capable of cleaving sugar chains with N-acetyl-lactosamine repeating units.

Spectrophotometric and fluorimetric assays of galactocerebrosidase activity have been used in the diagnosis of Krabbe's disease.[161] Derivatives of galactocerebroside were prepared containing coloured (ω-2,4,6-trinitrophenylaminolauric acid) or fluorescent (11-(9-anthroxy)undecanoic acid) fatty acid moieties. These cerebrosides were used as substrates for galactocerebrosidase activity. By overcoming problems associated with the radioactively labelled substrates normally used, yet retaining good enzyme-substrate specificity, these derivatives provided useful and reliable alternative substrates for galactocerebrosidase activity. Enzyme activities in whole extracts of brain, liver, fibroblasts, and cultured amniotic fluid cells were compared, using substrates for the novel cerebrosides as well as {^{3}H}galactocerebroside. Good correlation of activities was obtained. In extracts derived from patients with Krabbe's disease marked deficiency of galactocerebrosidase activity was observed as a partial reduction in enzyme activity. The results show that these coloured and fluorescent galactocerebrosides may be used with confidence in the diagnosis and carrier detection of Krabbe's disease.

8 α- and β-D-Glucosidases

The neutral isoenzymes of human α-D-glucosidase have been characterized by differences in substrate specificity, molecular weight, and electrophoretic mobility.[162] Neutral α-D-glucosidase C was precipitable in 40-60% ammonium sulphate, had a molecular weight of 92,000, an isoelectric point of 5.5, and released D-glucose from glycogen as well as from low-molecular-weight artificial and natural substrates which contained (1 → 4)-α-D-glucosidic linkages. Neutral α-D-glucosidase AB precipitated at 0-40% ammonium sulphate, bound to concanavalin A, had a molecular weight of greater than 150,000, and did not utilize (1 → 4)-linked α-D-glucose substrates larger than the disaccharide. Neutral α-D-glucosidase AB migrated more rapidly to the anode than α-D-glucosidase C when agarose, Cellogel, acrylamide, or starch was used as support medium. Both isoenzymes were equally inhibited by Zn^{2+}.

A study has been performed on α-D-glucosidase in four human colon malignant tumours developed into nude mice.[163] Both acid and neutral α-D-glucosidases were characterized in four human adenocarcinoma tumours obtained from cell lines of different glycogen content, by using either 4-methylumbelliferyl α-D-glucoside or glycogen and maltose as substrate. No obvious linear relationship between α-D-glucosidase activity and glycogen level could be found. However, the lowest and highest activities coincided with the lowest and highest glycogen contents, respectively. The hydrolytic activity of neutral α-D-glucosidase investigated in the presence of turanose, an inhibitor of the acid form, was found to be high towards 4-methylumbelliferyl-α-D-glucoside or maltose but weak towards glycogen. Kinetic parameters for both enzymes were found to be similar in the four tumours under investigation and close to what was reported in normal animal tissues.

The formation of monoclonal antibodies against human acid α-D-glucosidase has been reported.[164] Acid α-D-glucosidase purified from human placenta was used to immunize a mouse (strain Balb/cHeA) according to a procedure described earlier (Stahli, C., Staehlin, T., Miggiano, V., Schmidt, J., and Haring, P. (1980) J. Immunol. Methods, 32, 297-304). After fusion of spleen cells with myeloma cells, about 10% of the hybrid clones obtained produced antibodies against acid α-D-glucosidase. Finally, eight stable clones producing antibodies against the enzyme were obtained. When purified acid β-D-glucosidase was analysed by polyacrylamide gel electrophoresis in the presence of sodium dodecyl sulphate, two major protein bands (mol. wt. 76,000 and 70,000), a minor band of mol. wt. 96,000, and several minor bands with a mol. wt. of 67,000 or lower were seen. Since all these components react with the monoclonal antibodies, they must have at least one antigenic determinant in common.

The amounts of β-D-galactosidase, sucrose β-D-glucohydrolase, α-D-glucosidase, microvillus aminopeptidase, and dipeptidyl peptidase IV in tangentially sectioned biopsies fom jejunum have been studied by quantitative immunoelectrophoresis and enzymic assays.[129] For further details see initial citation of ref.129.

Apparent normal leukocyte α-D-glucosidase activity in glycogen storage disease type II (Pompe's Disease) has been reported.[165] An α-D-glucosidase determination in the peripheral leukocytes revealed normal activity. However, α-D-glucosidase activity was completely absent in a pre-mortem skeletal muscle biopsy. Post-mortem studies showed α-D-glucosidase activity to be absent in all tissues examined, including cultured skin fibroblasts. Massive glycogen deposition corresponded to the localization of the enzymic deficiency, except in the brain, where glycogen content was within the normal range. The α-D-glucosidase activity in mixed peripheral leukocytes was due to an isoenzyme of acid maltase in the granulocyte series. Ante-natal diagnosis was accurate in a subsequent pregnancy, but discordance between enzyme activity in different cell lines in an individual with a genetic disease is a conceivable source of error in both pre-natal and post-natal diagnoses.

It has been found that reduced glutathione is not required for measurement of α-D-glucosidase in human seminal plasma.[166]

Properties of the molecular forms of α-D-glucosidase and β-D-glucocerebrosidase from normal human and Gaucher disease spleen have been reported.[167] Comparative studies of β-D-glucosidases and β-D-glucocerebrosidases from normal human and Gaucher disease spleen have allowed the identification of three enzymatic groups on the basis of their subcellular localization (soluble or membrane-bound), enzymatic, and genetic properties. β-D-Glucocerebrosidase cleaves both the natural substrates, β-D-glucocerebrosidases, and the artificial substrate, methylumbelliferyl(MeUmb) β-D-glucopyranoside. A minor part (25%) is soluble in water whereas 75% of the total activity is membrane bound and can be solubilized by 0.25% Triton X-100. Using preparative electrofocusing, the soluble β-D-glucocerebrosidase shows two molecular forms (pI 5.2 ± 0.2 and 6.4 ± 0.2). Exactly the same forms are found for the membrane-bound β-D-glucocerebrosidases solubilized by Triton X-100. All these molecular forms display the same enzymatic properties: optimum pH of 5.4, heat stability at 42°C, K_m values of 3.7 ± 0.5 mM for the artificial substrate in the presence of sodium taurocholate and 0.06 ± 0.01 mM for the natural substrate. In addition they are activated to the same extent (600%) by sodium taurocholate. Furthermore, these two molecular forms are coded by the same gene since both are deficient in type I non-neuropathic Gaucher disease. The soluble, non-specific MeUmb-β-D-glucosidase catalyses the hydrolysis of the artificial substrate but not of the natural substrate and is focused in one single peak (pI 4.7 ± 0.2) different from the two forms of β-D-glucocerebrosidase discussed above. It is strongly inhibited by sodium taurocholate, weakly by Triton X-100, and is heat labile at 42°C. No deficiency is detected in the case of type I Gaucher disease reported here. It is thus suggested that this enzyme corresponds to a gene different from that coding for β-D-glucocerebrosidase. The membrane-bound, non-specific MeUmb-β-D-glucosidase also hydrolyses the artificial substrate but not the natural substrate. This enzyme differs from the soluble one with regard to its heat stability at 42°C and its strong inhibition by Triton X-100. On the other hand, the membrane-bound non-specific MeUmb-β-D-glucosidase differs from membrane-bound β-D-glucocerebrosidase in its heat stability at 50°C, its substrate specificity, and the sodium taurocholate effect. This enzyme activity is about normal in Gaucher disease and is thus coded by a gene different from the β-D-glucocerebrosidase one.

Significant α-D-glucosidase activity has been identified in human amniotic fluid during mid-pregnancy.[168] This enzyme is distinguishable from the lysosmal acid α-D-glucosidase, deficiency of which is associated with Pompe's disease. The two enzymes differ in optimum pH, thermal stability, electrophoretic migration, isoelectric point, molecular mass, and immunological response. Amniotic α-D-glucosidase is also different from the classical neutral form. Immuno-cross-reactions suggest that the amniotic fluid enzyme has a double foetal origin: renal and intestinal. It seems that α-D-glucosidase in amniotic fluid is linked to lipids.

A comparative study of α-D-glucosidase activities in rat and trout tissues has been performed.[169] All investigated tissues of rat and trout contained acid and neutral α-D-glucosidases which could be distinguished from one another and from the digestive α-D-glucosidase bound to the brush border of the intestine by their biochemical properties. In both species, the neutral α-D-glucosidase was found to be very active in liver and kidney and, to a lesser extent, in spleen, brain, and pancreas.

The skeletal muscle of cattle suffering from generalized glycogenesis type II has been shown to lack acid α-D-glucosidase activity.[170] Furthermore, there was no evidence of enzymically inactive proteins that cross-reacted with antibodies raised against acid α-D-glucosidase from the muscle of normal animals.

The activities of various glycosidases in homogenates of the small intestinal mucosa of two adult and 18 suckling tammar wallabies (M. eurenii) aged from 6 to 50 weeks have been investigated.[38] For further details see initial citation of ref.38.

Lysosomal α-D-glucosidase has been purified from a lysosome-enriched fraction of rat liver using an improved procedure.[171] A purification factor of 2900-fold was reached, with a yield of 35%. Polyacrylamide gel electrophoresis of the purified enzyme in non-denaturing conditions or in the presence of sodium dodecyl sulphate showed only one band. However, a microheterogeneity among enzyme subunits was detected by high-resolution two-dimensional electrophoresis.

Brush-border-membrane α-D-glucosidase has been purified from rat kidney by a procedure that included papain hydrolysis, ammonium sulphate fractionation, gel filtration, and affinity chromatography in which Tris, an inhibitor of α-D-glucosidase, served as the ligand.[172] The pure enzyme had a specific activity which represented a 1200-fold purification relative to the activity in the cortex homogenate. The enzyme was found to be a glycoprotein, containing 17.5% hexose. The native enzyme was an aggregate of four monomors or subunits, each with a molecular weight of 335,000 daltons, as judged by SDS-polyacrylamide gel electrophoresis. The K_m for α-D-glucosidase was 1.6 mM, and the K_i for Tris was 3.4mM.

Monospecific antibody produced in rabbits against the rat enzyme did not cross-react with dog kidney α-D-glucosidase. Unlike intestinal enzyme, the renal α-D-glucosidase was specific for maltose and did not hydrolyse starch or other disaccharides.

Further work on microsomal D-glucosidases of rat liver has confirmed that at least two enzymes are involved in the removal of D-glucose from the D-glucose-containing oligosaccharide.[173] One acted on the oligosaccharide containing three D-glucose residues and another on the oligosaccharide which has one or two D-glucoses. The D-glucosidase which acts on $(D\text{-Glc})_2(D\text{-Man})_9(D\text{-GlcNAc})_2$ could be purified with a concanavalin A-Sepharose column followed by electrofocusing. The purified preparation was active on the oligosaccharide containing one or two D-glucose residues. Heat inactivation and inhibition by disaccharides was parallel for both activities. Inhibition of the D-glucosidase active on $(D\text{-Glc})_3(D\text{-Man})_9(D\text{-GlcNAc})_2$ was obtained with kojibiose which has a (1 → 2) linkage, while the D-glucosidase acting on $(D\text{-Glc})_3(D\text{-Man})_9(D\text{-GlcNAc})_2$ was inhibited by nigerose {(1 → 3) linkage}, maltose {(1 → 4) linkage}, and D-glucose at a higher concentration. None of the β-anomers inhibited. These results were consistent with a configuration of the three D-glucoses of the dolichyl diphosphate-linked oligosaccharide. Kojibiose was found to inhibit D-glucosidase action not only on the free oligosaccharide but also on protein-bound one.

Fifteen glycosidases were assayed in lymphomyeloid and digestive tissues of *Ginglymostoma cirratum*, *Heterodontus francisci*, and *Etmopterus apinax*. Activities of α-D-mannosidase, β-D-galactosidase, β-D-2-acetamido-2-deoxygalactosidase, β-D-glucuronidase, and α- and β-D-glucosidase usually were higher in lymphomyeloid tissues than in digestive tissues.

Trehalose has been found to behave as a mixed inhibitor of the α-D-glucosidase activity of honey-bee haemolymph.[174] In samples with initially Michaelian kinetics, the maximum velocity V_m is decreased and the Michaelis constant K_m is increased by the same factor, equal to 1.7 in emerging bees and 1.3 in foraging bees. The dissociation constants K_i and K'_i, respectively, corresponding to trehalose binding to the enzyme and to the (enzyme-artificial substrate) complex, are equal to 62.5 mM and 168.7 mM in emerging bees and 303.9 and 754.5 mM in foraging bees. The ratio K'_i/K_i remains constant. The Hill coefficient for trehalose binding to the enzyme varies from 0.91 in emerging bees to 1.38 in foraging bees, which is consistent with the frequently non-Michaelian tendencies of this enzyme. The value of I_{50} rises from 98.3 mM to 358.3 mM. This increase, together with those of $\underline{K}_i$ and $\underline{K}'_i$, is correlated with the increase in trehalosemia from emerging to foraging bees in the samples studied.

Sucrose density gradients have been used to characterize lysosome-containing fractions from 0 h white prepupae of Stomoxys calcitrans.[42] The acidic glycosidases (α-D-glucosidase, α-D-galactosidase, α-D-mannosidase, β-D-glucosidase, β-D-galactosidase, β-D-glucuronidase, and β-D-2-acetamido-2-deoxyglucosidase) equilibrate at the same density as does acid phosphatase.

A comparison of α-D-glucosidase and α-D-mannosidase has been made in bloodstream forms of Trypanosoma brucei brucei.[175] The physicochemical and kinetic properties of the two major trypanosomal glycosidases, α-D-glucosidase and α-D-mannosidase, were compared in bloodstream forms of T. brucei brucei.[42] Both enzymes are membrane-bound and located intracellularly. The results are discussed in relation to the possible role of α-D-glucosidase and α-D-mannosidase in the processing or catabolism of trypanosomal glycoproteins.

Nine glycosidases in bloodstream forms of T. brucei brucei S42 have been partially characterized.[20] α-D-Glucosidase had similar physicochemical and enzymic properties to those of α-D-galactosidase, β-D-glucosidase, β-D-2-acetamido-2-deoxyglucosidase, and β-D-2-acetamido-2-deoxygalactosidase.

Purification and substrate specificity of sweet-corn α-D-glucosidase have been described.[176] An α-D-glucosidase was purified from sweet-corn seeds by fractionation with ammonium sulphate, chromatographies on Sepharose, and gel filtrations on Sephadex G-100. The enzyme was homogeneous in disc-electrophoretic analysis. The molecular weight was estimated to be about 96,000 by SDS-disc electrophoresis. The enzyme showed high activities toward maltose, nigerose, phenyl α-maltoside, and malto-oligosaccharides. The ratios of maximum velocity for maltose, nigerose, kojibiose, isomaltose, phenyl α-glucoside, phenyl α-maltoside, panose, turanose, and soluble starch were estimated to be 100 : 78 : 17 : 11 : 28 : 100 : 31 : 3.4 : 126, and the $\underline{K}_m$ values for these substrates 1.5 mM, 1.4 mM, 0.48 mM, 14 mM, 4.2 mM, 1.1 mM, 5.0 mM, 0.28 mM, and 52 mg ml^{-1}, respectively. The maximum velocity for soluble starch was high, but this α-D-glucan was not a favourable substrate because the $\underline{K}_m$ value was also very high. The $\underline{V}_{max}$ values for malto-oligosaccharides were somewhat dependent on the degree of polymerization (*n*). The K_m values for substrates having four or more D-glucose units increased with the increase in *n*.

Two forms of α-D-glucosidase have been isolated from rice seeds at the milky stage by a procedure that included fractionation with ammonium sulphate, column chromatography, and preparative disc gel electrophoresis.[177] The isolated α-D-glucosidases (α-D-glucosidase I and α-D-glucosidase II) were glycoproteins containing 5.5% and 3.4% carbohydrate, respectively. The molecular weight and isoelectric point were: 78,900 and pH 9.3 for α-D-glucosidase I, and 47,300 and pH 9.3 for α-D-glucosidase II. They were found to be homogeneous on polyacrylamide disc gel electrophoresis. The two enzymes hydrolysed maltose, maltotriose, phenyl α-maltoside, amylose, and soluble starch to liberate D-glucose but did not act on sucrose. The enzymes produced maltotriose and 4-α-D-nigerosyl-glucose from maltose as main α-D-glucosyl-transfer products, and they hydrolysed amylose to liberate α-D-glucose.

Three forms of α-D-glucosidase have been isolated from 5-day-old green gram (*Phaseolus vidissimus*) seedlings, by a procedure including fractionation with ammonium sulphate and polyethylene glycol 6000, column chromatography, preparative gel electrofocusing, and preparative disc gel electrophoresis.[178] The α-D-glucosidases isolated (α-D-glucosidase I, α-D-glucosidase II-1, and α-D-glucosidase II-2) were homogeneous on polyacrylamide disc gel electrophoresis. Their molecular weights were 145,000, 105,000, and 65,000, respectively. The three enzymes hydrolysed maltose, maltotriose, phenyl α-maltoside, amylose, and soluble-starch liberating D-glucose but did not act on sucrose. The enzymes hydrolysed phenyl α-maltoside into D-glucose and phenyl α-D-glucoside, and they hydrolysed amylose liberating α-D-glucose. Maltotriose was the main α-D-glucosyl-transfer product formed from maltose by the three α-D-glucosidases.

An α-D-galactosidase from tubers of *S. affinis* has been purified about 130-fold by ammonium sulphate fractionation, chromatography, and gel filtration.[179] The purified enzyme showed a single protein band on disc gel electrophoresis. The molecular weight of the enzyme was determined to be approximately 42,000 by gel filtration and 44,000 by SDS disc gel electrophoresis. The optimum reaction pH was 5.2. The enzyme hydrolysed raffinose more rapidly than planteose. The activation energy of raffinose and planteose by the enzyme was estimated to be 7.89 and 11.4 kcal

mol^{-1}, respectively. The enzyme activity was inhibited by various D-galactosidases and structural analogues of D-galactose. Besides hydrolytic activity, the enzyme also catalysed the transfer reaction of D-galactosyl residue from raffinose to methanol.

The secretion of lysosomal enzymes has been studied in the cellular slime mould Dictyostelium discoideum. One group of enzymes, including β-D-2-acetamido-2-deoxyglucosidase, α-D-mannosidase, β-D-glucosidase, β-D-galactosidase-1, and α-D-glucosidase-1, are very efficiently secreted, with 10 to 20% of the total cellular activity becoming extracellular within a few hours. All of these enzymes have similar or identical secretion kinetics and may come from the same lysosomal vesicles. For further details see initial citation of ref.46.

Active-site-directed irreversible inhibition of glycosidases by the corresponding glycosylmethyl-(4-nitrophenyl)triazines has been investigated.[72] β-D-Glucopyranosylmethyl-(4-nitrophenyl)triazine is an active-site-directed irreversible inhibitor (ASDIN) of both β-D-glucosidase and β-D-galactosidase activities of sweet-almond β-D-glucosidase B. It has no effect on yeast α-D-glucosidase, the lacZ β-D-galactosidase of E. coli, or glucoamylase from Aspergillus niger. For further details see initial citation of ref.72.

The induction of α-D-glucosidases in Streptomyces venezuelae has been investigated.[180] S. venezuelae contains intracellular α-D-glucosidases that are induced during growth on maltose, isomaltose, maltotriose, dextrin, starch, and other α-D-glucosides. Induction was prevented by rifampicin and inhibited by chloramphenicol or streptomycin, indicating that de novo synthesis of messenger ribonucleic acid and protein was required. D-Glucose and other readily utilizable sugars did not repress induction of α-D-glucosidase activity, whereas certain organic acids and amino acids effectively reduced enzyme synthesis. Extracts of mycelium grown in the presence of maltose as an inducer hydrolysed maltose and isomaltose rapidly. Sucrose and other α-D-glucosides were less suitable substrates, whereas trehalose and starch were not hydrolysed. No activity was observed with β-D-glucosides, α-D-galactosides, or methyl α-D-mannopyranoside.

The production of 4-nitrophenyl α-D-glucopyranoside-hydrolysing α-D-glucosidase by Bacillus cereus ATCC 7064 has been investigated.[181] Out of 17 strains of 14 different Bacillus species, B. cereus ATCC 7064 produced the highest levels of 4-nitrophenyl α-D-glucopyranoside-hydrolysing α-D-glucosidase.

Maximal enzyme yield was achieved after 6 h of cultivation, at an initial pH of 6.5 - 7.0, on a medium containing 2% soluble starch, 2% peptone, 0.3% yeast extract, 0.05% meat extract, 0.3% K_2HPO_4, and 0.1% KH_2PO_4. When grown on 2% starch, B. cereus synthesized α-D-glucosidase in the cytoplasm. At a starch concentration of less than 1%, however, the enzyme appeared in the culture broth. This accumulation was maximal at 0.4% starch when the extracellular enzyme amounted to 54% of the total enzyme produced.

Six glycoside hydrolases in the culture medium of Bacteroides fragilis - α-D-glucosidase, β-D-glucosidase, α-D-galactosidase, β-D-galactosidase, β-D-2-acetamido-2-deoxyglucosidase, and α-L-fucosidase - were systematically purified by ammonium sulphate precipitation, gel-filtration chromatography, and density-gradient isoelectric focusing, which resolved the glycosidases into distinct, well separated fractions.[49] α-D-Glucosidase was shown to possess dual affinity for the respective D-galactoside substrate. For further details see initial citation of ref.49.

A novel α-D-glucosidase produced by Bacillus amylolyticus has been reported.[182] Bacillus amylolyticus produces α-amylase, pullulanase, and α-D-glucosidase. By selection of carbon source in the growth medium, α-D-glucosidase was produced preferentially and with exclusion of the other two activities. The α-D-glucosidase was highly specific for maltose and to a lesser extent maltotriose but was inactive towards a range of other substrates including 4-nitrophenyl α-D-glucoside and isomaltose. Optima for activity were recorded at pH 7.0 and 40°C and the enzyme was insensitive to H_4 edta.

The isolation and certain properties of Saccharomyces cerevisiae α-D-glucosidase II have been described.[183] The enzyme was purified by ion-exchange chromatography on DEAE-Sephadex A-50 and isoelectric focusing. As a result, two forms of α-D-glucosidase were obtained, with pIs of 5.35 and 5.30. The isolated isoforms of the enzyme did not differ in specific activity, but their molecular weights, according to the data of SDS polyacrylamide gel electrophoresis, is equal to 60,000, while according to the data from gel filtration through Sephadex G-100 it is 55,000. Both forms exhibited electrophoretic polydispersion. It was shown that the isoforms of α-D-glucosidase are glycoproteins containing 1.5-2% 2-amino-2-deoxy-D-glucose and 5-8% (isoform A) and 2-3% (isoform B) neutral sugars. It was shown by paper chromatography and D-glucose oxidase reaction that D-glucose is the basic constituent of the neutral sugars in both isoforms of α-D-glucosidase.

A quantitative in situ assay of yeast α-D-glucosidase involving permeabilization of the cells by freezing and thawing has been described.[184] The assay was applied to different strains in different physiological states and was shown to give results comparable to those obtained with total cell homogenates. The primary advantage of the in situ assay was the possibility of analysing large numbers of samples from the same culture during a growth curve using a very reduced cell mass.

A new trisaccharide, 6^F-α-D-glucosylsucrose, has been synthesized by transglucosylation reaction of brewer's yeast α-D-glucosidase.[185] Enzymatic synthesis of oligosaccharide through the transglucosylation reaction of brewer's yeast α-glucosidase was attempted in the reaction system containing only sucrose as substrate. As the transglucosylation product, a new non-reducing trisaccharide was chromatographically isolated in crystalline form (monohydrate), giving m.p. 139 ~ 142°C and $\{\alpha\}_D^{22}$ +102° (in water). The trisaccharide was confirmed to be 6^F-α-D-glucosylsucrose, that is O-α-D-glucopyranosyl-(1 → 6)-O-β-D-fructofuranosyl-(2 → 1)-O-α-D-glucopyranoside, of which undeca-acetate gave m.p. 148°C and $\{\alpha\}_D^{22}$ +119° (in chloroform), and this sugar was named isomelezitose.

The coordinated regulation of riboflavin permease and α-D-glucosidase synthesis in the yeast has been reported.[186] Riboflavin permease of the yeast Pichia guilliermondii is a regulatable transport system. Its synthesis is induced by sucrose, maltose, methyl α-D-glucopyranoside, melezitose, and raffinose, but not D-glucose, trehalose, or cellobiose. The synthesis of riboflavin permease in the presence of sucrose or maltose is suppressed by cycloheximide, actinomycin D, and 8-hydroxyquinoline and thus is regulated at the level of transcription. Inducers of riboflavin permease are also capable of inducing α-D-glucosidase synthesis. Mutants were selected with a constitutive formation of riboflavin permease - the synthesis of α-D-glucosidase in such strains also occurs constitutively. Culturing of the yeasts in a medium with a high D-glucose content leads to a parallel decrease in riboflavin permease and α-D-glucosidase activity. The data obtained are evidence of co-ordinated regulation of riboflavin permease and α-D-glucosidase in P. guilliermondii. An insufficient or excess content of vitamin B_2 in the medium has no effect on the level of riboflavin permease in this yeast species.

The synthesis of α-D-glucoside in yeast cells depleted of intramitochondrial ATP has been investigated.[187] It was found that bongkrekic acid did not prevent the maltose-induced synthesis of α-D-glucosidase in derepressed cells of the wild-type and corresponding respiratory-deficient mutant of Saccharomyces cerevisiae. The results suggested that expression of nuclear genes specifying α-D-glucosidase and maltose catabolism in yeast is apparently not dependent on the proper function of mitochondrial adenine nucleotide translocase and does not even require the presence of normal levels of ATP in mitochondria.

To overcome the difficulties encountered when kinetic parameters of an enzyme are extrapolated to high initial substrate concentrations of an enzyme reactor, effective and fast methods to obtain significant data for technical applications have been developed based on on-line rate determinations.[188] Such extensive treatment has proved necessary for the following enzymes: alanine dehydrogenase, formate dehydrogenase, and α-D-glucosidase.

A thermostable activator of β-D-glucosidase in normal human saliva has been characterized.[189] The authors demonstrate the presence of a thermostable factor in normal human saliva which activates the β-D-glucosidase of saliva. The activator is specific for β-D-glucosidase. It is dialysable and susceptible to pronase digestion. Sephadex G-25 gel filtration indicates that the activator is of low molecular weight, and polyacrylamide gel electrophoresis shows the low-molecular-weight fraction to migrate as two closely moving protein bands. A major proportion of the activator does not bind to concanavalin A-Sepharose. The activator appears to act by binding to or modifying the salivary β-D-glucosidase in a time-dependent manner.

The conditions for maximal activity (pH, buffer, saturating substrate concentration, range of linear relationships between enzyme activity *versus* incubation time and *versus* enzyme concentration) in the fluorimetric assay of several glycohydrolases of lysosomal origin in human plasma and serum have been established.[26] The following enzymes were studied: α-D-galactosidase, β-D-galactosidase, β-D-glucosidase, β-D-glucuronidase, α-D-mannosidase, α-L-fucosidase. All examined enzymes turned out to be more or less unstable upon storage at 37°C, 4°C, and -20°C in both serum and plasma. For further details see initial citation of ref.26.

An improved fluorimetric leukocyte β-D-glucosidase assay has been developed for Gaucher's disease.[190] Three fluorimetric leukocyte β-D-glucosidase assays were compared for their ability to diagnose Gaucher's disease and identify carriers of the disorder: the acid β-D-glucosidase assay of Beutler and Kuhl (Beutler, E. and Kuhl, W. (1970) J. Lab. Clin. Med., 76, 747-755), a pH 5.5-sodium taurocholate-dependent assay, and a new procedure which employs conduritol B epoxide, an active-site-specific inhibitor of glucocerebrosidase. All three assays unambiguously identified patients wth Gaucher's disease. With regard to identifying carriers the bile salt-dependent assay of Peters et al. and the conduritol B epoxide-dependent procedure gave the greatest discrimination between the mean β-D-glucosidase values for the control and heterozygote samples when evaluated using Student's t test. The most reliable assay for the identification of the carrier state was the conduritol B epoxide-dependent procedure which can be expected to provide the fewest false negative results when classifying heterozygotes (5%). However, the fact that none of these methods completely separated control and heterozygote samples indicated that their use in screening programs will result in a significant number of incorrect assignments.

β-D-Glucosidase-stimulating proteins (co-β-D-glucosidase) have been isolated from bovine spleen by acidification of homogenized spleen, heat denaturation, and column chromatography.[191] Gel electrophoresis of the product revealed a trace of inert protein and two fast-moving bands, a major diffuse band and a minor, faster-moving band. The latter two bands could be eluted from the gel and shown to stimulate a D-glucosidase preparation from bovine spleen. They both stained with Stains All and Fast Green, but poorly with Coomassie Blue. The bands could also be visualized by ultraviolet scanning, and periodate-Schiff stain was positive for the major band. The M_r of the co-β-D-glucosidase was about 20,400 as measured with the gel-permeation column, but 4,900 as measured with a Sephacryl S-200 column containing guanidium chloride and roughly 6,200 as measured by SDS gel electrophoresis. A pI of 4.3-4.4 was indicated by isoelectric focusing. Neutral sugar was found to be present, but no sialic acid. The protein was destroyed by pronase, but not by lyophilization, N-ethylmaleimide, or alkaline phosphatase. Stimulation of the basal activity (1 nmol h^{-1} assayed with methylumbelliferyl D-glucopyranoside) was induced 50% when 0.15 μg ml^{-1} of coglucosidase was in the incubation. The activating protein raised the V values and lowered the K_m values when both D-glucosylceramide and the artificial substrate were used. In contrast, phosphatidyl serine raised both the *V* and the K_m for cerebroside hydrolysis. The activator protein was found to occur in the soluble part of spleen as well as in the mitochondrial and lysosomal fractions.

Phlorizin, labelled with tritium only in the D-glucose moiety, was used as substrate for the β-D-glucosidase present in brush-border membranes from hamster intestine in order to study, simultaneously, the kinetics of hydrolysis and the rate of the D-{^{3}H}glucose liberated by the enzyme.[192] The D-{^{3}H}glucose seems to experience the same hydrolase-related transport into the intestinal villi as the hexoses liberated from the common disaccharides by their respective hydrolases. The released D-{^{3}H}glucose accumulation rate is only partially inhibited by unlabelled D-glucose added to the medium either as the free sugar or as the precursors sucrose, lactose, or D-glucose 1-phosphate, and then only when these sugars are present at very high levels. Furthermore, D-glucose oxidase, added to the medium as a D-glucose scavenger, has no effect on the uptake rate of the phlorizin hydrolase-liberated sugar. These and other findings are presented as evidence that, under conditions where the Na^+-dependent D-glucose carrier is more than 97% inhibited by phlorizin, the D-glucose derived from the inhibitor, like the hexoses from disaccharides, has a kinetic advantage for transfer into the intestinal tissue.

The α-L-fucosidase, β-D-2-acetamido-2-deoxyglucosidase, α-D-mannosidase, β-D-glucuronidase, β-D-xylosidase, β-D-glucosidase, and α-L-arabinofuranosidase activities of normal and atrophic skeletal muscle of developing and adult rats have been collectively investigated.[36]

The immunohistochemical localization of β-D-glucosidases in lignin and isoflavone metabolism in chick-pea (*Cicer arietinum*) seedlings has been reported.[193] Coniferin-specific and isoflavone 7-β-D-glucoside-specific β-D-glucosidases have been localized in stem and root sections of chick-pea seedlings by the indirect immunofluorimetrical method. The coniferin-specific β-D-glucosidase has been found in the cell walls of the tracheary elements and of the endo-, epi-, and exo-dermis. All these tissues are known to contain either lignin or polymers, like subrin and cutin, which consist partially of phenylpropanoid elements. The localization of this β-D-glucosidase is therefore in agreement with its postulated relationship to the phenylpropanoid metabolism. The isoflavone 7-β-

D-glucoside-specific β-D-glucosidase, on the other hand, is predominantly located in the parenchymatic cortex cells, and obviously in the cytoplasm. These cells are known to contain the isoflavone formononetin, which has been shown to undergo turnover in chick-pea seedlings. The authors therefore assert that this β-D-glucosidase is involved in the metabolism of the 7-β-D-glucoside of this isoflavone.

The kinetic principles of the formation of D-glucose and cellobiose in the hydrolysis of microcrystalline cellulose under the action of cellulase complexes from eight different sources have been studied experimentally.[194] By successive addition of individual components of the cellulase complex {endo-(1 → 4)-β-D-glucanase and β-D-glucosidase} to the reaction system, the steps limiting the rate of enzymatic hydrolysis of microcrystalline cellulose were revealed. It was shown that in most of the cases studied the step determining the rate of formation of glucose with the participation of intermediate cellobiose is the action of β-D-glucosidase. Only in one case (complex from Aspergillus foetidus, enriched with β-D-glucosidase) is the rate of formation of D-glucose from microcrystalline cellulose limited by the action of the endo-(1 → 4)-β-D-glucanase of the cellulase complex. In accordance with the kinetic principles developed, it was shown that when an excess of β-D-glucosidase is added to the reaction system, the step limiting the rate of hydrolysis of microcrystalline cellulose under the action of all the cellulase complexes studied becomes the attack on the initial insoluble substrate by endo-(1 → 4)-β-D-glucanase. Under the same conditions, a linear correlation was found between the steady-state rate of formation of D-glucose from microcrystalline cellulose under the action of all the cellulase complexes studied, on the one hand, and the activity of endo-(1 → 4)-β-D-glucanase in these complexes, on the other. It was shown that the action of all the cellulase complexes studied (selected rather arbitrarily) is described by essentially the same kinetic principles, which is evidence of the same mechanisms of the hydrolysis of insoluble cellulose by cellulase preparations of different origin.

The substrate specificities of three cellulases and a β-D-glucosidase purified from Thermoascus aurantiacus have been examined.[195] All three cellulases partially degraded native cellulose. Cellulase I, but not cellulase II and cellulase III, readily hydrolysed the mixed (1 → 3)(1 → 6)-linked polysaccharides such as carboxymethylpachyman, yeast glucan, and laminarin. Both cellulase I and the β-D-glucosidase degraded xylan, and it was proposed that the xylanase activity was an inherent feature of these two enzymes; lichenin, a (1 → 3)(1 → 4)-β-D-glucan, was degraded by all three cellulases. Cellulase II cannot degrade carboxymethylcellulose, and with filter paper as substrate the end-product was cellobiose, which indicated that cellulase II was an exo-(1 → 4)-β-D-glucan cellobiosylhydrolase. Degradation of cellulose (filter paper) could be catalysed independently by each of the three cellulases. There was no synergistic effect between any of the cellulases, and cellobiose was the principal product of degradation. The mode of action of one cellulase (cellulase III) was examined by using reduced cellodextrins. The central linkages of the cellodextrins were the preferred points of cleavage, which, with the rapid decrease in viscosity of carboxymethylcellulose, confirmed that cellulase III was an endocellulase. The rate of hydrolysis increased with chain length of the reduced cellodextrins, and these kinetic data indicated that the specificity region of cellulase III was five or six D-glucose units in length.

Optimization experiments with response surface statistical analysis have been performed with *Schizophyllum commune* to obtain high β-D-glucosidase yields.[196] The factors in the optimization experiment were the concentrations of cellulose, peptone, and KH_2PO_4. Their optimal values were 3.2, 3.0, and 0.2 g (100 ml)$^{-1}$, respectively. Enzyme assays revealed very high β-D-glucosidase (22.2 U ml^{-1}) and cellobiase (68.9 U ml^{-1}) yields. The avicelase yield was low as compared with that from *Trichoderma reesei*. Mixtures of *S. commune* and *T. reesei* culture filtrates caused faster and more extensive saccharification of Avicel than could be achieved by either filtrate alone. A β-D-glucosidase was isolated and purified from the optimized culture filtrate of *S. commune*. The electrophoretic mobility of the purified β-D-glucosidase indicated a molecular weight of 97,000. The amino acid composition was similar to that of β-D-glucosidase from *T. reesei*. The acidic (L-aspartate and L-glutamate) residues or their amides or both made up approximately 20% of the protein. The NH_2-terminal amino acid of the enzyme was L-histidine.

A new ultrasonic method has been developed for determining the composition and properties of individual components of polyenzyme systems without their preliminary separation.[197] The method is based on a determination of the pH dependence of the rate constants of inactivation of individual components of the enzyme system under the action of cavitational ultrasound. The method was used to study the cellulase complex from the fungus Geotrichum candidum. It was shown that this complex contains at least four cellulolytic enzymes - endo-(1 → 4)-β-D-glucanase, exo-(1 → 4)-β-D-glucanase, β-D-glucosidase, and aryl-β-D-glucosidase. These enzymes differ in values of pK of the ionogenic groups, controlling the pH profiles of ultrasonic inactivation, as well as in values of the rate constants of inactivation of the form of the active site most stable to the action of ultrasound.

The relationship between carbohydrate moiety and thermostability of β-D-glucosidase from Mucor meihei YH-10 has been investigated.[48] Two forms of thermostable β-D-glucosidase (a more thermostable G-a and less thermostable G-b) were purified to homogeneity from the culture filtrate. G-a and the G-b were glycoenzymes, and they contained 23.4% and 13.0% carbohydrate residues, respectively. For further information see initial citation of ref.48.

The components of the cellulolytic system of Talaromyces emersonii grown on cellulosic material have been identified.[198] The enzyme system was found to be comprised of four to five forms of exo-(1 → 4)-β-D-glucanase, at least two forms of endo-(1 → 4)-β-D-glucanase, and three enzymes exhibiting β-D-glucosidase activity. One of the latter, termed β-D-glucosidase III, is induced concurrently with the cellulases but disappears from the medium because of the low pH that develops during growth. The cellulases by contrast are more acid stable. Later in the growth cycle β-D-glucosidase I accumulates. An intracellular β-D-glucosidase, β-D-glucosidase IV, was also detected. The possible functions of these enzymes are discussed.

Purification and characterization of the extracellular and intracellular β-D-glucosidases of the termophyllic fungus Talaromyces emersonii have been described.[199] The organism produces three extracellular and one intracellular enzymes exhibiting β-D-glucosidase activity. Two of the extracellular forms, β-D-glucosidase I and β-D-glucosidase III, have been purified, as has the intracellular β-D-glucosidase IV. The pH and temperature optima, stability kinetic parameters, and substrate specificity of each have been determined. It was concluded that β-D-glucosidase I and β-D-glucosidase IV are true β-D-glucosidases while β-D-glucosidase III is an exo-(1 → 4)-β-D-glucanase.

An enzyme from Rhizobium melitoti has been shown to be a β-D-glucosidase activity.[135] An earlier study had shown the existence of two enzymes, A and B, with β-D-galactosidase activity in R. meliloti 2011 Str3. A mutant lacking β-D-galactosidase A allowed a study to be made of the role of enzyme B. It is demonstrated that enzyme B, whose synthesis was inducible by cellobiose, is able to hydrolyse 4-nitrophenyl β-D-glucopyranoside and 4-nitrophenyl β-D-galactopyranoside with, respectively, K_m values of 0.135 and 6.4 mM. Enzyme B is actually a β-D-glucosidase with β-D-galactosidase activity.

The thermal-stability characteristics of the cellulase enzymes present in culture filtrates of the thermophilic fungus Sporotrichum thermophile have been investigated at different temperatures and at different times of exposure.[200] Maximum enzymic activities under assay conditions were found at 68°C for the filter-paper activity (FPA) and the Cx activity (carboxymethylcellulose), while the maxima for the C_1 activity (cotton) and β-D-glucosidase activity (cellobiose) were found to be at 55°C and 72°C, respectively. Culture filtrates were exposed to a given constant temperature for varying lengths of time to a maximum of 48 hours and then analysed for residual enzymic activities under assay conditions. The exposure temperatures studied were 50°C, 60°C, and 65°C. After 48 hours' exposure time at 50°C the residual activities for the FPA, Cx, and β-D-glucosidase were found to be 88%, 98%, and 93% of the original activities, respectively.

Lysosomal enzymes have been found to possess a common antigenic determinant in the cellular slime mould Dictyostelium discoideum, and antisera have been prepared against two such lysosomal enzymes.[47] Two purified enzyme preparations used for immunization, β-D-2-acetamido-2-deoxyglucosidase and β-D-glucosidase-1, showed no cross-contamination with each other and no significant contamination by other lysosomal enzymes. For further details see initial citation of ref.47.

Separation and some of the properties of two β-D-glucosidases of Sporotrichum (Chrysoporium) thermophile have been described.[201] Intracellular, inducible β-D-glucosidase was fractionated by gel chromatography or isoelectric focusing into components A and B. Enzyme A (molecular weight 440,000) had only aryl β-D-glucosidase activity whereas enzyme B (molecular weight 40,000) hydrolysed several β-D-glucosides but had only low activity against 2-nitrophenyl β-D-glucopyranoside. Both enzymes had temperature optima of about 50°C. The pH optimum was 5.6 for enzyme A and 6.3 for enzyme B, respectively. The K_m 2-nitrophenyl β-D-glucopyranoside value for enzyme A was 0.5 mM, and the corresponding values for enzyme B were 0.18 mM (2-nitrophenyl β-D-glucopyranoside) and 0.28 mM (cellobiose). Enzyme B, when tested with 2-nitrophenyl β-D-glucopyranoside, showed substrate inhibition at a substrate concentration above 0.4 mM which could be released by cellobiitol and other alditols. Enzyme A was isoelectric at pH 4.48, and enzyme B was isoelectric at pH 4.64. Several inhibitors were tested for their action on the activity of enzymes A and B. Both enzymes were found to be concomitantly induced in cultures with either cellobiose or cellulose as carbon source.

Two forms of β-D-glucosidase from the culture filtrate of Macrophomina phaseolina have been separated and partially purified by $(NH_4)_2SO_4$ precipitation, ion-exchange chromatography, and gel filtration.[202] The final preparation was purified 103-fold and 88-fold for β-D-glucosidase I and β-D-glucosidase II, respectively. Polyacrylamide gel electrophoresis of the purified enzymes imparted a single band at pH 8.3. The two forms differ from each other with respect to molecular weight (323,600 for β-D-glucosidase I and 220,000 for β-D-glucosidase II), pH optima, temperature optima, electrophoretic mobility, and substrate specificity. The two forms of β-D-glucosidase may also be differentiated by inhibition experiments using as inhibitors D-glucono-1,4-lactone and nojirimycin. Nojirimycin is more potent for β-D-glucosidase I than for β-D-glucosidase II. The energy of activation for the two enzymes was also different.

The purification, characterization, and properties of β-D-glucosidase enzymes from Sclerotium rolfsii have been described.[203] Four β-D-glucosidase enzymes were extensively purified from the culture filtrates of S. rolfsii and some of their physicochemical properties studied. All the enzymes showed a single protein band in SDS gel electrophoresis and in disc gel electrophoresis at pH 8.9 and 4.3. The purified β-D-glucosidases were free from endo-D-glucanase (carboxymethylcellulose viscosity-lowering activity). All the enzymes are glycoproteins and are composed of one polypeptide chain. The molecular weight of the four β-D-glucosidases varies between 90,000 and 107,000. The pH and temperature optima of the four β-D-glucosidases are 4.2 and 68°C with cellulose as substrate. The isoelectric points for the enzymes are 4.10, 4.55, 5.10, and 5.55, respectively. The specific activities of the enzymes with cellobiose as substrate are 55, 78, 175, and 51 μmol D-glucose released per minute per milligram protein, respectively. The enzymes are inhibited by the reaction product D-glucose, and by D-glucono-1,4-lactone and nojirimycin. A carboxylate group is implicated in the catalysis of β-D-glucosidase.

The purification and properties of β-D-glucosidase from Humicola insolens have been described.[204] This thermophilic fungus exhibited high β-D-glucosidase activity when grown in solid wheat-bran medium. The β-D-glucosidase was purified from the culture extract by consecutive column chromatographies and found to be homogeneous on polyacrylamide gel disc electrophoresis. The molecular weight was estimated to be 250,000 by SDS gel electrophoresis, and the isoelectric point was at pH 4.23. The enzyme had an optimum pH of 5.0, an optimum temperature of 50°C, and showed significant resistance to urea, dimethyl sulphoxide, and ethyl alcohol.

The kinetics of the hydrolyses of cellotriose and of cellotetraose by β-D-glucosidase have been studied using a convenient integral technique.[205] Reaction mechanisms and mathematical models were postulated to describe the reactions. The end-products of the reaction were found to be inhibitory toward hydrolysis in a competitive mode. Hydrolysis of cellotetraose produces cellobiose and hydrolysis of cellotriose produced cellobiose and D-glucose. Both sugars inhibited the enzyme, with cellobiose being a stronger inhibitor.

Oxygen-18 leaving-group kinetic isotope effects (KIEs) have been measured for a set of glycosyl-transfer reactions with 4-nitrophenyl β-D-glycosides as substrates.[206] Acid-catalysed hydrolysis and alkaline hydrolysis exhibit KIEs of k_{16}/k_{18} = 1.0355 ± 0.0015 and 1.386 ± 0.0032, respectively. Lysozyme and β-D-glucosidase A show KIEs on V_{max}/K_m (V/K) of $(V/K)_{16}/(V/K)_{18}$ = 1.0467 ± 0.0015 and 1.0377 ± 0.0061, respectively. The large magnitude of these KIEs requires that carbon-oxygen bond scission be far advanced in the transition states for these reactions; therefore, in the transition states for the first irreversible steps in these reaction sequences, scission of the glycosidic bond must be essentially complete for the reactions catalysed by lysozyme and β-D-glucosidase A, which are thought to proceed *via* S_N1 and S_N2 mechanisms, respectively. Acid-catalysed hydrolysis is shown to proceed through a transition state involving at least 80% C-O bond cleavage and only partial proton transfer to the leaving 4-nitrophenyl oxygen atom.

2-Deoxy-D-glucopyranose has been used in the selective

isolation of mutants of Trichoderma reesei with enhanced β-D-glucosidase production.[89] When 2-deoxy-D-glucose was applied to T. reesei QM 9414 in a similar way it permitted the selective isolation in very high plate yields of mutants derepressed and hyperproductive with respect to β-D-glucosidase.

Extracellular cellulase activities of Clostridium thermocellum LQRI and Trichoderma reesei QM9414 have been compared.[207] The crude extracellular cellulase of C. thermocellum LQRI (virgin strain) was very active and solubilized microcrystalline cellulose at one-half the rate observed for the extracellular cellulase of T. reesei QM9414 (mutant strain). C. thermocellum cellulase activity differed considerably from that of T. reesei as follows: higher endo-D-glucanase/exo-D-glucanase activity ratio, absence of extracellular β-D-glucosidase or β-D-xylosidase activity, long-chain oligosaccharides instead of short-chain oligosaccharides as initial (15 min) hydrolytic products on microcrystalline cellulose, mainly cellobiose or xylobiose as long-term (24 hour) hydrolysis products of Avicel and MN300 or xylan, and high activity and stability at 60 to 70°C. Under optimized reaction conditions, the kinetic properties (V_{max}, 0.4 μmol min^{-1} mg^{-1} of protein, energy of activation 33 kJ, temperature coefficient 1.8) of C. thermocellum cellulose-solubilizing activity were comparable to those reported for T. reesei, except that the dyed Avicel concentration at half-maximal velocity was two-fold higher (182 μM). The cellulose-solubilizing activity of the two crude cellulases differed considerably in response to various enzyme inhibitors. Most notably, Ag^{2+} and Hg^{2+} effectively inhibited C. thermocellum but not T. reesei cellulase at 20 μM, whereas Ca^{2+}, Mg^{2+}, and Mn^{2+} inhibited T. reesei but not C. thermocellum cellulase at 10 mM. Both enzymes were inhibited by Cu^{2+} (20 mM), Zn^{2+} (1.0 mM), and ethylene glycol-bis-(β-aminoethyl ether)-N,N-tetra-acetic acid (10 mM). T. reesei but not C. thermocellum cellulose-solubilizing activity was 20% inhibited by D-glucose (73 mM) and cellobiose (29 mM). Both cellulases preferentially cleaved the internal glycosidic bonds of cello-oligosaccharides. The overall rates of cello-oligosaccharide degradation were higher for T. reesei than for C. thermocellum cellulase, except that the rates of conversion of cellohexaose to cellotriose were equivalent. The production of cellulases and hemicellulases has been studied with Trichoderma reesei Rut C-30.[208] This organism produced, together with high cellulase activities, considerable amounts of xylanases and β-D-glucosidase. Three cellulose concentrations (1.0, 2.5, and 5.0%) were examined to determine the maximum levels of cellulase activity obtainable in submerged culture. Temperature and pH profiling were used to increase cell mass to maximum levels within two days, thereby enhancing fermenter productivity at higher substrate levels. The effects of temperature, pH, Tween-80 concentration, carbon source, and substrate concentration on the ratio of mycelial growth and extracellular enzyme production were described.

A mutant strain with increased production of cellulolytic enzymes has been induced from the good cellulase producer *Trichoderma reesei* QM 9414.[209] Cellulase activities of the mutant in fermenter cultivations were increased two- to three-fold and β-D-glucosidase activity up to six-fold when compared to the corresponding activities produced by QM 9414.

Cellulolytic enzyme complexes obtained from mutants of *Trichoderma reesei* with enhanced cellulase production have been characterized.[210] The cellulolytic enzyme complexes secreted by *T. reesei* QM 9414 and its mutants M 5, M 6, MHC 15, and MHC 22 were characterized by determining their specific filter paper, carboxymethylcellulase, and β-D-glucosidase activities. They were characterized further by measuring their carboxymethylcellulase and β-D-glucosidase profiles after separation on an isoelectric-focusing column over the pH range 3-10. While the overall filter-paper activity was roughly equal in all preparations, the specific β-D-glucosidase activity was highest in mutants MHC 15 and MHC 22, which are distinguished morphologically from the parent strain, QM 9414, by a higher degree of branching of their hyphae. Two peaks of β-D-glucosidase activity were detected by isoelectric focusing in preparations from QM 9414 and M 6, none in the enzyme from the mutant M 5, while 3 and 4 peaks, respectively, were found in preparations from morphological mutants MHC 15 and MHC 22. The higher β-D-glucosidase activity in these last two preparations was also reflected in the higher D-glucose to cellobiose ratio in the initial stages of cellulose hydrolysis by the individual enzyme preparations.

Production of extracellular cellulase by an isolate of a *Fusarium* sp. has been studied in shake cultures, and sequential appearance of cellulase components (β-D-glucosidase on the first day, *endo*-(1 → 4)-β-D-glucanase on the second day, and *exo*-(1 → 4)-β-D-glucanase on the third day of growth on insoluble cellulose) was observed.[211] Maximum production of all these components was achieved on the fifth day. The Fusarium produced significantly higher β-D-glucosidase within a short period of time as compared with Trichoderma reesei. The influence of different nitrogen and carbon sources on cellulase has been investigated. Crude cellulolytic enzyme was used for hydrolysis of common agricultural wastes both with and without sodium hydroxide pretreatment. Analysis of hydrolysates indicated glucose as the major constituent (about 83% of total reducing sugar).

The regulation of the cellulolytic system in Trichoderma reesei by sophorose has been investigated through measurements of induction of cellulase and repression of β-D-glucosidase.[212] Sophorose has two regulatory roles in the production of cellulase enzymes in T. reesei: β-D-glucosidase repression and cellulase induction. Sophorose also is hydrolysed by the mycelial-associated β-D-glucosidase. Repression of β-D-glucosidase reduces sophorose hydrolysis and thus may increase cellulase induction.

The enzymatic hydrolysis of cellulose to D-glucose using immobilized β-D-glucosidase has been investigated, because cellulase systems frequently contain inefficient β-D-glucosidase to prevent the accumulation of inhibitory amounts of cellobiose.[213] This research has investigated the use of supplemental immobilized β-D-glucosidase to increase yields of D-glucose. Immobilized β-D-glucosidase from Aspergillus phoenicis was produced by soption on controlled-pore alumina with about 90% activity retention. The product lost only about 10% of the original activity during an on-stream reaction period of 500 hours with cellobiose as substrate. Maximum activity occurred near pH 3.5 and the apparent activation energy with Trichoderma reesei cellulase to hydrolyse cellulosic materials, such as Solka Floc, corn stover, and exploded wood. Increased yields of D-glucose and greater conversions of cellobiose to D-glucose were observed when the reaction systems contained supplemental immobilized β-D-glucosidase.

The inhibition of β-D-glucosidase in Trichoderma reesei C30 cellulase by D-glucose, its isomers, and derivatives has been studied since using cellobiose and 4-nitrophenyl β-D-glucopyranoside as substrates for determining enzyme activity.[214] The enzymatic hydrolysis of both substrates was inhibited competitively by D-glucose with approximate K_i values of 0.5 mM and 8.7 mM for cellobiose and 4-nitrophenyl β-D-glucopyranoside as substrate, respectively. This inhibition by D-glucose was maximal at pH 4.8, and no inhibition was observed at pH 6.5 and above. The α-anomer of

D-glucose inhibited β-D-glucosidase to a greater extent than did the β-form. Compared with D-glucose, L-glucose, D-glucose 6-phosphate, and D-glucose 1-phosphate inhibited the enzyme to a much lesser extent, unlike D-glucose-L-cysteine which was almost as inhibitory as D-glucose itself when cellobiose was used as substrate. D-Fructose (2-100 mM) was found to be a poor inhibitor of the enzyme. It was suggested that high rates of cellobiose hydrolysis catalysed by β-D-glucosidase may be prolonged by converting the reaction product D-glucose to D-fructose using a suitable preparation of D-glucose isomerase.

The hydrolysis of cellobiose to D-glucose by β-D-glucosidase has been studied with emphasis upon susceptibility to product and substrate inhibition.[215] This work reports a model which combines both product and substrate inhibition effects for β-D-glucosidase isolated from a commercial preparation of *Trichoderma viride*. An integrated rate equation was presented which predicted the trends of time courses for hydrolyses of cellobiose at concentrations ranging from 14.6-146.0 mM cellobiose. The constants used in the model (determined from initial rate data) were compared to those reported for β-D-glucosidase obtained from other sources of *T. viride*. Most notable in this comparison was the apparently higher activity and reduced inhibition of this enzyme compared to other sources of β-D-glucosidase.

The role of sugar hydroxy groups in glycoside hydrolysis has been investigated.[216] The contribution of the hydroxy groups at C-2 and C-4 and of the hydroxymethyl group at C-5 of β-D-glucopyranosides to their hydrolysis by β-D-glucosidase A_3 from *Aspergillus wentii* was investigated with 4-methylumbelliferyl-β-D-glucosides with appropriate structural modifications. Relative hydrolysis rates expressed by k_{cat}/k_{cat} (D-glucoside) were: 2-deoxy 4×10^{-6}, 2-deoxy-2-amino 2.4×10^{-4}, 2-deoxy-2-amino 1×10^{-6}, 4-deoxy 1.8×10^{-4}, xyloside 6.3×10^{-4}, galactoside 1×10^{-6}. Binding to the active site as measured by the K_m value of these substrates or by the K_i value of appropriate inhibitors was only moderately decreased by the above modifications. A temperature study with the 2-deoxy-D-glucoside showed that the decrease in k_{cat} was not due to a change in ΔH^* but to a more negative ΔS. The steady-state hydrolysis of the 2-deoxy-D-glucoside was approached with a burst (rate constant 0.13 min^{-1}) at pH 6 and 1 mM substrate. Deglycosylation of the enzyme was partially rate limiting. Rate constants for glycosylation and deglycosylation calculated from pre-steady-state kinetics were in good agreement with the constants calculated from experiments where the 2-deoxy-D-glucoside was used as an inhibitor for the hydrolysis of the D-glucoside and where a slow approach to the steady state of the inhibited reaction was observed.

A radioactive glycopeptide with a molecular weight of 13,200 has been isolated from β-D-glucosidase A_3 after labelling the active site with {^{3}H}conduritol B epoxide and cleavage with trypsin.[217] The glycopeptide consisted of 63 amino acids and 29 ± 1 sugar residues. Its amino acid sequence was derived from the results of sequence analysis of peptic and cyanogen bromide peptides. The radioactive inhibitor was bound to L-aspartic acid 12 of the sequence; the sugar residues were probably bound as N-glycosides to L-asparagine 48 and L-asparagine 56.

β-D-Glucosidase production by Aspergillus wentii has been investigated in stirred-tank bioreactors.[218] Batch cultivation of A. wentii was carried out on Mandels and Reese (1957) medium containing 3% cellulose and 0.3% peptone at 30°C in 14 and 30 litre bioreactors. Use of a 48-hours-old inoculum (10% v/v), 0.5 v.v.m. aeration at 400-800 r.p.m., and maintenance of dissolved oxygen and pH above 50% saturation level and between 5.2 to 4.0, respectively, produced 10.58 g cells l^{-1} and 61.9 IU e^{-1} h^{-1} cellobiase productivity.

β-D-Glucosidase production by Aspergillus phoenicis has been investigated in stirred-tank fermenters.[219] This article reports on the isolated production of β-D-glucosidase to provide a supplementary addition of the enzyme during the fermentation of cellulose to D-glucose by Trichoderma reesei.

The purification and properties of β-D-glucosidases of Aspergillus aculeatus have been described.[220] Three distinct β-D-glucosidases were isolated and purified. Their substrate specificities are described. Two of the purified enzymes appear to play an important role in producing D-glucose from cellulose, and it is suggested that these should be classified as a new type of exo-β-D-glucosidase.

The effect of tunicamycin on secreted glycosidases of Aspergillus niger has been described.[50] The main glycosidases secreted into the culture medium during growth of A. niger were β-D-glucosidase, α-D-galactosidase, and β-D-2-acetamido-2-deoxyglucosidase. In the presence of tunicamycin, the activities of these enzymes in the culture medium were considerably decreased, whereas fungal growth and total mycelial protein content were not diminished. For further details see initial citation of ref.50.

D-Glucosidase activities capable of removing the three D-glucose residues from D-Glc_3D-Man_9D-$GlcNAc_2$ oligosaccharide have been detected in a cell-free preparation of Saccharomyces cerevisiae X-2180.[221] The D-glucosidase which cleaves the D-glucose residue at the non-reducing terminus (D-Glc_3D-Man_9D-$GlcNAc_2$ oligosaccharide D-glucosidase) was equally distributed between the particulate and the supernatant fractions obtained after centrifugation of the yeast homogenate at 27,000 x g for 30 min. The membrane-bound activity was stimulated by Triton X-100, whereas the supernatant activity was not affected. The soluble D-Glc_3D-Man_9D-$GlcNAC_2$ oligosaccharide D-glucosidase was partially purified from the supernatant by ammonium sulphate fractionation followed by DEAE-Sephadex chromatography. It was clearly separated from α-D-glucosidase, which acts on 4-nitrophenyl α-D-glucopyranoside, but still contained β-D-glucosidase and α-D-mannosidase acting on 4-nitrophenyl β-D-glucopyranoside and α-D-mannopyranoside, respectively. The D-Glc_3D-Man_9D-$GlcNAc_2$ oligosaccharide glucosidase has a pH optimum of 6.8 and showed no requirement for divalent cations. The enzyme was very active with D-glucose-labelled D-Glc_3D-Man_9D-$GlcNAc_2$, was slightly active with D-Glc_2D-Man_9D-$GlcNAc_2$, and showed no activity with D-Glc_1D-Man_9D-$GlcNAc_2$. These properties suggest that this enzyme is involved in the first step of processing of oligosaccharides after transfer from dolichyl pyrophosphate to proteins.

The formation and location of (1 → 4)-β-D-glucanases and (1 → 4)-β-D-glucosidases have been studied in cultures of Penicillium janthinellum grown on Avicel, sodium carboxymethyl cellulose, cellobiose, D-glucose, D-mannose, and maltose.[222] endo-(1 → 4)-β-D-Glucanases were found to be cell free, and their formation was induced by cellobiose. (1 → 4)-β-D-Glucosidases, on the other hand, were formed constitutively and were primarily cell free, but with a small amount strongly associated with the cell wall. Low (1 → 4)-β-D-glucosidase activities of periplasmic or intracellular origin were also found. A rotational viscosimetric method was developed to measure the total endo-(1 → 4)-β-D-glucanase activity of the culture (broth and solids). By this method it was possible to determine the endo-(1 → 4)-β-D-glucanase activity not only in the supernatant of the culture but also on the surface of the mycelium or absorbed on residual Avicel. During a 70 litre batch cultivation of P. janthinellum, the adsorption of endo-(1 → 4)-β-D-glucanases by residual and newly added 10% Avicel was measured. The adsorption of soluble protein and endo-(1 → 4)-β-D-glucanases by Avicel was found to be largely independent of the pH value but dependent on temperature.

The importance of cellulase and xylanase release from Bacteroides succinogenes in the rumen environment has been investigated.[223] During growth of B. succinogenes in a liquid medium with cellulose as the source of carbohydrate, greater than 80% of the endo-(1 → 4)-β-D-glucanase, xylanase, and aryl-β-D-xylosidase and 50% of the aryl-β-D-glucosidase was released from cells into the culture fluid. Less than 25% of the β-D-xylosidase activity was detected in the culture fluid. Approximately 50% of each of the released enzymes measured was associated with sedimentable subcellular membrane vesicles. The vesicles appeared to be released from the outer membrane of intact cells by bleb formation, primarily in pockets between the cells and the cellulose, although a few unattached cells with blebs were seen. Many vesicles were seen adhering to cellulose, and they were also seen free in the culture fluid. These data suggest that B. succinogenes releases hydrolytic enzymes in non-sedimentable and particulate forms during growth by a mechanism which has until now received little attention. Cellulose incubated in a porous nylon bag in the rumen was colonized by bacteria resembling B. succinogenes, and subcellular vesicles were seen penetrating channels and fractures in the cellulose. It was suggested that B. succinogenes cells in the rumen contribute to an extracellular population of subcellular vesicles that possess cellulolytic and hemicellulolytic activities which probably enhance polymer digestion and provide a source of sugars for microbes lacking polymer-degrading activity, thereby contributing to a stable heterogeneous microbial population.

Polyacrylamide gel electrophoresis of the cellulolytic system from culture supernates of Acetivibrio cellulolyticus has shown the presence of four major enzymes: a β-D-glucosidase, an exo-D-glucanase, and two endo-D-glucanases.[224] The relative proportions of these enzymes in the culture supernate were affected by the nature of the cellulosic substrate and by the length of the incubation period. The molecular weights of the cellulolytic enzymes were β-D-glucosidase 81,000, exo-D-glucanase 38,000, endo-D-glucanase 38,000, endo-D-glucanase C2 33,000, and endo-D-glucanase C3 10,400, as estimated by their electrophoretic mobilities relative to proteins of known molecular weight. Treatment of the high-molecular-weight endo-D-glucanase with sodium dodecyl sulphate-mercaptoethanol led to reversible dissociation of the enzyme into polypeptide subunits similar to the low-molecular-weight *endo*-D-glucanase. *endo*-D-Glucanase activity could be assayed for directly using a novel method of incorporating carboxymethyl cellulose in the polyacrylamide gels. The molecular weight and functions of these enzymes are compared with those detected in culture filtrates of various fungi.

The production of carboxymethyl cellulase and β-D-glucosidase by *Clostridium acetobutylicum* has been investigated in an industrial fermentation medium.[225] In liquid medium the carboxymethyl cellulase was induced by molasses, and it was not repressed by D-glucose. Optimum carboxymethyl cellulase activity occurred at pH 4.6 and 37°C.

The mechanism of the action of β-D-glucosidase from *Botryodiplodia theobromae* has been investigated.[226] In the presence of a high concentration of 4-nitrophenyl β-D-glucopyranoside (donor) the rates of production of 4-nitrophenyl and a transglucosylation product (1-glyceryl β-D-glucopyranoside) increased, whereas the rate of production of D-glucose decreased with increasing concentration of glycerol in reactions catalysed by the high-molecular-weight β-D-glucosidase obtained from culture filtrates of *B. theobromae*. When {donor} was greater than K_m, the rate of production of 4-nitrophenol was higher in the presence of glycerol than in its absence, whereas when {donor} was less than K_m the rate of production of 4-nitrophenol was lower in the presence of glycerol than in its absence. Glycerol increased both K_m and V_{max}, whereas dioxan increased K_m but decreased V_{max}. Up to 1 mM $AgNO_3$ had no effect on enzyme activity. A ^2H solvent-isotope-effect $\{V_{max}\cdot(H_2O)/V_{max}\cdot(^2H_2O)\}$ value of 1.40 ± 0.05 was found at pH (or p^{2}H) 5.8. α-^{2}H kinetic isotope effect (k_H/k_{2H}) values of 1.03 ± 0.01 and 1.05 ± 0.01 were found in the absence and presence of glycerol, respectively. Although maltose was a non-competitive inhibitor of β-D-glucosidase activity, the ratio of velocity in the presence of glycerol to that in its absence increased, after an initial decline, with increasing concentration of maltose. A mechanism is suggested which involves a solvent-separated glycosyl cation-carboxylate ion pair, which has greater affinity for alcoholic D-glucosyl acceptors, and an intimate ion pair, which has greater affinity for water as a D-glucosyl acceptor and which could collapse reversibly and rapidly into a preponderance of an unreactive covalent D-glucosyl enzyme.

Trichoderma viride secretes a cellulase complex that is rich in β-D-glucosidase and therefore well suited for the saccharification of cellulosic materials.[227] The cellulase was investigated with respect to optimum conditions of reaction and enzyme stability. Avicelase, carboxymethylcellulase, and β-D-glucosidase differed considerably in their physicochemical properties. At higher temperatures the β-D-glucosidase was not very stable.

The effect of β-D-glucosidase on three assays for cellulolytic enzymes, *i.e.* the activities against dyed Avicel, hydroxyethylcellulose, and filter paper, has been studied using cellulase enzyme derived from *Trichoderma reesei* VTT-D-80133.[228] The dyed Avicel and hydroxyethylcellulose assays were only slightly affected by β-D-glucosidase, whereas the filter-paper assay was linearly dependent on the level of β-D-glucosidase over a wide range of activity of this enzyme.

9 β-D-Glucuronidases

Immobilized β-D-glucuronidase has been characterized in aqueous and mixed-solvent systems.[229] β-D-Glucuronidase was covalently attached to alkylamine-controlled pore glass *via* a glutaraldehyde immobilization scheme. The activity of this immobilized β-D-glucuronidase was studied with respect to several kinetic parameters in comparison with the behaviour of the soluble enzyme. K_m values for 4-nitrophenyl oestriol-3- and oestriol-16α-glucosides of D-glucopyranuronic acid were determined. For each substrate the K_m was essentially the same, 0.2 mM, and this value did not change when the enzyme was immobilized. The soluble and immobilized enzymes both displayed a relatively broad pH maximum centred at pH 6.8 for all substrates. Several organic-aqueous mixtures including methanol, ethanol, acetonitrile, and ethylene glycol were tested, and their effects on the activity of immobilized β-D-glucuronidase were similar to those found for the soluble enzyme. Long-term (1 year) storage stability tests of the immobilized enzyme were carried out. The immobilized enzyme retained 40% of its initial activity after 1 year and was very robust towards most of the organic solvents tested.

Human fibroblasts with a genetic deficiency of a single lysosomal enzyme and fibroblasts from a patient with I-cell disease with a multiple deficiency of lysosomal hydrolases have been used as recipient cells in studies on recognition and uptake of β-D-acetamido-2-deoxyhexosidase, β-D-glucuronidase, and β-D-galactosidase.[58] For further details see initial citation of ref.58.

The conditions for maximal activity (pH, buffer, saturating substrate concentration, range of linear relationships between enzyme activity *versus* incubation time and *versus* enzyme concentration) in the fluorimetric assay of several glycohydrolases of lysosomal origin in human plasma and serum have been established.[26] The following enzymes were studied: α-D-galactosidase, β-D-galactosidase, β-D-2-acetamido-2-deoxyglucosidase, β-D-glucosidase, β-D-glucuronidase, α-D-mannosidase, α-L-fucosidase. All examined enzymes turned out to be more or less unstable upon storage at 37°C, 4°C, and -20°C in both serum and plasma. The only exceptions were β-D-glucuronidase, which was stable in plasma and serum. For further details see initial citation of ref.26.

The influence of cyproterone acetate on the activity of β-D-glucuronidase in the hypothalamus and cerebral cortex of the male mouse has been investigated.[230] Daily subcutaneous injections of cyproterone acetate greatly decrease the protein content and β-D-glucuronidase activity in the mouse hypothalamus. These effects are reversible and the recovery capacity of the animal seems to be inversely related to the duration of antiandrogenic treatment.

A commercial preparation of bovine liver β-D-glucuronidase has been found to contain two distinct enzyme species, both of which catalyse the hydrolysis of 4-methylumbelliferyl α-L-idopyranosyluronic acid.[231] The species with a molecular weight of about 290,000 was active towards phenyl α-L-idopyranosyluronic acid but lacked the latter activity. Studies of the kinetics of inhibition and heat inactivation suggested that the hydrolysis of 4-methylumbelliferyl α-L-idopyranosyluronic acid is due to the β-D-glucuronidase in the case of the 290,000 dalton species. The highly purified β-D-glucuronidase preparations derived from rat preputial gland and liver lysosomes also exhibited 4-methylumbelliferyl α-L-iduronidase activity. These findings support the view that β-D-glucuronidase can hydrolyse certain α-L-idopyranosyluronic acid bonds and raise the possibility that β-D-glucuronidase may play a role in the catabolism of L-iduronic acid-containing glycosaminoglycans.

A comparative study has been performed upon hepatic and intestinal β-D-glucuronidase from adult male fischer rats.[232] The subcellular distribution patterns of β-D-glucuronidase indicated that 80% of hepatic β-D-glucuronidase activity is present in the lysosomal-mitochondrial pellet and 60% of intestinal β-D-glucuronidase activity is in the soluble fraction (105,000 g supernate). A 55-fold purification with a yield of 45% of β-D-glucuronidase activity from each tissue was obtained by ammonium sulphate fractionation, gel filtration, and column chromatography. Isoelectric-focusing gel electrophoresis reveals a unique β-D-glucuronidase isenozyme for the intestine (pI 5.4) and two for the liver (pI 5.6 and 6.7).

Levels of collagenolytic activity, β-D-glucuronidase, and collagen prolyl hydroxylase in paws from rats with developing adjuvant arthritis have been reported.[233] The collagenolytic activity associated with insoluble collagen fibres separated from homogenates of inflamed paws from rats with adjuvant arthritis was quantitated using H_4 edta-sensitive solubilization of hydroxyproline as a measure of activity. Approximately 60% of the solubilized hydroxyproline was associated with dialysable products. The level of collagenolytic activity in the paws increased with time after the induction of adjuvant arthritis and paralleled to a large extent the development of inflammation in both the adjuvant injected (right) hind paw and in the non-injected contralateral paw. By day 26, the level of free collagenolytic activity in the injected paw had increased to a level 30 times normal while that in the contralateral paw had increased to a level 10 times normal. Treatment of the residues from the injected paws with trypsin resulted in the activation of a latent collagenolytic activity which, on day 26, accounted for approximately 50% of the total activity. The elevated level of collagen prolyl hydroxylase in the inflamed paw suggested that the rate of collagen synthesis was also increased. The activity of β-D-glucuronidase increased in the inflamed paw with time after the induction of adjuvant arthritis. The inflamed paw of the adjuvant rat may represent a useful system in which to study the role of collagenolytic enzymes in the destruction of connective tissue by inflammatory lesions.

The changes of sulphated glycosaminoglycans and of enzymes active upon them during bovine foetal development have been analysed.[35] β-D-Glucuronidase and β-D-2-acetamido-2-deoxyglucosidase remains without significant changes during the whole period. For further details see initial citation of ref.35.

Adjuvant-induced arthritis in rats has been studied by the

changes in serum and urinary protein-bound carbohydrate metabolites, changes in serum and tissue lysosomal glycohydrolases, and lysosomal fragility.[30] There is no change in the total activity of lysosomal glycohydrolases, including β-D-glucuronidase. The free activities of lysosomal glycohydrolases investigated, including β-D-glucuronidase, are increased in liver and spleen in the acute phase. The free activities of β-D-glucuronidase, β-D-2-acetamido-2-deoxyglucosidase, and cathepsin D of kidney showed no change. For further details see initial citation of ref.30.

The α-L-fucosidase, β-D-2-acetamido-2-deoxyglucosidase, α-D-mannosidase, β-D-glucuronidase, β-D-xylosidase, β-D-glucosidase, and α-L-arabinofuronosidase activities of normal and atrophic skeletal muscle of developing and adult rats have been collectively investigated.[36]

Diphenylhydantoin induction of intracellular β-D-glucuronidase and alkaline phosphate activity has been reported in cultured bovine dental pulp cells.[234] Diphenylhydantoin, a drug commonly used in the treatment of grande-mal epileptic seizures, is responsible for side effects which may result from the induction of enzymes, particularly those found in the microsomal fraction of the liver. This study describes an effect of this drug upon bovine dental pulp cells, the induction of β-D-glucuronidase and alkaline phosphatase activity.

The presence of a precursor form of β-D-glucuronidase, with a subunit molecular weight of 75,000, has been demonstrated in mouse kidney.[235] This was later processed to the mature form, with subunit molecular weight of 71,500. Tissue fractionation revealed that the precursor was associated with the microsomes whereas the mature form was associated with the lysosomes. In mice lacking egasyn both forms of β-D-glucuronidase were present, but the rate of processing was elevated compared to normal.

Comparative studies have been made of an L-asparagine-linked oligosaccharide structure of rat liver microsomal and lysosomal β-D-glucuronidases.[236] The carbohydrate chains of microsomal and lysosomal β-D-glucuronidases of rat liver were studied by *endo*-β-D-2-acetamido-2-deoxyglucanase H digestion and by hydrazinolysis. Only a part of the oligosaccharides released from microsomal β-D-glucuronidase was an acidic component. The acidic component was not hydrolysed either by sialidase or by calf intestinal and *Escherichia coli* alkaline phosphatases, but was converted to a neutral component by phosphatase digestion after mild acid treatment indicating the presence of a phosphodiester group. The neutral oligosaccharide portion of microsomal enzyme was a mixture of five high D-mannose-type sugar chains.

Fifteen glycosidases were assayed in lymphomyeloid and digestive tissues of Ginglymostoma cirratum, Heterodontus francisci, and Etmopterus spinax.[19] Activities of α-D-mannosidase, β-D-galactosidase, β-D-2-acetamido-2-deoxygalactosidase, β-D-glucuronidase, and α- and β-D-glucosidase usually were higher in lymphomyeloid tissues than in digestive tissues.

Sucrose density gradients have been used to characterize lysosome-containing fractions from 0 h white prepupae of Stomoxys calcitrans.[42] The acidic glycosidases α-D-glucosidase, α-D-galactosidase, α-D-mannosidase, β-D-glucosidase, β-D-galactosidase, β-D-glucuronidase, and β-D-2-acetamido-2-deoxyglucosidase equilibrate at the same density as does acid phosphatase.

Several mollusc glycosidases have been studied for their activities towards natural substrates.[70] β-D-Glucuronidase from Littorina littorea hydrolyses hyaluronic acid, chondroitin 4-sulphate, and heparin with a very low activity. However, it is much more active on oligosaccharides (from the above-mentioned macromolecules) containing non-reducing terminal glucosyluronic acid residues.

10 α-L-Iduronidases

Maturation of α-L-iduronidase has been studied in cultured human diploid fibroblasts.[237] {^{3}H}Leucine was administered to the cells, and the enzyme was isolated by immunoprecipitation with antiserum to human kidney α-L-iduronidase. Radioactive polypeptides were separated by polyacrylamide gel electrophoresis and visualized by fluorography. Three radioactive, immunoprecipitable polypeptides of apparent molecular weight of 75,000, 72,000, and 6,000 were found intracellularly and one of 76,000 in the medium. Pulse-chase experiments showed the 75,000 intracellular species to be a precursor which was trimmed to the intermediate form in a few hours and the latter to the smaller form over a period of 4 to 5 days. The extracellular form was barely detectable unless the fibroblasts had been labelled in the presence of 10 mM NH_4Cl or had been derived from patients with I-cell disease (two conditions known to cause enhanced secretion of lysosomal enzymes). The identification of the four radioactive polypeptides as different forms of α-L-iduronidase was established not only by their temporal relationship but also by their absence from fibroblasts of Hurler patients (genetically deficient in α-L-iduronidase activity) and by the similarity in patterns generated by partial proteolysis with V8 protease from Staphylococcus aureus. All species of α-L-iduronidase were phosphorylated, as shown by incorporation of ^{33}P into the four polypeptides. The phosphate label could be removed by treatment with endo-β-D-2-acetamido-2-deoxyglucanase H. There appeared to be no difference in catalytic activity between the large and small forms of the enzyme. On the other hand, there was a difference in their uptake. The 66,000 species appeared to be of low-uptake form, in contrast to the 76,000 secreted form which was readily taken up by the D-mannose 6-phosphate receptor.

11 α- and β-D-Mannosidases

The conditions for maximal activity (pH, buffer, saturating substrate concentration, range of linear relationships between enzyme activity *versus* incubation time and *versus* enzyme concentration) in the fluorimetric assay of several glycohydrolases of lysosomal origin in human plasma and serum have been established.[26] The following enzymes were studied: α-D-galactosidase, β-D-galactosidase, β-D-glucosidase, β-D-glucuronidase, α-D-mannosidase, α-L-fucosidase. For further details see initial citation of ref.26.

A study has been made of lysosomal enzyme activities in serum and leukocytes in chronic hepatic disease.[60] Five lysosomal enzyme activities (arylsulphatase A, α-D-mannosidase, α-L-fucosidase, β-D-2-acetamidodeoxyhexosidase, and β-D-galactosidase) were determined in serum and leukocytes of controls and patients suffering from various liver diseases. For further details see initial citation of ref.60.

An unusual case of mannosidosis with severe deficiency of acid α-D-mannosidase in leukocytes and high residual enzymatic activity in skin fibroblasts has been reported.[238] The mutant enzyme was thermostable and has an almost normal K_m value. The patient had mild clinical manifestations, but excretion of mannose-rich oligosaccharides was very high. This case of mannosidosis is compared with those reported in literature.

Intracellular and extracellular α-D-mannosidase activity of cultured skin fibroblasts has been investigated with respect to a relation to cystic fibrosis.[239] Fibroblast intracellular α-D-mannosidase was inactivated by 30-50% after 120 min at 50°C. Separation of α-D-mannosidase by isoelectrofocusing and wheat-germ lectin-Sepharose into neutral and acid components showed the former to be heat labile and the cause of the observed loss. Foetal-calf serum acid α-D-mannosidase could be activated at 60°C if kept at pH 9.0. Fibroblast extracellular α-D-mannosidase was inactivated by 50-70% after 120 min at 50°C, reflecting the greater level of the neutral component. Control and cystic-fibrosis α-D-mannosidase showed no difference.

The specific activity of α-D-mannosidase in serum of I-cell disease patients has been found to be considerably increased due to increased amounts of the component with optimal activity at pH 4.6 (acidic form).[240] The intermediate form with pH optimum of 6.0 remains unaltered. These conclusions were reached by using optimal conditions for differential assay of the α-D-mannosidases checked by partial separation of the components in serum by sucrose centrifugation.

Adjuvant-induced arthritis in rats has been studied by the changes in serum and urinary protein-bound carbohydrate metabolites, changes in serum and tissue lysosomal glycohydrolases, and lysosomal fragility.[30] The free activities of lysosomal glycohydrolases investigated including α-D-mannosidase are increased in liver and spleen in the acute phase. For further details see initial citation of ref.30.

The α-L-fucosidase, β-D-acetamido-2-deoxyglycosidase, α-D-mannosidase, β-D-glucuronidase, β-D-xylosidase, β-D-glucosidase, and α-L-arabinofuranosidase activities of normal and atrophic skeletal muscle of developing and adult rats have been collectively investigated.[36]

An α-D-mannosidase specific for the hydrolysis of 2-O-α-D-mannosyl linkages to D-mannose has been solubilized and partially purified from rabbit liver microsomes.[241] The enzyme is inhibited by H_4 edta and has optimal activity in the presence of calcium ions. The purified enzyme has a requirement for non-ionic detergents or for specific phospholipids. At detergent concentrations appreciably below the critical micelle concentration, the enzyme is active in the presence of phosphatidylcholine or phosphatidylethanolamine but not with phosphatidylinositol, phosphatidylglycerol, or phosphatidic acid. At concentrations of phosphatidylcholine which provide optimal activity, the enzyme is strongly inhibited by phosphatidylinositol or phosphatidylglycerol. The substrate specificity of the α-D-mannosidase toward oligosaccharide substrates suggests that the enzyme may be involved in the processing of the oligosaccharide chains of mammalian glycoproteins.

Evidence has been presented for the biochemical diagnosis of the first case of feline mannosidosis.[242] A marked deficiency of acidic α-D-mannosidase in the brain, kidney, and liver and excessive excretion of D-mannose-rich oligosaccharides in the urine were found in a kitten suffering from a nervous disorder. Residual acidic α-D-mannosidase, ranging from 2.0 to 5.5% of the normal activity, was observed in the tissues of the affected kitten. It had similar kinetic and physicochemical properties to the normal activity. The amount of D-mannose in the urine of the affected kitten was 19-fold greater than in a comparable control, and the molar ratio of D-mannose to 2-acetamido-2-deoxy-D-glucose was approx. 6:1. High concentrations of neutral oligosaccharides were detected in the urine. The predominant oligosaccharide appeared to be a hexasaccharide. The biochemical features of bovine, feline, and human mannosidosis were compared, and it was concluded that feline mannosidosis may be a useful animal model for studying the human disease.

Kinetics of α-D-mannosidase action on various α-D-mannopyranosyl linkages in hen ovalbumin glycopeptides have been monitored by carbon nuclear magnetic resonance spectroscopy.[243] The authors suggest that ^{13}C n.m.r. spectroscopy is a practical method for following the kinetics of enzymatic digestion of individual carbohydrate residues of glycopeptides and for determining the structures of the products of partial digestions. The reported rule that jack-bean α-D-mannosidase hydrolyses 2- and 6-O-α-D-mannosyl linkages to D-mannose at least 15 times faster than the corresponding 3-O-α-linkages (Tai, T., Yamashita, K., Ogata-Arakawa, M., Koide, N., Muramatsu, T., Iwashita, S., Inoue, Y., and Kobata, A. (1975) J. Biol. Chem., 250, 8569-8575) is not of general validity.

Fifteen glycosidases have been assayed in lymphomyeloid and digestive tissues of *Ginglymostoma cirratum*, *Heterodontus*, and *Etmopterus spinax*.[19] Activities of α-D-mannosidase, β-D-galactosidase, β-D-2-acetamido-2-deoxygalactosidase, β-D-glucuronidase, and α- and β-D-glucosidase usually were higher in lymphomyeloid tissues than in digestive tissues.

Sucrose density gradients have been used to characterize lysosome-containing fractions from 0 h white prepupae of Stomoxys calcitrans.[42] The acidic glycosidases α-D-glucosidase, α-D-galactosidase, α-D-mannosidase, β-D-glucosidase, β-D-galactosidase, β-D-glucuronidase, and β-D-2-acetamido-2-deoxyglucosidase equilibrate at the same density as does acid phosphatase.

Nine glycosidases in bloodstream forms of Trypanosoma brucei brucei S42 have been partially characterized.[20] α-D-Mannosidase was clearly distinct from the other glycosidases with respect to pH optimum thermostability, specific activity during the course of parasiaemia, and subcellular location. It was suggested that glycosidases studied may play a role in the turnover of trypanosomal glycoproteins, in particular the variant-specific surface antigen.

The physicochemical and kinetic properties of the two major trypanosomal glycosidases α-D-glucosidase and α-D-mannosidase were compared in blood-stream forms of Trypanosoma brucei brucei S42.[175] For further details see initial citation of ref.175.

Attempts have been made to isolate mutants of Dictyostelium discoideum which express α-D-mannosidase-2 constitutively.[244] Evidence was sought for a gene-specific negative regulatory element of D. discoideum similar to the repressor/operator systems that control some operons in prokaryotes and their viruses. Mutants of D. discoideum that constitutively express the stage-specific enzyme α-D-mannosidase-2 could provide such evidence. Reconstruction experiments demonstrated that the screening technique developed would detect mutants of that type. Over 2 X 10^5 survivors of heavily mutagenized cell populations were grown and their progeny tested. No mutants of the desired type were obtained. The possible implications of this finding are discussed.

Acid hydrolases from Dictyostelium discoideum have been reported to contain phosphomannosyl recognition markers.[69] To test this hypothesis, the binding and endocytosis of purified β-D-glucosidase, β-D-2-acetamido-2-deoxyhexosidase, and α-D-mannosidase by human fibroblasts were investigated. These enzymes underwent endocytosis. For further details see initial citation of ref.69.

An acidic α-D-mannosidase has been isolated from culture filtrates of Aspergillus saitoi.[245] The extracellular α-D-mannosidase was homogeneous in polyacrylamide gel electrophoresis.

The molecular weight of the enzyme was 51,000 and the isoelectric point pH 4.5. The purified enzyme has a pH optimum of 5.0, a $\underline{K}_m$ of 0.45 mM with baker's yeast D-mannan, and has no activity towards 4-nitrophenyl α-D-mannopyranoside. The mode of action of the enzyme has been studied with baker's yeast D-mannan and sake yeast D-mannan. The enzyme cleaves specifically the (1 → 2)-linked side chain, producing free D-mannose.

12 Neuraminidases (Sialidases)

The extracellular neuraminidase production by Pseudomonas aeruginosa isolated form cystic fibrosis has been investigated.[246] The cellular localization of glycoprotein and ganglioside neuraminidases in normal and I-cell disease cultured fibroblasts has been investigated.[247] Cellular organelles have been separated on a colloidal silica gradient. The subcellular distribution of these enzymes indicated that the glycoprotein neuraminidase is mainly a lysosomal hydrolase whereas the ganglioside neuraminidase is primarily located in the plasma membranes. The latter isoenzyme is tightly bound to these membranes and thus could not be extracted by homogenization in the presence of Triton X-000.

The wheat-germ lectin binding of β-D-galactosidase has been studied in human and normal and sialidosis type II fibroblasts.[127] The authors suggest that neuraminidase deficiency may be the primary defect in sialidosis type II cells. For further details see initial citation of ref.127.

It has been reported that the combined deficiency of β-D-galactosidase and neuraminidase in human fibroblasts can be corrected to nearly normal values.[124] This can be accomplished by addition of concentrated culture medium obtained after NH_4Cl stimulation of different types of human fibroblasts. For further details see initial citation of ref.124.

The characteristics of the neuraminidase of human leukocytes, fibroblasts, and amniotic fluid cell cultures have been determined with a radioactive assay method utilizing neuramin-{^{3}H}lactitol as the enzyme substrate.[248] Fibroblast cultures from patients with the inherited neuraminidase deficiency diseases, including mucolipidosis I, sialidosis I, and sialidosis II, juvenile type, have less than 10% normal neuraminidase activity using either this substrate, 2-(3'-methoxyphenyl)-N-acetyl-α-neuraminic acid, or 2'-(4-methylumbelliferyl)-N-acetyl-α-neuraminic acid. The total sialic acid content of fibroblasts and leukocytes from mucolipidosis I and sialidosis I patients is greatly elevated. This parameter is useful in establishing a diagnosis of sialidase deficiency. The sialic acid content of sialidosis II, juvenile type, with coexistent neuraminidase and β-D-galactosidase deficiencies, is only slightly elevated above normal levels. A patient with mucolipidosis I has 16% of normal neuramin-{^{3}H}lactitol neuraminidase activity in his peripheral leukocytes. His parents were clearly distinguished from the normal range using leukocyte enzyme levels, and a maternal aunt was identified as a possible carrier. The presence of this enzyme in amniotic fluid cell cultures, both fibroblastic and mixed-cell type, makes possible the prenatal detection of these diseases. A pregnancy from a family at risk for having a child with mucolipidosis I was monitored by amniocentesis and subsequent neuraminidase measurement of the amniotic fluid cell culture.

Degradation of mucin oligosaccharides of human colae systems has been found to be associated with extracellular, but not with cell-bound β-D-galactosidase, β-D-2-acetamido-2-deoxyglucosidase, and neuraminidase.[29] The most probable numbers (MPN) of faecal bacteria producing extracellular β-D-galactosidase, β-D-2-acetamido-2-deoxyglucosidase, and neuraminidase were estimated. For further details see initial citation of ref.29.

An adverse effect of neuraminidase on the analysis of cell surfaces by borohydride tritiation has been reported.[249] Using the erythrocyte surface as a model it was found that neuraminidase treatment has a deleterious effect on the mobility and resolution of labelled glycoproteins in sodium dodecyl sulphate polyacrylamide gel electrophoresis. The authors suggest that this undesirable consequence of neuraminidase needs to be carefully considered when the enzyme is used to enhance D-galactose oxidase/borohydride tritiation of cell surfaces.

It has been shown that at least two components of neuraminidase can be distinguished in human leucocytes on the basis of pH optimum, thermolability at 30°C, and the effect of the detergent octyl-D-glucoside.[250] With 4-methylumbelliferyl α-D-N-acetylneuraminate as substrate, the A component had a pH optimum of 5.0, was labile at 30°C, and was unaffected by 0.2 M octyl β-D-glucopyranoside. The B component had a pH optimum of 4.0-4.2, was stable at 30°C, but lost most of its activity in the presence of 0.2 M octyl β-D-glucopyranoside. Both A and B components were membrane bound but only the A component was solubilized by octyl β-D-glucopyranoside in an active form. Molecular weights of neuraminidases by X-ray radiation inactivation were found to be 240,000 ± 19,000 for the B component, 203,000 ± 17,000 for the A component, and 238,000 ± 8,000 for the octyl β-D-glucopyranoside-solubilized A component. Gel filtration of soluble A component in the presence of octyl β-D-glucopyranoside showed a single peak of activity eluted at or near the void volume, which suggested that the enzyme was still in an aggregated form. Profound deficiency of neuraminidase activity was found for both A and B components in leucocytes of patients affected with sialidoses types 1 and 2 (less than 15% normal) and intermediate activity in obligate heterozygotes. It was suggested that the A and B components of leucocyte neuraminidase are closely related from the genetic point of view and that rapid diagnosis of sialidoses can be done by fluorimetric assay of neuraminidase in leucocytes.

The conditions of a linear assay for 4-methylumbelliferyl α-D-N-acetylneuraminic acid neuraminidase activity in human liver have been determined and an acidic liver neuraminidase has been characterized with a pH optimum between 4.4 and 4.8, and an apparent Michaelis constant of 0.11 ± 0.02 mM for 4-methylumbelliferyl α-D-N-acetylneuraminic acid.[251] The acidic neuraminidase is labile to freezing, heating, and various storage conditions. The great majority of human liver neuraminidase activity is found in resuspended pellets after homogenization and centrifugation. A small amount of this resuspended pellet activity can be solubilized with the non-ionic detergent Triton X-100. The properties of human liver neuraminidase are compared to those of other mammalian neuraminidases.

The activities of various glycosidases in homogenates of the small intestinal mucosa of two adult and 18 suckling tammar wallabies (*M. eurenii*) aged from 6 to 50 weeks have been investigated.[38] β-D-Galactosidase, β-D-2-acetamido-2-deoxyglucosidase, α-L-fucosidase, and neuraminidase activities were high during the first 34 weeks *post partum* and then declined to very low levels. For further details see initial citation of ref.38.

Oligosaccharides with a new type of core structure, β-D-GalNAc-(1 → 4)-β-D-Gal-(1 → 4)-{NGNA-(2 → 3)}-β-D-Gal-(1 → 3)-D-GalNAcol, have been isolated from trout egg glycoprotein.[252] The occurrence of N-glycolylneuraminic acid (NGNA) linked (2 → 3) to the internal 2-acetamido-2-deoxy-D-galactose residue is not known in any other glycoproteins and glycolipids. This NGNA residue was found to show some anomalous nature such as resistance to neuraminidases and unusual chemical-shift values of H-3 protons in NMR spectra. A new feature is also found in the asialo-oligosaccharide structure β-D-GalNAc-(1 → 4)-β-D-Gal-(1 → 4)-D-GalNAc, linked (1 → 3) to the linkage region disaccharide β-D-Gal-(1 → 3)-D-GalNAc.

Studies have been carried out on the effect of bile salts on the rates of hydrolysis of the N-acetylneuraminyl linkages of several sialic acid-containing compounds by the neuraminidase of Clostridium perfringens.[253] When G_{M3}-ganglioside, two glycolipids (glycophorin and orosomucoid), and neuraminyl-lactose were used as substrates, hydrolysis was obtained even in the absence of bile salts, but addition of these detergents, below its critical micellar concentration, increased the reaction rates. Above the critical micellar concentration of the detergent rates decreased again. When a second ganglioside, G_{M1}, was used as substrate, the requirement for bile salts was absolute; hydrolysis was not observed at all without this detergent. With increasing concentrations of bile salt and in the presence of high concentrations of enzyme, rates of hydrolysis increased, reaching maximal values at fixed ratios of bile salt to G_{M1}-ganglioside. Physical measurements showed that mixtures of bile salt and G_{M1}-ganglioside formed mixed micelles that had a higher critical micellar concentration, a lower molecular weight, and greater axial ratio than the corresponding micelles of pure G_{M1}-ganglioside.

Neuraminidase of Corynebacterium ulcerans has been purified by affinity chromatography using immobilized colominic acid preparations.[254] Neuraminidase appears to be a thermolabile protein, molecular mass 70,000. The pH optimum of 5.5 is independent of the substrate used, the optimal temperature is 37°C, the Michaelis constant towards N-acetylneuraminosyl-lactose is 5.2 x 10^{-4}M. Ca^{2+} and Ba^{2+} activated the enzyme, but Zn^{2+}, Fe^{2+}, and chelating agent H_4 edta were inhibitory. The enzyme did not hydrolyse the (2 → 6) or (2 → 8) bonds of submaxillary pig mucin and colominic acid, respectively, but it hydrolysed such substrates as fetuin, ovomucin, orosomucoid, and horse serum glycoproteins.

Partial purification and properties of neuraminidase from Bifidobacterium lactentis have been described.[255] B.lactentis 659, a Gram-positive non-spore-forming anaerobic bacterium originally isolated from the faeces of breast-fed infants, produces neuraminidase after enzyme induction with 2mM 2-acetamido-2-deoxy-D-mannose added to the culture medium. Bacteria were transferred and grown in a medium containing casein hydrolysate, lactose, sodium acetate, amino acids, vitamins, salts, and 2% human milk for 48 h at 37°C under N_2/CO_2 atmosphere. Two subcultures derived from the original strain B. lactentis 659 were investigated separately because of different growth characteristics. However, their taxonomic identity was not doubtful. Neuraminidase was liberated from the bacterial sediments by ultrasonic treatment of 0.15M NaCl and was isolated by 60% ammonium sulphate precipitation, dialysis, concentration, and column chromatography on Sepharose CL-6B and Sephadex G-100. The enzyme exhibits a molecular weight of 38,000 and a pH optimum in the range of pH 5-6 in barbital/acetate buffers.

The neuraminidase activity of Pasturella haemolytica isolates has been investigated.[256] Type isolates of the 12 established serotypes and bovine and ovine field isolates were included in the study. Neuraminidase activities ranged from 0 to 0.87 U per mg (dry weight) of P. haemolytica. Activity levels among the isolates studied were serotype associated.

The complete neuraminidase of influenza A/PR8/34 (H_0N_1) has not been detected in its recombinant virus.[257] The neuraminidase located on the influenza recombinant A/E(H_{eq1}) x A/PR8(N_1) was relatively a poor antigen (NI titre < 10) as compared with the enzyme present on the parent virus A/PR8/34 (H_0N_1) (NI titre 160). This difference in the antigenic behaviour of the neuraminidase of A/PR8/34 (H_0N_1) virus was explored. Results obtained showed that rabbit anti-A/PR8/34 (H_0N_1) serum contained two distinct types of antineuraminidase antibody, and only one type of antineuraminidase antibody was precipitable by the neuraminidase located on the recombinant virus. Evidence suggests that the parent enzyme(s) is not detectable in its entirety on its recombinant virus.

Neuraminidase has been purified from the influenza virus A/Hong Kong/68 (H_3N_2) by treatment of the purified virus with sodium laurylsarcosinate, centrifugation at 110,000 x g, and chromatography on DEAE-Sephadex and Sephadex G-200.[258] It migrated as a single component during electrophoresis on polyacrylamide gel, and its molecular weight was estimated about 270,000. The enzyme was thermolabile, the activity being reduced to 60% in 10 min at 50°C. The purified neuraminidase had an apparent K_m value of 4.1×10^{-3} M for 5-N-acetyl-2-O-(3-methoxyphenyl)-α-D-neuraminic acid and was able to release sialic acid with linkages (2 → 3), (2 → 6), and (2 → 8) (with very different efficiency) from fetuin, gangliosides, colominic acid, and bovine and porcine submaxillary mucins. The enzymic activity was measured by several procedures which are discussed.

13 β-D-Xylosidases

The α-L-fucosidase, β-D-2-acetamido-2-deoxyglucosidase, α-D-mannosidase, β-D-glucuronidase, β-D-xylosidase, β-D-glucosidase, and α-L-arabinofuranosidase activities of normal and atrophic skeletal muscle of developing and adult rats have been collectively investigated.[36]

Aspergillus niger, strain 110.42 (CBS), has been selected as a producer of high xylanolytic activities.[259] The time-course of D-xylanase and β-D-xylosidase production as well as the effect of pH and temperature on the activity of these enzymes were studied. H.p.l.c. analysis of the enzymatic degradation of L-arabino-D-xylan showed a nearly complete conversion to pentose sugars. The authors discuss the use of crude D-xylanase preparations for the saccharification of D-xylans.

Cultures of Streptomyces flavogriseus have been found to produce considerable amounts of D-xylanase when grown on D-xylan-containing media.[260] Comparatively lower yields of this enzyme were obtained when hay or avicel served as main carbon source. β-D-Xylosidase was synthesized intracellularly and appeared less dependent on the fermentation substrate. The strain produced simultaneously various enzymes of the cellulase complex and the D-xylose-induced D-glucose isomerase.

Active-site-directed irreversible inhibition of glycosidases by the corresponding glycosylmethyl-(4-nitrophenyl)triazines has been investigated.[72] β-D-Xylopyranosylmethyl-(4-nitrophenyl)triazine is an active-site-directed irreversible inhibitor (ASDIN) of β-D-xylosidase from Penicillium wortmanni, but has no effect of the β-D-xylosidase of Bacillus pumilus, except by virtue of its consumption of stabilizing dithiothreitol. For further details see initial citation of ref.72.

The importance of cellulase and D-xylanase release from Bacteroides succinogenes in the rumen environment has been investigated.[223] During growth of B. succinogenes in a liquid medium with cellulose as the source of carbohydrate, greater than 80% of the *endo*-(1 → 4)-β-D-glucanase, D-xylanase, and aryl-β-D-xylosidase and 50% of the aryl-β-D-glucosidase was released from the cells into the culture fluid. For further details see initial citation of ref.223.

In a study of the extracellular cellulase activities of *Clostridium thermocellum* LQRI and *Trichoderma reesi* QM9414 extracellular β-D-xylosidase activity was found to be absent.[207]

14 *endo*-β-D-2-Acetamido-2-deoxyglucanases

Two *endo*-β-D-2-acetamido-2-deoxyglucanases (type F-I and type F-II) have been isolated from fig latex.[261] At pH 7.0, type F-I was retained by the DEAE-Sephadex A-50 column, whereas F-II was not absorbed by the column. The optimum pH of type F-I was found to be pH 5.9 and type F-II pH 5.4. Type F-I enzyme hydrolyses the tri-D-mannosyl derivatives of di-2-acetamido-2-deoxy-D-glucosyl L-asparagine faster than the penta- or hexa-D-mannosyl compounds. Type F-II hydrolyses the penta- and hexa-D-mannosyl derivatives but not the tri-D-mannosyl compound.

The distribution of the different types of oligosaccharides in cathepsin D and in β-D-2-acetamido-2-deoxyhexosidase synthesized in cultured human fibroblasts has been studied by using *endo*-β-D-2-acetamido-2-deoxyglucanase H as a probe for high D-mannose oligosaccharides.[262] The enzymes were specifically labelled in the protein or the carbohydrate moiety. In both enzymes, resistant and cleavable oligosaccharides were found. The resistant oligosaccharides prevailed in the secreted enzymes. Precursor molecules of cathepsin D contained two oligosaccharide side chains. Multiple forms of the precursor are synthesized with both, one, or more oligosaccharides sensitive to the action of the *endo*-β-D-2-acetamido-2-deoxyglucanase H. In fibroblasts unable to phosphorylate lysosomal enzymes (mucolipidosis II) the excessively secreted lysosomal enzymes contained predominantly oligosaccharides resistant to *endo*-β-D-2-acetamido-2-deoxyglucanase H.

An autolytic *endo*-β-D-2-acetamido-2-deoxyglucanase, capable of cleaving the glycoside linkages of *N*-unsubstituted D-2-amino-2-deoxy-D-glucose in the glycan moiety of cell-wall peptidoglycan, has been purified 470-fold from a salt extract of the 20,000 g precipitate fraction obtained after sonication of a lysozyme-resistant strain of *Bacillus cereus*.[263] The purified enzyme preparation was also active towards the glycan chain of fully *N*-

acetylated cell-wall peptidoglycan. The *endo*-β-D-2-acetamido-2-deoxyglucanase was inactive towards the cell-wall peptidoglycan unless the peptide portion of this polymer was removed either by the action of *N*-acetylmuramyl-L-alanine amidase or by the treatment with alkali in aqueous dimethyl sulphoxide. Studies on the action of this enzyme towards chemically modified glycans revealed that the carboxy groups of muramic acid residues were indispensable to a substrate for this enzyme.

endo-β-D-2-Acetamido-2-deoxyglucanase has been demonstrated in human tissues and has been partially characterized.[264] The presence of enzyme activity towards oligosaccharides of the oligomannosidic and of the *N*-acetyl-lactosaminic type is demonstrated in human tissues. Furthermore, some properties of the enzyme (pH optimum, influence of thiol reagents, and stability) in human kidney are described.

A new method has been devised for assaying the *endo*-β-D-2-acetamido-2-deoxyglucanase activity by using the dansyl asparaginyl oligosaccharide $(\text{D-Man})_5(\text{D-GlcNAc})_2$-L-Asn-DNS as the substrate and analysing the product, D-GlcNAc-L-Asn-DNS, by reverse-phase high-pressure liquid chromatography using a silica-based chemically bonded octadecyl column.[265] The column was eluted with 8% acetonitrile in 25 mM sodium borate buffer, pH 7.5. The effluent was monitored by a u.v. monitor at 213 nm and a fluorescence monitor (excitation 313 nm, emission 540 nm). Under these conditions, D-GlcNAc-L-Asn-DNS was well separated from $(\text{D-Man})_5(\text{D-GlcNAc})_2$-L-Asn-DNS and the analysis could be completed in 5 min. Under the conditions described above, the lower limit of detection was 0.1 nmol of dansyl glycopeptides.

15 Agarases

Agarase has been concentrated and purified from culture filtrates of an agar-degrading *Pseudomonas*-like bacteria by affinity chromatography on divinyl sulphone cross-linked macroporous agarose.[266] By gel filtration three fractions were obtained, two of which, agarases I and II, had agarase activity. Agarase II was further purified by isoelectric focusing, and the main peak, agarose IIb, was isoelectric at pH 5.1. Molecular-weight determinations indicated agarase I to be a dimer with M_r 210,000. The M_r of agarase IIb was 63,000 as determined by analytical ultracentrifugation, polyacrylamide gel electrophoresis in sodium dodecyl sulphate, and molecular-sieve chromatography in 6 M guanidinium chloride. The amino acid compositions of the two proteins were very similar and both were found to be glycoproteins. The pH optimum of the enzymes was 6.7 and the optimal temperature was 38°C for agarase I and 43°C for agarase IIb. Melted agarose and agarose gel were used as substrates for the enzymes. The ratio of the activities towards the different substrates was 4.3 for agarase I and 1.0 for agarase IIb. Agarase I hydrolysed the glycosidic linkages in neoagaro-octaose so as to produce two moles of neoagarotetraose for one mole of neoagarohexaose and one mole of neoagarobiose. Agarase IIb hydrolysed only the central glycosidic linkage to form two moles of neoagarotetraose.

16 Alginases and Alginate Lyases

The properties of isoenzymes of alginate lyase in the mid-gut gland of *Turbo cornutus* have been investigated.[267] Alginate lyase SP2 was separated on an SP-Sephadex C-50 column from SP1, whose properties have already been reported, and purified according to the method employed for SP1, to obtain information on SP2. The profiles of the optimal pH, pH stability, thermal inactivation, and molecular size of SP2 were entirely the same as those of SP1. The isoelectric point of SP1 and SP2 was 7.5 and 7.7, respectively. The action of SP2 on alginate caused a rapid decrease in solution viscosity. Analysis of digestion products of alginate with SP2 showed that the enzyme had an affinity toward the D-mannuronate-rich domains of the alginate molecule and released unsaturated oligomers mostly composed of D-mannuronic acid as final product.

17 α-Amylases

A kinetic equation has been derived which represents a synergystic action of *endo*- and *exo*-enzyme upon polysaccharides.[268] Its adequacy was tested by experiments using immobilized α-amylase and glucoamylase for hydrolysis of soluble starch and amylose.

An assay on amylase activities using NAD-dependent maltose dehydrogenase has been described.[269] It was demonstrated that the activities of α-amylase and β-amylase could be measured with NAD-dependent maltose dehydrogenase by the end-point assay method and by the rate assay method. One unit of the activity of α-amylase or β-amylase was defined as the amount of the enzyme which produced one μmol of maltose equivalent product from substrate per minute at 25°C under the assay condition.

The effect of albumin on α-amylase assay using blue starch polymer has been investigated.[270] The binding of albumin to the starch polymer was responsible for the stimulation of α-amylase activity, since the stimulation was observed with the albumin-bound blue starch polymer as substrate irrespective of the presence or absence of free albumin in the reaction mixture. When the albumin-bound blue starch polymer was hydrolysed by α-amylase, the blue oligosaccharide fragments were released with the albumin attached. These oligosaccharide fragments were about 1.3 times larger in average molecular size than the fragments released in the absence of albumin, suggesting that the apparent stimulation of α-amylase by albumin (about 1.3 times) is due to the liberation of the larger oligosaccharide fragments.

The suitability of control materials for determination of α-amylase activity has been assessed in comparison with reference groups of authentic human serum specimens containing α-amylase of either pancreatic or salivary origin, specimens from patients with no pancreatic pathology, and normal specimens to which porcine pancreatic α-amylase was added.[271] After determination of α-amylase activity by 11 commonly used techniques (five different principles), the results were processed by both classical (linear representation, regression) and multivariate (correspondence analysis, principal-components analysis) statistical techniques. Specimens containing porcine pancreatic α-amylase did not behave like any of the other groups. It was concluded that porcine enzyme should not be used for interlaboratory quality-control surveys or inter-method comparison studies. Determination of human salivary and pancreatic α-amylase showed intermethod biases similar to those for authentic patients' specimens. Human salivary α-amylase, because of both its behaviour and its commercial availability, is a satisfactory source for α-amylase activity of quality-control specimens. The nature of the matrix (polyvinylpyrrolidone, albumin, delipidated serum, bovine serum, or human serum) little influenced the behaviour of the specimens for any of the methods studied.

A polarimetric method for the determination of α-amylase activity has been described.[272] In the study of the relatively slow amylase-catalysed hydrolysis of oligosaccharides, it is assumed that the subsequent mutarotation has essentially no effect on the kinetics of the total reaction. Relatively large samples and long measurement times (slow rate of change in angle of rotation) are necessary for the polarimetric determination of α-amylase activity. For these reasons, only urine and duodenal juice are suitable as sample sources. Owing to the absence of coupled reactions and auxilliary enzymes, the polarimetrically determined kinetics are generally more representative of the true kinetics of the amylase reaction than those obtained with fully enzymic methods. In addition, single determinations by the polarimetric method are less subject to interference by sample components. With measurement times in the order of minutes, the polarimetric method shows excellent linearity, very good proportionality between analytical response and quantity of enzyme, and good precision in series (*CV* = 1.3%). Comparison of the polarimetric with a chromogenic and a full enzymic method showed an acceptable correlation (*r* for each method was about 0.98).

A number of enzymes, including amylases, dehydrogenases, and proteases, have been shown to be renaturable after polyacrylamide gel electrophoresis in the presence of sodium dodecyl sulphate.[273] Enzyme activity was detected *in situ* by action on substrates introduced into the gel and subsequent staining of either the product or unreacted substrate. This technique combines the advantage of enzyme identification with the resolution and molecular-weight dependence of gel electrophoresis in the presence of sodium dodecyl sulphate. Enzymes appeared to recover activity as soon as the detergent diffused out of the gel. Most monomeric enzymes could be renatured even after disruption of their disulphide bonds, but several proteases, including trypsin, could not. Oligomeric enzymes composed of identical subunits were poorly renaturable. Renatured enzymes were retained in gels after electrophoresis longer than native enzymes which had been subjected to electrophoresis in the absence of detergent. Re-electrophoresis of the renatured enzymes showed that part of the retained activity was physically anchored to the gel, possibly by the folding of polypeptides around the gel matrix as the enzymes were renatured.

An inhibitor I-1, capable of acting on both α-amylase and trypsin, has been purified to homogeneity from ragi (finger-millet) grains.[274] The factor was found to be stable to heat treatment at 100°C for 1 h in the presence of NaCl and also was stable over a wide pH range 1-10. Pepsin and pronase treatment of inhibitor I-1 resulted in gradual loss of both the inhibitory activities. Formation of trypsin-inhibitor I-1 complex, amylase-inhibitor I-1 complex, and trypsin-inhibitor I-1-amylase trimer complex was demonstrated by chromatography on a Bio-Gel P-200 column. This indicated that the inhibitor was double-headed in nature. The inhibitor was retained on a trypsin-Sepharose 4B column at pH 7.0. Elution at acidic pH resulted in almost complete recovery of amylase-inhibitory and trypsin-inhibitory activities. α-Amylase was retained on a trypsin-Sepharose column to which inhibitor I-1 was bound, but not on trypsin-Sepharose alone. Modification of amino groups of the inhibitor with 2,4,6-trinitrobenzenesulphonic acid resulted in complete loss of amylase-inhibitory activity but only 40% loss in antitryptic activity. Modification of L-arginine residues by cyclohexane-1,2-dione led to 85% loss of antitryptic activity after 5 h but no effect on amylase-inhibitory activity. The results show that a single bifunctional protein factor was responsible for both amylase-inhibitory and trypsin-inhibitory activities with two different reactive sites.

α-Amylase inhibitors from wheat kernels have been investigated as factors in resistance to post-harvest insects.[275] α-Amylase inhibitors were extracted from kernels of five hard winter-wheat varieties that were grown at different locations in two crop years and that had been evaluated for susceptibility or resistance to the rice weevil, *Sitophilus oryzae*. Inhibitor activity was assayed with larval α-amylase from two species of stored grain pests, the rice weevil and the yellow mealworm, *Tenebrio molitor*. Correlation in some wheat varieties was observed between *in vivo* resistance to the insect and the extent of *in vitro* inhibition of the insect larval α-amylase by the extracted inhibitors. These results indicated that α-amylase inhibitors in wheat could be involved in the resistance of wheat to post-harvest infestation.

The chemical structure of an amylase inhibitor S-AI produced by *Streptomyces diastaticus* has been investigated.[276]

Some properties of amylase inhibitor A produced by *Streptomyces* sp. No. 280 have been reported.[277] *Streptomyces* sp. No. 280 produced several kinds of amylase inhibitors (amylase inhibitor A, B, B', and C). Two amylase inhibitors (designated as $AI\text{-}A_1$ and $AI\text{-}A_2$) were obtained from an amylase inhibitor A fraction by paper chromatography. $AI\text{-}A_1$ inhibited muscle phosphorylase *a* much more than $AI\text{-}A_2$ and was hydrolysed by sweet-potato α-amylase whereas $AI-A_2$ was not. Both amylase inhibitors had a carbohydrate moiety and were hydrolysed by some kinds of amylases or acids. They lost their inhibitory activity against phosphorylase a after treatment with acids or hog pancreatic α-amylase, but they showed increased inhibitory activity toward porcine small-intestinal sucrase. Both $AI-A_1$ and $AI-A_2$ were composed of D-glucose and a basic moiety which gave a positive ninhydrin reaction. The molecular weights of $AI-A_1$ and $AI-A_2$ were estimated to be approximately 1,300 - 1,500 by gel filtration on a Sephadex G-15 column. The nitrogen content of the amylase inhibitors was found to be about 1.3% by elemental analysis.

The sequence of the α-amylase inhibitor Hoe-467 from Streptomyces tendae 4158 has been elucidated.[278] An α-amylase inhibitor isolated from S. tendae, strain 4158, was re-purified by chromatography on CM- and DEAE-cellulose column. Two inhibitors could be characterized: α-amylase inhibitor Hoe-457 A (with L-aspartic acid as N-terminal residue) and α-amylase inhibitor Hoe-467 S (with L-serine as N-terminal residue) The preliminary structure was determined by automatic Edman-degradation procedures of the aziranized inhibitor and tryptic peptides, derived from digestions of the performic oxidized, aziranized, and maleylated inhibitor, respectively. The α-amylase inhibitor Hoe-467 A consists of 74 residues and has a calculated molecular weight of 7958. It is composed of all common amino acids except L-methionine and L-phenylalanine. Digestion with pepsin was carried out to determine the disulphide bonds. Two fractions could be isolated, containing one L-cystine each, giving information about the position of the disulphide bridges.

The complete amino acid sequence of one of the major wheat protein iso-inhibitors of α-amylase has been determined.[279] The sequence was deduced from analysis of fragments and peptides from the protein by cleavage with cyanogen bromide and by digestion with trypsin, chymotrypsin, thermolysin, and the Staphylococcus aureus V8 protease. The molecule consists of a single polypeptide chain of 123 residues. Both L-serine and L-alanine were found in position 65, and further minor examples of microheterogeneity were observed in four other residues.

The interaction of four purified α-amylase inhibitors from wheat flour with human salivary and pancreatic α-amylases has been investigated.[280] The inhibitory activity of the four proteins towards salivary α-amylase was significantly increased by preincubation of the enzyme with inhibitor before adding substrate. This effect was not observed with the inhibition of pancreatic α-amylase by inhibitors 1 and 2. Inhibition of both amylases was affected to different degrees by incubating starch with inhibitor prior to the addition of enzyme. Maltose, at concentrations which only slightly affected amylase activity, prevented the inhibition of both enzymes by all four inhibitors. Gel-filtration studies on salivary amylase-inhibitor mixtures showed the formation of EI complexes on a mole-to-mole ratio. A similar complex between pancreatic α-amylase and inhibitor 4 was observed, though complex formation between pancreatic α-amylase and the other inhibitors was not clearly demonstrated.

A search has been made for natural plant enzyme inhibitors of human salivary α-amylase.[281] Twelve varieties of sorghum (Sorghum vicolor), 14 varieties of pearl millet (Pennisetum typhpoideum), 12 varieties of setaria (Setaria intalica), 4 varieties of ragi (Eleucine coracana), 11 varieties of echinocloa millet (Echinocloa colona), 13 varieties of proso (Panicium miliaceum), 11 varieties of kodo (Paspalum scorbiculatum), and 11 varieties of miliare (Panicium miliare) were screened for inhibitory activity against human salivary amylase. Echinocloa, proso, kodo, and miliare had no detectable activity. Two strains of sorghum and one strain of pearl millet did not show α-amylase inhibitory activity. All other seeds had activity, the highest being observed in sorghum. Setaria had no action on human, bovine, and porcine pancreatic amylases. Sorghum inhibitor did not act on bovine and porcine pancreatic amylases. Pearl millet and ragi extracts inhibited all the four α-amylases. The inhibitors were non-dialysable and were inactivated by pepsin treatment. Setaria and sorghum inhibitors were relatively thermolabile compared to ragi and pearl millet inhibitors.

An α-amylase protein inhibitor from Arachis hypohaea seeds has been purified and its properties have been described.[282] The protein, which showed a highly specific inhibitory activity towards hog pancreatic and human salivary α-amylases but not towards plant and bacterial α-amylases, has been purified 197-fold from an aqueous extract of peanut cotyledons using heat treatment, ammonium sulphate precipitation, and ion-exchange chromatography on DEAE- and CM-cellulose. The purified inhibitor was homogeneous by polyacrylamide gel electrophoresis. Its molecular weight, as determined by Sephadex G-100 gel filtration, and its electrophoretic mobility at pH 8 relative to Bromophenol Blue were 25,000 and 0.14, respectively. The inhibitory activity was relatively resistant to thermal treatment and markedly increased when the inhibitor was preincubated with the enzyme before the addition of starch. Further, the inhibition was found to be pH dependent and non-competitive in nature.

Two proteinaceous α-amylase inhibitors, I-1 and I-2, have been isolated from ragi (*E. coracana*) grains by extraction with 0.15 M sodium chloride, ammonium sulphate fractionation, chromatography on CM-cellulose, gel filtration on Sephadex G-50, and rechromatography on CM-cellulose.[283] The inhibitors I-1 and I-2 were purified 50-fold and 12-fold, respectively, and the preparations were found to be homogeneous by cellulose acetate and polyacrylamide gel electrophoresis. The estimated molecular weights based on gel chromatography were around 9,100 for I-1 and 8,300 for I-2. The inhibitors were active against α-amylase from hog pancreas, human pancreas, and human saliva but had no action on *Bacillus subtilis* α-amylase. Both the inhibitors were basic proteins and were stable over a wide pH range of 1-10. Sodium chloride increased the heat stability and solubility of the inhibitors. Pepsin and pronase treatments cause inactivation of both the inhibitors. Trypsin inactivated I-2 alone. Both I-1 and I-2 were resistant to chymotrypsin. The inhibitory patterns observed with I-1 and I-2 were non-competitive when hog pancreatic amylase was used. Amylase inhibitory activity was lost completely on germination of ragi for 48 h.

Four α-amylase inhibitors have been isolated from an albumin fraction of wheat flour by ion-exchange and gel-filtration chromatography.[284] The purified inhibitors were characterized according to their electrophoretic mobilities, molecular weights, carbohydrate content, sulphydryl content, susceptibility to proteolytic digestion, and specificities in inhibiting human salivary and pancreatic α-amylases. The properties of these inhibitors were compared to similar proteins isolated by other workers.

A structural study has been made of the sugar chains of α-amylases produced ectopically in tumours.[285] About thirty per cent of two α-amylases produced from a serous papillary cystadenocarcinoma of the ovarium (case 1) and a bronchioloalveolar adenocarcinoma of the lung (case 2) was glycoproteins containing 1 mol of L-asparagine-linked sugar chain, respectively. The structures of the sugar moieties were found by sequential enzymatic degradation and methylation analysis. Structures of L-asparagine-linked sugar chains were the same in the tumour of cases 1 and 2 and were incomplete in comparison with those of the parotid amylase.

Amylases have been purified and characterized from three amylase-producing human tumours.[286] The relative molecular mass of the amylases was estimated to be 54,000 on sodium dodecyl sulphate polyacrylamide gel electrophoresis, different from human salivary amylase (61,000 and 64,000) and human pancreatic amylase (60,000). The tumour amylases had completely identical antigenicities with human salivary and pancreatic amylases against antibody to human salivary amylase, while the intensity of the tumour amylases was less than 20% of that of human salivary and pancreatic amylases on single radial immunodiffusion. On isoelectric focusing, two of the three tumour amylases showed a major peak at pH 6.4, which corresponded to a major peak of human salivary amylase; the other showed a major peak at pH 6.4, which corresponded to a minor peak of human salivary amylase. The three tumour amylases showed similar amino acid compositions, different from those of human salivary and pancreatic amylases. These findings suggest that tumour amylases have a tertiary structure similar to that of normal human amylases but differ from them in amino acid composition.

The preparation and characterization of materials containing human pancreatic and salivary α-amylases have been described and their relationship to endogenous amylase in human serum examined.[287] Amylase was purified from human pancreas and saliva by solvent and salt fractionation and column chromatography to specific activities of 63 and 279 kU g^{-1}, respectively. Four liquid pools, differing only in activity, were prepared from each source of amylase, each in a matrix containing, per litre, 30 g of human albumin, 50 mmol of sodium chloride, 1 mmol of calcium chloride, and 50 mmol of Tris hydrochloride buffer, pH 7.4. Characterization of the pools showed that the amylase activity in the materials was stable for at least six months at 25°C, among-vial variability of amylase activity was 0.5% (2 CV), and the pools were free from eight possible contaminating enzymes. Plots of salivary *versus* pancreatic amylase activity measured in materials with eight commercially available methods showed least-squares slopes ranging from 0.51 to 1.0. The intermethod commutability of the materials (*i.e.* how closely they mimic endogenous serum amylase) was examined in relationship to approximately 100 human sera.

An enzymatic assay for the determination of α-amylase in serum employed a soluble substrate, maltoheptaose, and a coupled enzymatic indicator reaction consisting of β-D-glucosidase and the hexokinase β-D-glucose 6-phosphate dehydrogenase system.[288] H.p.l.c. was used to establish the action pattern of maltoheptaose under the test conditions: (A) the action pattern of α-amylase alone, (B) that of the combined action of α-amylase and β-D-glucosidase. Conducive to the effort was: the availability of pure maltoheptaose and human pancreatic α-amylase, the development of an adequate procedure for sample pretreatment (partition chromatography on a mixed-bed ion exchanger) and of an h.p.l.c. system for separation of substrate and reaction products without interference from by-products of the assay (partition chromatography on a cation-exchange column with acetonitrile-water), and the use of a new, very sensitive refractometric detector revealing sugar amounts as low as 40 ng. The following stoichiometric equations were derived:

$$\text{maltoheptaose} \xrightarrow{\alpha\text{-amylase}} 0.10\ \text{maltopentaose} + 0.79\ \text{maltotetraose} + 0.87\ \text{maltotriose} + 0.29\ \text{maltose}$$

$$\text{maltoheptaose} \xrightarrow[\beta\text{-D-glucosidase}]{\alpha\text{-amylase}} 0.10\ \text{maltopentaose} + 0.77\ \text{maltotetraose} + 0.06\ \text{maltotriose} + 0.07\ \text{maltose} + 2.94\ \text{glucose}$$

The standard deviation of the rate coefficients is about 5%.

A study of the action of human α-amylases on reduced malto-oligosaccharides has been reported.[289] Kinetic studies on the reactions of human pancreatic and salivary α-amylases with a series of reduced malto-oligosaccharides (maltotetraoitol through maltoheptaoitol) were conducted. Maltotetraoitol was barely hydrolysed by either α-amylase under the conditions used, and the minimum size of substrate susceptible to hydrolysis was maltopentaoitol. Among these substrates, maltohexaoitol showed the highest reactivity with each α-amylase, especially with salivary α-amylase, compared with maltopentaoitol. For the combination of each enzyme with each substrate, the predominant point of cleavage of the substrates was found to be the third (1 → 4)-α-D-glucosidic linkage from the D-glucitol residue.

Antihuman parotid α-amylase antibodies raised in rabbits and horses were used as the primary antibodies in both the peroxidase-antiperoxidase and sandwich techniques for the localization of human α-amylase.[290] Immunoreactive enzyme was demonstrated in the normal acinar cells of salivary glands and pancreas. Malignant transformation, which has occasionally resulted in ectopic production of α-amylase by various tissues, actually caused a loss of α-amylase synthesis by the transformed acinar cells of salivary glands and did not result in elaboration of α-amylase by transformed ductal cells.

Biochemical and kinetic studies have been performed on recrystallized urinary α-amylase of normals and of patients with cancer of the bladder.[291] Measurement of the reaction rate at a range of pH values indicated that pH 6.8 was the optimum pH value for the enzyme of normal subjections, while the abnormal enzyme has its optimum at pH 8. The kinetic behaviour provides strong supporting evidence in favour of the presence of multiple interaction sites in the enzyme of patients with cancer of the bladder.

The multiple-attack model of the α-amylase mechanism has been analysed and the theoretical rate coefficients dependent on the degree of polymerization of substrate (n) have been applied.[292] They are: effectiveness of hydrolysis (Yn), which affects V values, and the average number of unitary movements (ξ_m), which affects K_m values. The model explains the dependence of V and K_m on the degree of polymerization of substrates with n values higher than the number of subsites in the enzyme-binding site. Distribution of the degree of polymerization of hog pancreas α-amylase products (n_p) and average n_p value has been found experimentally and applied to calculate the maximal number of unitary movements (z) of the enzyme during the single enzyme-substrate meeting. V and K_m values of hog pancreas α-amylase for amylose and different branched substrates have been determined and discussed in terms of the multiple-attack theory. Their dependence on theoretic Yn and ξ_n values has been found in support of the model.

The sequences of four cyanogen bromide-cleaved peptides of porcine pancreatic α-amylase have been determined.[293] The sequence strategy has involved the cleavage by CNBr and the separation of the 9 CNBr:peptides into 7 fractions by gel filtration. The complete purification and partial-sequence determination of 3 peptides (CNBr-4, CNBr-5c, and CNBr-5d) have previously been reported. In the present work, the purification of 4 additional CNBr peptides (CNBr-1, CNBr-3b2, CNBr-7b1, and CNBr-7b2) and of smaller tryptic peptides has been carried out. The complete sequence and the order of CNBr-

4, CNBr-5c, CNBr-3b2, CNBr-7b1, and CNBr-7b2 as well as the partial sequence of CNBr-1, CNBr-2, and CNBr-3b1 have been determined. The ordering has been possible by taking advantage of the highly homologous mouse pancreatic α-amylase just being deduced from the nucleotidic sequence of mouse α-amylase mRNA.

A single-step purification of rat pancreatic and salivary α-amylase by affinity chromatography has been described.[294] The α-amylases were purified 20- to 50-fold with a yield above 75% by a single-step affinity-chromatographic procedure. The affinity ligand was an α-D-glucohydrolase inhibitor coupled to 6-aminohexyl-Sepharose 4B. Pancreatic α-amylase was eluted as a single peak at pH 7.4 with 0.1% glycogen or at pH 5.8 without glycogen. Salivary amylase could be eluted only with 0.1% glycogen-containing buffer at pH 7.4. Pancreatic α-amylase migrated as a single homogeneous band on sodium dodecyl sulphate polyacrylamide gels, whereas salivary α-amylase migrated as a doublet.

α-Amylase activity has been detected in unprocessed preamylase produced in the cell-free translation of porcine pancreatic RNA.[295] Preamylases, synthesized in the RNA-dependent rabbit reticulocyte lysate translation system supplemented with porcine pancreatic RNA, were identified by their specific immunoprecipitation with anti-amylase. The preamylases have apparent $\underline{M}_r$ = 55,000 and 58,000 as compared to 52,000 and 55,000 for the purified, secreted α-amylase isoenzymes. In order to establish whether the unprocessed precursors may assume enzymatically active conformations, a highly sensitive activity gel electrophoresis technique, by which picogram quantities of enzyme can be detected, was explored. When standard α-amylase and translation products are subjected to electrophoresis on polyacrylamide gel containing 0.01% starch and $CaCl_2$, active amylase which binds tightly to starch can only migrate as the starch is hydrolysed. When the gel is subsequently stained with I_2, the appearance of clear tracks, the lengths of which are roughly proportional to the logarithm of amylase concentration, signifies the presence of amylase activity. By this approach, it was possible to detect amylase activity in a range corresponding to about 100 pg of pure amylase $(10\ \mu l)^{-1}$ of translation mixture. This value agrees well with an estimate from radioactivity incorporation of total preamylase in the translation mixture, and it was consequently concluded that unprocessed preamylase can assume the appropriate conformation to give enzymatic activity.

The characterization of the amino termini of mouse salivary and pancreatic amylases has been reported.[296] Mouse amylase mRNA sequences and information concerning signal sequences from other proteins have been used to locate potential cleavage points for removal of the signal peptides from the mouse amylases. That in turn allowed the authors to predict the biochemical characteristics of the N-terminal tryptic peptide for each enzyme and then use those characteristics to locate and purify the peptide. Analysis of the N-terminal peptides showed that the secreted mouse amylases have pyroglutamic (pyrrolidone carboxylic) acid residues at their N-termini.

The formation of a complex of proteinous animal α-amylase inhibitor (Haim) with hog pancreatic α-amylase has been reported.[297] Novel proteinous inhibitors (Haim I, II) of Streptomyces griseosporeus YM-25 inhibit animal α-amylase specifically. As Haim is a protein containing no sugar moiety, the inhibition of Haim against the amylase was considered to be attributed to the result of protein-protein interaction. On the other hand, S-AI, an amylase inhibitor produced by S. diastaticus subsp. amylostaticus, inhibits various α-amylases of animal, plant, and microbial origin. S-AI consists of 6 mol of glucose with unknown moiety tentatively named as S-AI-X. S-AI may be classified as a substrate analogue inhibitor. In this paper, the interactions among hog pancreatic α-amylase, proteinous inhibitor (Haim), and XG derived from S-AI, as a substrate analogue, are described. Inferring the results from the triple interaction among Haim, XG, and α-amylase, Haim should join at a neighbourhood of a catalytic site or at a situation like an allosteric site of α-amylase.

An antiserum against purified rat parotid α-amylase has been used to localize the protein in parotid glands of developing and adult rats.[298] The unlabelled antibody peroxidase-antiperoxidase method and the protein A-gold colloid technique were used at the light and electron-microscope levels, respectively. Immunoreactive α-amylase was detected in a few scattered cells in the glands of 2-day-old rats. During the following days the number of cells stained immunocytochemically for α-amylase increased rapidly. At 15 days of age all acinar cells revealed α-amylase, but the intensity of immunostaining varied from cell to cell. Electron microscopically, α-amylase was localized in the secretory granules and, by using a more concentrated antiserum, in the rough endoplasmic reticulum and Golgi complex. At early stages of development the acinar cells contained fewer and smaller secretor granules than in adult animals.

The gold particles indicative of amylase were randomly distributed over the secretory granules. In the glands of adult rats, amylase was distributed inhomogeneously within the secretor granules. In the majority of secretory granules gold colloid particles were located over the electron-dense portions of the granules. However, secretory granules in which an amylase-rich shell surrounded an α-amylase-poor or α-amylase-negative core were not infrequent.

A comparative study of carbohydrase activities in marine invertebrates has been performed.[299] 103 species of marine invertebrates, belonging to 7 types, Spongia, Coelenterata, Annelida, Arthropoda, Mollusca, Echinodermata, Chordata, were tested for laminarinase, cellulase, and amylase activities.

A nephelometric method for determining cerial α-amylase has been evaluated.[300] A nephelometric assay adapted from a clinical method was evaluated for suitability in determining α-amylase levels in sprouted soft white wheat. In the manual mode of machine operation, the velocity of decline in light scattering was directly proportional to the amount of enzyme assayed. The accurate limits of the assay performed in the automated mode were defined as 0-720 machine units of activity, with a commercial β-limit dextrin as substrate. This substrate was more reactive than amylopectin and less subject to β-amylase interference. The automated assay yielded data that correlated highly with falling number results.

The properties of two amylase activities which differ in their substrate specificity and subcellular location as well as a chloroplast-associated R-enzyme (debranching activity) have been reported.[301] An extrachloroplastic amylase was resolved by gel-filtration chromatography into two activities of 80,000 and 40,000 daltons. Both extrachloroplastic activities hydrolyse amylopectin and shellfish glycogen and only slowly hydrolyse rabbit liver glycogen, β-limit amylopectin, and amylose. In contrast, the major chloroplastic amylase attacks all of these glucans at comparable rates. Glucan hydrolysis by both the extrachloroplastic and chloroplastic amylase generates not only maltose but appreciable amounts of other oligosaccharides, whereas maltotetraose hydrolysis produces D-glucose, maltose, and maltotriose. The action patterns displayed by the amylase activities indicate that both are endoamylases, although they lack the typical Ca^{2+} requirement or heat stability of seed endosperm α-amylases. Dithiothreitol, glutathione (oxidized or reduced), ascorbate, dehydroascorbate, and dithiothreitol plus thioredoxin have no effect on either the chloroplastic or extrachloroplastic amylase activities. The chloroplastic R-enzyme debranches amylopectin, βlimit amylopectin, pullulan, and α-limit dextrins, but not rabbit liver glycogen. An increase in extinction coefficient and λ_{max} is detected when the debranched amylopectin and β-limit amylopectin form a complex with I_2-KI. Based on these properties, the chloroplastic R-enzyme is similar in enzymic activity to the R-enzyme observed in endosperm tissue'

Endogenous gibberellins and amylase activity have been investigated in tall and dwarf strains of rice (Oryza sativa).[302] Two tall and seven dwarf strains of rice were sprayed with GA_3. Four dwarf strains responded to exogenous GA_3 and showed a markedly low endogenous gibberellin content than the tall strains, while two dwarf strains did not respond to GA_3 application and had considerably higher endogenous gibberellin levels than the tall ones. Amylase activity in germinating seeds showed a significant negative correlation with the endogenous gibberellin content.

Hard red winter-wheat cultivars have been found to differ markedly in sprouting and in α-amylase synthesis in sprouted grain, which increased as time between maturity and exposure to simulated rain was increased.[303] Sprouting and α-amylase activity measured by falling number and labelled starch degradation were highly correlated. However, cultivar and temporal deviations from that relationship occurred. Hard winter-wheat lines are generally more susceptible than were hard red winter-wheat cultivars, but exceptions occurred and differences were not more quantitative. The simulated sprouting environment clearly identified sprouting-resistant phenotypes, and it would be useful for selecting desirable germoplasm in wheat-improvement programmes.

The properties of partially purified α-amylase of pearl millet have been investigated.[304] The purification method affected the physicochemical characteristics of purified millet α-amylases. Polyacrylamide gel electrophoresis showed three protein bands for α-amylase purified by glycogen complex formation and seven for α-amylase purified by starch column procedures. All protein bands exhibited α-amylase activity. α-Amylase isoenzymes had molecular weights, determined by sodium dodecyl sulphate gel electrophoresis, ranging from 22,000 to 53,000 and isoelectric points ranging from 4.8 to 6.2. The pH optima were between 4.4 and 4.8, and temperature optimum was 55°C. Many of the above characteristics are similar to those reported for amylases purified from immature cereal grains.

The distribution and effect of α-amylase in field-sprouted wheats has been tested in Japanese-type sponge cake and by related physical and chemical tests.[305] To clarify the effect of field sprouting on quality, amylase activity was determined (0 - 4.13 dextrinizing units {DU} per gram) in soft white winter-wheat composites with typical levels of field sprouting (0 - 36.2%) and in their mill fractions. Average yields of the corresponding mill fractions of seven wheat composites were: patent flour 45%, midpatent flour 10%, clear flour 15%, bran 25.8%, shorts 3.2%, and red dog 1.0%. Ash and protein contents of the fractions were typical. α-Amylase activity was relatively low for 45% patent, midpatent, and clear flours and relatively high for bran, shorts, and red dog. When α-amylase activity in DU per gram of each mill fraction was calculated as DU per gram of wheat, based on yield, activity increased linearly in each mill fraction with increasing wheat α-amylase. When expressed as per cent of wheat DU, bran accounted for 42% of the α-amylase activity, patent flour for 32%, shorts for 9%, clear flour for 8%, midpatent flour for 7%, and red dog for 2%. Sponge-cake volume increased from 1,280 cm^3 for the control to 1,315 cm^3 for the flour milled from wheat containing 0.35 DU g^{-1}. Thereafter, volume decreased rapidly to 908 cm^3 as DU per gram of wheat increased to 4.13. At least 0.2 DU g^{-1} of wheat (about 2.5% sprouted wheat) was considered a doubly safe level that would have no adverse effect on sponge-cake quality. Sponge-cake volume per cent of sprouted wheat, amylograph viscosity, falling number, and gas production were all functions of the α-amylase activity of wheat or flour. It was concluded that different levels of field sprouting should not be simulated by supplementing unsprouted wheat with a highly sprouted one, especially not with a highly malted barley.

A modified procedure has been reported for use of the Perkin-Elmer Model 191 grain amylase analyser in determining low levels of α-amylase in wheats and flours.[306] The analyser was employed at its highest possible sensitivity to determine the α-amylase levels. β-Limit dextrin (0.5%) was used as substrate, and the rate of decrease in turbidity effected by the enzyme was followed on a recorder attached to the grain amylase analyser. Straight-line relationships were found between increases in concentration of enzyme and rates of decrease in nephelos per minute. The method was found satisfactory for assaying α-amylase levels in wheats ranging in Hagberg falling numbers from 240 to 460 s and in amylograph viscosities from 125 to

925 BU. The method has some advantages over an automated fluorimetric procedure for α-amylase.

An interlaboratory investigation of the Perkin-Elmer model 191 grain amylase analyser has been conducted in the United Kingdom.[307] The instrument was calibrated against the Farrand method for bread flours on the amylase-2 scale only. The conversion factor was 17 amylase-2 units to 1 Farrand unit. The calibration was stable with time and between three laboratories. The reproducibility was excellent, with a standard deviation of 0.035P at P amylase-2 units. An α-amylase level could be obtained in 10 min compared to 2-5 h by the Farrand method.

Plant α-amylases have been found to be sensitive to high calcium-ion concentration *in vitro*.[308] At 250 mM nearly 70% of the activity of maize, wheat, and barley α-amylases was inhibited, but in contrast hog pancreatic and bacterial enzymes were not affected. Inhibition of maize α-amylase was competitive, reversed by dialysis, and markedly increased by the pre-incubation of the enzyme with Ca^{2+} before the addition of the substrate. The apparent K_i and the apparent number of Ca^{2+} bound per enzyme molecule were found to be 85 mM and 1.1, respectively. The latter value indicates the absence of a co-operativity phenomenon. The interaction between the maize α-amylase and Ca^{2+} was also studied as a function of temperature. The efficiency of Ca^{2+} interaction decreases with increasing temperature and as expected the overall reaction was exothermic with a ΔH value of -13.8 kcal mol^{-1}. The values for ΔG and ΔS for the formation of Ca-enzyme complex at 25°C were found to be 1.6 kcal mol^{-1} and -51.6 cal mol^{-1} deg^{-1}, respectively. The relatively large negative ΔS value suggested that the enzyme molecule undergoes marked conformational change during its interaction with Ca^{2+}.

A detailed study of the *in vitro* interation between Zn^{2+} and α-amylases from different origins has shown that under physiological conditions plant α-amylases are strongly inhibited by zinc ions.[309] The inhibition is pH dependent and can be reversed by extensive dialysis against buffer. In addition, it can also be removed by the treatment with H_4 edta, particularly at low pH values. However, H_4 edta itself does not have any effect on enzymatic activity. The inhibition for maize α-amylase at its optimum pH (5.5) was competitive and the K_i was 10 mM. The Hill coefficient value was 1, suggesting that Zn^{2+} interacts with α-amylase in a simple manner with no co-operativity.

The purification by affinity chromatography of α-amylase from the cotyledons of germinating *Vigna mungo* seeds has been reported.[310] In *V. mungo* cotyledons, the α-amylase activity increased markedly during germination at 27°C in the dark, while the activity of other amylases was very low. The α-amylase was purified from 4-day-old cotyledons by affinity chromatography on epoxy-activated Sepharose 6B substituted with β-cyclohepta-amylose and by column chromatography on Bio-Gel P-200. Gel filtration and polyacrylamide gel electrophoresis showed that the enzyme exists mostly as monomer (43,000 daltons), but partially aggregates to form dimer, trimer, and further multimers. Ca^{2+} protected the α-amylase against heat inactivation. Incubation of the enzyme with 5 mM H_4 edta or dialysis against 10 mM H_4 edta resulted in a 50-90% loss of activity. The inactivation was partially reversed by the addition of Ca^{2+}. Other properties, such as the amino acid composition, K_m value, pH optimum, and activation energy were similar to those of other plant α-amylases.

A *Bacillus subtilis* transformant producing thermostable α-amylase has been isolated using DNA from a thermophilic bacterium, Thermophile V2.[311] The extracellular α-amylase did not crossreact with a rabbit antiserum against *B. subtilis* α-amylase. The structural gene for the thermostable α-amylase was integrated at a different locus from *B. subtilis* α-amylase and was linked to pyrA. The transformant was not thermophilic, and its upper temperature for growth was similar to that of the host bacterium.

The N-terminal amino acid sequence of a saccharifying-type α-amylase from *Bacillus subtilis* var. *amylosacchariticus* has been investigated and a comparison made with that of liquefying-type α-amylase.[312]

The *in vitro* synthesis of α-amylase-like products in *Bacillus subtilis* RNA-directed heterologous protein-synthesizing system has been described.[313] α-Amylase-like proteins were synthesized in a heterologous cell-free protein-synthesizing system prepared from *Escherichia coli*. The proteins were precipitable with anti-α-amylase serum and detected only when RNA extracted from α-amylase producing *B. subtilis* cells was used as messenger. The *in vitro* α-amylase-like products seemed to consist of two components having molecular weights of 30,000 and 13,000.

By the use of 2-deoxy-D-glucose, a non-metabolizable catabolic repressor of certain yeast enzymes, mutants derepressed and hyperproductive with respect to α-amylase have been obtained in very high plate yields from the yeast and *Lipomyces kononenkoae*.[89]

A method for the automatic measurement of α-amylase and glucoamylase activities during fermentation has been developed.[314] Soluble starch dyed with Remazol Brilliant Orange was used as the substrate for α-amylase and 4-nitrophenyl α-D-glucopyranoside for glucoamylase. The same automatic analysis system could be used for both of these enzymes because the reaction products were measured at the same wavelength. Simultaneous pick-up of enzyme and the respective substrate was enabled by using two samplers. The presence of α-amylase did not interfere with the glucoamylase determination. Absolute values for α-amylase activity were obtained using a mathematical correction. Monitoring of these enzymes was accomplished during microbial fermentation.

Bacillus amylolyticus produced α-amylase, pullulanase, and α-D-glucosidase. By selection of carbon source in the growth medium, α-D-glucosidase was produced preferentially and with exclusion of the other two activities.[182]

The effect of D-glucose on the enzymes involved in the degradation of a reserve D-glucan in Polyporus circinatus has been studied.[315] The levels of phosphorylase activity, amylase, and amylo-1,6-D-glucosidase were found to be regulated by D-glucose concentration.

The coproduction of several exoenzymes in Bacillus subtilis has been described.[316] Mutants defective in many of these enzymes, particularly α-amylase and alkaline and neutral protease, were isolated in order to provide tools for a study of the mechanism of secretion of exoproteins.

The cloning of thermostable α-amylase gene using Bacillus subtilis phage ρ11 has been achieved using a method consisting of two repetitions of prophage transformation.[317]

The molecular weight of Aspergillus oryzae α-amylase (Taka-amylase A) has been estimated at 51,000 ± 500 by the combined use of high-pressure silica-gel chromatography and a low-angle laser-light-scattering technique.[318] The study was carried out partly to assess the performance of the combined technique, and results obtained indicate that it is promising as a method to determine protein molecular weight both accurately and quickly.

An α-amylase from Streptomyces praecox has been purified and its characteristic action, the conversion of maltotriose (G3) to maltose (G2) without appreciable formation of D-glucose (G1), was investigated.[319] Isoelectric focusing showed that the enzyme preparation after chromatographic separation comprises three isoenzymes. The preparation from the glycogen-adsorption procedure showed the highest specific activity of any preparation of this enzyme ever obtained. Product analysis with uniformly labelled G3 revealed that at a high concentration (18 mM) of G3 much more G2 is produced than G1 (the product ratio G2/G1 is over 20), while at a lower concentration (10 μM) the reaction mixture was composed of nearly equimolar amounts of D-glucose and maltose. Based on the product analysis of reducing end-labelled G3 in addition to the above findings, the following conversion mechanism was proposed: Streptomyces α-amylase catalyses transglycosylation to produce maltotetraose (G4) as a transient product which is immediately degraded into two molecules of G2 by a subsequent hydrolytic reaction, i.e. two molecules of G3 are converted into three molecules of maltose without appreciable formation of D-glucose.

The production of extracellular amylase by P. omnivorum has been studied under different cultural conditions.[320] Maximum mycelial growth and amylase synthesis were found in basal synthetic medium containing 2% glycogen as the carbon source. Significant amounts of amylase and growth were obtained in media containing other carbon compounds with (1 → 4)-α-D-glucosidic linkages. Optimum temperatures for mycelial growth and amylase synthesis were 28° and 30°C, respectively. Of all the nitrogenous compounds tested, ammonium nitrate supported the maximum mycelial growth and amylase production. Increasing the concentration of starch in the basal medium increased growth as well as the enzyme synthesis.

Schwanniomyces castellii and Endomycopsis fibuligero have been found to produce extracellular amylase(s) when grown on various carbon sources and at different pH values.[321] Both yeast species showed significant amylase synthesis in the presence of either maltose or soluble starch. Of the other substrates tested (D-glucose, cellobiose, sucrose, trehalose, melezitose, raffinose, ethanol, glycerol) differences were found regarding growth and amylase production. Free D-glucose in the culture medium apparently inhibited enzyme synthesis. The pH range allowing maximal growth and amylase production was 4.5 - 6.0 for E. fibuligera and 5.5 - 7.0 for S. castellii.

The production of α-amylase by a strain of Bacillus stearothermophilus isolated from leaf litter was investigated in a tryptone-maltose medium at 55°C in batch and chemostat culture.[322] α-Amylase production in chemostat culture was influenced by the growth rate throughout the dilution rate range used.

18 β-Amylases

Some properties of amylase inhibitor A produced by Streptomyces sp. 280 have been reported.[277] Streptomyces sp. 280 produced several kinds of amylase inhibitors (amylase inhibitor A, B, B', and C). Two amylase inhibitors (designated as AI-A_1 and AI-A_2) were obtained from an amylase inhibitor A fraction by paper chromatography. AI-A_1 inhibited muscle phosphorylase a much more than AI-A_2 and was hydrolysed by sweet-potato β-amylase whereas AI-A_2 was not. Both amylase inhibitors contained carbohydrate and were hydrolysed by some kinds of amylases or acids. They lost their inhibitory activity against phosphorylase a after treatment with acids or hog pancreatic α-amylase, but they showed increased inhibitory activity toward porcine small intestinal sucrase. Both AI-A_1 and AI-A_2 were composed of D-glucose and a basic moiety which gave a positive ninhydrin reaction (1.3% nitrogen). The molecular weights of AI-A_1 and AI-A_2 were estimated to be approximately 1,300 - 1,500.

An assay method for amylase activities using NAD-dependent maltose dehydrogenase has been described.[269] It was demonstrated that the activities of α-amylase and β-amylase could be measured with dependent maltose dehydrogenase by the end-point assay method and by the rate assay method. One unit of the activity of α-amylase or β-amylase was defined as the amount of the enzyme which produced one μmol of maltose equivalent product from substrate per minute at 25°C under the assay conditions.

A comparative study of carbohydrase activities in marine invertebrates has been performed.[299] 103 species of marine invertebrates, belonging to 7 types, Spongia, Coelenterata, Annelid, Arthropoda, Mollusca, Echinodermata, Chordata, were tested for laminarinase, cellulase, and amylase activities.

The modification of wheat β-amylase by proteolytic enzymes has been investigated.[323] Two β-amylase components, I and II, were separated from wheat flour by ion-exchange chromatography. Component I contained three main forms and component II contained one form when analysed by polyacrylamide slab electrofocusing. Papain or a malted wheat extract converted the three β-amylases in component I to five forms with more basic pI values and the one β-amylase in component II to three forms with pI values identical to those arising from degradation of component I. The new β-amylases resulting from proteolytic attack were identical in pI values to β-amylases in germinating wheat, which indicated that the mechanism of formation of such enzymes is via limited proteolytic degradation.

An immunochemical study has been made of two β-amylase-deficient inbred lines of rye (Secale cereale), the kernels of which displayed a very low level of β-amylase activity (1-3% of the levels generally found in rye) in comparison with a third normal line.[324] An anti-wheat β-amylase immune serum which cross-reacted with the rye enzyme anti-wheat β-amylase immune serum absorbed the β-amylase activity in the three lines. Comparably small amounts of enzymatic antigen corresponded to the small levels of activity detected in the enzyme-deficient lines. Neither the level of activity nor the amount of enzymatic antigen were notably changed upon germination. The results indicated that the reduced activity is due neither to the presence of an inhibitor nor to the production of inactive enzymes. Germination can proceed normally without later production of β-amylase.

The behaviour of sulphydryl groups of soybean β-amylase has been investigated by chemical modification.[325] Two sulphydryl groups out of a total of five in the native enzyme reacted with 5,5'-dithiobis-(2-nitrobenzoic acid), iodoacetamide, or monoiodoacetate at high ionic strength, accompanied by inactivation of the enzyme. At low ionic strength, only one sulphydryl group was accessible to 5,5'-dithiobis-(2-nitrobenzoic acid) without inactivation. By using the reaction with iodoacetamide, the dissociation constants of these two sulphydryl groups and rate constants for the reaction were determined. The most reactive sulphydryl group, which was independent of the inactivation, reacted with monoiodoacetate 185 times faster than the essential sulphydryl group. Next, selective modification of one sulphydryl group with monoiodoacetate was carried out. The modified enzyme, having activity equal to that of the native enzyme, was purified by ion-exchange column chromatography. It was successively treated with 5,5'-dithiobis-(2-nitrobenzoic acid) or iodoacetamide in order to achieve selective modification of the essential sulphydryl group. The enzyme modified further with 5,5'-dithiobis-(2-nitrobenzoic acid) was fully reactivated by 2-mercaptoethanol treatment. It also recovered 65% of the original enzymatic activity on treatment with cyanide but only 7% with sulphite. Maltose and cyclohexa-amylose protected the essential sulphydryl group from modification, the former being effective. The ultraviolet absorption spectrum of soybean was changed by modification of the essential sulphydryl group, and the spectrum was also changed by the binding of maltose. The change induced by maltose was influenced by modification of the essential sulphydryl group. It was concluded that the essential sulphydryl group of soybean β-amylase does not participate in the catalysis but is situated near the binding site of maltose.

The conformational properties of soybean β-amylase have been investigated by the circular-dichroism probe and measurement of enzyme activity.[326] The enzyme exhibited a positive circular-dichroism band at 192 nm, a negative band at 222 nm, and a shoulder near 210 nm. Analysis of the spectrum in the far-ultraviolet zone indicated the presence of approximately 30% of α-helix and 5-10% of β-pleated sheet, the rest of the polypeptide main chain possessing aperiodic structure. In the near-ultraviolet region, the enzyme protein showed at least six positive peaks at 259, 265, 273, 281, 292, and 297 nm. The positive bands at 292 and 297 nm remained unaltered on acetylation of the enzyme by N-acetylimidazole and were assigned to L-tryptophanyl chromophores. These bands were affected in intensity in the presence of maltose or cyclohepta-amylose, which indicates that some tryptophan residues are situated at the binding sites. The native conformation of soybean β-amylase was found to be sensitive to pH variation (below pH 5 and above pH 10), sodium dodecyl sulphate, guanidinium chloride, and heating to 50-55°C. Complete disorganization of the secondary structure was attained by 6 M guanidinium chloride. Sodium dodecyl sulphate was effective in disturbing the tertiary structure of the enzyme but did not affect significantly the secondary structure. Enzymatic inactivation was paralleled by the decrease of circular-dichroism bands in the near-ultraviolet region as produced by the denaturants. It was concluded that the uniquely folded structure of the enzyme contains some less rigid domains and a rigid core stabilized by hydrophobic interactions, electrostatic interactions, and hydrogen bonds.

Subsite affinities of soybean β-amylase have been evaluated by a new method to elucidate the effect of multiple attack.[327] Product analysis of the reducing-end-labelled malto-oligosaccharides showed that G_6^* and G_7^* were degraded through the multiple-attack pathway, in 0.02 M acetate buffer, pH 5.4 at 25°C. The apparent first-order rate constants, $(k_0/K_m)e_0$, for n-mer substrates (n = 3 - 7) were obtained from the decrease of substrate followed by paper chromatography under the condition of {S} < K_m. Using the constants and the cleavage distribution of G_3^*, the subsite affinities (A_1, A_4, and A_5) of β-amylase were evaluated. It was revealed that the first subsite of soybean β-amylase had a larger subsite affinity than any subsites of other amylases so far evaluated. Thus the first subsite seems to play an important role for the characteristic action pattern of this enzyme, namely exclusive maltose-forming action. The frequency of multiple attack decreased as either or both pH and temperature become unfavourable for enzymic activity.

Endogenous gibberellins and amylase activity have been investigated in tall and dwarf strains of rice (Oryza sativa).[302] Two tall and seven dwarf strains of rice were sprayed with 0 or 40 μg ml^{-1} GA_3. Four dwarf strains responded to exogenous GA_3 and showed a markedly lower endogenous gibberellin content than the tall strains, while two dwarf strains did not respond to GA_3 application and had considerably higher endogenous gibberellin levels than the tall ones. Amylase activity in germinating seeds showed a significant negative correlation with the endogenous gibberellin content.

β-Amylase-producing micro-organisms have been effectively isolated by using an amylase inhibitor (S-AI).[328] Among forty-two strains isolated, one was found to be the most potent β-amylase-producing micro-organism, and was identified as Bacillus polymyxa. Two kinds of β-amylase, I and II, obtained from culture broth as electrophoretically homogeneous forms (mol. wts. 44,000), were very similar to each other in their enzymatic properties except the small difference in isoelectric point (I pH 8.35, II pH 8.59). β-Amylases I and II were most active at pH 7.5 at 45°C, and stable between pH 4 and 9 for 15 h at 37°C. Both enzymes were strongly inhibited by 4-chloronescuribenzoate, and reactivated by L-cysteine. The amino acid compositions of the enzymes were studied.

The use of ultrafiltration membrane systems in stirred-cell and in thin-channel systems for immobilizing enzyme (sweet-potato intrinsic and crystalline β-amylase) in the hydrolysis of sweet potato through a continuous operation mode has been studied.[329] Both the filtration rate and reducing sugars, produced as the result of enzymatic hydrolysis, decreased with the filtration time. The immobilized enzymes in the thin-channel system showed a much better performance compared to that in the stirred-cell system. Addition of crystalline sweet-potato β-amylase to the sweet potato increased both the filtration rate and reducing-sugars content. Alcoholic fermentation of the filtrate resulted in an alcohol content of 4.2%.

This represented fermentation of 95% of the sugars with an efficiency of 88%.

The effect of D-glucose on the enzymes involved in the degradation of a reserve β-D-glucan in Polyprus circinatus has been studied.[315] The levels of phosphorylase activity, amylase, and amylo-1,6-D-glucosidase were found to be regulated by D-glucose concentration.

19 Amylo-1,6-D-glucosidases

Reaction of rabbit muscle amylo-1,6-D-glucosidase/4-α-D-glucanotransferase with an active-site-directed irreversible inhibitor, 1-S-dimethylarsino-1-thio-β-D-glucopyranoside, has been investigated.[330] The inhibitor reacts with the debranching enzyme to inactivate both the D-glucosidase and the transferase activities. The two activities are lost at different rates. The rate of inactivation of each activity is first order in both enzyme and inhibitor concentration. D-Glucosidase and combined D-glucosidase-transferase activities are lost at the same rate, indicating that the D-glucosidase is rate-limiting for the combined action on glycogen phosphorylase limit dextrin. The relative rate of inactivation of the D-glucosidase and transferase activities by 1-S-dimethylarsino-1-thio-β-D-glucopyranoside is a function of the protein concentration. At high enzyme concentration, the transferase is inactivated twice as rapidly as the D-glucosidase. At low enzyme concentration the transferase is inactivated only one-half as fast. The rate of inactivation of both activities at high enzyme concentration is much slower, indicating that aggregation of the protein at these concentrations affects the accessibility of 1-S-dimethylarsino-1-thio-β-D-glucopyranoside to the active sites. The sites of 1-S-dimethylarsino-1-thio-β-D-glucopyranoside reaction with the debrancher were deduced from the protective effect of a series of substrates and substrate analogues. Glycogen phosphorylase limit dextrin protects both the D-glucosidase and transferase activities from inactivation by 1-S-dimethylarsino-1-thio-β-D-glucopyranoside. Bis-Tris, (2,2-bis(hydroxymethyl)-2,2',2''-nitrilotriethanol), a reversible inhibitor which binds specifically at the D-glucosidase active site, protects the D-glucosidase but not the transferase against inactivation. Cyclohexa-amylose, which binds to the polymer site, has no protective effect against the action of 1-S-dimethylarsino-1-thio-β-D-glucopyranoside. The results indicate that 1-S-dimethylarsino-1-thio-β-D-glucopyranoside acts as an active-site-directed irreversible inhibitor at both the D-glucosidase and the transferase active sites on the debranching enzyme but not at the polymer-binding site. The temperature effect on the 1-S-dimethylarsino-1-thio-β-D-glucopyranoside inactivation rate is quite large, ΔH+ = 35 kcal mol^{-1}, ΔS+ = 54 e.u. These values suggest that 1-S-dimethylarsino-1-thio-β-D-glucopyranoside inactivation of the debrancher occurs with a significant conformation change.

20 Cellulases

A review of cellulases and their biosynthesis and applications has been produced.[331]

Two different cellulases from Trichoderma viride have been immobilized on Sepharose by derivatization of the enzyme with a bifunctional vinyl monomer (glycidylmethylacrylate) and successive graft copolymerization of vinyl enzyme on the support.[332] The best yields for the derivatization reaction have been obtained at pH 8.3, 18°C, with a vinyl monomer:enzyme ratio of 10.2×10^{-2} and a reaction time of 6 hours. Fe^{2+}-H_2O_2 was used as redox initiation system in the copolymerization reaction. The kinetic behaviour of the free and immobilized enzyme of the CM-cellulose hydrolysis has been compared (K_m values 30 mM), and the V_{max} values differed by a factor of 1.8. The immobilized enzyme has been reused in a series of hydrolysis cycles, and no significant activity decrease nor any inhibition by metabolites was observed.

The kinetic principles of the formation of D-glucose and cellobiose in the hydrolysis of microcrystalline cellulose under the action of cellulase complexes from eight different sources have been studied experimentally.[194] By successive addition of individual component of the cellulase complex (endo-(1 → 4)-β-D-glucanase and β-D-glucosidase) to the reaction system, the steps limiting the rate of enzymic hydrolysis of microcrystalline cellulose were revealed. It was shown that in most of the cases studied the step determining the rate of formation of D-glucose with the participation of intermediate cellobiose is the action of β-D-glucosidase. Only in one case (complex from Aspergillus foetidus, enriched with β-D-glucosidase) is the rate of formation of D-glucose from microcrystalline cellulose limited by the action of the endo-(1 → 4)-β-D-glucanase of the cellulase complex. In accordance with the kinetic principles developed, it was shown that when an excess of β-D-glucosidase is added to the reaction system the step limiting the rate of hydrolysis of microcrystalline cellulose under the action of all the cellulase complexes studied becomes the attack on the initial insoluble substrate by endo-(1 → 4)-β-D-glucanase. Under the same conditions, a linear correlation was found between the steady-state rate of formation of D-glucose from microcrystalline cellulose under the action of all the cellulase complexes studied, on the one hand, and the activity of endo-(1 → 4)-β-D-glucanase in these complexes, on the other. It was shown that the action of all the cellulase complexes studied (selected rather arbitrarily) is described by essentially the same kinetic principles, which is evidence of the same mechanisms of the hydrolysis of insoluble cellulose by cellulase preparations of different origin.

A simple procedure that uses a cellulose-enriched culture started from sewage sludge has been developed for producing cellulolytic enzymes and converting cellulose to acetic acid rather than to methane and carbon dioxide.[333] In this procedure, the culture which concerts cellulose to CH_4 and CO_2 was mixed with a synthetic medium and cellulose and heated to 80°C for 15 min before incubation. The end-products formed were acetic acid, propionic acid, CO_2, and traces of ethanol and H_2. Supernatants from 6- to 10-day-old cultures contained 16 to 36 mM acetic acid. Cellulolytic enzymes in the supernatant were stable at 2°C under aerobic conditions for up to 4 weeks and had the ability to hydrolyse carboxymethylcellulose, microcrystalline cellulose, cellobiose, xylan, and filter paper to reducing sugars.

The effects of the surfactant Tween 80 on the enzymic hydrolysis of newspaper have been tested.[334] By monitoring sugar production it was found that the surfactant increased the rate and extent of cellulose saccharification. The rate of enzyme usage in the hydrolysis reactor was improved by 33%. In addition, in the presence of surfactant the recovery of enzymes was higher. Analysis of the enzyme solution showed that with Tween 80 present larger fractions of enzyme remained in solution throughout hydrolysis. It appeared that the surfactant hindered the immobilization of the enzymes on the substrate by reducing the strength of adsorption.

The effect of mild sodium hydroxide treatments on sugar-cane cellulosic wastes (bagasse, pith, and straw) to increase their biological degradability has been studied.[335] At a level of 8% NaOH (on a dry-matter basis) 60% digestibility measured by the in vitro technique was achieved for all materials tested. Indirect methods to predict the digestibility of treated materials such as the bacterial degradability, enzymatic degradability, hot-water solubility, and chemical oxygen demand were tried as alternative methods to the rumen fluid technique. High correlation coefficients for all materials were obtained with all alternative techniques. An important reduction of time and reagents was achieved by the utilization of the solubility and chemical-oxygen-demand tests.

It has been shown theoretically and experimentally that polyenzyme cellulase complexes can catalyse reactions in kinetic systems that are characterized by the absence of any one definite step limiting the rate of the enzymatic hydrolysis of cellulose, even when the rates of the individual steps of the process differ substantially.[336] Such a peculiarity of the kinetic behaviour of cellulase complexes is due to the presence of shunting pathways in a series of successive-parallel reactions of enzymatic degradation of cellulose to D-glucose. In such systems the equilibrium rate of formation of the end-product is determined, as a rule, by the aggregate of the rates of several steps simultaneously. A consequence of this reaction mechanism is the wide variety of the kinetic principles of the accumulation of D-glucose (the end-product) when the composition of individual components of the polyenzyme cellulase system is varied. An illustration of the kinetic principles of the enzymatic hydrolysis of insoluble cellulose (cotton linter) is cited, using ten cellulase complexes from fungi of the genera Trichoderma, Geotrichum, and Aspergillus. It was shown that the value of the equilibrium rate of D-glucose formation is determined, as a rule, by the action of two or three cellulolytic components of the complex. The presence of a single mechanism of the hydrolysis of cellulose for all the cellulase complexes studied, regardless of their composition or origin, was demonstrated theoretically and experimentally.

Modes of action of exo- and endo-cellulases have been investigated in the degradation of celluloses I and II.[337] Cotton and Valonia cellulose (cellulose I) were readily attacked by endo-cellulase of a highly endowise-hydrolysis type, with a sharp decrease in the degree of polymerization, while the simultaneous production of reducing sugar was low. In contrast, viscose rayon and alkali cellulose (cellulose II) showed little lowering of the degree of polymerization by *endo*-cellulase, though the effect was slightly greater than with *exo*-cellulase, while the simultaneous production of reducing sugar was very high. The synergistic effect of *exo*- and *endo*-cellulases on the hydrolysis of cellulose was much higher with cellulose I than with cellulose II. An explanation of these results was offered in terms of differences in the polarity of cellulose chains in the crystallinity and/or ultrastructure of the cellulose fibres.

The kinetics of the hydrolysis of cotton lint (short-fibred cotton) by eight cellulase complexes from fungi of the genera *Trichoderma*, *Geotrichum*, and *Asperigullus* were studied under steady-state and non-steady-state conditions.[338] It was found that the steady-state rate of D-glucose formation is proportional to the content of *endo*-D-glucanase in the cellulase preparations (which was observed for seven of the complexes studied and also after enrichment of the cellulase complex with other components). A mechanism of the enzymatic hydrolysis of cellulose, according to which *endo*-glucanase is the first enzyme to act on the insoluble substrate, is proposed on the basis of a kinetic analysis of the multienzyme cellulase system and substantiated. The concept that under certain experimental conditions a linear correlation must be observed between the rate of hydrolysis of native cellulose, on the one hand, and its soluble polymeric derivatives (in particular, carboxymethylcellulose), on the other, is substantiated theoretically and experimentally. The data obtained refute the hypothesis discussed in the literature for the last 30 years concerning the presence in cellulase complexes of a non-hydrolytic C_1 enzyme, which presumably first attacks native cellulose. The results of the work show that the role of the hypothetical C_1 enzyme is in fact performed by an *endo*-D-glucanase of statistical action.

Cellulose from the Gram-negative bacterium *Acetobacter xylinum* has been used as a model substrate for visualizing the action of cellulase enzymes from the fungus *Trichoderma reesei*.[339] High-resolution electron microscopy reveals that *A. xylinum* normally produced a ribbon of cellulose that is a composite of bundles of crystalline microfibrils. Visual patterns of the process of cellulose degradation have been established. Enzymes are initially observed bound to the cellulose ribbon. Within 10 min, the ribbon is split along its axis into bundles of microfibrils which are subsequently thinned until they are completely dissolved within 30 minutes. Incubations with purified components of the cellulase enzyme system produced less dramatic changes in ribbon structure. Purified 1,4-β-D-glucan cellobiohydrolase I produced no visible change in cellulose structure. Purified endo-1,4-β-D-glucanase produced some splaying of ribbons into microfibril bundles. In both cases, whole ribbons were present even after 60 minutes of incubation, visually confirming the synergistic mode of action of these enzymes.

The influence of major structural features of cellulose on rate of enzymatic hydrolysis, has been investigated.[340] Cellulosic samples with a wide range of crystallinity indices and specific surface areas showed a relationship between their rates of hydrolysis, and the two structural parameters, crystallinity index and specific surface area, were examined.

The binding of cellulases to their substrate at low temperature has been used to purify these enzymes.[341] The cellulases are adsorbed at 0°C and then released from the enzyme-substrate complex at 50°C. For the adsorption step a great excess of enzyme is used to reduce binding of closely related enzymes to the cellulose. Non-bound proteins are removed by extensive washing. The adsorbed enzymes are released by their own catalytic action on the substrate. The small polysaccharides formed from cellulose in this reaction are easily separated from the enzyme. The purity of enzymes isolated by the adsorption method was assayed by immunoelectrophoresis and the course of enzyme action on substrate during the isolation procedure followed by electron microscopy.

A comparative investigation of various cellulase assay procedures has been performed.[342] The cellulolytic activity of crude enzyme preparations from different cellulolytic fungi was assayed comparatively with several common analytical procedures described in the literature. The investigation was carried out with the objective of evaluating, with raw culture filtrates, the different cellulase tests in relation to their specificity for endo- and exo-cellulase action as well as to allow comparisons to be made between results from different research groups using different methods.

A new ultrasonic method has been developed for determining the composition and properties of individual components of polyenzyme systems without their preliminary separation.[197] The method is based on a determination of the pH dependence of the rate constants of inactivation of individual components of the enzyme system under the action of cavitational ultrasound. The method was used to study the cellulase complex from the fungus Geotrichum candidum. It was shown that this complex contains at least four cellulolytic enzymes: *endo*-(1 → 4)-β-D-glucanase, *exo*-(1 → 4)-β-D-glucanase, β-D-glucanase, β-D-glucosidase, and aryl-β-D-glucosidase. These enzymes differ in values of p*K* of the ionogenic groups, controlling the pH profiles of ultrasonic inactivation, as well as in values of the rate constants of inactivation of the form of the active site most stable to the action of ultrasound.

There are many methods of determining the activity of cellulolytic enzymes and numerous ways of expressing this activity.[343] Since, however, the results obtained in different research centres are almost completely incomparable, an attempt has been made to introduce a uniform system for expressing a unit of enzyme activity. This system is based on the relationship between the degree of enzyme dilution and the enzyme activity. The slope coefficient *a*, determined experimentally, is constant for a given method of determining enzyme activities by reducing sugars irrespective of the substrate used, time of reaction, or composition of cellulolytic complex.

A radiometric microassay for cellulase activity measures the release of ^{14}C products from {U-^{14}C}cellulose.[344] The assay is simple and utilizes very small incubation volumes. Activity is evident within the first 30 minutes of reaction. This method should be particularly suited to activity surveys of culture filtrates from fermentations as well as the analysis of fractions from enzyme purification studies.

Extensively ball-milled cellulose fibres have been used as natural substrate for the determination of cellulase activity.[345] This physical treatment breaks the large cellulose fibres to small but insoluble particles yielding a substrate accessible for complete enzymatic breakdown. The activity of cellulases was estimated from the decrease in optical density of ball-milled suspension of fibres and simultaneous measurement of liberated sugars during hydrolysis. A good correlation was found between the initial rate of reaction and the amount of sugar released at given times.

A convenient zymogram stain for cellulases has been reported.[346] The release of dye from phospheric acid-swollen cellulose azure by *endo*-cellulases and *exo*-cellulases is the basis of a technique for locating these enzymes on gels.

An automatic on-line cellulase assay in computer-coupled pilot fermentation has been reported.[347] The measurement was based on the use of dyed Avicel cellulose as substrate. The computer was programmed to calculate the cellulase activity during fermentation. The measured activity values were corrected against a standard sample of known activity by means of a mathematical model for the calibration curve which was stored in the computer.

Four species of tropical earthworm have been found to differ with regard to their enzyme activity.[348] The maximum activity of protease and of cellulase occurred in the posterior region of the gut of the earthworms. On average Octochaetona surensis shows maximum activity and Drawida calebi shows minimum activity for all the enzymes studied.

In a comparative study of carbohydrase activities in marine invertebrates 103 species of marine invertebrates, belonging to 7 types, Spongia, Coelenterata, Annelida, Arthropoda, Mollusca, Echinodermata, Chordaya, were tested for laminarinase, cellulase, and amylase activities.[299]

The application of carbohydrases to the extraction of proteins from commercial full-fat bran using a wet alkaline procedure at pH 8.5 obtained an extract containing 30% of the total nitrogen originally present in the bran.[349] However, the yield of total nitrogen in the extract could be increased to 38.5% with a pretreatment containing carbohydrases (cellulase, xylanase, and poly-D-galacturonase) followed by the normal extraction procedure.

The purification of a cellulase isoenzyme with a pI of 9.5 from kidney-bean abscission zones has been described.[350] An important step in the purification involved the adsorption of the cellulase isoenzyme onto an affinity column of CF-11 cellulose and the subsequent elution with cellobiose. Native and SDS polyacrylamide gel electrophoresis established that there was only one component in the purified cellulase samples. Antibodies raised against the purified pI 9.5 cellulase precipitated this isoenzyme from crude or purified solutions but did not cross-react with pI 4.5 cellulase from 2,4-D-treated abscission zones. The antibody was shown to be monospecific by immunoelectrophoresis and by the fact that it precipated only a single ^{14}C-labelled protein from an adscission zone extract heavily labelled with ^{14}C amino acids.

The hydrolyses of cellulose substrates by cellulase culture filtrates from three mutant strains of Trichoderma reesei grown on lactose and on cellulose have been compared.[351] Cellulose culture filtrates contained five to six times as much cellulase as lactose culture filtrates. Unconcentrated cellulose culture filtrates produced up to 10% sugar solutions from 15% cellulose in 24 hours. Specific activity in enzyme assays and efficiency in saccharification tests were low for enzymes from all the mutants. Over a wide range the percent saccharification of a substrate in a given time was directly proportional to the logarithm of the ratio of initial concentrations of enzyme and substrate. As a result of this, dilute enzyme was more efficient than concentrated enzyme, but if high sugar concentrations are desired very large quantities of enzyme are required. Since the slopes of these plots varied, the relative activity of cellulase on different substrates may be affected by enzyme concentration.

A mutant strain with increased production of cellulolytic enzymes has been induced from the good cellulase producer *Trichoderma reesei* QM 9414.[209] Cellulase activities of the mutant in fermenter cultivations were increased two- to three-fold and β-D-glucosidase activity up to six-fold when compared to the corresponding activities produced by QM 9414.

Cellulolytic enzyme complexes obtained from mutants of *Trichoderma reesei* with enhanced cellulase production have been characterized.[210] The cellulolytic enzyme complexes secreted by the fungus *T. reesei* QM 9414 and its mutants M 5, M 6, MHC 15, and MHC 22 were characterized by determining their specific filter-paper, carboxymethylcellulase, and β-D-glucosidase activities. They were characterized further by measuring their carboxymethylcellulase and β-D-glucosidase profiles after separation on an isoelectrofocusing column over pH range 3-10. While the overall filter-paper activity was roughly equal in all preparations, the specific β-D-glucosidase activity was highest in mutants MHC 15 and MHC 22, which are distinguished morphologically from the parent strain, QM 9414, by a higher degree of branching of their hyphae. Two peaks of β-D-glucosidase activity were detected by isoelectrofocusing in preparations from QM 9414 and M 6, none in the enzyme from the mutant M 5, while 3 and 4 peaks, respectively, were found in preparations from morphological mutants MHC 15 and MHC 22. The higher β-D-glucosidase activity in these last two preparations was also reflected in the higher D-glucose to cellobiose in the initial stages of cellulose hydrolysis by the individual enzyme preparations.

The production of cellulases and hemicellulases has been studied with *Trichoderma reesei* Rut C-30.[208] This organism produced, together with high cellulase activities, considerable amounts of xylanases and β-D-glucosidase. Three cellulose concentrations (1.0, 2.5, and 5.0%) were examined to determine the maximum levels of cellulase activity obtainable in submerged culture. Temperature and pH profiling were used to increase cell mass to maximum levels within two days and thereby enhancing fermentor productivity at higher substrate levels. The effects of temperature, pH, Tween-80 concentration, carbon source, and substrate concentration on the ratio of mycelial growth and extracellular enzyme production were described.

The potential for increased production of cellulase of Trichoderma sp. by pH cycling and temperature profiling has been investigated.[352] Cultivation of T. QM 9414 on 3% cellulose medium (C/N ratio = 8.5) produced 4.5 IU ml^{-1} cellulase in 180 hours at a cell growth of 8.0 g $litre^{-1}$ (0.266 g cell g^{-1} cellulose). It corresponded to an average cellulase productivity of 25.0 IU $litre^{-1}$ h^{-1} (3.5 IU g^{-1} cell h^{-1}). In the same medium 9.5 g $litre^{-1}$ cell mass (0.316 g cell g^{-1} cellulose), 6.2 IU ml^{-1} cellulase, and 38.75 IU $litre^{-1}$ h^{-1} (4.0 IU g^{-1} cell h^{-1}) cellulase productivity could be obtained using pH-cycling conditions during cultivation. Cell mass, cellulase yield, and productivity were further increased to 10.0 g $litre^{-1}$, 7.2 IU ml^{-1}, and 44.0 IU $litre^{-1}$ h^{-1} (4.5 IU g^{-1} cell h^{-1}), respectively, by simultaneous pH cycling and temperature-profiling strategy.

An enzyme preparation from a mutant strain of cellulomonas CS1-17 has been found to act synergistically with low levels of Trichoderma reesei cellulase in saccharification of alkali-pretreated sugar-cane bagasse and in assays of filter-paper activity.[353] Supplementation of the Cellulomonas preparation with T. reesei cellulase provided a particularly active preparation.

The regulation of the cellulolytic system in Trichoderma reesei by sophorose has been investigated through measurement of induction of cellulase and repression of β-D-glucosidase.[212] Sophorose has two regulatory roles in the production of cellulase enzymes in T. reesei: β-D-glucosidase repression and cellulase induction. Sophorose also is hydrolysed by the mycelial-associated β-D-glucosidase. Repression of β-D-glucosidase reduces sophorose hydrolysis and thus may increase cellulase induction.

The inactivation of the cellulase of Trichoderma reesei by sheer forces is of sufficient magnitude to merit consideration in the design of equipment for the enzymatic hydrolysis of cellulose.[354] The inactivation constant, k_d, is a function of the flow rate of the enzyme solution through a fine capillary tube; $\underline{k}_d$ increased slowly at low shear stress, and much more rapidly when the shear stress was greater than 15 dynes cm^{-2}.

In the light of the potential application of Trichoderma reesei mutants in the commercial saccharification of cellulose, the biochemical nature of the cellulases from these mutants has been examined in an effort to define the enzyme complex in relation to its saccharification efficiency and to gain knowledge which will contribute toward a better understanding of how to isolate more efficient mutants for enzyme production.[355]

The extracellular cellulase activities of Clostridium thermocellum LQRI and Trichoderma reesei QM 9414 have been compared.[207] The crude extracellular cellulase of C. thermocellum LQRI (virgin strain) was very active and solubilized microcrystalline cellulose at one-half the rate observed for the extracellular cellulase of T. reesei QM 9414 (mutant strain). C. thermocellum cellulase activity differed considerably from that of T. reesei as follows: higher endo-D-glucanase/exo-D-glucanase activity ratio, absence of extracellular β-D-glucosidase or β-D-xylosidase activity, long-chain oligosaccharides instead of short-chain oligosaccharides as initial (15 min) hydrolytic products on microcrystalline cellulose, mainly cellobiose or xylobiose as long-term (24 h) hydrolysis products of Avicel and MN300 or xylan, and high activity and stability at 60 to 70°C. Under optimized reaction conditions, the kinetic properties ($\underline{V}_{max}$ 0.4 µmol min^{-1} per mg of protein, energy of activation 33 kJ, temperature coefficient 1.8) of C. thermocellum cellulose-solubilizing activity were comparable to those reported for T. reesei, except that the dyed Avicel concentration at half-maximal velocity was two-fold high (182 µM). The cellulose-solubilizing activity of the two crude cellulases differed considerably in response to various enzyme inhibitors. Most notably, Ag^{2+} and Hg^{2+} effectively inhibited C. thermocellum but not T. reesei cellulase at <20µM, whereas Ca^{2+}, Mg^{2+}, and Mn^{2+} inhibited T. reesei but not C.thermocellum cellulase at >10 mM. Both enzymes were inhibited by Cu^{2+} (>20 mM), Zn^{2+} (>1.0 mM), and ethylene glycol-bis(β-aminoethyl ether)-N,N-tetra-acetic acid (>10 mM). T. reesei but not C. thermocellum cellulose-solubilizing activity was 20% inhibited by D-glucose (73 mM) and cellobiose (29 mM). Both cellulases preferentially cleaved the internal glycosidic bonds of cello-oligosaccharides. The overall rates of cello-oligosaccharide degradation were higher for T. reesei than for

C. thermocellum cellulase, except that the rates of conversion of cellohexaose to cellotriose were equivalent.

A method has been developed for the selective purification of a low-molecular-weight endo-D-glucanase (M_r 13,000) of the cellulase complex from Trichoderma koningii.[356] The enzyme was obtained in the form of a lyophilized preparation completely free from cellobiase activity. It was found that cellobiose and methyl cellobioside activate (6-fold at a saturating concentration) the low-molecular-weight endo-D-glucanase and almost completely inhibit the activity of high-molecular-weight endo-D-glucanases. Cellobiose had a weak inhibitory effect on the unfractionated cellulase complex from T. koningii. A kinetic analysis of this phenomenon showed that the activating effect of cellobiose or its analogues on the endo-D-glucanase takes place by a transglycosylation mechanism. The multiple kinetic manifestations of cellobiose, which acts as an activator or inhibitor of endo-D-glucanases from various sources, are explained by the varying amounts of the low- and high-molecular-weight endo-D-glucanases as well as by their capacity for transglycosylation and the transfer of the reducing groups of the reaction products to effectors acting as additional nucleophilic agents in enzymatic catalysis.

The thermal-stability characteristics of the cellulase enzymes present in culture filtrates of the thermophilic fungus Sporotrichum thermophile have been investigated at different temperatures and at different times of exposure.[200] Maximum enzymic activities under assay conditions were found at 68°C for the filter-paper activity (FPA) and the Cx activity (carboxymethylcellulose), while the maxima for the C_1 activity (cotton) and β-D-glucosidase activity (cellobiose) were found to be at 55°C and 72°C, respectively. After 48 h exposure time of culture filtrates to 50°C the residual activities for the FPA, Cx, and β-D-glucosidase were found to be 88%, 98%, and 93% of the original activities, respectively. D-Glucose, cellobiose, Avicel, and Solka Floc have been utilized as substrates for growth of Thermomonospora sp. in order to study the induction-repression characteristics of its associated cellulase system.[357] While D-glucose proved to be an effective repressor of the cellulase enzymes, the other three substrates induced relatively high levels of enzyme activity as measured by the filter-paper assay. On a unit-cell-mass basis the highest values of cellulase activity were obtained when Avicel was utilized as the carbon energy source. The nature of the cellulosic material and its initial concentration were identified as two very important parameters of the induction process.

Humicola insolens, a thermophilic fungus isolated from manure and compost heaps, has been found to produce significant amounts of thermostable cellulases in cultures on wheat-bran medium (50°C, 4 days).[358] The mould bran extract hydrolysed Avicel, carboxymethylcellulose, and newsprint at 90%, 45%, and 35%, respectively, to D-glucose. Then Avicelase and carboxymethylcellulase were purified from the culture extract by adsorption onto Avicel, heat and acid treatment, and consecutive column chromatographies to a homogeneous state on polyacrylamide gel electrophoresis. The purified cellulases, especially carboxymethylcellulase, were found to be highly thermostable (temperature optima 50°C). Avicelase was stable after heating at 65°C for 5 min and carboxymethylcellulase retained 45% of the original activity after heating at 95°C for 5 min.

An inhibitor of cellulase has been isolated from Dictyostelium discoideum in the culmination stage and partially purified.[359] This inhibitor was shown to be a protein of molecular weight 20,000; at pH 6.5 it inhibited the two cellulases present in D. discoideum but not cellulases from other organisms. Inhibition was prevented At pH 8.0 but could be recovered by adjustment of the pH to 6.5. The inhibitor was present in all stages of development from amoebae to spores but disappeared as the spores aged and was absent from germinating cells, indicating that specific destruction of this inhibitor may be responsible for the activation of the cellulases.

It has been found that the white form of Talaromyces emersonii produces more cellulase activity than the green or red/brown variants when grown on cellulose-peptone media.[360] The maintenance of the level of enzyme production was achieved with concomitant reduction in medium costs.

The cellulose degradation and carboxymethylcellulase production by Sporocytophaga myxococcoides have been investigated by growing the organism in a 3l air-lift fermenter using a medium containing 2% w/v insoluble cellulose.[361] The cellulose content of the medium reduced the k_La of the fermenter but during growth the dissolved oxygen concentration did not fall below 75% saturation. Rates of cellulose degradation and extracellular enzyme production were similar to those reported for a stirred-tank fermenter.

A new cellulase has been purified from Aspergillus aculeatus.[362] The enzyme was less or only slightly active on conventional substrates such as Avicel, carboxymethylcellulose, cellobiose, and salicin. However, the enzyme was potently active on amorphous cellulose such as insoluble cello-oligosaccharides, phosphate-swollen cellulose, and alkali-swollen cellulose. Thus the enzyme was classified as a cellulase. It was suggested that this enzyme may play an important role in saccharification of cellulose.

A new endo-cellulase component of carboxymethylcellulase type (E_n-1) has been obtained by gel filtration and column chromatography from Driselase, a commercial enzyme preparation from Irpex lacteus.[363] The enzyme behaved as a single protein on SDS polyacrylamide disc electrophoresis (mol. wt. 15,000) and it contained only 0.73% carbohydrate as D-glucose. The pattern of its amino acid composition is similar to those of other cellulases in respect of high contents of L-glycine, L-serine, and L-threonine. The cellulase was most active at pH 4.0 and was very stable in the pH range of 3.0 to 6.0, but was completely inactivated by heating at 70°C for 10 min. A series of cello-oligosaccharides, including cellobiose, was formed by this enzyme from carboxymethylcellulose as well as from water-insoluble celluloses. In the hydrolysis of carboxymethylcellulose, the increase in the fluidity of the substrate was relatively large as compared with the simultaneous increase in reducing power. From this result and the pattern of hydrolysis products, E_n-1 was elucidated to be an endo-cellulase, and it showed the highest randomness among the cellulase components obtained so far from Irpex lacteus.

A cellulolytic enzyme system has been identified in Acetivibrio cellulolyticus, a newly isolated anaerobe.[364] Crude enzyme preparations from the organism converted ball-milled pulp, cotton batting, filter and tissue paper, a microcrystalline cellulose, carboxymethylcellulose, cellobiose, and xylan to reducing sugars. The preparations showed maximum activity between pH 5 and 6 and at a temperature between 37 and 50°C, depending on the substrate used. The enzyme activity was fairly stable at 2°C for 4 weeks. The saccharifying ability of the preparation was comparable to that of commercially available cellulase preparations from Aspergillus niger and Trichoderma viride.

The regulation of cellulase synthesis in Acetivibrio cellulolyticus isolated from a cellulose-enrichment culture degraded cellulose by the secretion of cellulolytic enzymes into the culture medium.[365] Both exo-D- and endo-D-glucanase activities were detected and shown to be regulated by induction and catabolite repression. endo-D-Glucanase synthesis was induced by cellulose, cellobiose, and salicin. Synthesis induced during growth on cellobiose was inhibited, although not completely repressed, by D-glucose. exo-D-Glucanase activity was enhanced when cells were grown on insoluble poly-β-D-glucoside polymers. Activity was progressively inhibited by supplementing microcrystalline substrates with increasing amounts of cellobiose.

The fermentation of various saccharides derived from cellulosic biomass to ethanol has been examined in mono- and co-cultures of Clostridium thermocellum strain LQRI and C. thermohydrosulphuricum strain 39E. C. thermohydrosulphuricum fermented D-glucose, cellobiose, and xylose, but not cellulose or xylan, and yielded ethanol/acetate ratios of 1.0.[366] At non-limiting cellulosic substrate concentrations (1%), C. thermocellum cellulase hydrolysis products accumulated during monoculture fermentation of Solka Floc cellulose and included D-glucose, cellobiose, xylose, and xylobiose. A stable co-culture that contained nearly equal numbers of C. thermocellum and C. thermohydrosulphuricum was established that fermented a variety of cellulosic substrates, and the ethanol yield observed was twofold higher than in C. thermocellum fermentations. The metabolic basis for the enhanced fermentation effectiveness of the co-culture on Solka Floc cellulose included: the ability of C. thermocellum cellulase to hydrolyse α-cellulose and hemicellulose, the enhanced utilization of mono- and di-saccharides by C.thermohydrosulphuricum increased cellulose consumption, threefold increase in the ethanol production rate, and twofold decrease in the acetate production rate. The co-culture actively fermented MN300 cellulose, Avicel, Solka Floc, SO_2-treated wood, and steam-exploded wood. The highest ethanol yield obtained was 1.8 mol of ethanol per mol of D-glucose unit in MN 300 cellulose.

The degradation of softwood, hardwood, and grass lignocelluloses by two Streptomyces strains has been investigated. S. viridosporus T7A and S. setonii 75V12 were grown on softwood, hardwood, and grass lignocelluloses, and the lignocellulose decomposition was followed by monitoring substrate weight, lignin, and carbohydrate losses over time.[367] Results showed that both Streptomyces strains substantially degraded both the lignin and the carbohydrate components of each lignocellulose. However, these actinomycetes were more efficient decomposers of grass lignocelluloses than of hardwood or softwood lignocelluloses.

A synthetic medium for the cellulolytic anaerobe

Ruminococcus albus has been described[368]

A study of cellulase and protein production by Chaetomium cellulolyticum strains grown on cellulosic substrates has been reported.[369] Fermentation with C. cellulyticum was carried out on media containing either Avicel cellulose or newspaper. Production of free cellulolytic enzymes, cellulose degradation, and the formation on cell protein were studied with the original and mutant srains.

The variation of bagasse crystallinity and cellulose activity during the fermentation of Cellulomonas bacteria has been investigated.[370] At the early stage of the fermentation the crystallinity index of bagasse increased, suggesting that the major metabolized fraction corresponded to the hemicellulose during this stage. Later the crystallinity achieved a steady state and then decreased, which indicated that the most complex structure of bagasse was being attacked. The analysis of the cellulolytic activity of extracellular enzyme in the medium showed a sharp increase followed by an abrupt levelling off and decline in activity. These results along with the reduction of crystallinity index and bagasse utilization (70%) led the authors to the conclusion that the C_1 component was present in cellulase complex synthesized by the bacteria.

An enzyme preparation from a Cellulomonas strain has been shown previously to be active in releasing reducing sugars from alkali pretreated sugar-cane bagasse.[371] This enzyme preparation has been demonstrated to be very resistant to end-product inhibition by D-xylose, D-glucose, cellobiose, and ethanol.

During growth of Bacteroides succinogenes in a liquid medium with cellulose as the carbohydrate source, greater than 80% of the endo-(1 → 4)-β-D-glucanase, xylanase, and aryl-β-D-xylosidase and 50% of the aryl-β-D-glucosidase was released from cells into the culture fluid.[227] Less than 25% of the β-D-glucosidase activity was detected in the culture fluid. Approximately 50% of each of the released enzymes measured was associated with sedimentable subcellular membrane vesicles. Many vesicles were seen adhering to cellulose, and they were also seen free in the culture fluid. These data suggested that B. succinogenes releases hydrolytic enzymes in non-sedimentable and particulate forms during growth by a mechanism which has until now received little attention. Cellulose incubated in a porous nylon bag in the rumen was colonized by bacteria resembling B. succinogenes, and subcellular vesicles were seen penetrating channels and fractures in the cellulose. It was suggested that B. succinogenes cells in the rumen contribute to an extracellular population of subcellular vesicles that possess cellulolytic and hemicellulolytic activities which probably enhance polyment digestion and provide a source of sugars for microbes lacking polymer-degrading activity, thereby contributing to a stable heterogeneous microbial population.

The production of carboxymethylcellulase and β-D-glucosidase by Clostridium acetobutylicum has been investigated in an industrial medium.[225] The cellulase was induced by molasses, and it was not repressed by D-glucose. Optimum carboxymethylcellulase activity occurred at pH 4.6 and 37°C.

21 Chitinases

An in vitro system described for the study of chitinase involves soluble enzyme protein(s) and an insoluble substrate preparation.[372] With insect molting fluid chitinase, it shows properties that parallel those observed during in vivo breakdown of cuticle during the molt. For example, molting fluid chitinase activity not previously exposed to chitin is strongly and specifically adsorbed to the substrate, in contrast to other enzymatic activities including β-D-2-acetamido-2-deoxyglucosidase present in molting fluid. This leads to partial purification of molting fluid chitinase activity reflected in increased specific activity of chitinase associated with the insoluble chitin substrate. Molting fluid chitinase activity may involve a number of polypeptides ranging in molecular weight from 145,000 to less than 20,000 daltons. The system described gives results consistent with a processive mechanism for molting fluid chitinase, i.e. data are given demonstrating that molting fluid chitinase continues to act on the same chitin particle(s) with which it initially associates rather than diffusing freely from substrate particle, and the product of its action appears to be a monosaccharide rather than a mixture of oligosaccharides. Processive behaviour for chitinase would be predicted from the known structure, and the in vivo measured rate of breakdown, of cuticle chitin during the molt. The preliminary nature of this conclusion, based on what is so far known about the structure of the substrate used in the in vitro system, is briefly discussed.

High chitinase and lysozyme activities have been found in Leydig's organ of Etmopterus spinax, Somniosus microcephalus, and Torpedo nobilians, in Leydig's and epigonal organs and spleen of Raja radiata, and in the epigonal organ of Rhinoptera bonasus.[19] Strong chitinase activity with little or no lysozyme activity was noted in Leydig's and epigonal organs of Scyliorhinus canicula and in the epigonal organ of Ginglymostoma cirratum and Heterodontus francisci.

Chitinase production by Conidiobolus lamprauges and other Conidiobolus species has been described.[373] The production of chitinase is reported even in a medium free of chitin. The substrate specificity of the enzyme was tested with chitin gel, colloidal chitin, chitin powder, and ethylene glycol chitin.

A chitinase-overproducing mutant of Serratia marcescens has been isolated.[374] After mutagenesis with ultraviolet light, ethyl methane sulphonate, or N-methyl-N'-nitro-N-nitrosoguanidine, 19,940 colonies were screened for production of enlarged zones of clearing (indicative of chitinase activity) on chitin-containing agar plates. Forty-four chitinase high producers were tested further in shake flask cultures. Mutant IMR-1E1 was isolated, which, depending on medium composition, produced two to three times more endo-chitinase activity and other components of the chitinolytic enzyme system than the wild type. After induction by chitin, endo-chitinase and chitobiase activity appeared at similar times for both IMR-1E1 and QMB1466, suggesting possible co-ordinate control of these enzymes. The results were consistent with IMR-1E1 containing a regulatory mutation which increased production of the components of the chitinolytic enzyme system and/or with IMR-1E1 containing a tandem duplication of the chitinase genes. The high rate of reversion of IMR-1E1 to decreased levels of chitinase production suggested that the overproduction of chitinase by IMR-1E1 is due to a tandem gene duplication.

22 Dextranases

A limit dextrinase has been purified 2,700-fold from ungerminated peas by affinity chromatography.[375] The enzyme hydrolyses (1 → 6)-α-D-glucosidic linkages in alpha-limit dextrins containing at least one (1 → 4)-linked α-D-glucose residue on either side of the susceptible linkage. The limit dextrinase also hydrolyses the polysaccharides amylopectin, amylopectin beta-limit dextrin, glycogen beta-limit dextrin, and pullulan, but has no activity towards glycogen.

A technique has been described for the screening of dextranase-producing micro-organisms.[376] The culture of fungi (556 strains) and streptomyces (115 strains) was performed by the koji method on wheat bran and submerged method in liquid media. The culture media were extracted and the aqueous extracts screened for endo-dextranolytic activity by incubation with a 5% dextran solution. The activity of the enzyme was measured by viscometry. Eight strains were selected and the pH activities and stabilities of their dextranases were described.

The general properties and specificity of a dextran (1 → 2)-debranching enzyme from Flavobacterium have been examined in order to apply this enzyme to the structural analysis of highly branched dextrans.[377] The optimum pH range and temperature were pH 5.5 - 6.5 and 45°C, respectively. The enzyme was stable up to 40°C on heating for 10 min, and over a pH range of 6.5 - 9.0 on incubation at 4°C for 24 h. The effects of various metal ions and chemical reagents have also been examined. The debranching enzyme has a strict specificity for the (1 → 2)-α-D-glucosidic linkage at branch points of dextrans and related branched oligosaccharides, and produces D-glucose as the only reducing sugar. The degree of hydrolysis of the dextrans by this enzyme and the K_m value (mg ml^{-1}) were as follows: B-1298 soluble, 25.2%, 0.21, B-1299 soluble, 31.5%, 0.27, and B-1397, 11.8%, 0.91. The debranching enzyme thus has a novel type of specificity as a dextranhydrolase. The authors have termed this enzyme a dextran α-(1 → 2)-debranching enzyme, and its systematic name is also discussed.

23 (1 → 3)-α-D-Glucanases

In a new colorimetric method for the detection and assay of (1 → 3)-α-D-glucanases, the enzyme substrate consists of Cibracron Blue F3GA complexed with a dextranase-treated streptococcal glucan.[378] The method is especially convenient for tests involving large numbers of samples, and can be adapted to quantitative as well as qualitative applications. The assay is sufficiently sensitive for screening bacterial samples as potential sources of (1 → 3)-α-D-glucanase.

The purification and properties of endo-(1→ 3)-α-D-glucanase from a Streptomyces chartreusis strain have been reported.[379] An enzyme hydrolysing the water-insoluble glucans produced from sucrose by Streptococcus mutans was purified 6.4-fold from the culture concentrate of S. chartreusis strain F2 by ion-exchange chromatography and gel filtration. Electrophoresis of the purified enzyme protein gave a single band on a sodium dodecyl sulphate polyacrylamide gel slab. Its molecular weight was estimated to be approximately 68,000, but there is a possibility that the native enzyme exists in an aggregated form or is an oligomer of the peptide subunits, having a molecular weight larger than 300,000. The pH optimum of the enzyme was 5.5 - 6.0, and its temperature optimun was 55°C. The enzyme lost activity on heating at 65°C for 10 min. The enzyme activity was completely inhibited by the presence of 1 mM Mn^{2+}, Hg^{2+}, Cu^{2+}, Ag^{2+}, or merthiolate. The K_m value for the water-insoluble glucan of S. mutans OMZ176 was an amount of glucan equivalent to 1.54 mM D-glucose, i.e. 0.89 mM in terms of the α-(1 → 3)-linked-D-glucose residue. The purified enzyme was specific for glucans containing an (1 → 3)-α-D-glucosidic linkage as the major bond. The enzyme hydrolysed the S. mutans water-insoluble glucans endolytically, and the products were oligosaccharides.

24 endo-(1 → 3)-β-D-Glucanases

Action patterns of (1 → 3)-β-D-glucanases from Oerskovia zanthineolytica have been investigated on laminaran, lichenan, and yeast glucan.[380] Four D-glucanases in the crude culture broth act synergistically to degrade the walls of viable yeast. Enzyme I (mol. wt. 18,000) does not lyse viable yeast cells, even though it displays high activity against laminaran, and will readily depolymerize lichenan. Enzyme II is also an endo-glucanase (mol. wt. 29,000) and is active against both viable yeast cells and laminaran, but it exhibits restricted activity against lichenan. Enzymes IIIa and IIIb have molecular weights of 27,000 and 29,000, respectively, and exhibit high lytic activity against viable yeast cells, but release few reducing groups from laminaran or yeast glucan. They exhibit higher specific activity against yeast glucan than against laminaran and yield an oligosaccharide of d.p. 5 as the eventual hydrolysis product. In the initial stages of hydrolysis, they act in an endo manner, but they effect cleavage of a d.p. 14 substrate in an apparent exo manner and have high specificity for straight-chain β-(1 → 3)-linked D-glucans. The proposed systematic name for enzymes IIIa and IIIb is (1 → 3)-β-D-glucan pentaosehydrolase. When yeast glucan was hydrolysed by enzyme IIIb, 40% of the glucan was recovered as the products of d.p. 5 and 9, 20% was solubilized but remained as oligosaccharides having d.p. of between 15 and 35, 24% had a d.p. >35, and 20% was insoluble material that resembled yeast-bud scars microscopically.

Cell-free extracts, membranous fractions, and cell-wall preparations from Schizosaccharomyces pombe were examined for the presence of (1 → 3)-α- and (1 → 6)-β-D-glucanase activities.[381] The various glucanases were assayed in cells at different growth stages. Only (1 → 3)-β-D-glucanase activity was found, and this was associated with the cell-wall fraction. Chromatographic fractionation of the crude enzyme revealed two endo-(1 → 3)-β-D-glucanases, designated as glucanase I and glucanase II. Glucanase I consisted of two subunits of molecular weights 78,500 and 82,000, and glucanase II was a single polypeptide of 75,000. Although both enzymes had similar substrate specificities and similar hydrolytic action on laminarin, glucanase II had much higher hydrolytic activity on isolated cell walls of S. pombe. On the basis of differential lytic activity on cell walls, glucanase II was shown to be present in conjugating cells and highest in sporulating cells. Glucanase II appeared to be specifically involved in conjugation and sporulation since vegetative cells and non-conjugating and non-sporulating cells did not contain this enzyme. The appearance of glucanase II in conjugating cells may be due to de novo enzyme synthesis since no activation could be demonstrated by combining extracts from vegetative and conjugating cells. Increased glucanase activity occurred when walls from conjugating cells were combined with walls from sporulating cells. Studies with trypsin and proteolytic inhibitors suggest that glucanase II exists as a zymogen in conjugating cells. A temperature-sensitive mutant of S. pombe was isolated which lysed at 37°C. Glucanase activity was higher in vegetative cells held at 37°C than cells held at 25°C. Unlike the wild-type strain, this mutant contained glucanase II activity during vegetative growth and may be a regulatory mutant.

25 exo-(1 → 3)-β-D-Glucanases

It has been reported that glycosylation is not necessary for the secretion of exo-(1 → 3)-β-D-glucanase by Saccharomyces cerevisiae protoplasts.[382] It was shown that the normal synthesis and secretion of the major exo-(1 → 3)-β-D-glucanase secreted by S. cerevisiae protoplasts were affected by tunicamycin, and a new, presumably non-glycosylated form is exported to the cellulose medium. A similar form was also secreted in the presence of 2-deoxy-D-glucose.

Partial purification of endo- and exo-(1 → 3)-β-D-glucanase enzymes from Zea mays seedlings and their involvement in cell-wall autohydrolysis have been investigated.[383] Molecular-sieve chromatography of the cell-wall protein resolved endo-β-D-glucanase and exo-(1 → 3)-β-D-glucanase activities when Avena glucan and laminaran, respectively, were employed as substrates. The exoenzyme (mol. wt. 60,000) was strongly inhibited by inorganic mercury at a concentration which suppressed the release of monosaccharide from autolytically active cell wall. The endo-β-D-glucanase (mol. wt. 26,000) showed a marked preference for substrates of mixed-linkage and exhibited features indicating that it initiates the autolytic solubilization of wall glucan.

The purification and some properties of an exo-(1 → 3)-β-D-glucanase from basidiomycete species have been described.[384] An exo-(1 → 3)-β-D-glucanase was purified from the commercial enzyme Kitalase, which is a yeast cell-wall lytic enzyme preparation, by ammonium sulphate fractionation, ion-exchange chromatography, and gel filtration. The optimum pH value was 5.8, and the optimum temperature was 55°C. The enzyme was stable in the pH range of 5.1 - 9.8 and at temperatures below 53°C. The isoelectric point and the molecular weight were estimated to be pH 9.3 and 73,000, respectively. The enzyme was shown to bypass (1 → 6)-linkaged branches to cleave (1 → 3)-linkages when scleroglucan was used as substrate. The K_m values for laminarin, laminaripentaose, laminaritetraose, and laminaritriose were 0.16, 2.01, 2.24, and 1.34 mM, respectively.

26 endo-(1 → 4)-β-D-Glucanases

The specific properties have been examined of the (1 → 4)-β-D-glucanase component of Trichoderma koningii that participates in an early and effective stage of random breakdown of native cellulose to short fibres.[385] The enzyme was purified and freed from associated components of the cellulase complex (particularly β-D-glucosidase) that interfere with and complicate interpretation of the action of such enzymes. Purification increased the specific activity 25-fold over culture filtrates. The enzyme hydrolysed CM-cellulose faster than the purified β-D-glucosidase from the same organism hydrolysed any of its substrates (cellobiose or cellodextrins). The specificity of the glucanase was directed towards soluble derivatives of cellulose. The glucanase (temperature optimum 60°C) attacked larger cellodextrins (cellohexaose to cellotetraose, in that order) much more readily than small dextrins (cellobiose and cellotriose) and released a mixture of products, D-glucose up to cellopentaose. Similar examination of hydrolysates of the reduced cellodextrins showed clearly the high specificity of the enzyme for the central bond of its natural substrates (the cellodextrins), whatever their chain length, and indicated the nature of the enzyme as an endo-glucanase.

A new ultrasonic method has been developed for determining the composition and properties of individual components of polyenzyme systems without their preliminary separation.[197] The method is based on determination of the pH dependence of the rate constants of inactivation of individual components of the enzyme system under the action of cavitational ultrasound. The method was used to study cellulase complex from the fungus Geotrichum candidum. This complex contains at least four cellulolytic enzymes, endo-(1 → 4)-β-D-glucanase, exo-(1 → 4)-β-D-glucanase, β-D-glucosidase, and aryl-β-D-glucosidase, which differ in values of pK of the ionogenic groups, controlling the pH profiles of ultrasonic inactivation, as well as in values of the rate constants of inactivation of the form of the active site most stable to the action of ultrasound.

endo-(1 → 4)-β-D-Glucanase was purified from Clostridium thermocellum by centrifugation, ultrafiltration, selective precipitation, ion-exchange, Sephadex chromatography, and preparative gel electrophoresis.[386] The 22-fold-purified enzyme behaved as a homogeneous protein under non-denaturing conditions. The enzyme represented a significant component (> 25%) of total extracellular endo-glucanase activity, but was purified in low yield by the procedures employed. The native molecular weight of the endo-(1 → 4)-β-D-glucanase was determined by ultracentrifugational analysis, amino acid composition, and polyacrylamide-gel electrophoresis and varied between 83,000 and 94,000. The enzyme contained 11.2% carbohydrate and was isoelectric at pH 6.72. The pH and temperature optima of the endo-glucanase were 5.2 and 62°C, respectively. The enzyme lacked β-cysteine and was low in sulphur-containing amino acids. The purified endo-(1 → 4)-β-D-glucanase displayed: high activity towards carboxymethylcellulose, celloheptaose, cellohexaose, and cellopentaose, low activity towards Avicel microcrystalline cellulose and cellotetraose, no detectable activity towards cellotriose or cellobiose, increased activity towards cello-oligosaccharides with increasing degree of polymerization. The internal glycosidic bonds of cello-oligosaccharides were cleaved by the enzyme in preference to external linkages. K_m and V_{max} for cellopentaose and cellohexaose hydrolysis were 2.30 mM and 39.3 μmol min^{-1} per mg of protein and 0.56 mM and 58.7 μmol min^{-1} per mg of protein, respectively.

The kinetic principles of the formation of D-glucose and cellobiose in the hydrolysis of microcrystalline cellulose under the action of cellulase complexes from eight different sources have been studied experimentally.[194] By successive addition of individual components of the cellulase complex (endo-(1 → 4)-β-D-glucanase and β-D-glucosidase) to the reaction system, the steps limiting the rate of enzymatic hydrolysis of microcrystalline cellulose were revealed. It was shown that in most of the cases studied the step determining the rate of formation of D-glucose with the participation of intermediate cellobiose is the action of β-D-glucosidase. Only in one case (complex from *Aspergillus foetidus*, enriched with β-D-glucosidase) is the rate of formation of D-glucose from microcrystalline cellulose limited by the action of the endo-(1 → 4)-β-D-glucanase of the cellulase complex. In accordance with the kinetic principles developed, it was shown that when an excess of β-D-glucosidase is added to the reaction system the step limiting the rate of hydrolysis of microcrystalline cellulose under the action of all the cellulase complexes studied becomes the attack on the initial insoluble substrate by endo-(1 → 4)-β-D-glucanase. Under the same conditions, a linear correlation was found between the steady-state rate of formation of D-glucose from microcrystalline cellulose under the action of all the cellulase complexes studied, on the one hand, and the activity of endo-(1 → 4)-β-D-glucanase in these complexes, on the other. It was shown that the action of all the cellulase complexes studied (selected rather arbitrarily) is described by essentially the same kinetic principles, which is evidence of the same mechanisms of the hydrolysis of insoluble cellulose by cellulase reparations of different origin.

During growth of Bacteriodes succinogenes in a liquid medium with cellulose as the source of carbohydrate, greater than 80% of the endo-(1 → 4)-β-D-glucanase, xylanase, and aryl-β-D-xylosidase and 50% of the aryl-β-D-glucosidase was released from cells into the culture fluid.[227]

By using QAE-Sephadex A50 chromatography in the presence of 6 M urea, it was possible to split the glucanase complex of Clostridium thermocellum into distinct protein fractions.[387] One of these fractions contained an endo-(1 → 4)-β-D-glucanase which was isolated at a high degree of purity and was identified by its ability to hydrolyse trinitrophenylated carboxymethylcellulose. The enzyme is of monomeric nature, with a molecular weight of 56,000. It has an isoelectric pH of 6.2 and an optimum pH of 6.0. It hydrolysed carboxymethylcellulose and, at a slower rate, cellulose powder. The major end-products of cellulose degradation were D-glucose, cellobiose, and cellotriose. Cellotetraose is formed as an intermediate product.

The formation and location of (1 → 4)-β-D-glucanases and (1 → 4)-β-D-glucosidases have been studied in cultures of Penicillium janthinellum grown on Avicel, sodium carboxymethyl cellulose, cellobiose, glucose, mannose, and maltose.[222] endo-(1 → 4)-β-D-Glucanases were cell free, and their formation was induced by cellobiose. (1 → 4)-β-D-Glucosidases, on the other hand, were formed constitutively and were primarily cell free, but with a small amount strongly associated with the cell wall. A rotational viscosimetric method was developed to measure the total endo-(1 → 4)-β-D-glucanase activity of the culture (broth and solids). By this method, it was possible to determine the endo-(1 → 4)-β-D-glucanase activity not only in the supernatant of the culture but also on the surface of the mycelium or absorbed on residual Avicel. During a 70 litre batch cultivation of P. janthinellum, the adsorption of endo-(1 → 4)-β-D-glucanases by residual and newly added 10% Avicel was measured. The adsorption of soluble protein and endo-(1 → 4)-β-D-glucanase by Avicel was found to be largely independent of the pH value but dependent on temperature.

The kinetics of growth and extracellular cellulase production by Talaromyces emersonii grown on cellulosic material were measured.[198] The enzyme system was found to be comprised of four to five forms of exo-(1 → 4)-β-D-glucanase and at least two forms of endo-(1 → 4)-β-D-glucanase and three β-D-glucosidases. One of the latter, termed β-D-glucosidase III, is induced concurrently with the cellulases but disappears from the medium because of the low pH that develops during growth. The cellulases by contrast are more acid stable. The possible functions of these enzymes are discussed.

Production of extracellular cellulase by an isolate of a Fusarium sp. has been studied in shake cultures, and sequential appearance of cellulase components (β-D-glucosidase on the first day, endo-(1 → 4)-β-D-glucanase and exo-(1 → 4)-β-D-glucanase on the third day of growth on insoluble cellulose) was observed.[211] Maximum production of all these components was achieved on the fifth day. The influence of different nitrogen and carbon sources on cellulase production was investigated. Crude cellulolytic enzyme was used for hydrolysis of common agricultural wastes both with and without sodium hydroxide pretreatment. Analysis of hydrolysates indicated glucose as the major constituent (about 83% of total reducing sugar).

Purification and characterization of the extracellular and intracellular β-D-glucosidases of the thermophillic fungus Talaromyces emersonii have been described. The authors conclude that β-D-glucosidase I and β-D-glucosidase IV are true β-D-glucosidases while β-D-glucosidase III is an exo-(1 → 4)-β-D-glucanase.[199]

27 endo-(1 → 6)-β-D-Glucanases

An endo-(1 → 6)-β-D-glucanase of Flavobacterium M64 hydrolyses the octasaccharide repeating unit of succinoglycan to two tetrasaccharides.[388] An endo-(1 → 6)-β-D-glucanase capable of hydrolysing the octasaccharide repeating unit of succinoglycan to two tetrasaccharides has been isolated from cells of Flavobacterium. One tetrasaccharide was composed of D-glucose, succinic acid, and pyruvic acid (4:1:1, molar ratio), and the other was composed of D-glucose and D-galactose (3:1, molar ratio). This enzyme hydrolysed the (1 → 6)-β-D-glucosidic linkage to the (1 → 6)-linked β-D-glucose residue in the octasaccharide repeating unit of succinoglycan and also hydrolysed the octasaccharide repeating units of similar polysaccharides produced by many strains of *Agrobacterium* and *Rhizobium* species.

The structure of the extracellular acidic polysaccharide succinoglycan from *Alcaligenes faecalis* has been elucidated by successive fragmentation of the polysaccharide with extracellular β-D-glycanase (succinoglycan depolymerase) and intracellular *endo*-(1 → 6)-β-D-glucanase from *Flavobacterium* sp. M64 into two tetrasaccharides *via* its octasaccharide repeating unit and then methylation analysis and enzymic hydrolysis of the products.[389]

28 D-Glucanases (Miscellaneous)

The use of dye-polysaccharide interactions has been investigated for a β-D-glucanase assay.[390] Certain polysaccharides, in particular cereal β-D-glucans and substituted celluloses, induce bathochromic shifts in the adsorption spectra of such direct dyes as Congo Red. As cello-oligosaccharides showed little or no interaction, it followed that formation of dye-polysaccharide complexes might be useful in studies of β-D-glucanase action. This possibility has been investigated by monitoring the digestion of a sample of oat β-D-glucan with purified β-D-glucan-endohydrolase from *Bacillus subtilis*. Solutions of oat β-D-glucan (0.5% w/v) (pH 6.5) were treated with enzyme and their viscosity was monitored, and aliquots were removed at intervals, heated, and tested for dye interaction and reducing sugar. The results clearly showed a relationship between loss in viscosity and loss in dye interaction, whereas measurements of reducing sugar were not useful until later stages of degradation.

Partial purification of *endo*- and *exo*-β-D-glucanase enzymes from *Zea mays* seedlings has been described and their involvement in cell-wall autohydrolysis investigated.[391] Molecular-sieve chromatography of the cell-wall protein resolved *endo*-β-D-glucanase and *exo*-β-D-glucanase activities when *Avena* glucan and laminaran, respectively, were employed as substrates. The exoenzyme (molecular weight 60,000) was strongly inhibited by Hg^{2+} at a concentration which suppressed the release of monosaccharide from autolytically active cell wall. The *endo*-β-D-glucanase (mol.wt. 26,000), which showed a marked preference for substrates of mixed linkage, exhibited features indicating that it initiates the autolytic solubilization of wall glucan.

(1 → 3),(1 → 4)-β-D-Glucanase has been purified 142-fold from Bacillus subtilis by chromatography on cellulose and SE-Sephadex C-50 followed by gel filtration to homogeneity through Acrylex P-60.[392] SDS electrophoresis and gel filtration gave molecular weights for the enzyme of 30,000 and 33,000, respectively. The enzyme (pI 4.5) does not contain tryptophan, free sulphydryl groups, or carbohydrates.

It has been shown that cellobiose, in contrast to D-glucose, is a regulator of the activity of endoglucanases of cellulase complexes from various sources.[393] Cellobiose is an activator of the low-molecular-weight endo-D-glucanase from T. koningii and an inhibitor of the high-molecular-weight endo-D-glucanases from the same source. Kinetic analysis of the reaction showed that the binding of cellobiose or methylcellobiose to the low-molecular-weight endo-D-glucanase leads to a sixfold increase in the rate of degradation of carboxymethylcellulose. The activating action of cellobiose is due to its ability to play the role of a supplementary nucleophilic agent in the reaction of transglycosylation of intermediate oligosaccharides in the enzymatic hydrolysis of cellulose.

A new D-glucanase produced by a marine Bacillus species has been reported.[394] An isolate of bacteria from marine mud was found to produce a new enzyme capable of hydrolysing insoluble D-glucans produced by oral streptococci. The substrate specificity of the enzyme indicated that the enzyme was a new α-D-glucanase.

The structure of the extracellular acidic polysaccharide succinoglycan from Alcaligenes faecalis var. myxogenes 10C3 has been elucidated by successive fragmentation of the polysaccharide with extracellular α-D-glycanase (succinoglycan depolymerase) and intracellular endo-β-(1 → 6)-D-glucanase of Flavobacterium sp. M64 into two tetrasaccharides via its octasaccharide repeating unit.[389]

Polyacrylamide gel electrophoresis of the cellulolytic system from the culture supernates of Acetivibrio cellulolyticus has shown the presence of four major enzymes: a β-D-glucosidase, an exo-D-glucanase, and two endo-D-glucanases.[224]

29 Glucoamylases

A method for the automatic measurement of α-amylase and glucoamylase activities during fermentation has been developed.[314] For further details see initial citation of ref.314.

A kinetic equation which represents a synergistic action of endo- and exo-enzyme upon polysaccharides' adequacy has been tested by experiments using immobilized α-amylase and glucoamylase.[268]

Physical entrapment has been used as an approach to achieve thermal stabilization of D-glucose oxidase and glucoamylase.[395] The t values for the thermoinactivation of D-glucose oxidase and glucoamylase were increased several-fold by their entrapment in polyacrylamide gels. In polyacrylate gels the individual enzymes behaved differently, probably owing to microenvironmental effects arising by the polyelectrolyte nature of the carrier.

Intrinsic kinetic constants for pore-diffusion-limited immobilized enzyme reactions have been estimated.[396] A simple method using the Aris-Bischoff modulus that establishes intrinsic rate parameters when slow pore diffusion of substrate limits immobilized enzyme reactions that obey Michaelis-Menten kinetics has been presented. Data at high substrate concentrations, where the enzyme would be saturated in the absence of diffusion limitation, and at low substrate concentrations, where effectiveness factors are inversely proportional to reaction modulus, were used to determine maximum rate and Michaelis constant, respectively. Because Michaelis-Menten and Langmuir-Hinshelwood kinetics are formally identical, this method may be used to estimate intrinsic rate parameters of many heterogeneous catalysts. The technique was demonstrated using experimental data from the hydrolysis of maize dextrin with diffusion-limited immobilized glucoamylase. This system yielded a Michaelis constant of 0.14%, compared to 0.11% for soluble glucoamylase and 0.14% for immobilized glucoamylase free of diffusional effects.

The entrapment of chemical derivatives of glucoamylase in calcium alginate gels has been investigated.[397] Glucoamylase was retained in calcium alginate gels, when the molecular weight of the enzyme was increased, e.g. by attachment to polymeric supports or by intermolecular cross-linking. Alternatively, the enzyme could be retained by covalent attachment to the alginate matrix. In the majority of the glucoamylase derivatives more than 50% of the initial activity determined with maltose or partially hydrolysed starch such as substrate was recovered, while upon entrapment in gel particles less than 40% and 6% of the initial activity towards maltose and partially hydrolysed starch, respectively, was recovered. Thus, diffusional and steric limitations on substrate access appeared to influence the apparent enzymic activity, in spite of the high permeability of calcium alginate gels. Dextran-coupled glucoamylase entrapped in calcium alginate gels has been used as a packed-bed reactor for continuous hydrolysis of oligodextrins during a period of two months with no change in activity.

The thermal stability of immobilized glucoamylase from *Rhizopus niveus* entrapped in polyacrylamide gels and bound to SP-Sephadex C-50 has been investigated.[398] The thermal stability of immobilized glucoamylase entrapped in polyacrylamide gels was enhanced slightly compared with glucoamylase in free solution, and was independent of the acrylamide monomer concentration and *N,N'*-methylene-bis(acrylamide) content. To explain this phenomenon, the cellular structure of polyacrylamide gel was taken into consideration in addition to interactions between glucoamylase and gel and a decrease in dielectric constant in the gel. On the other hand, immobilized glucoamylase bound to SP-Sephadex by ionic interaction showed lower stability than free glucoamylase, and much greater stability than glucoamylase in the presence of dextran sulphate, a constituent of SP-Sephadex. Thermal stabilities for the free and immobilized enzymes were also compared at the pH not in the bulk solution but in the SP-Sephadex.

The thermal stability of immobilized glucoamylase in the presence of a substrate has been investigated using a mass balance model.[399] The applicability of this method to a maltose-immobilized glucoamylase system was examined.

The pH-dependent unfolding of glucoamylase from rabbit small intestine by methanol has been described.[400] Glucoamylase from the brush borders of the rabbit small intestine was solubilized by papain and Triton X-100. The Triton-solubilized fraction, when freed of Triton, assumed a micelle-like structure. In the presence of methanol, the enzyme micelle as well as the papain-solubilized enzyme monomer unfolded, obeying the first-order rate law. Unfolding of the micelle was described by two time constants (k = 10^{-3} s^{-1} and k = 10^{-4} s^{-1}), whereas papain-solubilized enzyme had single-phase kinetics of unfolding (k = 10^{-4} s^{-1}). The rate was dependent on pH and the value of activation energy lies between 1-6 kcal mol^{-1} at various pHs. Methanol was devoid of any effect on the activity of papain-solubilized glucoamylase at various pHs, whereas both Triton-100-solubilized enzyme and enzyme micelle were inhibited totally by 30% methanol between pHs 4 and 8.

An unusual type of glycoprotein structure has been identified for a glucoamylase.[401] Glucoamylase occurs in two isoenzyme forms (glucoamylase I and glucoamylase II) in extracts from certain fungi. The isoenzymes from Aspergillus niger are glycoenzymes containing D-mannose, D-glucose, and D-galactose as integral structural components. New data from experiments on reductive alkaline β-elimination and from methylation analysis show that the carbohydrate chains of glucoamylase I are linked O-glycosidically from D-mannose residues to L-serine or L-threonine residues of the protein moiety. In this enzyme, the carbohydrate residues are present as 20 single D-mannose residues, 11 disaccharide components having the structure 2-O-D-mannopyranosyl-D-mannose, 8 trisaccharides, and 5 tetrasaccharides composed of various combinations of D-mannose, D-glucose, and D-galactose residues joined by (1 → 3) and (1 → 6) glycosidic linkages. Such an array of carbohydrate chains in a glycoprotein is unusual, and may account for some of the unique properties exhibited by glucoamylase.

Amylolytic enzymes produced by a strain of Aspergillus niger cultivated on cassava starch in liquid or solid culture were found to be mainly glucoamylases.[402] For the same initial amount of substrate, the glucoamylase activity increased even after 60 h of culture on solid medium whereas it decreased in liquid culture. The pH optima and thermostabilities for enzymes produced in solid and liquid cultures were different. The K_m values expressed as mg soluble starch $(100\ ml)^{-1}$ were 0.1% for crude enzyme from solid culture and 0.057% for crude enzyme from liquid culture. Studies of active-site-directed irreversible inhibition of glycosidases by the corresponding glycosylmethyl-(4-nitrophenyl)triazines show that they have no action on glucoamylase from Aspergillus niger.[72]

A major glucoamylase of Aspergillus niger has been purified 23-fold (21% yield) by ultrafiltration followed by successive chromatography on DEAE-Sephadex, Ultrogel AcA 44, and SP-Sephadex.[403] The purified enzyme was proved homogeneous as judged by polyacrylamide gel electrophoresis, isoelectric focusing, ultracentrifugation, and also from the absence of the glycosidase activities. The enzyme was a glycoprotein containing neutral sugar (18%) and 2-amino-2-deoxy-D-glucose (0.77%), and its molecular weight (90,000) was estimated by SDS polyacrylamide gel electrophoresis and amino acid composition. The N-terminal amino acid was L-alanine. The pH optimum of the purified enzyme was 4.5 with soluble starch as a substrate. The enzyme was stable between pH 2.5 and 7.5 and retained full activity at temperatures up to 50°C. The enzyme activity was inhibited by Hg^{2+} and, to a lesser extent, by Pb^{2+} and Mn^{2+}. The K_m value for malto-oligomer markedly decreased with increasing chain length of substrate in glucose unit (*n*) and the V_{max} value increased with *n*, thus resulting in the increase in the V_{max}/K_m value with *n*. The kinetic parameters for other substrates such as soluble starch, glycogen, and isomaltose as well as the K_i values for some saccharides were also determined.

A thermostable glucoamylase has been purified from the culture filtrates of Thermomyces lanuginosus and has been established to be homogeneous by a number of criteria.[404] The enzyme was a glycoprotein (mol. wt. 57,000) with a carbohydrate content of 10.12%. The enzyme hydrolysed successive D-glucose residues from the non-reducing ends of the starch molecule but did not exhibit any glucosyltransferase activity. The enzyme appeared to hydrolyse maltotriose by the multi-chain mechanism. The enzyme (temperature optimum 70°C) was unable to hydrolyse (1 → 6)-α-D-glucosidic linkages of isomaltose and dextran, and the enzyme exhibited increase in the V_{max} and decrease in K_m values with increasing chain length of the substrate molecule. The enzyme was inhibited by the substrate analogue D-glucono-1,4-lactone in a non-competitive manner. The enzyme exhibited remarkable resistance towards chemical and thermal denaturation.

The thermal stability and kinetics of glucoamylase from Rhizopus niveus has been investigated in the presence of dextran, DEAE-dextran, dextran sulphate, polyethylene glycol, and Ficoll in a temperature range from 52.0 to 60.0°C.[405] Protective effects of these polymers against thermal inactivation of the enzyme increased in the order polyethylene glycol < Ficoll < dextran. The order was the same as that of affinities between glucoamylase and each of polymers determined by partition of glucoamylase in liquid-liquid two-phase systems consisting of these polymers. On the other hand, glucoamylase with dextran sulphate exhibited an interesting phenomenon which was a typical electrostatic interaction explicable in terms of an electrical double layer.

A Flavobacterium species which produces cyclodextrin-degrading glucoamylase has been isolated.[406] The inducible, cell-bound enzyme was purified about 10-fold to 75% purity in 57% yield. The action of the enzyme with cyclohexa-amylose, cyclohepta-amylose, and cyclo-octa-amylose and typical glucoamylase substrates always gave D-glucose as the final degradation product. Small amounts of maltose, which could be detected in the course of cycloamylose degradation, were hydrolysed at a lower rate. Apparently the enzyme preferred shorter α-D-glucopyranosyl chains, and amylopectin and glycogen proved to be very poor substrates.

30 Glycanases (Miscellaneous)

The occurrence of high and low M_r forms of glycogen phosphorylase in extracts of human brain has been reported.[407] Accorinding to gel-exclusion chromatography and sucrose density gradient sedimentation both phosphorylases a and b appear to exist in a high M_r form of 400,000 and a low M_r form whose apparent size differs with the method of determination. The low M_r form is probably an equilibrium mixture of dimer and monomer which gives different apparent M_r values depending upon the position of equilibrium.

The assignment of the human gene for liver-type 6-phosphofructokinase isoenzyme to chromosome 21 has been achieved by using somatic cell hybrids and monoclonal anti-L antibody.[408]

Substrate specificty of the glycanase activity associated with particles of Klebsiella bacteriophage No.6 has been investigated.[409] The glycanase catalyses cleavage of O-β-D-glucopyranosyl-(1 → 3)-4,6-O-(1-carboxyethylidene)-β-D-mannopyranose linkages in Klebsiella serotype-6 capsular polysaccharide. Of 74 heterologous Klebsiella polysaccharides and two derivatives of the type-6 glycan only the type-1 and type-57 polymers were addtionally degraded by the phage-6 enzyme. The repeating units in the three substrates have a 1ax-3eq, 1eq-eq-linked chain D-gluco- or D-galacto-pyranosyl residue in common (which constitutes the reducing end after glycanase action) and a carboxyl group on the next hexopyranosyl residue. Of the 72 polysaccharides not affected by the viral enzyme, at least the type-11 and type-21 glycans also contain the same homology of primary structure. This indicates that the conformation of the glycanase recognition site also constitutes an important feature of the substrates.

31 Heparin Hydrolases

Purification of heparinase and heparitinase by affinity chromatography on glycosaminolycan-bound AH-Sepharose 4B has been reported.[410] The recoveries of the purified preparations of heparinase and heparitinase were estimated to be 39 and 50%, respectively, from the crude enzyme fractions obtained by the first column chromatography on hydroxyapatite.

An *endo*-D-glycosidase (heparin lyase) acting on heparin and heparan sulphate was partially purified (~300 times) from human platelets by affinity chromatography on heparan sulphate-substituted Sepharose.[411] Only heparin-like polysaccharides were degraded by the enzyme. The susceptibility of various biosythetic heparin intermediates indicated that the platelet heparitanase had a requirement for sulphamino but not ester sulphate groups. No activity toward other uronic acid-containing glycosaminoglycans could be demonstrated. Alternate linkages in heparin or heparan sulphate were attacked by the enzyme as shown by analysis of the reducing sugar moiety in oligosaccharide products. The anticoagulant activity of heparin, determined in an antithrombin III activation assay, was markedly reduced after treatment with the heparin lyase. The enzyme was released from its storage site in platelets after induction of the platelet-release reaction. The physiological function of platelet heparin lyase is not known but may be to modify extracellular heparin-like polysaccharides in the vascular system.

Heparin hydrolase production by *Flavobacterium heparinum* in complex protein digest medium, with heparin employed as the inducer, has been improved 156-fold.[412] Rapid deactivation of heparinase activity, both specific and total, was observed at the onset of the stationary phase. Nutritional studies on growth and heparinase production showed an obligate requirement for L-histidine and no vitamin requirement. L-Methionine partially relieved the L-histidine requirement. A defined medium containing D-glucose, ammonium sulphate, basal salts, L-methionine, and L-histidine was developed for growth and heparinase production. The growth rate in this medium was 0.21 h^{-1}, which is 40% higher than that in complex medium. The maximum volumetric productivity of heparinase in the defined medium was increased to 1,475 U l^{-1} h^{-1}, providing a 640-fold increase over that achieved by previously published methods.

No rapid deactivation was observed. An examination of alternate inducers for heparinase showed that heparin degradation products, hyaluronic acid, heparin monosulphate, β-D-2-acetamido-2-deoxy-glucopyranose, and maltose, induce heparinase in complex medium. An Azure A assay was developed to measure the heparin concentration during fermentation and the heparinase specific activity of crude extracts of F. heparinum obtained from sonication, thus negating the need for further purification to measure activity.

32 Hyaluronidases

Purified bovine testicular hyaluronidase is inactivated by butane-2,3-dione in either borate or Hepes buffer, pH 8.3.[413] The presence of borate enhanced the inactivation process, which followed pseudo-first-order kinetics with a calculated second-order rate constant of 13.54 M^{-1} min^{-1}. Using kinetic data it was estimated that the modification of 1 mol L-arginine per mol enzyme was that sufficient for inactivation to occur, whereas amino acid analysis indicated that 4 mol L-arginine had been modified. The inactivation process was partially prevented by using either competitive inhibitors or substrates of the enzyme, thus indicating that the essential L-arginine residue is close to the active site of hyaluronidase. A full kinetic analysis of the enzyme with either hyaluronic acid or chondroitin 6-sulphate as substrate showed that the activity of hyaluronidase was uncompetitively activated by either H^+ or NaCl. The product obtained by reduction of the carboxyl groups of hyaluronic acid to the corresponding alcohol groups was a competitive inhibitor. The possibility that the microenvironment of hyaluronic acid was responsible for the observed kinetic effects of pH and ionic strength was dispelled. The authors conclude that these data are compatible with a mechanism that involves an ionic interaction between a carboxyl group on the substrate and an L-arginine residue on the enzyme.

A procedure has been described for the assay of bovine testicular hyaluronidase in human blood following intravenous adminstration of the enzyme.[414] Inhibitation of hyaluronidase by the reported non-specific serum inhibitor is minimal. However, the presence of human serum does alter the pH profile of hyaluronidase and enhances the activity of the enzyme at low pH values. Preliminary data indicate the effects caused by serum on the pH optimum and the presence of endogenous serum hyaluronidase. The activation effect is not specific for any particular blood type and is independent of whether serum of citrated plasma is used. A similar effect to that of serum on hyaluronidase activity is produced by different buffer mixtures or increased NaCl concentration. It is recommended that bovine testicular hyaluronidase be measured at pH 4.0 in 0.1 M sodium citrate buffer containing 0.15 M NaCl, as under these conditions the addition of human serum or citrated plasma does not alter the pH optimum of the enzyme.

Comparative kinetic studies have been performed on mouse and bovine testicular hyaluronidase.[415] The mouse enzyme has an optimum pH of 4.3 with a broad pH spectrum similar to that of bovine testicular hyaluronidase. While bovine and mouse testicular hyaluronidases show maximum activity at 37°C, different values for energy of activation are obtained frm Arrhenius plots for degradation of hyaluronic acid. Mouse hyaluronidase has a K_m value of 0.9 mg ml^{-1} as compared to 2.1 mg ml^{-1} for the bovine enzyme. Thermal-denaturation experiments reveal that the mouse testicular enzyme is more stable than the bovine enzyme at 55°C. The most potent inhibitors for mouse testicular hyaluronidase are $CuSO_4$ and $HgCl_2$ at 10^{-3} M concentration. L-Cysteine appeared to protect the enzyme inhibition caused by $HgCl_2$.

A rapid purification of bovine testicular hyaluronidase by chromatography on dermatan sulphate-substituted AH-Sepharose 4B followed by chromatography on acetylated AH-Sepharose 4B has been reported.[416] The procedure yielded a purified hyaluronidase with a specific activity of 19.1 units μmol min^{-1} mg^{-1} in high yield. SDS-polyacrylamide gel electrophoresis of the purified material revealed the presence of two close bands with approximate molecular weights of 61,000 and 67,200.

Characterization of β-D-2-acetamido-2-deoxyglucosidase from mouse testes showed that the enzyme does not cross-react in immunodiffusion plates with anti-mouse testicular hyaluronidase serum, suggesting that β-D-2-acetamido-2-deoxyglucosidase does not have common antigenic determinants in hyaluronidase or hyalurono-glucosaminidase complex of acrosome or testes.[37]

Hyaluronidase has been purified 94-fold from mouse testes by ion-exchange chromatography, gel filtration, and affinity chromatography.[417] The enzyme was relatively heat stable and lost 10-20% of its activity at 50-55°C for 10 min. E_a for heat denaturation of enzyme was 42-45 kcal between 45 and 63°C. The Michaelis constant of mouse testicular hyaluronidase was 1.1 mg ml^{-1} hyaluronic acid. Antibodies to the purified enzyme showed a single precipitin line by Ouchterlony gel diffusion. Antiserum to hyaluronidase inhibited enzyme activity by 25%. Immunologically, mouse testicular hyaluronidase was species specific. Tissue extracts of mouse vital organs, except testes and epididymis, did not react with the antisera, though non-specific precipitation did occur between intestinal extracts and anti-hyaluronidase serum. Hyaluronidase was localized in testis sections by indirect immunofluorescence. A specific dark-green fluorescence was localized on cell boundaries extending from spermatogonia to spermatids and appeared on the sperm acrosome. Cytoplasm of spermatogonia and spermatocytes showed light-green fluorescence whereas interstitial tissue was devoid of fluorescence.

In a simple plate assay for hyaluronidase activity hyaluronic acid is incorporated into agarose gels and the enzyme is allowed to diffuse from punched wells.[418] The undigested hyaluronic acid is then precipitated with cetylpyridinium chloride and the diameters of the clear circles are proportional to the logarithm of the enzyme concentration applied to the well. The assay was utilized to examine commercially available hymenoptera venoms, manufactured for use in allergy diagnosis and treatment, for their content of hyaluronidase as a measure of lot-to-lot consistency. The assay permits the analyses of a large number of samples with good reproducibility, without the need for any special instrumentation. Based on the quantity of purified hyaluronidase reported in honey-bee venom (T.P. King, A.K. Sobotka, L. Kochoumain, and L.M. Lichenstein, 1976, Arch. Biochem. Biophys., 172, 661-671), the assay is estimated to detect 70 ng ml^{-1} of purified honey-bee venom hyaluronidase.

The degradation process of hyaluronic acid by *Streptomyces* hyaluronidase has been investigated.[419] Saturated and unsaturated hyaluronate oligosaccharides were prepared from human umbilical-cord hyaluronic acid by partial digestion with bovine testicular and *Streptomyces* hyaluronidase, respectively. After treatment with *Streptomyces* hyaluronidase, the distribution of degradation products from these oligosaccharides was determined. Octasaccharides, either saturated or unsaturated, were substrates of the minimum size. Tetra- and hexa-saccharides were not degraded further and remained as final products. This indicates that at least four succeeding *N*-

acetylhyalobiuronosyl residues were required for this enzymatic degradation. Since unsaturated oligosaccharides were more susceptible to this enzyme than saturated ones of the same polymerization degree, and inner linkages were cleaved in preference to those at the outermost sites, some groups adjacent to this segment seemed to influence the susceptibility of the oligosaccharide to this enzyme.

Some properties of bovine foetal brain hyaluronidase have been described.[35] The possible role of chondroitin sulphate C and hyaluronidase in the processes of differentiation and division is discussed.

33 Inulinases

The separation and purification of β-D-fructofuranosidase and an inulinase from germinating garlic (*Allium sativum*) bulbs have been described.[80] Both the enzymes (mol. wts. 76,000 by gel chromatography) were inactivated by 4-chloromercuribenzoate, 5,5'-dithiobis(2-nitrobenzoic acid), and heat treatment at 55°C for 5 min at pH 7.0. The inulinase hydrolysed inulin, sucrose, and raffinose. The K_m values for inulin and sucrose were 10 mM and 25 mM, respectively. While in the early stages of plant growth (up to 6 days) the inulinase and β-D-fructofuranosidase activities increased in a parallel fashion in the bulbs; at later stages the increase in activity was more than that of inulinase activity.

The production of alcohol from fermentable extracts of Jerusalem artichoke has been investigated using yeasts with inulinase activity.[420]

34 Isoamylases

Immunoreactive trypsin and pancreatic isoamylase activity in serum of patients with chronic renal failure or hepatic cirrhosis has been investigated.[421] In 121 patients with either liver cirrhosis or chronic renal failure pancreatic enzymes in serum were a frequent finding. In renal insufficiency a decreased rate of enzyme elimination is the most likely cause of the above-normal values observed for serum and pancreatic isoamylase.

35 Laminaranases

A comparative study of carbohydrase activities in marine invertebrates has been performed.[299] 103 species of marine invertebrates belonging to 7 types, Spongia, Coelenterata, Annelida, Arhropoda, Mollusca, Echinodermata, Chordata, were tested for laminarinase, cellulase, and amylase activities.

36 Lysozymes

A simple method for the ultrasensitive quantitation of lysozyme has been developed.[422] The lytic activity against Micrococcus lysodeikticus was measured spectrophotometrically after an 18 h incubation period. The method is capable of quantitating lysozyme at concentrations as low as 5 μg ml^{-1} and is applicable to determinations of the enzyme in complex biological mixtures.

A radioimmunoassay developed for serum and urinary lysozyme involves use of an antibody, raised in rabbits, to lysozyme from urine from a patient with monocytic leukemia.[423] This antiserum, diluted 4000-fold, is incubated with radiolabelled lysozyme for 2 h and antibody-bound lysozyme is separated from free lysozyme with dextran-coated charcoal. Validation of the assay included precision and parallelism studies with serum and urine samples from patients with monocytic leukemia and analgesic nephropathy. Results of the radioimmunoassay and those obtained with a Micrococcus lysodeikticus lytic assay correlated well (r = 0.94). The main advantages of the assay are its good precision and reproducibility and its high sensitivity.

A new lysozyme assay based on fluorescence polarization or fluorescence intensity utilizing a peptidoglycan substrate from Micrococcus lysodeikticus subsequently labelled with fluorescein isothiocyanate (FITC) at the amino group of the peptide has been reported.[424] When the FITC-labelled substrate was subjected to lysozyme digestion, an increase of fluorescence intensity or a decrease of fluorescence polarization value (P value) was apparent in five minutes at a lysozyme concentration as low as 0.1 or 0.01 μg ml^{-1}, respectively. The effect of other hydrolytic enzymes, including α-D-mannosidase, proteases, and ribonuclease, on the P value was found to be negligible. The measured values represented the specificity and dose of lysozyme added. Apparent $\underline{V}_{max}$ and $\underline{K}_m$ values for two different lysozymes, chicken egg-white and human, could be determined by this method.

A study has been made of the rates of structural fluctuations of lysozyme in the range of thermal-unfolding transition.[425] The hydrogen-deuterium exchange reaction for the L-tryptophan residues in lysozyme has been followed in 4.5M LiBr at pH 7.2 in the temperature range of the unfolding transition by measuring the transmittance change at 293 nm. The exchange reaction proceeded in three relatively exposed tryptophan residues on the molecular surface. The third phase corresponded to the H-D exchange reaction of the three buried L-tryptophan residues. The H-D exchanges of three L-tryptophan residues buried in folded molecules were caused by fluctuation between the folded and unfolded structure of the protein molecule. The rates of such a fluctuation were determined from the rates of the exchange reaction at various temperatures. These rates agreed very well with those determined from the temperature-jump method. This means that a protein molecule in solution fluctuates between the N- and D-states at every temperature within the transition region, where the N-form is the tightly folded native structure and the D-form the randomly coiled chain. From measurements of thermal unfolding of ester-108-lysozyme and the binding constant of $(D\text{-GlcNAc})_3$, to ester-108-lysozyme, it was found that almost all cross-linked molecules are in the folded state near 50°C and pH 7.2 in 4.5M LiBr, where intact molecules are unfolded. A study was also made of the H-D exchange reaction of ester-108-lysozyme. In the temperature region of 43-50°C, about 70% of the exchangeable tryptophan residues of ester-108-lysozyme were exchanged within 1 s immediately after the mixing of D_2O, in spite of the fact that almost all molecules are in the folded state. This was considered the premelting of the surface of a cross-linked molecule.

The interaction of lysozyme with mixed 1,2-dipalmitoyl-D-phosphatidic acid/1,2-dimyristoyl-L-phosphatidylcholine liposomes has been investigated by laser Raman spectroscopy.[426] Substantial changes were observed in the spectra of both the lipid and protein in the mixed liposomes over the range 10-62°C. At temperatures below 27°C, interaction with lipid appears to increase slightly the amount of helical structure in lysozyme at the expense of random conformation. At temperatures above 30°C, considerable β-sheet is irreversibly formed. Onset of β-formation appears to coincide with the formation of disordered lipid side chains in the acidic component of the lipid. At all temperatures, the O-P-O diester stretching mode at 782 cm^{-1} is much more intense in the lipid-protein mixture than in lipid alone. It is observed that the dimyristoyl phosphatidylcholine chain-disorder transition is lowered by 3°C, while that of the phosphatidic acid is lowered by 12°C, yet the post-transition conformation contains a significantly higher proportion of trans segments in the presence of lysozyme. These results are interpreted in terms of (1) a polar interaction between acidic phospholipid and lysozyme at temperatures below either chain-disorder transition, in which lysozyme is essentially excluded from the hydrophobic portion of the lipid, and (2) an interaction at higher temperatures which involves the lipid side chains of dipalmitoyl phosphatidic acid in the disordered state and is manifested by a substantial conformational change.

The effects of detergents or fatty acids on proteolysis of lysozymes have been studied.[427] Low concentrations of ionic detergents with 8 or more carbons in the alkyl chain were predicted to shift the native-denatured transition in the lysozyme to the denatured state. Fatty acids were found to exert a similar effect, and their participation was proposed in proteolysis of alimentary-canal digestion and intracellular catabolism.

Lysinoalanine formation in lysozyme has been found to be dependent on alkali concentration, pH, temperature, and exposure time.[428] The upper limits of lysinoalanine formation in lysozyme and α-lactalbumin were related to the number of L-lysine residues with a cystine disulphide bond in the adjacent position rather than to the individual contents of these residues.

Preparations and immunological characterizations of two lysozyme derivatives dinitrophenylated at L-lysine-33 and L-lysine-96, respectively, have been reported.[429] Two derivatives of hen egg-white lysozyme with single substitutions of a 2,4-dinitrophenyl (DNP) residue were prepared. The reaction of lysozyme with a 10-fold molar excess of 2,4-dinitrobenzene sulphonic acid provided one mono-DNP-substituted lysozyme (DNP_1-33 lysozyme), which was purified by ion-exchange chromatographies. The other one (DNP_1-96 lysozyme) was prepared as follows. After maleylation of lysozyme in the presence of a 7-fold molar excess of maleic anhydride, the derivative with one free amino group was purified on DE-52 cellulose. This material was dinitrophenylated with 2,4-dinitrobenzene sulphonic acid and the mono-DNP-substituted derivative was purified on DE-52. DNP_1-96 lysozyme was finally purified on SE-Sephadex C-25, after demaleylation at pH 3.5, at 37°C for 5 days. DNP_1-33 lysozyme and DNP_1-96 lysozyme both migrated as a single band with slower mobility than that of native lysozyme. On reduction, carboxymethylation, and chymotrypsin digestion, both mono-DNP-substituted lysozymes yielded a single yellow peptide. The amino acid compositions or partial sequence of these peptides indicated that L-lysine-33 and L-lysine-96 were the only dinitrophenylated residues in DNP_1-33 lysozyme and DNP_1-96 lysozyme, respectively. DNP_1-33 lysozyme and DNP_1-96 lysozyme showed antigenic reactivities equal to that of native lysozyme with antisera to lysozyme. The DNP residues on the protein were accessible to anti-DNP antibodies, but the affinities of DNP_1-33 lysozyme to anti-DNP antibodies were lower than those of DNP_1-96 lysozyme. This result is discussed with respect to the local environments of the DNP residues in these proteins.

Lysozymes from hen, duck II, and goose egg-white and human lysozyme were compared at the level of their L-tryptophyl residues using the solvent perturbation technique and fluorescence emission in addition to two chemical modfications.[430] The three L-tryptophyl residues of goose lysozyme were found completely exposed to the solvent. The study of the situation of the aromatic residues of hen, duck II, and human lysozymes provided additional evidence that the three-dimensional structures of these enzymes are very similar, although an unexpected reactivity of L-tryptophan-28 of human lysozyme towards N-bromosuccinimide could be evidenced.

As part of a study of the whey proteins of various mammals a comparison has been made of the α-lactalbumins and lysozymes of the kangaroo and horse.[431] In the milk of the red kangaroo (Megaleia rufa) there is only one α-lactalbumin and it occurs throughout lactation, but no lysozyme has been detected. There are two α-lactalbumins in the milk of the grey kangaroo (Macropus giganteus): one, designated α-lactalbumin Zone B, is present throughout lactation, and the second, designated α-lactalbumin Zone A, is present only in late lactation. One lysozyme is also present. The milk of the horse (Equus caballus) contains one α-lactalbumin and at least one lysozyme. Partial amino acid sequences are proposed from sequence determination and from analyses of tryptic peptides compared with the known sequences of other α-lactalbumins and lysozymes.

Multiple forms of lysozyme found in the rat liver have been isolated and characterized from cellular organelles.[432] Isolation of the enzymes was achieved by Sephadex gel filtration and chromatography on carboxymethylcellulose. The purity of the preparations was examined by electrophoresis on polyacrylamide gel. The nuclear lysozyme moved as a single band, indicating homogeneity, whereas other subcellular lysozymes appeared heterogeneous due to the presence of more than one band, thus showing partial purity. Although the subcellular lysozymes were similar with respect to enzymatic properties, pH, buffer molarity optima, and electrophoretic mobility, differences were observed in elution patterns, responses to nuclear inhibitor, and heat sensitivity. Nuclear lysozyme was distinctly different by these criteria as compared to other subcellular lysozymes.

Evolutionary implications have been made about the relation between hen egg-white lysozyme and bacteriophage T4 lysozyme.[433] Hen egg-white lysozyme and T4 bacteriophage lysozyme have the same catalytic function, but have non-homologous amino acid sequences. Notwithstanding the differences in their primary structures, the two lysozymes have similarities in their overall backbone conformations, in their modes of binding substrates, and probably in their mechanisms of action. By different criteria, the similarity between the folding of the two enzymes can be shown to be statistically significant. Also the transformation which optimizes the agreement between the backbones of the two molecules is shown to align accurately their active-site clefts, so that saccharide units bound in the A, B, C, and D subsites of hen egg-white lysozyme coincide within 0.1 to 0.2 nm with analogous saccharides bound to phage lysozyme. Furthermore, a number of the specific interactions between enzyme and substrate which were observed for hen egg-white lysozyme, and thought to be important for catalysis, are found to occur in a structurally analogous way in the phage enzyme. Fifty-four atoms from the respective active sites which appear to be equivalent, including saccharides bound in the B and C sites, superimpose with a root-mean-square discrepancy of 1.35 nm. These structural and functional similarities suggest that the two lysozymes have arisen by divergent evolution from a common precursor. This is the first case in which two proteins of completely different amino acid sequence have been shown, with high probability, to have evolved by divergent rather than convergent evolution.

The binding mode of sodium dodecyl sulphate to lysozyme and the accompanying structural change of lysozyme by binding have been investigated by means of the binding isotherm, the precipitation curve, and the CD spectra in pure water, NaCl, and borate buffer solutions.[434] The precipitation phenomena could be explained in terms of the neutralization of the net charge of lysozyme due to dodecyl sulphate-ion binding. The analysis of the binding isotherms by the use of the BET equation gave the site number of the first layer corresponding to the positively charged residues at the pH studied. The conformational change from the β-structure to the α-helix has been observed in the second-layer binding. The environmental change of the side-chain residues has also been observed.

Fluorescence polarization studies have been made of saccharide binding to wheat-germ agglutinin and lysozyme.[435] 2-Acetamido-2-deoxy-D-glucopyranose has been shown to cause only slight increases in the polarization of wheat-germ agglutinin fluorescence at saturating levels whereas the disaccharide and trisaccharide produced increases in the polarization value from 0.116 to 0.151 and 0.154, respectively. These increases have suggested that rotational motions of the L-tryptophan residue at the binding sites were being restricted by an interaction between these tryptophans and the bound sugars. A model of the nature and location of these interactions was discussed. Comparable results were obtained with lysozyme, which showed a larger effect upon the binding of 2-acetamido-2-deoxy-D-glucopyranose but a maximal increase in polarization upon binding the corresponding disaccharide or trisaccharide.

Bindings of calcium to lysozyme and its derivatives have been studied by UV difference spectroscopy at various pH's.[436] The binding constant was 40 M^{-1} at neutral pH. The binding caused proton release from lysozyme and did not inhibit the binding of tri-2-acetamido-2-deoxy-β-D-glucopyranose to lysozyme. In the presence of 0.2M Ca^{2+}, lysozyme showed 26% of the activity of the free enzyme toward hexa-2-acetamido-2-deoxy-β-D-glucopyranose but the cleavage pattern was similar to that of the free enzyme. It was predicted that calcium bound near the catalytic carboxyls caused inhibition of lysozyme activity. It was found from the results of protease digestion that calcium binding shifted the native-denatured transition in lysozyme toward the native state.

Oxindolealanine-62 lysozyme equilibrium, calorimetric, and kinetics of its reaction with β-D-acetamido-2-deoxyglucopyranose oligosaccharides have been studied.[437] The enzyme, formed by reaction with N-bromosuccinimide and purified by using affinity chromatography, was examined in its binding of homologous oligosaccharides of β-D-acetamido-2-deoxyglucopyranose and in its catalysis of hydrolysis and of transglycosylation of the hexasaccharide of β-D-acetamido-2-deoxyglucopyranose. Equilibrium binding constants were determined by changes in absorbance or fluorescence associated with ligand binding. Enthalpies of binding were measured calorimetrically. The pattern of variation of ΔG° and ΔH° of binding with ligand chain length and pH was different for oxindolealanine-62 lysozyme compared with native lysozyme. These results indicate that the interactions of the ABC region of the active site with substrates are substantially altered by L-tryptophan-62 oxidation, more than expected for loss only of the L-tryptophan-62 interactions. The partitioning of the glycosyl enzyme intermediate between reaction with a saccharide acceptor (transglycosylation) and reaction with water is unaffected by oxidation of L-tryptophan-62. Similarly, the pattern of cleavage of the hexasaccharide, predominantly to tetra- and di-saccharide, is unaffected by oxidation of L-tryptophan-62. The 2000-fold slow rate of catalytic reaction (relative to free enzyme and substrate) apparently reflects a sterically hindered fit of the substrate into the active site of the modified enzyme and not a special catalytic importance of L-tryptophan-62. The geometry of the transition state for oligosaccharide hydrolysis is inferred to be the same for native and oxidized enzymes.

Oxygen-18 leaving-group kinetic isotope effects (KIEs) have been measured for a set of glycosyl transfer reactions with 4-nitrophenyl β-D-glycosides as substrates.[206] Acid-catalysed hydrolysis and alkaline hydrolysis exhibit KIEs of k_{16}/k_{18} = 1.0355 ± 0.0015 and 1.0386 ± 0.0032, respectively. Lysozyme and β-D-glucosidase A show KIEs on $V_{max}/K_m(V/K)$ of $(V/K)_{16}/(V/K)_{18}$ = 1.0467 ± 0.0015 and 1.0377 ± 0.0061, respectively. The large magnitude of these KIEs requires that carbon-oxygen bond scission be far advanced in the transition states for these reactions. Therefore, in the transition states for the first irreversible steps in these reaction sequences, scission of the glycosidic bond must be essentially complete for the reactions catalysed by lysozyme and β-D-glucosidase A, which are thought to proceed *via* S_N1 and S_N2 mechanisms, respectively. Acid-catalysed hydrolysis is shown to proceed through a transition state involving at least 80% C-O bond cleavage and only partial proton transfer to the leaving 4-nitrophenyl oxygen atom.

Binding of 4-methylumbelliferyl chitotetraoside to hen lysozyme has been studied by measuring changes in fluorescence at 375 nm.[438] Hydrolysis of 4-methylumbelliferyl N-acetyl-chitotetraoside catalysed by lysozyme was studied by measuring the release of 4-methylumbelliferone from 4-methylumbelliferyl N-acetyl-chitotetraoside fluorimetrically, and the kinetic constants were determined in the pH range of 2 to 8 at 0.1 ionic strength and 42°C. The binding and kinetic data showed that 4-methylumbelliferyl N-acetyl-chitotetraoside binds to subsites A to E (productive binding) and subsites A to D with the non-reducing sugar residue extending beyond subsite A (non-productive binding). The fraction of the productive complex was 0.77 at pH 8.5. The pH dependence of the kinetic constants was analysed assuming that the molecular species with ionized L-aspartic acid-52 and protonated L-glutamic acid-35 is active and L-aspartic acid-101 participates in the binding, and the p*K* values of these groups were determined. The p*K* values of L-aspartic-52, L-glutamic-35, and L-aspartic-101 acids were 3.6_0, 6.2_0, and 4.2_0, respectively, for free lysozyme, 3.4_0, 6.5_5, and 3.4_0, respectively, for the productive complex, and 3.9_5, 6.5_5, and 3.3_0, respectively, for the non-productive complex. The p*K* values for free lysozyme were in excellent agreement with those obtained by analysis of the kinetic constants for 4-methylumbelliferyl N-acetyl-chitotrioside. The free energy of activation was 24 kcal mol^{-1} at pH 5.2. Comparison with the corresponding value obtained for hydrolysis of chitohexaoside suggests that the interactions of 2-acetamido-2-deoxy-D-glucopyranose residues with subsites E and F in the transition state are important in lysozyme catalysis.

Binding of 4-methylumbelliferyl chitotrioside and 4-methylumbelliferyl chitotetraoside to human lysozyme has been studied by fluorescence measurement.[439] Hydrolysis of 4-methylumbelliferyl chitotrioside and 4-methylumbelliferyl chitotetraoside catalysed by human lysozyme was studied by measuring the release of 4-methylumbelliferone fluorimetrically, and the kinetic constants were determined in the pH range of 2 to 8 at 0.1 ionic strength and 42°C. On the basis of binding and kinetic data, it was shown that 4-methylumbelliferyl chitotrioside binds mainly to subsites A to D with the terminal methylumbelliferyl group bound to subsite D (non-productive binding) and that 4-methylumbelliferyl chitotetraoside binds to subsites A to E (productive binding) and subsites A to D with the non-reducing sugar residue extending beyond subsite A (non-productive binding). The pH dependences of the kinetic constants for hydrolysis of 4-methylumbelliferyl chitotrioside and 4-methylumbelliferyl chitotetraoside were analysed, assuming that non-productive binding occurs competitively, that an ionizable group in addition to the catalytic groups (L-aspartic-52 and L-glutamic-35) participates in the catalysis, and that the molecular species with ionized L-aspartic acid-52 and protonated L-glutamic acid-35 is active. Analyses of the kinetic constants for 4-methylumbelliferyl chitotrioside and 4-methylumbelliferyl chitotetraoside both gave the same p*K* values of the catalytic groups (pK_{52} = 3.6_3 and pK_{35} = 6.6_8 at 0.1 ionic strength and 42°C). These p*K* values were very close to the values determined previously by spectroscopic methods.

A lysozyme-catalysed reaction of chito-oligosaccharides has been described.[440] The time-courses of substrate consumption and product formation in the lysozyme-catalysed reaction were determined with chitotetraose and chitopentaose as substrate to accumulate data suitable for the estimation of rate constants by numerical analysis. The lysozyme-catalysed reactions were followed by t.l.c. or h.p.l.c. Chitotetraose decomposed apparently to small oligosaccharides within 5 h, and chitopentaose decomposed within 15 min at pH 5.0 and 50°C. The temperature dependence of the rate of disappearance of the initial substrate showed a different profile from that observed with glycol chitin as substrate by the reducing power method. The order (or distribution) of the amount of product formed from chitopentaose in the reaction time-course determined by t.l.c. differed from that determined by h.p.l.c. The relative error in h.p.l.c. was much less than that in t.l.c., and the time-course determined by h.p.l.c. was thought to be of sufficient accuracy for the estimation of rate constants by computer analysis.

High lysozyme and chitinase activities have been found in Leydig's organ of *Etmopterus spinax*, *Somniosus microcephalus*, and *Tompedo nobiliana*, in Leydig's and epigonal organs and spleen of *Raja radiata*, and in the epigonal organ of *Rhinoptera bonasus*.[19]

An enzyme actively lysing the cell walls of a haemolytic streptococcus of group A was shown to be a lysozyme mucopeptide-*N*-acetylmuranoylhydrolase.[441] It was isolated from the culture liquid of *Actinomyces levoris* by ammonium sulphate precipitation and further purification by gel filtration and ion-exchange chromatography. The molecular weight (12,500), isoelectric point (10.6), and amino acid composition of the enzyme were determined. A peptidoglycan obtained from the cell walls of a group A streptococcus was used as the substrate in a determination of the specificity of the enzyme.

Lysozyme activity has been observed in bacteriophage T4 ghosts.[442] This enzyme is probably responsible for the lysis from without, observed at high multiplicity of infection, a process independent of the presence of the e gene product which is also a lysozyme. The ghost lysozyme and e lysozyme differed with respect to their requirements for maximal catalytic activity and to some extent in substrate specificity. The ghost lysozyme was released from phage particle by the action of Triton X-100.

Bacillus cereus peptidoglycan with *N*-unsubstituted glucosamine residues has been found to be insensitive to treatment with bacteriophage T4 lysozyme.[443] After *N*-acetylation with acetic anhydride, T4 lysozyme-cleared solutions of the peptidoglycan and reducing sugars were liberated. The digestion products were mainly of high molecular weight, since the peptidoglycan is peptide cross-linked to a great extent. *N*-Propylation did not convert the partially *N*-unsubstituted peptidoglycan to a sensitive form. It is concluded that the acetamido groups are required for binding and/or catalysis by T4 lysozyme.

37 Oligo-1,6-D-Glucosidases

Detergent-solubilized pig intestinal sucrose-α-D-glucohydrolase-oligo-(1 → 6)-D-glucosidase has been purified 40 to 100 times with a yield of 10 to 20% by a rapid immunoadsorbent technique.[444] The purified enzyme was shown to be homogeneous by immunoelectrophoresis and was essentially free from other known brush-border peptidases and disaccharides. It consisted of two polypeptide chains with apparent molecular weights of 140,000 and 150,000, respectively. In contrast, the enzyme isolated from pigs in which the pancreas was completely disconnected from the duodenum 3 days before killing migrated in polyacrylamide gel electrophoresis in dodecyl sulphate as a single polypeptide chain with an apparent molecular weight of 260,000. Treatment with pancreatic proteases *in vitro* converted the large polypeptide chain into bands with molecular weights equal to or somewhat larger than those of sucrose-α-D-glucohydrolase-oligo-(1 → 6)-D-glucosidase from normal pigs. It was suggested that the single chain represents a precursor, which is converted to the final sucrose-α-D-glucohydrolase-oligo-(1 → 6)-D-glucosidase *in vivo* by pancreatic proteolytic enzymes. This is one of the few examples in vertebrates of a single polypeptide chain carrying two enzymatically active sites. The significance of the result for the mechanism of the biosynthesis of sucrose-α-D-glucohydrolase-oligo-(1 → 6)-D-glucosidase is discussed.

38 Pectate, Pectin, and Poly-D-Galacturonate Lyases

Pectate lyase has been purified to a nearly homogeneous state from the culture filtrate of *Streptomyces nitrosporeus*.[445] The molecular weight was estimated to be about 41,000. Isoelectric point was pH 4.6. The enzyme was most active at pH 10.0 and 50°C, and was relatively stable at a pH range of 4-11 (at 2°C for 48 h) and below 40°C (at pH 7.0 for 10 min). Ca^{2+} was required for maximum activity. The enzyme was an endopectate lyase which was more active on low methoxyl pectin than on poly-D-galacturonic acid and had macerating activity on potato tissue and Ganpi bark.

The degradation of poly-D-galacturonic acid by rumen ciliate protozoa has been investigated.[446] The depolymerase activity of cell-free extracts of nine species of rumen ciliate protozoa and two mixed protozoal preparations, grown *in vivo* and *in vitro*, towards poly-D-galacturonic acid was examined. The highest activity was found with *Eremoplastron bovis* and *Ostracodinium obtusum bilobum* while there was none in the spined or spineless forms of *Entodinium caudatum* and little in *Polyplastron multivesiculatum*. On the basis of the rapid drop in viscosity, inhibition by EDTA, and the production of u.v.-absorbing material, the enzymes from all active species were designated as endopectate lyases although some poly-D-galacturonase may be present. Neither pectin nor poly-D-galacturonic acid supported the survival or growth of any of the protozoal species tested.

High-performance liquid chromatography has been used to investigate pectic enzymes.[447] Technical pectic multienzyme preparations Pectinex Ultra and Rohament P were chromatographed on an analytical scale using medium-pressure liquid chromatography on a glycol methacrylate rigid macroreticular gel, Spheron 1000 and its ion-exchange derivatives. A combination of isocratic and linear-gradient elution (with gradients in ionic strength or pH) was employed and fractions were monitored by measurements of absorbance (A_{285} and A_{254}), conductivity, pH, and enzyme activity. Conditions for rapid separations of pectic enzymes are elaborated. The results indicate the possibilities of separating the technically undesirable pectin-esterase activity from the other enzyme activities, and of a more detailed biochemical investigation of these enzymes, important for the food industry.

The simultaneous synthesis of pectin lyase and carotovoricin has been induced by mitomycin C, nalidixic acid, or ultraviolet light irradiation in Erwinia carotovora.[448] When Erwinia carotovora Er, a bacteriocinogenic strain, was induced after irradiation by ultraviolet (UV) light or inhibitors of DNA synthesis, such as mitomycin C or nalidixic acid, pectin lyase and bacteriocin (designated carotovoricin) activity appeared in the culture fluid. The optimal dose of each of these agents for producing the enzyme or bacteriocin was identical, and the time-courses for both were essentially the same. The syntheses of the enzyme and bacteriocin were assumed to be regulated by the same mechanism, in which a repressor inactivated by UV light, mitomycin C, or nalidixic acid was involved. The other three bacteriocinogenic strains of E. carotovora also formed pectin lyase, in addition to carotovoricin in the presence of mitomycin C, indicating that simultaneous syntheses of pectic lyase and carotovoricin were widespread phenomena in bacteriocinogenic strains of E. carotovora.

The pectic enzyme activities of bacteria associated with rotted onions have been investigated.[449] The aerobic bacteria associated with soft rot in onions (Allium cepa) were isolated and identified as a Vibrio sp., Micrococcus epidermidis, Pseudomonas cepacia, and Acinetobacter sp., a Xanthomonas sp., Bacillus polymyxa, and Bacillus megaterium. With the cup-plate assay method, no pectin hydrolase could be detected from any of these isolates when they were cultured in pectin medium, but lyase and pectinesterases were detectable. Onion tissue cultures showed pectin hydrolase activity for P. cepacia and B. polymyxa and lyase and pectinesterase activities for all of the isolates, usually at higher levels of activity than those of the pectin medium culture filtrates. In both culture media, Vibrio sp. showed the highest lyase and pectinesterase activities. In the viscometric test, all of the isolates achieved at least a 50% decrease in viscosity for lyase enzyme, with M. epidermidis and Vibrio sp. recording viscosity decreases as high as 83%. The ability to cause soft rot in onion bulbs was demonstrated by P. cepacia and Xanthomonas sp. Benzoic acid at a concentration of 0.8 mg ml^{-1} caused total suppression of enzyme production, whereas sodium benzoate at this concentration reduced pectinesterase production by 71% and lyase prodution by 72%. The possible use of these preservatives in the control of soft rot in onions is noted.

39 Poly-D-Galacturonases

The degradation of poly-D-galacturonic acid by rumen ciliate protozoa has been investigated.[446] The depolymerase activity of cell-free extracts of nine species of rumen ciliate protozoa and two mixed protozoal preparations, grown in vivo and in vitro, towards poly-D-galacturonic acid was examined. The highest activity was found with Eremoplastron bovis and Ostracodinium obtusum bilobum, while there was none in the spined or spineless forms of Entodinium caudatum and little in Polypastron multicesiculatum. On the basis of the rapid drop in viscosity, inhibition by H_4 edta, and the production of u.v.-adsorbing material, the enzymes from all active species were designated as endopectate lyases, although some poly-D-galacturonase may be present. Neither pectin nor poly-D-galacturonic acid supported the survival or growth of any of the protozoal species tested.

The application of carbohydrases including poly-D-galacturonase to the extraction of proteins from wheat bran has been investigated.[349] For further details see initial citation of ref.349.

The effect of immobilization of Aspergillus niger extracellular poly-D-galacturonase on kinetics and action pattern has been investigated.[450] The extracellular poly-D-galacturonase of A. niger was covalently bound to 4-aminobutanoic acid, glycyl-glycine, and 6-aminohexanoic acid, respectively, as spacers. Immobilization of the enzyme invariably led to decreased activity, the extent of the decrease being inversely proportional to the chain length of the spacer. This fact, as well as apparent kinetic parameters of the immobilized enzyme, indicated that steric hindrance and, probably, diffusion effects are responsible for the decrease in activity. The covalent binding also caused an alteration of the action pattern of the enzyme on polymeric and oligomeric substrates. In the case of the polymers the randomness in degradation was lowered because of restriction of the enzyme action to peripheral areas of the substrate molecule. Among oligomeric substrates, the most important change was observed in tetra-(D-galactosiduronic acid), in which (2

+ 2) degradation occurred as well as the (1 + 3) degradation characteristic of the soluble enzyme.

The changes in poly-D-galacturonase activity which occur during ripening of normal and mutant tomato fruit have been investigated.[451] There is a sequential appearance of two isoenzymes, poly-D-galacturonases 1 and 2, during ripening. These isoenzymes have been purified and their properties compared. Poly-D-galacturonase 1 has a M_r of 1,000,000, is 50% inactivated at 78°C and has a density of 1.343 g cm^{-3} in caesium chloride. Poly-D-galacturonase 2 has a M_r of 42,000, is 50% inactivated at 57°C, and has a density of 1.300 g cm^{-3} in caesium chloride. Fruits from isogenic lines homozygous for the ever-ripe (Nr) mutation do not ripen normally and contain reduced amounts of poly-D-galacturonase. Only poly D-galacturonase 1 was produced in Nr fruit. Tomatoes from isogenic lines homozygous for the ripening inhibitor (rin) mutation do not ripen normally and produce very little detectable poly-D-galacturonase. Although poly-D-galacturonases 1 and 2 had different properties, they both gave rise to a single polypeptide on electrophoresis in polyacryamide gels in the presence of sodium dodecylsuphate (M_r - 46,000). A comparison of the major fragments produced by limited proteolysis of poly-D-galacturonases 1 and 2 with chymotrypsin suggests that the polypeptides from the two isoenzymes were similar. The same conclusion was reached from a comparison of poly-D-galacturonases 1 and 2 by radioimmunoassay, using antibody prepared against poly-D-galacturonase 2 and ^{125}I-labelled poly-D-galacturonase 2.

Some properties of the poly D-galacturonase-elicitor from the filtrates of Rhizopus stolonifer cultures have been examined in an attempt to understand its mode of action as an elicitor of casbene synthetase activity in castor-bean seedlings.[452] Both the poly-D-galacturonase activity and the elicitor activity are heat labile with similar heat-sensitivity profiles. Also, the catalytic activity of the enzyme is lost on treatment with sodium periodate, as had been shown previously for the elicitor activity. The pH optimum of the enzyme activity with poly-D-galacturonic acid as the substrate is 4.9. Exposures of germinating castor-bean seedlings to the elicitor for short-term periods of 1 to 10 minutes followed by washing and incubation in sterile, distilled water are partially effective in elicitation in comparison with the continuous exposure of the seedlings over 11 hours to the same amount of the elicitor. The initial rate of reaction catalysed by the enzyme is about 3 times faster with poly-D-galacturonic acid as a substrate than with partially (50%) methylated poly-D-galacturonic acid (pectin). The K_m value of the enzyme for poly-D-galacturonic acid is about 4.2 millimolar in terms of monomeric units and about 0.07 millimolar in terms of polymer concentration. Examination of the types of products formed by the action of the enzyme suggests that it is an endo-hydrolase. The amino acid composition of this enzyme is similar to those of other extracellular fungal proteins reported. The carbohydrate moiety of the glycoprotein poly-D-galacturonase elicitor is composed of 92% D-mannose and 8% D-glucosamine by gas chromatography-mass spectrometry analysis. The linkage-group analysis of the carbohydrate moiety showed that D-mannosyl residues which are 1,2-linked comprise about 70% of the total glycosyl residues and demonstrated the presence of some 1,3,6- and 1,2,6-linked branching D-mannosyl residues.

Apparently homogeneous poly-D-galacturonase elicitor purified from the filtrates of *Rhizopus stolonifer* cultures has been found to stimulate germinating castor-bean seedlings to produce greatly increased levels of casbene synthetase activity.[453] The purification procedure involved gel-filtration chromatography on Sephadex G-25 and G-75 columns followed by cation-exchange chromatography on a Sephadex CM C-50 column. Homogeneity of the purified preparation was indicated by the results of cationic polyacrylamide disc gel electrophoresis and isoelectric focusing (pI = 8.0). The identities of the casbene elicitor activity and poly-D-galacturonase were indicated by the coincidence of the two activities at all stages of purification, the coincidence of both activities with the single protein-staining band detected on a cationic polyacrylamide disc gel and an isoelectric-focusing gel, and the identical behaviour of both activities on an agarose gel affinity column. The purified poly-D-galacturonase elicitor is a glycoprotein with approximately 20% carbohydrate content and an estimated molecular weight of 32,000 by polyacrylamide disc gel electrophoresis.

The variability of poly-D-galacturonase and protein isoelectric-focusing patterns has been investigated in *Botrytis cinerea* isolates.[454] Acetone precipitates from culture filtrates of three isolates of *B. cinerea* were resolved by isoelectric focusing on polyacrylamide gels to detect differences in the poly-D-galacturonase and protein patterns. Only a few bands - four in the protein patterns and two in the poly-D-galacturonase patterns - were common to all the isolates. Differences were also detected in poly-D-galacturonase and protein patterns of all the same isolate at different ages of culture (7, 14, and 21 d). Identical poly-D-galacturonase patterns were obtained when isoelectric focusing was applied to an acetone precipitate either directly or after further purification by ion-exchange chromatography.

40 exo-Poly-D-Galacturonate Lyases

An exo-poly-D-galacturonate lyase has been purified to a homogeneous state from the culture filtrate of Streptomyces massasporeus.[455] The molecular weight was estimated to be about 54,000 and the isoelectric point was pH 5.5. The enzyme was most active at pH 9.5 and 40°C, and was relatively stable at pH 3.0 for 10 min. Ca^{2+} was required for maximum activity. The enzyme was most active on tri-D-galacturonic acid and had higher activity on low methoxyl pectins rather than on poly-D-galacturonic acid. The enzyme removed terminal unsaturated di-D-galacturonate units from the D-galacturonide chains.

41 Pullulanases

The properties of two amylases which differ in their substrate specificity and subcellular location as well as a chloroplast-associated R-enzyme (debranching activity) have been reported.[301]

The conversion of active and inactive debranching enzymes in rice seeds has been reported.[456] Debranching-enzyme activity in rice seeds increased during the early stage of ripening and then decreased, and increased again during germination. The inactive enzyme accumulated rapidly in ripening from the twentieth day after flowering. Radioactive amino acids were readily incorporated into the active debranching enzyme after their absorption into immature rice seeds. Subsequently, the radioactivity increased in the inactive enzyme, accompanying a decrease in the active enzymes. It was concluded that the debranching enzyme in rice seeds is synthesized during ripening in active form and that it accumulates in inactive form, which can be reactivated during germination.

The identification of an inactive debranching enzyme in rice seeds and its activation have been reported.[457] Incubation of rice flour with 1,4-dithiothreitol (dithiothreitol) generated high pullulanase activity. Maximum activity occurred after 24 h incubation with 6mM thiol. All of the thiols at 10mM activated the debranching enzyme, suggesting that the process involved cleavage of the disulphide bonds. This view was confirmed by the effect of other reductants. Sodium sulphite and sodium dithionite were very effective, but sodium borohydride was moderately effective in spite of the pH (~10) which was unfavourable for enzyme stability. The debranching enzyme was also activated to similar extents by proteases, e.g. papain, α-chymotrypsin, and trypsin.

Debranching enzyme has been purified from mature rice seeds.[458] The enzyme was extracted from a supernatant solution of rice flour following its homogenization in 10mM sodium dithionite solution. DEAE-cellulose was used as the extraction medium from which the enzyme protein was eluted with sodium phosphate buffer. Purification was achieved by the use of a Sephadex G-100 column and a CM-cellulose column. The purified enzyme had no amylase, maltase, and Q-enzyme activities and was homogeneous in gel filtration, disc electrophoresis, and isoelectric focusing. The molecular weight (ca. 70,000) determined by gel filtration and the optimum pH of 5.6 were identical to those of the milky-stage enzyme. Substrate specificities of the enzyme were also similar to those of the milky-stage enzyme. The enzyme rapidly hydrolysed pullulan but scarcely hydrolysed phytoglycogen, similar to other higher-plant debranching enzymes. Kinetic data similar to those of the milky-stage enzyme were also obtained.

Bacillus amylolyticus produces α-amylase, pullulanase, and α-D-glucosidase. By selection of carbon source in the growth medium α-D-glucosidase was produced preferentially and with exclusion of the other two activities.[182]

42 Sucrose-α-D-Glucohydrolase

The activities of various glycosidases etc. including sucrose-α-D-glucohydrolase in homogenates of the small-intestinal mucosa of two adult and 18 suckling tammar wallabies (*M. eugenii*) aged from 6 to 50 weeks have been investigated.[38] For further details see initial citation of ref.38.

The amounts of sucrose-α-D-glucohydrolase in tangentially

sectioned biopsies from jejunum have been studied by quantitative immunoelectrophoresis and enzymic assays.[129] For further details see initial citation of ref.129.

Sucrose-α-D-glucohydrolase-oligo-(1 → 6)-α-D-glucosidase has been purified from rat intestinal microvillus membranes after injection of D-(2-^{3}H)mannose and L-(6-^{3}H)fucose, using a column of monoclonal antibody protein A-Sepharose.[459] After pronase digestion and gel filtration of the glycopeptides labelled from both precursors, a major part of the radioactivity was recovered in asparagine-linked complex oligosaccharides, and a smaller amount in partially alkali-labile high-molecular-weight glycopeptides. Only a small amount of (^{3}H)mannose was found in *endo*-β-D-2-acetamido-2-deoxyglucosidase H-sensitive high-D-mannose oligosaccharides.

Detergent-solubilized pig intestinal sucrose-α-D-glucohydrolase oligo-(1 → 6)-D-glucosidase has been purified 40 to 100 times with a yield of 10 to 20% by a rapid immunoadsorbent technique.[444] For further details see initial citation of ref.444.

43 αα- and ββ-Trehalases

The activities of various glycosidases in homogenates of the small-intestinal mucosa of two adult and 18 suckling tammar wallabies (*M. eugenii*) aged from 6 to 50 weeks have been investigated.[38] αα-Trehalase activity was very low or absent during the first 34 weeks, and then increased.

In order to establish the specificity of αα-trehalases and their action on monomodified asymmetrical derivatives of trehalose, the purification of may-bug (*Melolontha vulgaris coleopterae*) *exo*-trehalase has been purified 16,000-fold.[460] The specific activity was 307 units mg^{-1} of proteins, approximatively 78 times higher than previously reported. Some of the properties of the enzyme have been studied: molecular weight, isoelectric point, action of various effectors and inhibitors.

Effects of acrylonitrile on trehalase activity in *Tribolium castaneum* and *Trogoderma granarium* everts has been studied.[461] Acrylonitrile inhibited trehalase and phosphorylases in larvae and adults of *Tribolium castaneum*. In *Trogoderma granarium* larvae phosphorylases alone were inhibited.

44 *endo*-(1 → 4)-β-D-Xylanases

The mode of action of an *endo*-(1 → 4)-β-D-xylanase from a Basidiomycete *Sporotrichum dimorphosporum* has been reported and its utility for the structural investigation including substitution pattern of the branched water-soluble L-*arabino*-(4-O-methylglucurono)-D-xylan from redwood (*Sequoia sempervirens*) has been studied.[462] The action of the D-xylanase appears to involve a region of the polysaccharide backbone having three D-xylosyl residues. A mode of action that requires unsubstituted hydroxyl groups at C-2, C-3, and C-2' of a xylobiosyl residue was proposed. The binding site seems to correspond to a shallow cavity. The composition and structure of the final residue of attack shows that the enzyme has no action when the D-xylosyl residues branched through O-2 are separated by only one unsubstituted D-xylose residue. This pattern of action, the nature of the dialysable products, and the production of a final residue in which the substituents are accumulated suggest that the L-arabinosyl and D-glucosyluronic groups are irregularly distributed on the main chain of the D-xylan from redwood and that in some regions they are in close vicinity when not actually on adjacent D-xylosyl residues.

The extracellular *endo*-(1 → 4)-β-D-xylanase of the yeast *Cryptococcus albidus* has been found to catalyse the degradation of aryl β-D-xylosides by reactions other than simple hydrolytic cleavage.[463] Liberation of phenol or 4-nitrophenol from the corresponding β-D-xylosides was accompanied by formation of D-xylose oligosaccharides and only small amounts of D-xylose. With the aid of phenyl β-D-{U-^{14}C}-xyloside synthesized from D-{U-^{14}C}-xylose, it was established that the reaction followed a complex pattern with the rate of phenyl β-D-xyloside digestion and appearance of various products varying markedly with time. The reaction involved multiple transglycosylic reactions leading first to phenyl-D-glycosides of D-xylo-oligosaccharides, which are subsequently hydrolysed mainly to D-xylobiose and D-xylotriose.

The action pattern and reaction mechanism of the *endo*-(1 → 4)-β-D-xylanase of the yeast *Cryptococcus albidus* have been investigated using reducing-end {1-^{3}H}-labelled and {U-^{14}C}-labelled (1 → 4)-β-D-xylo-oligosaccharides up to xylopentaose.[464] The enzyme was found to catalyse degradation of oligosaccharides also by pathways other than a simple hydrolytic cleavage. Bond-cleavage frequency of xylotriose, xylotetraose, and xylopentaose was found to be concentration dependent. At high substrate concentration reactions such as xylosyl, xylobiosyl, and xylotriosyl transfer occur and result in the formation of products larger than the starting substrate. D-Xylose and xylobiose to a significant extent enter the reaction pathways as glycosyl acceptors. None of the trans-glycosylic reactions observed with reducing-end-labelled substrates or acceptors was accompanied by a significant label redistribution from the reducing-end unit, suggesting that the enzyme-glycosyl intermediates effective in the transfer reactions can be formed from the non-reducing-end units of oligosaccharides. Evidence for the formation of a termolecular shifted complex of β-D-xylanase with xylotriose has also been obtained. All features of the degradation of oligosaccharides by β-D-xylanase were found to be consistent with the lysozyme-type reaction mechanism.

The substrate-binding site of *endo*-(1 → 4)-β-D-xylanase of the yeast *Cryptococcus albidus* has been investigated using (1 → 4)-β-D-xylo-oligosaccharides {1-^{3}H}-labelled at the reducing end.[465] Evaluation of the affinities of ten imaginary subsites pointed out that the substrate-binding site of the enzyme is composed of four subsites and that the catalytic groups are localized in the centre. The imaginary subsites on the left-hand side of the binding site (non-reducing-end side) showed little or no affinity to bind D-xylosyl residues. For the subsites on the right-hand side of the binding site ('reducing-end' side) negative values of affinity were obtained, which means this region of the enzyme is unfavourable for complexing with D-xylosyl residues. As a consequence of the asymmetric distribution of negative values of affinity around the binding site, the enzyme displays a strong preference for attaching near the reducing end of the substrate. Regardless of the length of {1-^{3}H}-xylo-oligosaccharides, {1-^{3}H}-xylobiose was the prevailing reaction product at an early stage of hydrolysis, and frequency distribution of bond cleavage decreased from the second glycosidic bond towards the non-reducing end. Additional information on the substrate-binding site of *C. albidus* β-D-xylanase was obtained by evaluating the efficiency of D-xylose, xylobiose, methyl β-D-xylopyranoside, and phenyl β-D-xylopyranoside to serve as glycosyl acceptors in the transglycosylic reaction proceeding at high concentrations of xylotriose.

45 Xylanases (Miscellaneous)

The accuracy of a D-xylanase assay has been tested and improved.[466] The assay was used to monitor D-xylanase production by a Cellulomonas isolate and to demonstrate that this activity is distinct from the β-D-xylosidase activity of the organism.

The application of carbohydrases to the extraction of proteins from wheat bran has included the efficacious pretreatment with xylanase.[349]

An extracellular D-xylanase from a soil fungus (Fusarium sp.) grown on a medium containing groundnut hemicellulose B was purified 76-fold by ammonium sulphate fractionation, ion-exchange chromatography, and gel filtration.[467] The enzyme was homogeneous by disc gel electrophoresis at pH 8 and showed optimal activity at pH 5.6 and 37°C. It was observed that groundnut and sesame hemicellulose B were degraded considerably (~80 and 58%, respectively) by the purified D-xylanase, whereas D-gluco-D-mannan and D-xylan from groundnut were comparatively poorly hydrolysed (~30-40%).

The production and characterization of thermostable D-xylanase from Talaromyces byssochlamydoides YH-50 have been described.[468] This strain isolated from compost heaps produced the highest amount of thermostable D-xylanase among 180 isolates tested. When it was cultivated in solid wheat-bran medium containing xylan as an additive carbon source, the maximal amount of thermostable xylanase was produced after 3 days at 50°C. This culture filtrate hydrolysed 90% of D-xylan to D-xylose. The enzyme (pH optimum 5.5, temperature optimum 70°C) was quite stable after heating at 65° for 5 min and retained 55% of original activity after heating at 95°C for 5 min.

Trichoderma reesei Rut C-30 organism produced, together with high cellulase activities, considerable amounts of D-xylanases and β-D-glucosidase. The effect of temperature, pH, Tween-80 concentration, carbon source, and substrate concentration on the ratio of mycelial growth and extracellulose enzyme production was described.[208]

Aspergillus niger strain 110.42 has been selected as a producer of high D-xylanolytic activities.[469] The time-course of xylanase and β-D-xylosidase production as well as the effect of pH and temperature on the activity of these enzymes were studied. H.p.l.c. analysis of the enzymatic degradation of arabinoxylan showed a nearly complete conversion to pentose sugars. The authors discuss the use of crude D-xylanase preparations for the saccharification of D-xylans.

The enhanced cellulolytic activity of a Cellulomonas mutant has been shown to apply also to its ability to hydrolyse D-xylan-containing hemicelluloses.[470] The hydrolytic activity was directly proportional to the D-xylose content in the hemicellulose substrate.

A Cellulomonas strain, with potential for saccharification of sugar-cane bagasse, has been found to possess a range of carbohydrase activities which could be necessary for effective hydrolysis of the hemicellulose fraction of bagasse.[471]

Inducible β-xyloside permease has been reported as a constituent of the D-xylan-degrading enzyme system of the yeast Cryptococcus albidus.[472] The yeast, depending on whether it is grown on D-xylan or D-glucose, differs remarkably in the ability to take up inducers of extracellular endo-(1 → 4)-β-D-xylanase synthesis. In washed, D-glucose-grown cells the initially low ability to take up xylobiose or methyl β-D-xylopyranoside increases during incubation with these compounds after a lag phase shorter than the induction time of the extracellular β-D-xylanase. Using methyl β-D-{U-^{14}C}-xylopyranoside as a very slow metabolizable inducer of β-D-xylanase it has been established that the increase in the rate of xylobiose or methyl xyloside uptake is due to induction of an active transport system for methyl β-D-xyloside and (1 → 4)-β-D-xylo-oligosaccharides. The system is called β-D-xylosidase permease. The permease activity of induced cells decreases in the absence of β-D-xylanase inducers. The induction of permease as well as its inactivation (degradation) can be prevented with cycloheximide; thus both events appear to be dependent on de novo protein synthesis. In analogy with other active transport systems, β-D-xyloside permease function can be blocked effectively by inhibitors of energy metabolism in the cells.

The importance of cellulase and D-xylanase release from Bacteroides succinogenes in the rumen environment has been investigated.[223] During growth of B. succinogenes in a liquid medium with cellulose as the source of carbohydrate, greater than 80% of the xylanase was released from cells into the culture fluid. For further details see initial citation of ref.223.

Cultures of Streptomyces flavogriseus have been found to produce considerable amounts of D-xylanase when grown on D-xylan-containing media.[260] Comparatively lower yields of this enzyme were obtained when hay or Avicel served as main carbon source.

D-Xylanase induction by β-D-xyloside was investigated in non-growing conditions using non-induced mycelia of Streptomyces species harvested from D-glucose medium.[473] The mycelia started to produce D-xylanase without lag time when β-D-xyloside was added. The rate of D-xylanase synthesis was dependent on the concentration of β-D-xyloside added to the inducing culture medium. The induction constants of various β-D-xylopyranosides were calculated from Lineweaver-Burk plots, and those of methyl, isopropyl, butyl, and ethylenecyanohydrin β-D-xylosides were 10.53 mM, 3.83 mM, 0.55 mM, and 0.25 mM, respectively. Some α-D-xylosides repressed D-xylanase synthesis. The rate of D-xylanase synthesis decreased suddenly after the addition of α-D-xyloside. The inhibition constants of methyl, ethyl, and isopropyl α-D-xylopyranosides were 8.80 mM, 12.50 mM, and 33.33 mM, respectively. The xylanase induction was also repressed by D-glucose until consumption of the additional D-glucose was complete.

The production of xylanase by a Streptomyces species using non-metabolizable inducer has been described.[474] It was found that a variety of non-metabolizable β-D-xylosides possessed a marked inducing ability in comparison to D-xylan or its related materials. In the production of D-xylanase using such a synthetic inducer, the difficulty in preparing the inducer is considered on an economical basis. A suitable culture condition for D-xylanase production by methyl β-D-xyloside in Streptomyces species No. 3137 and a method for re-using the inducer were investigated.

46 Carbohydrate Isomerases

D-Xylose Isomerases (D-Glucose Isomerase). -- The inhibitory effect of some polyols (D-mannitol, D-arabitol, D-glucitol, ribitol, and xylitol) on the activity of D-xylose isomerase from Actinomyces olivocinereus has been studied.[475] All the polyols studied are purely competitive reversible inhibitors. The values of K_i for D-mannitol, D-arabitol, D-glucitol, ribitol, and xylitol are 0.200, 0.140, 0.030, 0.024, and 0.020 M, respectively. The inhibition is partially eliminated by an increase in the Mg^{2+} and Co^{2+} concentration to 5×10^{-2} and 1×10^{-3} M, respectively.

The kinetic properties of immobilized and non-immobilized D-xylose isomerase containing Arthrobacter species cells have been investigated.[476] In both cases the kinetics could be described by a modified Michaelis-Menten expression. It appeared that immobilization caused no deactivation. Furthermore, it was shown that the permeability of the cell membrane was increased by heat and toluene treatments.

D-Xylulose, an intermediate of D-xylose catabolism, has been observed to be fermentable to ethanol and carbon dioxide in a yield of greater than 80% by yeasts (including industrial baker's yeast) under fermentative conditions.[477] This conversion appears to be carried out by many yeasts known for D-glucose fermentation. In some yeasts, xylitol, in addition to ethanol, was produced from D-xylulose. Fermenting yeasts are also able to produce ethanol from D-xylose when D-xylose-isomerizing enzyme is present. The results indicated that ethanol could be produced from D-xylose in a yield of greater than 80% by a two-step process. First, D-xylose is converted to D-xylulose by D-xylose isomerase. D-Xylulose is then fermented to ethanol by yeasts.

Cultures of Streptomyces flavogriseus has been found to produce inter alia D-xylose isomerase when induced by D-xylose.[260]

The effect of gamma-irradiation on D-xylose isomerase purified from the cells of Streptomyces phoeochromogenus in dilute solution has been investigated.[478] The activity of the enzyme decreased exponentially with the dose under all conditions investigated. The inactivation yields (G_0 value) in neutral solution were 0.069 in air and 0.115 in nitrogen. The role of the radicals produced by water radiolysis in the inactivation of D-glucose isomerase was estimated by using nitrous oxide or tert-butanol as selective radical scavengers. Under these conditions, the hydroxyl radical and the hydrogen atom were found to be important in the enzyme inactivation, and the hydrated electron contributed very little.

The radiosensitivity of D-xylose isomerase has been investigated under various irradiation conditions.[479] The inactivation of D-glucose isomerase irradiated in a cell-bound state was exponential, and an increase in inactivation was recognized in an oxygenated condition. The cell-free enzyme was highly radiosensitive and had a small oxygen effect compared to that in a cell-bound state. The oxygen enhancement ratio (OER) decreased with a degree in enzyme purification. Released D-glucose isomerase was protected by the addition of glutathione, and the inactivation curve in nitrogen almost agreed with that in the cell. The protective effect of glutathione in oxygen decreased at higher doses because glutathione in oxygen was easily decomposed by irradiation.

The subunit structure has been detected in D-xylose isomerase from Streptomyces griseofuscus by using denaturants.[480] The enzyme was quite stable to sodium dodecyl sulphate under mild conditions, and dissociation into smaller subunits was dependent on pH and temperature. At acidic pH and high temperature, the enzyme showed rapid dissociation and inactivation. The dissociation into constituent subunits paralleled the loss of enzymatic activity. The enzyme was also dissociated into a singular type of subunit by 6 M guanidinium chloride which gave a single symmetrical elution peak on Sepharose 6B column chromatography in the presence of 6 M guanidine hydrochloride, and the eluate showed no activity. Molecular-weight estimations of the subunit, using both sodium dodecyl sulphate polyacrylamide gel electrophoresis and gel filtration, were the same (43,000). The N-terminal amino acid was identified as L-serine and was estimated to be 43 per cent from analyses of PTH and DNP derivatives. The sequence from the N-terminal was found to be L-Ser-L-Asp-L-Gln.... Considering the molecular weight of the native enzyme as 180,000 and the results obtained in the experiments here, it was concluded that native D-glucose isomerase was composed of four identical or very similar subunits with equal molecular weights and dimensions.

Some physicochemical properties of purified D-xylose isomerase from Streptomyces griseofuscus have been examined.[481] The extinction coefficient ($E^{1\%}_{1cm}$) and isoelectric point (pI) were determined to be 11.4 and 4.0, respectively. The values for sedimentation coefficient (S^{0}_{20w}), partial specific volume ($\underline{v}$), diffusion coefficient (D_{20w}), and intrinsic viscosity ({N}) were 8.50 S, 0.73 $cm^3\ g^{-1}$, 4.6 x $10^7\ cm^2\ S^{-1}$, and 3.7 ml g^{-1}, respectively. The molecular weight was estimated to be 180,000 by the sedimentation equilibrium method and 185,000 by a gel electrophoretical method.[481] The enzyme was found to contain four cobalt ions per molecule in atomic-absorption spectrophotometrical analysis. The circular-dichroism spectrum in the far-ultraviolet region showed a minimum at 280 nm and a maximum at 198 nm. Mean residue ellipticity ({O}) at 220 nm was estimated to be -11,200 deg.cm^2/d mol. The computed values for the contents of the secondary structure were as follows: α-helix 40%, β-form 36%, and random coil 24%. The enzyme showed a visible circular-dichroism spectrum having negative peaks at 530 nm and 430 nm. This suggested the formation of a co-ordinated metal compound, Co-amino acid complex. The molecular dimension of the enzyme was examined by measuring the hydrodynamic parameters. The frictional ratio (a/b) was 5.0 for prolate ellipsoid and 0.2 for oblate ellipsoid. Stokes' radius was calculated to be 47 Å for a hydrated hypothetical sphere.

The purification and enzymatic properties of D-xylose isomerase from Streptomyces griseofuscus have been described.[482] The enzyme was purified 4.3-fold and obtained in crystalline form, by adding ammonium sulphate, was homogeneous on polyacrylamide gel electrophoresis and ultracentrifugation. The optimum pH for activity was 8.5 and optimum temperature 85°C. The enzyme was quite inactive in the absence of metal ions but was remarkably activated by the addition of magnesium or cobalt ions. The enzyme catalysed the isomerization of D-xylose, D-glucose, and D-ribose. The K_m values for D-glucose and D-xylose were 2.2×10^{-1} M and 5.4×10^{-2} M, respectively, while V_{max} values were 17.6 μmol min^{-1} mg^{-1}, respectively. The ratio of D-fructose content to D-glucose content was approximately 1.0 at reaction equilibrium. The enzyme activity was inhibited by sugars, sugar alcohols, and tris(hydroxymethyl)aminomethane in a competitive manner. The K_i values were as follows: D-xylitol 1.2×10^{3} M, D-sorbitol 1.1×10^{-2} M, D-mannitol 8.3×10^{-2} M, D-mannose 3.4×10^{-1} M, D-galactose 3.2×10^{-1} M, L-arabinose 2.3×10^{-1} M, and Tris 6.2×10^{-2} M. 4-Chloromercuribenzoate, iodoacetic acid, sodium azide, potassium cyanide, sodium fluoride, and 2-mercaptoethanol had no inhibitor effect on the activity of D-glucose isomerase, while H_4 edta caused a significant loss of activity.

The investigation by ^{1}H n.m.r. spectroscopy of the site of proton exchange catalysed by poly(D-mannuronic acid) C-5 epimerase, which catalyses the conversion of D-mannuronic acid residues into L-guluronic acid residues provided the substrate is a glycuronan of at least 10 units, has been reported.[483] Information is lacking as to the enzyme's specificity for a particular local arrangement of units in the chain.

L-Ribose Isomerases. -- Induction of L-ribose isomerase by L-ribose in Mycobacterium smegmatis has been investigated.[484] L-Ribose, which unlike D-ribose was not a substrate of the enzyme and also not a growth substrate for the organism, induced the same D-ribose isomerase possibly due to the characteristic configuration and conformation of the ribose molecule.

47 Carbohydrate Oxidases

D-Glucose Oxidases. -- D-Glucose oxidase and an oxidized version of the enzyme have been immobilized covalently to two basic types of sorbents - glycidylmethacrylate copolymers and bead cellulose.[485] The properties of the samples thus obtained were compared with those of immobilized D-glucose oxidase bound on to some common carriers. Samples which possessed not only a high absolute activity but also adequate mechanical and flow properties were characterized in greater detail with respect to the immobilization efficiency and kinetic properties of bound D-glucose oxidase.

New findings in the oxidation of glucose by means of immobilized D-glucose oxidase have been reported.[486] Based on proceeding publications the paper shows recent aspects, which seem to be important for an economical and technical use of the immobilized D-glucose oxidase-catalase system. The results of trials in continuous processing were negative, since the conversion rates were sufficient and technical difficulties arose in keeping back the enzyme particles in the reaction vessel. Batchwise processing with repeated enzyme application showed the major importance of optimal oxygen supply and keeping low temperatures of about 2°C. Different types of stirred-tank reactors were tested and classified concerning their suitability. The chances for large-scale application of the immobilized D-glucose oxidase-catalase system are judged to be good.

A modified lectin matrix has been used to study the three-dimensional structure of a dimeric glycoprotein, D-glucose oxidase, which was allowed to react with L-lysine-specific cross-linkers, both when immobilized on a succinoylated lectin matrix at a critically low density and also at a high density in solution.[487] Analysis of the cross-linked complexes thus obtained led to the following inferences with regard to the structure of this protein. Of the 15 L-lysine residues on each D-glucose oxidase protomer, none is available on the non-interfacial surfaces. Assuming that this protein possesses C_2 symmetry with isologous bonding between subunits, it may be inferred that on each protomer there are at least two L-lysine clusters along or close to the interprotomeric interface. These interfacial L-lysine residues on each protomer are so oriented that the ε-amino groups of L-lysine residues a and b on protomer 1 face, and are very close to, the ε-amino groups of L-lysine residues b' and a', respectively, on protomer 2. General inferences on the geometry of dimeric proteins derivable from an analysis of the cross-linked complexes obtained (as well as those not seen) by using this low-density matrix cross-linking approach were enumerated. It was suggested that modified lectin matrices may prove useful in studying the three-dimensional structure of glycoproteins, particularly non-crystallizable oligomers.

The enhancement of oxygen absorption by magnetite-containing beads of immobilized D-glucose oxidase has been investigated.[488] Rates of oxygen absorption of D-glucose solutions were measured using an immobilized-enzyme reactor, in which magnetite-containing beads of immobilized D-glucose oxidase were moved by a revolving magnetic field to reduce the mass-transfer resistances at the gas-liquid interface and around the bead. Data were also obtained for oxygen absorption into D-glucose solutions containing soluble or immobilized glucose oxidase (without magnetite), as well as for physical absorption for the runs with the magnetite-containing beads increased because of mechanical stirring caused by spinning of the beads at the gas-liquid interface. In this case the experimental enhancement factors were found to be larger than those predicted on the basis of the film theory for gas absorption with a pseudo-first-order reaction.

Physical entrapment has been used as an approach to achieve thermal stabilization of D-glucose oxidase.[395] The $t_{0.5}$ value for the thermoinactivation of D-glucose oxidase was increased several-fold by entrapment in polyacrylamide gels. In polyacrylate gels the enzyme behaved differently, probably owing to a microenvironmental effect of the polyelectrolyte nature of the carrier.

It has been reported found that in addition to oxygen six redox indicators can serve as substrates for D-glucose oxidase from *Pencillium vitale*.[489] The pH dependence of the rate of the reaction catalysed by D-glucose oxidase was investigated in the presence of oxygen and artificial electron acceptors. In contrast to the reaction with oxygen, whose optimum pH is 5.6, the optimum pH of reactions with phenazine methosulphate, basic dark blue 2K, polyviologen, and the ion-radical salt of *N*,*N*,*N*',*N*'-tetramethyl-4-phenylenediamine was observed at pH 7.5. The results indicated that the reduced enzyme can exist in two forms (protonated and deprotonated) differing in rate of oxidation by oxygen and in ability to bind and reduce artificial electron acceptors with the formation of the semiquinone form of the cofactor.

The use of D-glucose oxidase-catalase system in measuring

aeration capacity of fermenters has been investigated and a comparison made of the dynamic and steady-state methods of $\underline{k}_1\underline{a}$ measurement.[490] The conditions have been specified where spontaneous hydrolysis of lactone was sufficiently rapid, thus eliminating inhibitory action of lactone on the oxidation. In D-glucose oxidase-free batches, the $\underline{k}_1\underline{a}$ values were determined using various modifications of the dynamic method. The dynamic methods in which gas interchange was effected without interrupting aeration and agitation of the batch yielded erroneously lower $\underline{k}_1\underline{a}$ values as compared to the results of steady-state methods if the measured $\underline{k}_1\underline{a}$ value was higher than 0.03 s^{-1}. The value yielded by the dynamic method in which the gas interchange was effected at the same time with turning on aeration and agitation of the batch agreed with values resulting from the steady-state method, provided that the measured $\underline{k}_1\underline{a}$ values were lower than 0.08 s^{-1} and the simultaneous interfacial transport of nitrogen and oxygen had been taken into account in the evaluation. At $\underline{k}_1\underline{a}$ values higher than 0.08 s^{-1} this modification of the dynamic method also yielded lower $\underline{k}_1\underline{a}$ values as compared with the outcome of the steady-state method. The experiments performed did not, however, allow one to decide unambiguously on whether these lower $\underline{k}_1\underline{a}$ values were due to failure of the adopted model to descibe adequately the dynamic behaviour of the system or whether they were true values differing from those yielded by the steady-state method on account of different physical properties of compared batches.

The kinetics of oxidation of D-glucose solution by dissolved oxygen in the presence of D-glucose oxidase and excess catalase have been studied both theoretically and experimentally.[491] The kinetic model used was based on the detailed mechanism of the reaction. The description of the model was given by the Michaelis-Menten equation for the case of two substrates. The experiments were carried out in a closed reactor at a temperature of 25°C and pH 5.55 using commercial enzymes. Kinetic constants of the reaction were evaluated by fitting the oxygen concentration profiles calculated from the model to those found experimentally, by the method of non-linear regression. The rate of D-glucose mutarotation and distortion of data caused by the dynamics of the applied oxygen probe were taken into account. The determined kinetic constants of the reaction were in good agreement with those given in the literature for purified enzymes. Adoption of literature kinetic data of this reaction to the determination of aeration capacity of fermenters by the D-glucose oxidase system proved possible.

At pH 7.4, in 0.1 M phosphate buffer, D-glucose oxidase and Blue Dextran interact strongly with each other.[492] Under these conditions a soluble complex is formed, as shown by ultrafiltration experiments using a Diaflo XM-300 membrane (nominal molecular weight cut-off 300,000). In this complex, the enzyme is about 40% more active than when free in solution, and is also more stable towards acid denaturation. Comparative studies with a high-molecular-weight dextran devoid of dye residues (Dextran T-2000) suggested that the dye moiety of Blue Dextran is directly involved in the binding of the protein, and possibly also in the enhancement of its enzymatic activity.

D-Glucose oxidase purified from Aspergillus niger on subjection to isoelectric focusing and gel electrophoresis was found to be composed of at least six component enzymes (pI 3.9 to 4.3).[493] They all possess an identical protein moiety, since the amino acid compositions, C-terminal sequences, catalytic parameters, quantitative and qualitative immunological properties, and the electrophoretic patterns of the peptide fragments obtained by CNBr cleavage were practically the same. On the other hand, the carbohydrate contents of the enzymes were found to be different, and these differences were associated in the main with a particular peptide fragment.

L-Galactonolactone Oxidases. -- L-Galactonolactone oxidase has been found to be inactivated by various sulphydryl reagents: the order of inactivation rate was $HgCl_2$, 4-chloromercuribenzoate, 4,4'-dipyridyldisulphide, 2,2'-dipyridyldisulphide, 5,5'-dithio-bis-(2-nitrobenzoate), and N-ethylmaleimide iodoacetamide.[494] The inactivation by 4,4'-dipyridyldisulphide was studied in detail, and it was found that the maximum degree of inactivation attained with increasing reagent concentration was 93% and that the kinetics of inactivation were first order with respect to reagent concentration. The pH dependence of the second-order rate constant of the inactivation revealed that a sulphydryl group with a pK_a of 9.8 was involved in the inactivation process. The value of pK_a is high compared with that of low-molecular-weight thiols, indicating that the ionization of the sulphydryl group is affected by the electric field of a negatively charged group located in its vicinity. The values of K_m and V_{max} for the L-galactonolactone oxidase reaction did not change greatly around pH 9.8. This is consistent with the view that the sulphydryl group does not participate in the catalytic process. The velocity of inactivation in the presence of substrate was considerably greater than that observed in its absence. It appears that the sulphydryl group becomes more reactive during the catalytic cycle of the enzyme. It was also noted that the absorption spectrum of the flavin prosthetic group (the oxidized form) was modified by adding 4-chloromercuribenzoate.

L-Gulonolactone Oxidases. -- The activity of L-gulonolactone oxidase in livers of 49 species of eutherian mammals has been found to vary intraspecifically among individuals: coefficients of variation were 0.2 to 0.4 in many species.[495]

Cellobiose Oxidases. -- A species of the imperfect fungus Monilia produced cellobiose oxidase extracellularly when grown on cellulose.[496] The inducible enzyme was both bound to the mycelium and released into the growth medium and showed a high degree of specificity for cellobiose, but also oxidized lactose and 4-O-β-D-glucopyranosyl-D-mannose. The specificity of the electron acceptor was restricted to those compounds having a redox potential of +0.22 V. 4-Benzoquinone and several other quinones, however, were not reduced. Oxygen was not consumed nor was hydrogen peroxide produced by cellobiose oxidation of cellobiose. The enzyme had a molecular weight of 48,000 and a pI of 5.4. A new zymogram technique was developed for the detection of cellobiose oxidase in polyacrylamide gels following electrophoresis and isoelectric focusing.

L-Fructose Oxidases. -- An experimental procedure has been proposed for the study of liquid-phase reactions catalysed by enzymes immobilized on porous supports.[497] The kinetic behaviour can be assessed by experimental runs which allow a separate evaluation of the internal diffusion, the surface kinetics, and the fluid-particle mass transfer. The method has been applied in the study of sucrose inversion by β-L-fructose oxidase immobilized on IRA-93 resin.

D-Galactose Oxidases. -- The steric factors involved in the action of glycosidases and D-galactose oxidase have been investigated.[109] D-Galactose oxidase is sterically hindered by certain types of branching in substrate oligosaccharide chains and will not oxidize the D-galactose residue in trisaccharide (5) but will in (6). D-Galactose as in (7) is not susceptible to oxidation with D-galactose oxidase until after the application of β-D-2-acetamido-2-deoxygalactosidase.

```
                          α-NeuNGL
                              2
                              |
                              6
              β-D-Gal-(1→3)-D-GalNAcol
                         (5)

              β-D-Gal-(1→3)-D-GalNAcol
                    2
                    |
                    1
              ?-L-Fuc
                         (6)

   β-D-GalNAc-(1→3)-β-D-Gal-(1→3)-D-GalNAcol
                             2
                             |
                             1
                        α-L-Fuc
                          (7)
```

48 Carbohydrate Transferases

A review of the affinity chromatography of glycosyltransferases summarizes the use of biospecific chromatography techniques in the purification of mammalian glycosyltransferases.[498] Ligands that are analogues of donor or acceptor substrates have been linked to cyanogen bromide-activated agarose for use as affinity adsorbents. Immobilized lectins have been employed to recognize the carbohydrate moieties of glycosyltransferase and remove them from complex mixtures. The application of these methods has permitted extensive purification of many membrane-bound glycosyltransferases, some to homogeneity.

The possibility that murein glycosyltransferase of Escherichia coli may function as an exoenzyme to cleave the jurein sacculus in a systematic fashion has been investigated.[499] Two molecular species of this hydrolytic enzyme have been isolated and characterized: one is associated with the soluble fraction and the other with the envelope fraction of ruptured E. coli cells. The soluble enzyme was employed to digest murein sacculi that had been uniformly labelled with {^{3}H}diaminopimelic acid. The analysis of the reaction product indicated that the enzyme did not cleave the glycan chains randomly. To determine whether glycosyltransferase released muropeptide first from the 2-acetamido-2-deoxy-D-glucosyl or the 1,6-anhydromuramyl ends of the glycan chains, the {^{3}H}-diaminopimelate-labelled sacculi were further radiolabelled at their 2-acetamido-2-deoxy-D-glucosyl ends with {^{14}}-D-galactose by a D-galactosyltransferase reaction. The glycosyltransferase released D-galactose-labelled X + X' muropeptides (where X = 2-acetamido-2-deoxy-D-glucosyl-1,6-anhydro-N-acetylmuramyl-L-alanyl-D-α-glutamyl-meso-diaminopimelic acid, X' = X-D-alanine) early during the course of digestion, suggesting exoenzymatic cleavage of the glycan chains preferentially from the 2-acetamido-2-deoxy-D-glycosyl ends. The kinetics of the activity of the membrane-bound enzyme were found to be identical to those of the soluble enzyme, indicating that both molecular species of glycosyltransferase function as exoenzymes in vitro.

Inactivation of D-glucosyltransferases from oral Streptococcus mutans and S. sanguis by photochemical oxidation has been reported.[500] Cell-free D-glucosyltransferase of D-glucose-grown S. mutans AHT was completely inactivated in the presence of 0.002% of Methylene Blue at 25°C and pH 7.0 after illumination with a 150 W incandescent lamp. The rate of inactivation was decreased at pH values <7.0. L-Histidine was the only amino acid residue modified to a significant extent, and the rates of oxidation of L-histidine and loss of enzyme activity closely agreed. Production of both water-insoluble and -soluble L-glucan fractions from sucrose by the oxidized D-glucosyltransferase preparations was significantly inhibited. Photo-oxidation with 0.002% Rose Bengal at pH 7.0 also induced complete inactivation of the D-glucosyltransferase. These results strongly suggest that the imidazole portion of L-histidine may function as part of the active sites of both D-glucosyltransferase isoenzymes of S. mutans AHT, which are responsible for the synthesis of (1 → 3)- and (1 → 6)-α-D-glucosidic linkages.

It has been reported that the glucosyltransferases of

Streptococcus mutans have been resolved into two components essential to water-insoluble glucan synthesis.[501]

Collagen-D-galactotransferase and collagen-D-glucosyltransferase activities have been studied in cultured human foetal lung WI-38 and IMR-90 diploid fibroblasts.[502] These enzymes functioned in concert to synthesize D-glucosyl-D-galactosyl-L-hydroxylysine units as found naturally in collagens, basement membranes, and certain serum glycoproteins. The transferases used UDP-D-Gal and UDP-D-Glc as glycose donors, collagens and collagen-derived peptides or glycopeptides as glycose acceptors, and worked best in the presence of manganese as a required divalent cation. Two pH optima, between pH 6 and 6.5 and between pH 7.5 and 8, were noted for each type of transferase, and these optima, particularly in the case of D-glucosyltransferase, were evident regardless of size of acceptor employed in the assay. About 35% of the total activity of each enzyme was found in the soluble fractions of cell homogenates, and about 50% of the particulate fraction activities could be released by mild sonication or by treatment with Triton X-100. Assessment of transferase activities as a function of cellular ageing in culture revealed that significant decreases in enzyme levels occurred as the cell approached senescence (late phase II), and these effects were reversed when the cell attained senescence (phase III). Addition of ascorbic acid to young cultures, under conditions known to increase endogenous collagen-peptide hydroxylation, caused no effects on the activities of the glycosyltransferases toward exogenous acceptors. These results suggested that the activities of collagen-hydroxylases and glycosyltransferases might not be co-ordinately regulated and that, regardless of the hydroxylation events, glycosylation of the peptide might be limited to a specific fraction of L-hydroxylysine residues during the post-translational modification of collagen.

Bovine glycoprotein β-D-galactosyltransferase has been shown to have two metal-binding sites, the functional properties of which have been established by using kinetic, spectroscopic, and affinity-chromatographic approaches.[503] Metal site I, which is involved in maintaining the structural integrity of the protein, must be liganded prior to other substrates binding and prior to a second metal binding to site II, which is shown to be associated with UDP-D-galactose binding. Both metal sites can bind a variety of metals. However, calcium and its fluorescent analogue europium bind only to site II. Fluorescent resonance-energy transfer measurements between europium in site II and cobalt in site I indicate a distance of 18 ± 3 Å between the two sites. Chemical-modification studies with *S*-mercuric-*N*-dansylcysteine indicate that one (of a total of three exposed sulphydryl groups) can be specifically dansylated and that this sulphydryl group is in or near the UDP-D-galactose-binding site. Resonance-energy transfer measurements between this introduced sulphydryl group and cobalt in metal site I give a distance of 19 ± 3 Å between these points, consistent with the interpretation that the UDP-D-galactose-binding site, which is associated with metal site II, is located some distance from the structural metal site (site I).

A D-galactose D-galactosyltransferase (lactose synthetase A protein), which transfers D-galactose from UDP-D-galactose to 2-acetamido-2-deoxy-D-glucose, was purified 286,000-fold to homogeneity with 40% yield from human plasma by repeated affinity chromatography on α-lactalbumin-Sepharose.[504] Sodium dodecyl sulphate polyacrylamide gel electrophoresis of the purified enzyme showed a single protein band with a molecular weight of 49,000. The enzyme is a glycoprotein with 11% by weight carbohydrate which has only 2-acetamido-2-deoxy-glucose-L-asparagine glycopeptide linkages. The enzyme showed characteristic changes in activity at different α-lactalbumin concentrations, indicating that the enzyme is the A protein of lactose synthetase. K_m values for the substrates were found to be 0.056 mM for UDP-D-galactose and 0.20 mM for Mn^{2+}. The activity of the enzyme was neutralized by anti-enzyme antibody, but the antibody did not neutralize the bovine milk D-galactosyltransferase (A protein) activity.

Bovine α-lactalbumin has been dansylated to give an enzymatically fully active, highly fluorescent derivative labelled on the N-terminal L-glutamic acid residue.[505] This fluorescent derivative of α-lactalbumin has been covalently crosslinked to glycoprotein β-D-galactosyltransferase by using dimethyl pimelimidate. Resonance-energy transfer measurements using cobalt bound to the transferase as the acceptor of energy transfer from the dansyl group on the α-lactalbumin indicate that the dansyl group is 32 Å from the cobalt on the transferase. A model of the active site of the transferase and its interaction with α-lactalbumin is proposed on the basis of these and previous studies. These studies have allowed an extension of the active-site mapping of lactose synthetase and give greater insight into the interaction of α-lactalbumin with the transferase and its role in lactose synthetase.

Glycoprotein:L-fucosyltransferase from sheep brain, solubilized with Triton X-100 and prepurified by hydrophobic chromatography on ethyl-agarose, has been further purified by chromatofocalization.[506] This procedure yielded four isoenzymes.

The soluble L-fucosyltransferase of the rat small-intestinal mucosa purified by gel filtration on Sephadex G-100 has been resolved into two isoenzymes F_1 (pI = 4.47) and F_2 (pI = 4.96) by isoelectric focusing.[507] The use of DEAE-cellulose instead of Sephadex G-100 leads to another component F_3 (pI = 8.70). Kinetic parameters (K_m and V_{max}) for the three isoenzymes were determined. The enzymatic mechanism seems to be a bi-bi random type for the three isoenzymes, and F_3 appears as the most active species.

The isoelectrofocusing patterns of 2-α-L-, 3-α-L-, and 4-α-L-fucosyltransferases from human milk serum have been reported.[508]

The specificity of sialyltransferase has been investigated with reference to the sialylation of ovine submaxillary mucin in vitro which was {^{14}C}sialylated in vitro using a porcine liver cell-free preparation.[509] The oligosaccharide chains were cleaved from the product glycoprotein by β-elimination under reductive conditions, fractionated by gel filtration on Bio-Gel P-2, and characterized by thin-layer chromatography. The structure of the product chain was studied by periodate oxidation and analysis of the peeling products formed in the β-elimination step. It appeared that {^{14}C}sialic acid had been introduced exclusively to the D-galactose residues of D-Gal-(1 → 3)-D-GalNAc disaccharide units occurring on the mucin as minor chains. No indication for a transfer to 2-acetamido-2-deoxy-D-galactosyl residues on this glycoprotein was obtained. In agreement with this result sialyltransferase activities of porcine, rat, human, and canine liver with D-Gal-(1 → 3)-D-GalNAc-protein acceptors were invariably much higher than those with ovine submaxillary asialomucin. When the asialomucin had been {^{14}C}sialylated by an ovine submaxillary gland cell-free preparation, analysis of the product oligosaccharide chain revealed the introduction of {^{14}C}sialic acid to position C-6 on the 2-acetamido-2-deoxy-D-galactose residues. The specificity of this transfer was reflected by the very high sialyltransferase activities of gland preparations with D-Gal-(1 → 3)-D-GalNAc-protein as well as D-GalNAc-protein acceptors. Mixed-enzyme experiments indicated that the difference in liver and gland ovine submaxillary asialomucin sialyltransferase activities was not due to the presence of a specific inhibitor in the liver or an activator in the gland. It was concluded that livers of man, pig, rat, and dog contain a D-Gal-(1 → 3)-D-GalNAc-protein sialyltransferase which is involved in the sialylation of O-glycosidically linked carbohydrate chains on serum glycoproteins. D-GalNAc-protein sialyltransferase activity, which richly occurs in ovine submaxillary gland, however, appears to be lacking from liver tissue.

A technique has been described by affinity chromatography of orotate phosphoribosyltransferase from *Escherichia coli* K-12.[510]

49 L-Iduronic Acid 2-Sulphate Sulphatases

A more sensitive assay has been developed for the L-iduronic 2-sulphate sulphatase which is deficient in cases of the Hunter syndrome.[511] The substrate is O-(α-L-idopyranosyluronic acid 2-sulphate)-(1 → 4)-2,5-anhydro-D-{^{3}H-1}mannitol 6-sulphate, and, after incubation, it is separated from the product by ion-exchange chromatography on a micro-column of Dowex 1 x 2 (Cl^-) rather than by high-voltage electrophoresis or ECTEOLA cellulose chromatography. Since the blank correction is then much smaller, a shorter incubation time can be used and conversion of the substrate reduced from approximately 50% down to levels where complications resulting from substrate depletion and product inhibition are minimal. Using whole serum the apparent K_m for the substrate is 0.2 nmol l^{-1}. With an incubation time of 20 min, sera from heterozygotes exhibited approximately 35% of the normal levels of L-iduronate 2-sulphate sulphatase (0.11-0.61, mean 0.34, nmol h^{-1} mg^{-1} protein for 21 carriers, 0.24-2.35, mean 0.94, nmol h^{-1} mg^{-1} protein for 37 normal females). Serum analyses can thus be used to supplement those on hair roots in the detection of carriers of the Hunter syndrome.

The L-iduronic acid 2-sulphate sulphatase of human placenta has been separated by ion-exchange and gel-filtration chromatography into two components, a less acidic form A and a more acidic form B.[512] The two forms have different mobilities on gel electrophoresis and different isoelectric points (A pI 5.0, B pI 4.5). They show the same pH optima in sodium acetate buffer and similar K_m values for {^{3}H}disulphated disaccharide substrate. The A form is more heat labile than the B form. Different molecular weights were found by gel filtration while similar values were estimated by sucrose gradient centrifugation. Neuraminidase treatment of the two forms gives evidence that these enzymes contain sialic acid residues.

Human urine has been shown to contain a novel sulphatase which is specific for 2-deoxy-2-sulphamido-D-glucopyranoside 3-sulphate.[513] Of the three isomeric sulphamate derivatives, 3-O-, 4-O-, and 6-O-sulphate esters, only the 3-O-ester was hydrolysed. The enzymatic activity required that the amino group be sulphated. Sulphate is not released if the amino group is free or acetylated. The enzyme was purified 70-fold. It had a pH optimum of 6.3 and was inhibited by inorganic sulphate and phosphate. The specificity of this enzyme suggested that a 3-O-sulphated 2-amino-2-deoxy-D-glucose moiety may have a role in the physiological activity of heparin or heparan sulphate.

50 Arylsulphatases

A study has been made of lysosomal-enzyme activities including arylsulphatase A in serum and leukocytes in chronic hepatic disease.[60] For further details see initial citation of ref.60.

A variant form of arylsulphatase A has been identified in human urine derived directly from the renal pelvis. The variant form of the enzyme was found in nephrostomic urine, in addition to the minor form, which is the sole component of arylsulphatase A in voided urine.[514] The nephrostomic enzyme differed from the voided-urine enzyme with respect to the kinetic parameters, the isoelectric point, heat stability, and immunological reactivity. The isoelectric points of the voided-urine and nephrostomic enzymes were 4.7 and 5.3, respectively. The nephrostomic enzyme was more heat labile at 62.5°C than the voided-urine enzyme. Although the molecular weights (130,000) and K_m values of the two enzymes with nitrocatechol sulphate as substrate were almost the same, the V value of the nephrostomic enzyme was 10% that of the voided-urine enzyme. It was demonstrated by various methods, using IgG antibody against the purified voided-urine enzyme, that the nephrostomic enzyme was antigenically distinct from the voided-urine enzyme.

Rabbit liver arylsulphatase A is a glycoprotein containing 4.6% carbohydrate comprising D-mannose (25), 2-acetamido-2-deoxy-D-glucose (7), and sialic acid (3 residues per enzyme monomer mol. wt. 140,000).[515] Each monomer consists of two equivalent polypeptide chains. The protein has a relatively high content of L-proline, L-glycine, and L-leucine, and the amino acid composition is similar to that of other known liver sulphatases. The arylsulphatase A catalyses the hydrolysis of a wide variety of sulphate esters, although it appears that cerebroside sulphate is a physiological substrate for the enzyme because the K_m is very low (0.06 mM). The turnover rate for hydrolysis of nitrocatechol sulphate or related synthetic substrates is much higher than the rate with most naturally occurring sulphate esters such as cerebroside sulphate, steroid sulphates, L-tyrosine sulphate, or β-D-glucopyranose 6-sulphate. However, the turnover rate with ascorbate 2-sulphate is comparable to the rates measured using most synthetic substrates. These results are discussed in relationship to several previously described sulphatase enzymes which were claimed to have unique specificities.

Arylsulphatase has been purified 219-fold from Pseudomonas species.[516] The final preparation was homogeneous by electrophoretic and immunological analysis. The enzyme is a monomer of molecular weight about 51,000, with a Stokes radius of 3.0×10^{-7} cm, a frictional ratio of 1.2, and a sedimentation coefficient of 4.1 S.

It has been reported that arylsulphatase activity is stimulated by arylamines and evidence for substrate activation has been produced.[517] A number of arylamines (including tyramine and tryptamine) increased the in vitro activity of arylsulphatase from Pseudomonas sp. strain $C_{12}B$. Amino acid analogues of these amines (e.g. L-tyrosine and L-tryptophan) failed to exert an effect. Stimulation of activity by tyramine could not be accounted for in terms of sulphotransferase activity for this phenol, and no shift in the pH optimum for the enzyme occurred in the presence of tryptamine. Increased V_{max} due to these amines was independent of enzyme concentration but varied significantly with substrate concentration. Evidence is presented which suggests that arylamines enhance arylsulphatase activity by forming a salt linkage with the substrate and rendering it more susceptible to enzymatic and acid-catalysed hydrolyses. The recrystallized tryptamine salt of the substrate exhibited a reduced affinity for the enzyme but was hydrolysed more rapidly than the potassium salt, which is normally employed as the assay substrate.

The stimulation of arylsulphatase synthesis in Pseudomonas aeruginosa by exogenous nucleotides has been reported.[518]

51 2-Acetamido-2-deoxy-D-glucose 6-Sulphate Sulphates

A deficiency of 2-acetamido-2-deoxy-D-glucose 6-sulphate required for heparan sulphate degradation has been reported in patients with Sanfilippo disease type D.[519] Skin fibroblasts from two patients who had symptoms of the Sanfilippo syndrome (mucopolysaccharidosis III) accumulated excessive amounts of heparan sulphate and were unable to release sulphate from 2-acetamido-2-deoxy-D-glucose 6-sulphate linkages in heparan sulphate-derived oligosaccharides. Keratan sulphate-derived oligosaccharides bearing the same residue at the non-reducing end and 4-nitrophenyl 2-acetamido-2-deoxy-D-glucopyranoside 6-sulphate were degraded normally. Kinetic differences between the sulphatase activities of normal fibroblasts were found. These observations suggest that 2-acetamido-2-deoxy-D-glucose 6-sulphate activities degrading heparan sulphate and keratan sulphate, respectively, can be distinguished. It is the activity directed towards heparan sulphate that is deficient in these patients. The authors propose that this deficiency causes Sanfilippo disease type D.

52 Miscellaneous Enzymes

Dextransucrase. -- Dextransucrase from Leuconostoc mesenteroides has been produced in a semicontinuous culture with slow addition of a concentrated sucrose solution.[520] The resulting high activity of the fermentation broth allowed a one-step purification method, by gel-permeation chromatography in 96.4% yield. This procedure resulted in 140-fold purification, with specific activity of 122 U mg^{-1}. The enzyme was immobilized on to an amino-Spherosil support activated with glutaraldehyde. Preparations with dextransucrase activities as high as 40.5 U g^{-1} of support were obtained when low specific area supports were used and maltose was added during the enzyme coupling. Diffusional limitations were found during enzyme reaction, as shown by a kinetic study. As a consequence of immobilization, the average molecular weight of dextrans appeared to increase. Immobilized dextransucrase could be promising for low-molecular-weight dextran production. Clinical dextran was synthesized when the polysaccharides produced in the presence of maltose were used as acceptors of a second synthesis reaction. The molecular-weight distribution of the resulting production was less dispersed than when clinical dextran was produced by acid hydrolysis of high-molecular-weight dextran.

Studies have been carried out on the acceptor specificity of dextransucrase which had been isolated from Streptococcus sanguis.[521] Radioactive acceptors were employed in reactions with cold sucrose and the counts incorporated were taken as a measure of acceptor activity. An order of relative activity was found to be polysaccharide > oligosaccharide > glycoside > monosaccharide. An evaluation of the time-course of the reaction with methyl α-D-glucopyranoside, or maltose, showed that a homologous series of oligosaccharides was formed from each. This suggested that the individual members of the series were related as precursors and products. The kinetics of the reaction with different acceptors were studied. All acceptors studied caused an activation of the enzyme and changes in the K_m for sucrose. The kinetic constants obtained were also used to compare the various acceptors.

It has been demonstrated that α-D-1-fluoroglucose is a glucosyl donor for dextransucrase isolated from Streptococcus sanguis.[522] The details of the reaction have been established as well as proving that the mechanism of D-fluoroglucose transfer is comparable to that of D-glucosyl transfer from sucrose. A new procedure for monitoring the reaction is reported, and is based on the measurement of proton formation using the pH indicator Bromcresol Purple. Production of F^- was found to be stoichiometric with proton production. Rate studies with the substrate indicate that α-D-1-fluoroglucose undergoes spontaneous hydrolysis, which is greatly increased in the presence of nucleophilic buffers. When {^{14}C}maltose and α-D-1-fluoroglucose or α-1-D-{^{14}C}fluoroglucose and maltose were incubated with dextransucrase, a series of oligosaccharide products was observed. The results indicate that the D-glucosyl moiety of α-D-1-fluoroglucose transferred to the acceptor. The nature of formation of the products is consistent with a series of precursor-product reactions. Product analysis of the saccharides by borohydride reduction analysis demonstrated that the D-glucosyl unit was added to the non-reducing end of maltose. When either D-{^{14}C}fructose or α-D-1-{^{14}C}fluoroglucose was incubated with enzyme, a reaction was observed which was analogous to the isotopic-exchange reaction catalysed by the enzyme in the presence of D-{^{14}C}fructose and sucrose.

Studies on donor-substrate specificity have been performed with dextransucrase.[523] Previous studies (D.S. Genghoff and E.J. Hehre, Proc. Soc. Exp. Biol. Med., 1972, 140, 1298) have shown that an α-linked fluorine atom at C-1 of glucose provided sufficient activation to permit this analogue to be a donor substrate for dextransucrase. In order to study the specificity at the donor-substrate-binding site, a series of 1-fluoro-α-D-sugars has been synthesized. In kinetic experiments, it has been determined that they served as competitive inhibitors of sucrose, the natural substrate. A comparison of the K_is provided information about the importance of specific changes in the D-glucose moiety with regard to binding to the enzyme. Similar kinetic studies were carried out with several 1-fluoro-β-D-sugars and the corresponding free monosaccharides. These were found to be non-competitive inhibitors, and to bind poorly. The 1-fluoro-α-D-sugars were also examined as donor substrates in reactions with known acceptors. With the exception of 1-fluoro-α-D-glucose, none of these analogues was active in this capacity.

Several carbohydrates, including sucrose analogues, have been designed, synthesized, and tested as dextransucrase inhibitors.[524] Certain carbohydrates containing an amino group were shown to be potent inhibitors of dextransucrase.

Multiple forms of Leuconostoc mesenteroides dextransucrase have been separated by gel filtration and electrophoretic analyses.[525] Two components of enzyme, having different affinities for dextran gel, were separated by a column of Sephadex G-100. The major component was treated with dextranase and purified to an electrophoretically homogeneous state. The purified enzyme had a molecular weight of 64,000-65,000, pI value of 4.1, and contained 17% of carbohydrate. H_4-edta showed a characteristic inhibition on the enzyme while stimulative effects were observed by the addition of exogenous dextran to the incubation mixture. The enzyme activity was stimulated by various dextrans and its K_m value was decreased with increasing concentration of dextran. The purified enzyme showed no affinity for a Sephadex G-100 gel and readily aggregated after preservation at 4°C in a concentrated solution.

Synthases. -- Concentrations of ADP-D-glucose:α-1,4-D-glucan-4-glucosyltransferase (starch synthase) and α-1,4-D-glucan:α-1,4-D-glucan-6-glycosyltransferase (branching enzyme) from developing seeds of Pisum sativum have been measured.[526] Primed starch synthase activity increased from 8 to 14 days after anthesis and decreased by 50% at 26 days. Citrate-stimulated starch synthase activity was highest at 10 days after anthesis, decreasing to low levels by 22 days. Branching-enzyme activity increased from 8 to 18 days after anthesis and decreased little by 26 days. Two fractions of starch synthase were recovered by gradient elution from DEAE-cellulose of extracts from 12- and 18-day-old seeds. The two fractions differed in primer specificity, $\underline{K}_m$ for ADP-$\underline{D}$-glucose, and relative amount of citrate-stimulated activity. A major and a minor fraction of branching enzyme were observed in extracts from both 12- and 18-day-old seeds. Marked differences in the relative abilities of the two branching-enzyme fractions to stimulate phosphorylase and to branch amylose as well as in pH optima were found. Although the content of the starch synthase and branching-enzyme fractions varied with seed age, little difference was seen in the properties of chromatographically similar fractions. Therefore, the changes in starch synthase and branching-enzyme activity during pea seed development resulted from changes in the concentrations of a few enzyme forms but not the appearance of different enzyme forms.

The hormonal regulation of glycogen synthase phosphorylation in perfused rat skeletal muscle has been investigated.[527] Using the rat hind-limb perfusion technique, kinetic evidence that glycogen synthase is substantially phosphorylated in control skeletal muscle and that epinephrine causes further phosphorylation was obtained. The phosphate content of the enzyme purified from control and hormone-treated muscle has been measured and a good correlation between the phosphorylation state of the enzyme and activity has been found.

The amino acid sequence of a region in rabbit skeletal muscle glycogen synthase phosphorylated by cyclic AMP-dependent protein kinase has been reported.[528] Analogues of the amino acid sequence surrounding site 1a demonstrated that sites 1a and 1b are separated by only 13 amino acids in the primary structure of glycogen synthase, which comprises 770 residues.

The inhibitory effects of papulacandin B and aculeacin A on (1 → 3)-β-$\underline{D}$-glucan synthase from Geotrichum lactis have been investigated.[529] The inhibition is specific and occurs not only in vitro but also in vivo since the addition of the antibiotics to a culture leads to cell-free extracts with a partially inactive synthase. Some characteristics of the inhibition are described.

Chitin synthase in the stipe of the basidiomycete Agaricus bisporus has been isolated and its properties have been investigated.[530]

Lyases. -- Heparan sulphate lyase from Flavobacterium heparinum has been purified by repeated hydroxyapatite column chromatography.[531] The enzyme was used to degrade heparan sulphate occurring on the surfaces of ascites hepatoma cells, AH66, and was found to be more effective than trypsin in removing heparan sulphate from the cells. Furthermore, on analysing glycosaminoglycans and glycopeptides from the enzyme-treated cells and control cells, it was concluded that heparan sulphate was exclusively present on the cell surface and accessible to the heparan sulphate lyase whereas other cell-surface complex carbohydrates remained intact.

The changes in the catalytic activity of N-acetylneuraminate lyase have been studied as a function of temperature in the presence or absence of its substrates.[532] The lyase was found to be effectively protected against heat inactivation by pyruvate, and it was suggested that this protective effect can be used in a new simple step in the purification of the enzyme.

A multifunctional enzyme, active on both polysaccharides and glycosides, has been isolated from the wood-decay fungus P. placenta.[533] A molecular weight of 185,000 was determined with lower-molecular-weight subunits characterized on SDS polyacrylamide gel electrophoresis. An unusual pI of 1.8 was established for this protein.

Dehydrogenases. -- The genetic and biochemical characterization of D-arabinose dehydrogenase from Neurospora crassa has been reported.[534] The enzyme was purified to homogeneity from wild-type N. crassa 74-A and from two colonial mutants, found to contain more of the enzyme. The enzymes were characterized by measurement of several kinetic and physicochemical parameters and were the same in all characteristics studied. Immunological studies performed with enzyme preparations from the three strains showed antigenic identity and indicated that those colonial strains contain more normal enzyme, rather than the usual amount of an altered, improved enzyme. Quantitation of the enzyme in crude extracts, performed by single radial immunmodiffusion, showed that the colonial strains have twice the level of enzyme as the wild-type strain. Genetic characterization, performed by analysis of meiotic products, heterokaryosis, and reversions, indicated that the difference in D-arabinose dehydrogenase activity detected among the three strains is probably determined by one gene.

D-Fructose dehydrogenase of Gluconobacter industrius has been isolated and characterized for the enzymatic microdetermination of D-fructose.[535] D-Fructose dehydrogenase was solubilized and purified from the membrane fraction of glycerol-grown G. industrius by solubilization of the enzyme with Triton X-100 and subsequent fractionation by ion-exchange chromatographies. The purified enzyme was tightly bound to a c-type cytochrome and another peptide existing as a dehydrogenase-cytochrome complex. The purified enzyme was deemed pure by analytical ultracentrifugation as well as by gel filtration on a Sephadex G-200 column. The enzyme complex (mol. wt. 140,000) on sodium dodecyl sulphate polyacrylamide gel electrophoresis showed the presence of three components having molecular weights of 67,000 (dehydrogenase), 50,800 (cytochrome c), and 19,700 (unknown function). Only D-fructose was readily oxidized by the enzyme in the presence of dyes such as ferricyanide, 2,6-dichlorophenolindophenol, or phazine methosulphate. Nicotinamide adenine dinucleotide, nicotinamide adenine dinucleotide phosphate, and oxygen did not function as electron acceptors. The optimum pH of D-fructose oxidation was 4.0. The enzyme (K_m 10^{-2}mM for D-fructose) was stable at pH 4.5 to 6.0. Stability of the purified enzyme was much enhanced by the presence of detergent in the enzyme solution. Removal of detergent from the enzyme solution facilitated the aggregation of the enzyme and caused its inactivation.

Levansucrase. -- Levansucrase, which catalyses the synthesis of the polysaccharide levan from the D-fructosyl residues of sucrose, was isolated from the culture fluid of Gluconobacter oxydans and purified to a homogeneous state by chromatography on hydroxyapatite and gel filtration.[536] According to the data of electrophoresis, the molecular weight of levansucrase is equal to 58,000. The enzyme molecule contains 25.86% acidic amino acids, 13.74% basic amino acids, and 12.77% aromatic amino acids, and one mole of methionine per mole. The ultraviolet absorption spectrum of purified levansucrase has a maximum at 280 nm and a minimum at 254 nm. By measuring the initial rates of the reactions catalysed by levansucrase of G. oxydans in the presence of {^{14}C}sucrose, it was established that this enzyme catalyses a reaction of hydrolysis of sucrose, a reaction of synthesis of levan, and a reaction of exchange of the labelled D-glucose half of sucrose with free D-glucose. The initial rates of liberation of D-glucose and D-fructose during the transfructosylation reaction were determined. It was shown that the addition of low-molecular-weight levan to the reaction mixture as a precursor of polysaccharide synthesis leads to a doubling of the initial rate of liberation of labelled D-fructose and an increase in the yield of levan from 41 to 69%.

Pectinesterases. -- Technical pectic multienzyme preparations, Pectinex Ultra and Rohament P, were chromatographed on an analytical scale using medium-pressure liquid chromatography on a glycol methacrylate rigid macroreticular gel, Spheron 1000, and its ion-exchange derivatives.[447] A combination of isocratic and linear-gradient elution (with gradients in ionic strength or pH) was employed and fractions were monitored by measurements of absorbance (A_{285} and A_{254}), conductivity, pH, and enzyme activity. Conditions for rapid separations of pectic enzymes are elaborated. The results indicate the possibilities of separating the technically undesirable pectinesterase activity from the other enzyme activities, and of a more detailed biochemical investigation of these enzymes, important for the food industry.

The aerobic bacteria associated with soft rot in onions (*Allium cepa*) were isolated and identified as a *Vibrio* species, *Micrococcus epidermidis*, *Eudomonas cepacia*, an *Acinetobacter* species, *Xanthomonas* species, *Bacillus polymyxa*, and *Bacillus megaterium*.[449] With the cup-plate assay method, no pectin hydrolase could be detected from any of these isolates when they were cultured in pectin medium, but lyase and pectinesterases were detectable. Onion tissue cultures showed pectin hydrolase activity for *P. cepacia* and *B. polymyxa* and lyase and pectinesterase activities for all of the isolates, usually at higher levels of activity than those of the pectin medium culture filtrates. In both culture media, *Vibrio* species showed the highest lyase and pectinesterase activities. In the viscometric test, all of the isolates achieved at least a 50% decrease in viscosity for lyase enzyme, with *M. epidermidis* and *Vibrio* species recording viscosity decreases as high as 83%. The ability to cause soft rot in onion bulbs was demonstrated by *P. cepacia* and *Xanthomonas* species. Benzoic acid at a concentration of 0.8 mg ml^{-1} caused total suppression of enzyme production whereas sodium benzoate at this concentration reduced pectinesterase production by 71% and lyase production by 72%. The possible use of these preservatives in the control of soft rot in onions is noted.

References

1 Enzymes, Ed. Herbert C. Friedmann, Hutchinson Ross Publishing Co., 1981, 1, 716.
2 Supplement 2 to Enzyme Nomenclature, Eur. J. Biochem., 1981, 116, 423-435.
3 Methods in Enzymology, Ed. S.P. Colowick and N.O. Kaplan, 73, Immunochemical Techniques, Pt.B, Ed. J.J. Langone and H. van Vunakis, 1981.
4 Affinity Chromatography and Related Techniques, Ed. T.C.J. Gribnau, J. Visser and R.J.F. Nivard, Analytical Chemistry Symposia Series, Vol 9, Amsterdam, 1981.
5 Immunochemical Techniques, Pt. C, Ed. J.J. Langone and H. van Vunakis, 1981.
6 Advances in Enzymology and Related Areas of Molecular Biology, 52, Ed. A. Meister, Wiley, 1981
7 I. Knack and K.-H. Rohm, Hoppe-Seyler's Z. Physiol. Chem., 1981, 362, 1119.
8 Enzyme Inhibitors ed. U. Brodbeck, 1980, 19.
9 Microbial Enzymes and Bioconversions, Economic Microbiology, 5, Ed. A.H. Rose, Academic Press, London, 1980.
10 Cellulose and Other Natural Polymer Systems, Biogenesis, Structure and Degradation, Ed. R.M. Brown, Jr., Plenum, 1982.
11 Birth Defects: Original Article Series, 16, 1, 2nd Symp. on Enzyme Therapy in Genetic Diseases, Hilton Head Isl., Ed. R.J. Desnick, A.R. Liss Inc., New York, 1980.
12 Immobilized Enzymes for Food Processing, W.H. Pitcher Jr., CRC Press, 1980.
13 Polymer Catalysts and Affinants - Polymers in Chromatography, Ed. B. Sedláček, C.G. Overberger and H.F. Mark. Wiley, 1981.
14 J.F. Kennedy, J.D. Humphreys and S.A. Barker, Enzyme Microb. Technol., 1981, 3, 129.
15 S.F. D'Souza and G.B. Nadkarni, Biotechnol. Bioeng., 1981, 23, 431.
16 P.R. Coulet and D.C. Bautheron, J. Chromatogr., 1981, 215, 65.
17 T. Izumi and K. Suzuki, Biochim. Biophys. Acta, 1980, 615, 402.
18 R. Vazquez Pernas, I. Ramos Martinez and M. Freire Rama, Comp. Biochem. Physiol., 1981, 69, 851.
19 R. Fange, G. Lundblad, K. Slettengren and J. Lind, Comp. Biochem. Physiol. Pt.B, 1980, 67, 527.
20 M. Carroll and P. McCrorie, Comp. Biochem. Physiol. Pt.B, 1980, 67, 685.
21 R. Salvayre, A. Negre, A. Maret, G. Lenoir and L. Douste-Blazy (Toulouse and Lyon, France), Biochim. Biophys. Acta, 1981, 659, 445.
22 A. Erzberger, E. Conzelmann and K. Sandhoff, Clin. Chim. Acta, 1981, 108, 361.
23 P.H. Whiting, A.J. Nicholls, G.R.D. Catto, N. Edward and J. Engeset, Clin. Chim. Acta, 1980, 180, 1.
24 U. Diener, E. Knoll, B. Langer, H. Rautenstrauch, D. Ratge and H. Wisser, Clin. Chim. Acta, 1981, 112, 149.
25 R. Gibey, J.L. Dupond, D. Alber, R. Leconte de Floris and J.C. Henry, Clin. Chim. Acta, 1981, 116, 25.
26 A. Lombardo, L. Caimi, S. Marchesini, G.C. Goi and G. Tettamanti, Clin. Chim. Acta, 1980, 108, 337.
27 E. Horak, S.M. Hopfer and F.W. Sunderman, Clin. Chem. (Winston-Salem N.C.), 1981, 27, 1180.
28 Y. Tachibana, K. Yamashita, M. Kawaguchi, S. Arashima and A. Kobata, J. Biochem. (Tokyo), 1981, 90, 1291.
29 L.C. Hoskins and E.T. Boulding, J. Clin. Invest., 1981, 67, 163.
30 S. Arumugham and S.M. Bose, Ital. J. Biochem., 1980, 29, 27.
31 D. Mahuran and J.S. Lowden, Can. J. Biochem., 1980, 58, 287.
32 S. Toma, G. Coppa, P.V. Donnelly, R. Ricci and N.D. Ferrante, Carbohydr. Res., 1981, 96, 271.
33 A. Orlacchio, C. Maffei, L. Binaglia and G. Porcellati, Biochem. J., 1981, 195, 383.
34 A. Reglero, Int. J. Biochem., 1979, 10, 285.

35 L.O. Sampaio and C.P. Dietrich, J. Biol. Chem., 1981, 256, 9205.
36 J.A. Khan and M.H.R. Lewis, Biochem. Soc. Trans., 1980, 8, 447.
37 G.S. Gupta and D.K. Kapur, Indian J. Biochem. Biophys., 1981, 18, 187.
38 P.J. Walcott and M. Messer, Aust. J. Biol. Sci., 1980, 33, 521.
39 A. Reglero, M. Esteban and J.A. Cabezas, Int. J. Biochem., 1981, 13, 837.
40 K. Hatakryama, M. Hiramatsu and N. Minami, J. Biochem. (Tokyo), 1980, 88, 613.
41 J.A. Khan and M.H.R. Lewis, Biochem. Soc. Trans., 1981, 9, 322.
42 J.R. DeLoach, H.H. Mollenhauer and R.T. Mayer, Comp. Biochem. Physiol. Pt.B., 1981, 69, 279.
43 P.F. Santoro and J.A. Dain, Comp. Biochem. Physiol. Pt.B, 1981, 69, 337.
44 C.K. Yi, Plant Physiol., 1981, 67, 68.
45 G.L. Gustapson and L.A. Milner, Biochem. Biophys. Res. Commun., 1980, 94, 1439.
46 R.L. Dimond, R.A. Burns and K.B. Jordan, J. Biol. Chem., 1981, 256, 6565.
47 D.A. Knecht and R.L. Dimond, J. Biol. Chem., 1981, 256, 3564.
48 H. Yoshioka and S. Hayashida, Agric. Biol. Chem., 1981, 45, 571.
49 J.O. Berg, L. Lindquist and C.E. Nord, App. Envir. Microbiol., 1980, 40, 40.
50 B.K. Speake, D.J. Malley and F.W. Hemming, Arch. Biochem. Biophys., 1981, 210, 110.
51 G. Lundblad, G. Huldt, M. Elander, J. Lind and K. Slettengren, Comp. Biochem. Physiol., 1981, 68, 71.
52 I.D.J. Burdett, J. Gen. Microbiol., 1980, 120, 35.
53 G. Constantopoulos, S. Rees, B.G. Cragg, J.A. Barranger and R.O. Brady, Biochem. Biophys. Res. Common., 1981, 101, 1345.
54 S.C. Li, Y. Hirabayashi and Y.T. Li, J. Biol. Chem., 1981, 256, 6234.
55 A. Frisch and E.F. Neufeld, J. Biol. Chem., 1981, 256, 8242.
56 E.A. Neuwelt, J.A. Barranger, R.O. Brady, M. Pagel, F.S. Furbish, J.M. Quirk, G.E. Mook and E. Frenkel, Proc. Natl. Acad. Sci. USA, 1981, 78, 5838.
57 A.L. Miller, B.C. Kress, R. Stein, C. Kinnon, H. Kern, J.A. Schneider and E. Harms, J. Biol. Chem., 1981, 256, 9352.
58 D.J.J. Halley, N. Sacchi, A. d'Azzo, A.J.J. Reuser and H. Galjaard, Exp. Cell Res., 1980, 129, 383.
59 A. Haslik and K. von Figura, Eur. J. Biochem., 1981, 121, 125.
60 Y. Deugnier, A. le Treut, D. Glaise, P. Brissot, J.Y. le Gall, Clin. Chim. Acta, 1980, 108, 385.
61 A.L. Miller, B.C. Kress, L. Lewis, R. Stein and C. Kinnon, Biochem. J., 1980, 186, 971.
62 T. Ludolph, E. Paschke, J. Glössl and H. Kresse, Biochem. J., 1981, 193, 811.
63 A.A. Farooqui and P.N. Srivastava, Int. J. Biochem., 1981, 13, 893.
64 D.H. Joziasse, D.H. Van den Eijnden, J.J.W. Lisman and G.J.M. Hooghwinkel, Biochim. Biophys. Acta, 1981, 660, 174.
65 P.M. Kimball, M.G. Brattain and W.E. White, Biochem. J., 1981, 193, 109.
66 D. Mahuran and J.A. Lowden, Can. J. Biochem., 1981, 59, 237.
67 S. Jacobs, E. Hazum and P. Cuatrecasas, Biochem. Biophys. Res. Commun., 1980, 94, 1066.
68 S. Kyosaka, S. Yamashiro and M. Tanaka, Chem. Pharm. Bull. (Japan), 1980, 28, 2658.
69 H.H. Freeze, A.L. Miller and A. Kaplan, J. Biol. Chem., 1980, 255, 11081.
70 J.A. Cabezas, A. Reglero, A. de Pedro, T. Diez and P. Calvo, Int. J. Biochem., 1981, 13, 389.
71 K. Yamashita, T. Ohkura, H. Yoshima and A. Kobata, Biochem. Biophys. Res. Commun., 1981, 100, 226.
72 P.J. Marshall, M.L. Sinnott, P.J. Smith and D. Widdows, J. Chem. Soc. Perkin Trans 1, 1981, 366.
73 J.A. Khan and M.H.R. Lewis, Biochem. Soc. Trans., 1980, 8, 447.
74 M. Tanaka, A. Abe and T. Uchida, Biochim. Biophys. Acta, 1981, 658, 377.
75 A. Kaji, M. Satp and Y. Tsutsui, Agric. Biol. Chem., 1981, 45, 925.
76 W. Hansen and D. Schreyer, J. Clin. Chem. Clin. Biochem., 1981, 19, 39.
77 A.M. Ugolev, L.F. Smirnova, N.N. Iezuitova, N.M. Timofeeva, N.M. Mityushova,

V.V. Egorova and E.M. Parshkov, FEBS Lett., 1979, 104, 35.
78 A. Mahmood and S. Ansari, Indian J. Biochem. Biophys., 1981, 18 198.
79 J. Synowiecki, Z.E. Sikorski and M. Naczk, Biotechnol. Bioeng., 1981, 23. 231.
80 P.G. Bhat and T.N. Battabiraman, Indian J. Biochem. Biophys., 1980, 17, 338.
81 M.A. Vattuone, F.E. Prado and A.R. Sampietro, Phytochemistry, 1981, 20, 189.
82 H. Masuda and S. Sugawara, Agric. Biol. Chem., 1978, 42, 479.
83 L. Faye, Anal. Biochem., 1981, 112, 90.
84 L. Faye, C. Berjenneau and P. Rollin, Plant Sci. Lett., 1981, 22, 77.
85 L. Faye and C. Lambert, Biochim. Biophys. Acta, 1980, 615, 392.
86 F.C. Largitte, Biochim. Biophys. Acta, 1981, 668, 481.
87 R.C.F. Baker and A.G. Dickerson, J. Gen. Microbiol., 1980, 121, 267.
88 L. Lehle, R.E. Cohen and C.E. Ballou, J. Biol. Chem., 1979, 254, 12209.
89 M. Beja da Costa and N. van Uden, Biotechnol. Bioeng., 1980, 22, 2429.
90 K. Marshall and H. Weigel, Carbohydr. Res., 1980, 83, 315.
91 H. Nakagawa, S. Ishigami, K. Sekiguchi, K. Kurata and N. Ogura, Phytochemistry, 1981, 20, 1229.
92 J.W.D. Groot Wassink and S.E. Fleming, Enzyme Microb. Technol., 1980, 2, 45.
93 E. Vanschaftingen and H.G. Hers, Proc. Natl. Acad. Sci., USA, 1981, 78, 2861.
94 A.L. Majunder, FEBS Lett., 1981, 133, 189.
95 Y. Fujita and E. Freese, J. Bacteriol., 1981, 145, 760-767.
96 M.M. Hosey and F. Marcus, Proc. Natl. Acad. Sci. USA, 1981, 78, 91.
97 G. Ya. Wiederschain, E.M. Beyer, B.A. Klyashchitasky and A.S. Shaskov, Biochim. Biophys. Acta, 1981, 659, 434.
98 B. Colas, Biochim. Biophys. Acta, 1981, 657, 535.
99 B. Colas, Biochim. Biophys. Acta, 1980, 613, 448.
100 P. Watkins and J.A. Alhadeff, Comp. Biochem. and Physiol., Pt.B, 1981, 68, 509.
101 E.M. Beier, B.A. Klyashchitshkii and G. Ya. Vindershain, Biochemistry, 1979, 44, 1533.
102 S.F. Chien and G. Dawson, Biochim. Biophys. Acta, 1980, 614, 476.
103 T.F. Scanlin and M.C. Glick, Clin. Chim. Acta, 1981, 114, 269.
104 T. Maler, M. Duthie, J.R. Riordan, FEBS Lett., 1980, 121, 153.
105 A. Haris and D. Swallow, Clin. Chim. Acta, 1981, 116, 171.
106 R. Thorpe and D. Robinson, Clin. Chim. Acta, 1978, 86, 21.
107 J.A. Alhadeff and F.L. Andrews-Smith, Biochim. Biophys. Acta, 1980, 614, 466.
108 M. Bernard, M.J. Foglietti and F. Percheron, Clin. Chim. Acta, 1981, 116, 91.
109 W.D. Gathmann and D. Aminoff, Biochem. Biophys. Res. Commun., 1981, 103, 68.
110 R.A. di Cioccio, C. Piskorz, G. Salamida, J.J. Barlow and K.L. Matta, Anal. Biochem., 1981, 111 176.
111 J.S. Mayes, J.B. Scheerer, R.N. Sifers and M.L. Donaldson, Clin. Chim. Acta, 1981, 112, 247.
112 R. Vazquez Pernas, I. Ramos Martinez, M. Rodriguez Lopex and M. Freire Rama, Comp. Biochem. and Physiol., Pt.B., 1981, 68, 141.
113 P.M. Dey and H. Kauss, Phytochemistry, 1981, 20, 45.
114 E. del Campillo, L.M. Shannon and C.N. Hankins, J. Biol. Chem., 1981, 256, 7177.
115 I. Chinen, T. Nakamura and N. Fukuda, J. Biochem. (Tokyo), 1981, 90, 1453.
116 P.M. Dey, Phytochemistry, 1981, 20, 1493.
117 E.L. Smart and D.M. Pharr, Plant Physiol., 1980, 66, 731.
118 D.W.M. Leung and J.D. Bewley, Planta, 1981, 152, 436.
119 D.W.M. Leung and J.D. Bewley, Nature (London), 1981, 289, 587.
120 I.G. Floréz, P.S. Lazo, A.G. Ochoa and S.Gascon, Biochim. Biophys. Acta, 1981, 674, 71.
121 C.M. Heyworth, E.F. Neumann and C.H. Wynn, Biochem. J., 1981, 193, 773.
122 G. Leray, L. Geunet, A. Le Treut and J-Yves le Gall, Biochem. Biophys. Res. Commun., 1981, 100, 1491.
123 M. Naoi and K. Yagi, Anal. Biochem., 1981, 116, 98.

124 A. Hoogeveen, A. d'Azzo, R. Brossmer and H. Galyaard, Biochem. Biophys. Res. Commun., 1981, 103, 292.
125 Y. Suzuki, H. Sakuraba, K. Hayasgum, K. Suzuki and K. Imahori, J. Biochem. (Tokyo), 1981, 90, 271.
126 A.L. Miller, Biochim. Biophys. Acta, 1978, 522, 174.
127 C.M. Heyworth, S. Kirkham and C.H. Wynn, Biochem. Soc. Trans., 1981, 9, 397.
128 B.K. Burton, Y. Ben-Yoseph and H.L. Nadler, Clin. Chim. Acta, 1978, 88, 483.
129 H. Skovbjerg, Biochem. J., 1981, 193, 887.
130 S. Taguchi, H. Kouyama and N. Yago, J. Biochem. (Tokyo), 1981, 89, 483.
131 R. Pascolini, A. Orlacchio and A.M. Gargiulo, Comp. Biochem. and Physiol., Pt.B, 1981, 69, 869.
132 M. Kamp and D. Hess, Phytochemistry, 1981, 20, 973.
133 J.L. Etcheberrigaray, M.A. Vattuone and A.R. Sampietro, Phytochemistry, 1981, 20, 49.
134 A. Mustranto, E. Karvonen, H. Ojamo and M. Linko, Biotechnol. Lett., 1981, 3, 333.
135 C. Niel, F. Delmaere and J.-B. Guillaume, Compt. Rend., Ser.D., 1980, 291, 771.
136 B.J. Macris and P. Markakis, Appl. Envir. Microbiol., 1981, 41, 956-958.
137 E.V. Letunova, A.S. Tikhomirova, S.D. Shiya, V.A. Markin, A. Ya. Khorlina and A.M. Bezborodov, Biochemistry, 1981, 46, 745.
138 K.M. Helena Nevalainen, App. and Envir. Microbiol., 1981, 41, 593-599.
139 S. Hjertén, Y. Kunquan and M. Ogunlesi, J. Chromatogr., 1981, 215, 25.
140 D.E. Kohlwey and M. Cheryan, Enzyme Microb. Technol., 1981, 3, 64.
141 C.M. Jorstad and D.R. Morris, Biochem. Biophys. Res. Commun., 1981, 98, 556.
142 S.K. Pal and R.K. Poddar, Biochem. Biophys. Res. Commun., 1980, 95, 1341.
143 I. Ya Kalachev, B.N. Gershanovich and G.I. Burd, Biochemistry, 1980, 45, 670.
144 S. Rosenberg and J.F. Kirsoh, Biochemistry, 1981, 20, 3189.
145 R.E. Huber, K.L. Hurlburt, C.L. Turner, Can. J. Biochem., 1981, 59, 100.
146 W. Mandecki, A.V. Fowler and I. Zabin, J. Bacteriol., 1981, 147, 694.
147 D. Foley, P. Dennis and J. Gallant, J. Bacteriol., 1981, 145, 641.
148 J.D. Noti and H.E. Umbarger, J. Bacteriol., 1981, 145, 673.
149 M. Rose, M.J. Casadaban, D. Botstein, Proc. Natl. Acad. Sci. USA - Biol. Sci., 1981, 78, 2460.
150 J.K. Welply, A.N. Fowler and I. Zabin, J. Biol. Chem., 1981, 256, 6804.
151 J.K. Welply, A.V. Fowler and I. Zabin, J. Biol. Chem., 1981, 256, 6811.
152 R.S. Accolla, R. Cina, E. Montesoro and F. Celada, Proc. Natl. Acad. Sci. USA - Biol. Sci., 1981, 78, 2478.
153 J.L. McKnight and V.A. Fried, J. Biol. Chem., 1981, 256, 9652.
154 R.E. Huber, R. Pisko-Dubienski and K.L. Hurlburt, Biochem. Biophys. Res. Commun., 1980, 96, 656.
155 N. Chaubet, V. Pétiard and A. Pareilleux, Plant Sci. Lett., 1981, 22, 369.
156 M.N. Fukuda, J. Biol. Chem., 1981, 256, 3900.
157 J. Sanchez, M.E. Arias and C. Hardisson, Experientia, 1981, 37, 469.
158 T.D. Thomas and K.W. Turner, Appl. Envir. Microbiol., 1981, 41, 1289.
159 J.F. Frank and G.A. Somkuti, App. Envir. Microbiol., 1979, 38, 534.
160 M. Kitamikado, M. Ito and Y.T. Li, J. Biol. Chem., 1981, 256, 3906.
161 G.T.N. Besley and S. Gatt, Clin. Chim. Acta, 1981, 110, 19.
162 F. Martiniuk and R. Hirschhorn, Biochim. Biophys. Acta, 1981, 658, 248.
163 C. Castilla, M. Rousset, M. Fraisse, A. Crespo, G. Chevalier, A. Zweibaum and J.C. Murat, Int. J. Biochem., 1981, 13, 381.
164 J. Hilkens, J.M. Tager, F. Buijs, B. Brouwer-Kelder, G.M. van Thienen, F.P.W. Tegelaers and J. Hilgers, Biochem. Biophys. Acta, 1981, 678, 7.
165 J.L. Potter, H.B. Robinson, J.D. Kramer and I.A. Schafer, Clin. Chem. (Winston-Salem N.C.), 1980, 26, 1914.
166 R.F. Parrish and W.R. Fair, Clin. Chem. (Winston-Salem N.C.), 1981, 27, 641.
167 A. Maret, R. Salvayre, A. Nègre and L. Dauste-Blazy, Eur. J. Biochem., 1981, 115, 455.
168 L. Poenaru, M.C. Vinet and J.C. Dreyfus, Clin. Chem. Acta, 1981, 117, 53.
169 C. Castilla, H. Paris, A. Crespo and J.C. Murat, Comp. Biochem. Physiol.,

Pt.B, 1980, 67, 653.
170 P.R. Dorling, J.M. Howell and J.M. Gawthorne, Biochem. J., 1981, 198, 409.
171 C. Dissous, J.F. Ansart, A. Cheron and J. Krembel, Anal. Biochem., 1981, 116, 35.
172 U. Reiss and B. Sacktor, Arch. Biochem. Biophys., 1981, 209, 342.
173 R.A. Ugalde, R.J. Staneloni and L.F. Leloir, Eur. J. Biochem., 1980, 113, 97.
174 M. Bounias and I. Kruk, Compt. Rend. Soc. Biol., 1981, 175, 46.
175 M. Carroll and P. McCrorie, Comp. Biochem. and Physiol., Pt.B., 1981, 70, 319.
176 H. Matsui, I. Yazawa and S. Chiba, Agric. & Biol. Chem., 1981, 45, 887.
177 Y. Yamasaki and Y. Suzuki, Agric. & Biol. Chem., 1980, 44, 707.
178 Y. Yamasaki and Y. Suzuki, Agric. & Biol. Chem., 1979, 43, 481.
179 Y. Ueno, T. Ikami, R. Yamauchi and K. Kato, Agric. & Biol. Chem., 1980, 44, 2623.
180 S. Chatterjee and L.C. Vining, Can. J. Microbiol., 1981, 27, 639.
181 Y. Suzuki and R. Tanaka, Eur. J. Appl. Microbiol. Biotechnol., 1981, 11, 161.
182 C.T. Kelly, M.E. Heffernan and W.M. Fogarty, Biotechnol. Lett., 1980, 2, 381.
183 A.A. Glemzha, R.P. Krakenaite and K.K. Yanulaitene, Biochemistry, 1981, 46, 363.
184 D.E. Oliveira, A. Lúcia, C. Santos Neto and A.D. Panek, Anal. Biochem., 1981, 113, 188.
185 S. Chiba, M. Murata, K. Matsusaka and T. Shimomura, Agric. & Biol. Chem., 1979, 43, 775.
186 A.A. Sibirnyi, G.M. Shavlovskii, G.P. Ksheminskaya and A.G. Orlovskaya, Biochemistry USSR, 1980, 44, 1226.
187 Z. Lešková, E. Dudíková and J. Subik, Biochim. Biophys. Acta, 1981, 673, 10.
188 M.R. Kula, R. Wichmann, U. Oden and C. Wandrey, Biochemie, 1980, 62, 523.
189 R. Cherian and A.S. Balasubramanian, Clin. Chim. Acta, 1981, 110, 169.
190 L.B. Daniels, R.H. Glen, W.F. Diven, R.E. Lee and N.S. Radin, Clin. Chim. Acta, 1981, 115, 369.
191 S.L. Berent and N.S. Radin, Arch. Biochem. Biophys, 1981, 208, 248.
192 D.W. Hanke, D.A. Warden, J.O. Evans, F.F. Fannin and D.F. Diedrich, Biochim. Biophys. Acta, 1980, 599, 652.
193 G. Burmeister and W. Hösel, Planta, 1981, 152, 578.
194 A.A. Klesov and I.V. Churilova,Biochemistry (Engl. Transl.), 1980, 45, 1282.
195 M.G. Shepherd, C.C. Tong and A.L. Cole, Biochem. J., 1981, 193, 67.
196 M. Desrochers, L. Jurasek and M.G. Paice, Appl. and Envir. Microbiol., 1981, 41, 222.
197 A.A. Klesov and L.V. Churilova, Biochemistry (Engl. Transl.), 1980, 45, 1.
198 A. McHale and M.P. Coughlan. Biochim. Biophys. Acta, 1981, 662, 145.
199 A. McHale and M.P. Coughlan, Biochim. Biophys. Acta, 1981, 662, 152.
200 A. Margaritis and E. Creese, Biotechnol. Lett., 1981, 3, 471.
201 H.-P. Meyer and G. Canevascini, Appl. Environ. Microbiol., 1981, 41, 924-931.
202 S.C. Saha, A. Sanyal, R.K. Kundu, S. Dube and D.K. Dube, Biochim. Biophys. Acta, 1981, 662, 22.
203 J.G. Shewale and J. Sadana, Arch. Biochem. Biophys., 1981, 207, 185.
204 H. Yoshida and S. Hayashida, Agric. Biol. Chem., 1980, 44, 1729.
205 T.A. Hsu, C.S. Gong and G.T. Tsao, Biotechnol. Bioeng., 1980, 22, 2305.
206 S. Rosenberg and J.F. Kirsch, Biochemistry, 1981, 20, 3196.
207 T.K. Ng and J.G. Zeikus, Appl. Envir. Microbiol., 1981, 42, 231.
208 S.K. Tangnu, H.W. Blanch and C.R. Wilke, Biotechnol. Bioeng., 1981, 23, 1837.
209 K.M.H. Nevalainen, E.T. Palva and M.J. Bailey, Enzyme Microb. Technol., 1980, 2, 59.
210 I. Labudová, V. Farkaš, S. Bauer, N. Kolarová, A. Brányik, Eur. J. Appl. Microbiol. and Biotechnol., 1981, 12, 16.
211 L.S. Trivedi and K.K. Rao, Biotechol. Lett., 1981, 3, 281.

212 D. Sternberg and G.R. Mandels, J. Bacteriol., 1980, 144, 1197-1199.
213 D.W. Sundstrom, H.C. Klei, R.W. Coughlin, G.J. Biederman and C.A. Brouwer, Biotechnol. Bioeng., 1981, 23, 473.
214 J. Woodward and S.L. Arnold, Biotechnol. Bioeng., 1981, 23, 1553.
215 J. Hong, M.R. Ladisch, G.S. Gong, P.C. Wankat and G.T. Tsao, Biotechnol. Bioeng., 1981, 23, 2779.
216 K.R. Roeser and G. Legler, Biochim. Biophys. Acta, 1981, 657, 321.
217 E. Bause and G. Legler, Biochim. Biophys. Acta, 1980626, 459.
218 S.K. Srivastava, K.B. Ramachandran and K. Gopalkrishnan, Biotechnol. Lett., 1981, 3, 477.
219 A. Allen and D. Sternberg, Biotech. Bioeng. Symp., 1980, 10, 189.
220 S. Muroa and R. Sakamoto, Agric. & Biol. Chem., 1979, 43, 1791.
221 R.D. Kilker, B. Saunier, J.S. Tracz and A. Herscovics, J. Biol. Chem., 1981, 256, 5299.
222 P. Rapp, E. Grote and F. Wagner, Appl. Envir. Microbiol., 1981, 41, 857.
223 C.W. Forsberg, T.J. Beveridge and A. Hellstrom, Appl. Envir. Microbiol., 1981, 41, 886.
224 J.N. Saddler and A.W. Khan, Can. J. Microbiol., 1981, 27, 288.
225 E.R. Allcock and D.R. Woods, Appl. Envir. Microbiol., 1981, 41, 539.
226 G.M. Umezurike, Biochem. J., 1981, 199, 203.
227 D.Herr, Biotechnol. Bioeng., 1980, 22, 1601.
228 M.J. Bailey, Biotechnol. Lett., 1981, 3, 695.
229 L.D. Bowers and P.R. Johnson, Biochim. Biophys. Acta, 1981, 661, 100.
230 M. Milone, R.K. Rastogi, F. Delsorbo, A. Corvino, M.F. Caliendo, Experientia, 1981, 37, 1115.
231 H. Kosaka, M. Isemura, T. Ono, Y. Nishimura and K. Kato, J. Biochem. (Tokyo), 1980, 88, 69.
232 W.H. Curley, G.W. Lucier and C.M. Schiller, Comp. Biochem. and Physiol. Pt.B, 1981, 68, 1.
233 J.W. Coffey and R.A. Salvador, Biochim. Biophys. Acta, 1981, 677, 243.
234 M.M. Everett andW.A. Miller, Biochem. Pharmacol., 1980, 29, 2143.
235 J.A. Brown, G.P. Jahreis and R.T. Swank, Biochem. Biophys. Res. Commun., 1981, 99, 691.
236 T. Mizuochi, Y. Nishimura, K. Kato and A. Kobata, Arch. Biochem. Biophys., 1981, 209, 298.
237 R. Myerowitz and E.F. Neufeld, J. Biol. Chem., 1981, 256, 3044.
238 I.V. Tavetkova, E.L. Rosenfeld and I.G. Prigozina, Clin. Chim. Acta, 1980, 107, 37.
239 J. Butterworth, Clin. Chim. Acta, 1980, 108, 347.
240 A.F. Van Elsen and J.G. Leroy, Clin. Chim. Acta, 1981, 112, 159.
241 W.T. Forsee and J.S. Schutzbach, J. Biol. Chem., 1981, 256, 6577.
242 L.J. Burditt, K. Chotai, S. Hirani, P.G. Nugent, B.G. Winchester and W.F. Blakemore, Biochem. J., 1980, 189, 467.
243 E. Berman and A. Allerhand, J. Biol. Chem., 1981, 256, 6657
244 J.L. Macleod and W.F. Loomis, J. Gen. Microbiol., 1980, 120, 273.
245 E. Ichishima, M. Arai, Y. Shigematsu, H. Kumagai and R. Sumida-Tanaka, Biochim. Biophys. Acta, 1981, 658, 45.
246 R. Leprat and Y. Michel-Briand, Annales de Microbiologie, 1980, B131, 209.
247 M. Zeigler and G. Bach, Biochem. J., 1981, 198, 505.
248 O.T. Mueller and D.A. Wenger, Clin. Chim. Acta, 1981, 109, 313.
249 S. Thompson and A.H. Maddy, FEBS Lett., 1981, 133, 165.
250 N. Hong, G. Beauregard, M. Potier, M. Belisle, L. Mameli, R. Gatti and P. Durand, Biochim. Biophys. Acta, 1980, 616, 259.
251 J.A. Alhadeff and S. Wolfe, Int. J. Biochem., 1981, 13, 975.
252 S. Inoue, M. Iwasaki and G. Matsumura, Biochem. Biophys. Res. Commun., 1981, 102, 1295.
253 S. Gatt, B. Gazit and Y. Barenholz, Biochem. J., 1981, 193, 267.
254 Yu. V. Vertiev and Yu. V. Ezepchuk, Hoppe-Seyler's Z. Physiol. Chem., 1981, 362, 1339.
255 H. von Nicolai, P. Esser and E. Lauer, Hoppe-Seyler's Z. Physiol. Chem., 1981, 362, 153.

256 G.H. Frank and L.B. Tabatabai, Infection and Immunity, 1981, 32, 1119.
257 D.J.S. Arora and J. Jill-Schubert, Can. J. Microbiol., 1980, 26, 1369.
258 J.A. Cabezas, P. Calvo, P. Eid, J. Martin, N. Perez, A. Reglero and C. Hannoun, Biochim. Biophys. Acta, 1980, 616, 228.
259 D. Conrad, Biotechnol. Lett., 1981, 3, 345.
260 M. Ishaque and D. Kluepfel, Biotechnol. Lett., 1981, 3, 481.
261 S.-C. Li, M. Asakawa, Y. Hirabayashi and Y.-T. Li (USA), Biochim. Biophys. Acta, 1981, 660, 278.
262 A. Hasilik and K. von Figura, Eur. J. Biochem., 1981, 121, 125.
263 S. Kawagishi, Y. Apaki and E. Ito, Eur. J. Biochem., 1980, 112, 273.
264 B. Overdijk, W.M.J. van der Kroef, J.J.W. Lisman, R.J. Pierce, J. Montreuil, G. Spik, FEBS. Lett., 1981, 128, 364.
265 H. Iwase, T. Morinaga, Y.-T. Li and S.-C. Li, Anal. Biochem., 1981, 113, 93.
266 H. Malmqvist, Biochem. Biophys. Acta, 1978, 537, 31.
267 T. Muramatsu and K. Egawa, Agric. Biol. Chem., 1980, 44, 2587.
268 M. Fujii, S. Murakami, Y. Yamada, Y. Ona and T. Nakamuua, Biotechnol. Bioeng., 1981, 23, 1393.
269 Y. Kobayashi and K. Horikoshi, Agric. Biol. Chem., 1980, 44, 2991.
270 H. Akedo, K. Yoshioka, M. Ehara and T. Kitamura, Clin. Chim. Acta, 1981, 115, 333.
271 J.P. Bretaudiere, R. Reej, P. Drake, A. Vassault and M. Bailly, Clin. Chem., 1981, 27, 806.
272 W. Honig, E. Moshudis and K. Oette, J. Clin. Chem. Clin. Biochem., 1981, 19, 1057.
273 S.A. Lacks and S.S. Springhorn, J. Biol. Chem., 1980, 255, 7467.
274 B. Shivaraj and T.N. Pattabiraman, Biochem. J., 1981, 193, 29.
275 M.A. Yetter, R.M. Saunders and H.P. Boles, Cereal Chem., 1981, 56, 243.
276 S. Murao and K. Ohyama, Agric. & Biol. Chem., 1979, 43, 679.
277 H. Nakano, T. Tajiri, Y. Koba and S. Ueda, Agric. Biol. Chem., 1981, 45, 1053.
278 H. Aschauer, L. Vertesy and G. Braunitzer, Hoppe-Seyler's Z. Physiol. Chem., 1981, 362, 465.
279 N. Kashlan and M. Richardson, Phytochemistry, 1981, 20, 1781.
280 C.M. O'Connor and K.F. McGeeney, Biochim. Biophys. Acta, 1981, 658, 397.
281 G. Chandrasekher, D.S. Rajn and T.N. Pattabiraman, J. Sci. Food Agric., 1981, 32, 9.
282 M. Irshad and C.B. Sharma, Biochim. Biophys. Acta, 1981, 659, 326.
283 B. Shivaraj and T.N. Pattabiraman, Indian J. Biochem. Biophys., 1980, 17, 181.
284 C.M. O'Connor and K.F. McGeeney, Biochim. Biophys. Acta, 1981, 658, 387.
285 K. Yamashita, Y. Tachibana, T. Takeuchi and A. Kobata, J. Biochem. (Tokyo), 90, 1281.
286 T. Takauchi, H. Fujki and T. Kameya, Clin. Chem., 1981, 27, 556.
287 E.J. Sampson, P.H. Duncan, D.M. Fast, V.S. Whitner, S.S. McNeally, M.A. Baird, M.L. Macniel and D.D. Bayse, Clin. Chem., 1981, 27, 714.
288 E.-O. Haegele, E. Schiach, E. Rauscher, P. Lehmann and M. Grassl, J. Chromatogr., 1981, 223, 69.
289 N. Saito and T. Horiuchi, Carbohydr. Res., 1981, 96, 138.
290 J.E. Hodes, M.T. Hull, R.C Karn, A.D. Merritt, M.E. Hodes, Experientia, 1981, 37, 184.
291 O. Awad, E. Amin and N. Mohamed, Int. J. Biol. Macromol., 1981, 3, 272.
292 J. Hutney and M. Ugorski, Arch. Biochem. Biophys., 1981, 206, 29.
293 L. Pasero, B. Abadie, Y. Chicheportiche, Y. Mazzei, D. Moinier, J.P. Bizzozero, M. Fougereau and G. Marchis-Mouren, Biochimie, 1981, 63, 71.
294 P.H. Burrill, P.M. Brannon and N. Kretchmer, Anal. Biochem., 1981, 117, 402.
295 T.L. Brown and F. Wold, J. Biol. Chem., 1981, 256, 10743.
296 R.C. Karn, T.E. Petersen, J.P. Hjorth, J.T. Nielsen, P. Roepstorff, FEBS. Lett., 1981, 126, 292.
297 S. Murao, A. Goro and M. Arai, Agric. Biol. Chem., 1980, 44, 1683.
298 T.Tanaka, E.W. Gresik and T. Barka, J. Histochem. Cytochem., 1981, 29, 1189.

299 L.A. Elyakova, N.M. Shevchenko and S.M. Avaeva, Comp. Biochem. and Physiol., 1981, 69, 905.
300 B.T. O'Connell, G.L. Rubenthaler and N.L. Murbach, Cereal Chem., 1980, 57, 411.
301 T.W. Okita and J. Preiss, Plant Physiol., 1980, 66, 870.
302 B.D. Singh, R.P. Singh and R.B. Singh, Experentia, 1981, 37, 363.
303 A.J. McCrate, M.T. Nielsen, G.M. Paulsen and E.G. Heyne, Cereal Chem., 1981, 58, 424.
304 A. Beleia and E. Varriano-Marston, Cereal Chem., 1981, 58, 433.
305 K.F. Finney, O. Natsuaki, L.C. Bolte, P.R. Mathewson and Y. Pomeranz, Cereal Chem., 1981, 58, 355.
306 J.E. Kruger and K.H. Tipples, Cereal Chem., 1981, 58, 271.
307 B.G. Osborne, S. Douglas, T. Fearn, C. Moorhouse and M.J. Heckley, Cereal Chem., 1981, 58, 474.
308 M. Irshad and C.B. Sharma, Phytochemistry, 1981, 20, 1205.
309 M. Irshad, M. Goel and C.B. Sharma, Phytochemistry, 1981, 20, 2123.
310 T. Koshiba and T. Minamikawa, Plant and Cell Physiol., 1981, 22, 979.
311 S. Shinomiya, K. Yamane and T. Oshima, Biochem. Biophys. Res. Commun., 1980, 96, 175.
312 Y. Nagata, S. Suga, A. Kado and B. Maruo, Agric. Biol. Chem., 1980, 44, 215.
313 H. Yamauchi, Y. Nagata and B. Maruo, Agric. Biol. Chem., 1980, 44, 1291.
314 M. Leisola, H. Ojamo, V. Kauppinen, M. Linko and J. Virkkunen, Enzyme Microb. Technol., 1980, 2, 121.
315 M.B.M. Oliveira and G.T. Zancan, J. Bacteriol., 1981, 145, 171.
316 M. Sippola and P. Mäntsälä, FEMS Microbiol. Lett., 1981, 10, 303.
317 S. Shinomiya, K. Yamane, T. Mizukami, F. Kawamura and H. Saito, Agric. Biol. Chem., 1981, 45, 1733.
318 T. Takagi, J. Biochem. (Tokyo), 1981, 89, 363.
319 T. Suganuma, T. Mizukami K. Moori, M. Ohnishi and K. Hiromi, J. Biochem. (Tokyo), 1980, 88, 131.
320 M. Gunasekaran, Microbios, 1980, 29, 37.
321 F. Clementi, J. Rossi, L. Costamagna and J. Rosi, J. Microbiol. Serol., 1980, 46, 399.
322 P.E. Davis, D.L. Cohen and A. Whitaker, J. Microbiol. Serol., 1980, 46, 391.
323 J.E. Kruger, Cereal Chem., 1981, 56, 298.
324 J. Daussant, B. Zbaszyniak, J. Sadowski and I. Wiatroszak, Planta, 1981, 151, 176.
325 B. Mikami, S. Airbara and Y. Morita, J. Biochem. (Tokyo), 1980, 88, 103.
326 E. Mori, B. Mikami, Y. Morita and B. Jirgensons, Arch. Biochem. Biophys., 1981, 211, 382.
327 T. Suganuma, M. Ohnishi, K. Hiromi and Y. Morita, Agric. Biol. Chem., 1980, 44, 1111.
328 S. Murao, K. Ohyama and M. Arai, Agric. Biol. Chem., 1979, 43, 719.
329 A. Azhar and M.K. Hamdy, Biotechnol. Bioeng., 1981, 23, 1297.
330 B.K. Gillard, R.C. White, R.A. Zingaro and T.E. Nelson, J. Biol. Chem., 1980, 255, 8451.
331 D.D.Y. Ryu and M. Mandels, Enzyme Microb. Technol., 1980, 1, 91.
332 A. Marzetti, B. Focher, L. D'Angiuro and V. Sarto, Cellul. Chem. Technol., 1981, 15, 3.
333 A.W. Khan, D. Wall and L. van den Berg, Appl. Environ. Microbiol., 1981, 41, 1214.
334 M. Castanon and C.R. Wilke, Biotechnol. Bioeng., 1981, 23, 1365.
335 A. Cabello, J. Conde and M.A. Otero, Biotechnol. Bioeng., 1981, 23, 2737.
336 A.A. Klesov and S. Yu. Grigorash, Biochemistry, 1981, 46, 86.
337 T. Kanda, K. Wakabayashi and K. Nisizawa, J. Biochem. (Tokyo), 1980, 87, 1635.
338 A.A. Klesov and S. Yu. Grigorash, Biochemistry (USSR), 1980, 45, 166.
339 A.R. White, R.M. Brown, Proc. Natl. Acad. Sci. USA, 1981, 78, 1047.
340 L.T. Fan, Y.H. Lee and D.R. Beardmore, Biotechnol. Bioeng., 1981, 23, 419.
341 M. Nummi, M.L. Niku-Paavola, T.M. Enari and V. Raunio, Anal. Biochem., 1981, 116, 137.

342 G. Canevascini and C. Gattlen, Biotechnol. Bioeng., 1981, 23, 1573.
343 E. Galas, R. Pyć, K. Pyć and K. Siwińska, Eur. J. Appl. Microbiol. and Biotechnol., 1981, 11, 229.
344 A. Vardanis and M. Finkelman, Anal. Biochem., 1981, 115, 78.
345 M. Nummi, P.C. Fox, M.L. Niku-Paavola and T.M. Enari, Anal. Biochem., 1981, 116, 133.
346 A. McHale and M.P. Coughlan, Biochem. J., 1981, 199, 267.
347 M. Leisola, J. Virkkunen, E. Karvonen and A. Meskanen, Enzyme Microb. Technol., 1979, 1, 117.
348 P.C. Mishra and M.C. Dash, Experientia, 1980, 36, 1156.
349 N. Waszczynskys, C.S. Rao and R.S.F. Da Silva, Cereal Chem., 1981, 58, 264.
350 D.E. Koehler, L.N. Lewis, L.M. Shannon and M.L. Durbin, Phytochemistry, 1981, 20, 409.
351 M. Mandels, J.E. Medeiros, R.E. Andreotti and F.H. Bissett, Biotechnol. Bioeng., 1981, 23, 2009.
352 S.N. Mukhopadhyay and R.K. Malik, Biotechnol. Bioeng., 1980, 22, 2237.
353 N. Choudhury, N.W. Dunn and P.P. Gray, Biotechnol. Lett., 1981, 3, 493.
354 E.T. Reese and D.Y. Ryu, Enzyme Microb. Technol., 1980, 2, 239.
355 B.S. Montenecourt, T.J. Kelleher, D.E. Eveleigh and L.G. Pettersson, Biotechnol. Bioeng. Symp. No.10, 1980, 15.
356 I.V. Churilova, V.I. Maksimov and A.A. Klesov, Biochemistry (USSR), 1980, 45, 516.
357 A.R. Moreira, J.A. Phillips and A.E. Humphrey, Biotechnol. Bioeng., 1981, 23, 1339.
358 S. Hayashida and H. Yoshioka, Agric. Biol. Chem., 1980, 44, 1721.
359 T.H.D.Jones and M. Gupta, Biochem. Biophys. Res. Commun., 1981, 102, 1310.
360 M.A. Folan and M.P. Coughlan, Int. J. Biochem., 1981, 13, 243.
361 I. Vance, S.L. Blayden and J. Tampion, Biotechnol. Lett., 1981, 3, 45.
362 S. Murao and R. Sakamoto, Agric. & Biol. Chem., 1979, 43, 1789.
363 T. Kanda, K. Wakabayashi and K. Nisizawa, J. Biochem. (Tokyo), 1980, 87, 1625.
364 A.W. Khan, J. Gen. Microbiol., 1980, 121, 499.
365 J.N. Saddler, A.W. Khan and S.M. Martin, Microbios, 1980, 28, 97.
366 T.K. Ng, A. Ben-Bassat and J.G. Zeikus, Appl. Environ Microbiol., 1981, 41, 1337.
367 S.P. Antai and D.L. Crawford, Appl. Envir. Microbiol., 1981, 42, 378.
368 M. Taya, T. Kobayashi and S. Shimizu, Agric. Biol. Chem., 1980, 44, 2225.
369 P. Fahnrich and K. Irrgang, Biotechnol. Lett., 1981, 3, 201.
370 A. Enríquez, R. Montalvo and M. Canales, Biotechnol. Bioeng., 1981, 23, 1431.
371 N. Choudhury, P.P. Gray and N.W. Dunn, Biotechnol. Lett., 1980, 2, 427.
372 M.L. Bade and A. Stinson, Arch. Biochem. Biophys., 1981, 206, 213.
373 F. Ishikawa, K. Oishi and K. Aida, Agric. Biol. Chem., 1981, 45, 2361.
374 J.D. Reid and D.M. Orgrydziak, Appl. Envir. Microbiol., 1981, 41, 664.
375 D. Yellowlees, Carbohydr. Res., 1980, 83, 109.
376 A. Hattori and K. Ishibashi, Agric. Biol. Chem., 1981, 45, 2347.
377 Y. Mitsuishi, M. Kobayashi and K. Matsuda, Carbohydr. Res., 1980, 83, 303.
378 B.L. Lamberts, L.G. Simonson, E.D. Pederson and R.W. Gaugler, Anal. Biochem., 1981, 117, 320.
379 T. Takehara, M. Inoue, T. Morioka and K. Yokogawa, J. Bacteriol., 1981, 145, 729.
380 T.W. Jeffries and J.D. Macmillan, Carbohydr. Res., 1981, 95, 87.
381 B.Y. Reichelt and G.H. Fleet, J. Bacteriol., 1981, 147, 1085.
382 A. Sanchex, G. Larriba, J.R. Villanueva and T.G. Villa, FEBS. Lett., 1980, 121, 23.
383 D.J. Huber and D.J. Nevins, Planta, 1981, 151, 206.
384 Y. Tsujisaka, N. Hamada and R. Kobayashi, Agric. Biol. Chem., 1981, 45, 1201.
385 G. Halliwell and R. Vincent, Biochem. J., 1981, 199, 409.
386 T.K. Ng and T.G. Zeikus, Biochem. J., 1981, 199, 341.
387 J. Petre, R. Longin and J. Millet, Biochimie, 1981, 63, 629.

388 J. Abe, A. Amemura and T. Harada, Agric. Biol. Chem., 1980, 44, 1877.
389 M. Hisamatsu, J. Abe, A. Amemura and T. Harada, Agric. Biol. Chem., 1980, 44, 1049.
390 P.J. Wood, Carbohydr. Res., 1981, 94, C19.
391 D.J. Huber and D.J. Nevins, Planta., 1981, 151, 206.
392 R.K. Firantene, V. Yu. Avizhenis and N.A. Tiunova, Biochemistry, 1981, 46, 502.
393 I.V. Churilova, V.I. Maksimov and A.A. Klesov, Biochemistry Engl. Transl.), 1979, 44, 1663.
394 Y. Okami, S. Kurasawa and Y. Hirose, Agric. Biol. Chem., 1980, 44, 1191.
395 V. Basaveswara Rao, N.V.S.Sastri, P.V. Subba Rao, Biochem. J., 1981, 193, 389.
396 G.K. Lee, R.A. Lesch and P.J. Reilly, Biotechnol. Bioeng., 1981, 23, 487.
397 B. Svenson and M. Otteson, Carlsberg Res. Commun., 1981, 46, 13.
398 S. Moriyama, A. Noda, K. Nakanishi, R. Matsuno and T. Kamikubo, Agric. Biol. Chem., 1980, 44, 2047.
399 S. Moriyama, S. Kataoka, K. Nakanishi, R. Matsuno and T. Kamikubo, Agric. Biol. Chem., 1980, 44, 2737.
400 S. Sivakami and D. Chatterji, Indian J. Biochem. Biophys., 1980, 17, 327.
401 J.H. Pazur, Y. Tominaga, L.S. Forsberg and D.L. Simpson, Carbohydr. Res., 1980, 84, 103.
402 D. Alazard, M. Raimbault, Eur. J. Appl. Microbiol. Biotechnol., 1981, 12, 113.
403 T. Takahashi, N. Inokuchi and M. Irie, J. Biochem. (Tokyo), 1981, 89, 125.
404 V. Basaveswara Rao, N.V.S.Sastri and P.V. Subba Rao, Biochem. J., 1981, 193, 379.
405 S. Moriyama, R. Matsuno, K. Nakanishi and T. Kamikubo, Agric. Biol. Chem., 1980, 44, 2763.
406 H. Bender, Eur. J. Biochem. 1981, 115, 287.
407 C. Ponzetto-Zimmerman and A.M. Gold, FEBS. Lett., 1981, 132, 179.
408 S. Voara and U. Francke, Proc. Natl. Acad. Sci. USA, 1981, 78, 3738.
409 U. Elsässer-Beile and S. Stirm, Carbohydr. Res., 1981, 88, 315.
410 N. Ototani, M. Kikuchi and Z. Yosizawa, Carbohydr. Res., 1981, 88, 291.
411 Å. Oldberg, C.-H. Heldin, Å. Wasteson, C. Busch and M. Höök, Biochemistry, 1980, 19, 5755.
412 P.M. Galliher, C.L. Cooney, R. Langer and R.J. Linhardt, Appl. Envir. Microbiol., 1981, 41, 360.
413 P. Gacesa, M.J. Savitsky, K.S. Dodgson and A.H. Olavesen, Biochim. Biophys. Acta, 1981, 661, 205.
414 P. Gacesa, M.J. Savitsky, K.S. Dodgson and A.H. Olavesen, Anal. Biochem., 1981, 18, 76.
415 F.N. Kolisis, T.G. Sotiroudis, A.E. Evangelopoulos, FEBS. Lett., 1980, 121, 280.
416 M. Lyon and C.F. Phelps, Biochem. J., 1981, 199, 419.
417 G.S. Gupta and E. Goldberg, Biochim. Biophys. Acta, 1981, 657, 364.
418 P.G. Richman and H. Baer, Anal. Biochem., 1980, 109, 376.
419 E. Shimada and G. Matsumura, J. Biochem. (Tokyo), 1980, 88, 1015.
420 J.P. Guirand, J. Daurelles and P. Galzy, Biotechnol. Bioeng., 1981, 23, 1461.
421 J. Fahrenkrug, P. Staun-Olsen and E. Magid, Clin. Chem., 1981, 27, 1655.
422 M.E. Selsted and R.J. Martinez, Anal. Biochem., 1980, 109, 67.
423 M.J. Thomas, A. Russo, P. Craswell, M. Ward and I. Steinhardt, Clin. Chem., 1981, 27, 1223.
424 H. Maeda, J. Biochem. (Tokyo), 1980, 88, 1185.
425 S.-I. Segawa, M. Nakayama and M. Sakane, Biopolymers, 1981, 20, 1691.
426 J.L. Lippert, R.M. Lindsay and R. Schultz, Biochim. Biophys. Acta, 1980, 599, 32.
427 T. Imoto, S. Sumi and K. Yagishita, Agric. Biol. Chem., 1980, 44, 1031.
428 K. Hasegawa, N. Okamoto, H. Ozawa, S. Kitajima and Y. Takado, Agric. Biol. Chem., 1981, 45, 1645.
429 A. Hirayama, H. Fujio, Y. Kohi, Y. Takagaki, K. Kondo and T. Amano,

J. Biochem. (Tokyo), 1981, 89, 963.
430 J.Perraudin, Y.Looze, J.Vincentelli, and J.Leonis, Comp. Biochem. Physiol., Pt B, 1980, 65, 127.
431 K.Bell, H.A.McKensie, V.Muller, and D.C.Shaw, Mol. Cell. Biochem., 1980, 29, 3.
432 V. Sidhan and S. Gurnani, Agric. Biol. Chem., 1981, 45, 1817.
433 B.W. Matthews, S.J. Remington, M.G. Grütter and W.F. Anderson, J. Mol. Biol., 1981, 147, 545.
434 K. Fukushima, Y. Murata, N. Nishikido, G. Sugihara and M.Tanaka, Bull. Chem. Sec. Jpn., 1981, 54, 3122.
435 P.J. Stein and K. Craig Heehn, Biochem. Biophys. Res. Commun., 1980, 95, 547.
436 T. Imoto, T. Ono and H. Yamada, J. Biochem. (Tokyo), 1981, 90, 335.
437 A. Shrake and J.A. Rupley, Biochemistry, 1980, 19, 4044.
438 Y. Yang and K. Hamaguchi, J. Biochem. (Tokyo), 1980, 88, 829.
439 Y. Yang, S. Kuramitsu and K. Hamaguchi, J. Biochem. (Tokyo), 1981, 89, 1357.
440 A. Masaki, T. Fukamizo, A. Otakara, T. Torikata, K. Hayashi and T. Imoto, J. Biochem. (Tokyo), 1981, 90, 527.
441 E.P. Savel'ev, G.I. Petrov, Z.F. Shmakova and S.A. Bitko, Biochemistry (Engl. Transl.), 1980, 45, 247.
442 B. Szewczyk and T. Skorko, Biochim. Biophys. Acta, 1981, 662, 131.
443 G. Kleppe, E. Vasstrand and H.B. Jensen, Eur. J. Biochem., 1981, 119, 589.
444 H. Sjöström, O. Norén, L. Christiansen, H. Wacker and G. Semenza, J. Biol. Chem., 1980, 255, 11332.
445 M. Sato and A. Kaji, Agric. Biol. Chem., 1980, 44, 1345.
446 G.S. Coleman, D.C. Sandford, S. Beahon, J. Gen. Microbiol., 1980, 120, 295.
447 D. Mikeš, J. Sedláčkova, L. Rexová-Benková and J. Omelková, J. Chromatogr., 1981, 207, 99.
448 Y. Itoh, K. Izaki and H. Takahashi, Agric. Biol. Chem., 1980, 44, 1135.
449 S.K.C. Obi and G.M. Umezurike, Appl. Envir. Microbiol., 1981, 42, 585-589.
450 Ľ. Rexová-Benková and M. Mračková-Dobrotová, Carbohydr. Res., 1981, 98, 115.
451 G.A. Tucker, N.G. Robertson and D. Grierson, Eur. J. Biochem., 1980, 112, 119.
452 S.C. Lee and C.A. West, Plant Physiol., 1981, 67, 640.
453 S.C. Lee and C.A. West, Plant Physiol., 1981, 67, 633.
454 P. Magro, P. Dilenna, P. Marciano and C. Pallavicini, J. Gen. Microbiol., 1980, 120, 105.
455 M. Sato and A. Kaji, Agric. Biol. Chem., 1980, 44, 717.
456 J. Yamada, Agric. Biol. Chem., 1981, 45, 747.
457 J. Yamada, Carbohydr. Res., 1981, 90, 153.
458 J. Yamada, Agric. Biol. Chem., 1981, 45, 1269.
459 A. Herscovics, A. Quaroni, B. Bugge and K. Kirsch, Biochem. J., 1981, 197, 511.
460 E. Bar-Guilloux, D. Robic and J.-E. Courtois, Biochimie, 1980, 62, 719.
461 S. Rajendran and M. Muthu, Experientia, 1981, 37, 886.
462 J. Comtat and J.P. Joseleau, Carbohydr. Res., 1981, 95, 101.
463 P. Biely, M. Vršanská and Z. Krátký, Eur. J. Biochem., 1980, 112, 375.
464 P. Biely, M. Vršanová and Z. Krátký, Eur. J. Biochem., 1981, 119, 565.
465 P. Biely, Z. Krátký and M. Vršanská, Eur. J. Biochem., 1981, 119, 559.
466 P.A.D. Rickard and T.A. Laughlin, Biotechnol. Lett., 1980, 2, 363.
467 D.B. Wankhede, K.R. Vijayalakshmi and M.R.R. Rao, Carbohydr. Res., 1981, 96, 249.
468 H. Yoshioka, S. Chavanich, N. Nilubol and S. Hayashida, Agric. Biol. Chem., 1981, 45, 579.
469 D. Conrad, Biotechnol. Lett., 1981, 3, 345.
470 P.A.D. Rickard and S.P. Peiris, Biotechnol. Lett., 1981, 3, 39.
471 P.A.D. Rickard, M.I. Rajoka and J.A. Ide, Biotechnol. Lett., 1981, 3, 487.
472 Z. Krátký and P. Biely, Eur. J. Biochem., 1980, 112, 367.
473 K. Nakanishi and T. Yasui, Agric. Biol. Chem., 1980, 44, 1885.
474 K. Nakanishi and T. Yasui, Agric. Biol. Chem., 1980, 44, 2729.
475 A.V. Ananichev, I.V. Ulezlo, A.M. Egorov, A.M. Bezborodov, E.V. Berezin,

Biochemistry, 1980, 45, 753.
476 M.A. Van Keulen, K. Vellenga and G.E.M. Joosten, Biotechnol. Bioeng., 1981, 23, 1437.
477 C.-S. Gong, L-F. Chen, M.C. Flickinger, L.-C. Chiang and G.T. Tsao, Appl. Envir. Microbiol., 1981, 41, 430.
478 T. Kume, H. Watanabe, S. Aoki and T. Sato, Agric. Biol. Chem., 1981, 45, 1311.
479 T. Kume, H. Watanabe, M. Takehisa and T. Sato, Agric. Biol. Chem., 1981, 45, 1351.
480 T. Kasumi, K. Hayashi, N. Tsumura and T. Takagi, Agric. Biol. Chem., 1981, 45, 1097.
481 T. Kasumi, K. Hayashi, N. Tsumura and T. Takagi, Agric. Biol. Chem., 1981, 45, 1087.
482 T. Kasumi, K. Hayashi and N. Tsumura, Agric. Biol. Chem., 1981, 45, 619.
483 B. Larsen and H. Grasdalen, Carbohydr. Res., 1981, 92, 163.
484 K. Izumori, S. Sugimoto and B. Kraska, Agric. Biol. Chem., 1980, 44(1), 223.
485 O. Valentová, M. Marek, F. Švec, J. Štamberg and Z. Vodrážka, Biotechnol. Bioeng., 1981, 23, 2093.
486 W. Hartmeier, Starch, 1981, 33, 97.
487 S. Pillai, Biochem. J., 1981, 193, 825.
488 E. Sada, S. Katoh and M. Terashima, Biotechnol. Bioeng., 1981, 23, 1037.
489 Ya.A. Aleksandrovskii, L.V. Bezhikina and Yu.V. Rodionov, Biochemistry, 1981, 46, 593.
490 V. Linek, P. Beneš, F. Hovorka and O. Holeček, Biotechnol. Bioeng., 1981, 23, 1467.
491 V. Linek, P. Beneš, J. Sinkule, O. Holeček and V. Malý, Biotechnol. Bioeng., 1980, 22, 2515.
492 B. Solomon, N. Lotan and E. Katchalski-Katzir, J. Chromatogr., 1981, 215, 121.
493 S. Hayashi and S. Nakamura, Biochim. Biophys. Acta, 1981, 657, 40.
494 E. Noguchi, M. Nishikimi and K. Yagi, J. Biochem. (Tokyo), 1981, 90, 33.
495 R. Jenness, E.C. Birney and K.L. Ayaz, Comp. Biochem. Physiol., 1980, 67, 195.
496 R.F.H.Dekker, J. Gen. Microbiol., 1980, 120, 309.
497 L. Toro and D. Gaudioso, Can. J. Biochem., 1980, 58, 667.
498 J.E. Sadler, T.A. Beyer and R.L. Hill, J. Chromatogr., 1981, 215, 181.
499 E.H. Beachey, W. Keck, M.A. De Pedro and U. Schwarz, Eur. J. Biochem., 1981, 116, 355.
500 T. Koga and M. Inoue, Carbohydr. Res., 1981, 93, 125.
501 K. Fukushima, R. Motoda, K. Takada and T. Ikeda, FEBS Lett., 1981, 128, 213.
502 H.H. Carnicero, A.M. Adaamany and S. Englard, Arch. Biochem. Biophys., 1981, 210, 678.
503 E.T. O'Keeffe, R.L. Hill and J.E. Bell, Biochemistry, 1980, 19, 4954.
504 Y. Fujita-Yamaguchi and A. Yoshida, J. Biol. Chem., 1981, 256, 2701.
505 E.T. O'Keeffe, T. Mordick and J.E. Bell, Biochemistry, 1980, 19, 4962.
506 P. Broquet and P. Louisot, Biochimie, 1981, 63, 803.
507 A. Martin, M.C. Biol, E. Arrambide, M. Richard and P. Louisot, Biochimie, 1981, 63, 241.
508 C. C-Mulet, J. Badet, J.P. Cartron, FEBS. Lett., 1981, 126, 123.
509 D.H. van den Eijnden, M.L.E. Bergh, B. Dieleman and W.E.C.M. Schiphorst, Hoppe-Seyler's Z. Physiol. Chem., 1981, 362, 113.
510 G. Dodin, FEBS. Lett., 1981, 134, 20.
511 I.M. Archer, P.S. Harper and F.S. Wusteman, Clin. Chim. Acta, 1981, 112, 107.
512 I.G. Leder, Biochem. Biophys. Res. Commun., 1980, 94, 1183.
513 P. Di Natale and L. Ronsisvalle, Biochim. Biophys. Acta, 1981, 661, 106.
514 T. Ishibashi, A. Maru, Y. Imai, A. Makita and I. Tsuji, Biochim. Biophys. Acta, 1980, 616, 218.
515 A. Waheed and R.L. van Etten, Biochim. Biophys. Acta, 1980, 614, 92.

516 J.R. George and J.W. Fitzgerald, J. Bacteriol., 1981, 145, 1428.
517 J.R. George and J.W. Fitzgerald, J. Bacteriol., 1981, 147, 69.
518 J.W. Fitzgerald, R.B. Kellogg and G.J. Stewart, FEMS Microbiol. Lett., 1981, 11, 93.
519 H. Kresse, E. Paschke, K.V. Figura, W. Gilberg and W. Fuchs, Proc. Natl. Acad. Sci. USA - Biol. Sci., 1980, 77, 6822.
520 P. Monsan and A. Lopez, Biotechnol. Bioeng., 1981, 23, 2027.
521 M.M. Matthews, C.L. Futerman, V.K. Parnaik and S.M. Jung, Arch. Biochem. Biophys., 1981, 208, 278.
522 S.M. Jung and R.M. Mayer, Arch. Biochem. Biophys., 1981, 208, 288.
523 T.J. Greir and R.M. Mayer, Arch. Biochem. Biophys., 1981, 212, 651.
524 S. Thaniyavaran, S. Singh, C.M. Maynard, K.G. Taylor and R.J. Doyle, Carbohydr. Res., 1981, 96, 134.
525 M. Kobayashi and K. Matsuda, Biochim. Biophys. Acta, 1980, 614, 46.
526 G.L. Matters and C.D. Boyer, Phytochemistry, 1980, 20, 1805.
527 J.-L. Chiasson, J.H. Aylward, H. Shikama and J.H. Exton, FEBS. Lett., 1981, 127, 97.
528 P.J. Parker, A. Aitken, T. Bilham, N. Embi and P. Cohen., FEBS. Lett., 1981, 123, 332.
529 P. Pérez, R. Varona, I. Garcia-Acha, A. Durán, FEBS. Lett, 1981, 129, 249.
530 G.D. Craig, D.A. Wood and K. Gull, FEMS Microbiol. Lett., 1981, 10, 43.
531 Y. Ohkubo, I. Funakoshi and I. Yamashina, J. Biochem. (Tokyo), 1981, 89, 161.
532 F.N. Kolisis, T.G. Sotiroudis and A.E. Evangelopoulos, FEBS. Lett., 1980, 121.
533 K.E. Wolter, T.L. Highly and F.J. Evans, Biochem. Biophys. Res. Commun., 1980, 97, 1499.
534 A. Carrasco, G. Pincheira and T. Ureta, J. Bacteriol., 1981, 145, 164.
535 M. Ameyama, E. Shinagawa, K. Matsushita and O. Adachi, J. Bacteriol., 1981, 145, 814.
536 V.E. Elisashvili, Biochemistry (Engl. Transl.), 1980, 45, 14.

7
Glycolipids and Gangliosides

BY I. M. MORRISON

1 Introduction

Various aspects of glycolipids are reviewed in the second of a two-volume book on cell membranes and viral envelopes.[1] The structure and function of plant lipids occupy Volume 4 of the comprehensive treatise on the biochemistry of plants.[2] A review on the role of lipids in biological membranes has appeared.[3]

2 Analytical and General Methods

I.r. photoacoustic spectroscopy has been used to study skin lipids.[4] Using $CHCl_3$-MeOH (2:1) extracts, the results were similar to those obtained by i.r. transmission spectroscopy, but only 10^{-2} times the amount of material was required.

The phenol-sulphuric acid method for the estimation of sugars in lipids has been modified by keeping the products at 100°C for 5 min.[5] More reproducible results were obtained.

A dry-column method for the quantitative extraction and simultaneous separation of lipid classes has been developed.[6] The tissue sample, anhydrous sodium sulphate, and diatomaceous earth are ground together and packed in a column. Total lipid or lipid classes are then eluted by isocratic or sequential elution, respectively. A two-stage, one-dimensional t.l.c. method has been developed for the separation of lipid classes.[7] Lecithin was found to influence the chromatography of steroid glucosiduronates.[8] It was concluded that hydrogen bonds were formed between the phosphodiester group of lecithin and the hydroxy groups of the steroid conjugates.

D-Galactose oxidase is able to oxidize the hydroxymethyl group of lipids containing D-galactose or 2-acetamido-2-deoxy-D-galactose.[9] These sugar residues can then be labelled by reduction with radiolabelled borohydride. Conditions are described for the most efficient production of product.

Methylation techniques for the structural analysis of

glycolipids have been reviewed.[10]

It has been shown that particles, previously observed in freeze-fracture replicas of mixed phospholipid dispersions, occur in dispersions of mixed sn-3-D-galactopyranosyldiacyl-glycerols.[11] These particles correspond to lipid micelles sandwiched within a membrane bilayer.

3 Gangliosides

The gangliosides G_{M_1} and $G_{D_{1a}}$ have been specifically oxidized at the 3-position of the sphingosine moiety with 2,3-dichloro-5,6-dicyanobenzoquinone in the presence of Triton X-100.[12] Subsequent reduction with borohydride restored the ganglioside to its original form, and if a radiolabel was used 99% of the radioactivity was incorporated. Proof of the reaction was obtained by reoxidation when the label was removed.

The interaction of gangliosides with Ca^{2+} ions and some polar head-group requirements for the establishment of particular interactions with phosphatidylcholine have been studied in monolayers at the air-0.145M NaCl interface.[13] The ganglioside-Ca^{2+} interaction depended on the position occupied by the neuraminosyl residues in the oligosaccharide chain. Favoured interactions of polyneuraminosylgangliosides with phosphatidylcholine may result from a configuration which allows a partial matching of two oppositely orientated electrical vectors contributed by the zwitterionic phosphocholine group and particular neuraminosyl groups.

The gangliosides G_{D_3} and $G_{T_{1a}}$ were biosynthesized as secondary products from lactosylceramide and G_{M_1}, respectively.[14] By a series of reactions involving periodate oxidation, borohydride reduction, and hydrolysis, the linkage between the two neuraminic acid residues was identified as (2→8).

^1H n.m.r. spectroscopy has demonstrated that N-acetyl-α-D-neuraminic acid is the first product released by the neuraminidases of both Clostridium perfringens and Arthrobacter ureafaciens when hydrolysing neuraminic acid-containing carbohydrate chains.[15]

Spectrofluorimetric techniques have been used to study various ganglioside and ganglioside-lecithin dispersions.[16] The studies indicate that the ultimate and/or penultimate carbohydrate moieties of the neutral sugar backbone of gangliosides and the topographical difference in the locations of the neuraminic acid linkage influence the integrity of the membranes including the hydrophobic region.

The micellar properties of gangliosides in aqueous solutions have been investigated by quasi-elastic light – scattering measurements.[17] The scattered intensity data allowed the derivation of an upper limit to the critical micelle concentration and the evaluation of the molecular weight of the micelle from G_{M_1} and $G_{D_{1a}}$. Aqueous dispersions of G_{M_1} and Triton X-100, in different proportions, have been analysed for some physicochemical properties and their susceptibility to D-galactose oxidase.[18] Differences in the molar ratio gave well defined transitions between different micellar species as measured by the physicochemical methods, and these transitions were also evident in the kinetics of the D-galactose oxidase. The properties of these micelles were further studied using light-scattering methods.[19] The D-galactose oxidase acted very poorly on homogeneous G_{M_1} micelles, but at fixed G_{M_1} concentrations the rate was increased by increasing Triton X-100 concentrations. The formation of statistical micelle aggregates was followed by inhibition of the D-galactose oxidase activity.

^1H and ^{13}C n.m.r. spectroscopy, laser-light scattering, and differential scanning microcalorimetry have been used to study the influence of G_{M_1} and $G_{D_{1a}}$ on the structural and thermotropic properties of sonicated small 1,2-dipalmitoyl-α-L-phosphatidylcholine vesicles.[20] Since the average dimensions of all vesicles were the same, differences in the temperature dependence of heat capacity in the presence of the gangliosides were attributed to differences in interaction of the hydrophobic parts of the molecules. The higher content of C_{20} sphingosines in $G_{D_{1a}}$ as compared to G_{M_1} was considered to be one possible source of difference. Trineuraminosylgangliosides are spontaneously incorporated into 1,2-dipalmitoyl-α-L-phosphatidylcholine vesicles and, since the ganglioside, on incorporation, is susceptible to neuraminidase action, it is concluded that incorporation occurs only on the outer face of the bilayer.[21] Calorimetric studies have indicated that the ganglioside stabilizes the vesicular structures by inhibiting the fusion of small vesicles that occurs below the phase-transition temperature.

Cholesterol enhanced the binding of thyrotropin to 1,2-dipalmitoyl-α-D,L-phosphatidylcholine liposomes containing G_{T_1}

ganglioside but not those containing the gangliosides G_{M_1} and G_{D1a}.[22] The role of gangliosides in the binding and action of thyrotropin has been re-evaluated by comparing thyrotropin binding with cholera-toxin binding in a cloned line of normal rat thyroid cells and neuraminidase-treated cells.[23] The results indicate that more complex gangliosides do not serve as a component of the thyrotropin receptors nor are they involved in the transmission of the hormone signal across the cell membrane of these cells.

Gangliosides that bind cholera toxin have been detected by the direct binding of ^{125}I-labelled toxin on t.l.c.[24] The method is sensitive to 70 fmol G_{M_1} and should be applicable to other carbohydrate-binding materials. The preparation of a receptor-specific affinity chromatographic matrix for the large-scale purification of cholera toxin has been described.[25] The ganglioside G_{M_1} is hydrolysed to lyso G_{M_1}, which is coupled, via stabilized Schiff's bases, to porous silica beads. The amount of toxin bound was proportional to the amount of lyso G_{M_1} used.

Ganglioside affinity filters have been used to identify toxigenic strains of Clostridium botulinum types C and D but could be used on a general procedure for identifying botulinal toxin and toxigenic strains.[26]

Antibodies to the gangliosides G_{D_3} and G_{M_1} have been prepared and their properties determined.[27] The antibodies to G_{D_3} were highly specific while those to G_{M_1} reacted with many other gangliosides and their derivatives. A periodate-oxidized derivative of G_{M_1} cross-reacted almost as readily as G_{M_1} itself.

The specificity of limulin and wheat-germ agglutinin towards neuraminic acids has been studied using lipid vesicles containing G_{M_3}.[28] Limulin binds specifically to N-glycoloyl derivatives of G_{M_3}, but the hydroxy group at C-4 and the carboxy group must also be free. The side chain is not involved in binding. Wheat-germ agglutinin only binds when the hydrophilic tail is removed. The acetamido group but not the carboxy group is involved.

The ganglioside G_{M_1}, incorporated in a planar lipid bilayer, interacts specifically with Ricinus toxin.[29] A fourteen-fold increase in conductance was observed.

An interspecies comparison of the brain ganglioside patterns from fish, amphibians, birds, and mammals has been studied by two-dimensional t.l.c.[30] Over 30 components were detected in mammalian neural samples. Differences were observed between the vertebrate classes, but greater similarity exists between the lower and higher classes than had previously been noted.

Various surfactants have been shown to be able to replace the requirements for the activator protein in the hydrolysis of G_{M_2} ganglioside by the β-D-2-acetamido-2-deoxy-hexosidase A from human liver.[31] In the presence of saturating concentrations of the activator, the enzymatic hydrolysis of the substrate was further stimulated. The surfactants had no effect on the enzymatic hydrolysis of 4-methylumbelliferyl 2-acetamido-2-deoxy-β-D-glucose. A model for this reaction is discussed.

A mononeuraminosylganglioside has been shown to be the antigen of a monoclonal antibody that is specific for cells from carcinoma of the human colon.[32]

The surface distribution of the ganglioside G_{M_1} on human blood cells and the effect of endogenous G_{M_1} and neuraminidase on cholera-toxin surface labelling have been demonstrated by an immunocytochemical study.[33] Neutrophils were the most heavily labelled blood cells while lymphocytes, erythrocytes, and platelets were only labelled to a limited extent. The labelling was significantly altered on neuraminidase treatment.

The lymphocytes of human peripheral mononuclear cells have been activated by oxidation with periodate.[34] The neuraminosyl group was converted to a 7-carbon analogue, but the L-fucose and D-galactose residues were also oxidized. The response to periodate was reduced ca. 50% by pretreatment with neuraminidase. The gangliosides from human peripheral blood lymphoctyes and neutrophils have been isolated and their structures determined by mass spectrometry and glycosidase hydrolysis results.[35] The three structures (1)-(3) were found in both types with (1) being the major ganglioside in lymphocytes and (3) the major ganglioside in neutrophils. Structure (2) may be a leukocyte-specific glycosphingolipid. Large gangliosides with the general structure (4) were also isolated. The gangliosides in human leukocytes have been characterized and differences were observed between types of leukocyte.[36] More ganglioside was present in normal leukemic cells than in granulocytes, while the complex gangliosides were more abundant in myelogenous cells than lymphocyte cells.

Concanavalin A was shown to increase the incorporation of labelled 2-amino-2-deoxy-D-glucose into the gangliosides.[37] Thus, ganglioside synthesis is not only associated with cellular division but also occurs within a short time of lymphocyte activation.

α-Neu*p*5Ac-(2→3)-β-D-Gal*p*-(1→4)-β-D-Glc*p*-(1→1)Cer

(1)

α-Neu*p*5Ac-(2→?)-β-D-Glc*p*NAc-(1→3)-β-D-Gal*p*-(1→4)-β-D-Glc*p*(1→1)Cer

(2)

α-Neu*p*5Ac-(2→?)-β-D-Gal*p*-(1→3,4)-β-D-Glc*p*NAc-(1→3)-β-D-Gal*p*(1→4)-β-D-Glc*p*-(1→1)Cer

(3)

α-Neu*p*5Ac-(2→?)[-β-D-Gal*p*-(1→3,4)-β-D-Glc*p*NAc-(1→3)]$_n$-β-D-Gal*p*(1→4)-β-D-Glc*p*-(1→1)Cer

(4)

Anti-asialoganglioside sera has been shown to cross-react with human blood-group N antigen.[38]

An immunocytochemical study of the surface ganglioside G_{M_1} in haemic cell lines of normal human bone marrow has been carried out similarly to ref. 33.[39] The extent of incorporation was related to specific cell types and to their stage of maturation.

The ganglioside patterns of horse, donkey, and mule brains have been determined.[40] Forebrains contained the highest concentration with the brain stem of the mule having the lowest value. The N-glycoloylneuraminic acid content was <3% of the total neuraminic acid. G_{M_1}, $G_{D_{1a}}$, and $G_{D_{1b}}$ were the major gangliosides. The gangliosides from equine kidney and spleen have been characterized as haematoside, G_{M_2}, G_{M_1}, and $G_{D_{1a}}$.[41] G_{D_3} was found in the kidney and $G_{D_{1b}}$ in the spleen. The relative proportions of N-acetyl and N-glycoloyl neuraminic acids were also determined.

Liposomes, composed of gangliosides from bovine brain and dipalmitoyllecithin, have been studied by e.s.r. spectroscopy.[42] At temperatures above the transition temperature (47°C) the gangliosides diffuse freely in the lipid matrix, but at 19°C, in the gel state, the gangliosides cluster together. Gangliosides from bovine brain have been modified by attaching biotin hydrazide to the aldehydo group after periodate oxidation.[43] The modified gangliosides were incorporated into mature rat thymocytes, and the results suggested that gangliosides may be involved in transmembrane communication during lymphocyte stimulation.

A sulphated ganglioside has been isolated from bovine gastric mucosa, and its structure (5) was elucidated by partial acid hydrolysis, specific glycosidase degradation, periodate oxidation, and methylation analyses.[44] 71% of the neuraminic acid was the N-acetyl derivative. A sulphated ganglioside from bovine gastric mucosa had the structure (6) confirmed by the usual procedures.[45] 65% of its neuraminic acid was the N-acetyl derivative.

A non-specific lipid-transfer protein, purified from bovine liver, will stimulate the transfer of the ganglioside G_{M_1} and neutral glycosphingolipids from phosphatidylcholine vesicles to erythrocyte ghosts.[46]

N-Glycoloylhaematoside and a ganglioside, containing N-glycoloyl-neuraminic acid and having an oligosaccharide chain identical to that of erythrocyte neuraminosylparagloboside, have been isolated from both normal and leukemic bovine blood lymphocytes.[47] Three other gangliosides were identified in α-(SO_3H→8)-Neu*p*-(2→3)-β-D-Gal*p*-(1→4)-β-D-Glc*p*-(1→1)Cer

4

↑

1

β-D-Gal*p*NAc

(5)

α-(SO_3H→8)-Neu*p*-(2→8)-α-Neu*p*-(2→3)-β-D-Gal*p*-(1→4)-β-D-Glc*p*-(1→1)Cer

(6)

leukemic lymphocytes.

The gangliosides have been isolated from bovine optic nerve and were found to be G_{M_1} (129), $G_{D_{1a}}$ (98), $G_{D_{1b}}$ (97), $G_{T_{1b}}$ (80), G_{D_3} (31), and G_{Q_1} (12), the concentrations being $\mu g\ g^{-1}$ wet tissue.[48] Together they accounted for 97% of the total neuraminic acid present.

The three L-fucogangliosides (7)-(9) have been isolated from porcine nervous tissue.[49] All three were found in dorsal-root ganglia and spinal cord but only (7) and (9) in the forebrain. The chemical analyses were confirmed by mass-spectral analysis.

The specificity of the neuraminosyltransferase from bovine liver microsomes has been determined.[50] The transfer was solely to C-3 of D-galactose residues with no transfer to 2-acetamido-2-deoxy-D-galactose residues.

Eight gangliosides have been identified as constituents of ovine testis, with G_{M_3}, G_{D_3}, G_{M_1}, and $G_{D_{1a}}$, respectively, being the most abundant.[51]

A new solvent system has been developed for t.l.c. of the gangliosides of rat cerebellum.[52] The separation takes 1 h and virtually no tailing occurs. The considerable heterogeneity of this material was demonstrated by identifying 28 separate gangliosides.

2-Acetamino-2-deoxy-D-glucose has been shown to be incorporated 2-3 times higher into undernourished rats than normal rats.[53] This enhanced incorporation was found in all brain areas. The turnover of the gangliosides was slower in young undernourished rats, relative to the controls, but greater in older rats. Microsomal gangliosides in rat brain have a higher initial incorporation of 2-acetamido-2-deoxy-D-mannose than synaptosomal gangliosides.[54] The gangliosides $G_{T_{1b}}$ and $G_{Q_{1b}}$ were more highly labelled than other gangliosides, but a decrease in incorporation was observed on treatment with the convulsant pentylenetetrazol. The role of neuraminosyl compounds on choline uptake by synaptosomes has been studied.[55] Neuraminidase treatment of synaptosomal fractions brought about a reduction in the uptake of choline.

The activation of (Na^+, K^+) ATPase by the ganglioside G_{M1} has been shown to follow biphasic kinetics.[56] The break is at 50nM G_{M_1} with a stable complex below that value and a loosely bound one above it. It was suggested that the activation was a specific phenomenon not related to the amphiphilic and ionic properties of the ganglioside but due to a modification of the membrane lipid

```
α-L-Fucp-(1→2)-β-D-Galp-(1→3)-β-D-GalpNAc-(1→4)-β-D-Galp-(1→4)-β-D-Glcp-(1→1)Cer
                                                     3
                                                     ↑
                                                     2
                                                α-Neup5Ac
```

(7)

```
α-L-Fucp-(1→2)-β-D-Galp-(1→3)-β-D-GalpNAc-(1→4)-β-D-Galp-(1→4)-β-D-Glcp-(1→1)Cer
                                                     3
                                                     ↑
                                                     2
                                                α-Neup5Gl
```

(8)

```
α-L-Fucp-(1→2)-β-D-Galp-(1→3)-β-D-GalpNAc-(1→4)-β-D-Galp-(1→4)-β-D-Glcp-(1→1)Cer
                                                     3
                                                     ↑
                                                     2
                               α-Neup5Ac-(2→8)-α-Neup5Ac
```

(9)

environment around the enzyme. The highest degree of labelling in homogenized neuronal perikarya from rat brain with [^{3}H]-5-acetylneuraminic acid has been found in subcellular fractions that also showed the highest specific activities for several ganglioside glycosyltransferases.[57] The neuraminic acid-containing glycoconjugates remained associated with the membrane after all treatments tried except sodium deoxycholate extraction. Some of the results suggested that the gangliosides change their accessibility to added enzymes in the process of transport through the axon and incorporation into the synaptic membrane.

An increased threshold for electrical stimulation to induce somatosensory-evoked potentials in peripheral nerves followed the transection and rejoining of the nerve in rats.[58] The amount of this increase was diminished by treatment with gangliosides so it was concluded that gangliosides enhance the rate of nervous regeneration.

G_{M2}-Synthase (UDP-2-acetamido-2-deoxy-D-galactose-G_{M3}-2-acetamido-2-deoxy-D-galactosyltransferase) has been studied in Golgi-rich fraction from rat liver.[59] Many requirements of this enzyme were demonstrated along with its physical properties.

Methods for the chemical detection of gangliotetraosylceramide, the rat T lympohocyte macrophage-associated antigen, have been reported along with its cellular distribution.[60]

The glycolipids of murine lymphocyte subpopulations have been investigated. A neuraminosyl derivative of asialo G_{M1}, which is sensitive to neuraminidase treatment, is increased in concentration in splenic lymphocytes.[61] Asialo G_{M1} is a normally occurring lipid and not an acid-hydrolysed product of G_{M1}.

The total glycolipids from thymocytes of leukemic AKR/J murine thymus have been shown to be greater than those of non-leukemic thymocytes.[62] Of the individual gangliosides, G_{M3}, G_{M2}, G_{D1a}, and G_{D1b} contents were higher while G_{M1} and G_{T1} were lower. Parallel with these findings were the observations that two neuraminosyltransferases were also at an elevated level. During the differentiation of leukemia M1 cells from mice into macrophages, the ganglioside content has been shown to change markedly.[63] A several-fold increase in G_{M1b} content (confirmed by neuraminidase treatment) was noted with a concomitant decrease in the dineuraminosylganglioside fraction. The glycolipid content of a murine cell line (JLS-V_9) and its ouabain-resistant mutant clone (JLS-V_9 OR) have been compared.[64] The ganglioside, G_{M3}, content of the mutant cells was 1.14 times that of the parent cells but, whereas 98% of the neuraminic acid in the parent cells was in the N-acetyl form, only 69% of that in the mutants was in the same form.

The ganglioside pattern of rabbit brain has been studied by two-dimensional t.l.c. and autoradiography.[65] Numerous minor components, representing previously uncharacterized gangliosides, were identified, several of which were novel L-fucose-containing gangliosides.

The gangliosides of rabbit thymus have been analysed by a ganglioside mapping procedure.[66] The G_M gangliosides and the combined G_D and G_T species accounted for 76.6 and 23.3%, respectively, of the lipid-bound neuraminic acid. The gangliosides (10) and (11) accounted for 38.4 and 31.3%, respectively, of the total fraction while G_{D_3} (11.8%) and G_{M_3} (17%) accounted for the remainder. The G_{D_3} and 62% of G_{M_3} were in the N-glycoloyl form. The lacto-N-neotetraose backbone of (10) and (11) has been shown to be tissue specific for thymus with 70% of all thymus gangliosides possessing it.[67] In addition the N-glycoloyl derivative of neuraminic acid represents 90% of that species.

The mononeuraminosyl gangliosides of rabbit skeletal muscle have been analysed with G_{M_3} representing 73% of the fraction.[68] A novel ganglioside, N-acetylneuraminosyl lacto-N-noroctaosyl-ceramide (12), constituted 5% of the fraction.

Addition of gangliosides has induced a statistically significant increase in the number of neurites in the outgrowth zone of guinea-pig spinal ganglia.[69] However, only a slight increase in neurite length was observed.

Three novel gangliosides (13)-(15) have been isolated from frog fat body and their structures confirmed by the usual methods.[70]

During phylogeny and early ontogeny of higher vertebrates the number of ganglioside fractions was reduced especially with regard to the more polar species.[71] Arranging the different ganglioside fractions subsequent to their paths of formation, it has become apparent that in elasmobranch fish and embryonic chicken the third pathway is used whereas in the adult chicken there is a switch to the first and second pathway. The results are taken as further evidence of the recapitulation principle according to Haeckel's biogenetic rate.

An unusual pattern of myelin gangliosides has been observed in avian central nervous system.[72] The total concentration was α-Neup5Gl-(2→3)-β-D-Galp-(1→4)-β-D-GlcpNAc-(1→3)-β-D-Galp-(1→4)-β-D-GlcpNAc-(1→3)-β-D-Galp-(1→4)-β-D-Glcp-(1→1)Cer

(10)

α-Neup5Gl-(2→3)-β-D-Galp-(1→4)-β-D-GlcpNAc-(1→3)-β-D-Galp-(1→4)-β-D-Glcp-(1→1)Cer

(11)

α-Neup5Ac-(2→3)-β-D-Galp-(1→4)-β-D-GlcpNAc-(1→3)-β-D-Galp-(1→4)-β-D-GlcpNAc-(1→3)-
β-D-Galp-(1→4)-β-D-GlcpNAc-(1→3)-β-D-Galp-(1→4)-β-D-Glcp-(1→1)Cer

(12)

```
α-D-Galp-(1→3)-β-D-Galp-(1→3)-β-D-GalpNAc-(1→4)-β-D-Galp-(1→4)-β-D-Glcp-(1→1)Cer
                                                     3
                                                     ↑
                                                     2
                                                 α-Neup5Ac
```

(13)

```
β-D-Galp-(1→3)-α-D-Galp-(1→3)-β-D-Galp-(1→3)-β-D-GalpNAc-(1→4)-β-D-Galp-(1→4)-β-D-Glcp-(1→1)Cer
                                                                      3
                                                                      ↑
                                                                      2
                                                                  α-Neup5Ac
```

(14)

```
α-D-Galp-(1→3)-β-D-Galp-(1→3)-α-D-Galp-(1→3)-β-D-Galp-(1→3)-β-D-GalpNAc-(1→4)-β-D-Galp-(1→4)-β-D-Glcp-(1→1)Cer
                                                                                       3
                                                                                       ↑
                                                                                       2
                                                                                   α-Neup5Ac
```

(15)

```
α-Neup5Gl-(2→6)-D-Glcp-(1→8)-Neup5Gl-(2→6)-D-Glcp-(1→1)Cer
```

(16)

considerably higher than in other mammalian species, with G_{M_3} being present in concentrations similar to G_{M_1}.

A new t.l.c. system has been used to separate complex gangliosides from embryonic chicken brain.[73] Ten fractions were detected, eight of them also being present in the brain of rays. Six of these were similar to the gangliosides G_{T_3}, G_{T_2}, $G_{T_{1c}}$, $G_{Q_{1c}}$, $G_{P_{1c}}$, and $G_{P_{1b}}$ identified in cod brain. Four of these have been partially characterized and they contain neuraminic acid: sphingosine ratios of 4:1, 5:1, 6:1, and 7:1.[74] Mild neuraminidase treatments yield $G_{T_{1b}}$ and $G_{D_{1b}}$ transiently but G_{M_1} finally. During brain development the amounts of the forms with ratios of 6 and 7 decreased.

The brain gangliosides of the ice fish Trematomus hansoni have an extremely high content of neuraminic acid which confers a high degree of polarity on the neuronal membrane and maintains their functionality at low temperatures.[75] At least four gangliosides were detected which were more highly substituted than the pentaneuraminosyl ganglioside G_{P_1}.

Two novel gangliosides have been isolated and characterized from the sea urchin Strongylocentrotus intermedius.[76] The first had the structure (16) and the second was similar except that it contained a sulphate ester group at C-8 of the terminal neuraminic acid residue.

The neuraminidation of gangliosides has been studied during maturation of cultured neurons.[77] During the first 4 h G_{M_3} and G_{D_3} are produced, after 12 h $G_{D_{1a}}$ and $G_{D_{1b}}$ are produced, and finally G_{M_1} and $G_{T_{1b}}$ are produced. G_{D_3} does not appear to be incorporated into the membranes formed during the period of synaptogenesis in the same way as $G_{D_{1a}}$ and polyneuraminosylgangliosides.

The ganglioside contents of SV40-transformed Balb/c3T3 cells (SV3TC cells) and concanavalin A-selected SV3T3 revertant cells have been determined and compared with untransformed cells.[78] There was a decrease in the concentration of higher gangliosides in transformed and revertant cells, but the content of G_{M_3} in revertant cells was higher than in transformed or original cells. G_{M_3} may have a role in growth control.

4 Animal Glycolipids

A new solvent system coupled with a porous silica-gel column has been developed for separating glycolipids containing mono- to tetrakaidecasaccharide chains in 60 min.[79] The separation and microdetection (10 pmol) of oligosaccharide chains of glycolipids have been achieved on high-performance cellulose t.l.c. and autoradiofluorography.[80]

Differential scanning calorimetry and X-ray diffraction of anhydrous and hydrated N-palmitoyl-D-galactosylsphingosine have provided evidence for a complex polymorphic behaviour and interconversions between stable and metastable structural forms.[81] The anhydrous glycolipid exhibits three lamellar crystal forms below 143°C and a liquid-crystal form between 143 and 180°C.

The 360 MHz ^1H and 90.5 MHz ^{13}C n.m.r. spectra of D-galactosylceramides have been recorded in DMSO and show a $^4C_1$ configuration of the sugar ring.[82] In contrast to an aqueous solution, the hydroxymethylene group at C-6 can freely rotate around the C-6 : C-5 bond. A series of glycolipids has been analysed by 360 MHz ^1H n.m.r. spectroscopy.[83] The resonance of all H-1 and H-2 as well as most H-3 and H-4 and some H-5 protons could be assigned with the aid of spin-decoupling difference spectroscopy. Regularities were observed in the type, anomeric configuration, site of glycoside linkage, and sugar residues which could be used for the structure of more complex glycolipids.

Mass spectrometry has been used for the sequencing of the oligosaccharide chain of a dodecaglycosylceramide.[84] This is the largest saccharide conclusively identified. Mass spectrometry, after t.l.c., has been used to identify the structures of glycolipid mixtures from a variety of species.[85]

The anomalous thermotropic phase-transition behaviour of 1,2-distearoyl-D-galactolipids has been investigated.[86]

The penetration of melittin and myelin basic protein into glycosphingolipid monolayers depends on the lipid polar head group, the protein concentration, and the initial surface pressure.[87] The interaction leads to a modification of the surface properties of both the glycosphingolipid and the protein.

Glycolipid synthetases have been measured in a two-phase scintillation assay carried out in a small vial.[88] The method should be suitable for many assays but the specific enzyme determined in this investigation was ceramide : UDP-D-glucose D-glucosyltransferase.

D-Galactosylcerebroside has been derivatized for use as the ligand in affinity chromatography by ozonolysis of the sphingosine double bond followed by oxidation.[89] The anti-D-galactosylcerebroside antibodies purified on this material were highly specific. Proteins can be covalently bound to liposomes after periodate oxidation of the glycosphingolipids in the vesicle membrane.[90] The method has potential for the antibody-mediated targeting of vesicles to cells.

The synthetic glycolipids lactosyl- and melibiosyl-phosphatidylethanolamine can be incorporated into unilamellar liposomes.[91] Both are agglutinated by Ricinus communis lectin but only the latter by Banderiaea simplicifolia isolectin I. Efficient fusion of phospholipid vesicles with monolayer cultures of eukaryotic cells has been accomplished by attaching glycolipid-containing vesicles to the cell surface by using a lectin displaying binding for both the cell surface and the glycolipid.[92]

Single bilayer vesicles of glycosphingolipids and phosphatidylcholine have been prepared and their physical properties determined.[93] Incubations with high-density lipoprotein-3 resulted in the transfer of the glycolipid, a process which did not occur in the absence of phosphatidylcholine. The complex has been further characterized, and only 69% of the lipid is transferred from the vesicles.[94]

The lipid compositions of axolemma-enriched fractions and myelin from human brains have been determined.[95] D-Galactolipid accounted for 25.8% of the lipid in the axolemma fractions, 83% of this being cerebrosides.

The D-galactosylceramidase of human brain is activated by phosphatidylserine but the G_{M1} ganglioside β-D-galactosidase is not activated.[96]

A review of sulpho-D-galactosylceramide has suggested that this glycolipid possesses multifunctional properties in biological membranes.[97]

The characteristic deposits of glycolipid in tissues from patients with Fabry's disease have been studied by electron microscopy.[98] Fourier transformation for the analysis of the distribution periodicity of the inclusions showed that, in synovial membrane and muscle, the inclusions had a monocrystalline character while in the other tissues they were amorphous or polycrystalline. The crystalline morphology of the trihexoside ceramide inclusions characteristic of Fabry's disease have been further examined especially in synovial tissue.[99] Image processing confirmed the membrane-like structure of trihexoside ceramide inclusions and corroborated the proposed classification.

D-Glucosylcerebrosides (Gaucher's storage material) have been shown to stimulate the release of lymphocyte activating factor from macrophages.[100] Other ceramides had no effect. The importance of these observations to Gaucher's disease was discussed.

The complete structure of a trisaccharide (17) from a patient with mannosidosis has been established by ^{13}C n.m.r. spectroscopy.[101] One of the features of a patient with mannosidosis has been excessive gingival hyperplasia with storage of D-mannose-linked oligosaccharides.[102] The trisaccharide (17) was the predominant material in the urine, but tetra- and pentasaccharides were more abundant in gingiva.

The neutral glycolipid compositions of human gastric and colonic cancer tissues have been compared with those from normal tissues.[103] The cancerous tissues had elevated levels of lactosylceramide and L-fucoglycolipids. The L-fucoglycolipids contained both 3-O- and 4-O-substituted residues of 2-acetamido-2-deoxy-D-glucose. Similar studies have confirmed these results, especially those of the L-fucoglycolipids, and further showed that the levels of sulpho-D-galactolipids were also elevated.[104] Blood group A-active glycolipids were present in cancer tissue from two patients with blood group O but not in the associated healthy tissue.

The liver and kidney of a patient with I-cell disease have been found to contain elevated levels of di- and tri-hexoside ceramide.[105] This was associated with a β-D-galactosidase deficiency. It was concluded that the β-D-galactosidase deficiency in I-cell disease is a more specific phenomenon rather than a secondary inhibition as found in mucopolysaccharidoses.

Evidence has been presented that Niemann-Pick disease, type C, is caused by the deficiency of an activating factor stimulating sphingomyelin and D-glucocerebroside degradation in the spleen of these patients.[106]

The glycolipids of the meconium of a human O Le (a-b+) secretor have been characterized by structural and immunological methods.[107] The dominating species were D-glucosyl- (15%) and D-galactosyl-ceramide (30%). Of the higher saccharides, the lacto series (30% of total weight) was represented by lactotetraosylceramide and H-active, Le^a-active, and Le^b-active glycolipids of the lacto series. Similar studies, involving more detailed separations and characterization methods, have suggested α-D-Manp-(1→3)-β-D-Manp-(1→4)-D-GlcpNAc

(17)

```
β-D-Galp-(1→3)-β-D-GlcpNAc-(1→3)-β-D-Galp-(1→4)-β-D-Glcp-(1→1)-Cer
                        4
                        ↑
                        1
                   α-L-Fucp
```

(18)

```
α-L-Fucp-(1→2)-β-D-Galp-(1→3)-β-D-GlcpNAc-(1→3)-β-D-Galp-(1→4)-β-D-Glcp-(1→1)-Cer
                                        4
                                        ↑
                                        1
                                   α-L-Fucp
```

(19)

α-L-Fucp-(1→2)-β-D-Galp-(1→3)-β-D-GlcpNAc-(1→3)-β-D-Galp-(1→4)-β-D-Glcp-(1→1)-Cer

(20)

that meconium is a rich source of glycolipids which allows the chemical fingerprinting of the blood-group type in individuals.[108]

The non-acidic glycosphingolipids of human cord blood erythrocytes from blood groups O and A have been characterized.[109] Lactosylceramide and globoside were the major components. Increased amounts of penta- and hexa-glycosylceramides, which lacked L-fucose substituents, were present and could be biosynthetic precursors of ordinary blood-group A and H glycosphingolipids.

The glycosphingolipids of human plasma lipoproteins have been reviewed with specific interest in the way the different glycolipids exchange between the lipoproteins and erythrocytes.[110] The identification of a blood-group Le^b-active glycosphingolipid in plasma by m.s. and n.m.r. spectroscopy has suggested a new method of chemical blood-typing of human individuals.[111]

A series of papers has been published on the immunochemistry of the Lewis blood-group system. In the first, the complete structures of the oligosaccharide chains of the Le^a-, Le^b-, and H-type 1-active glycolipids have been established as (18)-(20), respectively.[112] The second and third papers consider the m.s. analysis[113] and ^{1}H n.m.r. spectroscopy,[114] respectively, which have been used to establish these structures.

Neutral glycosphingolipids have been isolated from the hairy cells of a patient with hairy-cell leukemia and comprise D-glucosylceramide, lactosylceramide, and two higher glycolipids of the globo series which are similar to those found in human lymphocytes and chronic lymphocytic leukemia cells.[115] The glycolipid pattern may be useful in classifying leukemias of uncertain origin.

A high incidence of an autoantibody against the neutral glycolipid asialo G_{M_1} has been observed in patients with systemic lupus erythematosus.[116] Since no activity was shown against other glycolipids or gangliosides it was suggested that the antibody played a role in the pathogenesis of such neurological disorders.

The glycosphingolipids of normal human lymphocytes have been separated by h.p.l.c. The major component was lactosylceramide with others being detected in smaller amounts.[117] Purified B- and T-cell fractions had a similar complement of glycolipid but the B-cells contained significantly more of each component. Similar studies have been carried out on human peripheral blood lymphocytes and the lymphoid cells of a patient with B-cell chronic lymphocytic leukemia.[118] The glycolipids were the same as those in ref. 114.

A blood-group A-specific tetrasaccharide (21) has been isolated and characterized from human urine.[119] The sugar had A activity almost as great as that of the most active oligosaccharide isolated from soluble blood-group A substance.

The alkyl-D-galactolipids from calf brain have been separated by h.p.l.c.[120]

The two oligosaccharides β-D-mannopyranosyl-(1→4)-2-acetamido-2-deoxy-D-glucose and 4^2-O-β-D-mannopyranosylchitobiose accumulated in the kidney of a goat with mannosidosis.[121] The amounts represented three times those found in the brain, but while lesions of the brain caused profound neurological disorders functional impairment of the kidney was not evident.

Using an epithelial cell line derived from monkey kidney, it has been shown that D-galactosylcerebroside is present in the cytoplasm in close association with colchicine-sensitive microtubule-like subcellular structures.[122]

The distribution of glycero-D-glucolipids in the mucous barrier of dog stomach fundus, body, and antrum has been investigated.[123] The content of neutral glycero-D-glucolipid in the antral portion was higher than that in the other two areas, but the content of sulphated glycolipid was even higher.

The localization of D-galactosylcerebroside in distal tubules, ascending limbs of Henle's loops and collecting tubules of kidney, periportal bile ducts and hepatic parenchyma of liver, and bronchioles and alveoli of lungs has been demonstrated by an immunological method in these organs from hamsters.[124] The existence was further confirmed by chemical analysis.

The presence of SV40-specific glycolipid antigens in the plasma membrane, the nuclear membrane, and probably other inner membranes of SV40-transformed cells from hamsters has been indicated by immunofluorescence methods.[125]

Incubation of NIL 2cl cells, a cloned hamster cell line, with an equilibrium mixture of radioactive UDP-D-glucose and UDP-D-galactose has suggested that this is a useful method for determining the glycolipid composition of cultured cells.[126] Using the same cell line, the sequential glycosylations of endogenous glycosphingolipids have been studied using nucleotide sugars.[127] Differences were noted in the sequential glycosylation between cells harvested from confluent cultures and sparse cultures.

```
β-D-GlcpNAc-(1→3)-β-D-Galp-(1→4)-D-Glc
                      2
                      ↑
                      1
                  α-L-Fucp
```

(21)

SO_3H→3-D-GalpNAc-(1→4)-D-Glcp-(1→1)-Cer

(22)

Furthermore, there was an indication that two distinct pools of each glycolipid were present. Precursor-product relationships do not exist among the main pool of each glycolipid.

The specific localization of UDP-D-galactosylceramide: D-galactosyltransferase in heavy-myelin and membrane fractions of rat brain at an early age when myelination is just beginning has been suggested as a site for the myelination process.[128] The turnover of glycolipids in the adult rat brain has been studied _in vivo_ by determining the half-life of the carbohydrate portion.[129] These half-lives were much shorter than that of the total carbon pool. Since lactosylceramide had the shortest half-life, it was suggested that this glycolipid may serve as a branch point for the biosynthesis of cerebral gangliosides _in vivo_ and not as a degradation product of more complex molecules.

In various rat neural tumours, the contents of lactosylceramide were up to 80 g kg^{-1} compared with normal neural tissue values of 10 g kg^{-1}.[130]

The changes in lipid-bound surface carbohydrates during cell differentiation have been studied in the small intestine of rats.[131] The D-glucosylceramide content increased 2-3 fold. There was little change in many of the complex glycolipids. In similar studies, the content of trihexosylceramide has been shown to increase up to about 13 days after parturition.[132]

A novel sulphoglycosphingolipid (22) has been isolated from rat kidney and characterized by the usual methods.[133]

The clearance of D-glucocerebrosidase by rat liver has shown the presence of two distinct enzyme forms with different recognition characteristics.[134] Neuraminosyl, D-galactosyl, and 2-acetamido-2-deoxy-D-glucosyl residues are important in the uptake of the enzyme. Modification of the enzyme to expose some of these residues results in increased delivery to specific cell types. The disaccharides α-D-mannopyranosyl-(1→2)-D-mannose and α-D-mannopyranosyl-(1→3)-D-mannose have been synthesized from dolichyl-D-mannosylphosphate and D-mannose by rat liver membrane systems.[135] The proportion of each product was dependent on the assay conditions.

The formation of dolichylphosphate-D-glucose and dolichyl diphosphate oligosaccharides in rat spleen lymphocytes has been studied.[136] It was observed that the phospho-oligosaccharide populations contain far fewer products which had been D-glucosylated than the dolichyldiphosphate oligosaccharides from which they are derived, suggesting that the D-glucosyl units inhibit the reaction.

Sulpho-D-galactosylacylalkyl glycerol is enriched in a plasma membrane fraction from adult rat testis.[137] By comparing 11 separate tissues from rats, the presence of this lipid has been shown to be restricted to the testis.[138] Several novel sulpholipids were detected during this study.

L-Fucose-terminated glycoconjugates are recognized by pinocytosis receptors on macrophages where the uptake of β-D-glucuronidase is blocked.[139]

An anti-glycosphingolipid antibody has been prepared and purified and has been used to demonstrate the localization of asialo G_{M1} on the tips of the surface membrane of rat ascites hepatomas.[140]

Analogues of ceramides which inhibit D-glucocerebroside synthetase in mouse brain have been investigated.[141] The most potent compound tested was 2-decanoylamino-3-morpholino-1-phenyl-propanol. An examination of the effects suggested that the active region of the enzyme contains four active sites, one of them being an anionic moiety that binds the D-glucose residue in an activated form. The concentrations of myelin-characteristic D-galactolipid in organotype cultures of newborn mouse cerebellum have been determined by h.p.l.c.[142] These lipids are detectable at 9 days and increase steadily in concentration until the explants are more than 3 weeks old. The D-galactosylceramide sulphotransferase activity of the cerebrum and cerebellum of normal and jimpy mice has been measured during postnatal development.[143] The activity was related to net sulphatide synthesis and was 25-50% lower in the brain parts of the mutants when myelination was taking place in the controls.

The effective antibody in an antiserum having anti-natural killer cell activity and raised against mouse brain tissue is directed against the cell-surface glycolipid asialo G_{M1}.[144]

The distribution of the same glycolipid in the small intestine of the mouse has been determined using an immunofluorescence staining method.[145] Clear staining of the brush border and basolateral membranes of epithelial cells, cell membranes of cryptic cells, and some secretory granules was observed.

D-Glucosylceramide, on intravenous injection in mice, is stored predominantly in the liver and has a half-life of 3.5 days.[146] High levels of this lipid were found in the bile of one patient and the liver of two patients with a biliary obstruction, which suggests a relationship between the biliary excretion of this lipid and Gaucher's disease.

The glycolipid on the surface of natural killer cells from mouse spleen has been identified as asialo G_{M_1}.[147]

A method has been developed for the extraction of carbohydrate-defined Ia antigens from murine spleen and serum cells.[148] The antigen appeared to have a glycolipid structure with the antigenic activity residing in an oligosaccharide unit containing more than seven monosaccharide residues.

Two marker glycolipids of alloantigen-activated murine T-lymphocytes have been partially characterized as α-D-galactosyl-β-D-galactosyl-D-glucosylceramide and the same ceramide with a 2-acetamido-2-deoxy-α-D-glucosyl residue at the non-reducing terminal position.[149]

A monoclonal antibody, directed to the stage-specific embryonic antigen in mice, has been shown to react with a branched glycosphingolipid similar in structure to Ii antigen.[150]

Two related sublines of the mouse lymphoma L5178Y, which were selected for their different susceptibility to natural killer cell-mediated lysis, have different glycolipid patterns.[151] The sensitive line contains a high level of asialo G_{M2} which was not detected in the resistant line. However, there was no correlation between asialo G_{M_2} content and sensitivity in sublines.

Modification of the surface of distearoylphosphatidylcholine vesicles with synthetic glycolipids has dramatically affected the rate of uptake of these vesicles by murine peritoneal macrophage.[152] The high rate of uptake of 6-amino-6-deoxy-D-mannose-modified vesicles is inhibited by cytochalin B and chloroquine but not by colchicine, indicating that the mechanism is phagocytosis. These vesicles remain intact after phagocytosis as aggregates of fused and intact vesicles surrounded by a single bilayer membrane structure.

Several blood-group-type glycolipids from the small intestine of an individual rabbit have been identified by select-ion monitoring.[153] The dominating species was a hexaglycosylceramide with a blood-group B-type sequence, while a second hexaglycosylceramide having a blood-group A-type sequence was also found.

The incorporation of [2-^{3}H]-D-mannose into lipid-linked oligosaccharides of rabbit mammary gland is stimulated by hormones.[154]

An I-active ceramide decasaccharide (23) has been isolated and chacterized from rabbit erythrocyte membranes.[155]

The homologous series isomaltose to isomaltoheptaose can be coupled to stearylamine by reductive amination using cyanoborohydride ion.[156] The stearyl oligosaccharides from isomaltotriose to the heptasaccharide, when incorporated into rabbit liposomes, were agglutinated in the presence of specific antibodies and were lysed if complement was also present. As the disaccharide conjugate was not agglutinated, it may not protrude far enough from the surface to react with the antibody. The specificities and the sizes and shapes of the antibody combining sites in the above systems have been investigated by quantitative precipitin and precipitin-inhibition studies.[157] The sizes of 15 antisera have been determined and all had combining sites in the range isomaltotriose to isomaltohexaose. One of the antisera to stearylisomaltopentaose has been separated into IgM and IgG fractions, and the antibody binding sites were studied.[158] Both were similar to unfractionated antibodies with sites complementary to isomaltotetraose. Another antiserum had two site sizes, one for isomaltopentaose and the other for isomaltotetraose.

The specific antibodies elicited in rabbits against D-glucosylceramide did not cross-react with D-galactosylceramide or sulphatide.[159] They did react with the natural hapten present in red-blood-cell membranes of sheep.

Very low levels of glycolipid have been found in chick embryonic brain yet nearly adult levels were reached by the 5th day.[160] The deposition of mono-D-galactosyldiglyceride in the cerebrum was unique in that all the secretion occurred between the 16th day of embryonic life and the 5th day after hatching, a period characterized by a high rate of myelination.

The spermatozoa of the fresh-water bivalve Hyriopsis schlegelii have a complex spectrum of neutral glycolipids.[161] Two of the minor components (24) and (25), which accounted for <3% of the total neutral glycolipid, were characterized. This is the first report of D-xylose residues being present in a ceramide.

Vesicular stomatitis virus-infected BHK cells when deprived of D-glucose are unable to incorporate D-glucose into the oligosaccharide-lipid intermediates and, consequently, the G protein.[162] A unique late effect of rubella virus infection of BHK21/13s cells was the formation of a glycolipid not detected in

```
α-D-Galp-(1→3)-β-D-Galp-(1→4)-β-D-GlcpNAc
                                  1
                                  ↓
                                  3
                              β-D-Galp-(1→4)-β-D-GlcpNAc-(1→3)-β-D-Galp-(1→4)-β-D-Glcp-(1→1)-Cer
                                  6
                                  ↑
                                  1
α-D-Galp-(1→3)-β-D-Galp-(1→4)-β-D-GlcpNAc
```

(23)

```
α-D-Manp-(1→3)-β-D-Manp-(1→4)-β-D-Glcp-(1→1)-Cer
```

(24)

```
α-D-Manp-(1→3)-β-D-Manp-(1→4)-β-D-Glcp-(1→1)-Cer
                    2
                    ↑
                    1
                β-D-Xylp
```

(25)

control cells or virus.[163] The glycolipid contained L-fucose and glycerol in a 2:1 molar ratio.

The differentiation of M1 cells into macrophages has been induced by lymphokine.[164] During differentiation, asialo G_{M1} appeared on the surface of the M1 cells.

5 Plant Glycolipids

A rapid procedure has been reported for the purification of glycolipids free from surface-active contaminants and suitable for the measurement of surface-pressure isotherms.[165] A major plant glycolipid, mono-D-galactosyldiglyceride, gave very reproducible results with the method.

The motional properties of four mono-D-galactosyldiacylglycerols have been determined using ^{13}C n.m.r. spectroscopy.[166] The longitudinal relaxation times of the D-galactosyl carbon atoms were essentially constant.

The use of stearoyl lactylates[167] and sucrose mono- and diesters[168] in breadmaking has been described.

The structure of a tri-D-galactosyldiglyceride from alfalfa leaves has been partially determined.[169] The non-reducing terminal and penultimate D-galactose residues are linked to the next residue, one by a 1→4 bond and the other by a 1→6 bond. The innermost D-galactose residue is linked to the glycerol residue by a 1→3 bond.

The chemical compositions of cerebrosides in Adzuki bean seeds have been determined.[170] 3-*sn*-Mono- and 3-*sn*-di-D-galactosyldilinolenoylglycerol have both been isolated from bean leaves and the corresponding distearoyl derivatives prepared by hydrogenation.[171] The unsaturated lipids formed stable monomolecular films at an air/water interface while the saturated glycolipids formed condensed monolayers which were relatively unstable. The saturated mono-D-galactosyl derivatives did not form dispersions suitable for fluorescence probe studies of the phase transition.

An alkylether linkage-containing glycero-oligolipid has been isolated from germinating mung beans and the structure (26) confirmed.[172]

It has been established that there are at least two sites for desaturation of 18:2 fatty acid to 18:3, one of them being associated with mono-D-galactosyldiglyceride and which is

1-O-hexadecanoyl-2-O-hexadecyl-3-O-[α-L-Rhap-(1→4)-β-D-Galp-(1→4)-β-D-Galp-(1→4)-β-D-Galp]-glycerol

(26)

inhibited by the pyridazinone herbicide San 9785.[173]

Membrane preparations from developing soybean cotyledons, at the time of glycoprotein synthesis, have been shown to catalyse the assembly of lipid-linked oligosaccharides from UDP-2-acetamido-2-deoxy-D-glucose and GDP-D-mannose with chain lengths of up to 10 sugar residues.[174] The glycolipids do not appear to turn over in pulse-chase experiments and do not appear to produce completed storage glycoproteins. A lipid-bound oligosaccharide has been isolated from pea cotyledons incubated with D-mannose.[175] It appeared to be identical to one obtained from rat liver which is known to contain three D-glucose, nine D-mannose, and two 2-acetamido-2-deoxy-D-glucose residues. Enzymes were present in the same tissue and soybean roots which catalysed the transfer of part of the oligosaccharide chain to endogenous protein.

Young cassava (Manihot esculenta) leaves have a low lipid content, 25% of which is glycolipid.[176] This mixture consists of steryl glycosides, mono- and di-D-galactosyldiglycerides, and cerebrosides.

Mono- and di-D-galactosyldiglycerides were the major glycolipids of ungerminated pollen from Crotalaria juncea.[177] In growing pollen tubes, glycolipid biosynthesis from acetate has been enhanced by the presence of cAMP but not by boric acid.

Spinach thylakoids have been fractionated by detergent treatment and more than 80% of the D-galactolipid was separated from the chlorophyll-protein complexes.[178] Since the integrity of these complexes was maintained on detergent treatment, the free D-galactolipid apparently originates from the lipid matrix of the thylakoid membrane. Further studies have shown that specific associations between endogenous D-galactolipid and chlorophyll-protein complexes do not occur after membrane solubilization. Other studies have been carried out on the same tissue. Alterations in the pH of isolation had no effect on peptide patterns but a great effect on D-galactolipid composition.[179] The changes could be explained by interlipid D-galactosyl transfer during isolation at pH 7.2, since there was a reduction in the concentration of the mono-derivative with corresponding increases in the tri- and tetra-derivatives. The rates and pattern of D-galactolipid synthesis have been studied at different pH values.[180] The highest rates of interlipid D-galactosyl transferase are seen at pH 6.0 in diacylglycerol-poor envelopes and at pH 7.2 in diacylglycerol-rich envelopes. Intact chloroplasts behave essentially as isolated envelopes.

Using spinach and tobacco membranes, lipid-bound D-glucose-containing oligosaccharides have been synthesized which are similar to those found in animals and yeasts.[181] These oligosaccharides were based on $(Glc)_n(Man)_9(GlcNAc)_2$ where n is 2 or 3.

Most of the polar lipids in tobacco leaves vary in concentration with the chlorophyll content during leaf development and senescence.[182] Of the individual polar lipids the greatest changes were in the lipids associated with the chloroplast membranes, namely mono- and di-D-galactosyl, and sulphoquinovosyl-diglycerides.

The mono- and di-D-galactosyldiglyceride contents of wheat and barley chloroplasts are reduced under water stress.[183] The reduction was greater in the more water-requiring cultivars with the less water-requiring cultivars showing a good recovery on irrigation.

Two D-glycosylsterols, β-D-glucopyranosyl-(1→3)-sitosterol and β-cellotriosyl-(1→3)-sitosterol, have been isolated from the leafy stem of rice.[184] A new type of monoglycosylceramide has been isolated from the alkali-stable lipid fraction of rice bran.[185] D-Glucose was the major sugar residue present.

The pollen of Oenothera missouriensis is rich in monoglycosylceramides and also in acidic glycosphingolipids.[186] Compatible pollination leads to a large increase in monoglycosylceramides whereas incompatible self-pollination leads to a decrease in the triglycosylceramide content, irrespective of the genotype of the flower.

Chloroplast lipids have been extracted from varieties of common groundsel (Senecio vulgaris), common lambsquarter (Chenopodium album), and redroot pigweed (Amaranthus retroflexus) and analysed for differences between triazine-resistant and -sensitive biotypes.[187] The resistant varieties had higher mono-D-galactosyldiglyceride contents and lower di-D-galactosyl-diglyceride contents.

It has been shown that in Volvox carteri the antibiotic tunicamycin inhibits the UDP-2-acetamido-2-deoxy-D-glucose dolichylphosphate:2-acetamido-2-deoxy-D-glucose-1-phosphate transferase which is the first enzyme in the dolichol cycle.[188] The antibiotic showdomycin inhibits the same enzyme and dolichol-phospho-D-glucose synthetase.

6 Microbial Glycolipids

Mono- and di-D-glucosyldiglycerides are the dominant lipids of the Acholeplasma laidlawii membrane.[189] The di-D-glucosyl derivative forms a lamellar liquid-crystalline phase while the mono-derivative forms a reversed hexagonal phase. In a mixture, a reversed cubic phase can be formed at physiological temperatures. The structure of the cubic phase is composed of aggregates where the lipids can diffuse over microscopical distances. These glycolipids can be extracted by different detergents and the results indicated that the mono-D-glucosyl-diglyceride was involed in anchoring the protein D_{12} in the membrane.[190]

The membrane-associated lipoglycan from A.granularum has been shown to possess 12 of the repeating units (27) in a linear form attached to a diacylglycerol.[191]

The composition of Bdellovibrio bacteriovorus lipopolysaccharide has been determined in cells grown axenically and intraperiplasmically on Escherichia coli or Pseudomonas putida.[192] The axenically grown cells contained D-glucose and 2-amino-2,6-dideoxy-D-galactose as the sole sugar residues, while the intraperiplasmically grown cells contained the above components plus some derived from the substrate cells. However, since D-galactose was not present, it was suggested that only the lipid A component was transferred.

The major sulphated glycolipid of the halophile Halobacterium salinarium has been characterized by m.s. and n.m.r. spectroscopy as a trihexosyl glycerol C_{20}-diether with the sulphate group being on the non-reducing terminal hexose residue.[193] It may be identical to a similar lipid characterized previously from H.cutirubrum. Three minor glycolipids have been isolated from the halophile H.cutirubrum and their structures characterized as (28), the desulphated product from (28) and (29).[194]

The carbohydrate contents of the lipopolysaccharide antigens from three strains of Klebsiella pneumoniae have been determined, but the presence of capsular antigens has not been demonstrated.[195]

Two unusual phosphoglycolipids, which account for 64% of the total cellular lipids, have been isolated from the methanogenic archaebacterium Methanospirillum hungatei and are derivatives of 1-dibiphytanyl diglycerol tetraether.[196] One of the free hydroxy groups is glycosidically linked to a disaccharide, the two β-D-Glcp-(1→2)-α-D-Glcp-(1→4)-α-D-Glcp-(1→3,4)-β-L-FucpNAc-(1→3)-α-D-Galp-(1→3)-

α-D-Galp-(1→3)-α-D-Galp-(1→3,4)-β-D-GlcpNAc-(1→3,4)-β-D-GlcpNAc

(27)

1-O-[(HSO_3→3)-β-D-Galp-(1→6)-α-D-Manp-(1→2)-α-D-Glcp]-2,3-di-O-phytanyl-sn-glycerol

3
↑
1
α-D-Galf

(28)

1-O-[β-D-Galp-(1→6)-α-D-Manp-(1→2)-α-D-Glcp]-2,3-di-O-phytanyl-sn-glycerol

(29)

α-D-Glcp-(1→2)-β-D-Galf

(30)

α-D-Glcp-(1→6)-β-D-Galf

(31)

disaccharide residues being identified as (30) and (31).[197]

A new synthesis of cord factors and analogues has been reported involving benzylation of 6,6'-di-O-trityl-trehalose.[198] The benzyl derivative is converted via the 6,6'-di-O-methane-sulphonyl derivative to the cord factors.

The C-mycosidic glycopeptidolipid typing antigens from all serovars in the Mycobacterium avium/M. intracellulare/ M.scrofulaceum complex have been examined to varying extent, and some of them consist of short acetylated oligosaccharides linked to a common fatty acyl-peptidyl-O-(3,4-di-O-methyl-L-rhamnose) 'core'.[199] A basal rhamnosyl-6-deoxytalosyl disaccharide is always linked to allo-threonine. The glycopeptidolipid antigens in their structural principals, cellular location, and physiological role bear a striking miniscular resemblance to cell-wall components of other bacteria such as the O-antigenic and R-antigenic lipopolysaccharides.

A novel phenolic glycolipid has been isolated from M.leprae which is distinct from, although related to, that from M. kansasii.[200] The methylated glycolipid contained residues of 2,3-di-O-methylrhamnose, 3-O-methylrhamnose, and 3,6-di-O-methyl-glucose.

The endotoxic glycolipid from Salmonella minnesota, when injected into mice, has produced increased levels of aminopyrine-N-demethylase and aniline hydroxylase activities in the liver.[201]

A glycolipid has been isolated from the cellular slime mould Dictyostelium discoideum and characterized as Δ^{22}-stigmasteryl-D-glucoside.[202] The amount present was higher in the vegetative-stage cells, late aggregation-stage cells, and 1-day sorocarps than in cells of other stages.

A new glycolipid from a strain of Nocardia caviae has been characterized as a 2',3"-di-O-acyl-α-D-glucopyranosyl-(1→2)-D-glyceric acid with myristic, palmitic, and stearic acids being the principal acyl substituents.[203]

A β-D-mannosylceramide has been identified in the hepatopancreas of the fresh-water bivalve Hyriopsis schlegelii and represents 5% of the monohexosylceramide fraction.[204]

(References begin overleaf)

References

1 'Cell Membranes and Viral Envelopes', ed. H.S. Blough and J.M. Tiffany, Academic Press, London, 1980, Vol. 2.
2 'Biochemistry of Plants, a Comprehensive Treatise', ed. P.K. Stumpf, Academic Press, New York, 1980, Vol. 4 (Lipids: Structure and Function).
3 N. Tsukagoshi, J. Agric. Chem. Soc. (Jpn.), 1980, 54, 769.
4 S.O. Kanstad, P.-E. Nordal, L. Hellgren, and J. Vincent, Naturwissenschaften, 1981, 68, 47.
5 S.C. Kushwaha and M. Kates, Lipids, 1981, 16, 372.
6 W.N. Marmer and R.J. Maxwell, Lipids, 1981, 16, 365.
7 J. Bitman, D.L. Wood, and J.M. Ruth, J. Liq. Chromatogr., 1981, 4, 1007.
8 V.R. Mattox and R.D. Litwiller, Lipids, 1980, 15, 999.
9 N.S. Radin and G.P. Evangelatos, J. Lipid Res., 1981, 22, 536.
10 H. Rauvala, J. Finne, T. Krusius, J. Karkkainen, and J. Jarnefelt, Adv. Carbohydr. Chem. Biochem., 1981, 38, 389.
11 A. Sen, W.P. Williams, A.P.R. Brain, M.J. Dickens, and P.J. Quinn, Nature (London), 1981, 293, 488.
12 R. Ghidoni, S. Sonnino, M. Masserini, P. Orlando, and G. Tettamanti, J. Lipid Res., 1981, 22, 1286.
13 B. Maggio, F.A. Cumar, and R. Caputto, Biochem. J., 1980, 189, 435.
14 H.C. Yoshe and R.K. Yu, Carbohydr. Res., 1981, 93, 1.
15 H. Friebolin, R. Brossmer, G. Keilich, D. Ziegler, and M. Supp, Z. Physiol. Chem., 1980, 361, 697.
16 T. Uchida, Y. Nagai, Y. Kawasaki, and N. Wakayama, Biochemistry, 1981, 20, 162.
17 M. Corti, V. Degiorgio, R. Ghidoni, S. Sonnino, and G. Tettamanti, Chem. Phys. Lipids, 1980, 26, 225.
18 M. Masserini, S. Sonnino, R. Ghidoni, and G. Tettamanti, Biochim. Biophys. Acta, 1980, 601, 282.
19 M. Corti, V. Degiorgio, S. Sonnino, R. Ghidoni, M. Masserini, and G. Tettamanti, Chem. Phys. Lipids, 1981, 28, 197.
20 H.J. Hinz, O. Korner, and C. Nicolau, Biochim. Biophys. Acta, 1981, 643, 557.
21 P.L. Felgner, E. Freire, Y. Barenholz, and T.E. Thompson, Biochemistry, 1981, 20, 2168.
22 S.P. Dufrane, A. Poss, E. Goormaghtigh, and J.M. Ruysschaert, Experientia, 1981, 37, 524.
23 S.K. Beckner, R.O. Brady, and P.H. Fishman, Proc. Natl. Acad. Sci. USA, 1981, 78, 4848.
24 J.L. Magnani, D.F. Smith, and V. Ginsburg, Anal. Biochem., 1980, 109, 399.
25 J.L. Tayot, J. Holmgren, L. Svennerholm, M. Lindblad, and M. Tardy, Eur. J. Biochem., 1981, 113, 249.
26 S. Hayes, Infect. Immun., 1979, 26, 150.
27 S.K. Kundu, D.M. Marcus, and R.W. Veh, J. Neurochem., 1980, 34, 184.
28 R. Maget-Dana, R.W. Veh, M. Sander, A.-C. Roche, R. Schauer, and M. Monsigny, Eur. J. Biochem., 1981, 114, 11.
29 G. Kayser, E. Goormaghtigh, M. Vandenbranden, and J.M. Ruysschaert, FEBS Lett., 1981, 127, 207.
30 G.D. Hunter, V.M. Wiegant, and A.J. Dunn, J. Neurochem., 1981, 37, 1025.
31 P. Hechtman and Z. Kachra, Biochem. J., 1980, 185, 583.
32 J.L. Mangnani, M. Brockhaus, D.F. Smith, V. Ginsburg, M. Blaszczyk, K.F. Mitchell, Z. Steplewski, and H. Koprowski, Science, 1981, 212, 55.
33 G.A. Ackerman, K.W. Wolken, and F.B. Gelder, J. Histochem. Cytochem., 1980, 28, 1100.
34 J.F.J. Banchereau, D.M. Danois, M. Guenounou, G.M.S. Durand, and J.C.L. Agnery, Biochim. Biophys. Acta, 1981, 678, 98.

35 B.A. Macher, J.C. Klock, M.N. Fukuda, and M. Fukuda, J. Biol. Chem., 1981, 256, 1968.
36 A. Tsuboyama, F. Takaku, S. Sakamoto, Y. Kano, T. Ariga, and T. Miyatake, Brit. J. Cancer, 1980, 42, 908.
37 A.J. Yates, S.L. Mattison, and R.L. Whisler, Biochem. Biophys. Res. Commun., 1980, 96, 211.
38 G.F. Springer and H. Tegtmeyer, Naturwissenschaften, 1981, 68, 152.
39 G.A. Ackerman, K.W. Wolken, and F.B. Gelder, J. Histochem. Cytochem., 1980, 28, 1334.
40 A. Reglero, J. Garcia-Alonso, and J.A. Cabezas, J. Neurochem., 1980, 34, 744.
41 S. Gasa and A. Makita, J. Biochem. (Tokyo), 1980, 88, 1119.
42 M. Delmelle, S.P. Dufrane, R. Brasseur, and J.M. Ruysschaert, FEBS Lett., 1980, 121, 11.
43 S. Spiegel and M. Wilchek, J. Immunol., 1981, 127, 572.
44 B.L. Slomiany, K. Kojima, Z. Banas-Gruszka, V.L.N. Murty, N.I. Galicki, and A. Slomiany, Eur. J. Biochem., 1981, 119, 647.
45 A. Slomiany, K. Kojima, Z. Banas-Gruszka, and B.L. Slomiany, Biochem. Biophys. Res. Commun., 1981, 100, 778.
46 B. Bloj and D.B. Zilversmit, J. Biol. Chem., 1981, 256, 5988.
47 E.V. Dyatlovitskaya, A.E. Zablotskaya, Yu. V. Volgin, Yu. M. Azizov, and L.D. Bergel'son, Biochemistry (Engl. Transl.), 1980, 45, 791.
48 S.K. Das and M.S. McCullough, Lipids, 1980, 15, 932.
49 P. Fredmam, J.-E. Mansson, L. Svennerholm, B. Samuelsson, I. Pascher, W. Pimlott, K.-A. Karlsson, and G.W. Klinghardt, Eur. J. Biochem., 1981, 116, 553.
50 M.L.E. Bergh, G.J.M. Hooghwinkel, and D.H. Van den Eijnden, Biochim. Biophys. Acta, 1981, 660, 161.
51 A.A. Bushway, B.G. Bouchard, G. Engdahl, T.W. Keenan, and R.J. Bushway, Comp. Biochem. Physiol., 1981, 68, 245.
52 J.-P. Zanetta, F. Vitiello, and G. Vincendon, Lipids, 1980, 15, 1055.
53 H.K.M. Yusuf and J.W.T. Dickerson, Indian J. Biochem. Biophys., 1980, 17, 32.
54 H.C. Yohe, K. Ueno, N.C. Chang, G.H. Glaser, and R.K. Yu, J. Neurochem., 1980, 34, 560.
55 T.Y. Wong, H. Dreyfus, S. Harth, J.-C. Louis, P. Mandel, and R. Massarelli, C.R. Hebd. Seances Acad. Sci., Ser. III, 1981, 293, 31.
56 A. Leon, L. Facci, G. Toffano, S. Sonnino, and G. Tettamanti, J. Neurochem., 1981, 37, 350.
57 C.A. Landa, S.S. Defilpo, H.J.F. Maccioni, and R. Caputto, J. Neurochem., 1981, 37, 813.
58 F. Norido, R. Canella, and F. Aporti, Experientia, 1981, 37, 301.
59 H.-J. Senn, C. Cooper, P.C. Warnke, M. Wagner, and K. Decker, Eur. J. Biochem., 1981, 120, 59.
60 T. Momoi, H. Wiegandt, R.Arndt, and H.G. Thiele, J. Immunol., 1980, 125, 2496.
61 G.A. Schwarting and A. Gajewski, J. Immunol., 1981, 126, 2403.
62 Y.A. Saat, R. Krishnaraj, and R.G. Kemp, Biochim. Biophys. Acta, 1981, 678, 213.
63 M. Saito, H. Nojiri, and M. Yamada, Biochem. Biophys. Res. Commun., 1980, 97, 452.
64 Y. Niimura and I. Ishizuka, Biochim. Biophys. Acta, 1981, 663, 266.
65 R.W. Ledeen, J.E. Haley, and J.A. Skrivanek, Anal. Biochem., 1981, 112, 135.
66 M. Iwamori and Y. Nagai, Biochim. Biophys. Acta, 1981, 665, 205.
67 M. Iwamori and Y. Nagai, Biochim. Biophys. Acta, 1981, 665, 214.
68 M. Iwamori and Y. Nagai, J. Biochem. (Tokyo), 1981, 89, 1253.
69 J.-J. Hauw, S. Fenelon, J.-M. Boutry, and R. Escourolle, C.R. Hebd. Seances Acad. Sci., Ser. III, 1981, 292, 569.
70 M. Ohashi, J. Biochem. (Tokyo), 1980, 88, 583.

71 R. Hilbig, H. Rosner, and H. Rahmann, Comp. Biochem. Physiol., 1981, 68, 301.
72 F.B. Cochran, R.K. Yu, S. Ando, and R.W. Ledeen, J. Neurochem., 1981, 36, 696.
73 H. Rösner, Anal. Biochem., 1980, 109, 437.
74 H. Rösner, J. Neurochem., 1981, 37, 993.
75 H. Rahmann and R. Hilbig, Naturwissenschaften, 1980, 67, 259.
76 N.V. Prokazova, A.T. Mikhatlov, S.L. Kocharov, L.A. Malchenko, N.D. Zvezdina, G. Buznikov, and L.D. Bergelson, Eur. J. Biochem., 1981, 115, 671.
77 H. Dreyfus, S. Harth, A. Gomez de Gracia, B. Hoflack, P. Mandel, and J.-C. Louis, Biochem. Soc. Trans., 1981, 9, 102.
78 G. Mugnai, R. Coppini, D. Tombaccini, A. Fallani, and S. Ruggieri, Biochem. J., 1981, 193, 1025.
79 K. Watanabe and Y. Arao, J. Lipid Res., 1981, 22, 1020.
80 T. Momoi and H. Wiegandt, Z. Physiol. Chem., 1980, 361, 1201.
81 M.J. Ruocco, D. Atkinson, D.M. Small, R.P. Skarjune, E. Oldfield, and G.G. Shipley, Biochemistry, 1981, 20, 5957.
82 J. Dabrowski, H. Egge, and P. Hanfland, Chem. Phys. Lipids, 1980, 26, 187.
83 J. Dabrowski, P. Hanfland, and H. Egge, Biochemistry, 1980, 19, 5652.
84 M.E. Breimer, G.C. Hansson, K.-A. Karlsson, H. Leffler, W. Pimlott, and B.E. Samuelsson, FEBS Lett., 1981, 124, 299.
85 M.E. Breimer, G.C. Hansson, K.-A. Karlsson, and H. Leffler, J. Biochem. (Tokyo), 1981, 90, 589.
86 A. Sen, D.A. Mannock, W.P. Williams, and P.J. Quinn, Biochem. Soc. Trans., 1981, 9, 134.
87 G.D. Fidelio, B. Maggio, F.A. Cumar, and R. Caputto, Biochem. J., 1981, 193, 643.
88 A.V. Hospattankar and N.S. Radin, Lipids, 1981, 16, 764.
89 T. Uchida and Y. Nagai, J. Biochem. (Tokyo), 1980, 87, 1829.
90 T.D. Heath, B.A. Macher, and D. Papahadjopoulos, Biochim. Biophys. Acta, 1981, 640, 66.
91 P. Ghosh, B.K. Bachhawat, and A. Surolia, Arch. Biochem. Biophys., 1981, 206, 454.
92 F. Szoka, K.E. Magnusson, J. Wojcieszyn, Y. Hou, Z. Derzko, and K. Jacobson, Proc. Natl. Acad. Sci. USA, 1981, 78, 1685.
93 B.C.P. Kwok, B.W. Shen, and G. Dawson, J. Biol. Chem., 1981, 256, 9698.
94 B.W. Shen, B.C.P. Kwok, and G. Dawson, J. Biol. Chem., 1981, 256, 9005.
95 G.H. DeVries, W.J. Zetusky, C. Zmachinski, and V.P. Calabrese, J. Lipid Res., 1981, 22, 208.
96 E. Hanada and K. Suzuki, Biochim. Biophys. Acta, 1980, 619, 396.
97 B. Zalc, F.B. Craves, M. Monge, L. Leybin, H. Loh, and N. Baumann, C. R. Seances Soc. Biol., 1980, 174, 457.
98 H. Dupoisot, A. Constans, G. Daury, F. Delbarre, and S. Laoussadi, C.R. Hebd. Seances Acad. Sci., Ser. D, 1979, 288, 783.
99 S. Laoussadi, H. Dupoisot, A. Constans, G. Daury, and F. Delbarre, C.R. Seances Soc. Biol., 1981, 175, 55.
100 I. Gery, J.S. Zigler, R.O. Brady, and J.A. Barranger, J. Clin. Invest., 1981, 68, 1182.
101 H.A. Nunez, F. Matsuura, and C.C. Sweeley, Arch. Biochem. Biophys., 1981, 212, 638.
102 P.F. Daniels, D.F. Defeudis, and I.T. Lott, Eur. J. Biochem., 1981, 114, 235.
103 T. Pacuszka, W. Jozwiak, H. Miller-Podraza, and J. Koscielak, Cancer Biochem. Biophys., 1980, 5, 1.
104 H. Hattori, K. Uemura, and T. Taketomi, Biochim. Biophys. Acta, 1981, 666, 361.
105 Y. Eto, M. Owada, T. Kitagawa, Y. Kokubun, and O.M. Rennert, J. Neurochem., 1979, 32, 397.

106 H. Christomanou, Z. Physiol. Chem., 1980, 361, 1489.
107 K.A. Karlsson and G. Larson, J. Biol. Chem., 1981, 256, 3512.
108 K.A. Karlsson and G. Larson, FEBS Lett., 1981, 128, 71.
109 G. Larson and B.E. Samuelsson, J. Biochem. (Tokyo), 1980, 88, 647.
110 J.T.R. Clarke, Can. J. Biochem., 1981, 59, 412.
111 K.E. Falk, K.A. Karlsson, and B.E. Samuelsson, FEBS Lett., 1981, 124, 173.
112 P. Hanfland and H.A. Graham, Arch. Biochem. Biophys., 1981, 210, 383.
113 H. Egge and P. Hanfland, Arch. Biochem. Biophys., 1981, 210, 396.
114 J. Dabrowski, P. Hanfland, H. Egge, and U. Dabrowski, Arch. Biochem. Biophys., 1981, 210, 405.
115 W.M.F. Lee, J.C. Klock, and B.A. Macher, Biochemistry, 1981, 20, 6505.
116 T. Hirano, H. Hashimoto, Y. Shiokawa, M. Iwamori, Y. Nagai, M. Kasai, Y. Ochiai, and K. Okumura, J. Clin. Invest., 1980, 66, 1437.
117 G.A. Schwarting, Biochem. J., 1980, 189, 407.
118 W.M.F. Lee, J.C. Klock, and B.A. Macher, Biochemistry, 1981, 20, 3810.
119 C. Derappe, A. Lundblad, L. Messeter, and S. Svensson, FEBS Lett., 1980, 119, 177.
120 S. Yahara and Y. Kishimoto, J. Neurochem., 1981, 36, 190.
121 F. Matsuura, R.A. Laine, and M.Z. Jones, Arch. Biochem. Biophys., 1981, 211, 485.
122 K. Sakakibara, T. Momoi, T. Uchida, and Y. Nagai, Nature (London), 1981, 293, 76.
123 A. Slomiany, N.I. Galicki, K. Kojima, Z. Banas-Gruszka, and B.L. Slomiany, Biochim. Biophys. Acta, 1981, 665, 88.
124 K. Sakakibara, M. Iwamori, T. Uchida, and Y. Nagai, Experientia, 1981, 37, 712.
125 S. Ansel and C. Huet, Int. J. Cancer, 1980, 25, 797.
126 T. Sasaki, Biochim. Biophys. Acta, 1981, 666, 418.
127 T. Sasaki, Biochim. Biophys. Acta, 1981, 666, 426.
128 O. Kuul, K.H. Chou, and F.B. Jungalwala, Biochem. J., 1980, 186, 959.
129 A.I. Barkai, J. Neurochem., 1981, 36, 317.
130 K.H. Chou and F.B. Jungalwala, J. Neurochem., 1981, 36, 394.
131 M.E. Breimer, G.C. Hansson, K.A. Karlsson, and H. Leffler, Experimental Cell Res., 1981, 135, 1.
132 D. Bouhours and J.-F. Bouhours, Biochem. Biophys. Res. Commun., 1981, 99, 1384.
133 K. Tadano and I. Ishizuka, Biochem. Biophys. Res. Commun., 1980, 97, 126.
134 F.S. Furbish, C.J. Steer, N.L. Krett, and J.A. Barranger, Biochim. Biophys. Acta, 1981, 673, 425.
135 H. Quill and G. Wolf, Biochim. Biophys. Acta, 1981, 674, 327.
136 B. Hoflack, R. Cacan, and A. Verbert, Eur. J. Biochem., 1981, 117, 285.
137 M.A. Shirley and H. Schachter, Can. J. Biochem., 1980, 58, 1230.
138 C. Lingwood, G. Hay, and H. Schachter, Can. J. Biochem., 1981, 59, 556.
139 V.L. Shepherd, Y.C. Lee, P.H. Schlesinger, and P.D. Stahl, Proc. Natl. Acad. Sci. USA, 1981, 78, 1019.
140 T. Taki, Y. Hirabayashi, K. Takagi, R. Kamada, K. Kojima, and M. Matsumoto, J. Biochem. (Tokyo), 1981, 89, 503.
141 R.R. Vunnam and N.S. Radin, Chem. Phys. Lipids, 1980, 26, 265.
142 D. Satomi and Y. Kishimoto, Biochim. Biophys. Acta, 1981, 666, 446.
143 T. Burkart, U.N. Wiesmann, H.P. Siegrist, and N.N. Herschkowitz, Biochim. Biophys. Acta, 1981, 673, 351.
144 M. Kasai, T. Yoneda, S. Habu, Y. Maruyama, K. Okumura, and T. Tokunaga, Nature (London), 181, 291, 334.
145 A. Suzuki and T. Yamakawa, J. Biochem. (Tokyo), 1981, 90, 1541.
146 P.G. Pentchev, A.E. Gal, R. Wong, S. Morrone, B. Neumeyer, J. Massey, R. Kanter, A. Sawitsky, and R.O. Brady, Biochim. Biophys. Acta, 1981, 665, 615.
147 M. Kasai, M. Iwamori, Y. Nagai, K. Okumura, and T. Tada, Eur. J. Immunol., 1980, 10, 175.

148 T.J. Higgins and C.R. Parish, Mol. Immunol., 1980, 17, 1065.
149 K.R. Gruner, R.V.W. Van Eijk, and P.F. Mühlradt, Biochemistry, 1981, 20, 4518.
150 E. Nudelman, S.-I. Hakomori, B.B. Knowles, D. Solter, R.C. Nowinski, M.R. Tam, and W.W. Young, jun., Biochem. Biophys. Res. Commun., 1980, 97, 443.
151 W.W. Young, J.M. Durdik, D. Urdal, S.I. Hakomori, and C.S. Henney, J. Immunol., 1981, 126, 1.
152 P.S. Wu, G.W. Tin, and J.D. Baldeschweiler, Proc. Natl. Acad. Sci. USA, 1981, 78, 2033.
153 M.E. Breimer, G.C. Hansson, K.A. Karlsson, H. Leffler, W. Pimlott, and B.E. Samuelsson, Biomed. Mass Spectrom., 1979, 6, 231.
154 J.P. Bradshaw and D.A. White, Biochem. Soc. Trans., 1981, 9, 66.
155 P. Hanfland, H. Egge, U. Dabrowski, S. Kuhn, D. Roelcke, and J. Dabrowski, Biochemistry, 1981, 20, 5310.
156 C. Wood and E.A. Kabat, J. Exp. Med., 1981, 154, 432.
157 C. Wood and E.A. Kabat, Arch. Biochem. Biophys., 1981, 212, 262.
158 C. Wood and E.A. Kabat, Arch. Biochem. Biophys., 1981, 212, 277.
159 B. Zalc, P. Dupouey, M.J. Coulon-Morelec, and N.A. Baumann, Mol. Immunol., 1979, 16, 297.
160 K.N. Singh and K.S. Rao, Indian J. Biochem. Biophys., 1979, 16, 349.
161 M. Sugita, T. Yamamoto, S. Masuda, O. Itasaka, and T. Hori, J. Biochem. (Tokyo), 1981, 90, 1529.
162 S.J. Turco, Arch. Biochem. Biophys., 1980, 205, 330.
163 G. Bardeletti and A. Voiland, Arch. Virol., 1981, 68, 285.
164 K.S. Akagawa, T. Momo, Y. Nagai, and T. Tokunaga, FEBS Lett., 1981, 130, 80.
165 P. Tancrède, G. Chauvette, and R.M. Leblanc, J. Chromatogr., 1981, 207, 387.
166 J.M. Coddington, S.R. Johns, D.R. Leslie, R.I. Willing, and D.G. Bishop, Biochim. Biophys. Acta, 1981, 663, 653.
167 B. Thewlis, J. Sci. Food Agric., 1981, 32, 125.
168 H. Chung, P.A. Seib, K.F. Finney, and C.D. Magoffin, Cereal Chem., 1981, 58, 164.
169 S. Ito and Y. Fujino, Agric. Biol. Chem., 1980, 44, 1181.
170 M. Ohnishi and Y. Fujino, Agric. Biol. Chem., 1981, 45, 1283.
171 A. Sen, W.P. Williams, and P.J. Quinn, Biochim. Biophys. Acta, 1981, 663, 380.
172 Y. Kondo, Biochim. Biophys. Acta, 1981, 665, 471.
173 N.W. Lem and J.P. Williams, Plant Physiol., 1981, 68, 944.
174 D.S. Bailey, V. Deluca, M. Durr, D.P.S. Verma, and G.A. MacLachlan, Plant Physiol., 1980, 66, 1113.
175 R.J. Staneloni, M.E. Tolmasky, C. Petriella, and L.F. Leloir, Plant Physiol., 1981, 68, 1175.
176 H.T. Khor and H.L. Tan, J. Sci. Food Agric., 1981, 31, 399.
177 I.S. Bhandal and C.P. Malik, Phytochemistry, 1981, 20, 429.
178 E. Heinz and D. Siefermann-Harms, FEBS Lett., 1981, 124, 105.
179 J.F.G.M. Wintermans, A. Van Besouw, and G. Bögemann, Biochim. Biophys. Acta, 1981, 663, 99.
180 A. Van Besouw, J.F.G.M. Wintermans, and G. Bögemann, Biochim. Biophys. Acta, 1981, 663, 108.
181 L. Lehle, FEBS Lett., 1981, 123, 63.
182 A. Koiwai, T. Matsuzaki, F. Suzuki, and N. Kawashima, Plant Cell Physiol., 1981, 22, 1059.
183 S. Chetal, D.S. Wagle, and H.S. Nainawatee, Plant Sci. Lett., 1981, 20, 225.
184 M. Ohnishi and Y. Fujino, Phytochemistry, 1981, 20, 1357.
185 M. Ohnishi, D. Park, and Y. Fujino, Agric. Biol. Chem., 1981, 45, 755.
186 B. Bris, C. Delbart, D. Coustaut, and R. Linder, Phytochemistry, 1981, 20, 1255.
187 P. Pillai and J.B. St. John, Plant Physiol., 1981, 68, 585.

188 Th. Müller, E. Bause, and L. Jaenicke, FEBS Lett., 1981, 128, 208.
189 Å. Wieslander, L. Rilfors, L.B.-A. Johansson, and G. Lindblom, Biochemistry, 1981, 20, 730.
190 K.-E. Johansson, C. Jagersten, A. Christiansson, and Å. Wieslander, Biochemistry, 1981, 20, 6073.
191 P.F. Smith, Biochim. Biophys. Acta, 1981, 665, 92.
192 D.R. Nelson and S.C. Rittenberg, J. Bacteriol., 1981, 147, 860.
193 K.E. Falk, K.-A. Karlsson, and B.E. Samuelsson, Chem. Phys. Lipids, 1980, 27, 9.
194 B.W. Smallbone and M. Kates, Biochim. Biophys. Acta, 1981, 65, 551.
195 P. Bienvenu, B. Jouanneteau, D. Robert, R. Guilluy, G. Normier, L. Dussourd d'Hinterland, and R. Fontanges, C.R. Hebd. Seances Acad. Sci., Ser. D, 1981, 292, 1113.
196 S.C. Kushwaha, M. Kates, G.D. Sprott, and I.C.P. Smith, Science, 1981, 211, 1163.
197 S.C. Kushwaha, M. Kates, G.D. Sprott, and I.C.P. Smith, Biochim. Biophys. Acta, 1981, 664, 156.
198 A. Liav and M.B. Goren, Chem. Phys. Lipids, 1980, 27, 345.
199 P.J. Brennan, H. Mayer, G.O. Aspinall, and J.E. Nam Shim, Eur. J. Biochemistry, 1981, 115, 7.
200 S.W. Hunter and P.J. Brennan, J. Bacteriol., 1981, 147, 728.
201 K. Egawa and N. Kasai, Microbiol. Immunol., 1979, 23, 87.
202 A. Hase, Arch. Biochem. Biophys., 1981, 210, 280.
203 M.-T. Pommier and G. Michel, Eur. J. Biochem., 1981, 118, 329.
204 T. Hori, M. Sugita, and H. Shimizu, Biochim. Biophys. Acta, 1981, 665, 170.

8
Chemical Synthesis and Modification of Oligosaccharides, Polysaccharides, Glycoproteins, Enzymes, and Glycolipids

BY C. M. STURGEON

1 Synthesis of Polysaccharides, Oligosaccharides, Glycoproteins, Glycopeptides, and Glycolipids

Polysaccharides. -- A review of the synthesis and polymerization of anhydro sugars has appeared.[1]

1,2-Anhydro-3,4,6-tri-O-benzyl-α-D-glucopyranose has been polymerized with a number of Lewis acids.[2] Phosphorus pentafluoride at -60°C caused polymerization to a product rich in β-linkages, while other Lewis acids at higher temperatures gave perbenzylated polysaccharides of lower molecular weight with less stereoselectivity. Debenzylation of the most regular derivative yielded a polysaccharide with properties similar to those of natural (1 → 2)-β-D-glucopyranans.

The influence of 4-bromo substitution on the polymerization of 1,6-anhydro-2,3,4-tri-O-benzyl-β-D-glucopyranose has been investigated and compared with effects in the polymerization and co-polymerization of other 2,3,4-tri-O-substituted 1,6-anhydro sugars.[3] Polymerization of 1,2-anhydro-3,4,6-tri-O-benzyl-β-D-mannopyranose proceeds in the presence of Lewis acids, cationic co-ordination catalysts, and strong bases.[4] Debenzylation of the products yields oligomeric saccharides or low polymers which have been physically characterized.

The regular heteropolysaccharide glucorhamnan {→ 6)-D-Glc*p*-(1 → 4)-L-Rha*p*-(1 →}$_n$ has been obtained by polycondensation of 3-O-acetyl-4-O-(2,3,4-tri-O-acetyl-6-O-trityl-β-D-glucopyranosyl)-1,2-O-(1-*exo*-cyanoethylidene)-β-L-rhamnopyranoside using a triphenylmethylium perchlorate catalyst.[5]

Stereospecific polycondensation of trityl ethers of trisaccharide 1,2-O-cyanoethylidene derivatives followed by deprotection has afforded regular heteropolysaccharides built up of repeating trisaccharide units, *i.e.* → 6)-β-D-Man*p*-(1 → 4)-α-L-Rha*p*-

(1 → 3)-β-D-Galp-(1 → and → 6)-α-D-Manp-(1 → 4)-α-L-Rhap-(1 → 3)-β-D-Galp-(1 →.[6]

Oligosaccharides. -- The use of partially benzylated oxazoline derivatives of 2-acetamido-2-deoxy-D-glucopyranoside as standardized intermediates for oligosaccharide synthesis has been reported.[7] Conditions were found for the successful coupling of O-acetyl-di-O-benzyl derivatives (1,2) of 2-methyl-(1,2-dideoxy-β-D-glucopyrano)-{2,1-d}-2-oxazoline to three benzyl tri-O-benzyl-1-thio-β-D-galactopyranosides (3-5) and to a partially protected glycoside (6) of 2-acetamido-2-deoxy-D-glucose. The products were fully substituted disaccharides of 2-acetamido-2-deoxy-β-D-glucose linked (1 → 3), (1 → 6), (1 → 4), and (1 → 4), respectively, and capable of chain extension from position 3, 3', 4, or 4' following removal of the single temporary O-acetyl protecting group from each.

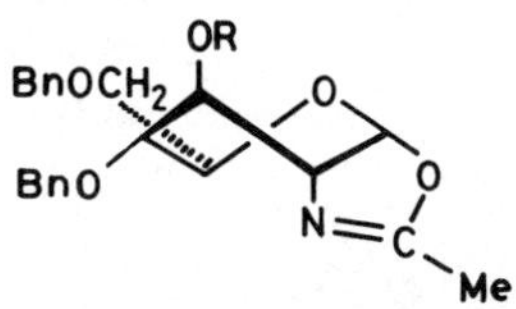

(1) R = Ac
(2) R = Bn

2-Methyl-{2-acetamido-4-O-acetyl-6-O-benzyl-3-O-(2-butenyl)-1,2-dideoxy-β-D-glucopyrano}-{2,1-d}-2-oxazoline, a glycosylating reagent in which the three hydroxyl groups are blocked with protecting groups of differing reactivity, has been synthesized in a ten-step sequence from 2-acetamido-2-deoxy-D-glucose.[8] The compound is a versatile intermediate for the synthesis of complex oligosaccharides of bacterial cell-wall, human-milk, and blood-group substances.

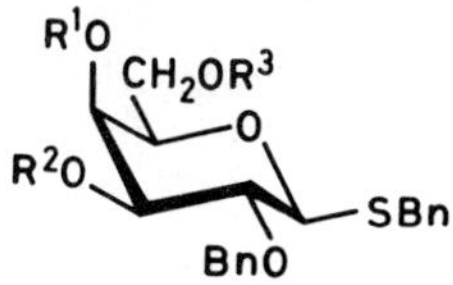

(3) $R^1 = R^3 = Bn$, $R^2 = H$
(4) $R^1 = H$, $R^2 = R^3 = Bn$
(5) $R^1 = R^2 = Bn$, $R^3 = H$

Allyl 6-O-benzyl-3,4-O-isopropylidene-α-D-galactopyranoside has been used to prepare a series of 2-O-benzoyl-3,6-di-O-benzyl-α-D-galactopyranosyl halides carrying either a second benzoyl group or a selectively removable temporary protecting group at position 4.[9] These compounds are designed to serve as precursors of β-linked interior D-galactopyranose units having chain extension at position 4 and branching if desired at position 2.

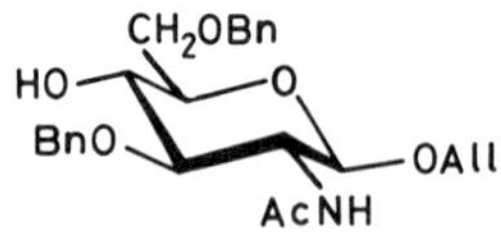

(6) All = allyl = $CH_2{=}CH{-}CH_2{-}$

3-O-Acetyl-2,4,6-tri-O-benzyl-α-D-mannopyranosyl bromide has been synthesized for use as a key intermediate in the preparation of branched-chain D-mannosyl oligosaccharides.[10]

Nitroxide spin-labelled α-D-glycopyranosides have been synthesized in good yield and in a highly stereoselective manner by reaction of per-O-glycopyranosyl bromides with 2,2,6,6-tetramethyl-4-piperidinol, followed by hydrogenation and oxidation.[11] 2-Amino-2-deoxy-D-glucose and -D-galactose derivatives, having a spin label at C-2, and the spin-labelled compound 1-{4-(β-D-galactopyranosyloxy)phenyl}-3-(2,2,6,6-tetramethylpiperidin-1-oxyl-4-yl)-2-thiourea were also synthesized.

Efficient formation of 1,2-*trans*-glycosidic linkages has been described using trimethylsilyl trifluoromethanesulphonate as catalyst.[12] The catalyst was successfully employed in reactions involving orthoester, oxazoline, or 1,2-*trans*-diacetate starting materials.

Crystalline 4-methylumbelliferyl 3,4,6-tri-O-acetyl-2-deoxy-2-oximino-α-D-*arabino*-hexopyranoside has been prepared.[13] Acetylation of this compound followed by reduction of the resulting crude O-acetyloxime with borane in oxolane, and acetylation, yields the 3,4,6-tri-O-acetyl derivative of 4-methylumbelliferyl 2-acetamido-2-deoxy-α-D-glucopyranoside, the substrate in a new sensitive fluorogenic assay of 2-acetamido-2-deoxy-α-D-glucosidase.

The syntheses of methyl 2-O-β-D-galactopyranosyl-β-D-galactopyranoside and methyl 2-O-(2-O-β-D-galactopyranosyl)-β-D-galactopyranoside have been reported using 2-O-benzoyl-3,4,6-tri-O-benzyl-

1-O-(4-tolylsulphonyl)-D-galactopyranoside as starting material.[14] An efficient synthesis of β-D-mannopyranosides has been described.[15] By acetalization under kinetic control D-mannose is converted into crystalline 2,3:4,6-di-O-cyclohexylidene-α-D-mannopyranose, which reacts in the presence of triethylamine with methanesulphonyl chloride to yield 2,3:4,6-di-O-cyclohexylidene-α-D-mannopyranosyl chloride for use in Koenigs-Knorr syntheses. 3-O-(3,6-Dideoxy-α-D-*xylo*-hexopyranosyl)-α-D-mannopyranosides of 8-methoxycarbonyloctan-1-ol and of 4-trifluoroacetamidophenol have been synthesized by reaction of 2,4-di-O-benzyl-3,6-dideoxy-α-D-*xylo*-hexopyranosyl bromide with suitably protected mannosides.[16]

The syntheses have been reported of isomers of the terminal disaccharides of the saccharide cytostatics olivomycin A and mithramycin.[17] As an approach to the synthesis of the lipid A component of the bacterial lipopolysaccharide, a facile preparation of β-D-(1→6)-linked disaccharides of N-fatty acylated 2-amino-2-deoxy-D-glucose has been reported using the oxazoline method.[18]

A convenient synthesis of 2-acetamido-2-deoxy-4-O-β-D-galactopyranosyl-D-glucose (N-acetyl-lactosamine) has been reported by addition of hydrogen cyanide to 1-N-benzyl-3-O-β-D-galactopyranosyl-D-arabinose.[19] The method allows preparation of 10-20 gram amounts of the disaccharide without any chromatographic purification.

The potential enzyme inducer 4-S-β-D-galactopyranosyl-4-thio-D-glucopyranose (thiolactose) has been synthesized *via* two routes.[20]

The T antigenic determinant 2-acetamido-2-deoxy-3-O-β-D-galactopyranosyl-α-D-galactopyranoside has been synthesized with an 8-methoxycarbonyl-octyl side chain[21] and with a water-soluble amide side chain.[22] Both side chains are suitable for further derivatization.

The chemical synthesis of 2-O-(4-O-methyl-α-D-glucopyranosyluronic)-D-xylose, a constituent of gymnosperm xylans, has been reported.[23]

The synthesis of disaccharides containing 2-deoxy-2-iodo-α-D-talopyranosyl groups has been reported using a novel glycosylating system previously described for the synthesis of disaccharides containing 2-deoxy-2-iodo-α-D-mannopyranosyl groups.[24] 3,4,6-Tri-O-acetyl-D-galactal was treated with silver imidazolate, mercury(II) chloride, and iodine in acetonitrile to yield a ~5:1 mixture of 3,4,6-tri-O-acetyl-2-deoxy-2-iodo-α-D-talo- and -D-galactopyranosyl chlorides, which was then treated with suitably protected alcohols to provide the required disaccharide.

The reaction of octa-O-acetylcellobiononitrile (7) with sodium methoxide, and acetylation of the resulting mixture, has afforded the anomeric peracetates of O-β-D-glucopyranosyl-(1→3)-D-arabinopyranose and -furanose.[25]

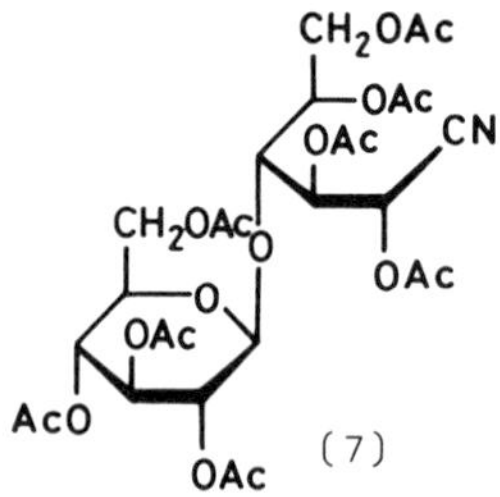

(7)

Methyl 2,3,6-trideoxy-4-O-(3,4-di-O-acetyl-2,6-dideoxy-α-L-lyxo-hexopyranosyl)-3-(trifluoroacetamido)-β-L-lyxo-hexopyranoside, a disaccharide having daunosamine and 2-deoxy-L-fucose in the sequence found naturally in rhodomycins, has been synthesized.[26] 4-Trifluoroacetamidophenyl 3-O-(α-D-glucopyranosyl)-α-D-mannopyranoside, of which the disaccharide moiety has been suggested to be the immunodominant part of O-antigen 14 in Salmonella bacteria (serogroup C_1), has been synthesized.[27]

4-Nitrophenyl 6-O-(2-acetamido-2-deoxy-β-D-glucopyranosyl)-α-D-mannopyranoside has been synthesized from 4-nitrophenyl 2,3-O-isopropylidene-α-D-mannopyranoside and 2-methyl-(3,4,6-tri-O-acetyl-1,2-dideoxy-α-D-glucopyrano)-{2,1-d-}-2-oxazoline via a crystalline 6-O-substituted disaccharide derivative.[28] Synthesis of methyl 6-O-(2-acetamido-2-deoxy-β-D-glucopyranosyl)-α-D-mannopyranoside was also accomplished by a similar reaction sequence. Acetal migration during mercuric cyanide-catalysed glycosylation has been shown to be responsible for the synthesis of 4-nitrophenyl 6-O-β-D-galactopyranosyl-β-D-galactopyranoside via 4-nitrophenyl 3-O-benzoyl-4,6-O-isopropylidene-β-D-galactopyranoside, instead of the expected product, the (1→3)-linked isomer.[29] A rapid synthesis has been described for 4-nitrophenyl 2-O-α-L-fucopyranosyl-β-D-galactopyranoside, a substrate for human (1→2)-α-L-fucosyltransferase.[30]

Phenyl 2-acetamido-2-deoxy-4,6-O-(4-methoxybenzylidene)-β-D-glucopyranoside has been reacted with 2,3,4-tri-O-benzyl-α-L-fucopyranosyl bromide under halide ion-catalysed conditions to yield phenyl 2-acetamido-2-deoxy-4,6-O-(4-methoxybenzylidene)-3-O-(2,3,4-tri-O-benzyl-α-L-fucopyranosyl)-β-D-glucopyranoside (8).[31] Mild treatment of (8) with acid, followed by hydrogenolysis, yielded the disaccharide phenyl 2-acetamido-2-deoxy-3-O-α-L-fucopyranosyl-β-D-glucopyranoside. Using a similar reaction sequence the synthesis of 6-(trifluoroacetamido)hexyl 2-acetamido-2-deoxy-3-O-α-L-fucopyranosyl-β-D-glucopyranoside was also reported.

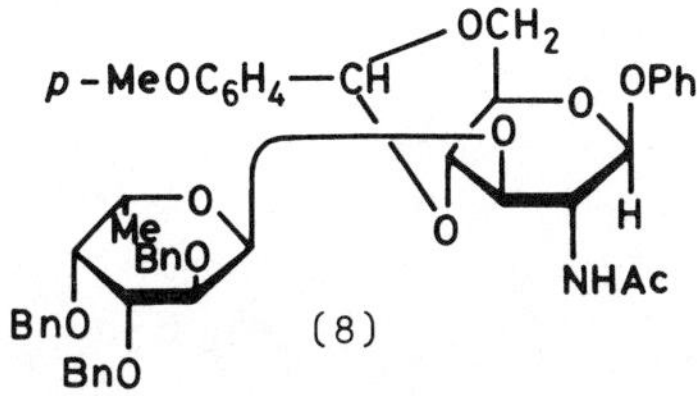

(8)

1,2,3-Tri-O-acetyl-4-O-benzyl-β-D-xylopyranose has been used for the stepwise synthesis of (1→4)-β-D-xylo-oligosaccharides.[32] The syntheses of α- and β-cellotriose hendecaacetates and of several 6,6',6'-tri-substituted derivatives of methyl β-cellotrioside have been described.[33] O-α-D-Glucopyranosyl-(1→2)-O-α-D-glucopyranosyl-(1→2)-D-glucopyranose (kojitriose) has been synthesized by condensation of 1,3,4,6-tetra-O-acetyl-α-D-glucopyranose with hepta-O-acetyl-α-kojibiosyl bromide under halide ion-catalysed conditions, followed by fractional recrystallization and deprotection.[34]

The synthesis of an aldotriuronic acid derivative related to (4-O-methylglucurono)-D-xylans has been reported.[35] Acetylation of methyl 4-O-(2-O-benzyl-β-D-xylopyranosyl)-β-D-xylopyranoside followed by catalytic debenzylation of the product yielded methyl 2,3-di-O-acetyl-4-O-(3,4-di-O-acetyl-β-D-xylopyranosyl)-β-D-xylopyranoside. Reaction of this nucleophile with methyl 2,3-di-O-benzyl-1-chloro-1-deoxy-4-O-methyl-α,β-D-glucopyranuronate in the presence of silver perchlorate and sym-collidine afforded the 4-O-methyl-α-D-glucuronic acid-containing trisaccharide derivative (9) as the major product, from which the desired methyl ester methyl glycoside (10) was readily obtained.

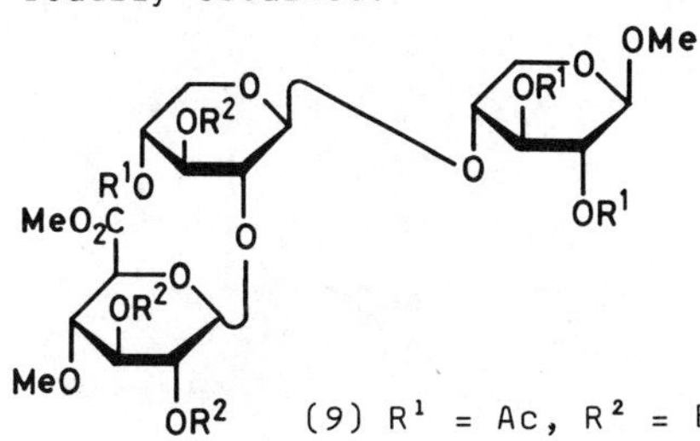

(9) R^1 = Ac, R^2 = Bn
(10) R^1 = R^2 = H

A key intermediate for solid-phase synthesis of β-(1→6)-linked D-galactopyranans, 2,3,4-tri-O-acetyl-6-O-(chloroacetyl)-α-D-galactopyranosyl bromide (11), has been prepared and condensed with 1,2,3,4-tetra-O-acetyl-D-galactopyranose to yield 1,2,3,4-tetra-O-acetyl-6-O-2,3,4-tri-O-acetyl-6-O-(chloroacetyl)-β-D-galactosyl-D-galactopyranose.[36] After selective removal of the chloroacetyl group from the adduct by thiourea treatment, the product was condensed again with (11) to yield, after deprotection, 6-O-(6-O-β-D-galactopyranosyl-β-D-galactopyranosyl)-D-galactopyranose.

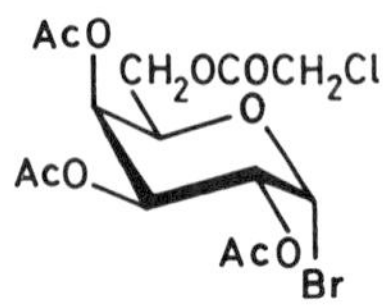

(11)

2,3,4,6-Tetra-O-acetyl-α-D-mannopyranosyl bromide has been reacted with methyl 3-O-benzyl-4,6-O-benzylidene-α-D-mannopyranoside in the presence of mercuric cyanide to yield methyl 3-O-benzyl-4,6-O-benzylidene-2-O-(2,3,4,6-tetra-O-acetyl-α-D-mannopyranosyl)-α-D-mannopyranoside.[37] Condensation of the debenzylidenated disaccharide with 2,3,4,6-tetra-O-acetyl-α-D-mannopyranosyl bromide afforded the corresponding trisaccharide, which is reported to be the repeating unit of an alkali-soluble polysaccharide (S-Iawe) isolated from the mycelia of Epidermophyton floccosum.

O-α-L-Rhamnopyranosyl-(1→3)-O-α-L-rhamnopyranosyl-(1→2)- and (1→3)-O-α-L-rhamnopyranose, two constituents of bacterial cell-wall polysaccharides, have been synthesized.[38] Utilizing 1,2:3,4-di-O-isopropylidene α-D-galactopyranose as starting material, convenient syntheses have been reported for O-α-L-rhamnopyranosyl-(1→3) and O-α-L-rhamnopyranosyl-(1→2)-O-α-L-rhamnopyranosyl-(1→6)-D-galactose, which are components of flavonol triosides.[39]

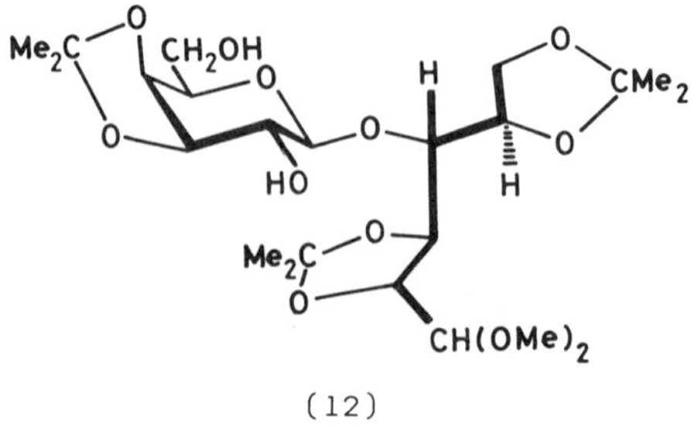

(12)

The synthesis of 2'-*O*-α-*L*-fucopyranosyl-lactose has been reported, using 2,3:5,6:3',4'-tri-*O*-isopropylidene-lactose dimethyl acetal (12) as starting material (Scheme 1).[40] A synthesis of 2'-*O*-methyl-lactose was also described.

The mercury salt-catalysed reaction of substituted α-*D*-galactosyl halides with the reactive 4-hydroxy groups of several rhamnosides has been studied and the effect of substituents on halide reactivity determined.[41] Various trisaccharides including part of the repeating unit (13) of the lipopolysaccharide from *Escherichia coli* O75 were synthesized from 2-acetamido-2-deoxy-*D*-glucose, *D*-galactose, and *L*-rhamnose. The full synthesis of the *E. coli* tetrasaccharide (13) was reported separately.[42] A new method has been developed for the stereoselective synthesis of oligosaccharides containing a β-linked *D*-mannose unit and previously not available by direct synthesis.[43] A silver silicate catalyst precipitated on aluminium oxide is used to promote the glycoside synthesis. The syntheses of several *D*-mannose-containing oligosaccharides are reported using this method.

β-*D*-Man*p*-(1→4)-α-*D*-Gal*p*-(1→4)-*L*-Rha
1
|
3
β-*D*-Glc*p*NAc
(13)

The syntheses of the terminal disaccharide and trisaccharide antigenic determinants of Groups A and C *Streptococci* and of the tetrasaccharide sequence thought to constitute the variant-A polysaccharide have been reported.[44] The synthesis has been reported of the branched pentasaccharide repeating unit of the O-specific side chain belonging to the lipopolysaccharide obtained from *Shigella dysenteriae* of serotype 2.[45,46] A trisaccharide halide block, obtained from three 2-azido-sugars, was combined with a disaccharide block to give the pentasaccharide. Each combination step of the synthesis gave α-glycosides stereoselectively.

The trisaccharide sequence 2-acetamido-2-deoxy-α-*D*-glucosyl-(1→4)-β-*D*-galactosyl-(1→4)-2-acetamido-2-deoxy-*D*-glucopyranose of blood-group-active substances has been obtained using a directly stereoselective α-glycoside synthetic approach.[47] A postulated trisaccharide of human erythrocyte membrane sialoglycoproteins, *O*-β-

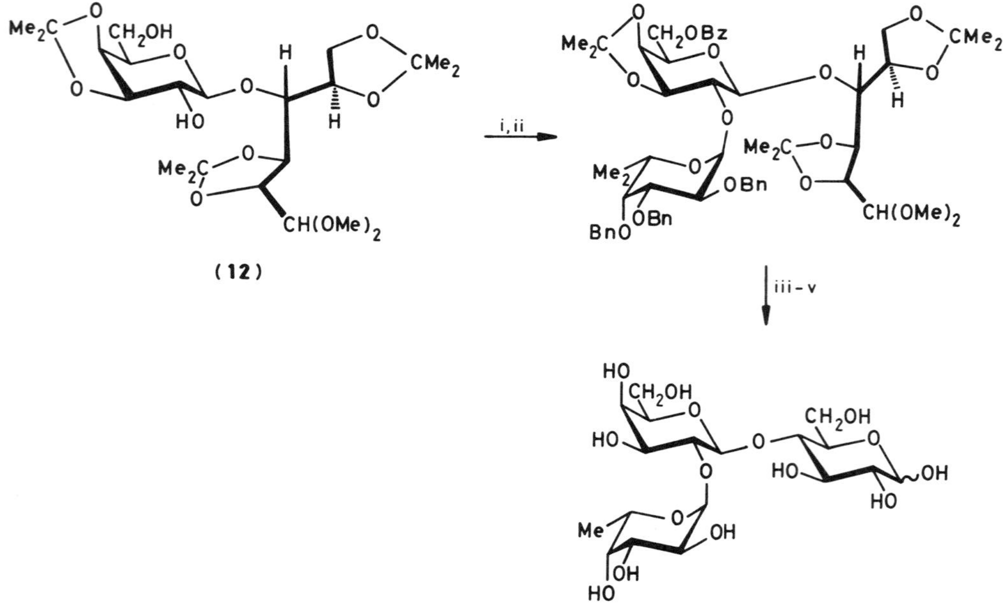

(12)

Reagents: i, benzoyl chloride-triethylamine-CH_2Cl_2; ii, Br^--2,3,4-tri-O-benzyl-α-L-fucopyranosyl bromide; iii, Na^+OMe^--MeOH; iv, AcOH (aq.), 80°C; v, H/Pd-AcOH (glac.)

Scheme 1

D-galactopyranosyl-(1→4)-O-2-acetamido-2-deoxy-β-D-glucopyranosyl-(1→3)-D-mannopyranose, has been synthesized.[48] Benzyl 6-O-allyl-2,4-di-O-benzyl-α-D-mannopyranoside was prepared and then condensed with an oxazoline derived from D-lactosamine to yield a protected form of the required trisaccharide.

Reaction of 2,3,4,6-tetra-O-acetyl-α-D-galactopyranosyl bromide with benzyl 2-acetamido-3,6-di-O-benzyl-2-deoxy-α-D-glucopyranoside in 1,2-dichloroethane has provided after deacetylation crystalline benzyl 2-acetamido-3,6-di-O-benzyl-2-deoxy-4-O-(β-D-galactopyranosyl)-α-D-glucopyranoside.[49] Acetonation followed by benzylation and mild acid hydrolysis yields crystalline benzyl 2-acetamido-4-O-(2,6-di-O-benzyl-β-D-galactopyranosyl)-3,6-di-O-benzyl-2-deoxy-α-D-glucopyranoside. Reaction of this diol with 1-O-(N-methyl)acetimidyl-2,3,4,6-tetra-O-benzyl-β-D-galactopyranose regiospecifically yielded a derivative from which the blood-group-active trisaccharide O-α-D-galactopyranosyl-(1→3)-O-β-D-galactopyranosyl-(1→4)-2-acetamido-2-deoxy-D-glucopyranose was obtained. The synthesis of the trisaccharide O-β-D-galactopyranosyl-(1→4)-O-(2-acetamido-2-deoxy-β-D-glucopyranosyl)-(1→6)-D-mannose, a blood-group Ii-active oligosaccharide, has been reported.[50] Benzyl 2,3,4-tri-O-benzyl-α-D-mannopyranoside was condensed with a disaccharide oxazoline, 2-methyl-{3,6-di-O-acetyl-1,2-dideoxy-4-O-(2,3,4,6-tetra-O-acetyl-β-D-galactopyranosyl)-α-D-glucopyrano}-{2,1-d}-2-oxazoline, to yield the required product.

The tetrasaccharide O-β-D-galactopyranosyl-(1→4)-2-acetamido-2-deoxy-β-D-glucopyranosyl-(1→3)-β-D-galactopyranosyl-(1→4)-2-acetamido-2-deoxy-D-glucopyranose has been synthesized.[51] This structure has been found in glycolipids isolated from erythrocyte membranes and is recognized by several anti-I antisera. As part of the synthetic route to the Lewis b blood-group antigenic determinant, the rapid synthesis of benzyl 2-acetamido-2-deoxy-3-O-β-D-galactopyranosyl-β-D-glucopyranoside has been reported.[52] This compound was derivatized to yield (14), which following condensation with 2,3,4-tri-O-benzyl-α-L-fucopyranosyl bromide and deprotection afforded the required tetrasaccharide, O-α-L-fucopyranosyl-(1→2)-β-D-galactopyranosyl-(1→3)-O-{α-L-fucopyranosyl-(1→4)}-2-acetamido-2-deoxy-D-glucose.

Treatment of benzyl 2-acetamido-3,6-di-O-benzyl-2-deoxy-4-O-β-D-galactopyranosyl-α-D-glucopyranoside with benzaldehyde in the presence of zinc chloride followed by regioselective benzoylation has provided crystalline benzyl 2-acetamido-4-O-(3-O-benzoyl-4,6-O-benzylidene-β-D-galactopyranosyl)-3,6-di-O-benzyl-2-deoxy-α-D-glucopyranoside.[53] Condensation of this disaccharide with 2,3,4-tri-O-benzyl-1-O-(N-methylacetimidyl)-β-L-fucopyranose yielded benzyl O-(2,3,4-tri-O-benzyl-α-L-fucopyranosyl)-(1→2)-O-(3-O-benzyl-4,6-O-benzylidene-β-D-galactopyranosyl)-(1→4)-2-acetamido-3,6-di-O-benzyl-2-deoxy-α-D-glucopyranoside. Debenzoylation, followed by condensation with 2,3,4,6-tetra-O-benzyl-1-O-(N-methylacetimidyl)-β-D-galactopyranose, provided a tetrasaccharide derivative that was catalytically hydrogenolysed to (15), the antigenic determinant of human blood group B-type 2.

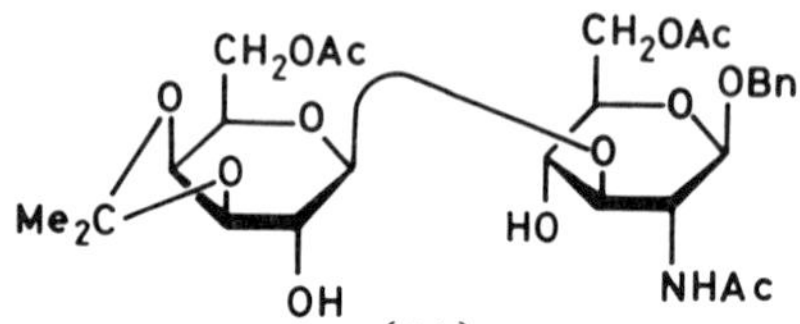

(14)

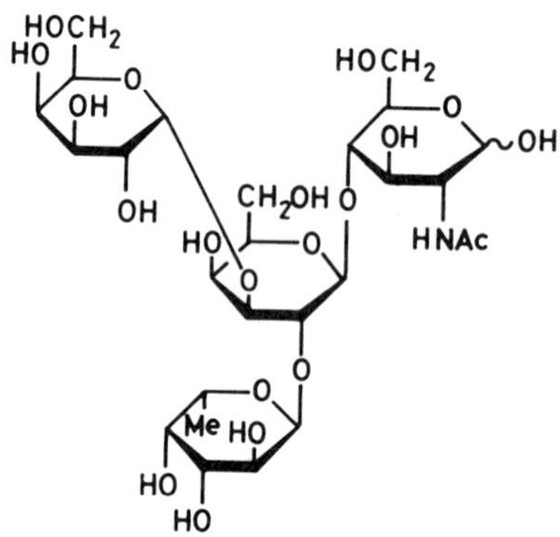

(15)

The antigenic determinants of type 2 of blood-group substances A and B as well as blood-group substance H have been synthesized by selective glycosidation reactions combined with a series of blocking and deblocking steps.[54] The pentasaccharide chain of the Forssman antigen has been synthesized using a novel block synthesis method involving catalysed reaction of suitably protected disaccharide and trisaccharide derivatives.[55]

Regio- and stereo-controlled synthetic routes to the branching pentasaccharides (16) and (17) have been reported.[56] Similar synthetic routes to the branched mannopentaoside (18) and mannohexa-

α-D-Manp-(1→2)-α-D-Manp-(1→6)-α-D-Manp-(1→OMe
3
↑
1
α-D-Manp
2
↑
1
α-D-Manp

(16)

β-D-GlcpNAc-(1→2)-α-D-Manp-(1→6)-α-D-Manp-(1→OMe
3
↑
1
α-D-Manp
2
↑
1
β-D-GlcpNAc

(17)

α-D-Manp-(1→6)-α-D-Manp-(1→6)-α-D-Manp-(1→OMe
3 3
↑ ↑
1 1
α-D-Manp α-D-Manp
2
↑
1
R

(18) R = H

(19) R = α-D-Manp

oside (19) have also been described.[57] Methyl 6-O-(2,6-di-O-α-D-mannopyranosyl-α-D-mannopyranosyl)-3-O-α-D-mannopyranosyl-α-D-mannopyranoside and methyl 6-O-(2,6-di-O-α-D-mannopyranosyl)-α-D-mannopyranoside have been synthesized.[58] 3-O-(3,6-Di-O-α-D-mannopyranosyl-α-D-mannopyranosyl)-6-O-α-D-mannopyranosyl-α-D-mannopyranoside and methyl 3-O-(3,6-O-α-D-mannopyranosyl-α-D-mannopyranosyl)-6-O-(2-O-α-D-mannopyranosyl-α-D-mannopyranosyl)-α-D-mannopyranoside, which provide models of the outer chain of the glycan of soybean agglutinin, have been prepared.[59] To provide a model of an inner chain of cell-wall proteoglycans isolated from Piricularia oryzae, efficient syntheses have been described for methyl 2,6-di-O-(2-O-α-D-mannopyranosyl-α-D-mannopyranosyl)-α-D-mannopyranoside and methyl 2,4-di-O-(2-O-α-D-mannopyranosyl-α-D-mannopyranosyl)-α-D-mannopyranoside.[60] The glycosyl acceptors methyl 3,4-di-O-benzyl-α-D-mannopyranoside and methyl 3,6-di-O-benzyl-α-D-mannopyranoside were reacted with suitably protected D-mannotriosides.

Silver trifluoromethanesulphonate-promoted condensation of 3,6-di-O-acetyl-4-O-(2,3,4,6-tetra-O-acetyl-β-D-galactopyranosyl)-2-deoxy-2-phthalimido-β-D-glucopyranosyl bromide with benzyl 3-O-benzyl-4,6-O-benzylidene-α-D-mannopyranoside, benzyl 3,6-di-O-benzyl-α-D-mannopyranoside, and benzyl 3,6-di-O-(3,4,6-tri-O-benzyl-α-D-mannopyranoside has yielded protected tri-, penta-, and heptasaccharide derivatives, all containing 2-acetamido-2-deoxy-β-D-lactosaminyl residues.[61] The free oligosaccharides, parts of the complex-type carbohydrate moiety of glycoproteins, were obtained after exchanging the 2-deoxy-2-phthalimido groups for 2-acetamido-2-deoxy groups and deblocking.

Regio- and stereo-controlled synthesis of the hexasaccharide (20) has been achieved by employing a properly protected mannotrioside as a key intermediate.[62] The synthesis of O-β-D-galactopyranosyl-(1→4)-O-(2-acetamido-2-deoxy-β-D-glucopyranosyl)-(1→6)-D-mannopyranose, part of the most highly branched complex type of carbohydrate portions of glycoproteins, has been reported.[63]

Glycoproteins. -- The acyl hydrazide derivative of 8-methoxycarbonyl octyl 2-acetamido-2-deoxy-3-O-(β-D-galactopyranosyl)-α-D-galactopyranoside has been reacted with bovine serum albumin to yield an effective artificial T-antigen containing 22 equivalents of hapten per mole of protein.[21] 4-Trifluoroacetamidophenyl 3-O-(α-D-glucopyranosyl)-α-D-mannopyranoside has been converted to its isothiocyanate derivative and coupled to bovine serum albumin to afford an artificial Salmonella antigen.[27]

Pseudoglycoproteins have been synthesized by treating proteins with carbohydrates (see e.g. Tables 1-5), chiefly for use in affinity chromatography. The preparation, properties, and applications of carbohydrate conjugates of proteins have been reviewed.[64]

```
β-D-GlcpNAc-(1→4)-α-D-Manp-(1→6)-α-D-Manp-(1→OMe
                      2                  3
                      ↑                  ↑
                      1                  1
               β-D-GlcpNAc          α-D-Manp
                                         2
                                         ↑
                                         1
                                    β-D-GlcpNAc
```

(20)

Glycopeptides. -- N-(L-β-Aspartyl)-α-D-glucopyranosylamine has been prepared, as a model of a corresponding derivative possibly present in glomerular basement membrane of rats.[65] α-Ethyl benzyloxycarbonyl-L-aspartate was condensed with 2,3,4,6-tetra-O-acetyl-α-D-glucopyranosylamine in the presence of diethylphosphocyanidate. This was followed by hydrogenolysis, acetylation, and de-ethoxylation of the resulting 2,3,4,6-tetra-O-acetyl-N-(α-ethyl benzyloxycarbonyl-L-β-aspartyl)-α-D-glucopyranosylamine to remove the protecting groups.

Other components of the glomerular basement membrane, O-(2-O-α-D-glucopyranosyl)-β-D-galactopyranosides of optically pure δ-hydroxy-L-lysylglycine and δ-hydroxy-L-lysylglycyl-L-glutamyl-L-aspartylglycine, have also been synthesized.[66] The use of the levulinyl group as a temporary blocking group for the 2-hydroxy function of D-galactose proved to be very efficient for the formation of a β-linkage between D-galactose and δ-hydroxy-L-lysylglycine, and also for the linkage of D-galactose to D-glucose *via* an α-interglycosidic bond. Synthesis in solution and in solid phase has been used to prepare N-acetylmuramyl derivatives of a number of oligopeptides.[67] Immunoadjuvant activities were determined for these derivatives.

In order to clarify the structural requirements for the immunoadjuvant activity of the carbohydrate moiety in N-acetylmuramoyl-L-alanyl-D-isoglutamine, 2-acetamido-2-deoxy-4- and -6-O-(D-2-propanoyl-L-alanyl-D-isoglutamine)-D-glucopyranose, 2-acetamido-2-deoxy-3-O-(D-2-propanoyl-L-alanyl-D-isoglutamine)-D-allopyranose, -D-gulopyranose, -D-galactopyranose, -D-mannopyranose, and -L-idopyranose, and 3-O-(D-2-propanoyl-L-alanyl-D-isoglutamine)-D- and -L-glucopyranose have been synthesized.[68] Immunoadjuvant activity of the N-acetylmuramoyl dipeptide analogues was examined in guinea pigs. The syntheses of novel N-acylmuramoyl dipeptides related to the lipid A constituent of the bacterial lipopolysaccharide have been reported.[69] All the synthetic muramoyl dipeptide analogues exhibited potent activity in inducing delayed-type hypersensitivity in guinea pigs.

The syntheses of three 4-deoxy analogues, a 4-chloro-4-deoxy- and a 4,6-dichloro-4,6-dideoxy-D-galacto analogue of N-acetylmuramoyl-L-alanyl-D-isoglutamine have been reported and their immunoadjuvant activities investigated.[70] Using solid-phase synthesis methods, 2,3,4,6-tetra-O-acetyl-1-N-{N-(tert-butyloxycarbonyl)-L-aspart-4-oyl}-D-glucopyranosylamine and 2-acetamido-3,4,6-tri-O-acetyl-1-N-{N-(tert-butyloxycarbonyl)-L-aspart-4-oyl}-2-deoxy-D-glucopyranosylamine have been introduced into the amino acid sequence 5 to 9 of somatostatin.[71] 3-O-(2-Acetamido-3,4,6-tri-O-acetyl-2-deoxy-β-D-glucopyranosyl)-N-(tert-butyloxycarbonyl)-L-serine has been prepared and condensed by the solid-phase procedure to give the model glycotripeptide (21) and a glycosomatostatin (22).[72]

Gly-{β-D-GlcpNAc-(1→3)-Ser}-Ala-OH

(21)

β-D-GlcpNAc-(1→3)-Ser-13-somatostatin

(22)

Glycolipids. -- Morprenyl phosphate has been treated with N,N'-sulphinyldi-imidazole and the resulting imidazolidate reacted with α-D-galactopyranosyl or α-D-glucopyranosyl phosphate to give the P^1-moraprenyl, P^2-glycosyl pyrophosphates (23) or (24).[73] The method was used to prepare the biologically active oligosaccharide polyprenyl derivatives (25) and (26), which are intermediates in the biosynthesis of O-specific polysaccharides of Salmonella serological group E.

Two disaccharide glycoside precursors of the palmitoyl analogues of lipid A and mono-dephospho lipid A have been synthesized.[74] Although palmitic acid was the sole fatty acid used, the synthetic scheme provides for the selective introduction of any desired saturated fatty acid at each of the five acylated positions in the final product. A convenient synthesis of 2-deoxy-2-D- and -L-(3-hydroxytetradecanoyl-amino)-D-glucose diastereoisomers of the monomeric lipid A component of the bacterial lipopolysaccharide has been reported, starting from diastereoisomeric benzyl 3,4,6-tri-O-acetyl-2-deoxy-2-(DL-3-hydroxytetradecanoylamino)-β-D-glucopyranoside.[75]

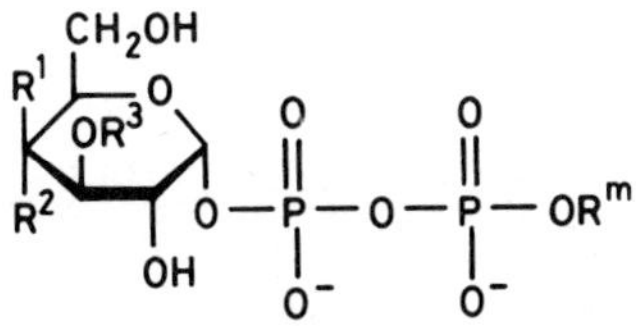

(23) R^1 = OH, R^2 = R^3 = H

(24) R^1 = R^3 = H, R^2 = OH

(25) R^1 = OH, R^2 = H, R^3 = α-L-Rhap-(1→

(26) R^1 = OH, R^2 = H , R^3 = β-D-Manp-(1→4)-α-L-Rhap-(1→

Cholesterol-containing β-D-galactosyl glycolipids having a new hydrophilic spacer group, 8-amino-3,6-dioxaoctanoic acid, have been synthesized.[76] The hydrophilic nature of this spacer group eliminates many of the problems inherent in the use of hydrophobic or charged spacer arms.

Condensation of N,N'-diacetylchitobiono-1,5-lactone, produced by oxidation of N,N'-diacetylchitobiose, with 1-aminoalkanes yields N,N'-diacetylchitobionic N-alkylamides.[77] The interaction of these glycolipids with wheat-germ agglutinin and Griffonia simplicifolia II lectin was described.

A convenient approach to the synthesis of triglycosyl-2,3-di-O-phytanyl-sn-glycerols has been reported (Scheme 2).[78]

2 Modification of Polysaccharides and Oligosaccharides and Uses of Modified Polysaccharides and Oligosaccharides

Introduction. -- Polysaccharide derivatives continue to be widely used in the preparation of matrices for affinity chromatography. A book has been published which presents a theoretical treatment of this technique together with detailed information concerning the preparation and use of affinity-chromatographic media.[79] The proceedings of the 4th International Symposium on Affinity

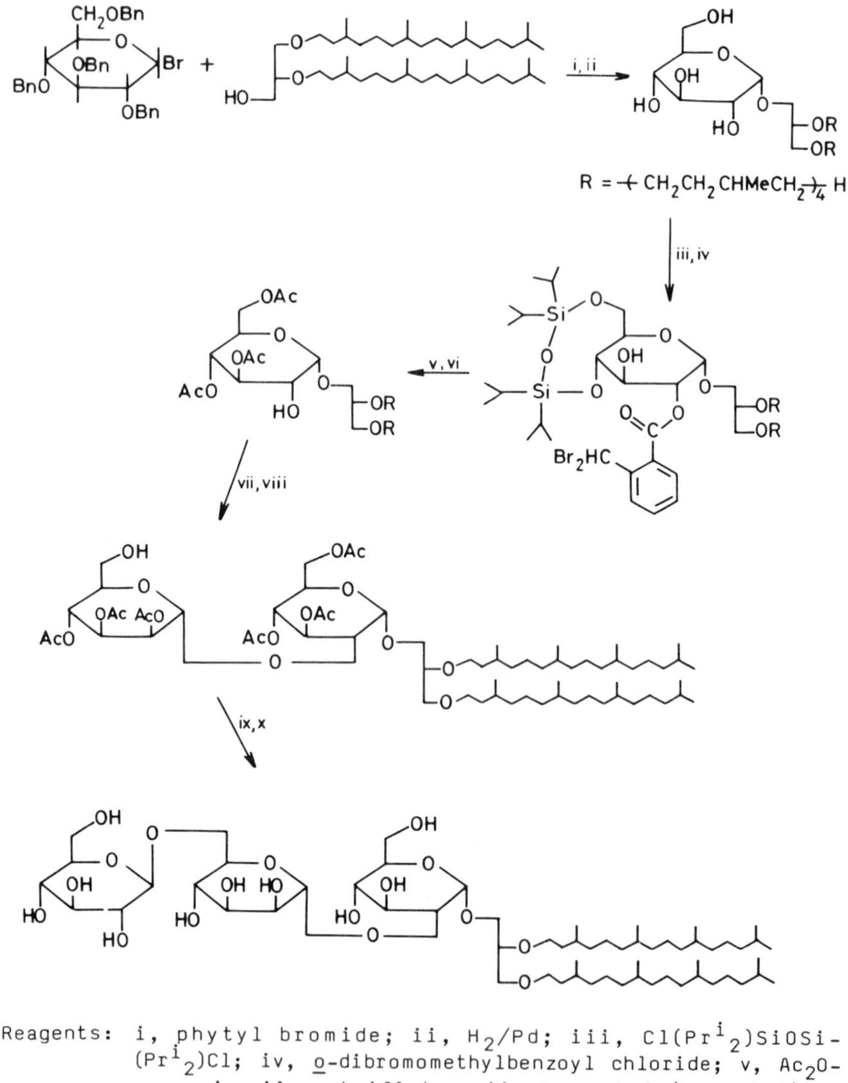

Reagents: i, phytyl bromide; ii, H_2/Pd; iii, Cl(Pr^i_2)SiOSi-(Pr^i_2)Cl; iv, o-dibromomethylbenzoyl chloride; v, Ac_2O-pyr; vi, silver triflate; vii, 2,3,4-tri-(O-acetyl)-6-(2,2,2-triethoxycarbonyl)-D-mannosyl bromide-acetonitrile/nitromethane; viii, Zn/THF-AcOH; ix, 2,3,4,6-tetra-O-acetyl-α-D-glucopyranosyl bromide, $HgBr_2$, $Hg(CN)_2$; x, Na^+OMe^-/MeOH/ether

Scheme 2

Chromatography, held in Veldhoven in June 1981, have been published.[80] They provide a survey of new developments in the field and illustrate the increasing importance of affinity techniques in industrial and biomedical/diagnostic applications. The use of polysaccharide polymers in chromatography has been described in the proceedings of the 18th Prague IUPAC Microsymposium on Macromolecules.[81] The use of polysaccharides as support polymers in the immobilization of macromolecules with biochemical and biomedical applications has been discussed.[82] Summaries of recent literature on molecular immobilization and bioaffinity phenomena, much of which involves use of polysaccharide matrices, are being published at regular intervals.[83-86]

The affinity chromatography of macromolecular substances on adsorbents bearing carbohydrate ligands has been reviewed.[87] The use of polysaccharide supports bearing isonitrile functional groups for covalent fixation of biologically active molecules has been described.[88] Reviews have appeared concerning the affinity-chromatographic purification of lectins utilizing polysaccharides, matrix-bound glycoproteins and glycopeptides, and matrix-bound mono- and di-saccharides,[89] and the use of affinity-chromatographic techniques in the purification of mammalian glycosyltransferases.[90] The application of affinity-chromatographic techniques for separations of and studies on molecular interactions of the components of the blood-coagulation system has been described.[91]

A unified treatment of systems containing immobilized biochemicals and chromatographic systems has been developed from basic thermodynamic considerations of partitioning in biphasic systems.[92] The theoretical treatment allows prediction of partition coefficients from readily obtainable experimental parameters, and simple procedures are suggested for the determination of binding constants by affinity or adsorption chromatography.

The technique of affinity electrophoresis has been reviewed.[93] In a theoretical treatment of the subject, evaluation of the effects of protein multivalency, immobilized ligand heterogeneity and microdistribution, and determination of effective concentrations of immobilized ligand has been considered.[94]

The use of affinity isoelectric focusing in polyacrylamide gels has been described for qualitative determination of ligand-binding proteins and for examining the functional homogeneity of purified protein preparations.[95] Affinity gels containing immobilized sugars or blue dextran were used to demonstrate the interaction of the immobilized ligands with lectins or lactate dehydrogenase.

Agarose. -- Agarose gels crosslinked to different extents have been used for the h.p.l.c. fractionation of human serum.[96] A method for running isoelectric-focusing separations under denaturing conditions in agarose has been described.[97] The method minimizes the risk of carbamylation of proteins when using urea in combination with agarose.

An improved rapid methodology for the isolation of nucleic acids from agarose gels has been described.[98] Hydroxy-apatite is used to bind the deoxyribonucleic acid after solubilization of agarose with potassium iodide or by heating, and the nucleic acid can then be eluted from the inorganic matrix free of agarose.

Living immobilized cells have been prepared by entrapment in agar gels.[99,100]

Agarose has been used without modification as an affinity-chromatographic support for the purification of anti-(Thy 1.2) monoclonal antibody covalently linked to ricin.[101]

Acid treatment of agarose to expose D-galactose units has provided an affinity matrix for the purification of D-galactose binding lectins from some Erythrina species.[102,103]

A mathematical model has been derived for the calculation of isothermal distribution coefficients for the immobilization of residues on agarose gels.[104] Similar coefficients have been obtained experimentally in chromatographic runs.

Agarose has been oxidized with periodate and then coupled with haemoglobin in the presence of sodium cyanoborohydride to yield an affinity matrix for the purification of haemoglobin-binding proteins.[105]

Gels prepared by immobilization of acriflavin on agarose have been successfully applied to the separation of nucleotides, oligonucleotides, and nucleic acids.[106] The use of flavin derivatives immobilized on agarose for affinity chromatography has been reviewed.[107]

Alkyl and aromatic derivatives of agarose continue to be used in preparative techniques involving hydrophobic chromatography and for enzyme immobilization. The relative contribution of hydrophobic and ionic interactions during chromatography of proteins on homologous series of hydrocarbon-coupled agaroses, obtained by coupling of alkylamines to agarose, has been assessed by comparing these columns with two similar series of columns devoid of charge.[108] Using a variety of pure proteins it was shown that columns of identical ligand densities have, qualitatively and quantitatively, similar adsorption and discrimination properties whether charged or not. From these results it was concluded that hydrophobic rather than ionic interations are of primary importance in chromatography on alkyl-agarose columns.

The ether-linked hydrophobic substituents of octyl-agarose and phenyl-agarose have been cleaved with boron tribromide and quantified by gas chromatography.[109] The procedure is sensitive, reproducible, and can easily be applied to other ether-linked ligands.

The binding of phytochrome from pea seedlings to alkyl-and ω-aminoalkyl-agaroses has been investigated.[110] ω-Aminobutyl-agarose has been applied to the isolation of a non-histone chromosomal high-mobility-group protein from mouse liver nuclei.[111] Decyl-agarose has been used for the purification of glycosyl-ceramidase from murine intestine[112] and for hydrophobic chromatography of _Staphylococcus aureus_ delta-toxin.[113] ω-Aminoethyl-agarose has been used for purification of glycogen-debranching enzyme from rabbit muscle.[114]

6-Aminohexyl-agarose has been applied to the purifications of nuclear androgen receptor,[115] phosphorylase b from liver of flounder (_Platichtys flesus_),[116] the pyruvate dehydrogenase component of the pyruvate dehydrogenase complex from _Escherichia coli_ K12,[117] amine oxidase from bovine serum,[118] a rat liver cyclic GMP-activated phosphodiesterase,[119] and a D-mannan-binding protein associated with the early chicken embryo.[120]

8-Amino-octyl-agarose has been used for the partial purification of cytochrome P-450 from rats.[121] Commercially available octyl-agarose has been used for the purification of the glycosyl-ceramidase from murine intestine,[112] of pig urinary kallikrein,[122] of a β-D-hydroxybutyrate dehydrogenase from _Zoogloea ramigera_ I-16-M,[123] of _Staphylococcus aureus_ delta-toxin,[113] of bovine brain hexosaminidase C,[124] of a topoisomerase from _Ustilago maydis_,[125] and of uroporphorinogen I synthase from human erythrocytes.[126] Octyl-agarose has also been applied successfully to the purification of a modified fluorescent heparin preparation.[127] A simple procedure for the preparation of human lymphoblastoid interferon in high yield involves chromatography on octyl-agarose as one step of the purification.[128] Sphingomyelin has been immobilized by adsorption onto octyl-agarose and then used as substrate in an enzyme assay for three bacterial phospholipases.[129] By immobilizing labelled lipid by the same method the performance of the same enzyme assay has been simplified and the sensitivity increased.[130]

Hydrophobic-interaction chromatography utilizing phenyl-agarose has been applied to the co-purification of alkaline phosphatase and 5'-AMP-specific nucleotidase from *Dictyostelium discoideum*,[131] and to the purification of vanillate hydroxylase (decarboxylating) from *Sporotrichum pulverulentum*,[132] of *Euglena gracilis* chloroplast elongation factor Ts,[133] and of human prolactin.[134] Membrane proteins from human brain have been chromatographed on phenyl-agarose after solubilization with Triton X-100.[135] Tyrosine hydroxylase,[136] *Escherichia coli* integration host factor,[137] *Staphylococcus aureus* delta-toxin,[113] rat uterine peroxidase,[138] bovine brain β-D-2-acetamido-2-deoxy-glucosidase,[124] and rat liver phenylalanine hydroxylase[139] have also been purified using phenyl-agarose. Subunits of pike-eel gonadotrophin have been separated by hydrophobic chromatography on phenyl-agarose.[140] A *Treponema* Reiter RNA antigen which precipitates with antibodies in human syphilitic sera has been purified on phenyl-agarose.[141] Phenyl-agarose has also been used for the purification of an ester hydrolase which hydrolyses phorbol diesters from murine liver,[142] and of a Ca^{2+}-activated neutral protease from rabbit skeletal muscle.[143]

10-Carboxydecylamino-agarose and *N*-(3-carboxypropionyl)amino-octyl-agarose have been prepared and used for chromatography of wheat-germ aspartate transcarbamoylase.[144] Under appropriate experimental conditions the enzyme can be specifically desorbed from either support using the substrate carbamoyl phosphate. The experimental evidence suggests that the enzyme is adsorbed at heterogeneous sites on each column only some of which are susceptible to substrate-specific desorption, that initial adsorption is not essentially biospecific, and that the desorption is biospecific. These results suggest that biospecific desorption from essentially nonbiospecific adsorbents may explain some published purifications currently regarded as being affinity chromatographic.

ω-Aminoalkyl-agarose derivatives of glutathione have been prepared by reaction of chloroacetyl derivatives with ω-aminoalkyl-agaroses.[145] The matrices were applied to the affinity chromatography of hepatic glutathione S-transferases.

Benzoquinone-activated agarose has been reacted with folic acid pre-coupled to bovine serum albumin to provide an affinity matrix for purification of folate-binding protein from porcine choroid plexus.[146]

Reactive carbonyl groups have been introduced into agarose gels by oxidation with aqueous bromine.[147] Proteins and other amines were coupled to the oxidized gels in high yield by reductive amination. The effect of the oxidation-reduction procedure on the chromatographic properties of the agarose gels was insignificant.

Benzoquinone-activated agarose has provided a satisfactory matrix for the immobilization of phosphotransferase.[148]

Bromoacetyl-1,6-diaminohexyl-agarose has been reacted with S-adenosylhomocysteine to provide an affinity-chromatographic matrix for purification of 5'-methylthioadenosine nucleosidase from *Lupinus luteus* seeds.[149]

The activation of agarose with 1,1'-carbonyldi-imidazole (CDI) has been further investigated and extended to other carbonylating reagents.[150] Results confirmed that use of CDI allows the facile preparation of activated matrices devoid of additional charge and suitable for affinity-chromatographic supports. 1,1'-Carbonyldi-1,2,4-triazole (CDT) gave a more reactive activated matrix, while 1,1'-carbonyldi-1,2,3-benzotriazole reacted only slowly and inefficiently. The introduction of imidazolyl carbamate groups onto cross-linked agarose by generating CDI *in situ* from phosgene and imidazole gave one-third of the level of activation of that obtained with pure CDI. CDI is concluded to be the most effective and convenient of the carbonylating reagents studied, but the CDT-activated matrix may be useful for the coupling of unstable protein ligands where short coupling times are essential.

Agarose matrices with a controlled degree of substitution can be synthesized using CDI, and a high level of activation can readily be achieved if required.[151] The CDI-activated agarose was found to have a half-life of more than fourteen weeks when stored in dioxane. Based on the results obtained, conditions suitable for the coupling of proteins were established. The N-alkylcarbamate linkage of the amino group of the ligands to the support was shown to possess good stability over a wide pH range. Agarose activated with CDI and subsequently coupled to aminophenyl boronic acid has been used successfully for the separation of glycosylated haemoglobins.[152]

Sulphate-agarose has been prepared by reaction of agarose with chlorosulphonic acid.[153] The matrix was applied to the chromatographic purification of bovine Factor VIII.

Agarose activated with cyanuric chloride has been reacted with colominic acid to provide an affinity-chromatographic matrix for purification of neuraminidase from Corynebacterium ulcerans.[154]

Agarose activated with cyanogen bromide continues to be used extensively for the immobilization of biologically active macromolecules and for affinity chromatography. Analytical methods have been developed for the determination of cyanate esters and imidocarbonates, the active species present on polysaccharide resins activated with cyanogen bromide.[155] A procedure for determining the coupling capacity of a resin is described: the amount of coupled ligand can be determined directly on the resin and without prior hydrolysis. Using these procedures the coupling capacity of activated resins towards small ligands was predicted by determining the amounts of cyanate esters and imidocarbonates present on the resin. Cyanate esters were found to be predominantly responsible for coupling of ligand to activated agarose. Alkaline treatment of cyanogen bromide-treated agarose both with or without coupled protein has been shown to release a thiol protease inactivator.[156] It was isolated and shown to be the cyanate ion. Experimental evidence suggests that the carbamate group in the activated polysaccharide is the most likely precursor of the cyanate ion.

Coupling of ligands to cyanogen bromide-activated agarose has hitherto been assumed to occur via nucleophilic attack of an amino group on the trans-cyclic imidocarbonate ring (see Vol.14, p.398). Recent results[155] suggest that cyanate esters are, in fact, predominantly responsible for coupling of ligand to the activated agarose and that the cyclic imidocarbonate is formed to a much lesser extent. Accordingly the term cyclic imidocarbonate previously used in these volumes to describe the product of reaction of agarose and other polysaccharides with cyanogen bromide is now replaced by the term cyanogen bromide activated. Recent references to the preparation of active immobilized forms of enzymes (for use as immobilized enzymes) and immunologically active macromolecules (for use as immunoadsorbents) and of various affinants (for use as affinity-chromatography matrices) by such methods are summarized in Tables 1,[157-182] 2,[183-205] and 3,[206-298] respectively.

Immobilized goat anti-(rabbit immunoglobulin) antibody has been used in column form for the separation of bound from free in a radioimmunoassay for human placental lactogen.[189] Using class-specific $F(ab')_2$ antibody fragments immobilized on cyanogen bromide-activated agarose immunoglobulin G has been isolated from human serum in a single chromatographic step with high yield and purity.[198]

Soluble immune complexes of bovine serum albumin—anti(bovine serum albumin) antibody (the model system) have been prepared and

Table 1 **Use of cyanogen bromide-activated agarose for the preparation of active immobilized enzymes**

Enzyme coupled to cyanogen bromide-activated agarose	E.C.No.	Agarose intermediate[a]	Ref.
Alcohol dehydrogenase		-	157
Chymotrypsin	3.4.21.4	-	158,159
Cystathione β-synthase		-	160
Deoxyribonuclease I	3.1.21.1	-	161,162
Elastase	3.4.21.11	-	163,164
D-Galactosyl hydroxy-L-lysyl glucosyltransferase		-	165
Glucoamylase	3.2.1.3.	-	166
Glyceraldehyde 3-phosphate dehydrogenase	1.2.1.12	-	167,168
Lactoperoxidase		-	169,170
Lysozyme	3.2.1.17	-	171
Neuraminidase	3.2.1.18	-	172,173
Papain	3.4.22.2	(27)	174
		-	175
Pepsin	3.4.23.1	(28)	176
Trypsin	3.4.21.4	-	177-180
		(29)	181
Tyrosinase	1.14.18.1	-	182

[a]See text at ref. 299 and Scheme 3

then bound to immobilized protein A.[202] The complexes could be desorbed from the protein A matrix and dissociated into antigen and

(Text continues on page 655)

Table 2 Use of cyanogen bromide-activated agarose for the preparation of immunoadsorbents

Immunologically active component coupled to cyanogen bromide-activated agarose	Use of product	Ref.
Anti-(human serum albumin) antibody	Purification of α_1-anti-chymotrypsin from human pleural fluid and serum	183
Anti-(α_1-antichymotrypsin) antibody	Purification of α_1-anti-chymotrypsin from human pleural fluid and serum	183
Anti-(human or rat α-feto-protein)antibody	Purification of human or rat α-fetoprotein	184
Anti-(human α-fetoprotein) monoclonal antibody	Purification of human α-fetoprotein	185
Anti-(blood-group I) antibody	Purification of sheep gastric mucins with blood-group Ii activity	186
Anti-(human chorionic gonadotrophin) antibody	Development of an enzyme immunoassay for human chorionic gonadotrophin	187
Anti-(human C5 component of complement) antibody	Purification of complement attack complexes from insulin-activated human serum	188
Anti-(rabbit immunoglobulin G) antibody	Radioimmunoassay for human placental lactogen	189
Anti-interferon antibody	Purification of human interferon	128
Anti-(urinary kallikrein) antibody	Purification of a prekallikrein from rat pancreas	190
Anti-renin antibody	Purification of human renin	191
Anti-(Rhizobium strain TA1 polysaccharide) antibody	Separation of pyruvylated polysaccharides into pyruvate-rich and -poor	192

Table 2 continued

Immunologically active component coupled to cyanogen bromide-activated agarose	Use of product	Ref.
	fractions	
Anti-(thermolabile antigen) antibody	Purification of thermolabile antigen from baker's yeast	193
Anti-(tumour-associated surface antigen) antibody	Isolation of polyoma virus-induced surface antigen from mouse cells	194
Human chorionic gonadotrophin carboxyl terminal tricosapeptide (synthetic)	Purification of anti-(human chorionic gonadotrophin) antibody	187
Immunoglobulin A, dimeric[a]	Purification of membrane receptor for polymer immunoglobulin	195
Immunoglobulin A, polymeric[b]	Purification of free secretory complex from rat bile	196
Immunoglobulin E	Purification of cell-surface receptor for IgE	197
Immunoglobulin $F(ab')_2$ antibody fragments	Single-step isolation of IgG from human serum	198
Intrinsic factor	Purification of a proteolytic derived ileal receptor for intrinsic factor-cobalamin	199
Protein A	Immobilization of monoclonal anti-(sucrase-isomaltase) antibody for purification of the enzyme	200
	Detection of complexes between IgG and inhibitory active part of the inter-α-trypsin inhibitor	201
	Isolation of soluble immune complexes prior to isoelec-	202

Table 2 continued

Immunologically active component coupled to cyanogen bromide-activated agarose	Use of product	Ref.
	tric focusing	
	Immobilization of anti-(NADP isocitrate dehydrogenase) antibody	203
	Study of differences in binding affinity of rat IgG subclasses to protein A	204
	Fractionation of normal human IgG proteins from Japanese	205

[a]See text and Scheme 3 (31)
[b]See text and Scheme 3 (28)

Table 3 Use of cyanogen bromide-activated agarose for the preparation of affinity-chromatography materials without linkage extension

Ligand or affinant coupled to cyanogen bromide-activated agarose	Use of product	Ref.
2-Acetamido-N-(6-aminohexanoyl)-2-deoxy-β-D-glucopyranosylamine	Affinity chromatography of β-D-2-acetamido-2-deoxyhexosidases A and B	206
2-Acetamido-2-deoxy-D-galactose	Affinity-chromatographic purification of a lectin from soybean (Glycine max)	207
	Affinity-chromatographic identification of blood-group A-active glycoproteins in human erythrocyte membrane	208
Acriflavin	Affinity chromatography of oligonucleotides	107
Actin, glutaraldehyde crosslinked	Affinity chromatography of light-chain isoenzymes of myosin head region	209
Actin, α-actinin	Affinity chromatography of proteolytic fragments of fibronectin	210
Albumin	Affinity chromatography of proteolytic fragments of fibronectin	210
8-(2-Aminoethylamino) cyclic AMP	Purification of a novel phosphodiesterase	211
N-(6-Aminohexanoyl)-β-L-fucosylamine	Purification of porcine thyroid fucosidase	212
N^6-(Aminohexyl)adenosine 2',5'-diphosphate	Affinity chromatography of glutamine synthetases from rice	213

Table 3 Continued

Ligand or affinant coupled to cyanogen bromide-activated agarose	Use of product	Ref.
	Purification of 3β-hydroxy-Δ^5-C_{27}-steroid oxidoreductase from rabbit liver	214
N[6]-(6-Aminohexyl)adenosine-5'-monophosphate	Affinity-chromatographic purification of lactate dehydrogenase isoenzyme C_4	215
	Purification of plant lactate dehydrogenase	216
	Purification of 3β-hydroxy-Δ^5-C_{27}-steroid oxidoreductase from rabbit liver	214
4-Aminophenyl mercuric acetate	Affinity chromatography of constituent enzyme activities of the multifunctional enzyme putrescine synthase	217
Antithrombin III	Affinity-chromatographic study of modified heparin	127
Asialofetuin	Purification of a phytohaemagglutinin from the Azolla-Anabaena symbiosis	218
	Purification of lactose-blocking lectin activity from fetal-calf skeletal muscle	219
Bandaeiraea simplicifolia I lectin	Affinity chromatography of blood-group ABH-active polyglycosyl peptides	220
Blue dextran	Affinity chromatography of L-threonine dehydrogenase from chicken liver	221
	Purification of glutamine synthetase from Escherichia coli	222
	Purification of acyl coenzyme A	223

Table 3 Continued

Ligand or affinant coupled to cyanogen bromide-activated agarose	Use of product	Ref.
	synthetase from *E. coli*	
	Affinity chromatography of glyceraldehyde 3-phosphate dehydrogenase	224
	Affinity chromatography of *Leuconostoc* phosphoglycerate mutase	225
	Affinity-chromatographic purification of rat Factor II	226
Calmodulin	Purification of smooth-muscle myosin light-chain kinase	227
Carboxypeptidase A inhibitor	Purification of carboxypeptidase A from rat skeletal muscle	228
Carcinoscorpius rotunda cauda (horseshoe crab) lectin	Affinity-chromatographic fractionation of sialoglycoproteins	229
Casein	Purification of a heparin-sensitive nuclear protein kinase	230
Citrulline	Affinity chromatography of putrescine synthase	217
Collagen	Purification of bovine dental sac collagenase	231
Concanavalin A	Purification of cathepsin D from pig myometrium	232
	Purification of an irreversible tissue inhibitor of collagenase in human amniotic fluid	233
	Affinity chromatography of *Trypanosoma brucei* antigens	234
	Purification of rat α-fetoprotein	235
	Purification of 53K glycoprotein from rabbit skeletal muscle	236
	Purification of acid phosphatase	237

Table 3 Continued

Ligand or affinant coupled to cyanogen bromide-activated agarose	Use of product	Ref.
	from rat prostatic adenocarcinoma	
	Purification of cholinesterase from chicken egg yolk	238
	Affinity-chromatographic fractionation of glycopeptides isolated from a rat liver biantennary glycan	239
	Purification of a phospholipid-dependent α-mannosidase from rabbit liver	240
	Purification of a concanavalin A receptor from boar spermatozoa	241
	Purification of human liver acid β-D-galactosidase	242
	Purification of two antigenic glycoproteins from rye-grass (*Lolium perenne*) pollen	243
	Purification of α_1-antitrypsin	244
	Affinity-chromatographic purification of human liver β-D-2-acetamido-2-deoxy-hexosidase	245
	Purification of glycoconjugates from *Spongia officinalis*	246
	Affinity-chromatographic fractionation of a mixture of 2-$\{^3H\}$-D-mannose-labelled glycopeptides derived from mouse lymphoma cell glycoproteins	247
	Isolation of two *endo*-β-D-2-acetamido-2-deoxy-glucanases from fig latex	248
	Purification of a neutral pro-	249

Table 3 Continued

Ligand or affinant coupled to cyanogen bromide-activated agarose	Use of product	Ref.
	tease cleaving type IV collagen	
	Purification of β-D-2-acetamido-2-deoxy-hexosidases A and B from human liver	250
	Purification of sciatin	251
	Purification of variant antigens of *Trypanosoma congolense*	252
	Purification of plasminogen activator secreted by human melanoma cells in culture	253
	Affinity chromatography of pepsin fragments of human placental and renal basement membrane laminin	254
	Purification of rat liver α-D-2-acetamido-2-deoxy-glucosyl phosphodiesterase	255
	Affinity chromatography of glycoproteins from adult-rat brain synaptic vesicles	256
	Affinity-chromatographic purification of bovine brain β-D-2-acetamido-2-deoxy-hexosidase C	124
	Copurification of alkaline phosphatase and 5'-AMP-specific nucleotidase from *Dictyostelium discoideum*	131
	Purification of glycosylceramidase from murine intestine	112
	Purification of rat uterine peroxidase	138
	Purification from murine liver	142

Table 3 Continued

Ligand or affinant coupled to cyanogen bromide-activated agarose	Use of product	Ref.
	of an ester hydrolase active on phorbol diesters	
	Purification of a root lectin from soybean (Glycine max)	207
Cytochrome $\underline{b}_5$	Purification of linoleoyl CoA desaturase from rat liver microsomes	257
Deoxyribonucleic acid	Affinity chromatography of porcine pancreas deoxyribonuclease I	258
Discoidin I	Purification of discoidin I-binding proteoglycan from axenic Dictyostelium discoideum	259
Dolichos biflorus lectin	Purification of receptors for the lectin from embryonal carcinoma cells	260
Factor X	Affinity-chromatographic purification of factor VIII	153
Fetuin	Purification of a neuraminic acid-binding lectin from horseshoe crab (Carcinoscorpius rotunda cauda)	261
Fibronectin	Purification of glycosaminoglycans	262
Hog gastric mucin (Smith degraded)	Purification of endo-β-D-galactosidase from Escherichia freundii	263
Gelatin	Affinity chromatography of proteolytic fragments of fibronectin	210
	Purification of a glycoprotein secreted by aortic endothelial cells in culture	264
	Purification of rat plasma fibro-	265

Table 3 Continued

Ligand or affinant coupled to cyanogen bromide-activated agarose	Use of product	Ref.
	nectin	
Guanosine diphosphate hexanolamine	Copurification of Lewis blood-group 2-acetamido-2-deoxy-D-glucosyl-α(1→4)-L-fucosyltransferase and a 2-acetamido-2-deoxy-D-glucosyl-α(1→3)-L-fucosyltransferase from human milk	266
Guanosine monophosphate	Purification of human hypoxanthine-guanine phosphoribosyl transferase	267
Helix pomatia lectin	Affinity-chromatographic study of subunit structure of enterokinase	268
Heparin	Purification of a heparin-sensitive nuclear protein kinase	230
	Purification of rat liver 2-acetamido-2-deoxy-D-glucosyl phosphodiesterase	255
	Purification of Xenopus laevis type 1 topoisomerase	269
	Purification of rat hepatic triglyceride lipase	270
	Affinity chromatography of native and chemically modified antithrombin III	271
	Purification of tyrosine hydroxylase	136
Histone	Purification of a catalytic subunit of a protein kinase from human erythrocyte membranes	272
Homoarginine	Affinity chromatography of	217

Table 3 Continued

Ligand or affinant coupled to cyanogen bromide-activated agarose	Use of product	Ref.
	putrescine synthase	
α-Lactalbumin	Purification of human serum D-galactosyltransferase	273
Lentil (Lens culinaris) lectin	Affinity-chromatographic study of binding of glycopeptides of different structures	274
Lichenan	Purification of (1→3),(1→4)-β-D-glucanase from Bacillus	275
{D-Leu^{6}} octapeptide (renin) inhibitor	Purification of human renin	191
Lotus tetragonolobus lectin	Affinity-chromatographic study of subunit structure of enterokinase	268
Low-density lipoprotein (human)	Affinity chromatography of various modified glycans	276
	Affinity chromatography of chondroitin 6-sulphate	277
L-Lysine	Affinity-chromatographic removal of plasminogen and plasmin from guinea-pig serum	278
	Large-scale isolation of functionally active components of the human complement system	279
	Purification of heparin	280
D-Mannan	Isolation of a D-mannan-binding protein from rat liver	281
Ovalbumin	Affinity chromatography of proteolytic fragments of fibronectin	210
Ovomucoid	Isolation of α-D-mannose : 1,2-(2-acetamido-2-deoxy-D-glucosyl) transferase from trachea mucosa	282
Pea (Pisum	Affinity-chromatographic study	274

Table 3 Continued

Ligand or affinant coupled to cyanogen bromide-activated agarose	Use of product	Ref.
sativum) lectin	of binding of glycopeptides of different structures	
	Affinity-chromatographic fractionation of a mixture of 2-{^{3}H}-*D*-mannose-labelled glycopeptides derived from mouse lymphoma cell glycoproteins	247
Phosvitin	Purification of a heparin-sensitive nuclear protein kinase	230
Plasminogen	Purification of α_2 plasmin inhibitor	283
Poly(inosine), poly(cytidylic acid)	Purification of human (2',5')(A)$_n$ synthetase	284
Poly(uridylic acid)	Purification of mammalian viral reverse transcriptases	285
Ricinus communis agglutinin	Purification of a glycoprotein from sera of cancer patients	286
	Affinity-chromatographic separation of cell-surface glycoproteins and glycolipids	287
Ricinus sanguinis agglutinin	Affinity chromatography of glycoproteins from adult-rat brain synaptic vesicles	256
Thyroid-stimulating hormone (TSH)	Purification of human TSH receptors from thyroid membrane	288
α-Toxin	Purification of nicotinic acetylcholine receptor proteins	289
Transferrin	Purification of autoantibodies to transferrin	290
Troponin T, dephosphorylated	Purification of dog cardiac troponin T kinase	291
Trypsin inhibitor	Purification of *Streptomyces*	292

Table 3 Continued

Ligand or affinant coupled to cyanogen bromide-activated agarose	Use of product	Ref.
(rice bran)	*griseus* trypsin	
Trypsin inhibitor (soybean)	Affinity chromatography of porcine pancreas deoxyribonuclease I	258
UDP-hexanolamine	Purification of rabbit liver oestrone and 4-nitrophenol UDP-glucuronyltransferases	293
Vicia cracca lectin	Affinity-chromatographic fractionation of blood-group ABH-active polyglycosyl peptides from human erythrocyte membranes	220
	Affinity chromatography of blood-group A-active glycoproteins from human erythrocyte membranes	294
von Willebrand factor	Isolation of human platelet receptors for von Willebrand factor	295
Wheat-germ agglutinin	Affinity chromatography of glycocon-conjugates from *Spongia officinalis*	246
	Purification of a glycoprotein from sera of cancer patients	286
	Purification of intracellular α-D-mannosidase from cultured skin fibroblasts	296
	Isolation of plasma membrane vesicles of human platelets	297
	Isolation of human platelet membrane fractions	298

antibody by isoelectric-focusing analysis. In a study of glycosminoglycan-lipoprotein interaction the effect of uronic acid composition and charge density of the glycan has been investigated using agarose substituted with low-density lipoprotein.[276]

Direct attachment of an enzyme or an affinant to agarose activated with cyanogen bromide has frequently been unrewarding, since approach to the bound ligand may be sterically impeded. This difficulty can be overcome by inserting a covalent bridge between matrix and ligand. Bridging is also used when functional groups on the ligand do not react with those on the matrix, since it allows a new type of functionality to be introduced. Modifications recently used to provide bridges in the preparation of affinity-chromatography matrices, etc., from agarose cyclic imidocarbonates are summarized in Schemes 3a and 3b and Table 4.[299-345] Where bridging has been used to immobilize enzymes or to prepare immunoadsorbents, details are given in Table 1 or 2, respectively. The products obtained by coupling various substances to agarose cyclic imidocarbonate for diverse uses are given in Table 5.[346-376]

Amine-substituted hormone analogues of tri-iodothyronine have been prepared and coupled to 6-aminohexyl-agarose in the development of support matrices for affinity chromatography of thyroid hormone receptors.[344]

2-Acetamido-2-deoxy-β-D-glucosyl hexanolamine coupled to agarose has been used in studies of the enzymatic synthesis and ^{13}C n.m.r. conformation of disaccharides containing β-D-galactose and β-D-{1-^{13}C}-galactose residues.[346] The solid-phase synthesis was achieved using soluble UDP-D-galactose and UDP-D-galactosyltransferase.

The use of immobilized lectins (concanavalin A and wheat-germ agglutinin) to prepare serum free of glycoprotein hormones, for use in radioimmunoassay, has been described.[363] A new method utilizing protein A-agarose has been developed for quantitation of the factor XIIIa-catalysed coupling of various structural proteins.[368] Using immobilized glutathione as an activator a new and very sensitive assay for thiol proteases (e.g. cathepsin D) has been reported.[369]

Immobilized heparan sulphate has been used in studies of self-association between various heparan sulphate species and oligosaccharides thereof.[370,371] Heparin immobilized on agarose has been used to assess the interaction among polyamines, nuclear protein kinase and heparin, as well as for purification of protein kinases inhibited by heparin.[372]

(Text continues on page 670)

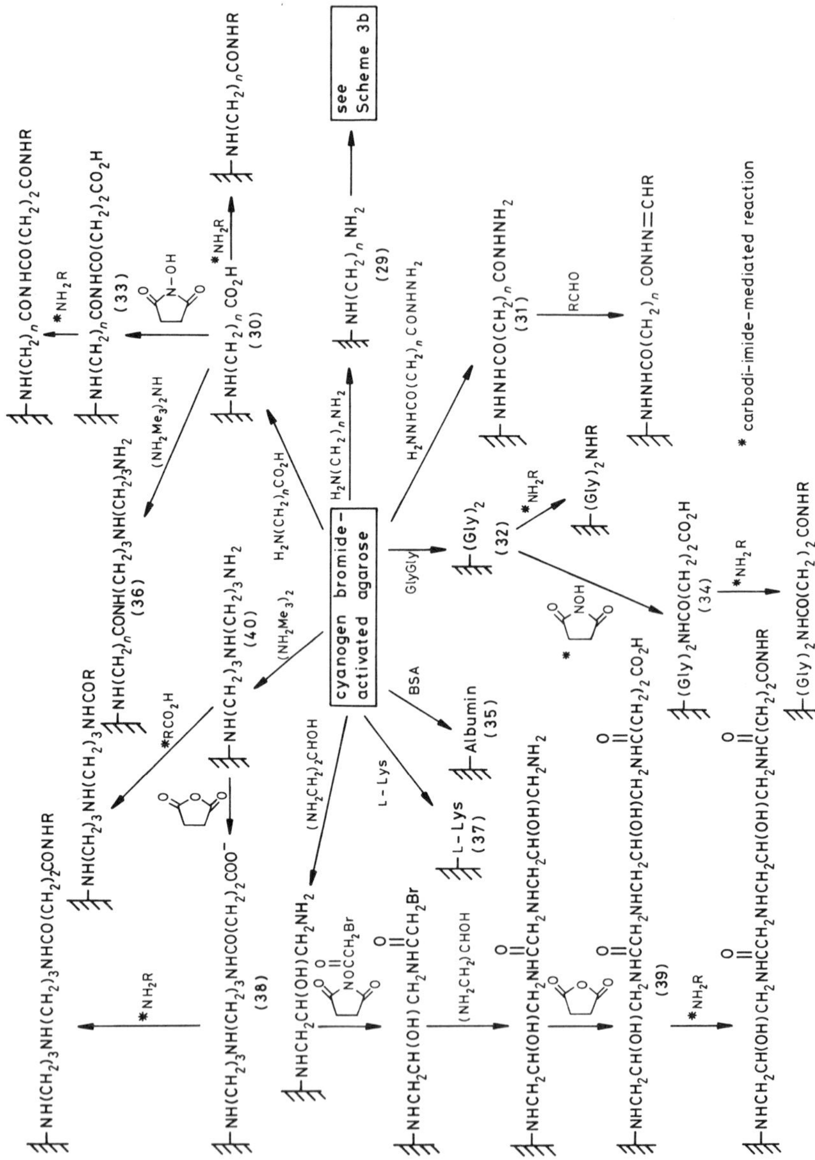
cyanogen bromide-activated agarose
see Scheme 3b
H2N(CH2)nNH2
NH(CH2)nNH2 (29)
H2N(CH2)nCO2H
NH(CH2)nCO2H (30)
NH(CH2)nCONHR
NH(CH2)nCONHCO(CH2)2CO2H (33)
NH(CH2)nCONHCO(CH2)2CONHR
H2NNHCO(CH2)nCONHNH2
NHNHCO(CH2)nCONHNH2 (31)
RCHO
NHNHCO(CH2)nCONHN=CHR
GlyGly
(Gly)2 (32)
(Gly)2NHR
(Gly)2NHCO(CH2)2CO2H (34)
(Gly)2NHCO(CH2)2CONHR
BSA
Albumin (35)
L-Lys
L-Lys (37)
(NH2Me3)2
NH(CH2)3NH(CH2)3NH2 (40)
(NH2Me3)2NH
NH(CH2)nCONH(CH2)3NH(CH2)3NH2 (36)
*RCO2H
NH(CH2)3NH(CH2)3NHCOR
NH(CH2)3NH(CH2)3NHCO(CH2)2COO− (38)
NH(CH2)3NH(CH2)3NHCO(CH2)2CONHR
(NH2CH2)2CHOH
NHCH2CH(OH)CH2NH2
NHCH2CH(OH)CH2NHCCH2Br
(NH2CH2)2CHOH
NHCH2CH(OH)CH2NHCCH2NHCH2CH(OH)CH2NH2
NHCH2CH(OH)CH2NHCCH2NHCH2CH(OH)CH2NHC(CH2)2CO2H (39)
NHCH2CH(OH)CH2NHCCH2NHCH2CH(OH)CH2NHC(CH2)2CONHR
*NH2R
* carbodi-imide-mediated reaction

Scheme 3a

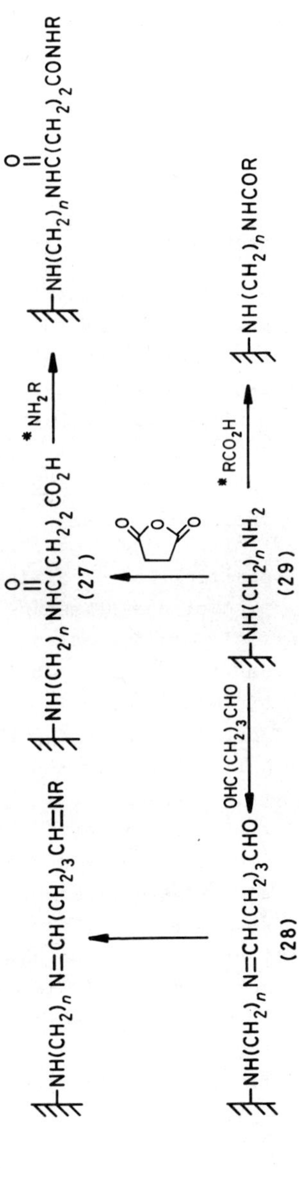

Scheme 3b

Table 4 Use of cyanogen bromide-activated agarose for the preparation of affinity-chromatography materials *via* linkage extension

Ligand or affinant coupled to cyanogen bromide-activated agarose	Agarose intermediate[a]	Use of product	Ref.
2-Acetamido-2-deoxy-β-D-galactose	30	Affinity-chromatographic purification of human liver β-D-2-acetamido-2-deoxy-hexosidase	245
Adenosine diphosphate	29	Affinity-chromatographic purification of glutamine synthetase from *Rhodospirillum rubrum*	299
Adenosine monophosphate	29	Affinity-chromatographic purification of Simian liver alcohol dehydrogenase	300
Adenosine monophosphate, periodate-oxidized	29	Affinity-chromatographic purification of two forms of RNAse U_2	301
	29	Affinity-chromatographic purification of RNAse T_2 from Taka-diastase	302
Adenosine triphosphate	31	Affinity-chromatographic purification of a heparin-sensitive nuclear protein kinase	230
		Affinity-chromatographic purification of flavokinase from *Phaseolus aureus* (mung bean)	303
4-Aminobenzamidine	32	Affinity-chromatographic	268

Table 4 continued

Ligand or affinant coupled to cyanogen bromide-activated agarose	Agarose intermediate[a]	Use of product	Ref.
		study of enterokinase	
2-Amino-2-deoxy-D-galactose	30	Affinity-chromatographic purification of a 2-amino-2-deoxy-D-galactose-binding protein from *Psophocarpus tetragonolobus* (winged bean)	304
6-Aminohexyl 2-acetamido-2-deoxy-1-thio-β-D-glucoside	30	Affinity-chromatographic purification of β-D-2-acetamido-2-deoxy-hexosidases A and B from human liver	250
6-Aminohexyl 1-thio-β-D-galactoside	30	Affinity-chromatographic removal of β-D-galactosidase during purification of hexosidases	250
4-Aminophenyl-2-acetamido-2-deoxy-β-D-glucoside	30	Affinity-chromatographic purification of bovine brain β-D-2-acetamido-2-deoxy-hexosidase C	124
4-Aminophenyl-1-thio-2-acetamido-2-deoxy-β-D-glucoside	30	Affinity-chromatographic purification of bovine brain β-D-2-acetamido-2-deoxy-hexosidase C	124
4-Aminophenyl-1-thio-β-D-glucoside	30	Affinity-chromatographic purification of a β-D-glucosyl hydrolase transferase present in walls of soybean cells	305
4-Aminosalicylic acid	33	Affinity chromatography of *Pseudomonas* salicylate	306

Table 4 continued

Ligand or affinant coupled to cyanogen bromide-activated agarose	Agarose intermediate[a]	Use of product	Ref.
		hydroxylase	
Aprotinin	30	Purification of pig urinary kallikrein	122
L-Argininal semicarbazone hydrofluoride	34	Affinity-chromatographic purification of trypsin	307
L-Arginine	30	Affinity-chromatographic purification of collagenase from *Clostridium histolyticum*	308
Bay g5421 amylase inhibitor	29	Affinity-chromatographic isolation of rat pancreatic and salivary amylases	309
4-Carboxybenzaldehyde	29	Rapid purification of human placental aldose reductase	310
2'3'-O-{1-(2-Carboxyethyl)ethylidene}-inosine	29	Quantitative affinity chromatography of adenosine deaminase	311
2'3'-O-{1-(2-Carboxyethyl)ethylidene}-7-methylguanosine 5'-diphosphate	29	Affinity-chromatographic purification of eukaryotic messenger RNA cap binding protein	312
2'3'-O-{1-(2-Carboxyethyl)ethylidene}-xanthosine	29	Affinity-chromatographic purification of guanine aminohydrolase	313
7-(5-Carboxypentyl)-guanosine 5'-diphosphate	29	Affinity-chromatographic purification of eukaryotic messenger RNA cap binding protein	312
Cholic acid	29	Affinity-chromatographic purification of steroid 12α-mono-oxygenase	314
	29	Affinity-chromatographic	315

Table 4 continued

Ligand or affinant coupled to cyanogen bromide-activated agarose	Agarose intermediate[a]	Use of product	Ref.
		purification of glutathione S-transferases A and C from rat liver cytosol	
Cortisol	35	Affinity-chromatographic purification of human transcortin	316
Creatine	29	Affinity-chromatographic purification of sarcosine oxidase from *Corynebacterium* species	317
Cyanocobalamin	36	Large-scale affinity-chromatographic purification of human holo-transcobalamin II	318
D-Cysteinyl-L-proline	29	Affinity-chromatographic purification of human serum angiotensin 1-converting enzyme	319
Dermatan sulphate	29	Purification of bovine testicular hyaluronidase	320
Dextran sulphate, cyanogen bromide-activated	37	Affinity-chromatographic purification of acid β-glucosidase from human placenta	321
Dihydrolipoic acid	29	Affinity-chromatographic purification of pyruvate dehydrogenase complex from *Escherichia coli* K12	322
Ethyl N-(3-carboxyphenyl) carbamate	29	Affinity-chromatographic isolation of plant tubulin from Azuki bean epicotyls	323
Folic acid	29	Affinity-chromatographic	324

Table 4 continued

Ligand or affinant coupled to cyanogen bromide-activated agarose	Agarose intermediate[a]	Use of product	Ref.
		purification of two components of a multienzyme system from *Clostridium thermoaceticum*	
5-Formyl-5,6,7,8-tetrahydropteroylglutamic acid	29	Affinity-chromatographic study of folate-binding proteins of rat liver mitochondria	325
O-α-L-Fucopyranosyl-(1→2)-β-D-galactopyranosyl-(1→3)-{α-L-fucopyranosyl-(1→4)}-2-acetamido-2-deoxy-β-D-glucosyl-(1→3)-β-D-galactosyl-(1→4)-D-glucose	-[b]	Purification of goat anti-hexaose immunoglobulin G	326
D-Gluco-D-mannan, cyanogen bromide-activated	29	Affinity-chromatographic purification of β-D-mannanase	327
Glycosphingolipid	29	Affinity-chromatographic purification of anti-glycosphingolipid antibody	328
Glycosaminoglycans, periodate-oxidized	31	Affinity chromatography of basement membrane laminin	329
5'-Guanosine monophosphate, periodate-oxidized	29	Affinity-chromatographic purification of ribonuclease T_1	330
	29	Affinity-chromatographic purification of ribonuclease from *Fusarium moniliforme*	331

Table 4 continued

Ligand or affinant coupled to cyanogen bromide-activated agarose	Agarose intermediate[a]	Use of product	Ref.
Heparin	29	Affinity-chromatographic purification of bovine testicular hyaluronidase	320
2-Iminobiotin	29	Affinity-chromatographic purification of avidin	332
	29	Affinity-chromatographic purification of streptavidin	333
Insulin	38	Affinity-chromatographic purification of human placental insulin receptor	334
Methotrexate	29	Affinity-chromatographic purification of dihydrofolic acid reductase	335
10-Methyl-5,8-dideazafolate	29	Affinity-chromatographic purification of mouse liver thymidylate synthetase	336
4-Methyl-nicotinamide adenine dinucleotide	39	Affinity-chromatographic purification of NAD-glycohydrolase from <u>Neurospora crassa</u>	337
1,2-Naphthoquinone	29	Affinity-chromatographic isolation of human liver dihydropteridine reductase	338
Nicotinamide adenine dinucleotide		Affinity-chromatographic purification of fragment of diphtheria toxin	339
Nicotinamide adenine dinucleotide phosphate		Affinity-chromatographic purification of aldehyde reductase from rat brain and liver and ox brain	340
Pepstatin	39	Affinity-chromatographic	232

Table 4 continued

Ligand or affinant coupled to cyanogen bromide-activated agarose	Agarose intermediate[a]	Use of product	Ref.
		purification of cathepsin D from pig myometrium	
	29	Affinity-chromatographic purification of a proteinase from *Candida albicans* culture supernatants	341
	29	Affinity-chromatographic purification of human renin	191
	29	Large-scale affinity-chromatographic isolation of rennin from mouse submaxillary gland	342
Pteroglutamic acid	40	Affinity-chromatographic isolation of a folate receptor from human placenta	343
Putrescine	30	Affinity chromatography of constituent enzyme activities of the multifunctional enzyme putrescine synthase	217
Riboflavin	29	Affinity-chromatographic purification of flavokinase from *Phaseolus aureus* (mung bean)	303
3,5,3'-Tri-iodothyronine	29	Affinity chromatography of thyroid hormone receptors	344
Uridine diphosphate D-glucuronate	29	Affinity-chromatographic purification of guinea-pig liver microsomal UDP D-glucuronosyltransferase	345

[a]See text at citation of Table 4 and Scheme 3

[b]Reaction with 1-amino-2-(4-aminophenyl)ethane, followed by reaction with cyanogen bromide-activated agarose

Table 5 Use of cyanogen bromide-activated agarose for the preparation of miscellaneous derivatives

Species coupled to cyanogen bromide-activated agarose	Use of product	Ref.
2-Acetamido-2-deoxy-β-D-glucosyl hexanolamine	Solid-phase enzymatic synthesis of disaccharides	346
Adenosine monophosphate, periodate-oxidized[a]	Detection and isolation of small amounts of base-specific RNAses in crude cell extracts	301
4-Aminophenyl β-D-glucopyranoside	Affinity-chromatographic studies of the specific interaction of concanavalin A and saccharides	347
Asialo-G_{M1}, ozonolysed[a]	Purification of anti-glycosphingolipid antibody	348
2-Chloro-10-(3-aminopropyl)phenothiazine	Affinity-chromatographic study of the Ca^{2+}-dependent interaction of S100b, troponin C, and calmodulin with the drug	349
Chromaffin granule membranes	Isolation by Ca^{2+}-dependent affinity chromatography of proteins that bind to chromaffin granule membrane	350
Concanavalin A	Demonstration of glycoprotein characteristics of the sodium channel saxitoxin-binding component from mammalian sarcolemma	351
	Characterization of *Xenopus* oocyte subcellular fractions and incubation media	352
	Study of transformation-dependent quantitative changes in glycopeptide binding to the lectin	353

Table 5 continued

Species coupled to cyanogen bromide-activated agarose	Use of product	Ref.
	Reversible binding of antithrombin III for use in the fractionation of heparin	354
	Study of binding of insulin receptors to lectins	355
	Removal of haemagglutinating activity during protease purification	356
	Study of the glycosylation of the arginine vasopressin/neurophysin II common precursor	357
	Study of microheterogeneity of ovalbumin	358
	Study of fibrinogen chains during synthesis	359
	Affinity-chromatographic demonstration of two α_1-acid glycoprotein populations in purified protein preparations	360
	Affinity-chromatographic separation of right-side-out oriented subfractions from purified calf thymocyte plasma membranes	361
	Demonstration of two types of colony-stimulating factor in serum of lipopolysaccharide-treated mice	362
	Removal of lutrophin and choriogonadotrophin β-subunit during preparation of glycoprotein hormone-free serum	363
	Demonstration of glycoprotein	364

Table 5 continued

Species coupled to cyanogen bromide-activated agarose	Use of product	Ref.
	nature of chlorophyllase	
	Demonstration of heterogeneity of heterogalactan from fruit bodies of *Fomitopsis pinicola*	365
Concanavalin A, succinylated	Investigation of chemical cross-linking of D-glucose oxidase on a modified lectin matrix	366
Deoxyribonucleic acid	Model system for development of a method for cytochemical hybridization with fluorochrome-labelled RNA	367
Fibrin	Development of a new method of examining and quantitating the Factor XIIIa-catalysed coupling of various structural proteins	368
Glutathione	Development of a very sensitive assay for thiol proteases using immobilized glutathione as an activator	369
Heparan sulphate, periodate-oxidized[b]	Demonstration of self-association of bovine lung heparan sulphates	370, 371
Heparin	Assessment of interactions among polyamines, nuclear protein kinase, and heparin. Purification of protein kinases inhibited by heparin	372
	Modification of heparin-agarose procedure for determination of plasma lipolytic activity of normalipidemic and hyperlipidemic subjects after heparin	373

Table 5 continued

Species coupled to cyanogen bromide-activated agarose	Use of product	Ref.
	injection	
Lentil lectin	Study of the carbohydrate-binding specificity of lentil lectin	247
	Study of binding of insulin receptors to lectins	355
Myosin	Development of a new method of examining and quantitating the Factor XIIIa-catalysed coupling of various structural proteins	368
Pea lectin	Study of the carbohydrate-binding specificity of pea lectin	247
	Study of binding of insulin receptors to lectins	355
Peanut agglutinin	Isolation of lectin receptors from human erythrocytes without use of detergents	374
Phytohaemagglutinin E	Support for immobilization of erythropoietin for attachment of erythroid colonies *in vitro*	375
Protein A	Development of a new method of examining and quantitating the Factor XIIIa-catalysed coupling of various structural proteins	368
Ricin I and ricin II	Study of binding of insulin receptors to lectins	355
Ricinus communis agglutin	Demonstration of glycoprotein characteristics of the sodium channel saxitoxin-binding component from mammalian sarcolemma	351
	Affinity chromatography of angiotensin-converting enzyme from swine kidney, serum, and culture	172

Table 5 continued

Species coupled to cyanogen bromide-activated agarose	Use of product	Ref.
	media	
Soybean agglutinin	Isolation of lectin receptors from human erythrocytes without use of detergents	374
Thrombin	Intermediate gel in method for demonstration of thrombin-binding platelet proteins by crossed immunoelectrophoresis	376
Wheat-germ agglutinin	Demonstration of glycoprotein characteristics of the sodium channel saxitoxin-binding component from mammalian sarcolemma	351
	Removal of lutrophin, choriogonadotrophin β-subunit, follitrophin, and thyrotrophin during preparation of glycoprotein hormone-free serum	363
	Support for immobilization of erythropoietin for attachment of erythroid colonies *in vitro*	375

[a]See Scheme 3 (29)
[b]See Scheme 3 (31)

Columns of diethylaminoethyl-agarose beads of high maximum capacity have been examined for their suitability for protein separation in chromatofocusing systems.[377] Large sudden changes of ionic strength should be avoided when using this medium, and when an anion exchanger is used the components of the pre-equilibration buffer and the elution buffer should be of the cationic type, while the counter-ions should not be subject to association-dissociation equilibria in the pH range of operation. Experimental details for optimal use of the system are reported.

Isolation of electrolectin, a β-D-galactoside-binding lectin from Electrophorus electricus (electric eel), has been achieved using lactose bound to divinylsulphone-activated agarose.[378] Divinylsulphone-activated agarose has been used to immobilize azodicoumarol for the affinity-chromatographic purification of cytosolic,[379] mitochondrial,[380] and microsomal [380] DT-diaphorase from liver of control and 3-methyl-cholanthrene-treated rats.

Epoxy-activated agaroses continue to be used widely for preparation of affinity matrices. Affinants coupled to epoxy-activated agarose have included 3-aminophenylboronic acid for isolation of polypeptides containing arginine modified with cyclohexane 1,2-dione,[381] fluphenazine for the rapid purification of calmodulin in good yield,[382] cyclohexa-amylose for the purification of a debranching enzyme of germinating rice endosperm,[383] 2-hydroxy-3,5-di-iodobenzoic acid for affinity-chromatographic purification of auxin affinity proteins from corn membranes,[384] maltose for purification of maltose-binding protein from Saccharomyces cerevisiae,[385] and ovomucoid for purification of a trypsin-like enzyme from Streptomyces erythreus.[386] Using polyethylene glycol coupled to epoxy-activated agarose, the influence of mobile-phase composition on the chromatographic behaviour of human peripheral blood cells has been investigated.[387] Polyethyleneimine has been coupled to epoxy-activated agarose and its properties and suitability for use as an ion exchanger have been investigated.[388] The matrix exhibits high capacity, fast equilibration on acid-base titration, and no maxima in a pH gradient. Resolving power could be improved by converting some of the imino groups into guanidine groups by modification with O-methylisourea. The matrix was applied successfully to the purifications of myoglobin and acetylated papain. Other affinants coupled to epoxy-activated agarose include raffinose, for the purification of 6^G-fructosyltransferase from asparagus roots,[389] D-ribose 5-phosphate for purification of Candida utilis transketolase free from D-ribulose 5-phosphate-3-epimerase,[390] and tris(hydroxymethyl)aminomethane for the purification of brush border membrane maltase from rat kidney.[391]

Epoxy-activated agarose has been treated with ammonium hydroxide to yield an amino-agarose which was treated with succinic anhydride to afford an N-hydroxysuccinimide ester.[392] Chitobiose was coupled to this matrix by reductive amination in the presence of sodium cyanoborohydride, and the resulting affinity matrix was applied successfully to the purification of a Cytisus-type Ulex europaeus haemagglutinin II. The derivatization of epoxy-activated agarose with various carbohydrates in order to prepare stable and high-capacity affinity adsorbents has been described with particular relevance to the use of the matrices for affinity chromatography of carbohydrate-binding proteins.[393]

Supports for metal-chelate affinity chromatography have been prepared by coupling iminodiacetic acid with epoxy-activated agarose to yield a bis-carboxymethylamino-agarose, which is then equilibrated with a suitable metal ion. In this manner Cu^{2+} chelate chromatography has been applied to the partial purification of RNA polymerase B stimulatory factor from nonhistone proteins of rat liver nuclei.[394] The use of metal-chelate affinity chromatography for the separation of nucleotides according to their metal affinity has been investigated in depth using Cu^{2+} and Ni^{2+} ions.[395] The adsorption and desorption of proteins in Zn^{2+}-chelate affinity chromatography has been studied using purification of albumin as a model system.[396] Adsorption was shown to be dependent to a great extent on the type of buffer ions used: acetate and phosphate ions, with a low affinity for the chelated metal, facilitate adsorption while amines prevent adsorption or cause desorption of adsorbed protein. Using Zn^{2+}-chelate chromatography successful purifications reported have included those of plasminogen activator secreted by human melanoma cells in culture,[253] of Dolichos biflorus seed lectin,[397] and of human fibroblast interferon.[398] Co^{2+}, Cu^{2+}, and Zn^{2+} complexes with bis-carboxymethylamino-agarose have been used for immobilization of alkaline phosphatase, lactate dehydrogenase, and malate dehydrogenase.[399]

Commercially available Affi-gel supports, agarose beaded gel included in a polyacrylamide support having different spacer arm lengths, have been used for the preparation of a number of affinity matrices. The spacers are anchored by extremely stable ether linkages. Reaction conditions have been optimized for immobilization of concanavalin A (as model protein) to agarose bearing an uncharged N-hydroxysuccinimide ester spacer arm (Affi-gel 10) and to agarose derivatized with an N-hydroxysuccinimide group with a tertiary amine in the spacer (Affi-gel 15).[400] Results showed that the charge of the protein relative to the charge of the gel is an important factor affecting the amount of protein coupled to active ester gels. Ligands immobilized on Affi-gel 10 have included monoclonal anti-(human leucocyte interferon) antibody for the purification of recombinant human leucocyte interferon,[401] 2-chloro-10-(3-aminopropyl)phenothiazine for the purification of calmodulin from bovine brain,[402] insulin for purification of the insulin receptor protein from porcine liver membranes,[403] 4-phenylbutylamine for purification of a chymotrypsin-like enzyme from human[404] or bovine[405] granulocytes, and soybean agglutinin for affinity-chromatographic studies of the lectin properties of herpes simplex virus type 1 glycoprotein.[406]

Tetraiodofluorescein has been immobilized on Affi-gel 102 and used for separation of heterogeneous protein mixtures.[407] Simple and rapid purification of tryptophan 5-mono-oxygenase from rabbit brain has been achieved using 2-amino-4-hydroxy-6,7-dimethyltetrahydropteridine immobilized on Affi-gel 202.[408]

Ethylene glycolyl bis(succinimidylsuccinate) has provided a bifunctional reagent for crosslinking and reversible immobilization of proteins through their amine groups.[409] Trypsin immobilized on agarose using this reagent retains full specific activity and can be released from the matrix with hydroxylamine.

N-Succinimidyl 3-(2-pyridyldithio)propionate has been used for controlled coupling of lysozyme and bovine serum albumin to gelatin-agarose.[410] The protein to be coupled and the gelatin-agarose conjugate were separately treated with the compound to yield 3-(2-pyridyldithio)-propionyl derivatives. The soluble derivative was reduced to thiopropionyl-protein and then conjugated to the derivatized gelatin conjugate through sulphydryl-disulphide exchange. This new method permits controlled coupling of proteins to an insoluble matrix, and the bond through which reaction occurs is known precisely.

A method for grafting enzymes onto polysaccharides using a redox system as a radical initiator has been reported.[411] Glycidylmethacrylate was used as a vinylating reagent and the reaction product with enzymes was copolymerized with agarose.

Organomercurial-agarose has been applied to the affinity

chromatography of constituent enzyme activities of the multifunctional enzyme putrescine synthase.[217]

A new method for preparation of polyacrylhydrazide-agarose has been described, based on the periodate oxidation of agarose followed by reaction with polyacrylhydrazide.[412] Enzymes, antibodies, lectins, and small ligands were coupled to the support: high degrees of binding and biological activity were obtained. The carrier combines the advantages of both agarose (minimal nonspecific interactions with proteins, good flow rates) and acrylamide (large number of modifiable groups, absence of charged groups).

Differences in the interactions of the catalytic groups of the active centres of actinidin and papain have been studied by affinity chromatography on the 2-pyridyl disulphide derivative of mercaptohydroxypropyl ether agarose gel.[413]

Thiol-agarose has been applied to the purification of α_1-antichymotrypsin from human pleural fluid and human serum,[183] the partial purification of biologically active low-molecular-weight human anti-haemophilic factor free of von Willebrand factor,[414] the purification of a Ca^{2+}-dependent thiol protease from human platelets,[415] the isolation of k-casein-like fractions from human casein,[416] and the fractionation of nuclear proteins ADP-ribosylated <u>in vitro</u> into thiol- and non-thiol-containing fractions.[417]

Thiopropyl-agarose has been shown to provide a simple and rapid means of isolating thiol peptides from large proteins, using ceruloplasmin as a model protein.[418] The matrix has also been applied to the purifications, by thiol-disulphide interchange chromatography, of biologically active low-molecular-weight human anti-haemophilic factor,[414] and α_1-antichymotrypsin from human pleural fluid and human serum.[183]

Agarose has been treated with 4-toluenesulphonyl chloride (tosyl chloride) with formation of 4-toluenesulphonic ester (tosylation).[419] The reaction is rapid and can be controlled to give a range of substitutions up to very high levels. The tosylated agarose shows excellent long-term stability and has swelling properties similar to those of untreated agarose, indicating that no cross-linking of the support occurs during tosylation. Efficient displacement of the tosyl groups occurs on addition of nucleophiles, such as amino- or mercapto-group-containing compounds, allowing ready preparation of immobilized affinity ligands and proteins.

Triazinyl dyes (usually Cibacron Blue F3GA) react with agarose *via* the chlorine atoms on the triazine ring and have been used for the purification of numerous proteins. Dye-ligand chromatography has been examined as a method for the purification of alkaline phosphatase, and 46 dye-Matrex^R gels were assessed for their ability to bind alkaline phosphatase.[420] Cibacron Blue F3GA and Procion Red HE-3B, each immobilized on agarose, have been used in a systematic investigation of their interactions with multinucleotide-dependent enzymes.[421] The results were used to design specific elution techniques for purification of these enzymes by affinity chromatography.

Cibacron Blue F3GA immobilized on agarose has been successfully applied to the purifications of calf thymus ribonuclease H I,[422] a cytosolic broad specificity β-D-glucosidase from human liver,[423] nitrate reductase from *Ankistrodesmus braunii*,[424] rat liver asparagine synthetase,[425] human platelet-derived growth factor,[426] a soluble glycoprotein from lung in pulmonary alveolar proteinosis,[427] a lymphocyte ATP-dependent deoxyribonuclease,[428] β-hydroxyisobutyrate dehydrogenase from *Candida rugosa*,[429] potato (*Solanum tuberosum*) lactate dehydrogenase isoenzymes,[430] human fibroblast interferon,[431] pig heart nucleoside diphosphate kinase,[432] L-glutamine synthetase from the cyanobacterium *Anabaena* 7120,[433] NAD(P)H : nitrate reductase and NADH : nitrate reductase from corn roots,[434] rat α_1-antitrypsin,[435] 2-enoyl-CoA reductase from *Mycobacterium smegmatis*,[436] an NAD-linked acetoacetyl-CoA reductase from *Zoogloea ramigera*,[437] phytochrome from rye (*Secale cereale*),[438] flavokinase from mung bean (*Phaseolus aureus*),[303] carnosine synthetase from avian muscle,[439] and NAD glycohydrolase from *Bungarus fasciatus* venom.[440]

The Cibacron Blue F3GA dye binding properties of α-feto-protein from four mammalian species (human, rat, bovine, ovine) have been investigated using the dye bound to agarose.[441] The affinity chromatography of serum albumins from human and animal plasmas has been investigated using Cibacron Blue F3GA-agarose.[442] Affinity gel electrophoretic studies were used to examine the effect of ligand presaturation with bilirubin and fatty acids on the interaction of serum albumins with immobilized dye.[443] The effect of nucleotide presaturation on the chromatographic behaviour of 6-phosphogluconate dehydrogenase from *Bacillus stearothermophilus* on immobilized Cibacron Blue F3GA and Procion Red HE3B has also been investigated.[444] Cibacron Blue F3GA-agarose has been used to identify the major component of the oestrogen-induced protein of rat uterus as the BB isoenzyme of creatine kinase.[445] Immobilized Cibacron Blue F3GA and Procion Red HE3B have been used for studies on the interaction of the dyes with dopamine β-monooxygenase.[446]

Triazine-activated agarose has been treated with succinic dihydrazide and then coupled with cysteinyl-proline derivatives to provide a matrix for affinity-chromatographic purification of human serum angiotensin-converting enzyme.[447]

Cibacron Blue F3GA is also available immobilized on Affi-gel, and these preparations have been used to investigate the interaction between transcobalamin II and the dye,[448] to purify the bile salt-stimulated lipase from human milk whey,[449] to purify calf thymus casein kinase I,[450] and to probe the molecular structure of phytochrome.[451] Affi-gel Blue has also been used for purifications of an apurinic/apyridiminic endonuclease from HeLa cells,[452] CTP synthetase from Ehrlich ascites tumour cells,[453] phytochrome from oat seedlings,[454] methyltransferase I from Bacillus subtilis,[455] and a membrane-bound ATP diphosphohydrolase from Cicer arietinum (chickpea) roots.[456] Cibacron Blue F3GA immobilized on Matrex^R gel has been used for the purification of keratan sulphate-degrading endo-β-galactosidase from Flavobacterium keratolyticus.[457] Matrex ^{R} Gel Red has been used for purification of CTP synthetase from Ehrlich ascites tumour cells.[453] Procion Blue-agarose has been used to remove contaminants during purification of human serum angiotensin l-converting enzyme,[319] and 6-phosphogluconate dehydrogenase from Acer pseudoplatamus L has been purified on Procion Red HE-3B immobilized on agarose.[458]

Blue dextran (dextran coupled with Cibacron Blue F3GA) coupled to cyanogen bromide-activated agarose has been used for the purifications of chicken liver L-threonine dehydrogenase,[221] L-glutamine synthetase from Escherichia coli,[222] coenzyme A synthetase from E. coli,[223] D-glyceraldehyde 3-phosphate dehydrogenase from lactic acid bacteria,[224] Leuconostoc phosphoglycerate mutase,[225] and rat Factor II,[226] and to probe the molecular structure of phytochrome.[451]

Amylose. -- Amylose has been reacted with the methylsulfinyl carbanion in dimethyl sulphoxide to form the alkoxide of amylose, which reacts with allyl bromide to afford tri-O-allylamylose, the starting material for synthesis of amylose-polyester block copolymer.[459]

The structure of the hydrated amylose-iodine complex has been determined using *X*-ray diffraction and stereochemical packing analysis.[460] Iodide was found as an almost linear chain in the centre of the 6-fold left-handed amylose helix. Eight water molecules of hydration per unit cell were located in good hydrogen-bonding positions between the amylose helices.

Characterization of the changes in absorption and c.d. spectra of tri-iodide ions in amylose with changing DP has been reported.[461] Possible explanations for the spectral changes were discussed.

Cellulose. -- Eighty-two compounds have been identified as volatile products after pyrolysis of cellulose under various conditions.[462] Comparisons between volatile pyrolysis products and components of cellulose cigarette smoke proved that they were qualitatively similar, but that most pyrolysis products were obtained in much larger quantities than smoke components per weight of cellulose. Pyrene, 1-methylpyrene, and 1-hydroxypyrene have been isolated and identified as mutation-enhancing principles from cellulose pyrolysate.[463]

Cellulases from *Trichoderma reesei* have been isolated by biospecific sorption on amorphous cellulose.[464] Thermo-chemical investigation of cellulose acetates has been reported.[465] *Leuconostoc mesenteroides* cells have been immobilized by entrapment in agar on an acetylcellulose filter, and used for the rapid analytical determination of *L*-phenylalanine.[99]

4-Aminobenzamidine aminododecylcellulose has been applied to the affinity-chromatographic purification of a plasminogen activator from mouse cells transformed by an oncogenic virus.[466] Covalent attachment of papain to 4-aminobenzoyl cellulose has been achieved using terephthalic diazide as mediator.[467] β-Amylase has been immobilized on 4-aminobenzyl cellulose after diazotization of the matrix.[468] Aminohexyl-cellulose has been used as a support for immobilization of epoxy-activated tannin.[469]

Avidin immobilized on cellulose discs activated with sodium t-butoxide and 1-tosyloxy-3-isocyanopropane has been used to develop a solid-phase assay for α-biotin.[470] Benzoquinone-activated cellulose has been used for the immobilization of papain [471] and of potato trypsin inhibitor.[472] The latter was used successfully for affinity-chromatographic purification of chymotrypsin.

The 1,1'-carbonyldi-imidazole method of activation of agarose has been extended to cellulose derivatives (see first citation of ref. 151).[151]

Quantitative studies on the binding of a variety of enzymes to carboxymethyl-cellulose have been carried out, and the magnitude of the affinity elution effect in the presence of substrates of the enzymes has been determined.[473] Some calculations were made demonstrating the range of strengths of interactions between enzyme and adsorbent, and the energy involved per charge on the protein molecule. Cellulose exchangers with pyrogallol as anchor group have been applied successfully to the chromatographic separation of antimony(III).[474]

Cyanogen bromide-activated cellulose has been used to immobilize anti-(polyinosine-polycytosine) antibody for specific separation of double-stranded RNA's[475] and to co-immobilize dihydrolipoamide reductase and 3α-hydroxysteroid dehydrogenase for use in the continuous-flow analysis of 3α-hydroxysteroids.[476] Nuclease P_1 has been immobilized by reaction with cyanogen bromide-activated cellulose, by carbodi-imide-mediated reaction with anion-exchange cellulose, and by reaction with titanium-cellulose complexes.[477] The last method was found to be the most effective, yielding an active immobilized enzyme with good stability properties.

Activation of cellulose with butan-1,4-diol diglycidyl ether has provided a matrix for simple and efficient immobilization of deoxyribonucleic acid.[478] Development of the method was described, and the affinity matrix was then applied to the purification of messenger ribonucleic acid. Deoxyribonucleic acid coupled to cellulose has also been used for the purification of an ATP-dependent deoxyribonuclease from _Bacillus laterosporus_.[479] Activated glucocorticoid-receptor complex from rat liver cytosol has been adsorbed on deoxyribonucleic acid-cellulose.[480] The complex could be extracted in a dose-dependent manner by incubation with different concentrations of sodium tungstate.

Using diazotized paper blots, heterogeneous protein samples obtained from differentiated neuroblastoma cells have been electrophoretically transferred to the blots and used for affinity purification of a heterogeneous anti-microtubule protein.[481]

Kinetic reasons for the inhomogeneity and character of deformability of the spatial network of chemically crosslinked cellulose ethers have been discussed.[482]

The behaviour of methylolated carbamoylethylated cellulose towards different classes of dyestuffs has been investigated.[483] The dyeing of cotton graft copolymers with some direct, basic, and reactive dyes has also been described.[484]

Well defined low-molecular-weight polystyrene has been grafted onto cellulose acetate in homogeneous solution.[485] The grafting was performed by esterifying the free hydroxyls in the cellulose acetate with anionically prepared polystyrene having a carboxylic acid end group which was activated either by conversion to the corresponding acid chloride or by reaction with trifluoroacetic anhydride.

Graft copolymerization of methyl methacrylate onto cellulose has been studied by varying the monomer, initiator, thiourea and acid concentrations, and temperature.[486] The graft yield increased steadily with an increase in monomer concentration. The effects of different monomers and the nature of the substrates were also studied.

Photosensitized graft polymerization of a vinyl monomer onto cotton has been investigated.[487] Modification of cellulose with vinyl monomers using the xanthogenate-hydrogen peroxide redox system has also been described.[488] Grafting of methyl methacrylate on cellulose containing sulphonic acid groups has been achieved using an Fe^{2+}-hydrogen peroxide redox system.[489]

A method for grafting enzymes onto polysaccharides using a redox system as a radical initiator has been reported.[411] Glycidylmethacrylate was used as a vinylating reagent and the reaction product with enzymes was copolymerized with cellulose.

Cellulose nitrate has been used to transfer tissue-specific antigens from polyacrylamide gels following electrophoresis.[490] Cellulose nitrate blotting has also been used in a study of hepatitis B surface antigen electrophoresed in agarose gels.[491]

Poly(adenylic acid)-containing ribonucleoproteins have been separated by affinity chromatography on oligothymidylic acid-cellulose.[492]

The covalent binding of trypsin to cellulose beads after periodate oxidation has been examined.[493] The degree of cellulose derivative solubilization is directly related to cellulose oxidation and increases with the increasing pH of the reaction mixture. Dextransucrase from Leuconostoc mesenteroides has been immobilized on porous phenoxyacetyl cellulose beads.[494] NAD pyrophosphatase has been immobilized by adsorption on phosphocellulose.[495]

The position and degree of ester sulphation in natural cellulose sulphate have been determined by ^{13}C n.m.r. spectroscopy.[496]

Thiosulphate derivatives, which can be reduced with mercaptoacetic acid, are suitable intermediates for the preparation of thiol derivatives of polymers.[497] Thiosulphate derivatives of cellulose have been prepared via chlorodeoxy- or via 3-chloro-2-hydroxy-propylcellulose, while mercaptodeoxy-cellulose prepared via a chlorodeoxy derivative was more suitable for the immobilization of non-thiol enzymes.

4-Toluenesulphonyl chloride-treated cellulose has been treated separately with aniline, benzylamine, 1-aminobutane, and piperazine to yield four different aminocelluloses which were further treated with carbon disulphide to furnish four dithiocarbamate celluloses.[498] A comparative study was made of their performance as adsorbents for several kinds of metal ions.

Chitin. -- Chitin has been dissolved in sodium hydroxide to produce a highly viscous alkaline chitin solution which was then treated with 2-chloroethanol to yield water-soluble glycolchitin.[499] The degree of glycolation of chitin was shown to be proportional to the amount of 2-chloroethanol used, under the conditions employed. Lysozyme hydrolysed 8-30% glycolated chitin with almost constant efficiency, but at a low degree of glycolation the glycolchitin was much less efficiently hydrolysed and therefore is not a suitable substrate for measurement of the activity of lysozyme.

Periodate oxidation of the non-reducing end groups of chitin has been reported.[500] The rates of hydrolysis of the modified substrates by chitinase and by lysozyme are increased relative to those of unmodified chitin.

The adsorption of uranium on chitin phosphate has been investigated to obtain information about uranium recovery from aqueous systems, especially sea water and uranium mine waste water.[501]

Chitin, unmodified, has been used for affinity-chromatographic purification of lysozyme.[502]

The suitability of krill chitin, prepared using different concentrations of KOH and HCl for deproteinization and demineralization, respectively, as a support for enzyme immobilization has been investigated.[503] The activity of enzymes immobilized on such supports depends on the degree of deproteinization of chitin, the availability of amino groups, the content of minerals, the mesh size, the structure of the surface, and the conformation of the chitin molecules. The immobilization of β-D-fructofuranosidase on krill chitin has been described.[504]

Chitosan. -- Oligosaccharides have been prepared from chitosan by partial acidic hydrolysis followed by fractionation on Dowex 50 (H^+ form).[505]

The ultrastructures of chitosan and some gel-forming branched-chain chitosan derivatives have been studied.[506]

Using sodium cyanoborohydride, lactose has been coupled to the 2-amino functions of chitosan to afford a branched 1-deoxylactit-1-yl derivative.[507] The product exhibited interesting and altered solution properties: it was insoluble in ethanol, aqueous ethanol, and other common organic solvents, but it did not gel or precipitate when dilute aqueous solutions were mixed with acid or base or with aqueous solutions of Ca^{2+}, Cr^{2+}, or Sn^{2+} chlorides, potassium chromate, boric acid, or combinations thereof.

Five derivatives of partially N-acetylated chitosan have been prepared from chitosan by reaction with acetic anhydride.[508] Each derivative was exhaustively oxidized with periodate, reduced with borohydride, and hydrolysed with dilute acid before gel-chromatographic fractionation. The composition of the reaction products indicated that free amino groups were heterogeneously present in the partially N-acetylated chitosan. Chitosan has been reacted with intramolecular dicarboxylic anhydrides to form novel N-(carboxyacyl) derivatives.[509] Some properties of the gels and viscous solutions obtained were reported.

Periodate oxidation of the non-reducing end groups of N-acetylchitosan has been reported.[500] The rates of hydrolysis of the modified substrate by chitinase and by lysozyme were increased relative to those of the unmodified polysaccharide.

Partially N-succinylated derivatives of chitosan and glycolchitosan have been prepared and crosslinked using a water-soluble carbodi-imide to form gels which may provide useful model systems for biochemical reactions.[510]

The adsorption of uranium on chitosan phosphate has been investigated to obtain information about uranium recovery from aqueous systems, especially sea water and uranium mine waste water.[501]

A new immobilization method has been described using the ionotropic gelation of chitosan (a polycation) with different multivalent anionic counter-ions ($Fe(CN)_6{}^{4-}$, $Fe(CN)_6{}^{3-}$, polyphosphates, poly(aldehydo-carbonic acid), poly-(1-hydroxy-1-sulphonate-prop-2-ene)).[511] *Escherichia coli* cells have been immobilized successfully by this method and retained ~57% of their initial tryptophan synthetase activity. Pepsin has been immobilized on chitosan.[512]

Cycloamyloses. -- Malto-oligosaccharides of DP ranging from 6 to 8 have been prepared by partial hydrolysis of cyclohexa-, cyclohepta-, and cyclo-octa-amyloses (α-, β-, and γ-cyclodextrins) with sulphuric acid, followed by chromatographic separation on a column of hydrophilic vinyl polymer gel.[513]

Two hexahydrate modifications of cyclohexa-amylose have previously been reported in aqueous solution without added guest molecules. A third crystal modification, with five water molecules located outside the molecular cavity and 2.57 within the cavity as disordered water, has now been described.[514]

Cyclohepta-amylose has been crystallized with 2,5-di-iodobenzoic acid as a guest molecule.[515] The X-ray crystal structures of this and of complexes of other meta-substituted guests were reported. The functionally important carboxylic acid group appears to lie in the primary-hydroxyl end of the cyclohepta-amylose molecule, which suggests that the structure seen in the crystals described does not correspond to a catalytically active species.

Formation constants for binary complexes of cyclohexa-amylose with alkanoic/alkanoate, 3- and 4-substituted phenol/phenolate, and benzoic/benzoate base substrates have been determined by pH potentiometry and visible spectrophotometry.[516] Dipolar interactions appear to be the main binding force in these reactions.

Two newly synthesized cyclohepta-amylose-nicotinamides have a nicotinamide group on the secondary-hydroxy side of cyclohepta-amylose, one in an axial and one in an equatorial configuration.[517] The corresponding reduced forms can reduce ninhydrin with large rate enhancements (40- and 60-fold, respectively) relative to NADH. They also show enzyme-like saturation kinetics in the reduction of ninhydrin, indicating that the reaction involves complex formation.

The rate of hydrolysis of 4-nitrophenyl acetate catalysed by azobenzene-capped cyclohepta-amylose is accelerated by photo-irradiation.[518]

Cyclohepta-amylose polymers have been used for the inclusion chromatography of some indole alkaloids.[519]

An aqueous solution of orange II dye and cyclo-octa-amylose forms a lyotropic mesophase at room temperature.[520] N.m.r. and c.d. studies suggest that the mesophase is formed from the inclusion compound which has a 1:1 structure, and that orange II enters cyclo-octa-amylose mainly from the long axis side of the molecule, which then aligns perpendicularly to the magnetic field. The equilibrium constants and rate constants for the formation of 13 azobenzene and one phenylazo-naphthalene dye-cyclohexa-amylose complexes have been determined at 26.6°C and an ionic strength of 0.15.[521] The effect of changing the substituents on the dyes was explained in terms of the sizes of the substituents and the charges on the substituents.

The rates of hydrolysis of prostacyclin(PGI_2) and its methyl ester (PGI_2Me) in aqueous solutions have been shown to be significantly retarded by cyclohexa-, cyclohepta-, and cyclo-octa-amyloses, and showed characteristic saturation kinetics and competitive inhibition.[522] The deceleration effects of cycloamyloses on the hydrolysis of PGI_2Me were about 3 times larger than those on the hydrolysis of PGI_2. The importance of the spatial relationship between the host and guest molecules was reflected in the kinetically determined stability constant for these inclusion complexations.

The formation and stability of inclusion complexes of guaiazulene with cyclohexa- and cyclohepta-amyloses have been studied.[523] The modified drug exhibited increased photostability and stability towards acid, suggesting that the cycloamylose derivatives may be of great utility in the preparation of stable drugs.

Cyclohepta-amylose has been used successfully to debitter citrus fruit products, presumably by complexation with naringin and limonin, the bitter factors in the fruits.[524]

Cyclo-octa-amylose slightly enhances the fluorescence intensity of α-naphthyloxy-acetic acid in aqueous solution.[525] When both cyclo-octa-amylose and cyclohexanol are present the intensity is markedly enhanced, showing the role of cyclohexanol as a space regulator which narrows the cavity of cyclo-octa-amylose to allow the inclusion of the fluorophore.

The formation of a charge-transfer complex between sodium α-naphthylacetate and picric acid has been shown to be promoted markedly by cyclo-octa-amylose but not significantly by cyclohepta-amylose, presumably because only cyclo-octa-amylose can include both electron donor and acceptor in the same cavity.[526]

The interactions between cycloamylose polyurethane resins and various organic compounds have been estimated by gas-solid chromatography.[527] Cycloamylose polyurethane resins were obtained by polymerization of cycloamyloses with di-isocyanates in pyridine and/or d.m.f. The cycloamylose resins exhibit strong interactions with guest molecules containing π-electrons or hetero-atoms. The resins can be used to distinguish between the configurations of xylene isomers and pyridine derivatives. The chromatographic behaviour of cyclohexa- and cyclohepta-amylose polyurethane resins prepared by this method has been studied using six aromatic acids as model compounds.[528] Phenylglycine, tyrosine, tryptophan, and phenylalanine can be separated completely, although phenylalanine is very strongly retained. The interactions between the cycloamylose units in the resins and the amino acids were discussed.

The sorption behaviour of organic compounds on polyurethane resins containing 6-deoxy-cyclohexa- or 6-deoxy-cyclohepta-amyloses, whose primary-hydroxyl groups are all deoxygenated, has been investigated.[529] Results suggested that these resins would be useful as column packings for the concentration and separation of organic compounds.

Dextrans. -- A mixed-solute exclusion method for pore-size distribution analysis of gel substances has been developed using crosslinked dextran gel.[530] A mixed solution of polymer and oligomers covering a sufficiently wide molecular-weight range is brought into contact with a gel sample, and the resulting differential dilution of the solute fractions is determined as a function of the molecular size by means of gel-permeation chromatography. The method compared favourably with two other methods.

The gel-chromatographic behaviour of aliphatic alcohols on crosslinked dextran gels (Sephadex^R G-10 and G-15) has been investigated.[531] The mechanism of interaction in aqueous systems was explained in terms of hydrophobic interactions using thermodynamic functions for dissolution in water and for partition in gel chromatography.

Affinity chromatography on crosslinked-dextran-gel columns has been used to demonstrate two distinct α(1→3) specificities of rabbit anti-dextran B1355 antibodies.[532] Concanavalin A, modified by u.v. irradiation in the presence of chloroacetamide, has been purified by affinity chromatography on crosslinked-dextran-gel columns.[533]

Lipophilic gel-filtration dextran gels (Sephadex LH-20 and Sephasorb^R HP Ultrafine) have been applied to the fractionation of polar organic constituents in environmental samples.[534] The method allowed essentially quantitative recovery of the different compounds investigated and is sufficiently rapid and convenient for use in routine analysis. Application of the method to a weathered Ekofisk crude oil was described.

Dextran T10 has been oxidized with aqueous bromine at pH 7.0.[535] The resulting oxodextran and its methoximated derivative were analysed by ^{13}C n.m.r. spectroscopy and the total number of keto groups and their positions were established. In accord with the proposed mechanism for bromine oxidation of mono- and disaccharides, the glucopyranosyl residues of dextran were oxidized mainly at C-2 and C-4. Over-oxidation resulted in a small proportion of acidic ring-cleavage products.

Dextran derivatives have been successfully activated with 1,1'-carbonyldi-imidazole, as previously described for agarose (see first citation of this reference).[151] Alkaline treatment of cyanogen bromide-treated crosslinked dextran, with or without coupled protein, has been shown to release a thiol protease inactivator, the cyanate ion.[155] Experimental evidence suggested that the carbamate group in the activated polysaccharide is the most likely precursor of the cyanate.

Sephadex activated with cyanogen bromide has been used to immobilize trypsin.[536] DEAE-dextran has been used for the isolation of casein and study of its interaction with the ion-exchange matrix.[537] DEAE-dextran has been adsorbed onto Spherosil^{R} porous silica beads and then crosslinked with 1,4-butanediol diglycidyl ether.[538] Following activation of the matrix with periodate, lyso $G_{M}1$ ganglioside was immobilized onto the DEAE-dextran/Spherosil matrix and applied successfully to the large-scale affinity-chromatographic purification of cholera toxin from *Vibrio cholerae* cultures. Periodate-oxidized dextran (dialdehyde dextran) has been reacted with 1,2-diaminoethane to yield a polyamine dextran derivative which was then coated onto the inner walls of *O*-alkylated nylon tubes.[539] Following glutaraldehyde activation of the derivatized nylon tubes, urease was successfully immobilized. Lytic enzymes from *Cytophaga* have been modified by reaction with periodate-oxidized dextran to yield a soluble enzyme preparation.[540]

Dextran sulphate has been activated with cyanogen bromide and then coupled with lysyl-agarose to provide a matrix for affinity-chromatographic purification of acid β-glucosidase from human placenta.[321]

A series of dyes covalently linked to dextran T70 have been used in a study of the relationship between the structure of anthraquinone-triazine compounds and their affinity to rabbit muscle lactate dehydrogenase.[541] The degree of substitution of the polysaccharide was found to affect the affinity. The influence of the structure of the polysaccharide constituent on the affinity of Cibacron Blue-(α-D-glucans) was also discussed.

D-Glucans. -- Cibacron Blue F3GA has been coupled to α-(1→3)-D-glucan to provide the substrate in a new assay for α-1,3-D-glucanases.[542] Lichenan immobilized on agarose has been used for the affinity-chromatographic purification of β-glucanases from *Bacillus* sp.[275]

Glycosaminoglycans. -- The vacuum-u.v. c.d. spectra of chondroitins and some of their complexes with poly(L-arginine) have been studied.[543] Periodate-oxidized chains of glycosaminoglycans have been coupled to adipic acid dihydrazide-substituted agarose for use in affinity chromatography.[329]

Dermatan sulphate-bound 6-aminohexyl-agarose has been coated with dermatan sulphate and used for the affinity-chromatographic removal of sulphatase contaminants during purification of heparinase from *Flavobacterium heparinum* extracts.[544] Heparin-bound aminohexyl-agarose coated with heparin was used in the same study for the affinity-chromatographic purification of heparatinase.

Porcine and whale heparins have been deaminated and some of the resulting sulphated oligosaccharides separated and identified.[545] The interactions of sulphated heparin and human low-density lipoprotein have been investigated with particular respect to the effect of the uronic acid composition and charge density of the glycan.[276] Gel-filtration studies of methylene blue-heparin interactions in aqueous solutions have been reported.[546]

D-Mannans. -- Guaran has been coupled to Procion Blue HB and the modified polysaccharide used to develop a reagent for detection of lectins.[547] The method involves use of the precipitin reaction: following incubation of dyed guaran and lectin, the amount of dye left in the supernatant is determined.

Alkali treatment of ivory nut D-mannan has yielded 3-deoxy-2-C-hydroxymethylpentonic, 3,4-dideoxy-pentonic, and 1,4-anhydro-3-deoxypentitol-2-carboxylic acids as the most abundant polyhydroxy acids produced, indicating that 1,4-glycosidic linkages predominated.[548] Results demonstrate that a large proportion of the

D-mannan molecules was brought completely into solution and that, in addition to non-reducing mannose end groups, non-reducing D-galactose end groups were present.

Miscellaneous Polysaccharides, Oligosaccharides, and Glycolipids. -- k-Carrageenan beads containing sulphate groups have been applied to the chromatographic purification of urokinase and as a matrix for binding tanin via Fe^{3+} ions.[549] Crosslinked and desulphated k-carrageenan beads were used for gel-filtration chromatographic purification of urease, asparaginase, and bovine serum albumin.

Colominic acid, immobilized by reaction with cyanuric chloride-activated agarose, has been used for the affinity-chromatographic purification of neuraminidase from Corynebacterium ulcerans.[154]

After epoxy activation of crosslinked pectate, iminodiacetate has been coupled to the gel and then equilibrated with Fe^{3+} ions.[550] The activated gel was used as a support matrix for immobilization of cells, and the influence of pH, ionic strength, and cell strain on immobilization was studied.

The use of 4-nitrophenyl-2-O-(α-L-fucopyranosyl)-β-D-galactopyranoside for the rapid detection of substrate-specific α-L-fucosidases has been described.[551]

The preparation of several 6,6'-disubstituted derivatives of methyl β-laminaribioside has been reported.[552] The derivatives methyl 6- and 6'-deoxy-β-laminaribiosides were also prepared. Laminaribiose, cellobiose, and gentiobiose have been acetonated with 2,2-dimethoxy-propane under various conditions.[553] Laminaribiose yielded two isopropylidene acetals in which the reducing D-glucose residue had the furanoid form, while two in which the reducing D-glucose residue formed the acyclic dimethyl acetal were obtained from cellobiose. Gentiobiose gave both types of isopropylidene compound.

The selective 4-toluenesulphonylation of 1,6-anhydro-4',6'-O-benzylidene-β-maltose has been reported.[554] The five tosylates obtained were separated and identified.

Using the nitromethane-mercuric cyanide procedure, 7-bromo-3-hydroxy-2-naphth-o-anisidide (naphthol AS-BI) has been condensed with 2,3,4,6-tetra-O-acetyl-α-D-mannopyranosyl chloride.[555] The product obtained after deprotection, 7-bromo-3-(α-D-mannopyranosyloxy)-2-naphth-o-anisidide, has been used successfully for histochemical detection of α-mannosidase.

On treatment with sodium methylsulphinylmethanide, 2-O-(4-O-methyl-α-D-glucopyranosyluronic acid)-D-xylose has been shown to be rapidly degraded by β-elimination to form 2-O-(4-deoxy-β-L-threo-hex-4-enopyranosyluronic acid)-D-xylose.[556] The kinetics of acid hydrolysis of these two compounds have been studied. Glycosyl-polyacrylamide gel copolymers have been used for affinity electrophoresis of lectins and cytochrome c.[94]

Periodate-oxidized glycolipids in liposomes have been attached to immunoglobulin G using sodium borohydride or sodium cyanoborohydride.[557] The conjugates are of potential use for antibody-mediated targeting of vesicles of cells.

A glycosphingolipid obtained from asialo-GM_1 by ozonolysis and subsequent alkali treatment has been immobilized on amino-hexyl agarose by reductive amination.[328] Anti-(asialo-GM_1) antibody was successfully purified on the immunoadsorbent thus obtained.

The antigenicity and immunological properties of a series of glycolipid antigens prepared by coupling isomaltose oligosaccharides (isomaltose – isomaltoheptaose) to a lipid carrier (stearylamine) have been investigated.[558] The antigenicities of the different stearyl isomaltosyl oligosaccharides thus obtained were compared, either when injected alone or when incorporated into liposomes.

Glycosyl-polyacrylamide gel copolymers have been used for affinity electrophoresis of lectins and cytochrome c.[94]

Starch. -- Starch has been coupled to agarose and then reacted with butylamine to introduce hydrophobic groups onto the matrix.[559] The immobilized starch derivative was successfully applied to the affinity-chromatographic purification of chloroplast α-1,4-glucan phosphorylase from spinach leaves.

Starch dyed with Remazol brilliant orange has been used as substrate for measurement of α-amylase and glucoamylase activities produced during fermentation.[560]

The photochemical grafting of vinyl monomers onto starch has been reported.[561]

3 Modification of Glycoproteins and Uses of Modified Glycoproteins

The sensitivity of a fluorescence labelling method for structural analyses of asparagine-linked sugar moieties of glycoproteins has been increased by using h.p.l.c. with a fluorescence detector.[562] After separation of sugar moieties from polypeptide portions by hydrazinolysis, free amino groups thus exposed were acetylated and the reducing ends of sugar chains reductively aminated with a fluorescent reagent, 2-aminopyridine. As little as 0.1 pmol of pyridylamino derivatives could be detected.

Albumins. -- The immunochemistry of conjugates prepared from serum albumins and acridine nitrogen mustards has been investigated.[563]

Bovine serum albumin microspheres containing adriamycin have been prepared by heat solidification of albumin in an emulsion of adriamycin, albumin, and cottonseed oil.[564] Results suggest that albumin microspheres containing adriamycin may be applicable as drug carriers in the adjuvant chemotherapy of liver metastases.

The effect of modification of albumin on tolbutamide-serum albumin binding has been studied.[565]

Antibodies. -- Following periodate activation, anti-(human serum albumin) antibody has been covalently linked to amyloglucosidase and the conjugate quantitated using an enzyme-cycling assay.[566] Conjugation of enzymes to immunoglobulins using the maleimide and periodate methods of coupling has been compared.[567] Use of N,N'-(oxydimethylene)dimaleimide rather than periodate to effect coupling was found to be superior for coupling of peroxidase, D-glucose oxidase, penicillinase, and β-D-galactosidase. Horseradish peroxidase-conjugated goat anti-immunoglobulin G antibody has been applied to the determination of immune complexes in human serum.[568]

Anti-Thy 1.2 monoclonal antibody has been covalently linked to ricin and shown to be a potent cell-type-specific toxin.[101] Thiol groups have been introduced into antibody molecules using N-succinimidyl-3(3-pyridyldithio)propionate.[569] The resulting 2-pyridyldisulphide groups on the protein were reduced with dithiothreitol to yield free thiol groups through which oxidation-induced disulphide coupling with ricin A was then effected. The conjugates were applied successfully to the selective killing of normal and neoplastic B-cells.

A simple method for preparation of antibody targeted liposomes has been described.[570] $F(ab')_2$ antibody fragments were covalently coupled to phosphatidylethanolamine and subsequently associated with dimyristoylphosphatidylcholine liposomes. Side reactions during the coupling were minimized by prior citraconylation of the $F(ab')_2$ fragments.

Phytohaemagglutinins. -- Methotrexate and chlorambucil have both been covalently linked to abrus agglutinin, abrin, ricinus agglutinin, ricin, or concanavalin A.[571] When injected into sarcoma-bearing mice each conjugate showed a higher inhibitory effect on DNA biosynthesis of tumour cells than did an equivalent dose of the free chemotherapeutic agent.

Succinylated or acetylated concanavalin A, but not the native lectin, lyses sheep erythrocytes in the presence of guinea-pig complement.[572] The effect appeared to be specific since succinylated wheat-germ agglutin is inactive and haemolysis was inhibited selectively by methyl α-D-glucopyranoside. Succinylated concanavalin A has been immobilized on agarose and used for chemical crosslinking of a dimeric protein (D-glucose oxidase).[366]

The effects of chemical modification on the conformation and biological activity of peanut agglutinin have been investigated.[573] Free amino groups on the lectin were modified with succinic anhydride and with 1-isothiocyanato-4-benzenesulphonic acid. Although the extents of modification were respectively 95% and 85%, the derivatives showed no loss of sugar-binding capacity, but their agglutinating and mitogenic activities were reduced. Derivatization of peanut agglutinin with tetranitromethane, 4-aminophenyl α-D-glucopyranoside, and 2-O-(4-aminobenzyl)-α-D-neuraminic acid was investigated: agglutinating and mitogenic activities were not seriously diminished. The influence of these modifications on the conformation of the lectin was investigated by means of c.d. studies in the far- and near-u.v. regions.

A non-agglutinating derivative of wheat-germ agglutinin has been prepared by treatment of the lectin with formic acid and cyanogen bromide.[574] The derivative binds to platelets and precipitates an antibody to the lectin.

Transferrin. -- Production of autoantibodies to transferrin has been achieved by immunization of rabbits with transferrin modified at either its neuraminic acid or its neuraminic acid and D-galactose residues with methionine sulphone hydrazide.[290] Autoantibodies obtained reacted with the native rabbit transferrin in radioimmunoassay and gel-diffusion systems.

Miscellaneous Glycoproteins. -- The effect of the carbohydrate moiety of k-casein on complex formation with β-lactoglobulin has been studied.[575] The carbohydrate portion does influence complex formation, and Ca^{2+} has an inhibitory effect on the interaction.

Conjugates of fetuin and asialofetuin with diphtheria toxin fragment A have been prepared.[576] The conjugates inhibited protein synthesis on cultured rat hepatocytes and bound to asialoglycoprotein receptors on the hepatocytes.

Sulphydryl-agarose has been used to remove Hg^{2+} from Hg^{2+}-fibronectin for regeneration of its thiol groups.[577]

Asialo-(human chorionic gonadotrophin) has been modified at its D-galactose residues by treatment with D-galactose oxidase or with periodate and borohydride.[578] The effects of such modification on the biological properties of the hormone were studied. Covalent crosslinking of human chorionic gonadotrophin to its receptor in rat testes has been achieved using disuccinimidyl suberate or dithiobis(succinimidyl propionate).[579]

Desialylated ovine submaxillary mucin, immobilized on agarose, has been used as an acceptor for D-galactose transfer from UDP-D-galactose.[580]

Immobilized Derivatives of Glycoproteins. -- The modification of glycoproteins to yield biologically active immobilized derivatives continues to attract much attention. These derivatives and their uses are summarized in Table 6.[581-598]

The use of monoclonal antibodies in biochemical research, including their use in immunoadsorption chromatography, has been described.[599] Monoclonal antibodies to α-fetoprotein have been immobilized and applied to purification of human α-fetoprotein.[185] Anti-interferon monoclonal antibody immobilized on Affi-gel 10 has been used to purify recombinant human leucocyte interferon.[401]

Anti-(human immunoglobulin G) antibodies immobilized on Spheron gel have been used successfully for separation of human T and B cells.[587] Pregnancy-specific β_1-glycoprotein has been measured successfully by fluoroimmunoassay using antibody coupled to magnetizable particles.[588]

A review has appeared describing the purification of enzymes by chromatography on heparin-agarose matrices.[595] The interaction of heparin with enzymes is described, methods for covalent attachment of heparin to agarose are reviewed, and advantages and disadvantages of the purification method are discussed.

An immunoadsorbent has been prepared by coupling albumin to gelatin pre-coupled to agarose.[410] The method permits controlled

(Text continues on page 714)

Table 6 Modification of glycoproteins by coupling to insoluble matrices and other macromolecules

Glycoprotein	Macromolecule or matrix coupled and mode of coupling	Use of product	Ref.
Albumin	Reaction with cyanogen bromide-activated agarose	Affinity chromatography of proteolytic fragments of fibronectin	210
	Reaction with agarose activated by oxidation with Br_2	Demonstration of new method of protein immobilization	147
	N-Succinimidyl-3-(2-pyridyldithio)propionate-mediated reaction with gelatin-agarose	Development of a new method for preparation of immunoadsorbents	410
	Carbodi-imide-mediated reaction with nylon-co(acrylic acid) polymers	Study of use of graft copolymers as supports for protein immobilization	581
	Glutaraldehyde or carbodi-imide-mediated co-crosslinking with α-cobra toxin or acetylcholine receptor	Immunoadsorption of receptor-specific ligands	582
	Glutaraldehyde-mediated co-cross-linking with D-glucose oxidase	Development of a direct method for measurement of activity of	583

Table 6 continued

Glycoprotein	Macromolecule or matrix coupled and mode of coupling	Use of product	Ref.
	and peroxidase onto a membrane	immobilized enzymes	
	Reaction with 3-(2',3'-epoxypropoxy)propyl-glass, 3-(2',3'-epoxypropoxy)propyl-silica, epoxy-activated glycidyl methacrylate copolymer or oxirane-acrylic beads	Study of the effect of the nature of proteins on their coupling to different epoxide-containing supports	584
Anti-(human serum albumin) antibody	Reaction with cyanogen bromide-activated agarose	Purification of α_1-anti-chymotrypsin from human pleural fluid and serum	183
Anti-(α_1-antichymotrypsin) antibody	Reaction with cyanogen bromide-activated agarose	Purification of α_1-anti-chymotrypsin from human pleural fluid and serum	183
Anti-(human or rat α-fetoprotein) antibody		Purification of human or rat α-fetoprotein	184
Anti-(human α-fetoprotein) monoclonal antibody		Purification of human α-fetoprotein	185

Table 6 continued

Glycoprotein	Macromolecule or matrix coupled and mode of coupling	Use of product	Ref.
Anti-(blood group I) antibody		Purification of sheep gastric mucins with blood-group Ii activity	186
Anti-(human chorionic gonadotrophin) antibody		Development of an enzyme immunoassay for human chorionic gonadotrophin	187
Anti-(human C5 component of complement) antibody		Purification of complement attack complexes from insulin-activated human serum	188
Anti-(Factor VIII) antibody	Reaction with cyanogen bromide-activated Biogel[R] A	Study of the structure and function of the Factor VIII complex	585
Anti-(immunoglobulin A) antibody	Glutaraldehyde-mediated reaction with amino-derivatized controlled pore glass	Purification of polyclonal IgA by repetitive semi-automatic immunoadsorption chromatography	586
Anti-(human immunoglobulin G) antibody	Reaction with cyanogen bromide-activated Spheron[R]	Separation of human T and B cells	587
Anti-(immunoglobulin M) antibody	Glutaraldehyde-mediated reaction with amino-derivatized controlled pore glass	Purification of polyclonal IgM by repetitive semi-automatic immunoadsorption chromatography	586
Anti-interferon	Reaction with cyanogen bromide-	Purification of human lymphoblastoid	128

Table 6 continued

Glycoprotein	Macromolecule or matrix coupled and mode of coupling	Use of product	Ref.
antibody	activated agarose	interferon in high yield	
Anti-interferon monoclonal antibody	Reaction with Affi-gel^{R} 10	Purification of recombinant human leukocyte interferon	401
Anti-(rabbit immunoglobulin G) antibody	Reaction with cyanogen bromide-activated agarose	Radioimmunoassay for human placental lactogen	189
Anti-(urinary kallikrein) antibody	Reaction with cyanogen bromide-activated agarose	Purification of a prekallikrein from rat pancreas	190
Anti-(NADP-isocitrate dehydrogenase) antibody	Dimethylsuberimidate-mediated crosslinking onto protein A-agarose	Purification of *Escherichia coli* NADP-isocitrate dehydrogenase by immunoadsorption	203
Anti-(poly(I)-poly(C)) antibody	Reaction with cyanogen bromide-activated cellulose	Specific separation of double-stranded ribonucleic acids	475
Anti-renin antibody	Reaction with cyanogen bromide-activated agarose	Purification of human renin	191
Anti-(*Rhizobium* strain TA1 polysaccharide) antibody		Separation of pyruvylated polysaccharides into pyruvate-rich and -poor fractions	192
Anti-(SP_1) antibody, cyanogen bromide-	Reaction with magnetizable cellulose-iron oxide particles	Sequential fluoroimmunoassay for measurement of pregnancy-specific	588

Table 6 continued

Glycoprotein	Macromolecule or matrix coupled and mode of coupling	Use of product	Ref.
activated		β_1-glycoprotein	
Anti-(thermolabile antigen) antibody	Reaction with cyanogen bromide-activated agarose	Purification of thermolabile antigen from baker's yeast	193
Anti-(thrombin-associated surface antigen) antibody		Isolation of polyoma virus-induced surface antigen from mouse cells	194
Anti-thrombin III		Study of properties of modified (fluorescent) heparin	127
		Comparative study of structures of highly active and relatively inactive forms of whale heparin	589
Asialofetuin	Reaction with cyanogen bromide-activated agarose	Purification of a phytohaemagglutinin from the Azolla-Anabaena symbiosis	218
		Purification of lactose-blocking lectin activity from fetal-calf skeletal muscle	219
Asialo-G_{M1}, ozonolysed		Purification of anti-glycosphingolipid antibody	348

Table 6 continued

Glycoprotein	Macromolecule or matrix coupled and mode of coupling	Use of product	Ref.
Bandeiraea simplicifolia I lectin		Affinity chromatography of blood-group ABH-active polyglycosyl peptides	220
Carcinoscorpius rotunda cauda (horseshoe crab) lectin		Affinity-chromatographic fractionation of sialoglycoproteins	229
Casein		Purification of a heparin-sensitive nuclear protein kinase	230
Clq component of complement	Adsorption on inner surfaces of polystyrene tubes	Determination of immune complexes in human serum	568
Concanavalin A	Reaction with cyanogen bromide-activated agarose	Purification of glycosylceramidase from murine intestine	112
		Purification of rat uterine peroxidase	138
		Purification of bovine brain 2-acetamido-2-deoxy-hexosidase C	124
		Co-purification of alkaline phospha-	131

Table 6 continued

Glycoprotein	Macromolecule or matrix coupled and mode of coupling	Use of product	Ref.
		tase and 5'-AMP-specific nucleotidase from *Dictyostelium discoideum*	
		Purification of an ester hydrolase from murine liver	142
		Isolation of a root lectin from soybean (*Glycine max*)	207
		Purification of cathepsin D from porcine myometrium	232
		Purification of an irreversible tissue inhibitor of collagenase in human amniotic fluid	233
		Affinity chromatography of *Trypanosoma brucei* antigens	234
		Purification of rat α_1-fetoprotein	235
		Purification of 53K glycoprotein from rabbit skeletal muscle	236
		Purification of acid phosphatase from rat prostatic adenocarcinoma	237
		Purification of cholinesterase from	238

Table 6 continued

Glycoprotein	Macromolecule or matrix coupled and mode of coupling	Use of product	Ref.
		chicken egg yolk	
		Affinity-chromatographic fractionation of glycopeptides isolated from a rat liver biantennary glycan	239
		Purification of a phospholipid-dependent α-D-mannosidase from rabbit liver	240
		Purification of a concanavalin A receptor from boar spermatozoa	241
		Purification of human liver acid β-D-galactosidase	242
		Purification of two antigenic glycoproteins from rye-grass (*Lolium perenne*) pollen	243
		Purification of α_1-antitrypsin	244
		Affinity-chromatographic purification of human liver β-D-2-acetamido-2-deoxy-hexosidase	245
		Purification of glycoconjugates	246

Table 6 continued

Glycoprotein	Macromolecule or matrix coupled and mode of coupling	Use of product	Ref.
		from *Spongia officinalis*	
		Affinity-chromatographic fractionation of a mixture of 2-{^{3}H}-D-mannose-labelled glycopeptides derived from mouse lymphoma cell glycoproteins	247
		Isolation of two *endo*-β-D-2-acetamido-2-deoxy-glucanases from fig latex	248
		Purification of a neutral protease cleaving type IV collagen	249
		Purification of β-D-2-acetamido-2-deoxy-hexosidases A and B from human liver	250
		Purification of sciatin	251
		Purification of variant antigens of *Trypanosoma congolense*	252
		Purification of plasminogen activator secreted by human	253

Table 6 continued

Glycoprotein	Macromolecule or matrix coupled and mode of coupling	Use of product	Ref.
		melanoma cells in culture	
		Affinity chromatography of pepsin fragments of human placental and renal basement membrane laminin	254
		Purification of rat liver α-D-2-deoxy-glucosyl phosphodiesterase	255
		Affinity chromatography of glyco-proteins from adult-rat brain synaptic vesicles	256
		Demonstration of glycoprotein characteristics of the sodium channel saxitoxin-binding compo-nent from mammalian sarcolemma	351
		Characterization of *Xenopus* oocyte subcellular fractions and incubation media	352
		Study of transformation-dependent quantitative changes in glyco-peptide binding to the lectin	353

Table 6 continued

Glycoprotein	Macromolecule or matrix coupled and mode of coupling	Use of product	Ref.
		Reversible binding of antithrombin III for use in the fractionation of heparin	354
		Study of binding of insulin receptors to lectins	355
		Removal of haemagglutinating activity during protease purification	356
		Study of the glycosylation of the arginine vasopressin/neurophysin II common precursor	357
		Study of microheterogeneity of ovalbumin	358
		Study of fibrinogen chains during synthesis	359
		Affinity-chromatographic demonstration of two α_1-acid glycoprotein populations in purified protein preparations	360
		Affinity-chromatographic separation	361

Table 6 continued

Glycoprotein	Macromolecule or matrix coupled and mode of coupling	Use of product	Ref.
		of right-side-out oriented subfractions from purified calf thymocyte plasma membranes	
		Demonstration of two types of colony-stimulating factor in serum of lipopolysaccharide-treated mice	362
		Removal of lutrophin and chorio-gonadotrophin β-subunit during preparation of glycoprotein hormone-free serum	363
		Demonstration of glycoprotein nature of chlorophyllase	364
		Demonstration of heterogeneity of heterogalactan from fruit bodies of <u>Fomitopsis</u> <u>pinicola</u>	365
		Support for immobilization of β-<u>D</u>-2-acetamido-2-deoxy-hexosidase and aryl sulphatase A	590

Table 6 continued

Glycoprotein	Macromolecule or matrix coupled and mode of coupling	Use of product	Ref.
		Affinity chromatographic study of oligosaccharides from band 3 glycoprotein from human erythrocyte membranes	591
		Purification of proteinase inhibitors	592
	Reaction with Affigel^{R} 15	Model protein for study of effect of protein charge on immobilization	400
	Reaction with cyanogen bromide-agarose	Investigation of crosslinking of D-glucose oxidase onto the modified lectin matrix	366
Discoidin I		Purification of discoidin I-binding proteoglycan from axenic *Dictyostelium discoideum*	259
	Glutaraldehyde-mediated reaction with Biogel^{R} P300	Identification of endogenous binding proteins for the lectin in *Dictyostelium discoideum*	593
Dolichos biflorus lectin	Reaction with cyanogen bromide-activated agarose	Purification of receptors for the lectin from embryonal carcinoma cells	260

Table 6 continued

Glycoprotein	Macromolecule or matrix coupled and mode of coupling	Use of product	Ref.
Erythrocytes	Formalinized	Purification of chitin-binding haemagglutinin from *Conidiobolus lampranges*	594
Erythropoietin	Adsorption on agarose-bound wheat-germ agglutinin or phytohaemagglutinin	Stimulation of erythroid colony formation *in vitro*	375
Factor X		Affinity-chromatographic purification of Factor VIII	153
Fetuin		Purification of a neuraminic acid-binding lectin from horseshoe crab (*Carcinoscorpius rotunda cauda*)	261
Fibronectin		Purification of glycosaminoglycans	262
Hog gastric mucin, desialylated		Purification of *endo*-β-D-galactosidase from *Escherichia freundii*	263
Helix pomatia lectin		Affinity-chromatographic study of subunit structure of enterokinase	268
		Study of lectin properties of herpes simplex virus type 1 glycoprotein	406
Heparin		Purification of a heparin-sensitive	230

Table 6 continued

Glycoprotein	Macromolecule or matrix coupled and mode of coupling	Use of product	Ref.
		nuclear protein kinase	
		Purification of rat liver 2-acetamido-2-deoxy-D-glucosyl phosphodiesterase	255
		Purification of *Xenopus laevia* type 1 topoisomerase	269
		Purification of rat hepatic triglyceride lipase	270
		Affinity chromatography of native and chemically modified antithrombin III	271
	Carbodi-imide-mediated reaction with aminohexyl-agarose	Affinity-chromatographic purification of bovine testicular hyaluronidase	320
	Reaction with cyanogen bromide-activated agarose	Assessment of interactions between polyamines, nuclear protein kinase, and heparin. Purification of protein kinases inhibited by heparin	372
		Modification of heparin-agarose pro-	373

Table 6 continued

Glycoprotein	Macromolecule or matrix coupled and mode of coupling	Use of product	Ref.
		cedure for determination of plasma lipolytic activity of normal lipidemic and hyperlipidemic subjects after heparin injection	
		Purification of tyrosine hydroxylase	136
		Purification of the bile salt-stimulated lipase from human whey	449
		Purification of lysosomal hydrolases, RNAse, and DNA and RNA polymerases	595
	Reaction with Ultrogel[R]	One-step purification of eggwhite lysozyme	596
	Glutaraldehyde-mediated reaction with polyvinyl alcohol beads	Inactivation of thrombin by anti-thrombin III on heparinized bio-material	597
Immunoglobulin A, dimeric	Reaction with adipic dihydrazide-derivatized agarose	Purification of membrane receptor for polymer immunoglobulin	195
Immunoglobulin A, polymeric	Glutaraldehyde-mediated reaction with aminohexyl-agarose	Purification of free secretory complex from rat bile	196
Immunoglobulin E	Reaction with cyanogen bromide-	Purification of cell-surface	197

Table 6 continued

Glycoprotein	Macromolecule or matrix coupled and mode of coupling	Use of product	Ref.
	activated agarose	receptor for IgE	
Immunoglobulin $F(ab')_2$ antibody fragments		Single-step isolation of IgG from human serum	198
α-Lactalbumin		Purification of human serum _D_-galactosyltransferases	273
Lentil (_Lens culinaris_) lectin		Study of the carbohydrate-binding specificity of lentil lectin	247
		Affinity-chromatographic study of binding of glycopeptides of different structures	274
		Study of binding of insulin receptors to lectins	355
		Affinity-chromatographic study of oligosaccharides from band 3 glycoprotein from human erythrocyte membranes	591
Lotus tetragonolobus lectin		Affinity-chromatographic study of subunit structure of enterokinase	268
Ovalbumin		Study of interactions with anionic	171

Table 6 continued

Glycoprotein	Macromolecule or matrix coupled and mode of coupling	Use of product	Ref.
		detergents	
		Affinity chromatography of proteolytic fragments of fibronectin	210
Ovomucoid		Isolation of α-D-mannose:1,2-(2-acetamido-2-deoxy-D-glucosyl) transferase from trachea mucosa	282
	Reaction with cyanogen bromide-activated agarose or reaction with epoxy-activated agarose	Affinity chromatography of a trypsin-like enzyme from *Streptomyces erythreus*	386
Pea (*Pisum sativum*) lectin	Reaction with cyanogen bromide-activated agarose	Affinity chromatography of a mixture of 2-{^{3}H}-D-mannose-labelled glycopeptides derived from mouse lymphoma cell glycoproteins. Study of the carbohydrate-binding specificity of pea lectin	247
		Affinity—chromatographic study of binding of glycopeptides of different structures	274
		Study of binding of insulin recep-	355

Table 6 continued

Glycoprotein	Macromolecule or matrix coupled and mode of coupling	Use of product	Ref.
		tors to lectins	
Peanut agglutinin		Isolation of lectin receptors from human erythrocytes without use of detergents	374
Phytohaemagglutinin E		Support for immobilization of erythropoietin for attachment of erythroid colonies *in vitro*	375
Ricin I and ricin II		Study of binding of insulin receptors to lectins	355
Ricinus communis agglutinin		Affinity chromatography of angiotensin-converting enzyme from porcine kidney, serum, and cultured porcine aorta endothelial cells	172
		Purification of a glycoprotein from sera of cancer patients	286
		Affinity-chromatographic separation of cell-surface glycoproteins and glycolipids	287
		Demonstration of glycoprotein	351

Table 6 continued

Glycoprotein	Macromolecule or matrix coupled and mode of coupling	Use of product	Ref.
		characteristics of the sodium channel saxitoxin-binding component from mammalian sarcolemma	
Ricinus sanguinis agglutinin		Affinity chromatography of glycoproteins from adult-rat brain synaptic vesicles	256
Soybean agglutinin		Isolation of lectin receptors from human erythrocytes without use of detergents	374
		Study of lectin properties of herpes simplex virus type 1 glycoprotein	406
Submaxillary mucin (ovine), desialylated		Acceptor for D-galactose transfer catalysed by D-galactosyl transferase from human erythrocyte ghosts	580
Thrombin		Intermediate gel in method for demonstration of thrombin-binding platelet proteins by crossed immunoelectrophoresis	376

Table 6 continued

Glycoprotein	Macromolecule or matrix coupled and mode of coupling	Use of product	Ref.
Thyroid-stimulating hormone (TSH)		Purification of human TSH receptors from thyroid membrane	288
Transferrin		Purification of autoantibodies to transferrin	290
Trypsin inhibitor (potato)	Reaction with benzoquinone-activated cellulose	Affinity chromatography of chymotrypsin	472
Trypsin inhibitor (rice bran)	Reaction with cyanogen bromide-activated agarose	Purification of *Streptomyces griseus* trypsin	292
Trypsin inhibitor (soybean)		Affinity chromatography of porcine pancreatic deoxyribonuclease I	258
	Glutaraldehyde-mediated reaction with aminosilanized LiChrospher^R	Study of resolution achievable in high-performance liquid affinity chromatography	598
Vicia cracca lectin	Reaction with cyanogen bromide-activated agarose	Affinity-chromatographic fractionation of bloodgroup ABH-active polyglycosyl peptides from human erythrocyte membranes	220
		Affinity chromatography of blood-group A-active glycoproteins from	294

Table 6 continued

Glycoprotein	Macromolecule or matrix coupled and mode of coupling	Use of product	Ref.
		human erythrocyte membranes	
von Willebrand factor		Isolation of human platelet receptors for von Willebrand factor	295
Wheat-germ agglutinin		Affinity chromatography of glycoconjugates from *Spongia officinalis*	246
		Purification of a glycoprotein from sera of cancer patients	286
		Purification of intracellular α-D-mannosidase from cultured skin fibroblasts	296
		Isolation of plasma membrane vesicles of human platelets	297
		Isolation of human platelet membrane fractions	298
		Demonstration of glycoprotein characteristics of the sodium channel saxitoxin-binding component from mammalian sarcolemma	351
		Removal of lutrophin, choriogonado-	363

Table 6 continued

Glycoprotein	Macromolecule or matrix coupled and mode of coupling	Use of product	Ref.
		trophin, and thyrotrophin during preparation of glycoprotein hormone-free serum	

coupling of proteins to an insoluble matrix, and the bond through which reaction occurs is known precisely. Anti-albumin antibodies bound to the agarose-gelatin-albumin conjugate are bound less tightly than to agarose-albumin, permitting recovery of purified antibody from the column immunoadsorbent in improved yield.

Immobilized Cells. -- The immobilization of cellular material for use both in affinity chromatography and for the production of enzymatically active preparations continues to receive considerable attention.

Acetobacter aceti cells adsorbed on cylindrical monoliths of cordierite have been used in fixed-cell reactors for rapid production of acetic acid.[600] *Alcaligenes eutrophus* cells entrapped in alginate or k-carrageenan gel have been applied to the detritiation of water.[601] The kinetics of isomerization of D-glucose catalysed by immobilized and non-immobilized *Arthrobacter* cells have been compared.[602] *Azotobacter chroococcum* cells entrapped in agar gel have been used in a batch system for continuous nitrogen fixation.[603]

A hydrogen peroxide sensor system has been developed using *Beneckea harveyi* cells immobilized on photo-crosslinkable resins or within alginate gel.[604] *Brevibacterium ammoniagenes* cells entrapped in polyacrylamide gel have formed part of a reactor cell system for production of glutathione and NADP.[605] L-Glutamic acid has been produced in a recycle reactor employing *Brevibacterium flavum* cells immobilized on collagen.[606] The modified immobilization procedure devised allowed use of any given pH and ionic strength. *Brevibacterium fuscum* cells entrapped in k-carrageenan gel have been used for continuous production of 12-ketochenodeoxycholic acid.[607]

Caldariella acidophila cells have been immobilized by glutaraldehyde-mediated co-crosslinking in chicken eggwhite.[608] Magnetite could also be entrapped in the resin to afford a biocatalyst with magnetic properties. *Candida tropicalis* cells have been immobilized in epoxy-activated crosslinked pectate to demonstrate the utility of a new support for whole-cell immobilization.[550]

The effect of immobilization on various glass-forming monomers (hydrophilic and hydrophobic) on the stability of photosystem II activity of spinach chloroplasts has been investigated.[609,610] The thermostability of the chloroplasts was greatly increased by immobilization. Spinach chloroplasts and *Clostridium butyricum* cells have been entrapped in agar gel and used together in a photochemical-energy conversion system.[611] Immobilized Clostridium butyricum spores have been used in the development of a method for starting a fermentor aseptically.[612] Biophotolytic hydrogen production by immobilized Clostridium pasteurianum cells has been investigated.[613] The cells were immobilized by glutaraldehyde-mediated reaction with aminopropyl glass or amino-Spherosil^R. Coriolus versicolor fungal cells entrapped in alginate gel have been used for decolorization of a kraft mill effluent.[614] Daucus carota cells entrapped in alginate gel have been used for conversion of gitoxigenin to 5β-hydroxygitoxigenin.[615]

Neuraminidase-treated human erythrocytes have been bound to lectin-coated agarose.[374] Cells could be sheared off the beads by mechanical disruption, leaving lectin receptors still attached to the beads. Erythroid colonies have been attached to agarose-lectin-erythropoietin complexes and the mechanism of colony formation studied.[375]

Escherichia alcalescens cells entrapped in polyacrylamide gel have been used to correlate the fate of {^{14}C}-L-leucine-labelled bacteria with changes in the enzymic activity of the entrapped cells.[616] E. coli cells have been immobilized by entrapment in polyurethane foam,[617] by reaction with epoxy-activated crosslinked pectate,[550] by entrapment in polyacrylamide gel,[605] and by a new immobilization method involving ionotropic gelation of chitosan with a multivalent anionic counter-ion.[511] Kluyveromyces fragilis cells entrapped in alginate gel have been used for batch production of ethanol from lactose,[618] or from Jerusalem artichoke tubers.[619] Leuconostoc mesenteroides cells immobilized by entrapment in agar on an acetylcellulose filter have been used for rapid determination of L-phenylalanine using a lactate electrode.[99] Methanosarcina barkeri cells entrapped in alginate gel have been applied to the conversion of methanol to methane.[620] Nocardia erythropolis, N. opaca, and Mycobacterium phlei cells have each been immobilized by entrapment in polyacrylamide gel or by adsorption on DEAE-cellulose and silica.[621] The immobilized-cell preparations have 4-en-3-oxosteroid:(acceptor)-1-en-oxidoreductase activities and have been used in continuous column processes to transform 1,4-dien-3-oxo-steroids into 4-en-3-oxo-steroids. Pachysolen tannophilus cells entrapped in alginate gel have been used for production of ethanol from D-xylose.[622] The stability of the L-aspartate β-decarboxylase activity of Pseudomonas dacunhae cells entrapped in k-carrageenan gel has been investigated.[623] The performance of a hollow-fibre microbe reactor has been compared with a reactor containing Pseudomonas denitrificans cells entrapped in alginate gel.[624]

Rhizopus nigricans cells entrapped in polyacrylamide, alginate, and agar gels have been used for the 11α-hydroxylation of progesterone.[625] Rhodopseudomonas capsulata cells immobilized by glutaraldehyde-mediated crosslinking with albumin have been used in an investigation of the influence of immobilization on light-induced phosphorylation of ADP by the chromatophore system.[626] ATP production by R. capsulata cells immobilized by a number of methods has been compared.[627] Cells were immobilized by glutaraldehyde-mediated co-crosslinking with albumin or gelatin, by entrapment in alginate or k-carrageenan gel, or by copolymerization with urethane prepolymers. An electron-microscopic study of immobilized Rhodopseudomonas capsulata cells, entrapped in alginate gel, has been described.[628] Rhodotorula minuta cells entrapped within photocrosslinked or polyurethane resin gels have been used for stereoselective hydrolysis of DL-menthyl succinate.[629]

Saccharomyces cerevisiae cells entrapped in alginate gel have been used for the continuous production of ethanol from D-glucose.[630-634] Polycation treatment following entrapment in alginate gel resulted in a matrix with higher resistance to phosphate.[634] The treatment inhibited respiration of entrapped cells but did not reduce ethanol production. Ethanol production by S. cerevisiae cells immobilized by adsorption on different ion-exchange resins,[635] by entrapment in k-carrageenan gels,[636] and by entrapment in polyacrylamide gel[637] has also been reported. S. cerevisiae cells entrapped in polyacrylamide gel have been used as part of a reactor cell system, simulating the glycolytic pathway, for the production of glutathione and NADP.[605,638] S. cerevisiae cells have been used as a model system for investigation of cell immobilization methods involving entrapment in silica hydrogel,[639] reaction with epoxy-activated crosslinked pectate,[550] entrapment in gelatin,[640] and entrapment in alginate beads followed by treatment with polyethyleneimine and glutaraldehyde or reaction with periodate- or carbodi-imide-activated alginate beads followed by polyethyleneimine crosslinking.[641]

Continuous production of steroid glycoalkaloids has been achieved in a packed-column recycle reactor employing Solanum aviculare cells immobilized by glutaraldehyde-mediated reaction with poly(phenylene oxide).[642]

Staphylococcus aureus cells, heat killed and formalin fixed on a pleat membrane filtration system, have been shown to produce tumoricidal responses in spontaneous canine neoplasms after extracorporeal perfusion over the immobilized cells.[643] The tumoricidal effect was presumably due to protein A present in the cells removing immune complexes.

Stemphylium loti cells immobilized by treatment with polyelectrolyte flocculating agents have been used for batch-wise or continuous degradation of cyanide.[644]

Streptomyces aureofaciens cells have been immobilized by glutaraldehyde crosslinking into a gelatin matrix and applied to the biotransformation of daunomycinone.[645] A mild new method for cell immobilization involving glyoxal-, glutaraldehyde-, or periodate-oxidized polyvinyl alcohol-crosslinking onto polyacrylamide hydrazide gels has been developed using S. clavuligerus cells as the model system.[646] The immobilized cells were used for antibiotic (cephalosporin) production and yielded significantly higher amounts of product than did cells immobilized by direct polymerization of acrylamide monomer. S. fradiae cells entrapped in polyacrylamide gel have been used for protease production.[647] S. phaeochromogenes cells with D-glucose isomerase activity have been adsorbed on diatomaceous earth and used in a column form for a study of substrate pulse response in reversible and consecutive reactions.[648] Results obtained were in good agreement with those predicted theoretically. S. phaeochromogenes cells have also been immobilized by aggregation of chitosan and used in an enzyme reactor.[649] The 16α-hydroxylation of dehydroepiandrosterone has been achieved using S. roseochromogenes cells immobilized in a photocrosslinkable resin.[650]

A study of the immobilization of lettuce thylakoids by several methods has been reported.[651] The effect of different microenvironments on activity yield, storage stability, and functional stability was described. The fermentative phosphorylation of nucleotides has been achieved using yeast cells entrapped in polyethylene glycol hydroxyethylacrylate gels.[652] Zymomonas mobilis cells entrapped in alginate gel have been used for efficient production of ethanol in a bioreactor.[653]

The use of immobilized cells has recently been reviewed.[654] A mathematical treatment has been published which allows calculation of the theoretical maximum reactor productivities which can be achieved in gas-producing immobilized-cell systems.[655]

4 Modification of Enzymes and Uses of Modified Enzymes

Acid Phosphatase. -- The stabilization of acid phosphatase by linear-chain, commercially available polyacrylamides in ultrafiltration membrane reactors has been studied.[656] Enzyme stabilities were compared with that of the enzyme after immobilization by glutaraldehyde-mediated crosslinking with human serum albumin.

β-D-Fucosidase. -- β-D-Fucosidase has been inactivated by treatment with iodine, N-acetylimidazole, or tetranitromethane, providing evidence for the existence of essential tyrosine residues.[657] Effective protection against inactivation of the enzyme was achieved by including a substrate analogue (lactose) in the incubates.

β-D-Galactosidase. -- Prostaglandin $F_{2\alpha}$ has been labelled with β-D-galactosidase using the mixed-anhydride method of coupling.[658] The conjugate was used in the development of a sensitive and specific enzyme immunoassay for prostaglandin $F_{2\alpha}$.

Glucoamylase. -- Anti(human serum albumin) immunoglobulin G has been covalently linked to glucoamylase and quantitated in an enzyme-cycling assay.[566]

D-Glucose Isomerase. -- The effect of γ-irradiation on purified D-glucose isomerase in dilute solution has been investigated.[659]

Hyaluronidase. -- Functional arginine residues in purified bovine testicular hyaluronidase have been modified with butane 2,3-dione.[660]

Lysozyme. -- The binding of Ca^{2+} to lysozyme and its derivatives has been studied by u.v. difference spectroscopy at various pH's.[661] Binding caused proton release from lysozyme and did not inhibit the binding of chitotriose to lysozyme. Ca^{2+} was suggested to bind near the catalytic carboxyl groups in the enzyme, thereby causing inhibition of enzyme activity. Ion binding shifted the native-denatured transition in lysozyme toward the native state.

The preparation and immunological characterization of two lysozyme derivatives, dinitrophenylated at Lys-33 and Lys-96, respectively, have been reported.[662]

The specificity of sialyltransferases has been studied using glycosylated lysozyme derivatives as substrates.[663]

Miscellaneous Glycoenzymes. -- Conjugation of horse-radish peroxidase to *Staphylococcus aureus* protein A has been compared using benzoquinone, glutaraldehyde, or periodate as crosslinking reagents.[664] Benzoquinone and glutaraldehyde treatment gave high yields of coupled protein A in conjugates of low molecular size, whereas periodate treatment resulted in a high yield of polymeric conjugates of large molecular size.

Immobilized Derivatives of Enzymes. -- Considerable interest continues to be taken in the immobilization of enzymes. The proceedings of the 5th International Enzyme Engineering Conference have been published.[665] The application of immobilized enzymes in the food industry has been reviewed.[666]

The reversible immobilization of enzymes has been reviewed with special reference to their use in analytical procedures.[667] The possibility of using a low degree of substitution of the support permits high sensitivity in inhibitor assays and may encourage use of binding assays in combination with conventional enzyme-based analytical techniques. The application of enzyme electrodes in analysis[668] and the use of enzyme thermistors for process control[669] have been reviewed. Different methods for the immobilization of enzymes on liposomes have been reviewed.[670]

The use of polyethylene beads as supports for enzyme immobilization has been discussed.[671] Films of highly polymerized collagen prepared under industrial conditions have been used routinely after acyl azide activation for the covalent immobilization of numerous enzymes from different classes.[672] The stability of the resulting membranes to operational and storage conditions, their excellent mechanical strength and their resistance to bacterial degradation render them suitable for a number of applications.

A new amperometric analytical system based on biocatalysts has been described.[673]

A model has been proposed in which a deactivated enzyme protects an active enzyme from deactivation, for both immobilized and soluble enzymes.[674] Steric hindrances and the immobile nature of the enzyme-support link of immobilized enzymes appear to lessen the extent of this protection.

A theoretical treatment describing approximate expressions of effectiveness factors for immobilized biocatalysts has appeared.[675] The effect of intraparticle diffusion resistance on the apparent stability of immobilized enzymes subject to first-order deactivation has been quantitatively studied.[676] A general expression was derived to describe the relationship between the decreasing observed enzymatic reaction rate and the intrinsic enzyme deactivation rate. A method of estimating the intrinsic deactivation rate constant was also proposed. The theoretical treatment was validated experimentally. A simple method has been presented which establishes intrinsic rate parameters when slow pore diffusion of substrate limits immobilized-enzyme reactions that obey Michaelis-Menten kinetics.[677] In a theoretical treatment optimal operation and replacement policies have been proposed for reactors with immobilized enzymes subject to deactivation.[678]

A convenient apparatus has been described for the determination of immobilized enzyme activities by the stirred batch method.[679] Advantages of the system include rapid working, efficient equilibration of temperature, compactness of the reaction chamber, the possibility of assaying the sample repeatedly or only once, and the ease of removal of reagent solutions or immobilized preparations from the apparatus.

A graphical method has been developed for determining the Michaelis-Menten constant free of external mass-transfer resistance for immobilized enzymes in a packed bed reactor.[680] The validity of the theoretical treatment was illustrated with examples from three different enzyme reactions.

Some observations on the similarity of batch and plug flow systems as applied to enzyme reactors have been published with particular relevance to extraction of true values of the kinetic parameters, even when the external mass-transfer effects were not eliminated.[681]

A theoretical treatment giving rate expressions for enzyme poisoning which are consistent with a Michaelis-Menten main reaction has been published.[682] The expressions can be used to analyse the performance of a fixed bed reactor containing immobilized enzyme.

An operational scheme for using immobilized enzymes in packed bed reactors that permits operation at a constant throughput rate and with constant product quality has been described.[683] The method involves use of columns operated in series with continuous enzyme addition. The design of a non-uniformly distributed biocatalyst has been described.[684]

Many preparations of immobilized enzymes and of enzymes coupled to other macromolecules to give water-soluble or -insoluble forms have been reported during the past year. These derivatives and their uses are summarized in Table 7.[685-743] Immobilized enzymes are referred to in this Table, whether they are enzymatically active or not.

The following information on immobilized glycoside and polysaccharide hydrolases and carbohydrate isomerases and oxidases is worthy of note. Immobilization of dextransucrase appears to be favourable for low-molecular-weight dextran production.[707] The kinetic modelling of a multiple immobilized-enzyme system has been reported.[744] The validity of the model was successfully demonstrated using D-glucose oxidase and β-D-fructofuranosidase co-immobilized in poly-(2-hydroxymethacrylate gel).[745] The β-D-fructofuranosidase activity of immobilized yeast cells has been utilized in an enzyme thermistor method for controlling sucrose concentration during fermentation.[746] β-D-Fructofuranosidase has been successfully immobilized by adsorption on krill chitin.[504] It was observed that treatment with glutaraldehyde resulted in 10-40% reduction of the activity of the enzyme preparations. β-D-Fructofuranosidase immobilized by glutaraldehyde-mediated crosslinking in hen eggwhite retained about 25% of its original activity.[708] Properties of the immobilized enzyme were investigated. The enzyme has also been immobilized on molecular sieves[710] and on the inner surfaces of nylon tubes.[711] β-D-Fructofuranosidase has been adsorbed on bead cellulose containing weak, basic N,N-diethylamino-2-hydroxypropyl groups.[709] Conditions most suitable for adsorption were described and inversion of sucrose using the preparation was studied in a stirred reactor. Use of the immobilized enzyme for sucrose inversion under flow conditions was also investigated.[747] Under these conditions the inversion was affected by the concentration and flow rate of the substrate and by the reaction temperature. The half-life of the enzyme was 215 days. It was found that the immobilization was more effective under flow conditions than under the equilibrium conditions of a stirred reactor.

The immobilization of β-D-galactosidase for use in dairy processing has been reviewed.[748] β-D-Galactosidase has been immobilized by adsorption on DEAE-cellulose followed by glutaraldehyde crosslinking[713] and by carbodi-imide-mediated reaction with nylon-co(acrylic acid) polymers.[581] Adsorption of the

(Text continues on page 752)

Table 7 Modification of enzymes by coupling to insoluble matrices and other macromolecules

Enzyme	EC No.	Matrix or macromolecule coupled and mode of coupling	Use of product	Ref.
Acetylcholinesterase	3.1.1.7	Reaction with thiosulphate derivatives of cellulose	Development of a new method of enzyme immobilization	497
β-D-2-Acetamido-2-deoxy-hexosidase	3.2.1.52	Reaction with concanavalin A	Study of kinetic properties of the active immobilized enzyme	590
Acetobacter aceti cells	-	Adsorption on cordierite	Production of acetic acid in a fixed-cell reactor	600
Acid phosphatase	3.1.3.2	Carbodi-imide-mediated reaction with nylon-co(acrylic acid) polymer	Development of a new method of enzyme immobilization	581
		Reaction with commercially available water-soluble poly-acrylamides or glutaraldehyde-mediated crosslinking with albumin	Comparative study of enzyme stabilization by reaction with soluble and insoluble macromolecules	656
		Co-gelation or co-polymerization and gelation with albumin onto membrane reactor walls	Study of the kinetic properties of the enzymes	685
Adenosine (phosphate)	3.5.4.17	Reaction with ion-exchange	Active immobilized enzyme	686

Table 7 continued

Enzyme	EC No.	Matrix or macromolecule coupled and mode of coupling	Use of product	Ref.
deaminase		resins *via* Ti^{4+} complex method		
Alcaligenes eutrophus cells	-	Entrapment in alginate or k-carrageenan gels	Detritiation of water in batch tank or column reactor	601
Alcohol dehydrogenase	1.1.1.1	Glutaraldehyde-crosslinked in an albumin matrix	Study of kinetics and diffusional effects of immobilized enzyme	687
		Reaction with cyanogen bromide-activated agarose	Active immobilized enzyme for use in an enzyme electrode NADPH/NADP regeneration	688
		Glutaraldehyde-mediated reaction with porous glass, reaction with cyanogen bromide-activated agarose, adsorption on Amberlite^{R} IRA-94, entrapment in polyacrylamide gel	Comparison of properties of immobilized enzyme	157
Aldehyde dehydrogenase	1.2.1.3	Glutaraldehyde-mediated reaction with nylon-polyethyleneimine copolymer	Automated determination of cholesterol	689
Alkaline phosphatase	3.1.2.1	Reaction with *bis*(carboxymethyl-	Development of a new method of	399

Table 7 continued

Enzyme	EC No.	Matrix or macromolecule coupled and mode of coupling	Use of product	Ref.
		amino)agarose pre-equilibrated with Cu^{2+}, Zn^{2+}, or Co^{2+} ions	enzyme immobilization on regenerable metal-chelate carriers	
Aminoacylase	3.5.1.14	Adsorption on DEAE-cellulose	Continuous isolation of D-phenylglycine from its racemic mixture	690
		Reaction with metal (Co^{2+}, Ni^{2+}, or Fe^{2+}) complexes of crosslinked polyvinyl alcohol	Development of a new method of enzyme immobilization	691
α-Amylase	3.2.1.1	Adsorption on krill chitin	Development of a new method of enzyme immobilization	503
		Graft polymerization to different polysaccharides following vinylation of the enzyme	Development of a new method of enzyme immobilization	411
		Glutaraldehyde-mediated reaction with zirconia-coated glass	Study of properties of active immobilized enzyme	692
		Reaction with polymethacrylate esters, reaction with CM-cellulose	Active immobilized enzyme	693

Table 7 continued

Enzyme	EC No.	Matrix or macromolecule coupled and mode of coupling	Use of product	Ref.
β-Amylase	3.2.1.2	Reaction with diazotized 4-aminobenzyl cellulose	Active immobilized enzyme	468
Arginase	3.5.3.1	Encapsulation within nylon membrane microcapsules by interfacial polymerization	Study of properties of active immobilized enzyme, study of effect of conjugate on growth of L1210 murine leukaemia	694
Arthrobacter cells	-	Not given	Study of kinetics of D-glucose isomerase activity of cells	602
Aryl sulphatase	3.1.6.1	Reaction with concanavalin A	Study of kinetic properties of the active immobilized enzyme	590
Ascorbate oxidase	1.10.3.3	Glutaraldehyde-mediated reaction with nylon netting	Enzyme electrode for amperometric assay of vitamin C	695
Aspartate aminotransferase	2.6.1.1	Unspecified reaction with agarose	Development of an apparatus for measurement of immobilized enzyme activities	679
		Reaction with acyl azide-activated collagen membranes	Development of a new method of enzyme immobilization	696
Azotobacter chroococcum cells	-	Entrapment in agar gel	Continuous nitrogen fixation	603

Table 7 continued

Enzyme	EC No.	Matrix or macromolecule coupled and mode of coupling	Use of product	Ref.
Beneckea harveyi cells	-	Reaction with photo-crosslinkable resins: entrapment in alginate gel	Development of a sensor system for hydrogen peroxide	604
Brevibacterium ammoniagenes cells	-	Entrapment in polyacrylamide	Production of glutathione and NADP	605
Brevibacterium flavum cells	-	Co-casting with collagen followed by glutaraldehyde-treatment	Production of glutamic acid in a recycle reactor	606
Brevibacterium fuscum cells	-	Entrapment in k-carrageenan gel	Continuous production of 12-ketochenodeoxycholic acid	607
Butyrylcholine esterase	3.1.1.8	Reaction with thiosulphate derivatives of cellulose	Development of a new method of enzyme immobilization	497
Caldariella acidophila cells	-	Glutaraldehyde-mediated reaction with chicken eggwhite, with or without magnetite present	Living immobilized cells with β-D-galactosidase activity	608
Candida tropicalis cells	-	Reaction with epoxy-activated pectate pre-reacted with iminodiacetic acid and equilibrated with Fe^{3+} ions	Development of a new method for whole cell immobilization	550
Catalase	1.11.1.6	Adsorption on polyethylene-	Active immobilized enzyme	697

Table 7 continued

Enzyme	EC No.	Matrix or macromolecule coupled and mode of coupling	Use of product	Ref.
		imine-coated glass microbeads		
		Co-immobilized with D-glucose oxidase by glutaraldehyde-mediated reaction with Pt gauze	Oxygen-stabilized enzyme electrode for D-glucose analysis in fermentation broths	698
Cephalexin-synthesizing enzyme	-	Adsorption on kaolite, bentonite, celite, activated carbon or silica	Study of kinetics and stabilities of the active immobilized enzymes	699
Chloroplasts	-	γ-Ray-induced polymerization with glass-forming monomers	Study of stabilities of immobilized cells	609
		γ-Ray irradiation with vinyl monomers	Study of stability of immobilized cells	610
		Entrapment in agar gel	Photochemical-energy conversion system	611
Cholesterol oxidase	1.1.3.6	Glutaraldehyde-mediated reaction with cellulose beads	Immobilized enzyme sensor utilizing chemiluminescence for measurement of cholesterol in serum	700
Chymotrypsin	3.4.21.1	Glutaraldehyde-mediated reaction	Active immobilized enzyme for	701

Table 7 continued

Enzyme	EC No.	Matrix or macromolecule coupled and mode of coupling	Use of product	Ref.
		with Ni-NiO-albumin complex	casein hydrolysis in stirred tank reactors	
		Reaction with 3-(2',3'-epoxy-propoxy)propylglass, 3-(2',3'-epoxypropoxy)propyl-silica, epoxy-activated glycidyl methacrylate copolymer or oxirane-acrylic beads	Study of effect of protein nature on coupling to different epoxide-containing supports	584
		Reaction with cyanogen bromide-activated agarose or dextran or entrapment in polyacrylamide gel or sheet	Study of stability of the active immobilized enzyme	158
		Reaction with cyanogen bromide-activated agarose	Study of autodigestion of chymotrypsin	159
		Graft polymerization to different polysaccharides following vinylation of the enzyme	Development of a new method of enzyme immobilization	411
		Reaction with polymethacrylate esters or reaction with CM-	Active immobilized enzyme	693

Table 7 continued

Enzyme	EC No.	Matrix or macromolecule coupled and mode of coupling	Use of product	Ref.
		cellulose		
		Reaction with Chromagel[R] A-2	Partial purification of urinary thiol proteinase inhibitors	702, 703
		Glutaraldehyde-mediated reaction to form monomers, soluble polymers, and membranes	Comparative study of native and chemically modified chymotrypsin	704
		Reaction with derivatized poly(vinyl alcohol) or reaction with methacrylate copolymers	Study of new potential carriers for enzyme immobilization	705
<u>Clostridium</u> <u>butyricum</u> cells	-	Not given	Development of method for starting a fermentor aseptically	612
		Entrapment in agar gel	Photochemical-energy conversion system	611
<u>Clostridium</u> <u>pasteurianum</u> cells	-	Glutaraldehyde-mediated reaction with aminopropyl-glass or amino-Spherosil[R]	Biophotolytic hydrogen production	613
<u>Coriolus</u> <u>versicolor</u> cells	-	Entrapment in alginate gel	Decolorization of a kraft mill effluent	614

Table 7 continued

Enzyme	EC No.	Matrix or macromolecule coupled and mode of coupling	Use of product	Ref.
Cystathione β-synthase	-	Reaction with cyanogen bromide-activated agarose	Active immobilized enzyme	160
Daucus carota cells	-	Entrapment in alginate gel	Bioconversion of gitoxigenin	615
Deoxyribonuclease I	3.1.21.1	Immobilized on erythrocytes	Model system for patients with systemic lupus erythematodes	706
		Reaction with cyanogen bromide-activated agarose	Demonstration of a DNAase-binding protein distinct from actin	161
			Isolation of actin from human hair follicles	162
Desulfovibrio desulfuricans cells	-	Glutaraldehyde-mediated reaction with aminopropyl-glass or amino-Spherosil[R]	Biophotolytic hydrogen production	613
Dextransucrase	2.4.1.5	Glutaraldehyde-mediated reaction with amino-Spherosil[R]	Active immobilized enzyme for production of dextran	707
		Reaction with phenoxyacetyl-cellulose beads	Active immobilized enzyme	494
Dihydrolipoamide reductase	1.6.4.3	Reaction with cyanogen bromide-activated cellulose	Continuous flow analysis of 3α-hydroxysteroids	476

Table 7 continued

Enzyme	EC No.	Matrix or macromolecule coupled and mode of coupling	Use of product	Ref.
Elastase	3.4.21.11	Reaction with cyanogen bromide-activated agarose	Purification of an elastase-specific inhibitor from bronchial mucus	163
			Active immobilized enzyme for proteolysis of human plasminogen	164
Erythrocytes	-	Reaction with agarose-lectin-erythropoietin complex	Study of stimulation of erythroid colony formation *in vitro*	375
Escherichia alcalescens cells	-	Entrapment in polyacrylamide gel	Living immobilized cells	616
Escherichia coli cells	-	Entrapment in polyacrylamide gel	Production of glutathione and NADP	605
		Reaction with epoxy-activated pectate pre-reacted with iminodiacetic acid and equilibrated with Fe^{3+} ions	Development of a new method for whole cell immobilization	550
		Entrapment in polyurethane foam	Living immobilized cells with	617

Table 7 continued

Enzyme	EC No.	Matrix or macromolecule coupled and mode of coupling	Use of product	Ref.
			penicillin G acylase activity	
		Ionotropic gelation with chitosan	Living immobilized cells with tryptophan synthetase activity	511
Ficin	3.4.22.3	Immobilized on carboxymethylcellulose	Validation of modified method for determining the Michaelis-Menten constant	680
β-D-Fructofuranosidase	3.2.1.26	Adsorption on krill chitin	Development of a new method of enzyme immobilization	503, 504
		Adsorption on Amberlite[R] IRA-94	Experimental validation of theory concerning effect of intraparticle diffusion resistance on the apparent stability of immobilized enzymes	676
		Adsorption on an ion-exchanger	Validation of a modified method for determining the Michaelis-Menten constant	680
		Glutaraldehyde-mediated cross-linking with hen eggwhite	Active immobilized enzyme and assessment of suitability of hen eggwhite as support for	708

Table 7 continued

Enzyme	EC No.	Matrix or macromolecule coupled and mode of coupling	Use of product	Ref.
			enzyme immobilization	
		Adsorption on N,N-diethylamino-2-hydroxypropylcellulose	Inversion of sucrose in a stirred reactor	709
		Metal-link-mediated reaction with molecular sieves	Active immobilized enzyme	710
		Glutaraldehyde-mediated reaction with O-alkylated nylon tubes	Active immobilized enzyme	711
β-D-Galactosidase	3.2.1.23	Carbodi-imide-mediated reaction with nylon-co(acrylic acid) polymer	Development of a new method of enzyme immobilization	581
		Adsorption on brushite	Active immobilized enzyme	712
		Adsorption on DEAE-cellulose followed by glutaraldehyde-crosslinking	Study of properties of the active immobilized enzyme	713
D-Galactosyl hydroxy L-lysyl glucosyl-transferase	-	Reaction with cyanogen bromide-activated agarose	Immunoadsorption purification of anti-(chick-embryo galactosyl hydroxy-L-lysyl-glucosyltransferase) antibody	165

Table 7 continued

Enzyme	EC No.	Matrix or macromolecule coupled and mode of coupling	Use of product	Ref.
endo-D-Galacturonanase	3.2.1.15	Carbodi-imide-mediated reaction with Spheron^{R} pre-coupled to glycine, β-alanine, 4-amino-butanoic acid, glycylglycine, or 6-aminohexanoic acid	Investigation of effect of immobilization on the activity of the enzyme	714
Glucoamylase	3.2.1.3	Reaction with cyanogen bromide-activated agarose	Immunoadsorption purification of anti-glucoamylase antibody	166
		Adsorption on krill chitin	Development of a new method of enzyme immobilization	503
		Glutaraldehyde-mediated reaction with alkylamine glass	Experimental validation of a new method for estimating intrinsic kinetic constants for pore diffusion-limited immobilized-enzyme reactions	677
		Reaction with polymethacrylate esters or reaction with CM-cellulose	Active immobilized enzyme	693
		Adsorption onto a copolymer of 2-dimethylaminoethyl methacry-	Study of properties of active immobilized enzyme	715

Table 7 continued

Enzyme	EC No.	Matrix or macromolecule coupled and mode of coupling	Use of product	Ref.
		late and methyl methacrylate		
		Entrapment in polyacrylamide gel	Study of thermal stability of the active immobilized enzyme	716
		Reaction with Ti^{4+}-activated controlled pore glass followed by crosslinking with tannic acid, or glutaraldehyde or 1,6-diaminohexane	Study of stability of the active immobilized enzyme	717, 718
		Reaction with acyl azide-activated collagen membranes	Development of a new method of enzyme immobilization	696
Glucose dehydrogenase	1.1.1.47	Immobilization by metal-link method	Study of kinetic properties of the active immobilized enzyme	719
Glucose isomerase	5.3.1.18	Reaction with radiation-modified gelatin gel	Active immobilized enzyme	720
D-Glucose oxidase	1.1.3.4	Reaction with acyl azide-activated collagen membranes	Development of a new method of enzyme immobilization	696
		Graft polymerization to different polysaccharides following	Development of a new method of enzyme immobilization	411

Table 7 continued

Enzyme	EC No.	Matrix or macromolecule coupled and mode of coupling	Use of product	Ref.
		vinylation of the enzyme		
		Adsorption on polyethyleneimine-coated glass microbeads	Active immobilized enzyme	697
		Co-immobilized with catalase by glutaraldehyde-mediated reaction with Pt gauze	Oxygen-stabilized enzyme electrode for D-glucose analysis in fermentation broths	698
		Glutaraldehyde-mediated reaction with cellulose beads	Immobilized enzyme sensor utilizing chemiluminescence for measurement of D-glucose in serum	700
		Entrapment in polyacrylamide gel	Study of thermal stability of the active immobilized enzyme	716
		Glutaraldehyde-mediated reaction with nylon 66 to form a membrane	Membrane with enzyme activity only when mechanically stretched	721
		Adsorption on porous polyvinyl chloride membrane	Active immobilized enzyme, study of mechanical control of enzymic activity	722
		Glutaraldehyde-mediated reaction	Active immobilized enzyme,	723

Table 7 continued

Enzyme	EC No.	Matrix or macromolecule coupled and mode of coupling	Use of product	Ref.
		with the porous side of an asymmetric acetylcellulose membrane	enzyme electrode for measurement of D-glucose in human serum or whole blood	
β-D-Glucuronidase	3.2.1.31	Glutaraldehyde-mediated reaction with alkylamine glass	Study of properties and stability of active immobilized enzyme	724
			Analysis of glucuronide conjugates prior to analysis by h.p.l.c.	725
Glyceraldehyde 3-phosphate dehydrogenase	1.2.1.12	Reaction with cyanogen bromide-activated agarose	Study of properties of active immobilized enzyme	167, 168
3α-Hydroxysteroid dehydrogenase	1.1.1.50	Reaction with cyanogen bromide-activated cellulose	Continuous flow analysis of 3α-hydroxysteroids	476
Inulinase	3.2.1.7	Adsorption on DEAE-cellulose	Study of properties of active immobilized enzyme, continuous production of fructose syrup from inulin	726
Kluyveromyces fragilis cells	-	Entrapment in alginate gel	Continuous conversion of lactose to ethanol	618

Table 7 continued

Enzyme	EC No.	Matrix or macromolecule coupled and mode of coupling	Use of product	Ref.
			Repeated batch production of ethanol from Jerusalem artichoke tubers	619
Lactate dehydrogenase	1.1.1.27	Adsorption on alumina	Study of properties of active immobilized enzyme	727
		Adsorption on PTFE particles packed in h.p.l.c. columns	Study of new support for enzyme immobilization	728
Lactoperoxidase	-	Reaction with cyanogen bromide-activated agarose or with concanavalin A-agarose	Comparison of properties of active immobilized enzymes	170
Leuconostoc mesenteroides cells	-	Entrapment in agar on an acetylcellulose filter	Used in conjunction with a lactate electrode for analysis of L-phenylalanine in human serum	99
Luciferase	-	Glutaraldehyde-mediated cross-linking with albumin onto a dialysis membrane	Microscale analysis of creatine kinase activity	729
Lysozyme	3.2.1.17	Reaction with cyanogen bromide-activated agarose	Study of interaction with anionic detergents	171

Table 7 continued

Enzyme	EC No.	Matrix or macromolecule coupled and mode of coupling	Use of product	Ref.
		N-Succinimidyl-3-(2-pyridyldithio)propionate-mediated reaction with gelatin-agarose	Development of a new method for preparation of immunoadsorbents	410
Methanosarcina barkeri cells	-	Entrapment in alginate gel	Conversion of methanol to methane	620
Mycobacterium phlei cells	-	Adsorption on DEAE-cellulose or silica, entrapment in polyacrylamide gel	Living immobilized cells for steroid transformations	621
Naringinase	-	Glutaraldehyde-mediated reaction with alkylamine glass	Active immobilized enzyme	730
Neuraminidase	3.2.1.18	Reaction with cyanogen bromide-activated agarose	Desialylation of angiotensin-converting enzyme	172
			Desialylation of human platelets	173
NAD^+ pyrophosphatase	3.6.1.22	Adsorption on phosphocellulose	Enzyme reactor for production of nicotinamide mononucleotide	495
Nocardia erythropolis or *opaca* cells	-	Adsorption on DEAE-cellulose or silica entrapment in polyacrylamide gel	Living immobilized cells for steroid transformations	621

Table 7 continued

Enzyme	EC No.	Matrix or macromolecule coupled and mode of coupling	Use of product	Ref.
Nuclease P	-	Reaction with cyanogen bromide-activated cellulose, carbodiimide-mediated reaction with anion-exchange cellulose, or reaction with Ti^{4+}-cellulose complex	Active immobilized enzyme	477
		Reaction with ion-exchange resins _via_ Ti^{4+}-complexation	Active immobilized enzyme	731
Micrococcal _endo_-nuclease	3.1.31.1	Reaction with cyanogen bromide-activated agarose	Study of kinetics of nucleic acid hydrolysis by active immobilized enzyme	732
Pachysolen tannophilus cells	-	Entrapment in alginate gel	Production of ethanol from D-xylose	622
Papain	3.4.22.2	Reaction with succinylaminoethyl-agarose	Active immobilized enzyme	174
		Reaction with cyanogen bromide-activated agarose	Use of active immobilized enzyme in a column reactor	175
		Terephthalic diazide-mediated crosslinking with 4-amino-	Study of properties of active immobilized enzyme	467

Table 7 continued

Enzyme	EC No.	Matrix or macromolecule coupled and mode of coupling	Use of product	Ref.
		benzyl-cellulose		
		Reaction with 4-benzoquinone-activated cellulose	Active immobilized enzyme	471
		Reaction with reactive poly-methacrylate esters or reaction with CM-cellulose	Active immobilized enzyme	693
		Reaction with hydrazide derivatives of alginate-polyacrylamide graft copolymers	Study of new support for enzyme immobilization	733
Penicillin amidase	3.5.1.11	Entrapment in gelatin	Validation of modified method for determining the Michaelis-Menten constant	680
		Cyanuric chloride-mediated reaction with water-soluble poly-electrolyte complexes	Active immobilized enzyme	734
Penicillium duponti enzyme	-	Glutaraldehyde-mediated cross-linking onto collagen	Active immobilized enzyme	735
Pepsin	3.4.23.1	Glutaraldehyde-mediated reaction with 6-aminohexyl-agarose	Digestion of IgG to obtain $F(ab')_2$ fragments for	176

Table 7 continued

Enzyme	EC No.	Matrix or macromolecule coupled and mode of coupling	Use of product	Ref.
			intravenous use	
		Reaction with chitosan	Active immobilized enzyme	512
Peroxidase	1.11.1.7	Graft polymerization to different polysaccharides following vinylation of the enzyme	Development of a new method of enzyme immobilization	411
		Copolymerization with a photo-crosslinkable prepolymer into a film followed by glutaraldehyde-crosslinking	Immobilized enzyme sensor utilizing chemiluminescence for measurement of serum D-glucose	700
		Glutaraldehyde-mediated reaction with alkylamine glass beads	Active immobilized enzyme	736
Pseudomonas dacunhae cells	-	Entrapment in k-carrageenan gel	Living immobilized cells	623
Pseudomonas denitrificans cells	-	Entrapment in alginate gel	Living immobilized cells	624
Rhizopus nigricans cells	-	Entrapment in polyacrylamide or agar gel	11α-Hydroxylation of progesterone	625
Rhodopseudomonas capsulata chromatophores	-	Glutaraldehyde-mediated cross-linking with albumin and with/	Investigation of the influence of immobilization on light-	626

Table 7 continued

Enzyme	EC No.	Matrix or macromolecule coupled and mode of coupling	Use of product	Ref.
		without hexokinase present	induced phosphorylation of ADP by the chromatophore system	
		Glutaraldehyde-mediated cross-linking with albumin or gelatin, entrapment in alginate or k-carrageenan gel, or copolymerization with urethane prepolymers	Comparative stabilization of biological photosystems by several immobilization procedures. Production of ATP	627
		Entrapment in alginate gel	Electron microscopic study of an immobilized living system	628
<u>Rhodotorula minuta</u> cells	-	Entrapment within photocross-linked or polyurethane resin gels	Stereoselective hydrolysis of <u>dl</u>-menthyl succinate in organic solvent	629
Ribonuclease		Reaction with derivatized poly-(vinyl alcohol) or reaction with methacrylate copolymers	Study of new potential carriers for enzyme immobilization	705
<u>Saccharomyces cerevisiae</u> cells	-	Entrapment in alginate gel	Continuous production of ethanol	630-634

Table 7 continued

Enzyme	EC No.	Matrix or macromolecule coupled and mode of coupling	Use of product	Ref.
			Development of methods for preparing alginate gels stable in phosphate-containing media, for use in cell immobilization	641
		Adsorption on ion-exchange resins	Continuous production of ethanol	635
		Entrapment in k-carrageenan gel	Continuous production of ethanol	636
		Entrapment in polyacrylamide gel	Continuous production of ethanol	637
			Production of glutathione	638
		Entrapment in silica hydrogel	Living immobilized cells	639
		Entrapment in gel	Living immobilized cells with β-D-fructofuranosidase activity	640
		Reaction with epoxy-activated pectate pre-reacted with imino-diacetic acid and equilibrated	Development of a new method for whole cell immobilization	550

Table 7 continued

Enzyme	EC No.	Matrix or macromolecule coupled and mode of coupling	Use of product	Ref.
		with Fe^{3+} ions		
		Entrapment in polyacrylamide gel	Production of glutathione and NADP	605
Solanum aviculare cells	-	Glutaraldehyde-mediated reaction with poly(phenylene oxide)	Living immobilized cells for continuous production of steroid glycoalkaloids	642
Sorbitol dehydrogenase	1.1.1.14	Reaction with acyl azide-activated collagen membranes	Development of a method of enzyme immobilization	696
Staphylococcus aureus cells	-	Reaction with epoxy-activated pectate pre-reacted with iminodiacetic acid and equilibrated with Fe^{3+} ions	Development of a new method for whole cell immobilization	550
Stemphylium loti cells	-	Treatment with flocculating agents	Cyanide degradation	644
Streptomyces aureofaciens cells	-	Glutaraldehyde-crosslinking onto a gelatin matrix	Conversion of daunomycinone to 13-dihydrodaunomycinone	645
Streptomyces aureofaciens enzyme extract	-	Reaction with glutaraldehyde-activated bead cellulose	Conversion of daunomycinone to 13-dihydrodaunomycinone	645

Table 7 continued

Enzyme	EC No.	Matrix or macromolecule coupled and mode of coupling	Use of product	Ref.
Streptomyces clavuligerus cells	-	Crosslinking onto pre-formed linear water-soluble polyacrylamide gels	Development of a mild new method for immobilization of microbial cells and living immobilized cells for antibiotic production	646
Streptomyces fradiae cells	-	Entrapment in polyacrylamide gel	Living immobilized cells for protease production	647
Streptomyces phaeochromogenes cells	-	Adsorption onto diatomaceous earth	Living immobilized cells with D-glucose isomerase activity	648
		Reaction with aggregating chitosan	Living immobilized cells for use in an enzyme reactor	649
Streptomyces roseochromogenes cells	-	Entrapment in a photocross-linkable resin	Living immobilized cells for 16α-hydroxylation of dehydroepiandrosterone	650
Thermolysin	3.4.24.4		Synthesis of an aspartamine precursor	737
Thylakoids (lettuce)	-	Glutaraldehyde-mediated cross-	Study of properties and	651

Table 7 continued

Enzyme	EC No.	Matrix or macromolecule coupled and mode of coupling	Use of product	Ref.
		linking with albumin or gelatin, entrapment in alginate or i-carrageenan gels, copolymerization with urethane polymers	stability of biological photosystem	
Trypsin	3.4.21.4	Reaction with cyanogen bromide-activated agarose	Purification of proteinase inhibitors from *Acacia elata*	177
			Purification of protease inhibitors from developing embryos of *Hemileuca oliviae*	178
			Purification of duck ovomucoid	179
			Investigation of the characteristics of the covalent binding of complement protein C3 to trypsin	180
		Reaction with 6-aminohexyl agarose	Comparative study of inhibition properties of immobilized trypsin and a protease on the	181

Table 7 continued

Enzyme	EC No.	Matrix or macromolecule coupled and mode of coupling	Use of product	Ref.
			surface of Ehrlich ascites tumour cells	
		Reaction with agarose mediated by ethylene glycolyl bis-(succinimidylsuccinate)	Demonstration of potential of a new cleavable reagent for crosslinking and reversible immobilization of proteins	409
		Reaction with periodate-activated cellulose beads	Study of new method for enzyme immobilization	493
		Reaction with thiosulphate derivatives of cellulose	Development of a new method of enzyme immobilization	497
		Reaction with cyanogen bromide-actived agarose or cross-linked dextran	Investigation of the influence of effectors on the refolding of immobilized trypsin	536
		Reaction with derivatized poly-(vinyl alcohol) or reaction with methacrylate copolymers	Study of new potential carriers for enzyme immobilization	705
		Reaction with hydrazide derivatives of alginate-polyacryl-	Active immobilized enzyme	733

Table 7 continued

Enzyme	EC No.	Matrix or macromolecule coupled and mode of coupling	Use of product	Ref.
		amide graft copolymers		
Tyrosinase	1.14.18.1	Reaction with cyanogen bromide-activated agarose	Active immobilized enzyme and study of subunit interactions of tyrosinase	182
Urease	3.5.1.5.	Glutaraldehyde-mediated reaction with O-alkylated nylon tubes	Active immobilized enzyme and study of effects of substrate flow rate on immobilized-urease assays	738
		Glutaraldehyde-mediated reaction with O-alkylated nylon tubes coated with polyaminated derivatives of starch, dextran, dialdehyde dextran, or dialdehyde starch	Study of use of polysaccharide derivatives as coats for nylon tube-immobilized urease	539
		Reaction with derivatized poly-(vinyl alcohol) or reaction with methacrylate copolymers	Study of new potential carriers for enzyme immobilization	705
		Entrapment in polyacrylamide gel	Study of performance of	739

Table 7 continued

Enzyme	EC No.	Matrix or macromolecule coupled and mode of coupling	Use of product	Ref.
		in the presence of magnetite	fluidized-bed immobilized-enzyme reactors utilizing magnetic fields	
Uricase	1.7.3.3	Glutaraldehyde-mediated reaction with cellulose beads	Immobilized-enzyme sensor utilizing chemiluminescence	700
		Glutaraldehyde-mediated reaction with nylon tube surfaces	Kinetic study of the active immobilized enzyme	740
			Automated analysis of serum urate	741
		Glutaraldehyde-mediated reaction with silica filaments on the inner wall of glass tubes used as reactors in continuous-flow analysers	Active immobilized enzyme and determination of uric acid in human serum	742
Urokinase	3.4.21.31	Reaction with agarose activated by oxidation with Br_2	Demonstration of new method of protein immobilization	147
		Reaction with collagen-synthetic polymer composite material	Study of <u>in vitro</u> and <u>in vivo</u> behaviour of the active	743

Table 7 continued

Enzyme	EC No.	Matrix or macromolecule coupled and mode of coupling	Use of product	Ref.
			immobilized enzyme	
Xanthine oxidase	1.2.3.2	Copolymerization with a photo-crosslinkable prepolymer into a film, followed by glutaraldehyde crosslinking	Immobilized-enzyme sensor utilizing chemiluminescence for measurement of hypoxanthine in tuna muscle	700
Yeast cells	-	Entrapment in polyethyleneglycol hydroxyethylacrylate gel	Study of properties of an immobilized glycolysis system of yeast in fermentative phosphorylation of nucleotides	652
Zymomonas mobilis cells	-	Entrapment in alginate gel	Living immobilized cells for efficient production of ethanol in a bioreactor	653

enzyme onto brushite ($CaHPO_4.2H_2O$) has been reported to be advantageous as immobilization can be performed in almost any buffer even at high salt concentrations, provided high concentrations of phosphate are avoided.[712] Very little leakage of enzyme from the support was observed, and the enzyme could be desorbed by increasing the concentration of phosphate ions.

Glucoamylase has been immobilized on alkylamine glass by glutaraldehyde-mediated reaction,[677] or by the metal-link method.[718] In the latter study treatment of the immobilized enzyme with 1,6-diaminohexane or glutaraldehyde was found to yield superior products. The influence of coupling conditions on activity and operational stability of glucoamylase immobilized on Ti^{4+}-activated controlled pore glass has been discussed.[717] Bound enzyme was found to be stabilized very effectively by crosslinking with bifunctional reagents such as glutaraldehyde and/or tannic acid. The thermostabilities of glucoamylase and D-glucose oxidase have been increased several-fold by their entrapment in polyacrylamide gels.[716] In polyacrylate gels the enzymes behaved differently, probably owing to microenvironmental effects arising from the polyelectrolyte nature of the carrier: the thermo-stability of glucoamylase from one source (*Thermomyces lanuginosus*) increased 7-fold, while that from *Rhizopus* sp. decreased considerably. The thermostability of D-glucose oxidase from *Aspergillus niger* increased slightly on immobilization in polyacrylate gel. Lower thermostability was also observed after immobilizing glucoamylase by adsorption onto a copolymer of 2-dimethylaminoethyl methacrylate and methyl methacrylate.[715]

Commercially available immobilized D-glucose isomerase has been used in experiments to validate a theoretically derived optimum temperature control policy for an immobilized glucose isomerase reactor system.[749] The theoretical treatment allowed for enzyme deactivation during continuous reactor operation.

Rates of oxygen absorption into D-glucose solutions have been measured using D-glucose oxidase immobilized by entrapment with magnetite in polyacrylamide gel.[750] An immobilized enzyme reactor was employed in which the enzyme beads were moved by a revolving magnetic field to reduce the mass-transfer resistances at the gas-liquid interface and around the bead. A comparison of different methods of D-glucose oxidase immobilization has been published.[751]

Immobilized D-glucose oxidase has been used in an enzyme electrode for measurement of D-glucose in human serum or whole blood.[723] The preparation of a stress-sensitive D-glucose oxidase-nylon membrane has been reported.[721] The enzyme, immobilized on glutaraldehyde-treated nylon-66, had no activity under normal conditions, but only when the membrane was mechanically stretched. A linear relationship between stress strength and D-glucose oxidase activity was found. Mechanical control of the activity of D-glucose oxidase immobilized on porous polyvinyl chloride membranes has also been reported.[722] Anomalous binding kinetics have been reported for a high-yield immobilized-enzyme system employing D-glucose oxidase adsorbed on non-porous polyethyleneimine-coated glass microbeads.[697] The presence of several low-molecular-weight impurities which apparently competed with D-glucose oxidase for binding was found to be responsible for the atypical adsorption curves obtained for the system. When large excesses of protein were added to the beads the binding of impurities became significant and the amount of enzyme activity per unit of bead was reduced, explaining the adsorption isotherms observed.

β-D-Glucosidase adsorbed on controlled pore alumina has been used in conjunction with soluble *Trichoderma reesei* cellulase to hydrolyse cellulosic materials.[752] Increased yields of D-glucose and increased conversion of cellobiose to D-glucose were observed when supplemental immobilized β-D-glucosidase was used in fixed bed, fluidized bed, and hydrolysis reactors.

Inulinase adsorbed on DEAE-cellulose has been applied to the continuous production of fructose syrup from inulin, and properties of the immobilized enzyme have been studied.[726]

Immobilized neuraminidase has been applied to the desialylation of angiotensin-converting enzyme to determine the effects of this treatment on lectin binding,[172] and to the desialylation of human platelets for studies of the accumulation of 5-hydroxytryptamine by the platelets.[173]

(References begin overleaf)

References

1 C.Schuerch, Adv. Carbohydr. Chem. Biochem, 1981, 39, 157.
2 P.F. Sharkey, R. Eby and C. Schuerch, Carbohydr. Res., 1981, 96, 223.
3 H. Ito and C. Schuerch, J. Polym. Sci. Polym. Letts., 1981, 19, 43.
4 H. Yamaguchi and C. Schuerch, Biopolymers, 1980, 19, 297.
5 N.K. Kochetkov and E.M. Klimov, Tetrahedron Lett., 1981, 22, 337.
6 N.K. Kochetkov, V.I. Betaneli, M.V. Ovchinnikov and L.V. Backinowsky, Tetrahedron, 1981, 37 Suppl. 9, 149.
7 M.A. Nashed, M. Kiso, C.W. Slife and L. Anderson, Carbohydr. Res., 1981, 90, 71.
8 P.L. Durette and E.P. Meitzner, Carbohydr. Res., 1981, 89, 279.
9 C.W. Slife, M.A. Nashed and L. Anderson, Carbohydr. Res., 1981, 93, 219.
10 S.S. Rana, J.J. Barlow and K.L. Matta, Carbohydr. Res., 1981, 88, C20.
11 N.R. Plessas and I.J. Goldstein, Carbohydr. Res., 1981, 89, 211.
12 T. Ogawa, K. Beppu and S. Nakabayashi, Carbohydr. Res., 1981, 93, C6.
13 P. Chow and B. Weissmann, Carbohydr. Res., 1981, 96, 87.
14 R. Eby and C. Schuerch, Carbohydr. Res., 1981, 92, 149.
15 P.J. Garegg, T. Iversen and R. Johansson, Acta Chem. Scand., 1980, 34B, 505.
16 P.J. Garegg, H. Hultberg and T. Norberg, Carbohydr. Res., 1981, 96, 59.
17 J. Thiem and J. Elvers, Chem. Ber., 1981, 114, 1442.
18 M. Kiso, H. Nishiguchi, K. Nishihori, A. Hasegawa and I. Miura, Carbohydr. Res., 1981, 88, C10.
19 J. Alais and A. Veyrières, Carbohydr. Res., 1981, 93, 164.
20 L.A. Reed and L. Goodman, Carbohydr. Res., 1981, 94, 91.
21 R.M. Ratcliffe, D.A. Baker and R.U. Lemieux, Carbohydr. Res., 1981, 93, 35.
22 J.C. Jacquinet and H. Paulson, Tetrahedron Lett., 1981, 22, 1387.
23 P. Kováč, E. Petráková and P. Kočiš, Carbohydr. Res., 1981, 93, 144.
24 P.J. Garegg, B. Samuelsson, P. Konradsson, I. Kvarnström and S.C.T. Svensson, Carbohydr. Res., 1981, 92, 157.
25 M.E. Gelpi and R.A. Cadenas, Carbohydr. Res., 1981, 88, 277.
26 H.S. El Khadem and D. Matsuura, Carbohydr. Res., 1981, 88, 332.
27 P.J. Garegg and T. Norberg, Acta Chem. Scand., 1981, 35B, 305.
28 S.S. Rana, J.J. Barlow and K.L. Matta, Carbohydr. Res., 1981, 96, 79.
29 S.A. Abbas, J.J. Barlow and K.L. Matta, Carbohydr. Res., 1981, 98, 37.
30 K.L. Matta, C.F. Piskorz and J.J. Barlow, Carbohydr. Res., 1981, 90, C1.
31 S.S. Rana, J.J. Barlow, and K.L. Matta, Carbohydr. Res., 1981, 91, 149.
32 P. Kováč and J. Hirsch, Carbohydr. Res., 1981, 90, C5.
33 K. Takeo, T. Yasato, and T. Kuge, Carbohydr. Res., 1981, 93, 148.
34 K. Takeo, Carbohydr. Res., 1981, 88, 158.
35 J. Hirsch, P. Kováč, J. Alföldi, and V. Mihálov, Carbohydr. Res., 1981, 88, 146.
36 A.K. Bhattacharjee, E. Zissis, and C.P.J. Glaudemans, Carbohydr. Res., 1981, 89, 249.
37 T. Takeda, I. Kawarasaki, and Y. Ogihara, Carbohydr. Res., 1981, 89, 301.
38 V. Pozsgay, P.Nánási, and A. Neszmélyi, Carbohydr. Res., 1981, 90, 215.
39 A. Lipták, P. Nánási, A. Neszmélyi, I. Riess-Maurer, and H. Wagner, Carbohydr. Res., 1981, 93, 43.
40 S.A. Abbas, J.J. Barlow, and K.L. Matta, Carbohydr. Res., 1981, 88, 51.
41 H. Paulsen and O. Lockhoff, Chem. Ber., 1981, 114, 3079.
42 H. Paulsen and O. Lockhoff, Chem. Ber., 1981, 114, 3115.
43 H. Paulsen and O. Lockhoff, Chem. Ber., 1981, 114, 3102.
44 T. Iversen, S. Josephson, and D.R. Bundle, J. Chem. Soc. Perkin Trans. I, 1981, 2379.
45 H. Paulsen and H. Bünsch, Chem. Ber., 1981, 114, 3126.
46 H. Paulsen and H. Bünsch, Tetrahedron Letters, 1981, 22, 47.
47 H. Paulsen and D. Schnell, Chem. Ber., 1981, 114, 333.
48 J. Alais and A. Veyrières, J. Chem. Soc. Perkin Trans. I, 1981, 377.

49 J.C. Jacquinet, D. Duchet, M.L. Milat, and P. Sinay, J. Chem. Soc. Perkin Trans. I, 1981, 326.
50 J. Alais and A. Veyrières, Carbohydr. Res., 1981, 92, 310.
51 A. Veyrières, J. Chem. Soc. Perkin Trans. I, 1981, 1626.
52 S.S. Rana, J.J. Barlow, and K.L. Matta, Carbohydr. Res., 1981, 96, 231.
53 M.L. Milat and P. Sinay, Carbohydr. Res., 1981, 92, 183.
54 H. Paulsen and C. Kolář, Chem. Ber., 1981, 114, 306.
55 H. Paulsen and A. Bunsch, Angew. Chem. Internat. Edn., 1980, 19, 902.
56 T. Ogawa, K. Katano, K. Sasajima, and M. Matsui, Tetrahedron, 1981, 37, 2779.
57 T. Ogawa and K. Sasajima, Tetrahedron, 1981, 37, 2787.
58 T. Ogawa and K. Sasajima, Carbohydr. Res., 1981, 93. 231.
59 T. Ogawa and K. Sasajima, Carbohydr. Res., 1981, 93, 53.
60 T. Ogawa and K. Sasajima, Carbohydr. Res., 1981, 93, 67.
61 J. Arnarp and J. Lönngren, J. Chem. Soc. Perkin Trans. I, 1981, 2070.
62 T. Ogawa and S. Nakabayashi, Carbohydr. Res., 1981, 93, Cl.
63 J. Arnarp, J. Lönngren, and H. Ottosson, Carbohydr. Res. 1981, 98, 154.
64 J.D. Aplin and J.C. Wriston, C.R.C. Critical Rev., Biochem., 1981, 10, 259.
65 T. Takeda, Y. Sugiura, Y. Ogihara, and S. Shibata, Can. J. Chem., 1980, 58, 2600.
66 H.J. Koeners, C. Schattenkerk, J.J. Verhoeven, and J.H. Van Boom, Tetrahedron, 1981, 37, 1763.
67 M. Zadral, J. Jezek, V. Krchnák, and R. Straka, Coll. Czech. Chem. Comm., 1980, 45, 1424.
68 A. Hasegawa, Y. Kaneda, Y. Goh, K. Nishibori, M. Kiso, and I. Azuma, Carbohydr. Res., 1981, 94, 143.
69 M. Kiso, Y. Goh, E. Tanahashi, A. Hasegawa, H. Okumura, and I. Azuma, Carbohydr. Res., 1981, 90, C8.
70 A. Hasegawa, E. Tanahashi, Y. Goh, M. Kiso, and I. Azuma, Carbohydr. Res., 1981, 92, 75.
71 S. Lavielle, N.C. Ling, and R.C. Guillemin, Carbohydr. Res., 1981, 89, 221.
72 S. Lavielle, N.C. Ling, R. Saltman, and R.C. Guillemin, Carbohydr. Res., 1981, 89, 229.
73 L.L. Danilov, S.D. Maltsev, V.N. Shibaev and N.K. Kochetkov, Carbohydr. Res., 1981, 88, 203.
74 M.A. Nashed and L. Anderson, Carbohydr. Res., 1981, 92, C5.
75 M. Kiso, H. Nishiguchi, S. Murase, and A. Hasegawa, Carbohydr. Res., 1981, 88, C5.
76 J. Slama and R.R. Rando, Carbohydr. Res., 1981, 88, 213.
77 R. Kaifu and I.J. Goldstein, Carbohydr. Res., 1981, 96, 241.
78 C.A.A. Van Boeckel, P. Westerduin, and J.H. Van Boom, Tetrahedron Letters., 1981, 22, 819.
79 W.H. Scouten, 'Affinity Chromatography', John Wiley and Sons Ltd., Chichester, 1981.
80 'Affinity Chromatography and Related Techniques', eds. T.C.J. Gribnau, J. Visser, and R.J.F. Nivard, Elsevier, Amsterdam, 1981.
81 'Polymer Catalysts and Affinants--Polymers in Chromatography', ed. B. Sedláček, C.G. Overberger, and H.F. Mark, John Wiley and Sons, Ltd., Chichester, 1981.
82 J.F. Kennedy and C.A. White, Biomed. Appl. Polymers. 1981, 11. 1.
83 C.M. Sturgeon and J.F. Kennedy, Enzyme Microb. Technol., 1981, 3, 76.
84 C.M. Sturgeon and J.F. Kennedy, Enzyme Microb. Technol., 1981, 3, 160.
85 C.M. Sturgeon and J.F. Kennedy, Enzyme Microb. Technol., 1981, 3, 260.
86 C.M. Sturgeon and J.F. Kennedy, Enzyme Microb. Technol., 1981, 3, 367.
87 J.H. Pazur, Adv. Carbohydr. Chem. Biochem., 1981, 39, 405.
88 L. Goldstein, J. Chromatogr., 1981, 215, 31.
89 H. Lis and N. Sharon, J. Chromatogr., 1981, 215, 361.
90 J.E. Sadler, T.A. Beyer and R.L. Hill, J. Chromatogr., 1981, 215, 181.
91 L.O. Andersson, J. Chromatogr., 1981, 215, 153.
92 J.B. Taylor and H.E. Swaisgood, Biotechnol Bioeng., 1981, 23, 1349.
93 V. Hořejši, Anal. Biochem., 1981, 112, 1.

94 V. Hořejši and M. Tichá, J. Chromatogr., 1981, 216, 43.
95 V. Hořejši and M. Tichá, Anal. Biochem., 1981, 116, 22.
96 S. Hjertén and Y. Kunquan, J. Chromatogr., 1981, 215, 317.
97 I. Olsson and T. Låås, J. Chromatogr., 1981, 215, 373.
98 S. Tracy, Prep. Biochem., 1981, 11, 251.
99 T. Matsunaga, I. Karube, N. Teraoka, and S. Suzuki, Anal. Chim. Acta, 1981, 127, 245.
100 H.H. Weetall, B.P. Sharma, and C.C. Detar, Biotechnol. Bioeng., 1981, 23, 605.
101 R.J. Youle and D.M. Neville, Proc. Nat. Acad. Sci., U.S.A., 1980, 77, 5483.
102 L. Bhattacharyya, P.K. Das, and A. Sen, Arch. Biochem. Biophys, 1981, 211, 459.
103 T.K. Datta and P.S. Basu, Biochem. J., 1981, 197, 751.
104 H.P. Jennissen, J. Chromatogr., 1981, 215, 73.
105 F. Delers, C. Lombart, M. Domingo, and S. Musquera, Anal. Biochem., 1981, 118, 353.
106 J.M. Egly, J. Chromatogr., 1981, 215, 243.
107 A.H. Merrill and D.B. McCormick, Methods in Enzymol., 1980, 66, 338.
108 G. Halperin, M. Breitenbach, M. Tauber-Finkelstein, and S. Shaltiel, J. Chromatogr., 1981, 215, 211.
109 H.G. Genieser, D. Gabel, and B. Jastorff, J. Chromatogr., 1981, 215, 235.
110 K.T. Yamamoto and W.O. Smith, Biochim. Biophys. Acta, 1981, 668, 27.
111 B.J. Conner and D.E. Comings, J. Biol. Chem., 1981, 256, 3283.
112 T. Kobayashi and K. Suzuki, J. Biol. Chem., 1981, 256, 7768.
113 F.S. Nolte and F.A. Kapral, Infection and Immunity, 1981, 31, 1094.
114 R.C. White, C.J. Ruff, and T.E. Nelson, Anal. Biochem., 1981, 115, 388.
115 N. Bruchovsky, P.S. Rennie, and T. Comeau, Eur. J. Biochem., 1981, 120, 399.
116 I.M. Peterson, Comp. Biochem. Physiol., 1981, 69B, 47.
117 H. Saumweber, R. Binder, and H. Bisswanger, Eur. J. Biochem., 1981, 114, 407.
118 A. Svenson and P.A. Hynning, Prep. Biochem., 1981, 11, 99.
119 D. Couchie, C. Erneux, and J.E. Dumont, Biochem. J., 1981, 199, 441.
120 N.G. Rutherford and G.M.W. Cook, F.E.B.S. Letts., 1981, 136, 105.
121 T. Kamataki, K.Maeda, and Y. Yamazoe, Biochem. Biophys. Res. Comm., 1981, 103, 1.
122 M. Maier, E. Polivka, and B.R. Binder, Hoppe-Seyler's Z. Physiol. Chem., 1981, 362, 883.
123 T. Nakada, T. Fukui, T. Saito, K. Miki, C. Oji, S. Matsuda, A. Ushijima, and K. Tomita, J. Biochem. (Jpn), 1981, 89, 625.
124 B. Overdijk, W.M.J. Van der Kroef, G.E.J. Van Steijn, and J.J.W. Lisman, Biochim. Biophys. Acta, 1981, 659, 255.
125 T.C. Rowe, J.R. Rusche, M.J. Brougham, and W.K. Holloman, J. Biol. Chem., 1981, 256, 10354.
126 P.M. Anderson and R.J. Desnick, J. Biol. Chem., 1980, 255, 1993.
127 H. Uchiyama and K. Nagasawa, J. Biochem. (Jpn), 1981, 89, 185.
128 S. Yonehara, Y. Yanase, T. Sano, M. Imai, S. Nakasawa, and H. Mori, J. Biol. Chem., 1981, 256, 3770.
129 T. Malmqvist, M. Malmqvist, and R. Möllby, Acta Pathol. Microbiol. Scand., 1981, 89B, 357.
130 T. Malmqvist and R. Möllby, Acta Pathol. Microbiol. Scand., 1981, 89B, 363.
131 D.R. Armant and C.L. Rutherford, J. Biol. Chem., 1981, 256, 12710.
132 J.A. Buswell, K.E. Eriksson, and B. Pettersson, J. Chromatogr., 1981, 215, 99.
133 L. Fox, N. Brot, and H. Weissbach, J. Biol. Chem., 1981, 256, 7796.
134 S.C. Hodgkinson and P.J. Lowry, Biochem. J., 1981, 199, 619.
135 S.D. Carson and W.H. Konigsberg, Anal. Biochem., 1981, 116, 398.
136 T. Lloyd and M.A. Walega, Anal. Biochem., 1981, 116, 559.
137 H.A. Nash and C.A. Robertson, J. Biol. Chem., 1981, 256, 9246.
138 R.L. Olsen and C. Little, Acta Chem. Scand., 1981, 35B, 1.
139 M. Parniak, H. Hasegawa, H. Wilgus, and S. Kaufman, Biochem. Biophys. Res. Commun., 1981, 99, 707.
140 T.B. Lo, F.L. Huang, and G.D. Chang, J. Chromatogr., 1981, 215, 229.

141 C.S. Petersen, N.S. Pedersen, and N.H. Axelsen, Anal. Biochem., 1981, 117, 231.
142 M. Shoyab, T.C. Warren, and G.J. Todaro, J. Biol. Chem., 1981, 256, 12529.
143 M. Takahashi-Nakamura, S. Tsuji, K. Suzuki, and K. Imahori, J. Biochem. (Jpn), 1981, 90, 1583.
144 R.J. Yon, Anal. Biochem., 1981, 113, 219.
145 M. Inoue, M. Hara, F. Nagashima, S. Matsui, N. Mitsuyasu, and Y. Morino, Biochim. Biophys. Acta, 1981, 659, 362.
146 S.A. Suleiman and R. Spector, Arch. Biochem. Biophys., 1981, 208, 87.
147 M. Einarsson, B. Forsberg, O. Larm, M.E. Riquelme, and E. Scholander, J. Chromatogr., 1981, 215, 45.
148 J.I. Ademola and D.W. Hutchinson, Biotechnol. Bioeng., 1980 22, 2419.
149 A.B. Guranowski, P.K. Chiang, and G.L. Cantoni, Eur. J. Biochem., 1981, 114, 293.
150 G.S. Bethell, J.S. Ayers, M.T.W. Hearn, and W.S. Hancock, J. Chromatogr., 1981, 219, 353.
151 G.S. Bethell, J.S. Ayers, M.T.W. Hearn, and W.S. Hancock, J. Chromatogr., 1981, 219, 361.
152 A.K. Mallia, G.T. Hermanson, R.I. Krohn, F.K. Fujimoto, and P.K. Smith, Anal. Letters, 1981, 14, 649.
153 G.A. Vehar and E.W. Davie, Biochemistry, 1980, 19, 401.
154 Y.V. Vertiev and Y.V. Ezepchuk, Hoppe-Seyler's Z. Physiol. Chem., 1981, 362, 1339.
155 J. Kohn and M. Wilchek, Anal. Biochem., 1981, 115, 375.
156 T.S. Hseu, S.L. Lan, and M.D. Yang, Anal. Biochem., 1981, 116, 181.
157 H. Ooshima, Y. Genko, and Y. Harano, Biotechnol. Bioeng., 1981, 23, 2851.
158 Y. Kawamura, K. Nakanishi, R. Matsuno, and T. Kamikubo, Biotechnol. Bioeng., 1981, 23, 1219.
159 H.L. Wu, D.A. Lace, and M.L. Bender, Proc. Nat. Acad. Sci., U.S.A., 1981, 78, 4118.
160 S.V. Amontov, E.A. Tolosa, and E.V. Goryachenkova, Biochemistry (U.S.S.R), 1980, 45, 1484.
161 K. Kohama and H. Holtzer, J. Biochem. (Jpn), 1981, 89, 341.
162 A.J.M. Vermorken, P.J.J.M. Weterings, M.A. Kibbelaar, J.A. Lenstra, and H. Bloemendal, F.E.B.S. Lett., 1981, 127, 105.
163 K. Hochstrasser, G.J. Albrecht, O. Schönberger, B. Rasche, and K. Lempart, Hoppe-Seyler's Z. Physiol. Chem., 1981, 362, 1369.
164 H.R. Lijnen, B. Van Hoef, and D. Collen, Eur. J. Biochem., 1981, 120, 149.
165 R. Myllylä, Biochim. Biophys. Acta, 1981, 658, 299.
166 J.H. Pazur, K.R. Forry, Y. Tominaga, and E.M. Ball, Biochem. Biophys. Res. Commun., 1981, 100, 420.
167 L.I. Ashmarina, V.I. Muronetz, and N.K. Nagradova, F.E.B.S. Lett., 1981, 128, 22.
168 T.V. Cherednikova, V.I. Muronetz, and N.K. Nagradova, Biochim. Biophys. Acta, 1980, 631, 292.
169 N. Koch and D. Haustein, J. Immunol. Methods, 1981, 41, 163.
170 J. Tenovuo and K. Kurkijärvi, Arch. Oral Biol., 1981, 26, 309.
171 P.F. Burns, C.W. Campagnon, I.M. Chaiken, and A.T. Campagnon, Biochemistry, 1981, 20, 2463.
172 S.F. Ching, L.W. Hayes, and L.L. Slakey, Biochim. Biophys. Acta, 1981, 657, 222.
173 V.M. Masters, J. Webster, and G.M.W. Cook, Biochem. Pharmacol., 1980, 29, 3189.
174 A. Kilara, Process Biochem., 1981, 16, 25.
175 L.A.E. Sluyterman and J. Wijdenes, Biotechnol. Bioeng., 1981, 23, 1977.
176 T. Tomono, T. Suzuki, and E. Tokunaga, Biochim. Biophys. Acta, 1981, 660, 186.
177 A.A. Kortt and M.A. Jermyn, Eur. J. Biochem., 1981, 115, 551.
178 M. Kučera and R.B. Turner, Biochim. Biophys. Acta, 1981, 662, 72.
179 M.N. Shul'gin, T.A. Valueva, A.Y. Kestere, and V.V. Mosolov, Biochemistry (U.S.S.R.), 1981, 46, 390.

180 R.B. Sim, T.M. Twose, D.S. Paterson, and E. Sim, Biochem. J., 1981, 193, 115.
181 F.S. Steven, M.M. Griffin, S. Itzhaki, and A. Al-Habib, Biochim. Biophys. Acta, 1981, 660, 333.
182 J.L. Iborra, J.A. Ferragut, and J.A. Lozano, Biochem. J., 1981, 197, 581.
183 A. Laine and A. Hayem, Biochim. Biophys. Acta, 1981, 668, 429.
184 H. Hirai, Y. Tsukada, A. Hara, N. Hibi, S. Nishi, H.T. Wepsic, T. Koji, and N. Ishnii, J. Chromatogr., 1981, 215, 195.
185 U.H. Stenman, M.L. Sutinen, R.K. Selander, K. Tontti, and J. Schröder, J. Immunol. Methods, 1981, 46, 337.
186 E. Wood, E.P. Hounsell, and T. Feizi, Carbohydr. Res., 1981, 90, 269.
187 S. Iwasa, C. Kitada, I. Yoshida, K. Kondo, M. Hori, and M. Fujino, J. Biochem. (Jpn), 1981, 89, 1091.
188 C.F. Ware, R.A. Wetsel, and W.P. Kolb, Molec. Immunol., 1981, 18, 521.
189 P. Cornale, M. Bonazzi, C. Molinu, P. Romelli, L. Vancheri, and F. Pennisi, Clin. Chem., 1981, 27, 896.
190 R. Matsas, D. Proud, K. Nustad, and G.S. Bailey, Anal. Biochem., 1981, 113, 264.
191 E.E. Slater and H.V. Strout, J. Biol. Chem., 1981, 256, 8164.
192 I.W. Sutherland, J. Chromatogr., 1981, 213, 301.
193 Y. Tamai, H. Shinmoto, and M. Takakuwa, Agric. Biol. Chem., 1981, 45, 2713.
194 Y. Barra, O. De Lapeyrière, and J. Planche, J. Nat. Cancer Inst., 1979, 63, 1441.
195 L.C. Kühn and J.P. Kraehenbuhl, J. Biol. Chem., 1981, 256, 12490.
196 G.A. Altamirano, C. Barranco-Acosta, E. Van Roost, and J.P. Vaerman, Molec. Immunol., 1980, 17, 1525.
197 B.L. Hempstead, C.W. Parker, and A. Kulczycki, J. Biol. Chem., 1981, 256, 10717.
198 J.F. Kennedy and J.A. Barnes, J. Chromatogr., 1981, 212, 179.
199 B. Seetharam, S.S. Bagur, and D.H. Alpers, J. Biol. Chem., 1981, 257, 183.
200 A. Herscovics, A. Quaroni, B. Bugge, and K. Kirsch, Biochem. J., 1981, 197, 511.
201 K. Hochstrasser, O.L. Schönberger, K. Lempart, and M. Metzger, Hoppe-Seyler's Z. Physiol. Chem., 1981, 362, 1363.
202 B.W. Maidment, L.D. Papsidero, M. Gamarra, T. Nemoto, and T.M. Chu, Anal. Biochem., 1981, 111, 336.
203 H.C. Reeves, R. Heeren, and P. Malloy, Anal. Biochem., 1981, 115, 194.
204 J. Rousseaux, M.T. Picque, H. Bazin, and G. Biserte, Molec. Immunol., 1981, 18., 639.
205 Y. Tamaki, T. Kishida, K. Shibata, N. Takahashi, and M. Fukuda, J. Immunol. Methods, 1981, 45, 177.
206 T. Bearspark, S. Bouquelet, B. Fournet, J. Montreuil, G. Spik, J. Stirling, and G. Strecker, F.E.B.S. Lett., 1977, 84, 379.
207 W. Gade, M.A. Jack, J.B. Dahl, E.L. Schmidt, and F. Wold, J. Biol. Chem., 1981, 256, 12905.
208 K.K. Karhi and C.G. Gahmberg, Biochim. Biophys. Acta, 1980, 622, 344.
209 M.A. Winstanley, D.A.P. Small, and I.P. Trayer, Eur. J. Biochem., 1979, 98, 441.
210 J. Keski-Oja and K.M. Yamada, Biochem. J., 1981, 193, 615.
211 D.M. Helfman, M. Shoji, and J.F. Kuo, J. Biol. Chem, 1981, 256, 6327.
212 D.S. Grove and G.S. Serif, Biochim. Biophys. Acta, 1981, 662, 246.
213 R.K. Iyer, R. Tuli, and J. Thomas, Arch. Biochem. Biophys., 1981, 209, 628.
214 K. Wikvall, J. Biol. Chem, 1981, 256, 3376.
215 A.A. Ansari, Biochem. J., 1981, 199, 75.
216 J. Barthová, P. Plachá, and S. Leblová, Coll. Czech. Chem. Commun., 1980, 45, 1608.
217 K.S. Srivenugopal and P.R. Adiga, J Biol. Chem., 1981, 256, 9532.
218 R.B. Mellor, G.M. Gadd, P. Rowell, and W.D.P. Stewart, Biochem. Biophys. Res. Commun., 1981, 99, 1348.
219 G.T. Montelione, S. Callahan, and T.R. Podleski, Biochim. Biophys. Acta, 1981, 670, 110.

220 J. Viitala, K.K. Karhi, G.G. Gahmberg, J. Finne, J. Järnefelt, G. Myllylä, and T. Krusius, Eur. J. Biochem., 1981, 113, 259.
221 Y. Aoyama and Y. Motokawa, J. Biol. Chem.., 1981, 256, 12367.
222 Z.F. Burton, T. Sutherland, and D. Eisenberg, Arch. Biochem. Biophys., 1981, 211, 507.
223 K.Kameda and W.D. Nunn, J. Biol. Chem., 1981, 256, 5702.
224 K. Kawai, Y. Saito, and T. Eguchi, J. Ferment. Technol., 1980, 58, 569.
225 K. Kawai, M. Mori, T. Eguchi, H. Horitsu, and Y. Eguchi, Agric. Biol. Chem. Jpn., 1981, 45, 2781.
226 M.R. Owens, L.L. Miller, and C.D. Cimino, Biochim. Biophys. Acta, 1981, 676, 365.
227 R.S. Adelstein and C.B. Klee, J. Biol. Chem., 1981, 256, 7501.
228 J.E. Bodwell and W.L. Meyer, Biochemistry, 1981, 20, 2767.
229 D.T. Dorai, B.K. Bachhawat, and S. Bishayee, Anal. Biochem., 1981, 115, 130.
230 K.M. Rose, L.E. Bell, D.A. Siefken, and S.T. Jacob, J. Biol. Chem., 1981, 256, 7468.
231 K. Iijima, J. Kichi, and T. Hayakawa, J. Biochem. (Jpn), 1981, 89, 1101.
232 E.G. Afting and M.L. Becker, Biochem. J., 1981, 197, 519.
233 J. Aggeler, E. Engvall, and Z. Werb, Biochem. Biophys. Res. Commun., 1981, 100, 1195.
234 C.A. Auffret and M.J. Turner, Biochem. J., 1981, 193, 647.
235 B. Bayard and J.P. Kerckaert, Eur. J. Biochem., 1981, 113, 405.
236 K.P. Campbell and D.H. Maclennan, J. Biol. Chem., 1981, 256, 4626.
237 C. Lee, G.P. Murphy, and T.M. Chu, Cancer Res., 1980, 40, 1245.
238 A. Saeed, S. DeBoeck, I. Debruyne, J. Wouters, and J. Stocks, Biochim. Biophys. Acta, 1980, 614, 389.
239 H. Debray, B. Fournet, J. Montreuil, L. Dorland, and J.F.C. Vliegenthart, Eur. J. Biochem., 1981, 115, 559.
240 W.T. Forsee and J.S. Schutzbach, J. Biol. Chem., 1981, 256, 6577.
241 J. Hermann and B. Keil, Biochim. Biophys. Acta, 1981, 643, 30.
242 C.M. Heyworth, E.F. Neumann, and C.H. Wynn, Biochem. J., 1981, 193, 773.
243 B.J. Howlett and A.E. Clarke, Biochem. J., 1981, 197, 695.
244 T. Ikuta, H. Okubo, H. Ishibashi, K. Shibata, M. Nagamo, and J. Kudo, Biochem. Int., 1981, 2, 619.
245 D.H. Joziasse, D.H. Van den Eijnden, J.J.W. Lisman, and G.J.M. Hooghwinkel, Biochim. Biophys. Acta, 1981, 660, 174.
246 S. Junqua, M. Lemonnier, and L. Robert, Comp. Biochem. Physiol., 1981, 69B, 445.
247 K. Kornfeld, M.L. Reitman, and R. Kornfeld, J. Biol. Chem., 1981, 256, 6633.
248 S.C. Li, M. Asakawa, Y. Hirabayashi, and Y.T. Li, Biochim. Biophys. Acta, 1981, 660, 278.
249 L.A. Liotta, K.T. Ryggvason, S. Garbisa, P.G. Robey, and S. Abe, Biochemistry, 1981, 20, 100.
250 T. Ludolph, E. Paschke, J. Glössl, and H. Kresse, Biochem. J., 1981, 193, 811.
251 G.J. Markelonis and T.H. Oh, J. Neurochem., 1981, 37, 95.
252 E. Reinwald, P. Rautenberg, and H.J. Risse, Biochim. Biophys. Acta, 1981, 668, 119.
253 D.C. Rijken and D. Collen, J. Biol. Chem., 1981, 256, 7035.
254 L. Risteli and R. Timpl, Biochem. J., 1981, 193, 749.
255 A. Varki and S. Kornfeld, J. Biol. Chem., 1981, 256, 9937.
256 J.P. Zanetta, A. Reeber, and G. Vincendon, Biochim. Biophys. Acta, 1981, 670, 393.
257 T. Okayasu, M. Nagao, T. Ishibashi, and Y. Imai, Arch. Biochem. Biophys., 1981, 206, 21.
258 H. Tanaka, I. Sasaki, K. Yamashita, K. Miyazaki, Y. Matuo, J. Yamashita, and T. Horio, J. Biochem. (Jpn), 1980, 88, 797.
259 J.R. Bartles, B.C. Santoro, and W.A. Frazier, Biochim. Biophys. Acta, 1981, 674, 372.
260 T. Muramatsu, H. Muramatsu, and M. Ozawa, J. Biochem. (Jpn), 1981, 89, 473.
261 D.T. Dorai, B.K. Bachhawat, S. Bishayee, K. Kannan, and D.R. Rao, Arch.

Biochem. Biophys., 1981, 209, 325.
262 J. Laterra, R. Ansbacher, and L.A. Culp, Proc. Nat. Acad. Sci. (U.S.A.), 1980, 77, 6662.
263 M.N. Fukuda, J. Biol. Chem., 1981, 256, 3900.
264 J. McPherson, H. Sage, and P. Bornstein, J. Biol. Chem., 1981, 256, 11330.
265 R.E. Weiss and A.H. Reddi, Biochem. J., 1981, 197, 529.
266 J.P. Prieels, D. Monnom, M. Dolmans, T.A. Beyer, and R.L. Hill, J. Biol. Chem., 1981, 256, 10456.
267 J.M. Wilson, B.W. Baugher, L. Landa, and W.N. Kelley, J. Biol. Chem., 1981, 256, 10306.
268 A.G. Magee, D.A.W. Grant, and J. Herman-Taylor, Clin. Chim. Acta, 1981, 115, 241.
269 D.G. Attardi, A. De Paolis, and G.P. Tocchini-Valentini, J. Biol. Chem., 1981, 256, 3654.
270 G.L. Jensen and A. Bensadoun, Analyt. Biochem., 1981, 113, 246.
271 M.N. Blackburn and C.C. Sibley, J. Biol. Chem., 1980, 255, 824.
272 K. Suzuki, T. Terao, and T. Osawa, J. Biochem. (Jpn), 1981, 89, 1.
273 Y. Fujita-Yamaguchi and A. Yoshida, J. Biol. Chem., 1981, 256, 2701.
274 K. Kornfeld, M.L. Reitman, and R. Kornfeld, J. Biol. Chem., 1981, 256, 6633.
275 R. Borriss, Z. Allgem. Mikrobiol., 1981, 21, 7.
276 L.A. Fransson and B.G. Johansson, Int. J. Biol. Macromol., 1981, 3, 25.
277 P.A.S. Mourão, S. Pillai, and N. Di Ferrante, Biochim. Biophys. Acta, 1981, 674, 178.
278 J. Burge, A. Nicholson-Weller, and K.F. Austen, Molec. Immunol., 1981, 18, 47.
279 C.H. Hammer, G.H. Wirtz, L. Renfer, H.D. Gresham, and B.F. Tuck, J. Biol. Chem., 1981, 256, 3995.
280 M.W.C. Hatton, L.R. Berry, H. Kaur, A. Koj, and E. Regoceczi, Canad. J. Biochem., 1979, 57 1183.
281 Y. Mizuno, Y. Kozutsumi, T. Kawasaki, and I. Yamashina, J. Biol. Chem., 1981, 256, 4247.
282 J. Mendicino, E.V. Chandrasekaran, K.R. Anumula, and M. Davila, Biochemistry, 1981, 20, 967.
283 I. Clemmensen, S. Thorsen, S. Müllertz, and L.C. Petersen, Eur. J. Biochem., 1981, 120, 105.
284 K. Yang, H. Samanta, J. Dougherty, B. Jayaram, B. Broeze, and P. Lengyel, J. Biol. Chem.., 1981, 256, 9324.
285 M.G. Sarngadharan, V.S. Kalyanaraman, R. Rahman, and R.C. Gallo, J. Virol., 1980, 35, 555.
286 S.L. Bolmer and E.A. Davidson, Biochemistry, 1981, 20, 1047.
287 D. Tsao and Y.S. Kim, J. Biol. Chem., 1981, 256, 4947.
288 C. Rickards, P. Buckland, B.R. Smith, and R. Hall, F.E.B.S. Lett., 1981, 127, 17.
289 F. A. Stephenson, R. Harrison, and G.G. Lunt, Eur. J. Biochem., 1981, 115, 91.
290 A.L. Barsoum and E.A. Davidson, Molec. Immunol., 1981, 18, 367.
291 C. Villar-Palasi and A. Kumon, J. Biol. Chem., 1981, 256, 7409.
292 M. Tashiro, N. Sugihara, Z. Maki, and M. Kanamori, Agric. Biol. Chem. (Jpn), 1981, 45, 519.
293 R.H. Tukey and T.R. Tephly, Arch. Biochem. Biophys., 1981, 209, 565.
294 K.K. Karhi and C.G. Gahmberg, Biochim. Biophys. Acta, 1980, 622, 344.
295 L.P. Thuy, R.F. Baugh, J.E. Brown, and C. Houghie, Biochim. Biophys. Acta, 1981, 678, 187.
296 J. Butterworth, Clin. Chim. Acta, 1980, 108, 347.
297 J. Kambayashi, M. Sakon, H. Ohno, and G. Kosaki, Thromb. Res., 1981, 21, 129.
298 A. Rotman and S. Linder, Thromb. Res., 1981, 22, 227.
299 A. Soliman, S. Nordlund, B.S. Johansson, and H. Baltscheffsky, Acta Chem. Scand., 1981, 35B, 63.
300 W.P. Dafeldecker, P.E. Meadow, X. Parés, and B.L. Vallee, Biochemistry, 1981, 20, 6729.

301 T. Uchida and Y. Shibata, J. Biochem. (Jpn), 1981, 90, 463.
302 S. Kanaya and T. Uchida, J. Biochem. (Jpn), 1981, 90, 473.
303 J. Sobhanaditya and N. Appaji Rao, Biochem. J., 1981, 197, 227.
304 P.S. Appukattan and D. Basu, Anal. Biochem., 1981, 113, 253.
305 K. Cline and P. Albersheim, Plant Physiol., 1981, 68, 207.
306 K.S. You and C.R. Roe, Anal. Biochem., 1981, 114, 177.
307 M. Nishikata, K.I. Kasai, and S.I. Ishii, Biochim. Biophys. Acta, 1981, 660, 256.
308 I. Emöd, N.T. Tong, and B. Keil, Biochim. Biophys. Acta, 1981, 659, 283.
309 P.H. Burrill, P.M. Brannon, and N. Kretchmer, Anal. Biochem., 1981, 117, 402.
310 P.F. Kador, D. Carper, and J.H. Kinoshita, Anal. Biochem., 1981, 114, 53.
311 H. Rosemeyer and F. Seela, Anal. Biochem., 1981, 115, 339.
312 K.M. Rupprecht, N. Sonenberg, A.J. Shatkin, and S.M. Hecht, Biochemistry, 1981, 20, 6570.
313 H. Rosemeyer and F. Seela, Carbohydr. Res., 1981, 90, 53.
314 K. Murakami and K. Okuda, Biochem. Biophys. Res. Commun., 1981, 100, 91.
315 N. Pattinson, Anal. Biochem., 1981, 115, 424.
316 U.W. Mueller and J.M. Potter, Biochem. J., 1981, 197, 645.
317 M. Suzuki, J. Biochem. (Jpn), 1981, 89, 599.
318 J. Van Kapel, B.G. Loef, J. Lindemans, and J. Abels, Biochim. Biophys. Acta, 1981, 676, 307.
319 R.B. Harris, J.T. Ohlsson, and I.B. Wilson, Anal. Biochem., 1981, 111, 227.
320 M. Lyon and C.F. Phelps, Biochem. J., 1981, 199, 419.
321 B. Shafit-Zagardo and B.M. Turner, Biochim. Biophys. Acta, 1981, 659. 7.
322 E. Schmincke-Ott and H. Bisswanger, Eur. J. Biochem., 1981, 114, 413.
323 K. Mizuno, M. Koyama, and H. Shibaoka, J. Biochem. (Jpn), 1981, 89, 329.
324 H.L. Drake, S.I. Hu, and H.G. Wood, J. Biol. Chem., 1981, 256, 11137.
325 A.J. Wittwer and C. Wagner, J. Biol. Chem., 1981, 256, 4102.
326 C.V. Warner, D.F. Smith, D.A. Zopf, and V. Ginsberg, Molec. Immunol., 1980, 17, 1163.
327 B.V. McCleary, Phytochemistry, 1979, 18, 757.
328 T. Taki, Y. Hirabayashi, K. Takagi, R. Kamada, K. Kojima, and M. Matsumoto, J. Biochem. (Jpn), 1981, 89, 503.
329 M. DelRosso, R. Cappelletti, M. Viti, S. Vannucchi, and V. Chiarugi, Biochem. J., 1981, 199, 699.
330 S. Kanaya and T. Uchida, J. Biochem. (Jpn), 1981, 89, 591.
331 H. Yoshida, I. Fukuda, and M. Hashiguchi, J. Biochem. (Jpn), 1980, 88, 1813.
332 G. Heney and G.A. Orr, Anal. Biochem., 1981, 114, 92.
333 K. Hofmann, S.W. Wood, C.C. Brinton, J.A. Montibeller and F.M. Finn, Proc. Nat. Acad. Sci. (U.S.A.), 1980, 77, 4666.
334 T.W. Siefel, S. Ganguly, S. Jacobs, O.M. Rosen, and C.S. Rubin, J. Biol. Chem., 1981, 256, 9266.
335 C. Morandi and G. Attardi, J. Biol. Chem., 1981, 256, 10169.
336 D.G. Priest, M.T. Doig, and J.B. Hynes, Experientia, 1981, 37, 119.
337 F. Menegus and M. Pace, Eur. J. Biochem., 1981, 113, 485.
338 F.A. Firgaira, R.G.H. Cotton, and D.M. Banks, Biochem. J., 1981, 197, 31.
339 T.N. Oeltmann and J.T. Forbes, Arch. Biochem. Biophys., 1981, 209, 362.
340 A.J. Rivett, I.L. Smith, and K.F. Tipton, Biochem. J., 1981, 197, 473.
341 R. Rüchel, Biochim. Biophys. Acta, 1981, 659, 99.
342 F. Suzuki, Y. Nakamura, Y. Nagata, T. Ohsawa, and K. Murakami, J. Biochem. (Jpn), 1981, 89, 1107.
343 A.C. Antony, C. Utley, K.C. Van Horne, and J.F. Kolhouse, J. Biol. Chem., 1981, 256, 9684.
344 K.R. Latham, J.W. Apriletti, N.L. Eberhardt, and J.D. Baxter, J. Biol. Chem., 1981, 256, 12088.
345 O.M.P. Singh, A.B. Graham, and G.C. Wood, Eur. J. Biochem., 1981, 116, 311.
346 H.A. Nunez and R. Barker, Biochemistry, 1980, 19, 489.
347 Y. Oda, K.I. Kasai, and S.I. Ishii, J. Biochem. (Jpn), 1981, 89, 285.
348 T. Taki, Y. Hirabayashi, K. Takagi, R. Kamada, K. Kojima, and M. Matsumoto, J. Biochem. (Jpn), 1981, 89, 503.

349 D.R. Marshak, D.M. Watterson, and L.J. Vaneldik, Proc. Nat. Acad. Sci. (U.S.A.), 1981, 78, 6793.
350 C.E. Creutz, Biochem. Biophys. Res. Commun., 1981, 103, 1395.
351 S.A. Cohen and R.L. Barchi, Biochim. Biophys. Acta, 1981, 645, 253.
352 A. Colman, C.D. Lane, R. Craig, A. Boulton, T. Mohun, and J. Morser, Eur. J. Biochem., 1981, 113, 339.
353 P. Conradt, M. Emura, and S. Matthel, Cancer Letts., 1981, 14, 55.
354 J. Denton, W.E. Lewis, I.A. Nieduszynski, and C.F. Phelps, Anal. Biochem., 1981, 118, 388.
355 J.A. Hedo, L.C. Harrison, and J. Roth, Biochemistry, 1981, 20, 3385.
356 F. Ishikawa, T. Kameyama, A. Takenaka, K. Oishi, and K. Aida, Agric. Biol. Chem. (Jpn), 1981, 45, 2105.
357 R. Ivell, H. Schmale, and D. Richter, Biochem. Biophys. Res. Commun., 1981, 102, 1230.
358 H. Iwase, Y. Kato, and K. Hotta, J. Biol. Chem., 1981, 256, 5638.
359 J.M. Nickerson and G.M. Fuller, Biochemistry, 1981, 20, 2818.
360 I. Nicollet, J.P. Lebreton, M. Fontaine, and M. Hiron, Biochim. Biophys. Acta, 1981, 668, 235.
361 K. Resch, S. Schneider, and M. Szamel, Anal. Biochem., 1981, 117, 282.
362 F.G. Staber and A.W. Burgess, J. Cell. Physiol., 1980, 102, 1.
363 M.C. Stuart, S. Ellis, L. Gowlland, and S. Tuff, Clin. Chem., 1981, 27, 52.
364 W. Terpstra, F.E.B.S. Letts., 1981, 126, 231.
365 T. Usui, S. Hosokawa, T. Mizuno, T. Suzuki, and H. Meguro, J. Biochem. (Jpn), 1981, 89, 1029.
366 S. Pillai, Biochem. J., 1981, 193, 825.
367 J.G.J. Bauman, J. Wiegant, and P. Van Duijn, J. Histochem. Cytochem., 1981, 29, 227.
368 D.R. Kahn and I. Cohen, Biochim. Biophys. Acta, 1981, 668, 490.
369 R.M. Metrione, Anal. Biochem., 1981, 110, 117.
370 L.-A. Fransson, Eur. J. Biochem., 1981, 120, 251.
371 L.-A. Fransson, B. Havsmark, and J.K. Sheehan, J. Biol. Chem., 1981, 256, 13039.
372 T. Hara, K. Takahashi, and H. Endo, F.E.B.S. Letts., 1981, 128, 33.
373 C.S. Wang, H.B. Dass, D. Downs, and R.R. Witmer, Clin. Chem., 1981, 27, 663.
374 A. Jakobovits, Y. Eshdat, and N. Sharon, Biochem. Biophys. Res. Commun., 1981, 100, 1484.
375 G.D. Roodman, J.L. Spivak, and E.D. Zanjani, J. Lab. Clin. Med., 1981, 98, 684.
376 I. Hagen, F. Brosstad, N.O. Solum, and R. Korsmo, J. Lab. Clin. Med., 1981, 97, 213.
377 L.A.A.Sluyterman and J. Wijdenes, J. Chromatogr., 1981, 206, 429.
378 G. Levi and V.I. Teichberg, J. Biol. Chem., 1981, 256, 5735.
379 B. Höjeberg, K. Blomberg, S. Stenberg, and C. Lind, Arch. Biochem. Biophys., 1981, 207, 205.
380 C. Lind and B. Höjeberg, Arch. Biochem. Biophys., 1981, 207, 217.
381 K. Rose, J.D. Priddle, and R.E. Offord, J. Chromatogr., 1981, 210, 301.
382 S. Kakiuchi, K. Sobue, R. Yamazaki, J. Kambayashi, M. Sakon, and G. Kosaki, F.E.B.S. Letts., 1981, 126, 203.
383 K. Iwaki and H. Fuwa, Agric. Biol. Chem. (Jpn), 1981, 45, 2683.
384 B. Tappeser, D. Wellnitz, and D. Klämbt, Z. Pfla zenphysiol., 1981, 101, 295.
385 M.R. Siro, K. Viitanen, and T. Lövgren, Acta Chem. Scand., 1981, 35B, 29.
386 K. Nagata and N. Yoshida, J. Biochem. (Jpn), 1981, 89, 1121.
387 U. Matsumoto and Y. Shibusawa, J. Chromatogr., 1981, 206, 17.
388 L.A.A. Sluyterman and J. Wijdenes, J. Chromatogr., 1981, 206, 441.
389 N. Shiomi, Carbohyd. Res., 1981, 96, 281.
390 T. Wood, Biochim. Biophys. Acta, 1981, 659, 233.
391 U. Reiss and B. Sacktor, Arch. Biochem. Biophys., 1981, 209, 342.
392 Y. Konami, T. Tsuji, I. Matsumoto, and T. Osawa, Hoppe-Seyler's Z. Physiol. Chem., 1981, 362, 983.
393 I. Matsumoto, H. Kitagaki, Y. Akai, Y. Ito, and N. Seno, Anal. Biochem.,

1981, 116, 103.
394 H. Kikuchi and M. Watanabe, Anal. Biochem., 1981, 115, 109.
395 P. Hubert and J. Porath, J. Chromatogr., 1981, 206, 164.
396 H. Hansson and L. Kågedal, J. Chromatogr., 1981, 215, 333.
397 C.A.K. Borrebaeck, B. Lönnerdal, M.E. Etzler, F.E.B.S. Lett., 1981, 130, 194.
398 J.W. Heine, J. Vandamme, M. Deley, A. Billiau, and P. Desomer, J. Gen. Virol., 1981, 54, 47.
399 P.R. Coulet, J. Carlsson, and J. Porath, Biotechnol. Bioeng., 1981, 23, 663.
400 R.G. Frost, J.F. Monthony, S.C. Engelhorn, and C.J. Siebert, Biochim. Biophys. Acta, 1981, 670, 163.
401 T. Staehelin, D.S. Hobbs, H. Kung, C.Y. Lai, and S. Pestka, J. Biol. Chem., 1981, 256, 9750.
402 C.R. Caldwell and A. Haug, Anal. Biochem., 1981, 116, 325.
403 H.E. Meyer, H.J. Bubenzer, L. Herbertz, L. Kuehn, and H. Reinauer, Hoppe-Seyler's Z. Physiol. Chem., 1981, 362, 1621.
404 K. Marossy, Biochim. Biophys. Acta, 1981, 659, 351.
405 K. Marossy, M. Hauck, and P. Elödi, Biochim. Biophys. Acta, 1981, 662. 36.
406 S. Olofsson, S. Jeansson, and E. Lycke, J. Virol., 1981, 38, 564.
407 R.F. Tucker, J. Babul, and E. Stellwagen, J. Biol. Chem., 1981, 256, 10993.
408 H. Nakata and H. Fujisawa, J. Biochem. (Jpn), 1981, 90, 567.
409 P.M. Abdella, P.K. Smith, and G.P. Royer, Biochem. Biophys. Res. Commun., 1979, 87, 734.
410 A.F.S.A. Habeeb, Biochim. Biophys. Acta, 1981, 673, 527.
411 L. D'Angiuro, B. Focher, P. Cremonesi, G. Mazzola, and G. Vecchio, J. Polymer Sci., Polymer Symp., 1980, 68, 69.
412 T. Miron and M. Wilchek, J. Chromatogr., 1981, 215, 55.
413 K. Brocklehurst, B.S. Baines, and J.P.G. Malthouse, Biochem. J., 1981, 197, 739.
414 R.B. Harris, A.J. Johnson, and L.T. Hodgins, Biochim. Biophys. Acta, 1981, 668, 471.
415 J.A. Truglia and A. Stracher, Biochem. Biophys. Res. Commun., 1981, 100, 814.
416 J.M. Chobert, F. Addeo, and B. Ribadeau-Dumas, J. Dairy Res., 1981, 48, 429.
417 M.R. Purnell, P.R. Stone, and W.J.D. Whish, Biochem. Soc. Trans., 1981, 9, 321.
418 L. Rydén and H. Norder, J. Chromatogr., 1981, 215, 341.
419 K. Nilsson, O. Norrlöw, and K. Mosbach, Acta Chem. Scand., 1981, 35B, 19.
420 V. Bouriotis and P.D.G.Dean, J. Chromatogr., 1981, 206, 521.
421 Y.D. Clonis and C.R. Lowe, Biochim. Biophys. Acta, 1981, 659, 86.
422 W. Büsen, J. Biol. Chem., 1981, 255, 9434.
423 L.B. Daniels, P.J. Coyle, Y.B. Chiao, R.H. Glew, and R.S. Labow, J. Biol. Chem., 1981, 256, 13004.
424 M.A. De La Rosa, C. Gomez-Moreno, and J.M. Vega, Biochim. Biophys. Acta, 1981, 662, 77.
425 S. Hongo and T. Sato, Anal. Biochem., 1981, 114, 163.
426 T.F. Deuel, J. San Huang, R.T. Proffitt, J.U. Baenziger, D. Chang, and B.B. Kennedy, J. Biol. Chem., 1981, 256, 8896.
427 R.H. Glew, J.U. Balis, and S. Shelley, T. Kuelenschmidt, and R. Jaffe, Biochim. Biophys. Acta, 1981, 670, 101.
428 J. Graw, E.J. Schlaeger, and R. Knippers, J. Biol. Chem., 1981, 256, 13207.
429 J. Hasegawa, Agric. Biol. Chem. (Jpn), 1981, 45, 2805.
430 L. Jervis, Biochem. J., 1981, 197, 755.
431 E. Knight and D. Fahey, J. Biol. Chem., 1981, 256, 3609.
432 I. Lascu, M. Duc, and A. Cristea, Anal. Biochem., 1981, 113, 207.
433 J. Orr, L.M. Keefer, P. Keim, T.D. Nguyen, T. Wellems, R.L. Heinrikson, and R. Haselkorn, J. Biol. Chem., 1981, 256, 13091.
434 M.C. Redinbaugh and W.H. Campbell, Plant Physiol., 1981, 68, 115.
435 D.E. Roll and R.H. Glew, J. Biol. Chem., 1981, 256, 8190.
436 T. Shimakata and T. Kusaka, J. Biochem. (Jpn), 1981, 89, 1075.
437 H. Shutu, T. Fukui, T. Saito, Y. Shirakura, and K. Tomita, Eur. J. Biochem.,

1981, 118, 53.
438 W.O. Smith and S.M. Daniels, Plant Physiol., 1981, 68, 443.
439 M.R.G. Wood and P. Johnson, Biochim. Biophys. Acta, 1981, 662, 138.
440 D.A. Yost and B.M. Anderson, J. Biol. Chem., 1981, 256, 3647.
441 P.C.W.Lai, A.J. Luft, L.L. Huang, and F.L. Lorscheider, Comp. Biochem. Physiol., 1981, 68B, 107.
442 T. Mahany, B.S. Khirabadi, D.M. Gersten, P.Kurian, R.S. Ledley, and P.W.Ramwell, Comp. Biochem. Physiol., 1981, 68B, 319.
443 E.C. Metcalf, B. Crow, and P.D.G.Dean, Biochem. J., 1981, 199, 465.
444 F. Qadri and P.D.G. Dean, Biochem. J., 1981, 199, 471.
445 N.A. Reiss and A.M. Kaye, J. Biol. Chem., 1981, 256, 5741.
446 T. Skotland, Biochim. Biophys. Acta, 1981, 659, 312.
447 R.B. Harris, J.T. Ohlsson, and I.B. Wilson, Arch. Biochem. Biophys., 1981, 206, 105.
448 J.A.Begley, S.M.Heckman, and C.A.Hall, Arch. Biochem. Biophys., 1981, 208, 548.
449 L. Bläckberg and O. Hernell, Eur. J. Biochem., 1981, 16, 225.
450 M.E. Dahmus, J. Biol. Chem., 1981, 256, 3319.
451 W.O. Smith, Proc. Nat. Acad. Sci., U.S.A., 1981, 78, 2977.
452 C.M. Kane and S. Linn, J. Biol. Chem., 1981, 256, 3405.
453 H. Kizaki, T. Sakurada, and G. Weber, Biochim. Biophys. Acta, 1981, 662, 48.
454 P.S. Song, I.S. Kim, and T.R. Hahn, Anal. Biochem., 1981, 117, 32.
455 A.H.J. Ullah and G.W. Ordal, Biochem. J., 1981, 199, 795.
456 F. Vara and R. Serrano, Biochem. J., 1981, 197, 637.
457 M. Kitamikado, M. Ito, and Y.T. Li, J. Biol. Chem., 1981, 256, 3906.
458 W. Jessup and P.D.G.Dean, J. Chromatogr., 1981, 219, 419.
459 K.S. Lee and R.D. Gilbert, Carbohydr. Res., 1981, 88, 162.
460 T.L. Bluhm and P. Zugenmaier, Carbohydr.Res., 1981, 89, 1.
461 T. Handa and H. Yajima, Biopolymers, 1980, 19, 723.
462 H. Sakuma, S. Munakata, and S. Sugawara, Agric. Biol. Chem. (Jpn), 1981, 45, 443.
463 H. Okamoto and D. Yoshida, Agric. Biol. Chem. (Jpn), 1981, 45, 1291.
464 M. Nummi, M.L. Niku-Paavola, T.M. Enari, and V. Raunio, Anal. Biochem., 1981, 116, 137.
465 V.J. Shoshina, A.K. Enikeeva, G.V. Nikonovich, and K.U. Usmanov, Cell. Chem. Technol., 1981, 15, 567.
466 K. Danø, V. Moller, L. Ossowski, and L.S. Nielsen, Biochim. Biophys. Acta, 1980, 613, 542.
467 B. Pekin, A. Telefoncu, and A. Aslan, Biotechnol. Bioeng., 1981, 23, 1907.
468 R. Ohba, T. Shibata, and S. Ueda, J. Ferment. Technol., 1979, 57, 146.
469 T. Watanabe, T. Mori, T. Tosa, and I. Chibata, Agric. Biol. Chem. (Jpn), 1981, 45, 1001.
470 S.A. Yankofsky, R. Gurevitch, A. Niv, G. Cohen, and L. Goldstein, Anal. Biochem., 1981, 118, 307.
471 M.L.R.Vairo, W. Borzani, M.R.de M. Cruz, L. Perego, L.H. Koshimizu, and E.I.V. Gómez, Biotechnol. Bioeng., 1981, 23, 1669.
472 J. Kučera, J. Chromatogr., 1981, 213, 352.
473 R.K. Scopes, Anal. Biochem., 1981, 114, 8.
474 J. Tscholakowa, P. Burba, B. Gleitsmann, and K.H. Lieser, Z. Analyt. Chem., 1980, 300, 121.
475 Y. Kitagawa and E. Okuhara, Anal. Biochem., 1981, 115, 102.
476 R. Bovara, G. Carrea, P. Cremonesi, and G. Mazzola, Anal. Biochem., 1981, 112, 239.
477 K. Rokugawa, T. Fujishima, A. Kuninaka, and H. Yoshino, J. Ferment. Technol., 1979, 57, 570.
478 L.G. Moss, J.P.Moore, and L. Chan, J. Biol. Chem., 1981 256, 12655.
479 T. Fujiyoshi and M. Anai, J. Biochem. (Jpn), 1981, 89, 1129.
480 N. Murakami and V.K. Moudgil, Biochim. Biophys. Acta, 1981, 676, 386.
481 J.B. Olmsted, J. Biol. Chem., 1981, 256, 11955.
482 G.A. Petropavlovsky and G.G. Vasilieva, Cell. Chem. Technol., 1981, 14, 851.
483 N.Y. Abou-Zeid, W. Anwar, and A. Hebeish, Cell. Chem. Technol., 1981, 15,

321.
484 M.H. El-Rafie, M. El-Kashouti, F. El-Sisi, and A. Hebeish, Cell.Chem. Technol., 1981, 15, 427.
485 P. Mansson and L. Westfelt, J. Polym. Sci. Polym. Chem., 1981, 19, 1509.
486 S.K. Kar, P.L. Nayak, and G. Sanu, J. Poly. Sci. Polym. Chem., 1981, 19, 1581.
487 R. Herold, A. Merlin, and J.P. Fouassier, Compt. Rend. Acad. Sci. C, 1979, 288, 189.
488 K. Dimov, P. Pavlov, and N. Simeonov, Cell. Chem. Technol., 1980, 14, 457.
489 A. Hebeish, A. Waly, S. Abdel Bari, and S. Bedewi, Cell. Chem. Technol., 1981, 15, 505.
490 W.F. Glass, R.C. Briggs, and L.S. Hnilica, Science, 1981, 211, 70.
491 J.C. McMichael, L.M. Greisiger, and I. Millman, J. Immunol. Methods, 1981, 45, 79.
492 P. DeMeyer, E. DeHerdt, M. Kondo, and H. Siegers, J. Biochem. Biophys. Methods, 1980, 2, 311.
493 J. Turková, J. Vajčner, D. Vančurova, and J. Stamberg, Coll. Czech. Chem. Comm., 1979, 44, 3411.
494 H.N. Chang, Y.S. Ghim, Y.R. Cho, D.A. Landis, and P.J. Reilly, Biotechnol. Bioeng., 1981, 23, 2647.
495 J. Berghäuser, R. Jeck, and M. Pfeiffer, Biotechnol. Letts., 1981, 3, 339.
496 S. Hunt and T.N. Huckerby, Int. J. Biol. Macromol., 1980, 2, 389.
497 P. Gemeiner and J. Zemek, Coll. Czech. Chem. Comm., 1981, 46, 1693.
498 S. Imai, M. Muroi, A. Hamaguchi, R. Matsushita, and M. Koyama, Anal. Chim. Acta, 1980, 113 139.
499 H. Yamada and T. Imoto, Carbohydr. Res., 1981, 92, 160.
500 S. Hirano and Y. Yagi, Carbohydr. Res., 1981, 92, 319.
501 T. Sakaguchi, T. Horikoshi, and A. Nakajima, Agric. Biol. Chem. (Jpn)., 1981, 45, 2191.
502 E.N. Vasstrand and H.B. Jensen, Scand. J. Dental Res., 1980, 88, 219.
503 J. Synowiecki, Z.E. Sikorski, and M. Naczk, Biotechnol. Bioeng., 1981, 23, 2211.
504 J. Synowiecki, Z.E. Sikorski, and M. Naczk, Biochnol. Bioeng., 1981, 23, 231.
505 S. Tsukada and Y. Inoue, Carbohydr. Res., 1981, 88, 19.
506 L.D. Hall, M. Yalpani, and N. Yalpani, Biopolymers, 1981, 20, 1413.
507 L.D. Hall and M. Yalpani, J. Chem. Soc. Chem. Commun., 1980, 1153.
508 S. Hirano, S Tsuneyasu, and Y.Kondo, Agric. Biol. Chem. (Jpn), 1981, 45, 1335.
509 S. Hirano and T. Moriyasu, Carbohydr. Res., 1981, 92, 323.
510 R. Yamaguchi, Y. Arai, T. Itoh, and S. Hirano, Carbohydr. Res., 1981, 88, 172.
511 K.D. Vorlop and J. Klein, Biotechnol. Letts., 1981, 3, 9.
512 V.S. Nithianandam, K.S.V. Srinisvasan, K.T. Joseph, and M. Santappa, Ind. J. Biochem. Biophys., 1979, 16, 119.
513 H. Kondo, H. Nakatani, and K. Hiromi, Agric. Biol. Chem. (Jpn), 1981, 45, 2369.
514 K.K. Chacko and W. Saenger, J. Amer. Chem. Soc., 1981, 103, 1708.
515 J.A. Hamilton, M.N. Sabesan, and L.K. Steinrauf, Carbohydr. Res.., 1981, 89, 33.
516 R.I. Gelb, L.M. Schwartz, B. Cardelino, H.S. Fuhrman, R.F. Johnson, and D.A. Laufer, J. Amer. Chem. Soc., 1981, 103, 1750.
517 M. Kojima, F. Toda, and K. Hattori, J. Chem. Soc. Perkin I, 1981, 1647.
518 A. Ueno, K. Takahashi, and T. Osa, J. Chem. Soc. Common., 1981, 94.
519 B. Zsadon, M. Szilasi, F. Tüdös, and J. Szejtli, J. Chromatogr., 1981, 208, 109.
520 M. Suzuki and Y. Sasaki, Chem. Pharm. Bull. (Jpn), 1981, 29, 585.
521 R.P. Rohrbach and J.F. Wojcik, Carbohydr. Res., 1981, 92, 177.
522 K. Uekama, F. Hirayama, T. Wakuda, and M. Otagiri, Chem. Pharm. Bull. (Jpn), 1981, 29, 213.
523 Y. Yonezawa, S. Maruyama, and K. Takagi, Agric. Biol. Chem. (Jpn), 1981, 45,

505.
524 A. Konno, M. Miyawaki, M. Misaki, and K. Yasumatsu, Agric. Biol. Chem. (Jpn), 1981, 45, 2341.
525 A. Ueno, K. Takahashi, Y. Hino, and T. Osa, J. Chem. Soc. Chem. Commun., 1981, 194.
526 N. Kobayashi, A. Ueno, and T. Osa, J. Chem. Soc. Chem. Commun., 1981, 340.
527 Y. Mizobuchi, M. Tanaka, and T. Shono, J. Chromatogr., 1980, 194, 153.
528 Y. Mizobuchi, M. Tanaka, and T. Shono, J. Chromatogr., 1981, 208, 35.
529 M. Tanaka, Y. Mizobuchi, T. Kuroda, and T. Shono, J. Chromatogr., 1981, 219, 108.
530 S. Kuga, J. Chromatogr., 1981, 206, 449.
531 K. Ujimoto and H. Kurihara, J. Chromatogr., 1981, 208, 183.
532 M. Torii, B.P. Alberto, and S. Tanaka, J. Biochem. (Jpn), 1980, 88, 1855.
533 I. Tanaka, Y. Abe, T. Hamada, O. Yonemitsu, and S. Ishii, J. Biochem. (Jpn), 1981, 89, 1643.
534 T. Barth, K. Tjessem, and A. Aaberg, J. Chromatog., 1981, 214, 83.
535 O. Larm, K. Larsson, E. Scholander, B. Meyer, and J. Thiem, Carbohydr. Res., 1981, 91, 13.
536 V.V. Mozhaev and K. Martinek, Eur. J. Biochem., 1981, 115, 143.
537 A.N. Gurov, N.V. Lozinskaya, A.V. Poteshnikh, and V.B. Tolstoguzov, J. Dairy Sci., 1981, 64, 380.
538 J.L. Tayot, J. Holmgren, L. Svennerholm, M. Lindblad, and M. Tardy, Eur. J. Biochem., 1981, 113, 249.
539 F.N. Onyezili and A.C. Onitiri, Anal. Biochem., 1981, 117, 121.
540 J.A. Asenjo and P. Dunnhill, Biotechnol. Bioeng., 1981, 23, 1045.
541 P. Gemeiner, D. Mislovičová, J. Zemek, and L. Kuniak, Coll. Czech. Chem. Commun., 1981, 46, 419.
542 B.L. Lamberts, L.G. Smonson, E.D. Pederson, and R.W. Gaugler, Anal. Biochem., 1981, 117, 320.
543 A.J. Stipanovic and E.S. Stevens, Biopolymers, 1981, 20, 1565.
544 N. Ototani, M. Kikuchi, and Z. Yosizawa, Carbohydr. Res., 1981, 88, 291.
545 M. Kosakai and Z. Yosizawa, J. Biochem. (Jpn), 1981, 89, 1933.
546 S. Kadokura, T. Miyamoto, and H. Inagaki, Biopolymers, 1981, 20, 1113.
547 B.S. Rathaur, G.S. Khatri, K.C. Gupta, C.K. Narang, and N.K. Mathur, Anal. Biochem., 1981, 112, 55.
548 L. Löwendahl, L.A. Lindström, and O. Samuelson, Acta Chem. Scand., 1980, 34B, 623.
549 I. Chibata, T. Tosa, and T. Sato, J. Chromatog., 1981, 215, 93.
550 M.A. Vijayalakshmi, A. Marcipar, and N. Cochet, J. Polymer Sci., Polymer Symp., 1980, 68, 57.
551 R.A. Di Cioccio, C. Piskorz, G. Salamida, J.J. Barlow, and K.L. Matta, Anal. Biochem., 1981, 111, 176.
552 K. Takeo, Carbohydr. Res., 1981, 93, 157.
553 Y. Ueno, K. Hori, R. Yamauchi, M. Kiso, A. Hasegawa, and K. Kato, Carbohydr. Res., 1981, 96, 65.
554 M. Mori and S. Tejima, Chem. Pharm. Bull. (Jpn), 1981, 29, 71.
555 T. Yamazaki, J. Yoshikawa, D.A. Jeanloz, and R.W. Jeanloz, Carbohydr. Res., 1981, 89, 342.
556 K. Shimizu, Carbohydr. Res., 1981, 92, 219.
557 T.D. Heath, B.A. Macher, and D. Papahadjopoulos, Biochim. Biophys. Acta, 1981, 640, 66.
558 C. Wood and E.A. Kabat, J. Exp. Med., 1981, 154, 432.
559 M. Steup, Biochim. Biophys. Acta, 1981, 659, 123.
560 M. Leisola, H. Ojamo, V. Kauppinen, M. Linko, and J. Virkkonen, Enz. Microb. Technol., 1980, 2, 121.
561 R. Hérold and J.P. Fouassier, Starch, 1981, 33, 90.
562 S. Hase, T. Ikenaka, and Y. Matsushima, J. Biochem. (Jpn), 1981, 90, 407.
563 H.J. Creech and A.P. O'Connell, Cancer Res., 1981, 41, 3844.
564 Y. Morimoto, K. Sugibayashi, and Y. Kato, Chem. Pharm. Bull. (Jpn), 1981, 29, 1433.
565 H. Yoshitomi and S. Goto, Chem. Pharm. Bull. (Jpn), 1981, 29, 2374.

566 J.R. Harper and A. Orengo, Anal. Biochem., 1981, 113, 51.
567 P.D. Weston, J.A. DeVries, and R. Wrigglesworth, Biochim. Biophys. Acta, 1980, 612, 40.
568 C. Wehler, J.M. Andrews, and D.H. Bing, Mol. Immunol., 1981, 18, 157.
569 K.A. Krolick, C. Villemez, P. Isakson, J.W. Uhr, and E.S. Vitetta, Proc. Nat. Acad. Sci., U.S.A., 1980, 77, 5419.
570 V.K. Jansons and P.L. Mallett, Anal. Biochem., 1981, 111, 54.
571 J.Y. Lin, J.S. Li, and T.C. Tung, J. Nat. Cancer Inst., 1981, 66, 523.
572 J.L. Langone and R. Ejzemberg, Biochem. Biophys. Res. Commun., 1981, 99, 768.
573 D. Nonnenmacher and R. Brossmer, Biochim. Biophys. Acta, 1981, 668, 149.
574 P. Ganguly and N.G. Fossett, Biochem. Biophys. Res. Commun., 1981, 99, 176.
575 H. Doi, F. Ibuki, and M. Kanamori, Agric. Biol. Chem. (Jpn), 1981, 45, 2351.
576 D.B. Cawley, D.L. Simpson, and H.R. Herschman, Proc. Nat. Acad. Sci., U.S.A., 1981, 78, 3383.
577 S.R. Gonda and J.R. Shainoff, Thromb. Res., 1980, 20, 581.
578 H.M. Madnick, N.K. Kalyan, H.L. Segal, and O.P. Bahl, Arch. Biochem. Biophys., 1981, 212, 432.
579 R.V. Rebois, F. Omedeo-Sale, R.O. Brady, and P.H. Fishman, Proc. Nat. Acad. Sci., U.S.A., 1981, 78, 2086.
580 F.J. Hesford, E.G. Berger, and D.H. Van den Eijnden, Biochim. Biophys. Acta, 1981, 659, 302.
581 C.G. Beddows, J.T. Guthrie, and F.I. Abdel-Hay, Biotechnol. Bioeng., 1981, 23, 2885.
582 B.H. Yang, P.V. Sundaram, and A. Maelicke, Biochem. J., 1981, 199, 317,
583 Y. Malpiece, M. Sharan, J.N. Barbotin, P. Personne, and D. Thomas, J. Histochem. Cytochem., 1980, 28, 961.
584 I. Zemanová, J. Turková, M. Capka, L.A. Nakhapetyan, F. Svec, and J. Kálal, Enzyme Microb. Technol., 1981, 3, 229.
585 H.A. Cooper, D. Lee, M.A. Lamb, and R.H. Wagner, Brit. J. Haematol., 1980, 44, 149.
586 W. Romer and E.W. Rauterberg, J. Immunol. Methods, 1980, 38, 239.
587 H. Tlaskalová-Hogenová, J. Coupek, M. Pospíšil, L. Tučková, J. Kaminková, and P. Mančal, J. Polymer Sci., Polymer Symp., 1980, 68, 89.
588 A.T.M. Al-Ani, M.H.H.Al-Hakeim, and T. Chard, Clin. Chim. Acta, 1981, 112, 91.
589 N. Ototani, M. Kikuchi, and Z. Yosizawa, J. Biochem. (Jpn), 1981, 90, 241.
590 A.A. Farooqui and P.M. Srivastava, Int. J. Biochem., 1981, 13, 893.
591 T. Tsuji, T. Irimura, and T. Osawa, J. Biol. Chem., 1981, 256, 10497.
592 E. Wachter and K. Hochstrasser, Hoppe-Seyler's Z. Physiol. Chem., 1981, 362, 1351.
593 W. Brever and C.H. Siu, Proc. Nat. Sci., U.S.A., 1981, 78, 2115.
594 F. Ishikawa, K. Oishi, and K. Aida, Agric. Biol. Chem. (Jpn), 1981, 45, 557.
595 A.A. Farooqui, J. Chromatogr., 1980, 184, 335.
596 E. Boschetti, P. Girot, J.P. Secheresse, and J.S. Blancard, J. Chromatogr., 1981, 210, 469.
597 M.F.A. Goosen, M.V. Sefton, and M.W.C. Hatton, Thromb. Res., 1980, 20, 543.
598 V. Kasche, K. Buchholz, and B. Galunsky, J. Chromatogr., 1981, 216, 169.
599 G.S. Eisenbarth, Analyt. Biochem., 1981, 111, 1.
600 C. Ghommidh, J.M. Navarro, and G. Durand, Biotechnol. Lett., 1981, 3, 93.
601 A.M. Klibanov and J. Huber, Biotechnol. Bioeng., 1981, 23, 1537.
602 M.A. Van Keulen, K. Vellenga, and G.E.H.Joosten, Biotechnol. Bioeng., 1981, 23, 1437.
603 I. Karube, T. Matsunaga, Y. Otomine, and S. Suzuki, Enzyme Microb. Technol., 1981, 3, 309.
604 K. Saga, T. Kobayashi, and S. Shimizu, Agric. Biol. Chem. (Jpn), 1981, 45, 2895.
605 K. Murata, K. Tani, J. Kato, and I. Chibata, Enzyme Microb. Technol., 1981, 3, 233.
606 A. Constantinides, D. Bhatia, and W.R. Vieth, Biotechnol. Bioeng., 1981, 23, 899.

607 H. Sawada, S. Kinoshita, T. Yoshida, and H. Taguchi, J. Ferment. Technol., 1981, 59, 111.
608 M. De Rosa, A. Gambacorta, L. Lama, B. Nicolaus, and V. Buonocore, Biotechnol. Lett., 1981, 3, 183.
609 F. Yoshii, T. Fujimura, and I. Kaetsu, Biotechnol. Bioeng., 1981, 23, 833.
610 T. Fujimura, F. Yoshii, and I. Kaetsu, Plant Physiol., 1981, 67, 351.
611 H. Kayano, T. Matsunaga, I. Karube, and S. Suzuki, Biotechnol. Bioeng., 1981, 23, 2283.
612 P.G. Krouwel, W.F.M. Van der Laan, and N.W.F. Kossen, Biotechnol. Lett., 1981, 3, 158.
613 R.H.A. Plasterk, K.K. Rao, and D.O. Hall, Biotechnol. Lett., 1981, 3, 99.
614 D. Livernoche, L. Jurasek, M. Desrochers, and I.A. Veliky, Biotechnol. Letts., 1981, 3, 701.
615 I.A. Veliky and A. Jones, Biotechnol. Lett., 1981, 3, 551.
616 H. Skodová, J. Chaloupka, and J. Skoda, Biotechnol. Bioeng., 1981, 23, 2151.
617 K. Klein and M. Kluge, Biotechnol. Lett., 1981, 3, 65.
618 Y.Y. Linko, H. Jalanka, and P. Linko, Biotechnol. Lett., 1981, 3, 263.
619 A. Margaritis and P. Bajpai, Biotechnol. Lett., 1981, 3, 679.
620 P. Scherer, M. Kluge, J. Klein, and H. Sahm, Biotechnol. Bioeng., 1981, 23, 1057.
621 P. Atrat and H. Groh, Z. Allgem. Mikrobiol., 1981, 21, 3.
622 R. Maleszka, I.A. Veliky, and H. Schneider, Biotechnol. Lett., 1981, 3, 415.
623 S. Takamatsu, K. Yamamoto, T. Tosa, and I. Chibata, J. Ferm. Technol., 1981, 59, 489.
624 B. Mattiasson, M. Ramstorp, I. Nilsson, and B. Hahn-Hagerdal, Biotechnol. Lett., 1981, 3, 561.
625 I.S. Maddox, P. Dunnill, and M.D. Lilly, Biotechnol. Bioeng., 1981, 23, 345.
626 V. Larreta Garde, G. Gellf, and D. Thomas, Eur. J. Biochem., 1981, 116, 337.
627 V. Larreta Garde, B. Thomasset, A. Tanaka, G. Gellf, and D. Thomas, Eur. J. App. Microbiol. Biotechnol., 1981, 11, 133.
628 V. Larreta Garde, B. Thomasset, and J.N. Barbotin, Enzyme Microb. Technol., 1981, 3, 216.
629 T. Omata, N. Iwamoto, T. Kimura, A. Tanaka, and S. Fukui, Eur. J. App. Microbiol. Biotechnol., 1981, 11, 199.
630 D. Williams and D.M. Munnecke, Biotechnol. Bioeng., 1981, 23, 1813.
631 Y.Y. Linko and P. Linko, Biotechnol. Lett., 1981, 3, 21.
632 C.F. Mandenius, B. Danielsson, and B. Mattiasson, Biotechnol. Lett., 1981, 3, 629.
633 G.H. Cho, C.Y. Choi, Y.D. Choi, and M.H. Han, Biotechnol. Lett., 1981, 3, 667.
634 I.. Veliky and R.E. Williams, Biotechnol. Lett., 1981, 3, 275.
635 A.J.Daugulis, N.M. Brown, W.R. Cluett, and D.B. Dunlop, Biotechnol. Lett., 1981, 3, 651.
636 M. Wada, J. Kato, and I. Chibata, Eur. J. App. Microbiol. Biotechnol., 1981, 11, 67.
637 M.H. Siess and C. Divies, Eur. J. Microbiol. Biotechnol., 1981, 12, 10.
638 K. Murata, K. Tani, J. Kato, and I. Chibata, Eur. J. App. Microbiol. Biotechnol., 1981, 11, 72.
639 P.G. Rouxhet, J.L. Van Haecht, J. Didelez, P. Gerard, and M. Briquet, Enzyme Microb. Technol., 1981, 3, 49.
640 P. Parascandola and V. Scardi, Biotechnol. Lett., 1981, 3, 369.
641 S. Birnbaum, R. Pendleton, P.O. Larsson, and K. Mosbach, Biotechnol. Lett., 1981, 3, 393.
642 V. Jirků, T. Macek, T. Vaněk, V. Krumphanzl, and V. Kubánek, Biotechnol. Lett., 1981, 3, 447.
643 D.S. Terman, Fed. Proc., 1981, 40, 45.
644 N. Nazly and C.J. Knowles, Biotechnol. Lett., 1981, 3, 363.
645 M. Marek, O. Valentova, K. Demnerova, J. Jizba, M. Blumaurerova, and J. Kas, Biotechnol. Lett., 1981, 3, 327.
646 A. Freeman and Y. Aharonowitz, Biotechnol. Bioeng., 1981, 23, 2747.
647 T. Kokubu, I. Karube, and S. Suzuki, Biotechnol. Bioeng., 1981, 23, 29.

648 S. Adachi, K. Hashimoto, K. Miyai, H. Kurome, R. Matsuno, and T. Kamikubo, Biotechnol. Bioeng., 1981, 23, 1961.
649 K.C. Chen, T. Waki, K. Suga, H. Taguchi, and D. Ryu, J. Ferm. Technol., 1981, 59, 357.
650 Y.Y. Chun, M. Iida, and H. Iizuka, J. Gen. App. Microbiol., 1981, 27, 505.
651 M.F. Cocquempot, B. Thomasset, J.N. Barbotin, G. Gellf, and D. Thomas, Eur. J. App. Microbiol. Biotechnol., 1981, 11, 193.
652 A. Kimura, Y. Tatsutomi, R. Matsuno, A. Tanaka, and H. Fukuda, Eur. J. App. Microbiol. Biotechnol., 1981, 11, 78.
653 A. Margaritis, P. Bajpai, and B. Wallace, Biotechnol. Lett., 1981, 3, 619.
654 I. Chibata and T. Tosa, Annu. Rev. Biophys. Bioeng., 1981, 10, 197.
655 P.G Krouwel and N.W.F. Kossen, Biotechnol. Bioeng., 1981, 23, 651.
656 G. Greco and L. Gianfreda, Biotechnol. Bioeng., 1981, 23, 2199.
657 B. Colas, Biochim. Biophys. Acta, 1981, 657, 535.
658 T. Yano, Y. Hayashi, and S. Yamamoto, J. Biochem. (Jpn), 1981, 90, 773.
659 T. Kume, H. Watanabe, S. Aoki, and T. Sato, Agric. Biol. Chem. (Jpn), 1981, 45, 1311.
660 P. Gacesa, M.J. Savitsky, K.S. Dodgson, and A.H. Olavesen, Biochim. Biophys. Acta, 1981, 661, 205.
661 T. Imoto, T. Ono, and H. Yamada, J. Biochem. (Jpn), 1981, 90, 335.
662 A. Hirayama, H. Fujio, Y. Kohi, Y. Takagaki, K. Kondo, and T. Amano, J. Biochem. (Jpn), 1981, 89, 963.
663 R. Schauer, E. Moczar, and M. Wember, Hoppe Seyler's Z. Physiol. Chem., 1979, 360, 1587.
664 H. Nygren and H.A. Hansson, J. Histochem. Cytochem., 1981, 29, 266.
665 Enzyme Engineering, Vol. 5., eds. H.H. Weetall and G.P. Royer (Plenum Press, N.Y., 1980).
666 Immobilized Enzymes for Food Processing, W.H. Pitcher, CRC Press, Boca Raton, 1980).
667 B. Mattiasson, J. App. Biochem., 1981, 3, 183.
668 G.G. Guilbault, Ann. N.Y. Acad. Sci., 1981, 369, 285.
669 B. Mattiasson, B. Danielsson, C.F. Mandenius, and F. Winquist, Ann. N.Y. Acad. Sci., 1981, 369, 295.
670 V.P. Torchilin and A.L. Klibanov, Enzyme Microb. Technol., 1981, 3, 297.
671 T.T. Ngo, J. Ivy, and H.M. Lenhoff, Biotechnol. Lett., 1980, 2, 429.
672 P.R. Coulet and D.C. Gautheron, J. Chromatogr., 1981, 215, 65.
673 J.J. Kulys, Enzyme Microb. Technol., 1981, 3, 345.
674 A. Sadana, Enzyme Microb. Technol., 1981, 3, 357.
675 T. Yamane, J. Ferm. Technol., 1981, 59, 375.
676 H. Ooshima and Y. Harano, Biotechnol. Bioeng., 1981, 23, 1991.
677 G.K. Lee, R.A. Lesch, and P.J. Reilly, Biotechnol. Bioeng., 1981, 23, 487.
678 K.I. Suga, K.C. Chen, and H. Taguchi, J. Ferm. Technol., 1981, 59, 137.
679 K. Kurkijärvi and T. Korpela, Biotechnol. Bioeng., 1981, 23, 1389.
680 B.L. Seong, M.H. Han, J.M. Park, and C.Y. Choi, Biotechnol. Lett., 1981, 3, 607.
681 N.G. Karanth, Biotechnol. Bioeng., 1981, 23. 225.
682 D.D. Do. and R.H. Weiland, Biotechnol. Bioeng., 1981, 23, 691.
683 F.H. Verhoff and S.T. Schlager, Biotechnol. Bioeng., 1981, 23, 41.
684 S.H. Park, S.B. Lee, and D.D.Y. Ryu, Biotechnol. Bioeng., 1981, 23, 2591.
685 F. Alfani, D. Albanesi, M. Cantarella, and V. Scardi, Biotechnol. Lett., 1981, 3, 309.
686 K. Rokugawa, T. Fujishima, A. Kuninaka, and H. Yoshino, J. Ferment. Technol., 1980, 58, 583.
687 J.R. Millis and L.B. Wingard, Biotechnol. Bioeng., 1981, 23, 965.
688 R.J. Lamed, E. Keinan, and J.G. Zeikus, Enzyme Microb. Technol., 1981, 3, 145.
689 W. Hinsch, A. Antonijevic, and P.V. Sundaram, J. Clin. Chem. Biochem., 1981, 19, 307.
690 E. Szwajcer, A. Szewczuk, and M. Mordarski, Biotechnol. Bioeng., 1981, 23, 1675.
691 I.A. Yamskov, M.V. Budanov, L.S. Lobareva, and V.A. Davankov, Bioorg. Khim.,

1981, 7, 86.
692 V. Ramesh and C. Singh, Enzyme Microb. Technol., 1981, 3, 246.
693 J. Kučera and M. Kminkovā, Coll. Czech. Chem. Commun., 1980, 45, 298.
694 P. O'Grady and P.Joyce, Enzyme Microb. Technol., 1981, 3, 149.
695 P. Posádka and L. Macholálan, Coll. Czech. Chem. Commun., 1979, 44, 3395.
696 P.R. Coulet and D.C. Gautheron, J.Chromatogr., 1981, 215, 65.
697 B.P. Wasserman, B.S. Jacobson, and H.O. Hultin, Biochim. Biophys. Acta, 1981, 657, 52.
698 S.O. Enfors, Enzyme Microb. Technol., 1981, 3, 29.
699 W.G. Choi, S.B. Lee, and D.D.Y. Ryu, Biotechnol. Bioeng., 1981, 23, 361.
700 T. Kobayashi, K. Saga, S. Shimizu, and T. Goto, Agric. Biol. Chem. (Jpn), 1981, 45, 1403.
701 P.A. Munro, P. Dunnill, and M.D. Lilly, Biotechnol. Bioeng., 1981, 23, 677.
702 K. Taniguchi, J. Ito, and M. Sasaki, J. Biochem. (Jpn), 1981, 89, 179.
703 M. Sasaki, K. Taniguchi, and K. Minakata, J. Biochem. (Jpn)., 1981, 89, 169.
704 M.H. Remy, D. Guillochon, and D. Thomas, J. Chromatogr., 1981, 215, 87
705 G. Manecke and D. Polakowski, J. Chromatogr., 1981, 215, 13.
706 T. Kitao, Y. Kitao, and K. Hattori, Experientia, 1981, 37, 452.
707 P. Monsan and A. Lopez, Biotechnol. Bioeng., 1981, 23, 2027.
708 S.F. D'Souza and G.B. Nadkarni, Biotechnol. Bioeng., 1981, 23, 431.
709 J. Hradil and F. Svec, Enzyme Microb. Technol., 1981, 3, 331.
710 L. Iyengar and A.V.S. Prabhakara Rao, J. Gen. Appl. Microbiol., 1981, 27, 339.
711 F.N. Onyezili and A.C. Onitiri, Anal. Biochem., 1981, 113, 203.
712 S. Hjertén, Y. Kunquan, and M. Ogunlesi, J. Chromatogr., 1981, 215, 25.
713 Y.K. Park and G.M. Pastore, J. Ferment. Technol., 1981, 59, 165.
714 L. Rexová-Benková and M. Mračková-Dobrotová, Carbohydr. Res., 1981, 98, 115.
715 V.S. Nithianandam, K.S.V. Srinivasan, K.T. Joseph, and M. Santappa, Biotechnol. Bioeng., 1981, 23, 2273.
716 V. Basaveswara Rao, N.V.S. Sastri, and P.V. Subba Rao, Biochem. J., 1981, 193, 389.
717 J.M.S. Cabral, J.P. Cardoso, and J.M. Novais, Enzyme Microb. Technol., 1981, 3, 41.
718 J.M.S. Cabral, J.M. Novais, and J.P. Cardoso, Biotechnol. Bioeng., 1981, 23, 2083.
719 E. Bissé, A. Scholer, and D.J. Von Der Schmitt, J. Appl. Biochem., 1981, 3, 176.
720 S. Bachman, L. Gebicka, and Z. Gasyna, Starch, 1981, 33, 63.
721 Y. Ishimori, I. Karube, and S. Suzuki, Eur. J. App. Microbiol. Biotechnol., 1981, 13, 19.
722 Y. Ishimori, I. Karube, and S. Suzuki, Biotechnol. Bioeng., 1981, 23, 2601.
723 T. Tsuchida and K. Yoda, Enzyme Microb. Technol ., 1981, 3, 326.
724 L.D. Bowers and P.R. Johnson, Biochim. Biophys. Acta, 1981, 661, 100.
725 L.D. Bowers and P.R. Johnson, Anal. Biochem., 1981, 116, 111.
726 J.P. Guiraud, S. Demeulle, and P. Galzy, Biotechnol. Lett., 1981, 3, 683.
727 G.A. Kovalenko, N.B. Shitova, and V.D. Sokolovskii, Biotechnol. Bioeng., 1981, 23, 1721.
728 N.D. Danielson and R.W. Siergiej, Biotechnol. Bioeng., 1981, 23, 1913.
729 O. Rodriguez and G.G. Guilbault, Enzyme Microb. Technol., 1981, 3, 69.
730 N.H. Park and H.N. Chang, J. Ferment. Technol., 1979, 57, 310.
731 K. Rokugawa, T. Fujishima, A. Kuninaka, and H. Yoshino, J. Ferment. Technol., 1980, 58, 423.
732 J.M. Guisán and A. Ballesteros, Enzyme Microb. Technol., 1981, 3, 313.
733 M.D.K. Kumaraswamy, K. Panduranga Rao, K.T. Joseph, and M. Santappa, Biotechnol. Bioeng., 1981, 23, 1889.
734 A.L. Margolin, V.A. Izumrudov, V.K. Svedas, A.B. Zezin, V.A. Kabanov, and I.V. Berezin, Biochim. Biophys. Acta, 1981, 660, 359.
735 B. Adu-Amankwa, A. Constantinides, and W.R. Vieth, Biotechnol. Bioeng., 1981, 23, 2609.
736 J. Lobarzewski, Biotechnol. Bioeng., 1981, 23, 2161.

737 K. Oyama, S. Nishimura, Y. Nonaka, K. Kihara, and T. Hashimoto, J. Org. Chem., 1981, 46, 5241.
738 F.N. Onyezili and A.C. Onitiri, Biochim. Biophys. Acta, 1981, 659, 244.
739 E. Sada, S. Katoh, M. Shiozawa, and T. Fukui, Biotechnol. Bioeng., 1981, 23, 2561.
740 A.B. Salleh and W.M. Ledingham, Int. J. Biochem., 1981, 13, 1113.
741 A.B. Salleh and W.M. Ledingham, Anal. Biochem., 1981, 116, 40.
742 A. Iob and H.A. Mottola, Clin Chem., 1981, 27, 195.
743 S. Watanabe, Y. Shimizu, T. Teramatsu, and T. Murachi, J. Biomed. Mater. Res., 1981, 15, 553.
744 D.G. Mercer and K.F. O'Driscoll, Biotechnol. Bioeng., 1981, 23, 2447.
745 D.G. Mercer and K.F. O'Driscoll, Biotechnol. Bioeng., 1981, 23, 2465.
746 C.F. Mandenius, B. Danielsson, and B. Mattiasson, Acta Chem. Scand., 1980, 34B, 463.
747 J. Hradil and F. Svec, Enzyme Microb. Technol., 1981, 3, 336.
748 N.A. Greenberg and R.R. Mahoney, Process Biochem., 1981, 16, 2.
749 S.H. Park, S.B. Lee, and D.D.Y. Ryu, Biotechnol. Bioeng., 1981, 23, 1237.
750 E. Sada, S. Katoh, and M. Terashima, Biotechnol. Bioeng., 1981, 23, 1037.
751 O. Valentová, M. Marek, F. Svec, J. Štamberg, and Z. Vodrážka, Biotechnol. Bioeng., 1981, 23, 2093.
752 D.W. Sundstrom, H.E. Klei, R.W. Coughlin, G.J. Biederman, and C.A. Brouwer, Biotechnol. Bioeng., 1981, 23, 473.

Author Index

In this index the number given in parentheses is the chapter number, which is followed by the relevant reference number(s) within that chapter.